A. Blancher · J. Klein · W. W. Socha (Eds.)

Molecular Biology and Evolution of Blood Group and MHC Antigens in Primates

Springer-Verlag Berlin Heidelberg GmbH

A. Blancher · J. Klein · W. W. Socha (Eds.)

Molecular Biology and Evolution of Blood Group and MHC Antigens in Primates

With 103 Figures and 72 Tables

Springer

ANTOINE BLANCHER, MD, PhD
Laboratoire d'Immunogénétique moléculaire
Université Paul Sabatier, CHU de Toulouse
Hôpital Purpan
Place du Docteur Baylac
31059 Toulouse cedex · France

JAN KLEIN, PhD
Max-Planck-Institut für Biologie
Abteilung Immungenetik
Correnstr. 42
72076 Tübingen · Germany

WLADYSLAW SOCHA, MD
Laboratory for Experimental Medicine and Surgery in Primates (LEMSIP)
New York University Medical Center
10 Townsend Avenue
Upper Grandview, NY 10960-4830 · USA

ISBN 978-3-642-63843-5

Library of Congress Cataloging-in-Publication Data
Molecular biology and evolution of blood group and MHC antigens in primates / A. Blancher, J. Klein, S. W. Socha (eds.).
op, cm.
  Includes bibliographical references and Index.
  ISBN 978-3-642-63843-5   ISBN 978-3-642-59086-3 (eBook)
  DOI 10.1007/978-3-642-59086-3
1. Blood group antigens. 2. Blood group antigens-Evolution. 3. Major histocompatibility complex.
4. Major histocompatibility complex-Evolution. 5. Primates-Evolution. I. Blancher, A. (Antoine),
1954- . II. Klein, Jan, 1936- . III. Socha, Wladyslaw W., 1926- .
QP98.M653>h2>1997
571.9'645-dc21

© Springer-Verlag Berlin Heidelberg 1997
Originally published by Springer-Verlag Berlin Heidelberg New York in 1997
Softcover reprint of the hardcover 1st edition 1997

Typesetting: Hermann Hagedorn GmbH, D-68519 Viernheim
Cover design: design & production, D-69121 Heidelberg
Production: PRO EDIT GmbH, D-69126 Heidelberg

SPIN: 10482806 27/3136 - 5 4 3 2 1 0 - Printed on acid-free paper

Preface

Zoologists have categorized primates into a single order, and no one doubts today that they share a common ancestry. Humans and Old and New World non-human primate species, from the lemurs of Madagascar to the African anthropoid apes, represent diverging branches of an evolutionary common trunk. Along with species-specific characters, all primates have retained a number of ancestral traits, relics of their common origin. The comparative study of these species-specific and ancestral traits makes it possible to reconstruct the evolutionary pathways of humans and nonhuman primates.

The discovery of the human blood groups and, later, of the Major Histocompatibility Complex (MHC) had a seminal effect on the field of human genetics, providing the first sound examples of mendelian polymorphisms. The use of blood group and MHC alleles as genetic markers in biological anthropology generated a conceptual revolution and persuaded researchers to begin to think in terms of populations and not only in terms of typology. The counterparts of these human red and white cell antigens were found and studied in nunhuman primates, and progress in this field is summarized in this book.

Investigations of red cell antigens in apes and monkeys were initiated by Karl Landsteiner and Alexander S. Wiener, the co-discoverers of the Rhesus factor in human blood, and were continued by their collaborators and resulted in hundreds of publications. *Blood groups of Primates. Theory, Practice, Evolutionary meaning* published in 1983 by Wladyslaw W. Socha and Jacques Ruffié was the first book to summarize the status of knowledge of blood group serology of non-human primates and to discuss the relationships between human and nonhuman primate red cell antigens. The discovery of the human Major Histocompatibility Complex (MHC) by Jean Dausset was followed by an explosive progress in the comprehension of the structure and physiology of the MHC antigens. Most importantly, the role of MHC class I and II molecules in the presentation of peptides to the T lymphocyte receptors was recognized. The study of MHC was extended to subhuman primate species and results were compiled in *Evolution of Major Histocompatibility Complex* (Jan Klein and Dagmar Klein, Eds. Springer, 1991).

Natural selection in response to the pressure generated by the numerous environmental pathogens to which human and nonhuman primates have been exposed has affected, over time, the frequencies of alleles and haplotypes of the various MHC loci. Thus, the immunological profiles we observe today in humans and other primates must be the result of these darwinian selection processes. Is it therefore possible to reconstruct all the infectious episodes experienced by a species? Nothing is less certain. Some aggressors may have become extinct due to their rapid destruction by the immune defenses of their hosts. In other cases, the host disappeared and the fight ceased for lack of fighters. Only those pathogens that allowed the host to survive long enough to infect other individuals and that were sufficiently resistant to the host's immune system could be maintained in an endemic state. Moreover, not a single species is free of infections; rather, each carries its particular load of parasites, bacteria, viruses and symbionts. These potentially pathogenic companions may become rigorously adapted to their host, occur ubiquitously or locally, or be common to many species. Furthermore, all these pathogens will continue to evolve with their host species, which may eventually become a unique reservoir for this infectious agent.

In this book, A. Blancher, J. Klein, and W. W. Socha have summarized the results of their many years of research, and, as editors, invited the contributions of a number of eminent colleagues and collaborators. The volume contains a vast list of references and represents a summary of modern molecular immunogenetics of primates, including humans. For many years *Blood Groups*, by Robert Race and Ruth Sanger, was the reference textbook for immunohematologists. The book by Blancher, Klein and Socha, which centers on the immunogenetics of blood groups and the MHC of nonhuman primates but also provides abundant information on human immunohematology, will also undoubtedly find a wide circle of readers.

April 1997   Jacques Ruffié
                                                                  Luca Cavalli-Sforza

Contents

List of Contributors

Bergström, T.
Department of Medical Genetics
Uppsala University Biomedical Center
Box 589
751 23 Uppsala
Sweden

Blancher, A.
Laboratoire d'Immunogénétique moléculaire
Université Paul Sabatier, CHU de Toulouse
Hôpital Purpan
Place du Docteur Baylac
31059 Toulouse
France

Blumenfeld, O.
Department of Biochemistry
Albert Einstein College of Medicine
1300 Morris Park Avenue
Bronx, NY 10461
USA

Bontrop, R.
TNO Medical Biological Laboratory
TNO Healt Research
PO Box 5815
2280 HV Rijswijk
The Nederlands

Cadavid, L. F.
Wisconsin Regional Primate Research Center
University of Wisconsin-Madison
1220 Capitol Court
Madison, WI 53715
USA

Chaudhuri, A.
Laboratory of Cell Biology
Lindsley F. Kimball Research Institute
The New Work Blood Center
310 E. 67th St.
New York, NY 10021
USA

Dawkins, R. L.
Center for Molecular Immunology and Instrumentation
GPO Box X2213
Perth, WA 6001
Australia

Fratalli Eder, A.
Department of Pathology and Laboratory Medicine
University of Pennsylvania
3400 Spruce Street
Philadelphia, PA 19104-4283
USA

Erlich, H. A.
Roche Molecular Systems, Inc.
1145 Atlantic Avenue, Suite 100
Alameda, CA 94501
USA

Figueroa, F.
Max-Planck-Institut für Biologie
Abteilung Immungenetik
Correnstr. 42
72076 Tübingen
Germany

Galili, U.
Department of Microbiology
Medical College of Pennsylvania
2900 Queen Lane
Philadelphia PA 19129
USA

Gaudieri, S.
Department for Molecular Immunology and Instrumentation
GPO Box X2213
Perth, WA 6001
Australia

Gongora, R.
Max-Planck-Institut für Biologie
Abteilung Immungenetik
Correnstr. 42
72076 Tübingen
Germany

Grahovac, B.
Head of Department of Molecular Immunogenetics
Croatian Institute of Transfusion Medicine
Petrova 3, 10 000 Zagreb
Croatia

Groves, C.
Department of Archaeology and Anthropology
Australian National University
Canberra, A.C.T. 0200
Australia

Gyllensten, U.
Department of Medical Genetics
Uppsala University Biomedical Center
Box 589
751 23 Uppsala
Sweden

Henry, J.
CNRS/ERS 7101
Laboratoire de Biochimie Médicale
2, rue du Dr. Marcland
87025 Limoges
France

Huang, C.-H.
Laboratory of Cell Biology
Lindsley F. Kimball Research Institute
The New York Blood Center
310 E. 67th St.
New York, NY 10021
USA

Klein, J.
Max-Planck-Institut für Biologie
Abteilung Immungenetik
Correnstr. 42
72076 Tübingen
Germany
and
Department of Microbiology and Immunology
University of Miami School of Medicine
Miami, FL 33101
USA

Kulski, J. K.
Department of Microbiology and Infectious Diseases
Royal Perth Hospital
GPO Box X2213
Perth, WA 6001
Australia

O'hUigin, C.
Max-Planck-Institut für Biologie
Abteilung Immungenetik
Correnstr. 42
72076 Tübingen
Germany
and
Department of Microbiology and Immunology
University of Miami School of Medicine
Miami, FL 33101
USA

Pogo, O.
Laboratory of Cell Biology
Lindsley F. Kimball Research Institute
The New York Blood Center
310 E. 67th St.
New York, NY 10021
USA

Pontarotti, P.
INSERM U119
27, Bd Leï Roure
13009 Marseille Cedex
France

Satta, Y.
The Graduate University for Advanced Studies
Hayama
240-01 Kanagawa
Japan

Shufflebotham, C.
Wisconsin Regional Primate Research Center
University of Wisconsin-Madison
1220 Capitol Court
Madison, WI 53715
USA

Socha, W. W.
New York University Medical Center
Laboratory for Experimental Medicine and Surgery in Primates (LEMSIP)
10 Townsend Avenue
Upper Grandview, NY 10960-4830
USA

Spitalnik, S. L.
Department of Pathology and Laboratory Medicine
University of Pennsylvania
3400 Spruce Street
Philadelphia, PA 19104-4283
USA

Tazi Ahnini, R.
CIGH/CNRS
Hôpital Purpan
31300 Toulouse Cedex
France

Watkins, D. I.
Wisconsin Regional Primate Research Center and Department of Pathology
and Laboratory Medicine
University of Wisconsin-Madison
1220 Capitol Court
Madison, WI 53715
USA

Xie, S.
Dept. of Molecular Biology
Hunan Medical University
Chingsa, Hunan 410078
P.R. of China

I Taxonomy and Phylogeny of Primates

1 Taxonomy and Phylogeny of Primates

Colin P. Groves

Introduction

Taxonomy is defined by Ernst Mayr (1969) as "The theory and practice of classifying organisms"; as a shorthand, it is used ("a taxonomy") for an actual classification itself. Phylogeny was defined by Haeckel, who coined the term, as "the entire science of the changes in form through which the phyla or organic lineages pass through the entire time of their discrete existence". As in the case of the term "taxonomy", we commonly use "a phylogeny" to mean a given group's own evolutionary history.

What is, or should be, the relationship between these two concepts? Are they, or should they be, two separate but overlapping spheres of influence; or are they so closely intertwined that one can almost regard taxonomy as a subset of phylogeny?

Taxonomy Above the Species Level

Linnaeus, the inventor of the system of taxonomy that we use today, divided the animal kingdom into six classes. He divided each class into a number of orders; each order into genera; each genus into species. Since his day, knowledge of animal diversity has increased enormously; six classes are far from adequate, and the classes are grouped into phyla. Within each order, the genera are grouped into families. The taxonomic hierarchy now has seven levels: kingdom, phylum, class, order, family, genus and species. Every individual is classified using these

Molecular Biology and Evolution of Blood Group
and MHC Antigens in Primates
Blancher/Klein/Socha (Eds.)
© Springer-Verlag Berlin Heidelberg 1997

seven levels (or ranks). Human beings belong to the animal kingdom; with other backboned animals, to the phylum Chordata; within the chordates, to the class Mammalia; within the mammals, to the order Primates; within the primates, to the family Hominidae; within the hominids, to the genus *Homo*; and finally to the particular species *Homo sapiens*. Each species bears a generic and a specific name, both placed in italics, the genus commencing with a capital letter, the specific name with a small letter. Families end in -idae; other ranks have no standardised ending.

Nonetheless, the seven obligatory ranks are rarely sufficient to represent the fine degrees of relatedness. It is necessary to insert even more subdivisions: suborders, infraorders, superfamilies, as much as is needed in any given case. In some instances, the plethora of ad-hoc levels become so unwieldly that it has been suggested that, above the family level at least, groups (or taxa) be unranked, and just the names used, in sequence. As we shall see, the ranks are threatening to get a little out of hand even among the primates.

Hennig thought that the relationship between taxonomy and phylogeny is clear: "the sequence of subordination [of taxa in the taxonomic hierarchy] corresponds to the recency of common ancestry" (Hennig 1966 p. 83); the phylogenetic branching sequence, that is, determines taxonomic ranking. Hennig attempted further to link taxonomic ranks with time, e.g. if two groups split during the Mesozoic, they should be ranked as families, if during the Cenozoic as tribes, and so on: the Order Primates, therefore would be at most a family of the Order Mammalia! Some consistency within major groups, within classes perhaps, might be desirable, but to try this across the entire animal and plant kingdoms would simply not be possible.

The logical basis of Hennig's system, which he called phylogenetic systematics and is today usually designated cladistic taxonomy, is that branching order in phylogeny is the only testable (falsifiable in the Popperian sense) measure of relatedness. Authors such as Mayr (1969) have strenuously objected to this assessment: relatedness is a consequence not only of branching sequence but of subsequent divergence as well. This objection can now be answered: when total differentiation can be measured, as by various genetic distance measures or by DNA-DNA hybridisation, it turns out to be roughly proportional to time since divergence (as reviewed, for example, by Easteal et al. 1995). In other words morphological differentiation, which is what Mayr doubtless had in mind when he referred to "subsequent divergence", accounts for such a minuscule proportion of the total genome as to be undetectable in the context of total genetic change. Even under Mayr's definition of relatedness, therefore, Hennig's insistence on the primacy of branching order is still justified – ironically, for reasons not contemplated by either of those two authors!

Reconstructing a Phylogeny

Phylogeny reconstruction is performed as follows:

1. List the taxa, or the groups it is proposed to analyse. The groups designated for analysis are called the *operational taxonomic units* (OTUs). One may choose, for example, to analyse the chimpanzee, gorilla, human, orangutan and gibbon separately, but to keep the Old World monkeys together and the New World monkeys together as single groups for analysis; thus, there will now be seven OTUs.

2. List all the *characters* which differ among the OTUs. Number of premolar teeth and presence/absence of a tail will be among numerous characters that exhibit differences among the OTUs in the example given above.

3. List the character *states* – the differences, character by character – between the OTUS. Thus, the New World monkeys will be listed as having, under the character "number of premolars", the state "three" (in each half of each jaw); Old World monkeys, gibbons, etc., will be listed as having character state "two". New and Old World monkeys will be listed as having, in the character "tail", the state "present", while gibbons, humans, etc. have the state "absent".

4. Determine the *polarity* – the direction of evolutionary change – of each character state. In the present case, more distantly related primates (lemurs, tarsiers) have three premolars, so it is likely that "three" in this case is the *primitive* or *plesiomorphic* state, and that "two" is the *derived* or *apomorphic* state. These more distantly related primates all have tails, so it is most plausible that "tail present" is plesiomorphic, "tail absent" apomorphic.

5. Find out which pairs of taxa are linked by *shared derived* or *synapomorphic* character states. Old World monkeys, gibbons, humans, etc., are linked by the synapomorphic state of "two premolars"; on the other hand, the character state that links Old and New World Monkeys, "tail present", is *shared primitive* or *symplesiomorphic*.

6. Synapomorphic character states are the only ones which can constitute evidence for relatedness, i.e. for having a more recent common ancestor. Sometimes parallel evolution occurs; as parallelism is considered an unlikely event, it is generally reckoned that the two OTUs with the greatest number of shared derived character states are those most closely related. Another process that sometimes spoils a perfect cladogram is reversal to a more primitive state.

7. The OTUs are linked in pairs, more and more inclusively, until a complete branching diagram, a *cladogram*, is constructed. Pairs of closest relatives, those which branch off a common node of the cladogram (each being *monophyletic*, i.e. descended from an exclusive common ancestor), are called *sistergroups*.

A cladogram is not precisely a "family tree". To construct a family tree (evolutionary tree, phylogeny), we need to know time relationships as well and to make some working hypothesis about which (fossil) taxa might be ancestors.

A pair of sister-groups in a cladogram might be plausibly hypothesised, on the basis of further data, to be an ancestor-descendant pair.

If the aim of taxonomy is to depict degrees of relatedness, and if relatedness is dependent to all intents and purposes on phylogenetic branching sequence, i.e. on sister-group status, then the relation of taxonomy to phylogeny is, in theory, one to one. Groves (1989) has drawn attention to instances in which such a strict mapping of phylogeny onto taxonomy, even when the branching sequence is definitely known (which is far from being always the case!), might actually be undesirable; for example, when the implied distance between one branch and the next is very short, barely distinguishable from a trifurcation. In such a case, it would be wise simply to rank the three taxa equally. For example, the time between the separation of the gorilla from the chimpanzee/human stem and that of the chimpanzee and human lineages seems to be so short that its very existence is controversial, and three taxa of equal rank would seem to be the best taxonomic solution. The other instances in which it might be difficult to make taxonomy a precise map of phylogeny are when fossil taxa are in question, but this is outside the scope of the present discussion.

One question remains to be asked: to what extent do phylogenies reconstructed by the cladistic method agree with those generated by molecular methods, the "total genome" methods mentioned above? For a while it seemed that agreement might be unsatisfactory, but in recent years the employment of more and more complete datasets, in both morphological and molecular spheres, has brought the two approaches closer together. The latest attempt (Shoshani et al. 1996) has found excellent agreement in primate phylogeny.

Cladistic methodology has created a revolution in primate taxonomy, which has become (above the species level) a search for monophyletic groups accompanied by a rejection of traditionally recognised but unquestionably non-monophyletic taxa. There are two glaring instances of this:

1. What are the basic divisions within the primates? The dead hand of tradition has the primates divided into two suborders, Prosimii and Anthropoidea. Yet it is perfectly clear that one of these, Prosimii, cannot be justified cladistically: all the character states used to define it (small brains, for example!) are shared primitive. When we look for shared derived conditions, we find them between one of the groups traditionally assigned to the Prosimii (the Tarsiiformes) and the other traditional suborder, the Anthropoidea. These shared derived conditions include:
    - Rhinarium dry, and upper lip not split
    - Tapetum lucidum absent from retina
    - Placenta haemochorial
    - Orbit walled in behind by bone
    - Lack of endogenous manufacture of vitamin C.

It is clear that the Tarsiiformes form a clade with the Anthropoidea. Taxonomically, therefore, they should be linked with the latter in a suborder Haplorrhini, leaving the other "prosimians" (the lemurs and lorises) alone in a suborder called Strepsirrhini.

2. What is the correct taxonomy for humans and great apes? The superfamily Hominoidea, to which humans, great apes and lesser apes (gibbons) are always assigned, divides cladistically into two branches, the gibbons and the rest. The "rest" also divides into two branches: the orangutan and the rest. There is some indication, in addition, that the next branching is gorilla vs chimpanzee/human; but, as this is not absolutely certain, and all indications are that if it is exists this dichotomy is very closely succeeded by the chimpanzee vs human one, it is best to be conservative and retain a three-way split (see above). The traditional taxonomy of the Hominoidea is:
  - Family Hylobatidae (gibbons)
  - Family Pongidae (great apes, i.e. orangutan, chimpanzee, gorilla)
  - Family Hominidae (humans).

But this, as we have seen, bears no relation to actual phylogeny and tells us nothing beyond the fact that the great apes are to some extent evolutionarily conservative (they retain small brains and a lot of body hair). The phylogenetically informative taxonomy is:
  - Family Hylobatidae (gibbons)
  - Family Hominidae
  - Subfamily Ponginae (orangutan)
  - Subfamily Homininae (chimpanzee, gorilla, human).

Having established this taxonomy, we can proceed to make predictions from it. The combining of humans and great apes in a single family tells us that we may expect synapomorphies between them in spheres unrelated to those (molecular, morphological) on which the taxonomy was based, such as cognitive psychology, in which wide-ranging similarities are known, to the degree that it has recently been argued that there are moral, and perhaps legal, implications (Cavalieri and Singer 1993). It would be interesting to take this further: are there cognitive synapomorphies of the Homininae which are not shared by the Ponginae? Cladistic taxonomy tells us that we may expect them; but they have not, as yet, been sought.

In the annotated taxonomy given in the Appendix, I have in every case recognised monophyletic groups in which the evidence is good (Haplorrhini; Hominoidea), but have been conservative otherwise (suprafamilial groupings within the Strepsirrhini). I have even had in some cases to recognise taxa that are demonstrably nonmonophyletic, when it is not clear what the groupings really should be (as in the New World monkeys).

The Species Problem

For some taxonomists the rank of species is where the phylogenetic association ceases. For Mayr, "species are groups of interbreeding natural populations that are reproductively isolated from other such groups" (Mayr 1969 p. 26). Groves

(1989) discusses the meaning of "reproductively isolated": essentially, it means that species *do not* interbreed, not that they *cannot*. The very best criterion of species status is *sympatry*: that two OTUs overlap in their distributions, yet remain separate and distinguishable, is prima facie evidence that they are distinct species. Two well-known macaque species, the crab-eating (*Macaca fascicularis*) and pigtailed (*Macaca nemestrina*) macaques, are found throughout much of Southeast Asia, over the same areas and often in the same forests; they are reproductively isolated, hence distinct species, for all that they can be persuaded to interbreed in captivity. In this case the reproductive isolating mechanisms are presumably behavioural; but they are as effectively separated by this means as if they had chromosomal differences (which they do not).

This definition, the "biological species concept", is accepted in its original form or in its variant forms (recognition concept, cohesion concept) by the majority of biologists; but there are difficulties. Cases of total *allopatry* (two OTUs occupying separate distributions) obviously cannot be decided on such a basis. Sometimes there is local breakdown in reproductive isolating mechanisms, which nonetheless do not result in the two species merging. In fact, Bernstein (1966) has reported just such a case between *Macaca fascicularis* and *M. nemestrina*. Hybrid zones between two OTUs whose distributions meet, but do not overlap, may allow for unrestricted gene flow between them, or may actually inhibit gene flow, or be semipermeable. The assumption tends to be that only species that are sister-groups interbreed, but this is not necessarily the case. Groves et al. (1993) have documented just such a case in the genus *Colobus*, in which two species which are not sisters (i.e. not each others' closest relatives) have interbred, in fact very widely so that in one particular geographic area one has effectively "swamped" the other, but a few nontypical cranial and pelage features crop up to remind us that the population concerned is ultimately of hybrid origin.

To remedy this, Cracraft (1989 etc.) has proposed the "phylogenetic species concept": a species is "an irreducible, monophyletic cluster of organisms possessing at least one diagnostic character". Most taxonomists would consider that this draws the species boundaries much too narrowly; nonetheless, the idea that species, too, must be monophyletic, has been widely adopted. I recommend here a recent inclusive definition, "the theoretical (ontological) concept": a species is "a single lineage of ancestor-descendant sexual populations, genetically integrated by historically contingent events of interbreeding" (Christoffersen 1995, p. 447).

In the appended annotated classification, I have tended to follow the most recent published revision, or given my reasons why not. I have also indicated subspecies, where these are known. Subspecies are geographic segments of a species that are morphologically distinguishable, at least as a whole. (Note that a species either has two or more subspecies, or it has none. Subspecies have three names – generic, specific and subspecific; each species also has what is called a *nominotypical* subspecies, the one whose subspecific name simply repeats the specific name, although, be it noted, it is no more "typical" of the species than is any other of the subspecies!). Where subspecies are almost

100 % diagnosable, they come close to being full (allopatric) species; this accounts for most of the disagreements about how many species there should be in the order primates.

Appendix: An annotated classification of the primates with vernacular or "book-" names[1] (modified from Groves 1993)

Order: Primates

Suborder Strepsirrhini (for the correct spelling of this name, see Jenkins 1987)

Cheirogaleidae

Cheirogaleinae

Allocebus
Allocebus trichotis: hairy-eared dwarf lemur

Cheirogaleus
Cheirogaleus major: greater dwarf lemur
Cheirogaleus medius: lesser dwarf lemur

Microcebus
Microcebus coquereli: giant mouse lemur
Microcebus murinus: grey mouse-lemur
Microcebus rufus: red mouse lemur
Microcebus myoxinus: pygmy mouse lemur (formerly considered a synonym of *M.murinus*, but see Schmid and Kappeler 1994)

Phanerinae

Phaner
Phaner furcifer (subspecies are: *P.f.furcifer, pallescens, parienti, electromontis*)

Lemuridae (This family may not be monophyletic; *Varecia* may perhaps be a sister-group of the Indridae.)

Eulemur (synonym *Petterus*)
Eulemur coronatus: crowned lemur
Eulemur fulvus: brown lemur (subspecies are: *E.f.fulvus* (= *mayottensis*), *rufus, sanfordi, albifrons*)

---

[1] Note: Taxa ending in -oidea are ranked as superfamilies, those in -idae as families, in -inae as subfamilies, in -ini as tribes. Taxa above the family level are left unranked, unless otherwise indicated. Some species have no established vernacular names. Often an entire genus has a well-known vernacular name (e.g. sportive lemurs for *Lepilemur*). For distributional data, readers are referred to Groves (1993).

Eulemur collaris (formerly considered a subspecies of *E.fulvus*; but see Rumpler 1993)

Eulemur albocollaris (formerly considered a subspecies of *E.fulvus*; but see Rumpler 1993)

Eulemur macaco: black lemur (subspecies are: *E.m.macaco, flavifrons*)

Eulemur mongoz: mongoose lemur

Eulemur rubriventer. red-bellied lemur

Hapalemur

Hapalemur aureus: golden bamboo lemur

Hapalemur griseus: grey gentle lemur; lesser bamboo lemur (subspecies are: *H.g.griseus, alaotrensis, meriodionalis, occidentalis.* Groves 1989 suggested that *H.g.alaotrensis* may be a full species.)

Hapalemur simus: broad-nosed gentle lemur; greater bamboo lemur

Lemur

Lemur catta: ring-tailed lemur

Varecia

Varecia variegata: ruffed lemur
   (subspecies are: *V.v.variegata, rubra*)

Megaladapidae (Rumpler 1993 calls this Lepilemuridae)

Lepilemur: sportive lemurs; weasel lemurs

Lepilemur dorsalis

Lepilemur edwardsi

Lepilemur leucopus

Lepilemur microdon

Lepilemur mustelinus

Lepilemur ruficaudatus

Lepilemur septentrionalis

Indridae (for the correct spelling of this name, see Jenkins 1987)

Avahi: woolly indri; avahi (The name *Lichanotus* is sometimes, incorrectly, used for this genus.)

Avahi laniger

Avahi occidentalis (formerly considered a subspecies of *A.laniger*, but see Rumpler et al. 1990)

Indri

Indri indri Indri

Propithecus

Propithecus diadema: diademed sifaka (subspecies are: *P.d.diadema* (= *edwardsi, holomelas*), *candidus, perrieri*)

Propithecus tattersalli: Tattersall's sifaka
Propithecus verreauxi: Verreaux's sifaka (subspecies are: *P.v.verreauxi (= majori),
deckeni, coronatus, coquereli*)

Daubentoniidae (It is uncertain whether this family or the Loridae/Galagidae pair is sister to other strepsirrhines.)

Daubentonia
Daubentonia madagascariensis: aye-aye

Loridae (usually called Lorisidae, but see Jenkins 1987)

Arctocebus: angwantibo
Arctocebus aureus (usually considered a subspecies of *A.calabarensis,* but see Groves 1989)
Arctocebus calabarensis

Loris
Loris tardigradus: slender loris (subspecies are: *L.t.tardigradus, grandis (= nordicus), nycticeboides*)

Nycticebus
Nycticebus coucang: greater slow loris (subspecies are: *N.c.coucang, menagensis, javanicus, bengalensis*)
Nycticebus pygmaeus: lesser slow loris (*N.intermedius* is a synonym of *N.pygmaeus*)

Perodicticus
Perodicticus potto: potto (subspecies uncertain; in need of revision)

Pseudopotto
Pseudopotto martini (see Schwartz, 1996)

Galagonidae (Usually called Galagidae, but see Jenkins 1987. The entire family is under revision, with several new species to be described: see Bearder et al. 1994. This family is sister-group to Loridae, but is conveniently given separate family standing.)

Euoticus: needle-clawed bushbaby
Euoticus elegantulus
Euoticus pallidus (usually considered a subspecies of *E.elegantulus,* but see Groves 1989)

Galago
Galago alleni Allen's bushbaby (subspecies uncertain, in need of revision)
Galago gallarum (usually considered a subspecies of *G.senegalensis,* but see Nash et al. 1989)
Galago matschiei (often placed in *Euoticus;* usually called *G.inustus,* but see Nash et al. 1989)
Galago moholi: moholi bushbaby (usually considered a subspecies of *G.senegalensis,* but see Jenkins 1987)

Galago senegalensis: Senegal bushbaby; lesser bushbaby (subspecies uncertain, in need of revision)

Galagoides
Galagoides demidoff: pygmy bushbaby; Prince Demidoff's bushbaby (usually called *demidovii,* but see Jenkins 1987; subspecies uncertain, in need of revision)
Galagoides zanzibaricus (subspecies are: *G.z.zanzibaricus, granti*)

Otolemur: thick-tailed bushabies; greater galagos
Otolemur crassicaudatus
Otolemur garnettii

Suborder Haplorrhini (for the correct spelling of this name, see Jenkins 1987)

Infraorder Tarsiiformes

Tarsiidae

Tarsius
Tarsius bancanus: Horsfield's tarsier; western tarsier
Tarsius dianae
Tarsius pumilus
Tarsius sangirensis (usually considered a subspecies of *T.spectrum,* but see Feiler 1990)
Tarsius spectrum: spectral tarsier
Tarsius syrichta: Philippine tarsier

Infraorder Simiiformes (usually called Anthropoidea, but see Hoffstetter 1982)

Platyrrhini

Callitrichidae

Callimico
Callimico goeldii: Goeldi's marmoset; Goeldi's monkey

Callithrix (*C.argentata* group of de Vivo 1991)
Callithrix argentata: silvery marmoset
Callithrix chrysoleuca (usually considered a subspecies of *C.humeralifer,* but see de Vivo 1991; Rylands et al. 1993 do not accept it as a full species)
Callithrix emiliae (usually considered a synonym of *C.argentata melanura,* but see de Vivo 1991; Rylands et al. 1993 consider that de Vivo misidentified the taxon, but that it is still a distinct species)
Callithrix humeralifera: Santarem marmoset (usually called *humeralifer,* but see de Vivo 1991)
Callithrix intermedia (usually considered a subspecies of *C.humeralifer,* but see de Vivo 1991; Rylands et al. 1993 consider it a subspecies of *C.argentata*)

Callithrix leucippe (usually considered a subspecies of *C.argentata*, but see de Vivo 1991)

Callithrix mauesi (described by Mittermeier et al. 1992)

Callithrix melanura (usually considered a subspecies of *C.argentata*, but see de Vivo 1991)

Callithrix nigriceps (described by Ferrari and Lopes 1992; Rylands et al. 1993 suggest it is a subspecies of *C.argentata*)

(C.jacchus group of de Vivo 1991)

Callithrix aurita

Callithrix flaviceps: buffy-headed marmoset (considered a subspecies of *C.aurita* by Rylands et al. 1993)

Callithrix geoffroyi

Callithrix jacchus: common marmoset

Callithrix kuhlii (considered by de Vivo 1991 to be only an intergrade between *C.jacchus* and *C.penicillata*)

Callithrix penicillata (probably a subspecies of *C.jacchus*, with which it intergrades widely; de Vivo 1991)

(C.pygmaea group: often referred to a separate genus, *Cebuella*)

Callithrix pygmaea: pygmy marmoset

Leontopithecus

Leontopithecus caissara: Superagui lion tamarin

Leontopithecus chrysomela: golden-headed lion tamarin

Leontopithecus chrysopygus: golden-rumped lion tamarin

Leontopithecus rosalia: golden lion tamarin

Saguinus

(hairy-faced group of Hershkovitz 1966)

Saguinus fuscicollis: saddleback tamarin (subspecies are: *S.f.fuscicollis, acrensis, avilapiresi, crandalli, cruzlimai, fuscus, illigeri, lagonotus, leucogenys, melanoleucus, primitivus, weddelli*; Rylands et al. 1993 consider that *S.f.melanoleucus* is probably a distinct species, with *acrensis* and *crandalli* as subspecies)

Saguinus imperator: emperor tamarin (subspecies are: *S.i.imperator, subgrisescens*)

Saguinus labiatus: red-bellied tamarin (subspecies are: *S.l.labiatus, thomasi*)

Saguinus midas: golden-handed tamarin (subspecies are: *S.m.midas, niger*; but Rylands et al. 1993 consider that *S.m.niger* is probably a distinct species)

Saguinus mystax: moustached tamarin (subspecies are: *S.m.mystax, pluto*)

Saguinus nigricollis. black-and-red tamarin (subspecies are: *S.n.nigricollis, graellsi*)

Saguinus tripartitus

(bare-faced group of Hershkovitz 1966)

Saguinus bicolor: pied tamarin (subspecies are: *S.b.bicolor, ochraceus, martinsi*)

Saguinus geoffroyi

Saguinus inustus: mottle-faced tamarin

Saguinus leucopus: white-footed tamarin

Saguinus oedipus: cottontop tamarin; pinché

Cebidae (This family is certainly polyphyletic, but there are disagreements how to break it up.)

Alouattinae

Alouatta
Alouatta belzebul: red-handed howler monkey (subspecies are: *A.b.belzebul, discolor, nigerrima, ululata*)
Alouatta caraya: black howler monkey
Alouatta coibensis (subspecies are: *A.c.coibensis, trabeata*)
Alouatta fusca: brown howler monkey
Alouatta palliata: mantled howler monkey (subspecies are uncertain, in need of revision)
Alouatta pigra
Alouatta sara
Alouatta seniculus: red howler monkey(subspecies are uncertain, in need of revision)

Aotinae

Aotus: night monkeys; owl monkeys; douroucouli (All members of this genus have traditionally been assigned to one species, *A.trivirgatus*, but there are very clearly several. Hershokvitz 1983 has nine, which Ford 1994 reduces to "seven, probably five")
(Red-necked group of Hershkovitz 1983)
Aotus azarae (*A.a.boliviensis* is transferred by Ford (1994) to *A.infulatus*, but the latter may in any case be conspecific with *A.azarae:* Ford 1994.)
Aotus infulatus (subspecies are *A.i.infulatus, boliviensis*; but may be subspecies of *A.azarae:* Ford 1994)
Aotus miconax
Aotus nancymae (may be a subspecies of *A.miconax* according to Ford 1994)
Aotus nigriceps
(Grey-necked group of Hershkovitz 1983)
Aotus trivirgatus
Aotus vociferans (subspecies are: *A.v.vociferans, lemurinus, griseimembra, brumbacki,* according to Ford 1994, and, by inference, *hershkovitzi*)

Atelinae

Ateles
Ateles belzebuth: long-haired spider monkey (subspecies are: *A.b.belzebuth, hybridus*; Froehlich et al. 1991 suggest that *A.b.hybridus* may be a separate species)
Ateles chamek
Ateles fusciceps (subspecies are: *A.f.fusciceps, robustus*)
Ateles geoffroyi: Geoffroy's spider monkey (subspecies uncertain, in need of revision)
Ateles marginatus
Ateles paniscus: black spider monkey

Brachyteles

Brachyteles arachnoides: woolly spider monkey; muriqui (subspecies are: *B.a. arachnoides, hypoxanthus*. The differences are considerable, and they are possibly distinct species.)

Lagothrix

Lagothrix flavicauda: Hendee's woolly monkey; yellow-tailed woolly monkey

Lagothrix lagotricha: Humboldt's woolly monkey (subspecies are: *L.l.lagotricha, cana, lugens, poeppigii*)

Callicebinae

Callicebus

Callicebus brunneus (Groves 1992 suggests that this may be a subspecies of *C.moloch*)

Callicebus caligatus (Groves 1992 raises the possibility that this could be a colour morph of *C.cupreus cupreus*)

Callicebus cinerascens (Groves 1992, suggests that this may be a subspecies of *C.moloch*)

Callicebus cupreus (subspecies are: *C.c.cupreus, discolor, ornatus*)

Callicebus donacophilus (subspecies are: *C.d.donacophilus, pallescens*)

Callicebus dubius (Groves 1992 proposes that this could be a colour morph of *C.cupreus cupreus*)

Callicebus hoffmannsi (subspecies are: *C.h.hoffmannsi, baptista*; Groves 1992 suggests that these may actually be subspecies of *C.moloch*)

Callicebus modestus

Callicebus moloch: dusky titi

Callicebus oenanthe

Callicebus olallae

Callicebus personatus: masked titi (subspecies are: *C.p.personatus, melanochir, barbarabrownae, nigrifrons*)

Callicebus torquatus: widow monkey; collared titi (subspecies are: *C.t.torquatus, medemi, purinus, regulus, lugens, lucifer*)

Cebinae

Cebus

*Cebus:*white-fronted capuchin (subspecies uncertain, in need of revision)

Cebus apella: tufted capuchin (subspecies uncertain, in need of revision; *C.a.xanthosternos* is very distinct and may be a distinct species according to Mittermeier et al. 1988)

Cebus capucinus: white-throated capuchin (subspecies uncertain, in need of revision)

Cebus kaapori (described by Queiroz 1992)

Cebus olivaceus: weeper capuchin (often referred to as *C.griseus* or *C.nigrivittatus*; subspecies uncertain, in need of revision)

Saimiri
Saimiri boliviensis: black-capped squirrel S. monkey (subspecies are: *S. b.boliviensis, peruviensis, jaburuensis, pluvialis*)
Saimiri oerstedii (subspecies are: *S.o.oerstedii, citrinellus*)
Saimiri sciureus: common squirrel monkey (subspecies are: *S.s.sciureus, albigena, cassiquiarensis, macrodon*)
Saimiri ustus
Saimiri vanzolinii: black squirrel monkey

Pitheciinae

Cacajao
Cacajao calvus: bald uakari (subspecies are: *C.c.calvus, ucayalii, rubicundus, novaesi*)
Cacajao melanocephalus: black-headed uakari (subspecies are: *C.m.melanocephalus, ouakary*)

Chiropotes
Chiropotes albinasus: white-nosed saki
Chiropotes satanas: bearded saki (subspecies are: *C.s.satanas, chiropotes, utahickae*)

Pithecia
Pithecia aequatorialis
Pithecia albicans
Pithecia irrorata (subspecies are: *P.i.irrorata, vanzolinii*)
Pithecia monachus: monk saki (subspecies are: *P.m.monachus, milleri*)
Pithecia pithecia: white-faced saki (subspecies are: *P.p.pithecia, chrysocephala*)

Catarrhini

Cercopithecoidea

Cercopithecidae

Cercopithecinae

Allenopithecus
Allenopithecus nigroviridis: Allen's swamp monkey

Cercocebus (This genus does not include *"C."albigena* or *aterrimus,* which belong to *Lophocebus.*)
Cercocebus agilis: agile mangabey (subspecies are: *C.a.agilis, chrysogaster,* and *sanjei;* these may be distinct species)
Cercocebus galeritus: Tana river mangabey
Cercocebus torquatus: collared mangabey; red-capped mangabey (subspecies are: *C.t.torquatus, lunulatus, atys; C.atys* may be a distinct species (sooty mangabey), with *lunulatus* a subspecies)

Cercopithecus (This genus does not include *"C."talapoin, patas* or *aethiops.* Species are arranged into groups: (1) *cephus* group: *C.cephus, ascanius, petaurista, erythrogaster, erythrotis, sclateri;* (2) *mona* group: *C.mona, campbelli, pogonias, wolfi;* (3) *diana;* (4) *dryas;* (5) *hamlyni;* (6) *lhoesti* group: *C.lhoesti, preussi, solatus;* (7) *mitis* group: *C.mitis, nictitans;* (8) *neglectus.*)

Cercopithecus ascanius: redtailed monkey (subspecies are: *C.a.ascanius, atrinasus, katangae, whitesidei, schmidti*)

Cercopithecus campbelli: Campbell's mona monkey (subspecies are: *C.c.campbelli, lowei*)

Cercopithecus cephus: moustached monkey (subspecies are: *C.c.cephus, cephodes*)

Cercopithecus diana: diana monkey (subspecies are: *C.d.diana, roloway*)

Cercopithecus dryas: Salonga monkey

Cercopithecus erythrogaster: red-bellied guenon

Cercopithecus erythrotis: red-eared guenon (subspecies are: *C.e.erythrotis, camerunensis*)

Cercopithecus hamlyni: Hamlyn's monkey; owl-faced monkey (subspecies are: *C.h.hamlyni, kahuziensis*)

Cercopithecus lhoesti: L'Hoest's monkey

Cercopithecus mitis: blue monkey; Sykes monkey (This species consists of two subspecies-groups, as follows: (1) *mitis* group (blue monkeys); subspecies are: *C.m.mitis, boutourlinii, doggetti, elgonis, kandtii, heymansi, opisthostictus, schoutedeni, stuhlmanni.* (2) *albogularis* group (Sykes monkeys); subspecies uncertain, in need of revision.)

Cercopithecus mona: common mona monkey

Cercopithecus neglectus: De Brazza monkey

Cercopithecus nictitans: putty-nosed monkey; greater spot-nosed monkey (subspecies are: *C.n.nictitans, martini, stampflii*)

Cercopithecus petaurista: lesser spot-nosed monkey (subspecies are: *C.p.petaurista, buettikoferi*)

Cercopithecus pogonias: crested mona monkey (subspecies are: *C.p.pogonias, grayi, nigripes*)

Cercopithecus preussi: Preuss's monkey

Cercopithecus sclateri: Sclater's guenon

Cercopithecus solatus: sun-tailed monkey

Cercopithecus wolfi: Wolf's mona monkey (subspecies are: *C.w.wolfi, elegans, pyrogaster, denti*)

Chlorocebus (traditionally included in *Cercopithecus,* but more closely related to *Erythrocebus*)

Chlorocebus aethiops: vervet, grivet, tantalus, green monkeys (This species consists of five subspecies-groups, as follows: (1) *aethiops* group (vervets); (2) *djamdjamensis;* (3) *pygerythrus* group (grivets); (4) *tantalus* group (tantalus); (5) *sabaeus* (green monkey); these groups may be distinct species.)

Erythrocebus
Erythrocebus patas: Patas monkey

Lophocebus (often included in *Cercocebus*, but the two are not sister-groups)

Lophocebus albigena: grey-cheeked mangabey (subspecies are: *L.a.albigena, osmani, johnstoni*)

Lophocebus aterrimus: black mangabey (subspecies are: *L.a.aterrimus, opden-boschi*)

Macaca (This genus is perhaps nonmonophyletic.)

Macaca arctoides: stump-tailed macaque; bear macaque

Macaca assamensis: Assam macaque (subspecies are: *M.a.assamensis, pelops*)

Macaca cyclopis: Formosan rock macaque

Macaca fascicularis: crab-eating or long-tailed macaque; "cyno" (subspecies are: *fascicularis, aurea, philippinensis, umbrosa, fusca, lasiae, atriceps, condorensis, tua, karimondjawae;* see Fooden 1996)

Macaca fuscata: Japanese macaque (subspecies are: *M.f.fuscata, yakui*)

Macaca maura: moor macaque

Macaca mulatta: rhesus monkey

Macaca nemestrina: pigtailed macaque (subspecies are: *M.n.nemestrina, leonina*)

Macaca nigra: Sulawesi crested macaque; "Celebes black ape" (subspecies are: *M.n.nigra, nigrescens;* but *M.n.nigrescens* may be a distinct species)

Macaca ochreata: booted macaque; ashy-black monkey (subspecies are: *M.o. ochreata, brunnescens*)

Macaca pagensis: Mentawai Islands macaque (subspecies are *M.p.pagensis, siberu;* the latter inadvertently described by Fuentes and Olson 1995)

Macaca radiata: Bonnet macaque (subspecies are: *M.r.radiata, diluta;.* probably subspecies of *M.sinica*)

Macaca silenus: lion-tailed macaque

Macaca sinica: toque macaque (subspecies are: *M.s.sinica, aurifrons*)

Macaca sylvanus: Barbary macaque; Barbary "ape"

Macaca thibetana: Tibetan macaque

Macaca tonkeana. Tonkean macaque (subspecies are: *M.t.tonkeana, hecki;* but *M.t.hecki* may be a distinct species)

Mandrillus

Mandrillus leucophaeus: drill (subspecies are *M.l.leucophaeus, poensis*)

Mandrillus sphinx: mandrill

Miopithecus

Miopithecus talapoin: Angolan talapoin

(*M.species* undescribed: Gabon talapoin)

Papio

Papio hamadryas: sacred, mantled or hamadryas baboon

Papio papio: Guinea baboon

Papio anubis: olive baboon; anubis baboon

Papio cynocephalus: yellow baboon (subspecies are *P.c.cynocephalus, kindae*)

Papio ursinus: Chacma baboon (subspecies are *P.u.ursinus, griseipes, ruacana*)

Theropithecus
Theropithecus gelada: gelada (subspecies are *T.g.gelada, obscurus*)

Colobinae (Groves 1989 ranks this as a full family, Colobidae)

Colobus
Colobus angolensis: Angola colobus (subspecies are *C.a.angolensis, cottoni, cordieri, ruwenzorii, prigoginei* (according to Colyn 1991), *palliatus*)
Colobus guereza: mantled colobus; mantled guereza (subspecies are *C.g.guereza, gallarum, dodingae, percivali, occidentalis, matschiei, kikuyuensis, caudatus*; but these are in need of revision)
Colobus polykomos: king colobus
Colobus satanas: black colobus
Colobus vellerosus: ursine colobus (has an extensive hybrid zone with *C.polykomos*, but the two are not sister species: see Groves et al. 1993)

Nasalis
Nasalis concolor: simakobu (subspecies are *N.c.concolor, siberu*)
Nasalis larvatus: proboscis monkey

Presbytis
Presbytis comata: Javan surili (subspecies are *P.c.comata, frederici*)
Presbytis femoralis: banded leaf-monkey (subspecies are *P.f.femoralis, siamensis, robinsoni, chrysomelas, natunae, paenulata, percura, catemana, cana*)
Presbytis frontata: white-fronted langur
Presbytis hosei: Hose's langur; Bornean grey langur (subspecies are *P.h.hosei, sabana, canicrus*)
Presbytis melalophos: Sumatran surili (subspecies are *P.m.melalophos, mitrata, bicolor, sumatrana*)
Presbytis potenziani: Mentawai Island langur; joja (subspecies are *P.p.potenziani, siberu*)
Presbytis rubicunda: maroron leaf-monkey (subspecies are *P.r.rubicunda, ignita, rubida, carimatae*)
Presbytis thomasi: Thomas's leaf monkey

Procolobus
Procolobus badius: western red colobus (subspecies are *P.b.badius, temmincki, waldronae*)
Procolobus pennantii: central red colobus (subspecies are *P.p.pennantii, bouvieri, tholloni, oustaleti, ellioti, parmentierorum, foai, semlikiensis, tephrosceles, gordonorum, kirkii*; this could actually be broken into several species)
Procolobus preussi: Preuss's red colobus
Procolobus rufomitratus: Tana river red colobus
Procolobus verus: olive colobus

Pygathrix
Pygathrix nemaeus: red-legged douc langur
Pygathrix nigripes: black-legged douc langur

Rhinopithecus

Rhinopithecus avunculus: Tonkin snub-nosed monkey

Rhinopithecus bieti: black or Yunnan snub-nosed monkey

Rhinopithecus brelichi: grey or Gweizhou snub-nosed monkey

Rhinopithecus roxellana: golden snub-nosed monkey (subspecies are *R.r.roxellana, hubeiensis, qinlingensis*)

Semnopithecus

Semnopithecus entellus: Indian grey langur; sacred or hanuman langur (Subspecies can be placed in four groups, according to Napier 1985: (1) *entellus* group: *S.e.entellus, anchises;* (2) *schistaceus* group: *P.e.schistaceus, hector, ajax;* (3) *hypoleucos* group: *P.e.hypoleucos, aeneas, achates, iulus, dussumieri;* (4) *priam* group: *S.e.priam, elissa, thersites.*)

Trachypithecus (Brandon-Jones 1995 synonymises this with *Semnopithecus*)

Trachypithecus auratus: Javan lutung (subspecies are *T.a.auratus, mauritius*)

Trachypithecus cristatus: silvery leaf-monkey (subspecies are *T.c.cristatus, barbei, germaini, caudalis*)

Trachypithecus delacouri

Trachypithecus ebenus (recently described (but as a subspecies of *T.auratus*) by Brandon-Jones 1995)

Trachypithecus francoisi: François' leaf monkey (subspecies are *T.f.francoisi, hatinhensis, leucocephalus, poliocephalus, laotum;* some of these are distinct species or belong to other species according to Brandon-Jones 1995)

Trachypithecus geei: Gee's golden langur

Trachypithecus johnii: Nilgiri langur

Trachypithecus obscurus: dusky or spectacled langur (subspecies are *T.o.obscurus, flavicauda, halonifer, carbo, styx, sanctorum, seimundi*)

Trachypithecus phayrei: Phayre's langur (subspecies are *T.p.phayrei, crepuscula, shanica*)

Trachypithecus pileatus: (capped Langur) (subspecies are *T.p.pileatus, brahma, tenebrica, durga, shortridgei*)

Trachypithecus vetulus: purple-faced leaf monkey (often referred to as "*Presbytis senex*"; subspecies are *T.v.vetulus, nestor, monticola, philbricki*)

Hylobatidae

Hylobates (The four subgenera may be full genera)

Subgenus *Hylobates*

Hylobates agilis: agile gibbon (subspecies are *H.a.agilis, albibarbis*)

Hylobates klossii: Kloss gibbon; bilou

Hylobates lar: white-handed gibbon; Lar gibbon (subspecies are: *H.l.lar, vestitus, entelloides, carpenteri, yunnanensis*)

Hylobates moloch: silvery gibbon

Hylobates muelleri: Müller's Bornean gibbon (subspecies are: *H.m.muelleri, funereus, abbotti;* these may be subspecies of *H.agilis*)

Hylobates pileatus: pileated gibbon

Subgenus *Nomascus*
Hylobates concolor: concolor gibbon; black crested gibbon (subspecies are *H.c.concolor, hainanus, jingdongensis, furvogaster, lu*)
Hylobates leucogenys: white-cheeked gibbon (subspecies are *H.l.leucogenys, siki*)
Hylobates gabriellae: red-cheeked gibbon

Subgenus *Bunopithecus*
Hylobates hoolock: Hoolock gibbon (subspecies are *H.h.hoolock, leuconedys*)

Subgenus *Symphalangus*
Hylobates syndactylus: siamang

Hominidae

Homininae

Gorilla
Gorilla gorilla: gorilla (subspecies are *G.g.gorilla, graueri, beringei*; *G.g.beringei* may be a distinct species, with *graueri* a subspecies)

Homo
Homo sapiens: human

Pan
Pan paniscus: pygmy chimpanzee; bonobo
Pan troglodytes: common chimpanzee (subspecies are *P.t.troglodytes, schweinfurthii, verus*; but they are in need of revision)

Ponginae

Pongo
Pongo pygmaeus: orangutan (subspecies are *P.p.pgymaeus, wurmbii, abeli*; *P.p.abeli* may be a distinct species)

References

Bearder SK, Honess PE, Ambrose L. Species diversity among galagos with special reference to mate recognition. In: Alterman L, Doyle GA, Izard MK (eds) Creatures of the dark: the nocturnal prosimians. Plenum, New York, pp 1–21, 1994

Bernstein IS. Naturally occurring primate hybrid. Science 154:1559–1560, 1966

Brandon-Jones D. A revision of the Asian Pied Leaf-monkeys (Mammalia: Cercopithecidae: superspecies *Semnopithecus auratus*) with a description of a new subspecies. Raffles Bull Zool 43:3–43, 1995

Cavalieri P, Singer P (eds) The great ape project: equality beyond humanity. Fourth Estate, London, 1993

Christoffersen ML. Cladistic taxonomy, phylogenetic systematics, and evolutionary ranking. Syst Biol 44:440–454, 1995

Colyn M. L'importance zoogéographique du bassin du Fleuve Zaïre pour la spéciation. Annalen der koninklijk Museum voor Midden-Afrika, Zool Wetenschappen, Tervuren, Belgium 264:1–250, 1991

Cracraft J. Speciation and its ontology: the emprical consequences of alternative species concepts for understanding patterns and processes of differentiation. In: Otte D, Endler JA (eds) Speciation and its consequences, Sinauer, Sunderland, MA, pp 28–59, 1989

de Vivo M. Taxonomia de Callithrix Erxleben, 1777 (Callitrichidae, Primates). Fundaçåo Biodiversitas, Belo Horizonte, 1991

Easteal S, Collett C, Betty D. The mammalian molecular clock. Springer, Berlin Heidelberg New York, 1995

Feiler A. Über die Säugetiere der Sangihe- und Talaud-Inseln der Beitrag A.B.Meyers für ihre Erforschung. Zool Abh Staatl Mus Tierkd Dresden 46:75–94, 1990

Ferrari SF, Lopes MA. A new species of marmoset from Western Brazilian Amazonia. Goeldiana Zool 12:1–13, 1992

Fooden, J. Systematic review of Southeast Asian Longtail Macaques, *Macoca fascicularis* (Raffles, [1821]). Fieldiana Zool No. 81, vit 206 pp

Ford SM. Taxonomy and distribution of the owl monkey. In: DeBoer L (ed) *Aotus*: the owl monkey. Academic, New York, pp 1–57, 1994

Froehlich JW, Supriatna J, Froehlich PH. Morphometric analyses of *Ateles*: systematic and biogeographic implications. Am J Primatol 25:1–22, 1991

Fuentes A, Olson M. Preliminary observations and status of the Pagai macaque. Asia Primates 4 (4):1–4, 1995

Groves CP. A theory of human and primate evolution. Oxford University Press, Oxford, 1989

Groves CP. Book review: titis, new world monkeys of the genus *Callicebus*, by Philip Hershkovitz. Int J Primatol 13:111–112, 1992

Groves CP. Order primates. In: Wilson DE, Redder D-A (eds) Mammal species of the world: a taxonomic and geographic reference. Smithsonian Institution, Washington, pp 243–277, 1993

Groves CP, Angst R, Westwood C. The status of *Colobus polykomos dollmani* Schwarz. Int J Primatol 14:573–586, 1993

Hennig W. Phylogenetic systematics. University of Illinois Press, Urbana, 1966

Hershkovitz P. Taxonomic notes on tamarins, genus *Saguinus*, with descriptions of four new forms. Folia Primat 4:381–395, 1966

Hershkovitz P. Two new species of night monkeys, genus *Aotus* a preliminary report on Aotus taxonomy. Am J Primatol 4:209–243, 1983

Hoffstetter R. Les Primates Simiiformes (= Anthropoidea) (compréhension, phylogénie, histoire biogéographique). Ann Paleont (Vert Invert) 68:241–290, 1982

Jenkins PD. Catalogue of primates in the British Museum (natural history), part IV, 1987

Mayr E. Principles of systematic zoology. McGraw-Hill, New York, 1969

Mittermeier RA, Rylands AB, Coimbra-Filho AF, da Fonseca GAB. Ecology and behavior of neotropical primates. WWF, Washington DC, 1988

Mittermeier RA, Schwarz M, Ayres, JM. A new species of marmoset from the Rio Maués Region, Stare of Amazonas, Central Brazilian Amazonia. Goeldiana Zool 14: 1–17, 1992

Napier PH. Catalogue of primates in the British Museum (natural history), part III, 1985

Nash LT, Bearder SK, Olson TR. Synopsis of *Galago* species characteristics. Int J Primatol 10:57–79, 1989

Queiroz HL. A new species of Capuchin Monkey, Genus Cebus Erxleben, 1777 from Eastern Brazilian Amazonia. Goeldiana Zool 15: 1–13, 1992

Rumpler Y. Die Klassifikation der Lemuren. In:Ceska V, Hoffmann H-U, Winkelsrater KH (eds) Lemuren im Zoo, Paul Parey, Berlin, pp 71–77, 1993

Rumpler Y, Warter S, Rabarivola C, Petter J-J, Dutrillaux B. Chromosomal evolution in Malagasy lemurs. XII. Chromosomal banding study of *Avahi laniger occidentalis*. Am J Primatol 21:307–316, 1990

Rylands AB, Coimbra-Filho AF, Mittermeier RA. Systematics, geographic distribution, and some notes on the conservation status of the Callithrichidae. In: Rylands AB (ed) Marmosets and tamarins: systematics, behaviour, and ecology. Oxford University Press, Oxford, pp 11–77, 1993

Schmid J, Kappeler PM. Sympatric mouse lemurs (*Microcebus* spp.) in Western Madagascar. Folia Primatol 63:162–170, 1994

Schwartz, JH. *Pseudopotto martini*: a new genus and species of extant lorisiform primate. Anthrop Pap Amer Mus nat Hist no. 78, 1996

Shoshani J, Groves CP, Simons EL, Gunnell GF. Primates phylogeny: morphological vs. molecular results. Mol Biol Evol (5:102–154), 1996

II Red Blood Cell Antigens

1 Introduction

W.W. SOCHA AND A. BLANCHER

The discovery of the blood groups in humans is one of the most important achievements of modern biology. At the beginning of this century, Karl Landsteiner described the ABO system (Landsteiner 1901), the knowledge of which enabled blood transfusion to become current practice, thus opening the way to the use of modern surgical methods. Ten years later, the presence of a red cell antigen similar to that found in humans was, for the first time, confirmed in a nonhuman species, namely in the chimpanzee (von Dungern and Hirszfeld 1913). This work points to the very early interest in the comparative serology of blood groups in an attempt to trace their evolutionary pathways. The systematic study of the homologues of human ABO groups in primate animals was undertaken by Karl Landsteiner and his pupil Philip Miller and reported in a series of articles published in 1925 (Landsteiner and Miller 1925a–c). The discovery of another major human blood group system, the M-N system, by Landsteiner and Levine (1927a,b) prompted investigations on the distribution of these newly defined antigens in the blood of nonhuman primates. It was again Landsteiner, this time jointly with Alexander Wiener, who carried out extensive experiments with ape and monkey blood using rabbit anti-human anti-M and anti-N reagents. They found out, through direct titrations and absorption-titrations, that not only are M and N factors recognizable on the red cells of the nonhuman primates but that immunization of rabbits with monkey blood yielded anti-M reagents equal in strength to the most potent anti-human M sera (Landsteiner and Wiener 1937; Wiener 1938). Serological analysis of various antibody fractions contained in anti-rhesus M sera led to the identification of a new, hitherto unknown, antigen of human red cells, the Rhesus factor; a milestone in human blood group serology (Landsteiner and Wiener 1940). Definition of this "monkey-type" antigen in human blood gave a strong new impulse to the study of blood groups in nonhuman primates using not only reagents originally prepared for typing human blood but also antisera specifically directed against antigens on the red cells of apes and monkeys. Both these approaches were applied in investigations conducted in a few specialized laboratories and resulted in the definition of a number of antigenic specificities and complex blood group systems in various species of anthropoid apes and monkeys.

The study of the blood groups of nonhuman primates, particularly comparative immunological investigations, was greatly facilitated by the creation of the

Molecular Biology and Evolution of Blood Group
and MHC Antigens in Primates
Blancher/Klein/Socha (Eds.)
© Springer-Verlag Berlin Heidelberg 1997

Laboratory for Experimental Medicine and Surgery in Primates (LEMSIP), at the New York University School of Medicine, through the initiative of Jan Moor-Jankowski. This laboratory, originally devoted to erythrocyte immunology, was organized to have access to an important colony of animals, which provided priceless materials for the study of various species of monkeys and anthropoid apes.

With the accumulation of serological observations on red cell antigens of humans, apes, monkeys and prosimians, it became obvious that not only individual antigens but entire blood group systems are shared, at least in part, by all primates including humans (Socha and Ruffié 1983). When one compares the blood group systems in monkeys and in humans, a distinction can be made between two categories of antigenic structures which are the products of sequences of DNA that have remained unchanged since early stages of evolution, i.e. *paleosequences,* or are found only in a single species or in a few closely related species and thus represent more recent sequences, i.e. *neosequences* (Ruffié et al. 1982). This distinction is not an exclusive attribute of red blood cell antigens; it is also observed within the MHC system. The latter observations are the basis of the concept of *transpecies evolution of polymorphism* proposed by Klein (1987).

Until recently, knowledge of red cell antigens in humans, anthropoid apes, monkeys and prosimians was based on serological tests which allowed only superficial and one-sided characterization of the red cell membrane, even though the arrival of monoclonal reagents helped distinguish minute differences among cross-reactive antigens of various species. The progress of molecular genetics methodologies has allowed identification of the genes encoding the structures responsible for blood group antigens or the enzymes responsible for the synthesis of some of those structures. The comparison of these sequences in humans and nonhuman primates allowed, for the first time, the study of the evolution of blood group genes at the molecular level and the identification of some of the mechanisms of gene diversification in the course of evolution, including the mechanisms at work in their fixation in species and populations. Among those mechanisms is the role of infectious diseases as factors of selection, which will be discussed in a separate chapter.

Acknowledgements. The authors wish to thank Professor Jacques Ruffié, Director, Laboratory of Physical Anthropology, Collège de France, Paris, for his continuous interest in this work, his advice and his encouragement. We are deeply indebted to Ms. Sally Lasano for her devoted and invaluable help in organizing, performing, and documenting blood grouping tests on nonhuman primates. Without her contribution, the blood group serology of apes and monkeys would not have been the same.

References

Klein J. Origin of the major histocompatibility complex polymorphism: the transpecies hypothesis. Hum Immunol 19:155–162, 1987

Landsteiner K. Über Agglutinationserscheinungen normalen menschlichen Blutes. Wien Klin Wochenschr 14:1132–1134, 1901

Landsteiner K, Levine P. Further observations on individual differences of human blood. Proc Soc Exp Biol Med 24:941–942, 1927a

Landsteiner K, Levine P. A new agglutinable factor differentiating individual human bloods. Proc Soc Exp Biol Med 24:600–602, 1927b

Landsteiner K, Miller CP. Serological studies on the blood of primates. I. The differentiation of human anthropoid blood. J Exp Med 42:841–852, 1925a

Landsteiner K, Miller CP. Serological studies on the blood of primates. II. The blood groups in anthropoid apes. J Exp Med 42:853–862, 1925b

Landsteiner K, Miller CP. Serological studies on the blood of primates. III. Distribution of serological factors related to human isoaggluntinogens in the blood of lower monkeys. J Exp Med 42:863–870, 1925c

Landsteiner K, Wiener AS. On the presence of M agglutinogens in the blood of monkeys. J Immunol 33:19–25, 1937

Landsteiner K, Wiener AS. An agglutinable factor in human blood recognized by immune sera for rhesus blood. Proc Soc Exp Biol Med 43: 223–224, 1940

Ruffié J, Moor-Jankowski J, Socha WW. Immunogenetic evolution of primates. In: Chiarelli AB, Curriccini RS (eds) Advanced views in primate biology. Springer, Berlin Heidelberg New York, pp 28–34, 1982

Socha WW, Ruffié J. Blood groups of primates. Theory, practice, evolutionary meaning. Liss, New York, 1983

von Dungern E, Hirszfeld L. Über Vererbung gruppenspezifischer Strukturen des Blutes (II). Z Immunitätsforsch 8:526–530, 1913

Wiener AS. The agglutinogens M and N in anthropoid apes. J Immunol 3:11–18, 1938

2 The ABO, Hh and Lewis Blood Group in Humans and Nonhuman Primates

A. Blancher and W.W. Socha

Molecular Biology and Evolution of Blood Group
and MHC Antigens in Primates
Blancher/Klein/Socha (Eds.)
© Springer-Verlag Berlin Heidelberg 1997

Introduction: Discovery of the First Human Blood Groups

Karl Landsteiner (born in Vienna on June 14, 1868, and died in New York on June 25, 1943) discovered the first blood groups in humans at the age of 32 (Landsteiner 1901). Based on patterns of cross-agglutination observed when red blood cells of certain subjects were mixed with the sera of others, he distinguished two antigens that he called A and B. Depending on the presence or absence of those red cells traits, three types of blood, A, B, and O, were identified. The fourth type, the AB blood group, characterized by the simultaneous presence of both A and B factors, was described 1 year later by Decastello and von Sturli (1902). The sera were found to contain antibodies directed against antigen(s) absent from the individual's own red cells (Landsteiner's rule). The reciprocal relationships between the red cell antigens and the antibodies in the serum in the four main ABO blood groups are shown in Table 1.

Table 1. Serology and genetics of the A-B-O blood group system

Blood group	Reactions of the red cells with antiserum			Reactions of te serum with RBCs of group			Possible genotype(s)
	Anti-A	Anti-A_1	Anti-B	A	B	O	
O	−	−	−	+	+	−	00
A_1	+	+	−	−	+	−	$A^1A^1, A^1A^2,$ or A^1O
A_2	+	−	−	−	+	−	A^2A^2 or A^2O
B	−	−	+	+	−	−	BB or BO
A_1B	+	+	+	−	−	−	A^1B
A_2B	+	−	+	−	−	−	A_2B

Only the first three blood groups, A, B, and O were discovered by Landsteiner.
The fourth AB blood group was described one year later by Decastello and von Sturli (1902).

In 1908, Epstein and Ottenberg put forward the hypothesis that the A and B blood factors were hereditary, which soon thereafter was confirmed by von Dungern and Hirszfeld (1911). The model of heredity of these factors by a series of three allelic genes (two codominant, A and B, and a recessive, O) was proposed by Bernstein (1924, 1925) and subsequently confirmed by other investigators.

The discovery of ABO groups which, 30 years later (1930), earned Landsteiner the Nobel Prize, had important consequences. For the first time, traits had been discovered that were directly, or almost directly, controlled by clearly identified genes. Human genetics and physical anthropology were to be profoundly modified by these findings. Furthermore, the discovery of the human blood groups led to the comprehension of the compatibility rules important for the safety of blood transfusion.

With advancements in biochemistry, it became possible to characterize the biochemical structures responsible for the expression of ABO antigens on the red cell membrane. A and B antigens are oligosaccharides and their biosynthesis is related to other blood groups, such as Hh and Lewis (to be described later in this chapter). The red cell antigens ABH and Lewis, also detectable in secretions, are synthesized by the sequential addition of sugars to common precursor substances. They are also expressed in tissues and, therefore, have to be considered as histo-blood group antigens (Clausen and Hakomori 1989).

The ABO System in Humans

Following the description of the four main blood groups of humans (A, B, AB and O), several subgroups of the A and B antigens were discovered.

Human Subgroups of A: A_1 and A_2

The two principal subgroups of A are A_1 and A_2. Red blood cells of both these types are strongly agglutinated by anti-A reagents . However, only A_1 red blood cells are agglutinated with the anti-A_1 reagent prepared by absorption from group B (or group O) human serum or from the plant lectin *Dolichos biflorus*. The number of A determinants is much fewer on A_2 red blood cells than on A_1 cells. It was not clear whether the differences between the A determinants of A_1 and A_2 were of qualitative or quantitative nature. Only recent studies have resolved that question by showing that those differences are not only quantitative but, indeed, also qualitative: Type 3 (type 3 repetitive A) and type 4 (globo A) A antigens are present almost exclusively on A_1 red blood cells (Clausen and Hakomori 1989). Type 3 repetitive A is a glycolipid absent from A_2 red cell membrane but it constitutes more than one half of the total A glycolipid on A_1 erythrocytes. In the type 3 repetitive A, a terminal A trisaccharide built on a third type peripheral core structure [Gal(β1–3)GalNac(α)] is linked to an A trisaccharide built on a type 2 core structure [Gal(β1–4)GlcNac(β)]. The A determinant of globo A is carried on type 4 core sequence [Gal(β1–3)GalNac(β)]. Globo A is a minor component of the glycolipids of A_1 erythrocytes which was not detected in the glycolipid fraction extracted from A_2 erythrocytes. Another integral glycolipid of the red cell membrane of A_1 and A_2 phenotypes is built on a type 2 core [Gal(β1–4)GlcNac(β)] and a small quantity of A glycoplipids with type 1 peripheral core [Gal(β1–3)GlcNac(β)] are passively adsorbed on red cells. In their minimum energy conformations, the type 3 repetitive and type 4 globo A have a more bent shape than the type 1 and type 2 A active oligosaccharides (Nylhom et al. 1989). It was suggested that this difference in terms of conformation might explain the chain-type specificity observed for different monoclonal anti-A antibodies (Nylhom et al. 1989). In addition to the differences concerning the structure of antigen A carried by glycolipids, the A_1 red cells, unlike those of subgroup A_2, are very weakly, or not at all, agglutinated by anti-H reagents from *Ulex europaeus* or from eel serum. The red cells weakly agglutinated by anti-A_1 reagents and agglutinated by anti-H were called A_{int} (int for intermediate). Heredity of A_1 and A_2 phenotypes depends on two alleles at the ABO locus, A^1 and A^2, with A^1 being dominant over A^2.

Weak Subgroups of A and B in Humans

In addition to A_1 and A_2 subgroups, there are several weak A and B subtypes (Harmening 1993; Mollison et al. 1994). Description of all these variants can be found in Tables 2 and 3 (for more details, see Salmon et al. 1984).

The classification of these weak subgroups is based on three criteria:
- Degree of agglutination of red cells by anti-A or anti-B reagents
- Presence or absence of isoagglutinins in the serum
- Presence or absence of soluble A, B and H substances in the saliva.

Weak A phenotypes include A_3, A_x, A_{end}, A_m, A_y, and A_{el}. The A_3 phenotype red cells are characterized by a mixed-field pattern of agglutination with anti-A and anti A+B reagents. A_x red blood cells show weak agglutination with anti-A+B but they are not (or only weakly) agglutinated with anti-A reagent. The phenotype A_{end} corresponds to a true inherited mosaicism, with only 10% of A-positive cells. The presence of the A antigen on red cells of A_m, A_y, and A_{el} is demonstrable only by absorption-elution of anti-A. The saliva of A_m secretors contains normal quantities of substance A. Saliva of A_y secretors, by contrast, contains small quantities of substance A, while the saliva of A_{el} secretors contains only substance H. Although the distinction of these subclasses has some practical merits, they are genetically heterogeneous even within the same subgroup. For example, through kinetic assays of A-transferase, the group A_3 was split into three subtypes (Salmon et al. 1984).

A_3, A_x, A_{end}, A_m and A_{el} characters are inherited as ABO alleles. The mode of inheritance of the A_y phenotype implied that the developement of antigen A on red cells depended on the presence of a common gene Y (Wiener et al. 1957). The

Table 2. Characteristics of human weak A phenotypes

Phenotypes	Conventional determination of blood groups							Standard agglutination (% by Anti-A)	Soluble substances		Presence of α3N-acetylgalactosaminyl-T[a] in serum
	Red cell				Serum						
	Anti-B	Anti-A	Anti-A+B	Anti-H	A_1	A_2	B		A	H	
A_1	−	+++	+++	−	−	−	+++	95 ± 3	+	++	Optimum pH ≈ 6
A_2	−	+++	+++	+++	−	−	+++	90 ± 5	+	++	Optimum pH ≈ 7
A_3	−	Mixed population	Mixed population	+++	+ or −	−	+++	63 ± 10	+	++	Weak
A_x	−	(+)	++	+++	+ or −	−	+++	33 ± 10	(A_x)[b]	++	Not detectable
A_{end}	−	Mixed population	Mixed population	+++	+ or −	−	+++	10 ± 5	−	++	Not detectable
A_m	−	−	−	+++	−	−	+++	0	+	++	Yes, two types
A_y	−	−	−	+++	−	−	+++	0	Weak	++	Yes, very weak
A_{el}	−	−	−	+++	++	+ or −	+++	0	−	++	Not detectable

a α3N-acetylgalactosaminyltransferase.
b (A_x), substance A in saliva of A_x individuals is detected only by inhibition of the agglutination of A_x red cells.

Table 3. Characteristics of human weak B phenotypes

Phenotypes	Conventional determination of blood groups							Standard agglutination (% by Anti-B)	Soluble substances		Presence of α3-Dgalactosyl-T[a] in serum
	Red cell				Serum						
	Anti-B	Anti-A	Anti-A+B	Anti-H	A_1	A_2	B		B	H	
B_3	Mixed population	−	Mixed population	+++	+++	++	−	60 ± 15	+	++	Yes, 10%–100% of normal B
B_x	(+) or −	−	(+)	+++	+++	++	(+)	< 60	(B_x)[b]	++	Not detectable
B_m	−	−	−	+++	+++	++	−	0	+++	(+)	Yes, 50%–100% of normal B
B_{el}	−	−	−	+++	+++	++	Weak or absent	0	−	+++	Not detectable

a α3-Dgalactosyltransferase.
b (B_x)[c], substance B in saliva of B_x individuals is detected only by inhibition of the agglutination of B_x red cells.

presence of allele *y* in double dose is responsible for the A_y phenotype. *Y and y* genes segregate independently of the ABO locus.

Similar to weak variants of A, several weak B subgroups were described: B_3, B_x, B_m, and B_{el}. Although these weak B subgroups are less frequent than A subgroups, comparative studies of many samples led to a classification similar to that established for weak A. Thus, the subgroups B_3 B_x, B_m, and B_{el} are the counterparts of A_3, A_x, A_m, and A_{el}, respectively (Table 3).

Cis-AB

Simultaneous expression of A and B antigens that depends on a single genetic unit on one chromosome has been reported by Seyfried et al.(1964). The term *cis*-AB was introduced by Yamaguchi et al. (1965) to differentiate this genotype (*cis AB/O* or *A* or *B*), from the ordinary AB genotype, in which two alleles (*A* and *B*) are inherited in a *trans* manner. The term "*cis*-AB" is, therefore, restricted to those cases in which inheritance of the allele in the *cis* manner can be confirmed by a family study.

Bombay and Para-Bombay Phenotypes

The Bombay phenotype was first reported in 1952 (Bhende et al. 1952) and the story of this discovery was recalled by Bhatia (1987). The cells of this phenotype are not agglutinated by anti-A, anti-B, nor anti-H and are usually Le^a +. The sera of these individuals contain anti-A, anti-B and anti-H. Thus, they can be transfused only with red blood cells from Bombay donors. The Bombay individuals are classified as nonsecretors since their secretions lack H, A and B determinants. This rare phenotype led to the discovery of a new locus, independent of the ABO locus. The absence of the A, B and H antigens at the surface of the Bombay red cells is ascribed to the double inheritance of *h*, the recessive silent allele of *H*. The latter allele encodes for the fucosyltransferase responsible for the synthesis of the H substance, a precursor of the A or B determinants in erythroid cells (see below and Figs. 1, 2). Very rare individuals are known who do not express H or A and B on their red cells but show normal quantities of H (and, possibly, A and B substances) in their saliva (Solomon et al. 1965; Mollicone et al. 1988b; Le Pendu et al. 1983a, 1983b). Properties of this phenotype, called para-Bombay, suggest that expressions of H on the red cells and in saliva are, in fact, two independent traits, for which Oriol et al. (1981) proposed a two-locus model wherein *Se* and *H* loci correspond to two distinct structural genes, each encoding an $\alpha(1,2)$fucosyltransferase. The *H* gene is expressed in the erythroid lineage (Oriol et al. 1986), endothelial cells (Oriol et al. 1992), epidermis (Oriol et al. 1986; Clausen and Hakomori 1989) and in primary sensory neurons of the peripheral nervous system (Mollicone et al. 1986). In contrast, the expression of the *Se* gene is restricted to the epithelial cells of the digestive, respiratory and urinary tracts (Szulman 1977; Rouger et al. 1986). The segregation of the *H*

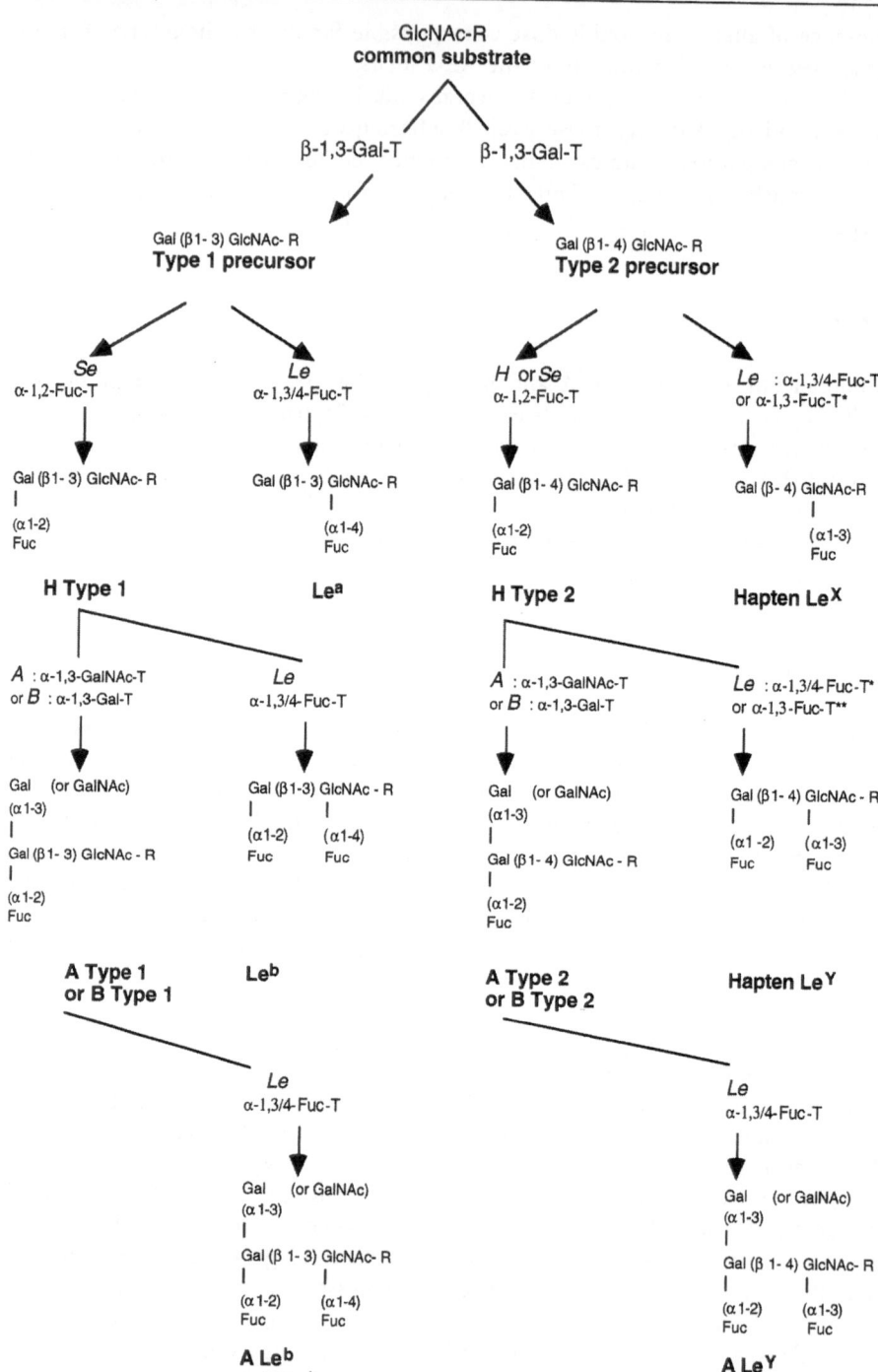

ABH of Type 2
Independent of *Se* & *Le*

ABH of Type 1 & 2
Controlled by *Se* & *Le*

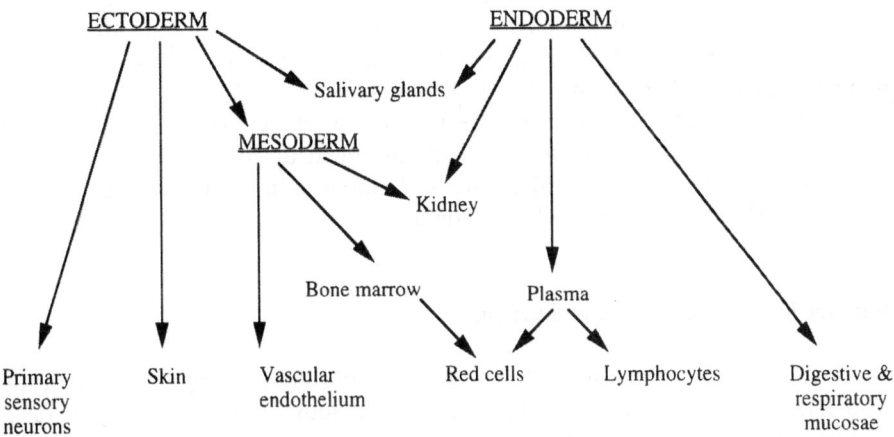

Fig. 2. Genetic control of the ABH and related antigens in human tissues. Origins from the three embryonic layers are indicated. There are some exceptions to this model, but most of the ABH molecules found in humans fit this general scheme. The exceptions include some of the deep glands of the stomach and the duodenum and some cells of the sweat and mammary glands

and *Se* genes in the H-deficient families of Reunion Island confirmed that the two genes are independent, though closely linked, on the long arm of chromosome 19 (Gerard et al. 1982; Ball et al. 1991). In conclusion, Bombay individuals are assumed to be of genotype *h/h; se/se* and are thus unable to express any α(1,2)fucosyltransferase.

Para-Bombay individuals have the genotype: *h/h; Se/se* (or *Se/Se*) and express a functional Se α(1,2)fucosyltransferase in some tissues, but not in the erythroid lineage. As a consequence, they do not express H at the surface of their red blood cells or endothelial cells but secrete the H substance in saliva.

Fig. 1. Biosynthetic pathways of ABH and Lewis antigens. *Gal*, D-galactose; *Fuc*, L-fucose; *GlcNAc*, N-acetyl-D-glucosamine; *GalNAc*, N-acetyl-D-galactose; *R*, remainder of molecule; *T*, transferase; *The enzyme encoded by *Le* (*FUT3*) has a much higher affinity for type 1 than for type 2 acceptor substrates. In plasma or at the surface of red cells, only type 1 Le^a and Le^b glycosphingolipids are found. Nevertheless, in epithelial cells of the digestive mucosa type 2 Le^X and Le^Y epitopes are synthesized under the control of *FUT3*. **The α(1,3)fucosyltransferase leading to the synthesis of hapten Le^X in leucocytes is encoded by FUT 4 or FUT7 genes (see text for more details)

ABO Blood Groups in Nonhuman Primates

Some 10 years after the discovery of the blood groups of humans, the presence of a red cell antigen similar to that found in humans was, for the first time, confirmed in a nonhuman species, namely, in the chimpanzee (von Dungern 1911). Systematic study of the homologues of human ABO groups in primate animals was undertaken by Landsteiner and Miller and reported in a series of articles published in 1925 (Landsteiner and Miller 1925a,b).

Most of the data presented here result from work carried out by Alexander Wiener until his death in 1976, and then continued by his collaborators at the Laboratory for Experimental Medicine and Surgery in Primates (LEMSIP) of the New York University School of Medicine.

Tests for ABO, H and Lewis Blood Groups

For details of blood grouping techniques in humans and nonhuman primates, the reader is referred to specialized articles and textbooks (Socha et al. 1972, 1984; Erskine and Socha 1978; Socha and Ruffié 1983).

Determination of the ABO Groups by Hemagglutination Tests of Nonhuman Primate Red Cells

The tests for the ABO groups are, by and large, carried out by the so-called *saline agglutination technique* in which the red cells to be blood-grouped are suspended in normal saline (or a low-ionic strength medium) and mixed with anti-A, anti-B and anti-H reagents to produce direct agglutination.

Anti-A and anti-B antibodies (isoagglutinins) normally occurring in the sera of most human and nonhuman primates (except those belonging to group AB) were used in the earliest tests for the ABO groups. The study of blood groups of apes and monkeys was initially directed toward the homologues of the human red cell antigens and employed reagents routinely used for the blood grouping of human blood. Since reagents of this type are, by and large, readily available commercially and do not require expensive, cumbersome, and time-consuming immunizations of primate animals, tests for homologues of human blood groups could be performed by any laboratory involved in human blood grouping work. In critically evaluating the results obtained it was particularly important to recognize possible interference of species-specific components present in most human sera and directed against red cells of nonhuman primates.

Elimination of heteroagglutinins by dilution is not practical with antisera of low titers and has the disadvantage of also diluting the group-specific antibody. A more reliable method consists of rendering specific the anti-A and anti-B sera by absorption with chimpanzee group O red cells. The thoroughly washed and packed chimpanzee red cells are mixed with two volumes of the undiluted anti-A or anti-B antiserum and incubated at room temperature for 60 min. After incubation, the tube is centrifuged and the supernatant serum removed immediately and tested with a fresh batch of washed chimpanzee group O cells to follow the progress of absorption. If necessary, absorption is repeated until all traces of heteroagglutinins are removed. The resulting reagent can be used full strength for ABO typing of chimpanzees, gibbons, orangutans, and gorillas. The specificity of the reactions of the so-prepared reagents with the red cells of apes can be ascertained by absorption experiments: for instance, absorption of the anti-A serum with group A chimpanzee (or gibbon or orangutan) red cells will render the reagent inactive against human group A red cells, while absorption of the serum with human A_1 cells will render the serum inactive against red cells of apes of group A.

Free of heteroagglutinins are various seed extracts (lectins) bestowed with type-specific hemagglutinating activity. The anti-A reagent made of saline extracts of lima beans (*Phaseolus vulgaris*) was used for ABO typing of chimpanzee blood (Wiener and Moor-Jankowski 1969). Another reagent of anti-A specificity, free of nonspecific heteroagglutinins and thus suitable for blood grouping primate red cells, was obtained from the snails *Helix pomatia* (Prokop et al. 1965) and *Achatina granulata* (Wiener et al. 1969).

Reagents derived from monoclonal antibodies are now available for blood typing. However, before using these reagents for tests on nonhuman primate blood, the specificity of each reagent has to be compared to that of reference polyclonal reagent tested against primate red cells of known phenotypes.

Testing of the Sera for Isoagglutinins Anti-A and Anti-B

As in human blood grouping practice, the results of tests for A and B agglutinogens on ape red cells are corroborated by tests on the animal's serum for the presence of anti-A and anti-B agglutinins, which, in most cases of nonhuman primate blood, follow the "Landsteiner's rule", i.e. are directed against agglutinogens that are absent from the red cells and from secretions.

This so-called reverse testing is done using, as test cells, human red cells of known types: O, A_1, A_2, and B. To eliminate possible interference of any nonspecific heteroagglutinins that might be present in ape serum, the serum is either tested diluted or, better, preabsorbed with human group O red cells.

The origin of the natural antibodies against antigens A and B was initially ascribed to the action of two pairs of linked genes, one leading to the expression of the A or B blood antigens, the other resulting in the production of the complementary anti-A and/or anti-B antibodies. At present, it is assumed that only the expression of ABO antigens is inherited while the serum isoagglutinins are of immune origin. The body's own antigens expressed on red cells or in secretions are tolerated by the immune system while it produces antibodies against the antigen(s) absent from the surface of red cells (and not secreted in saliva) (Wiener 1951). It was demonstrated that the appearance of these antibodies depends on exposure to exogenous antigens which mimic antigens A and B (Springer and Horton 1969). Various microorganisms or parasites express glycosaccharides which cross-react with anti-A and anti-B antibodies (Springer et al.

1969; Springer 1970; Smith et al. 1983). Exposure to these microorganisms after birth is responsible for the production of natural antibodies. Although all these antibodies are developed in response to external antigens, the functional properties of natural anti-A and anti-B antibodies produced by type O individuals differ from those produced by type A or type B individuals (Salmon et al. 1984). The serum of type O individuals contains antibodies reacting with A or B antigens (anti A + B). The natural antibodies differ from the "immune" type antibodies by their class and their affinities for the antigen (Filitti-Wurmser 1976; Salmon et al. 1984). The production of these immune antibodies is infrequent and can be induced by various immune stimulations (for a review, see Mollison 1979). The expected natural antibodies, complementary to the erythrocyte ABO antigen, are not always present in the serum (Springer and Tegtmeyer 1974). While it is relatively rare in humans, the absence of serum isoagglutinins is more frequently observed in nonhuman primates (see later in this chapter).

It was demonstrated that in humans the tolerance to the A and B histo-blood group antigens could not be broken by potent polyclonal lymphocyte B activation in vitro (Rieben et al. 1992), thus suggesting that the tolerance resulted either from a highly resistant anergic state or from clonal deletion of self-group-specific anti-A/B lymphocyte B cells. In another study, in vitro polyclonal activation of peripheral blood lymphocytes from type A and type B donors showed a significant number of clones producing anti-A or anti-B with apparent autoreactivity (Conger et al. 1993). However, although they were reactive against synthetic A or B trisaccharides, these pseudo autoantibodies were not absorbable with appropriate red cells. In certain pathological conditions, tolerance to A (or B) antigen can be broken (reviewed by Mollison 1979). It was possible to isolate in vitro a human hybridoma clone producing an auto-anti-A antibody (Inoue et al. 1989). This cell line was obtained by the fusion of lymphocytes from a lung cancer patient.

The tolerance to A/B antigens varies among species. For example, in the C57BL/10 mouse, the production of natural anti-A is sex-dependent. The pronounced production of these antibodies in females may have resulted from auto-immunization against an ovarian glycolipid carrying antigen A (Arend and Nijssen 1977).

Saliva Inhibition Tests for A, B, H and Lewis Group Substances

All apes (except for a single orangutan so far observed) and monkeys are secretors and their saliva, urine, seminal fluid, gastric juice, and amniotic fluid can be tested for the presence of A, B and H substances by the *hemagglutination inhibition technique*. In this method, the fluid investigated for the presence of A, B, H and Lewis is incubated with anti-A, anti-B and anti-H or anti-Lewis reagents, and then the indicator cells of group A, B and O (or Lewis-positive) are added to that mixture and their agglutination or absence of agglutination are observed. When the indicator cells of group A, for example, are not agglutinated by anti-A antiserum, it is concluded that the latter was inhibited by the A substance presumably present in the substance tested.

The saliva inhibition test is the basic technique for ABO grouping of the Old and New World monkeys, which have red cells that do not usually react with anti-A, anti-B, or anti-H reagents but which are all secretors of A-B-H substances. The tests are carried out by titration to determine the highest dilution of saliva capable of inhibiting agglutination of indicator red cells by appropriately prepared anti-A, anti-B, and anti-H reagents.

The anti-A and anti-B sera of human origin to be used in saliva inhibition tests must be titrated and then diluted to yield a titer of about 8 units (at room temperature) for human A_2 cells and B cells. The anti-H reagent is prepared by extracting seeds of *Ulex europaeus* and adjusting its titer to about 4–8 units for human group O cells, at refrigerator temperature.

Whenever possible, saliva inhibition tests are complemented and confirmed by testing the animal's serum for the presence of anti-B and anti-A agglutinins.

The ABO Blood Group System of Apes

In the past few decades of investigation into the nonhuman primate blood groups, most of the attention has concentrated on blood groups of the apes, the nonhuman primates which, by far, most closely resemble humans, both physically and physiologically. It appeared reasonable to expect that the study of blood groups of apes would be of interest not only as an end in itself and as a basis for multiple practical applications (transfusion, transplantation), but also that it would contribute to greater insight into the origins and nature of the human blood groups. For reasons of limited availability of blood specimens and specific typing reagents, not all ape species were equally thoroughly investigated for their blood groups. Understandably, the most abundant data relate to species that are most often employed in biomedical research and, therefore, constitute the bulk of primate animals maintained in captivity (Table 4).

Chimpanzee

Of the four groups of the human ABO blood group system, only two, A and O, were found in the common chimpanzee (*Pan troglodytes*) (Wiener and Gordon 1960; Wiener et al. 1963; Moor-Jankowski 1972a). Group A can be subdivided into three subgroups on the basis of the reactions observed with anti-A_1 reagents. Chimpanzee A_1 red cells are strongly agglutinated though reactions are less intense than those observed with human A_1 red cells. Chimpanzee red cells of subtype $A_{1,2}$ are weakly agglutinated by anti-A_1 while chimpanzee A_2 cells fail to react with anti-A_1. In contrast to human A_2, chimpanzee group A red cells, irrespective of subgroup, invariably do not react with anti-H lectin (Wiener and Moor-Jankowski 1972). As in humans, the A antigen of chimpanzee red cells is incompletely developed at birth (Table 4).

Table 4. A-B-O blood groups of anthropoid apes

Species	Blood group frequency (%)				Total number tested	Remarks
	O	A	B	AB		
Chimpanzees						Subgroups A_1, $A_{1,2}$ and A_2 observed, though not as sharply defined as in humans. No H detectable on the red cells. All secretors of A and/or H.
Pan troglodytes	10.1	89.9	0.0	0.0	972	
Pan paniscus	0.0	100.0	0.0	0.0	14	All subgroup A_1 indistinguishable from human A_1 and all secretors of A and H.
Gorillas						Red cells very weakly reacting with anti-B;
Gorilla g.gorilla	0.0	0.0	100.0	0.0	50	all secretors of B and H; differences of saliva inhibition titer point ot the presence of subgroups of B.
Gorilla g.beringei	0.0	0.0	100.0	0.0	6	All secretors of B and H; red cells very weakly reacting with anti-B.
Gibbon						Subgroups A_1, A_2, A_1B and A_2B well defined; only group B cells strongly reacting with anti-H; all secretors of A and/or B and H.
(Hylobates)	0.0	18.9	41.9	39.2	143	
Siamang						
(Symphalangus)	0.0	0.0	100.0	0.0	2	
Orangutan						Subgroups A_1, $A_{1,2}$, A_2, A_1B, $A_{1,2}B$ and A_2B ovserved; no H detectable on red cells; all but one secretors of A and/or B and H.
(Pongo pygmaeus)	0.0	56.1	16.9	27.0	91	

The reactions of group O red cells of *P. troglodytes* with anti-H are weaker than those of human group O adults and resemble the reactivity of human newborn group O red cells (Wiener et al. 1974).

All chimpanzees (over 500 animals) tested thus far for the expression of the A and H antigens in their saliva were found to be secretors (Wiener et al. 1974). As in humans, group O chimpanzee saliva contains H only, while group A saliva has both A and H substances (Wiener et al. 1974).

The frequencies of ABO types differ significantly in various subspecies of *Pan troglodytes*, with the frequency of group O varying from as low as 9.5 % in *P. t. verus* to as high as 39.5 % in *P.t. schweinfurthi* (Moor-Jankowski and Wiener 1967b). This is comparable to the situation in various human populations (Bernard and Ruffié 1966, 1972; Mourant et al. 1958). The rare pygmy chimpanzees (*Pan paniscus*), considered to be a distinct species and not a subspecies of common chimpanzee, are all type A₁ (Socha 1984). The earlier observations of group O pygmy chimpanzees (André et al. 1961; Schmitt 1968) were based on unreliable absorption or elution techniques and could not be confirmed in our tests. Reactions of pygmy chimpanzee red cells with anti-A₁ reagents are indistinguishable from those of human A₁ red cells (Moor-Jankowski and Wiener 1972; Socha 1984). The relative agglutinability of human and chimpanzee red cells by anti-A reagents is, therefore, as follows: human A_1 > *P. paniscus* A > *P. troglodytes* A_1 > *P. troglodytes* $A_{1,2}$ > *P. troglodytes* A_2 > human A_2. Note that in the case of *P. paniscus*, group A red cells are designated simply as A, even though they react like human subgroup A₁, because subgroup A₂ has so far not been found in that species. Not unlike the group A common chimpanzees, all dwarf chimpanzees so far studied were secretors of the A and H substances.

Serum anti-B isoagglutinins are regularly present in the sera of group A individuals, while anti-A, anti-B and anti-C are detected in the sera of blood group O individuals. The anti-C antibody reacts with both A and B antigens as well as with the so-called C factor which is always linked with A and B. This applies not only to human red cells but also to ape cells of group A, B and AB. Moreover, inhibition tests on salivas of humans, apes and monkeys show that whenever specificity A and/or B is present, so is specificity C (Socha and Wiener 1973a).

Gibbons and Siamangs

In gibbons, isoagglutination defines three blood groups homologous to the human groups A, B and AB (Wiener et al. 1968). Group O was not found among more than 140 animals investigated (Wiener et al. 1974). Gibbon group B red cells differ from their human homologues in that they are strongly agglutinated by anti-H lectin (*Ulex europaeus*); however, gibbon group A and AB red cells give negative, or weak reactions with anti-H.

All 57 gibbons tested by saliva inhibition were secretors of A and/or B substance, corresponding to their red cell ABO types. They were also all secretors of H substance. Tests on their sera showed no exception to Landsteiner's rule, and cross-matching of the serum and red cells of gibbons of various ABO types gave the reactions expected from their blood groups. Observations on the distribution of the ABO groups in gibbons are in accordance with a two-allele model including genes *A* and *B*, with the *O* allele absent, or extremely rare.

Only two siamangs (*Symphalangus*) have been tested and both were found to be group B (Wiener et al. 1968). More recently, red cells from one animal were tested with monoclonal and polyclonal antibodies. The red cells were agglutinated by anti-B and by an anti-H reagents (Rearden et al. 1984).

Orangutans

Agglutination tests carried out up to date have shown the presence of only three red cell blood groups homologous to the human groups, namely, A, B and AB. As in gibbons, group O appears to be absent in orangutans (Wiener et al. 1964b). All, but one, orangutans were found to be secretors of A and/or B group susbtances together with H. Discovery of a single nonsecretor orangutan, confirmed in multiple tests, was of special importance as the first and unique case of a nonsecretor primate animal. The serum of all orangutans thus far investigated contained isoagglutinins of specificities reciprocal to the antigen(s) expressed on red cells and in saliva.

Application of the Hardy Weinberg formula shows a marked deficiency of group AB animals which could be explained by the existence of three alleles A, B and O. However, not a single group O animal was encountered among the 91 orangutans studied so far (Socha et al. 1995), so the presence of O allele cannot be confirmed in this species.

Gorillas

The red cells of gorillas are not agglutinated (in saline method) by any anti-A or anti-H reagents. Weakly positive reactions are produced by some, but not all, polyclonal anti-B reagents. It is only after enzymatic treatment by ficin that it was possible to evidence, more convincingly, the presence of a B-like antigen on all red cells of all gorillas (Wiener et al. 1971; Socha et al. 1973b). Thus, all the gorillas tested appeared to have an extremely weak B-like antigen on their red cells. Despite enhancement of the reactions by pretreatment of the red cells by ficin, anti-H reagents gave no clumping of gorilla red cells. By contrast, the presence of B and H substances is always demonstrated in the saliva of all gorillas and their sera regularly contain anti-A isoagglutinins (Wiener et al. 1974).

In conclusion, as can be seen in Table 4, none of the ape species tested so far seem to have attained the level of the ABO polymorphism found in humans. Orangutans and gibbons display only three main groups, namely, A, B and AB; common chimpanzees only two, O and A; pygmy chimpanzees, siamangs and gorillas appear to be monomorphic with respect to the ABO groups. Unlike human populations, which are composed of 60 %–98 % ABH secretors and 2 %–40 % nonsecretors, practically all apes (with a notable exception of a single orangutan) proved to be secretors. It is of interest that gorillas, which are all group B, show only traces of B activity on their red cells but high concentrations of the B substance in their saliva and other secretions. Thus, in this respect, they resemble more the Old World monkeys rather than other anthropoid apes.

In human populations, about one third of the group A Caucasians and Africans are subgroup A_2, but no subgroup A_2 is found among Asians. While typical A_1 human cells give distinct and strong agglutination with anti-A_1 reagents, reactions of ape cells with these reagents differ from one species to another. Red cells of all pygmy chimpanzees (*Pan paniscus*) give reactions with anti-A_1 indistinguishable from those of human A_1 red blood cells. Gibbon A_1 and A_1B red cells are also, by and large, avidly agglutinated by anti-A_1. The A_1 and A_1B cells of orangutans as well as A_1 red cells of common chimpanzees (*Pan troglodytes*) give slightly weaker agglutination with anti-A_1 reagents so the distinction between A_1 and A_2 subgroups in these species is not as sharp as in human, pygmy chimpanzee or in gibbon. Rare chimpanzee and orangutan red cells have been observed that gave extremely weak but unequivocally positive reactions with anti-A_1, thus resembling human blood of the rare type A_{int}. Red cells of newborn and very young (up to the age of 2 months) chimpanzees give significantly weaker reactions with anti-A_1 reagents than do older and adult group A animals.

The ABO Blood Group System of Monkeys

The earliest attempt to define red cell antigens of lower primates was that by Landsteiner and Miller (1925c). Using eluates of normal human sera and immune rabbit anti-human reagents, they tested red cells of representatives of 30 species of monkeys and came to the conclusion that all New World monkeys (*Platyrrhini*) have in their blood an agglutinogen similar to, but not identical with, human B antigen. The red cells of Old World monkeys (*Cercopithecidae*) investigated in the same series of experiments failed to react with any of the reagents, and therefore, were presumed to be devoid of any ABO blood group antigens. However, Wiener et al. demonstrated that the secretions and glands of all lower monkeys regularly contained blood group substances, A or B and H and that these could be used for the ABO typing (Wiener et al. 1942). As proposed by Wiener and his coworkers (Wiener et al. 1942), saliva inhibition, together with the reverse testing of serum for anti-A and anti-B isoagglutination, became the standard procedure for ABO typing of primate species below the taxonomic level of anthropoid apes.

Old World Monkeys

Table 5 shows the updated information on the distribution of the ABO groups among Old World monkeys. The table gives pooled data for each species without, however, taking into account significant geographical and/or racial differences that were observed among various species of baboons (Brett et al. 1976; Downing

Table 5. Distribution of the A-B-O group in Old World monkeys (based on saliva inhibition tests)

Species	Blood group freqency (%)				Total number tested	Remarks
	O	A	B	AB		
Baboon *Papio anubis* (various sources)	0.0	14.5	47.5	38.0	750	Red cells give weak reactions with anti-A lectin (lma bean). This A-like RBC antigen is independent of the A-B-O type defined by saliva and serum tests.
P.hamadryas (various sources)	0.0	8.7	62.2	29.1	172	
Hybrids *hama-dryas anubis*	0.0	3.9	62.8	33.3	129	
P.cynocephalus	0.0	31.0	33.0	36.0	80	
P.ursinus (various sources)	0.0	22.0	43.5	34.5	168	
P.papio	1.0	14.4	49.5	35.1	188	
Papio (species unknown)	0.5	24.5	38.0	37.0	184	
Geladas (*Theropithecus gelada*)	100.0	0.0	0.0	0.0	20	Expected isoagglutinins are not always detectable in the serum.
Macaques *Macaca arctoides*	0.0	0.0	100.0	0.0	41	
M.fascicularis (Various sources)	1.4	26.7	37.9	34.0	985	
M.fuscata	0.0	0.0	100.0	0.0	14	Based on serum tests only; red blood cells stronlgy react with anti-B reagents.
M.maurus (tonkeana)	0.0	100.0	0.0	0.0	12	
M.mulatta	1.0	1.0	97.0	1.0	215	
M.nemestrina	74.2	15.7	7.9	2.2	89	
M.radiata (various sources)	0.0	41.5	36.4	22.0	77	
M.sylvanus	0.0	100.0	0.0	0.0	32	
Drill (*Mandrillus*)	0.0	100.0	0.0	0.0	4	
Vervet monkeys (*Cercopithecus*) (various sources)	0.5	70.1	9.4	20.0	562	Landsteiner's rule does not always hold; geographical dif-ferences are noticeable.
Langurs (*Presbytis entellus*)	0.0	0.0	88.9	11.1	18	
Celebes black apes (*Cynopithecus*)	3.8	88.5	7.7	0.0	26	Expected isoagglutinins not always detectable in serum.
Patas monkeys (*Erythrocebus*)	0.0	100.0	0.0	0.0	26	

Fig. 3. Map of the gene frequencies of the human-like ABH salivary antigens in vervet monkeys from Kenya, Ethiopia and South Africa. The radius of the circle is proportional to the sample size analyzed in each country (From Dracopoli and Jolly 1983)

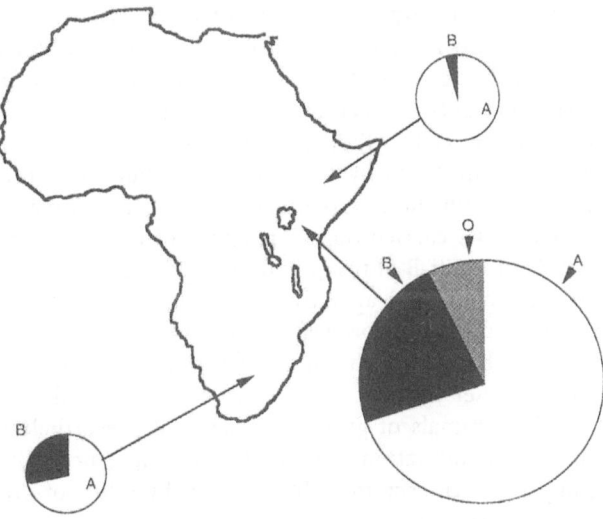

et al. 1975; Moor-Jankowski et al. 1964; Socha et al. 1977; Wiener et al. 1974), in crab-eating macaques of various origins (Terao et al. 1981), and in vervet monkeys (Downing et al. 1973; Dracopoli et al. 1983; Jolly et al. 1977). For example, as shown in Fig. 3, marked geographic variations of the *ABO* gene frequencies were observed among populations of vervet monkeys (*Cercopithecus aethiops*) (Dracopoli et al. 1983). The variations of gene frequencies are comparable to those observed in human populations (Mourant et al. 1958; Bernard and Ruffié 1966; Cavalli-Sforza et al. 1994). The processes that led to and maintained these interpopulation variations remain obscure (see the last section of this chapter). It was evoked that selection due to environmental factors (including endemic diseases) could modify the frequency of ABO alleles in given populations (see the chapter by Eder and Spitalnik). None of these factors seemed to be involved in the variations of *ABO* gene frequencies observed in *Cercopithecus aethiops* in Kenya (Dracopoli et al. 1983). However it is intriguing that the troupes of vervet monkeys with the highest frequency of allele *O* were encountered in a region Kimana (Kenya) adjacent to the territory where the highest frequency of allele *O* was observed in another African species of Cercopithecidae, namely, the baboon (*Papio* sp) (Byles and Sanders 1981). Since gene flow between these two species is impossible, one has to assume the role of some selective agent(s) acting in a similar way upon baboons and vervets.

Landsteiner's rule holds for most of the Old World monkeys, even though their ABO type is established by testing saliva for its contents of group substances, and not by hemagglutination tests (as it is the case in humans and apes). Rare animals occur with anti-A agglutinin in the serum despite the presence of blood group substance A in saliva. The anti-A in these group A and AB animals is peculiar in that, unlike regular anti-A from group B humans or monkeys, it is not inhibitable by saliva of nonhuman primate or human group

A secretors. To explain this phenomenon, Wiener et al. (1974) assumed the existences of anti-A agglutinins of two kinds: one, designated anti-A^c, reactive exclusively with the A substance on the red cells and not in secretions, and the other, designated anti-A^s, reactive with the blood group substances both in secretions and on red cells. Obviously, a group A baboon or group A vervet monkey would hardly be likely to form anti-A^s but could produce anti-A^c which would not react with the group A substance in the animal's own secretions.

Our tests carried out on red cells of the Old World monkeys using standard anti-A and anti-B antisera and anti-H lectin (*Ulex europeus*) failed to detect the presence of A or B agglutinogens, thus confirming early observations by Landsteiner and Miller. However, when tests were carried out with a potent anti-A lectin (lima bean), agglutination of baboon red cells resulted but at only very low titer (Wiener et al. 1969c). This reactivity with anti-A lectin was about the same for animals of group A, B and AB. Nevertheless it could be demonstrated that the agglutination was due to an A-like specificity since it was inhibited by group A or AB secretor saliva but not by saliva of group O or group B or saliva from human nonsecretors. Thus, all baboons have on their red cells an agglutinogen with a weak A-like specificity quite apart from their ABO groups. The presence of this A-like agglutinogen on the red cells apparently does not interfere with the occurrence of anti-A in the serum.

New World Monkeys

As in the Old World monkeys, the ABO blood groups of the New World monkeys are detected by saliva inhibition tests for the presence of the A, B and/or H blood group substances and by the reverse testing of sera for anti-A and anti-B isoagglutinins (Wiener et al. 1964c, 1967). It must be stressed, however, that the latter are not regularly present in the sera of American monkeys and, therefore, Landsteiner's rule does not always hold in these species. As shown in Table 6, the ABO polymorphism was confirmed in at least three of the five New World monkey species investigated thus far. Unlike the situation in humans and higher primates, in lower monkeys the H substance does not regularly accompany the A and B substances in saliva and other body secretions. Independent of monkeys' salivary ABO type is the B-like antigen detected on the red cells of all New World monkeys (Wiener et al. 1974; Froehlich et al. 1977), and the A-like antigen on the red cells of marmosets (Gengozian 1966). Absorption of human anti-B polyclonal antibodies on the red cells of New World monkeys left behind fractions of anti-B antibodies capable of agglutinating human B and AB and rabbit red cells. Other absorption experiments demonstrated that the anti-B polyclonal antibodies can be separated into at least four fractions: the first, specific for human B red cells; the second, cross-reacting with human and rabbit red cells; the third, reactive with rabbit and New World monkeys but not with human red cells; a fourth, cross-reactive with all B and B-like substances. The B-like substance expressed on the red cells of the New World monkey corresponds to the α-galactosyl epitope [Galα1–3Galβ1–4GlcNAc-R], which is a product of α1,3galactosyl-

Table 6. The A-B-O blood groups of New World monkeys and prosimians

Species	Blood group freqency (%)				Total number tested	Remarks
	O	A	B	AB		
Spider monkey (*Ateles*) various species	6.0	67.0	27.0	0.0	15	B-like agglutinogen regularly present on the red cells, independent of the A-B-O saliva type.
Squirrel monkey (*Saimiri sciureus*)	18.0	55.0	0.0	27.0	11	B-like agglutinogen regularly present on the red cells, independent of the A-B-O saliva type; expected isoagglutinins often weak or absent from the serum.
Capuchins (*Cebus*) various species	11.0	56.0	33.0	0.0	9	
Howler monkey (*Alouatta*)	0.0	0.0	100.0	0.0	52	B-like agglutinogen on the red cells.
Marmosets (*Callithrix*) various species	0.0	100.0	0.0	0.0	45	B-like agglutinogen regularly present on the red cells; H very weak or not detectable in saliva.
Galagos (*Galago crassicaudatus*)	0.0	100.0	0.0	0.0	17	B-like and H-like agglutinogens regularly present on RBCs; isoagglutinins erratic.
Lemurs (*Lemur catta*)	0.0	0.0	100.0	0.0	13	B-like and H-like agglutinogens reglularly present on RBCs; isoagglutinins in serum erratic.

tranferase (see the chapter by Galili, this volume). This enzyme is not expressed by the Old World monkeys nor by apes and humans because the appropriate genes are nonfunctional (see later in this chapter and the chapter by Galili, this volume).

Prosimians

Among prosimians only thick-tailed galagos (*Galago crassicaudatus*) and ring-tailed lemurs (*Lemur catta*) were investigated for their ABO groups. All 17 galagos tested so far were secretors of A and H group substances, while all 13 lemurs were secretors of B and H (Socha et al. 1995). Not unlike the New World monkeys, prosimians display a B-like antigen on the red cells irrespective of the animal's secretory ABO type. However, red cells of prosimians show some reactivity with selected anti-H reagents, which was not observed in tests on the red cells of

South American monkeys. As in New World monkeys, the isoagglutinins in Prosimian sera are generally weak or, often, absent.

ABO-H and Lewis Monoclonal Reagents in Tests with Red Cells of Nonhuman Primates

At the Second International Workshop on Monoclonal Antibodies Against Red Blood Cells and Related Antigens (Lund, Sweden, 1–4 April 1990) red cells of various species of anthropoid apes, Old World and New World monkeys and prosimians were tested with panels of monoclonal antibodies of anti-A, anti-B, anti-A+B specificities. Antibodies are listed in the proceedings of the workskop (1990). All anti-A, anti-B reagents were employed using the saline technique at room temperature. Initial tests with anti-H and anti-Lewis antibodies were carried out, in parallel, by saline, antiglobulin and ficinated red cell methods, as is customary in our practice when dealing with most primate immune antisera or other reagent not thoroughly tested with monkey blood. In order to assess their precise specificities the same antibodies were tested on synthetic glycosaccharides by Oriol et al. (1990).

Anti-A Antibodies

Reactions obtained with 13 anti-A monoclonal antibodies show a uniform pattern of reactivity, basically identical with patterns observed with most human polyclonal anti A reagents: strong agglutinations of A-positive gibbon, orangutan and chimpanzee red blood cells, and negative reactions with lower monkey red blood cells.

A notable exception was an antibody (W#001) that recognized an A-like epitope on the red blood cells of South American marmosets, thus resembling some of the most potent polyclonal anti-A reagents. Unlike strong polyclonal anti-A_1 sera and lectin, the single anti-A_1 monoclonal antibody tested failed to react with gibbon, orangutan or chimpanzee red cells of proper ABO types.

Anti-B Antibodies

Nine out of 10 anti-B monoclonals strongly agglutinated B-positive gibbon and orangutan red blood cells, but gave only very weak reactions with gorilla blood thus paralleling polyclonal anti-B. Unlike the latter, however, the monoclonals did not recognize the B-like antigen on red cells of all New World monkeys. This reinforces the notion (already evoked in this chapter) that the B-like antigen on New World monkey red cells corrresponds to the α-galactosyl epitope (Galα1–3Galβ1–4GlcNAc-R). One antibody (W#042) gave false positive reactions with group A chimpanzee blood.

Anti-A+B Antibodies

All five anti-A+B monoclonals strongly agglutinated all A- and/or B-bearing red cells of anthropoid apes, with the exception of gorillas. Only one antibody (W#052) reacted with red cells of some of the South American monkeys. It agglutinated all tamarins and detected individual differences among marmosets. It failed, however, to clump the red cells of another South American monkey species: squirrel monkeys. Yet A and/or B-like antigens are usually detectable on the red cells of all New World monkeys by human polyclonal anti-A and/or anti-B sera.

Agglutinations of human as well as primate red cells with anti-A, anti-B and anti-A + B monoclonals are all inhibitable with proper ABO-type saliva of anthropoid apes and Old World monkeys as well as by human secretor saliva.

Anti-H Antibodies

Patterns of reactions of the ten anti-H monoclonal antibodies tested varied from one reagent to another, depending on the species tested and the technique applied. With two exceptions (W# 058 and 059), positive reactions were obtained only with ape bloods, gibbon and chimpanzee blood being clumped by most of the antibodies. Unlike tests with anti-H lectin, reactions with monoclonal anti-H were, by and large, unrelated to ABO type. Hemagglutinations by monoclonal anti-H were strongly inhibited by saliva of most primates tested, irrespective of the saliva ABO type.

The Lewis Blood Group

The Lewis Blood Groups of Humans

The Lewis types of human blood are inherited by a pair of allelic genes, *Le* and *le*, independent of the genes *ABO*, *H-h*, and the secretor genes *Se-se*. Yet, the Lewis types are closely associated with the ABO blood groups and A-B-H secretor types (Grubb 1951) because the enzyme produced by the *Le* gene acts on the same substrates as do the ABO- and Se- and H-transferases (for a general review see Henry et al. 1995). The Lewis antigens appear to depend on the addition of a fucose monosaccharide, which is attached on the same blood group oligosaccharide chain as the fucose transferred by the *H* gene encoded enzyme. However, each fucose uses different points of attachment (Watkins 1970) (Fig. 1).

In contrast to the A-, B- and H-transferases, the Le transferase does not occur in plasma or in the cells of the hematopoietic system. The Lewis blood types are unique in that the transferase determined by the *Le* gene produces mainly soluble group substances: glycoproteins in saliva and glycosphingolipids in plasma (Fig. 2). The latter are secondarily adsorbed onto the red cell mem-

Table 7. Lewis types in adult humans

Genotypes	Saliva soluble antigens		Red cell phenotypes: reactions with		
	Le^a	Le^b	Anti-Le^a	Anti-Le^b	Anti-Le^x
sese LeLe *sese Lele*	+	−	+	−	+
SeSe LeLe *Sese LeLe* *SeSe Lele* *Sese Lele*	−	+	−	+	+
SeSe lele *Sese lele*	−	−	−	−	−
sese lele	−	−	−	−	−

brane as a result of a continuous exchange of glycosphingolipid between plasma and the red cell membrane. The same passive adsorption was demonstrated on lymphocytes (Oriol et al. 1981). The Lewis antigens are absent or extremely weak at birth. Red cells of very young infants do not react with anti-Lewis reagents, but become agglutinable during the first year of life (Le^a develops more rapidly than Le^b). The synthesis of Lewis antigens requires the interaction of two enzymes, one encoded by the *Le* gene and the other by the *Se* (secretor) gene (Grubb 1951) (see Fig. 1, and later in this chapter). Because Lewis and secretor genes are independent (Cepellini 1956), they can combine into four phenotypes, all detectable in tests on secretions such as saliva (see Table 7).

The Lewis phenotypes of red cells can be influenced by age, pregnancy and transfusion but usually correspond to the salivary phenotype (Salmon et al. 1984). The red cell ABO type influences the Lewis phenotyping. Two kinds of anti-Le^b were identified in tests carried out on Le^b+ red cells of various ABO phenotypes: anti-Le^{bH} which agglutinates A$_2$ or O Le^b+ red cells, and anti-Le^{bL} (true anti-Le^b) which reacts with A$_1$ Le^b+ red cells as well as with A$_2$ and O Le^b+ red cells (see Mollison 1979; Salmon et al. 1984).

The Lewis Blood Groups of Nonhuman Primates

Tests for the Lewis types of red cells of nonhuman primates have not been possible because of the unavailability of potent antisera free of nonspecific heteroagglutinins. However, as tests on the secretions are carried out by the inhibition technique, in which human red cells are used as test cells, the presence of heteroagglutinins in the reagents does not interfere.

More recently agglutination tests were performed on nonhuman primate red cells with monoclonal antibodies (see later in this chapter). As shown in Table 8, all 26 orangutans and eight out of 30 chimpanzees tested proved to be Le^a-posi-

Table 8. Lewis types of saliva of apes and monkeys

Species	Lewis types		Total
	Le^a	nL	
Anthropoid apes			
Chimpanzees *(Pan troglodytes)*	8	22	30
Gibbons *(Hylobates lar)*	0	8	8
Orangutans *(Pongo pygmaeus)*	26	0	26
Old World monkeys			
Baboons *(Papio cynocephalus)*	31	8	39
Drills *(Mandrillus leucophaeus)*	0	3	3
Celebes black apes *(Cynopithecus niger)*	26	0	26
Patas monkeys *(Erythrocebus patas)*	16	10	26
Vervet monkeys *(Cercopithecus pygerythrus)*	3	1	4
Rhesus monkeys *(Macaca mulatta)*	38	8	46
Crab-eating macaques *(Macaca fascicularis)*	8	0	8
Pig-tailed macaques *(Macaca nemestrina)*	10	0	10
New World monkeys			
Spider monkeys *(Ateles,* various species)	0	5	5
Cinnamon capuchins *(Cebus albifrons)*	0	4	4
Squirrel monkeys *(Saimiri sciureus)*	0	4	4
Marmosets (various species)	0	31	31

nL, absence of the Le^a substance in saliva.
Modified after Wiener *et al.* (1964a) and Moor-Jankowski and Wiener (1967a).
The presence of the substance Le^a was demonstrated by inhibition of agglutination.

tive, while all eight gibbons were Le^a-negative. Whereas most of the Old World monkeys tested were Le^a-positive, all 44 New World monkeys investigated thus far were Le^a-negative. These results suggest that the *Le*-encoded fucosyltransferase is probably not expressed (or not functional) in gibbons and the New World monkeys.

Secretion of Lewis substance is a polymorphic trait in most species investigated except orangutans, Celebes black apes, and crab-eating and pigtailed macaques, which, allowing for the limited number of animals tested and reported in Table 8, appeared to be monomorphically Le^a-positive.

Tests with Monoclonal Anti-Le^a and Anti-Le^b Antibodies

During the Second International Workshop on Monoclonal Antibodies Against Red Blood Cells and Related Antigens (Lund, Sweden, 1–4 April 1990) red cells of various species of anthropoid apes, Old World and New World monkeys and prosimians were tested with monoclonal antibodies of anti-Le^a, anti-Le^b specificities (for the list of monoclonal antibodies, see the proceedings of the workshop 1990).

All three anti-Le[a] monoclonal reagents reacted (by the antiglobulin technique), with gibbons and gorillas, but only one also agglutinated orangutan red blood cells. Tests with all other primate red blood cells and human controls were negative. Enzyme treatment rendered cells of chimpanzees and marmosets strongly agglutinable by two of the reagents while all three antibodies clumped ficinated, human, positive control red cells. The clumping of marmoset red cells was probably due to cross-reactions with other glycosaccharides because red cells of this species, as of those of other New World monkeys, are not agglutinated by polyclonal anti-Le[a].

Four anti-Le[b] monoclonal reagents gave uniformly positive results with ficinated red blood cells of orangutans, baboons and human positive controls. Some of the reagents consistently showed individual differences among gorillas, chimpanzees and bonnet macaques.

The Biosynthesis of the AB, H and Lewis Substances

At the molecular level, the antigenic expressions of the ABO blood groups correspond to the terminal oligosaccharaide epitopes of complex oligosaccharide chains belonging to either glycoproteins or glycolipids. The molecular entities which express ABO antigens (glycoproteins or glycosphingolipids, free oligosaccharides) are present in solution. Soluble glycoproteins (including salivary mucin) and some circulating serum glycoproteins contain ABH determinants displayed by O-linked oligosaccharides. At the red cell membrane, the ABO antigens are associated with glycolipids (reviewed by Watkins 1995; Lowe 1995) and integral membrane proteins. Biochemical analysis of the ABH determinants on human red cells indicates that 80 % of them are associated with band 3 (the anion transport protein of the erythrocyte) (Laine and Rush 1988). The ABH oligosacharides of band 3 correspond to a unique asparagine-linked chain (Tanner et al. 1988). The ABH determinants are displayed at the terminals of the branched poly-N-acetyllactosaminoglycan chains (Fukuda and Fukuda 1974). Other membrane glycoproteins, such as the Rh-50 associated glycoprotein, are also involved in erythrocyte expression of the ABH antigens (Moore and Green 1987). Red cell membrane glycolipids display the remaining red cell ABH determinants. They are referred to as polyglycosylceramides or glycolipids.

The synthesis of the A, B and H molecules is determined by several distinct genetic loci which encode glycosyltransferases. These enzymes catalyze the synthesis of the oligosaccharide chains which display the ABH antigens. The enzymatic activity of each transferase is defined in terms of the precursor chain to which a given sugar is transferred and according to the type of bond which is catalyzed.

The synthesis starts with the intervention of two α(1,2)fucosyltransferases which are encoded by the *H* and the *Se* genes. Both enzymes transfer a fucose to the disaccharide precursors. The fucose, from the GDP-fucose, is covalently

attached by the enzyme in the alpha anomeric linkage to the second carbon of
the galactose residue at the nonreducing end of the precursor.

Six different types of disaccharide core structures function as acceptor substrates for H or Se enzymes namely:

- Type 1: Galβ1–3GlcNAcβ1-R
- Type 2: Galβ1–4GlcNAcβ1-R
- Type 3: Galβ1–3GalNAcα1-R
- Type 4: Galβ1–3GalNAcβ1-R
- Type 5: Galβ1–3Galβ1-R
- Type 6: Galβ1–4Glcβ1-R

The core structure of type 5 corresponds to a synthetic disaccharide which
functions in vitro as an acceptor substrate for the enzyme H. The type 6 core
structure [Galβ1–4Glcβ1-R] was found in milk. Blood-group-active oligosaccharides based on this core structure were found on glycolipids extracted from small
intestine epithelial cells and from renal vein tissue (for references and more
details concerning the biochemistry of core structures see Watkins 1995). The
pathways of synthesis of the antigens H, A, B, and Le from type 1 and type 2
core structures are shown in Fig. 1.

The core structure of ABH and Lewis antigens varies from tissue to tissue
(Fig. 2). Type 1 precursors are generally synthesized in tissues derived from the
endoderm (Oriol et al. 1986). These include epithelial cells of the digestive, urinary, respiratory and reproductive tracts, and also some exocrine glands. Type 1
determinants are the major carriers of the ABH antigens on proteins and lipids
in secretions and other body fluids.

Type 2 molecules are synthesized in tissues derived from the mesoderm, such
as erythrocytes, or in tissues derived from the ectoderm, such as epidermis
(reviewed by Oriol et al. 1986). All the glycolipid A-B-H antigens produced by
erythrocytes are built on type 2 precursors and the small quantities of ABH glycolipids with a type 1 core sequence are adsorbed from plasma onto erythrocytes

Mono-fucosylated Le^a and di-fucosylated Le^b determinants occur only on a
type 1 core structure while monofucosylated Le^X and di-fucosylated Le^Y determinants only occur on type 2 core structure. Type 3 and type 4 core structures give
rise to ABH determinants but not to any Lewis determinants.

Type 3 ABH-associated oligosaccharides are mainly restricted to O-linked
chains of glycoproteins. These glycoproteins are expressed by epithelial cells.
On the red cells, type 3 oligosaccharides are carried only by glycolipids (Clausen
et al. 1985; Le Pendu et al. 1986).

Type 4 chains are carried by glycolipids of the red cell membrane (Clausen
et al. 1986) and of solid organs, such as the kidney (Samuelsson et al. 1994).

The H α(1,2)fucosyltransferase transfers a fucose to type 2 and type 4 precursors which give rise to type 2 and type 4 H substances. The enzyme encoded by
the *H* gene (fucosyltransferase 1, FUT1) is predominantly expressed by the erythroid lineage and endothelial cells. The Se α(1,2)fucosyltransferase uses, as preferential substrates, the type 1 and type 3 precursors to form type 1 and type 3 H
determinants. The enzyme encoded by the *Se* gene (fucosyltransferase 2, encoded

by the *FUT2* gene) is responsible for the synthesis of the H determinants in the exocrine secretions (Fig. 1, Table 2).

The synthesis of the Lewis blood group substances is the result of epistatic interactions between various fucosyltransferases (Wiener et al. 1964a). The pathway of production of Lewis substances in saliva is depicted in Fig. 1. The Lewis antigens have been divided into Le^a, Le^b, Le^X and Le^Y. It must be stressed that the Le^X antigen is unrelated to an antibody, originally called anti-Le^x, and able to react with both Le^a and Le^b substances (Arcilla and Sturgeon 1974).

The human Lewis system encompasses two alleles: *Le*, which encodes a $\alpha(1,3/4)$fucosyltransferase (FUT3) and is dominant over a null allele, *le*. The FUT3 enzyme catalyzes the transfer of a fucose (from the donor GDP-fucose) to the subterminal and/or internal N-acetylglucosamine residues on glycolipids and glycoproteins. The enzyme can use type 1 or type 2 chains as well as type 1 or type 2 sialyl N-acetyllactosamine as acceptors. However, the precise specificity of the human Lewis encoded enzyme remains to be clarified (Watkins 1995). The biochemical pathways using the type 1 molecules are described in Fig. 1. Individuals with the functional FUT3 (*Le/Le or Le/le*) and the functional FUT2, who are of genotypes *Se/Se or Se/se*, represent 70 % of the population and express ABH, Le^a and Le^b substances in their saliva. Individuals with a functional FUT3 (*Le/Le or Le/le*) but who are nonsecretors (genotype *se/se*) express only Le^a subtance in their saliva. Finally, individuals who do not possess the functional FUT3 (*le/le*), whether secretors or nonsecretors, do not express any Lewis substance in their saliva. When the FUT3 enzyme uses, as acceptor, type 2 chains, it leads to the production of Le^X and Le^Y antigens. However, since the enzyme encoded by the *FUT3* gene has a much greater affinity for type 1 than for type 2 acceptor substrates, in some tissues like plasma or at the surface of red cells, only type 1 Le^a and Le^b glycosphingolipids are found. Nevertheless, the expression of type 2 Le^X and Le^Y epitopes in epithelial cells of digestive mucosa is under the control of FUT3.

As mentioned earlier, the Lewis a/b determinants on human red cells are not endogeneously synthesized by the erythroid cells but are passively adsorbed from the plasma. These antigens correspond to neutral lipids most probably synthesized by mucosal cells of the digestive tract and shed into the plasma as glycosphingolipids. The oligosaccharide chains of these neutral lipids are built on type 1 precursor chains. These passively adsorbed Le antigens are also found on lymphocytes (Oriol et al. 1981).

Tissue Distribution of the ABH Antigens in Humans and Nonhuman Primates

Following their discovery on red cells, the ABH antigens were found to be secreted in saliva and expressed in many human tissues, especially on endothelial and epithelial cells where they are widely distributed (for a general review, see Salmon et al. 1984; Oriol et al. 1992). Research in this field was stimulated by the observation of modifications of the expression of the ABH and Le antigens in malignant cells (for a general review, see Hakomori 1985, 1989, 1991; Fukada 1985). The development of allotransplantations and, more recently, experiments with xenotranplantations gave a new impulse to the study of tissue expression of ABO and Le antigens in humans and in nonhuman primates. Panels of monoclonal antibodies specific for various oligossacharides, because of their narrow specificity, became useful and preferred tools in the study of the tissue distributions of these antigens (Feizi 1985). Although knowledge of the nonhuman primate ABH histo-blood groups seems of particular importance for the comprehension of their role in experimental allo- or xenografts, the study of the expression of ABH tissue antigens has been initially limited to a few nonhuman primate species (Stejskal et al. 1980; Oriol et al. 1984; Socha et al. 1987).

The distribution of the ABH and Le antigens in human tissues is the effect of complex processes because the synthesis of these antigens results from the epistatic interactions of numerous enzymes. Moreover, there are numerous fucosyltransferases (see special section in this chapter) which are encoded by various genes and display individual patterns of tissue expression. Figure 2 gives a brief overview of the genetic control of the expression of ABH and Le substances in human tissues. In humans, due to the existence of silent alleles at the *H, Se* and *Le* loci, the role of the three fucosyltransferases could be studied by comparison of individuals possessing or lacking these alleles. In this way, the function of the H, Se and Le enzymes in the biosynthesis of ABH and Le antigens in human tissues was assessed. Glycosyltransferases A or B, when expressed in tissues capable of producing the H substances, transform the latter into A or B antigens.

The tissue expression of histo-blood antigens varies during embryogenesis (Salmon et al. 1984; Oriol et al. 1995) (Fig. 2). For example, the ABH antigens are found very early in embryonic life on endothelial and epithelial cells of most organs while the development of the ABH antigens at the surface of red cells is not yet fully achieved at birth. In adults, ABH antigens are predominantly expressed by endothelial and epithelial cells as well as red blood cells. Their expression varies during differentiation and maturation of the cells. For example, in the epidermis, all the cells of the stratum spinosum express antigen H, but only the most mature cells of stratum granulosum, situated next to the stratum corneum, express A or B antigens (Oriol et al. 1992). The glycosylation pattern in oral mucosa and salivary gland epithelia follows cellular differentiation (Mandel 1992). In the course of hematopoiesis, the H-fucosyltransferase and ABO glycosyltransferases are expressed only in the red cell lineage and platelets but not in leucocytes (Mollicone et al. 1988a). As mentioned earlier, the ABH and Le antigens found on lymphocytes, red blood cells and platelets are passively acquired from the plasma (Mollicone et al. 1988a).

The expression of the ABH and Le antigens in tissues depends on the expression of the various glycosyltransferases. The difucosylated antigens, such as Le[b] and Le[y], depend on the successive intervention of two enzymes. For example, synthesis of Le[b] antigen in acini of salivary glands of Lewis-positive ABH secretor individuals requires the action of the $\alpha(1,2)$fucosyltransferase encoded by the *Se* gene and the subsequent transfer of a fucose on the subterminal GlcNAc of the type 1 precursor chain by the $\alpha(1,3/4)$fucosyltransferase encoded by the *Le* gene

(Fig. 1). Similarly, the difucosylated Le^Y antigen found in the gastric glands and Brunner's glands of the duodenal mucosa depends on epistatic interactions of the enzymes encoded by the genes *H* and *Le^X* (Molicone et al. 1985; Oriol et al. 1992). In individuals with either gene *A* or gene *B*, the Le^Y could theoretically be transformed in these locations into substances ALe^Y or BLe^Y. Surprisingly, only one half of the cells in gastric or Brunner glands are capable of transforming the subtance Le^Y into either antigen ALe^Y or antigen BLe^Y (Oriol et al. 1992). In humans, the synthesis of antigen H in the acinar cells of sweat glands is under the control of the fucosyltransferase encoded by the gene *Se*, while in the epidermis the H antigen is synthesized by the fucosyltransferase encoded by the gene *H*, so that the H substance and its derivatives (antigens A and/or B) are found in secretors as well as nonsecretors (Mollicone et al. 1988b). Fucosyltransferases, other than H and Se, are at work in some tissues. We will see later that the antigen Le^X on human leucocytes is under the control of a specific fucosyltransferase (most probably encoded by the *FUT7* gene). The expression of the Le^X in epithelial cells of proximal tubules and descending limbs of Henle's loop in the kidney is independent of *H*, *Le* and *Se* genes, but depends on the gene which controls the plasma α(1,3)fucosyltransferase (Oriol et al. 1992). In the same organ, the synthesis of antigen H in glomerular endothelial cells is under control of the *H* gene, but the synthesis of H in the tubular cells of collecting tubes depends on the *Se* gene.

Distribution of ABH epitopes was investigated in various organs and tissue samples obtained from cynomologus macaques and baboons used as donors in cardiac allo- and xenotransplantation experiments (Socha et al. 1987). Paraffin-embedded tissue sections were stained using monoclonal anti-A, and anti-B, and anti-H reagents and standard indirect, avidin-biotin complex immunoperoxidase technique. There were no significant differences between these two species in the distribution of tissue A, B and H epitopes and the intensity of their staining. Table 9 summarizes our findings which, by and large, are in agreement with

Table 9. Tests for A, B, H antigens in tissues of baboons and macaques: summary of findings

Positive	Negative
Endothelium of capillaries and larger blood vessels in heart, glomeruli and renal interstitium, trachea, bronchi, lung, liver, pancreas, gastrointestinal (GI) tract, spleen, bladder, skeletal muscles, and brain. A and/or B expressed according to ABO type, but H negative. Splenic sinusoidal lining stained in some animals only.	Myocardial, smooth and striated muscles; valvular endocardium, proximal renal tubular cells, pulmonary alveolar lining cells, hepatocytes, liver sinusoidal lining cells, bile ducts and pancreatic acinar, and endocrine cells.
Epithelium, surface and glandular, of the GI tract, pancreatic ducts, trachea, bronchi, uterus. A and/or B usually expressed in accordance with ABO type together with H. Deep gastric and duodenal Brunner's glands show H acitivity exclusively.	

earlier observations based on different staining techniques (Mollicone et al. 1986; Oriol et al. 1984; Stejskal et al. 1980; Alroy et al. 1982). As in humans, the dominant elements expressing the ABH epitopes are omnipresent on vascular endothelia and surface and glandular epithelia (Oriol et al. 1984, 1992) but the red cells are totally negative in Old World monkeys. It is interesting to note that in deep gastric and duodenal Brunner glands of macaque and baboons, only the H activity was detected, while in humans half the cells in these locations were found to express the A or B antigens (Oriol et al. 1992). In the epidermis of macaques, as in that of humans, all layers from stratum germinatum to stratum corneum were positive with goat anti-H reagent, whereas only a few scattered cells at the surface were positive with anti-A or anti-B (Oriol et al. 1984; Socha et al. 1987).

Salivary and sweat glands of all animals showed the presence of antigen H in addition to A or B (Stejkal et al. 1980). All vascular endothelial cells in liver biopsies of 12 baboons were positive for H, while antigens A and/or B were also demonstrated at the surface of endothelial cells (Oriol et al. 1984). Biliary ducts of baboons were found positive for H (Oriol et al. 1984).

The tissues of a New World monkey (marmoset) were studied in comparison with tissues from baboons (Mollicone et al. 1986). The paraffin-embedded sections of marmoset digestive tract revealed the presence of A and H antigens in mucus-secreting cells (all marmosets are known to have antigens A and H in their secretions) but their absence from erythrocytes. In addition, marmosets were found to lack ABH antigens on vascular endothelium, similar to rat and other mammals but unlike other members of the order of Primates (Mollicone et al. 1985). ABH antigens were detected, as in baboons and humans, in the posterior root ganglia of marmosets (Mollicone et al. 1986). Surprisingly enough, it was demonstrated that, in humans, expression of the ABH antigens in the posterior root ganglia was independent of the secretor gene. Although no systematic study of the central nervous systems has been performed, the neurons of humans, baboons, and marmosets have been found negative in all areas studied so far (grey and white matter of the cerebral cortex, and the cerebellum, olfatory tract, optic nerve and ventral horns of spinal cord) with the only exception of primary sensory neurons derived from placodes.

When one traces the variations of the expression of antigens ABH and Lewis from verterbrates phylogenetically distant from humans toward species closer to humans, one can conclude that the expression of ABH and Lewis antigens progressively expanded during the evolution of the species (Ruffié and Socha 1980; Oriol et al. 1992). The ABH and Lewis antigens were first expressed in tissues of endodermal origin (mucus secreting cells) then expanded into tissues of ectodermal origin (skin, nervous system) and finally into those of mesodermal origin (endothelial cells and erythrocytes). For example, amphibians express the ABH antigens only on epithelial cells derived from endoderm (Oriol et al. 1992) while endo- and ectodermal tissues are positive in mammals below the Primate order, such as mouse and rats (Oriol et al. 1992). Vascular endothelium is positive in Old World monkeys but negative in New World monkeys. Erythrocytes are the last cells in the processes of phylogenesis (and, as mentioned earlier, also during ontogenesis) to express ABH antigens; they do not become positive until the taxonomic level of higher primates, i.e. anthropoid apes and humans. However, this general picture, although by and large correct, is complicated by specific details. For example, in humans, expression of the H antigen on endothelial cells depends on the fucosyltransferase encoded by H (FUT1 gene), suggesting a concordance between the expression of H at the surface of red cells and that on the surface of endothelial cells. Unlike the expression of ABH on red cells, which is restricted to humans and some anthropoid apes (chimpanzee, gorilla, orangutan, gibbon), the expression of ABH antigen in other tissues is widely distributed throughout the Primate taxon, as it is, indeed, in other vertebrates.

Seemingly contradicting the above described phylogenetic pathway of the tissue expression of the ABH antigens is, for instance, the observation that pigs can have A-positive or A-negative red cell types. However, the porcine red cell (and leucocyte) A antigen is not part of the glycoproteins of the cell membrane but results from passive adsorption of circulating A glycosphingolipid. A-negative pigs, by contrast, have a circulating glycosphingolipid of type H (Oriol et al. 1994).

Although in tests using routine anti-A and anti-B reagents the red cells of lower monkeys appear to be lacking ABH antigens, weak A or B expressions were still demonstrated by the very sensitive absorption-elution techniques (Nakajima et al. 1993) or by very potent anti-A or anti-B reagents.

In the absence of silent alleles at the H and Se loci in almost all nonhuman primate species (only a single orangutan was found to be a nonsecretor among thousands of various primates tested and no equivalent of the Bombay phenotype was ever encountered in nonhuman primates) it has not been possible to study the role of the H- and Se-fucosyltransferases in the expression of histo-blood antigens in various tissues.

Practical Aspects of the Blood Group Serology of Apes and Monkeys

Serological affinities among blood groups of various members of the order Primates point not only to common pathways of molecular evolution but also to the similar biological significance of red cell antigens in the species of this taxonomic entity. If this is so, all progress made during the last 80 years in the detection and practical uses of the blood groups of humans should be equally applicable to nonhuman primates, at least those maintained in captivity.

Blood Transfusion

The utmost importance of blood groups, and, in fact, the main reason for their study, lies with the safety of blood transfusions between members of the same, or closely related, species. While in humans blood transfusion is one of the basic tools of modern medicine, there are still very few veterinarian-primatologists who resort to this therapeutic method in their everyday practice. Yet, numerous cases can be quoted in which sick animals or those incapacitated by experimentation were successfully treated and saved by administration of compatible homologous, or even heterologous, blood. The importance of compatibility of the transfused blood, so obvious in human medicine, is not yet generally appreciated by practitioners dealing with primate animals. There is no doubt, however, that, at least in most of the apes which show the ABO red cell polymorphism and simultaneous presence of anti-A and anti-B isoagglutinins in their sera, transfusion reactions following even the first infusion of ABO incompatible blood will be as severe and dramatic as those observed in human patients. Reactions due to red cell incompatibilities other than ABO will come into play during multiple transfusions, in which the agglutinating antibodies are formed as a result of previous nonmatched transfusions. The latter reactions will also occur when performing more than one transfusion in lower monkeys (macaques, baboons, etc.), which do not express ABH antigens on red cell antigens but which do have a number of highly immunogenic simian-type antigens on their red blood cells (see the chapter by Socha).

Transplantation

Since the ABO polymorphism is an exclusive attribute of the highest taxa in the animal kingdom, namely, simian primates and humans, apes and monkeys constitute a unique model for the in vivo study of ABO incompatibility in organ transplantation. Due to their relatively good availability and physical characteristics, the Old World monkeys, and particularly baboons and macaques, became experimental models of choice for allo- and xenotransplantations across the ABO barrier (Socha et al. 1987).

There are two features of ABO differentation that are of paramount importance for the fate of an ABO incompatible graft: (1) the presence of A, B and H substances in the transplanted organ or tissues that are directly involved in the engrafting process, and (2) the simultaneous existence of anti-A and/or anti-B antibodies in the recipient's serum that can directly react with homologous ABO epitopes in tissues of the transplant.

We have already mentioned the presence of ABH antigens at the surface of endothelial cells. The antibody-antigen reaction that takes place at the surface of these cells activates the complement cascade, severely damaging the cells (Platt and Bach 1991). This results in capillary disruption and interstitial hemorrhage causing rapid destruction of the transplanted tissues. However, some ABO incompatible allografts are tolerated through a mechanism called *accomodation* (Platt et al. 1990). This phenomenon was first observed in the transplantation of ABO-incompatible kidney allografts (Chopek et al. 1987; Alexandre et al. 1987). Although the ABO incompatible allograft constitues a paradigm of accomodation, similar phenomena were observed in HLA incompatibility and in xenotransplantations (reviewed by Platt 1994). The mechanism by which accomodation occurs remains unclear but several mechanisms have been evoked. One mechanism involves a decrease in the antibody-antigen interaction that gives rise to the xenograft rejection. It is noteworthy that the removal of natural anti-A and/or anti-B agglutinins from the plasma of the recipient promotes the appearance of accomodation in the ABO-incompatible allograft model (Cooper et al. 1994). However, accomodation is observed in the absence of any specific immune intervention in approximately one third of patients who received an ABO incompatible heart (Cooper 1990), kidney (Cook et al. 1987; Chopek et al. 1987), or liver (Gordon et al. 1987). Long-term graft survival in these groups of patients contrasted with the presence of anti-A and/or anti-B antibodies. A similar spontaneous accomodation was demonstrated after ABO-incompatible heart transplantation in the baboon. In this species spontaneous accomodation is more frequent than in humans (two thirds of baboon recipients vs one third of human recipients) (Cooper at al. 1988). In the former species, early hyperacute or acute rejection occurred only when ABO incompatibility was present (Cooper et al. 1988). Experimental studies were developed in the baboon model to search for immune manipulations capable of promoting accomodation (reviewed by Cooper et al. 1994). Intravenous injection of A or B trisaccharides efficiently inhibited the fixation of anti-A or anti-B to the A or B antigens expressed by cells of allo-incompatible organ grafts. In baboons which were hypersensitized against the antigen A or B, the mean survival time of ABO-incompatible cardiac allografts was 19 min. With continuous injection of trisaccharide associated with conventional immunosuppressive therapy (cyclosporin, azathioprine, or cyclophosphamide and corticosteroids), mean allograft survival was more than 28 days (Copper et al. 1993a,b). The experimental ABO-incompatible allograft constitutes a powerful model for comprehending the pathophysiology of acute rejection of xenografts. The importance of natural antibodies against α-Gal epitope in the rejection of transplanted organs was demonstrated when pig organs were transplanted to species which do not express the α-Gal epitope (Galα1–3Galβ1–4GlcNAc-R) (Samuelsson et al. 1994).

Carbohydrates as Ligands for Selectins

Selectins are molecules implicated in cell adhesion by their capacity to specifically bind carbohydrate ligands expressed by other cells and tissue components. The selectin family encompasses three molecules: L-selectin (expressed by leucocytes); E-selectin (expressed by endothelial cells); P-selectin (expressed by platelets and present in the Wiebel Palade bodies of endothelial cells). The structure of selectins was deduced from cDNA sequences (for general reviews see Bevilacqua et al. 1993; Carlos and Harlan 1994). Selectins are transmembrane glycoproteins with a short intracytoplasmic segment which represents the carboxylic end of the molecule, a transmembrane domain, and an extracellular part encompassing three domains. The first extracytoplasmic domain is represented by a discrete number of short consensus repeats (SCRs) (approximately 60 amino acids each) similar to those found in certain complement binding proteins. This segment is followed by a domain showing homology with epidermal growth factor (EGF) which, in turn, is followed by the NH_2-terminal segment, homologous to the hepatic C-type lectins (Drickamer 1988). The term "selectin" was chosen to recall the *selective* adhesion induced by these molecules between blood cells and endothelial cells, and the presence of a *lectin*-like domain at the NH_2-terminal part of these molecules (for general reviews see Feizi 1993; Kuijpers 1993). The presence of this region, related to carbohydrate recognition domains of previously described calcium-dependent animal lectins, inspired a search for carbohydrate ligands.

Selectin L was first defined in mouse as a lymphocyte receptor, recognized by the monoclonal antibody MEL-14. This membrane receptor was identified as the lymphocyte receptor which mediates the specific binding of lymphocytes to the high endothelial venule (HEV) in the peripheral lymph nodes. It was shown that adhesion of lymphocytes to endothelial cells was hindered by sialidase treatment of the endothelial cells. These results suggested that the counter-receptor of MEL-14 on HEV was carbohydrate in nature. Other studies demonstrated that MEL-14 antigen is expressed also by murine neutrophils and monocytes and suggested a role for this molecule in leukocyte adhesion to endothelium at sites of inflammation.

The L-selectin, identified independently on human lymphocytes, neutrophils and monocytes, was shown to be the human counterpart of the murine molecule MEL-14 (Cameniri et al. 1989; Tedder et al. 1990). Although selectin L was the first to be recognized, the exact nature of its carbohydrate ligand still remains elusive.

More details are known about carbohydrates which act as ligands for E- and P-selectins. In the early 1990s three groups reported independently that the sialyl-Le^X structure could specifically block E-selectin adhesion (the other name of this molecule is ELAM1 for endothelial-leucocyte-adhesion-molecule). Moreover, transfection of COS cells (African green monkey kidney cells transformed by SV40 virus, and unable to synthesize the Le^X structure) with an expression vector containing a gene encoding an $\alpha(1-3)$fucosyltranferase, resulted in acquisition by the transfected cells of an E-selectin-binding activity (Lowe et al. 1990; Goeltz et al. 1990). Later on, Le^X, Le^a, sialyl-Le^a, and sulfated Le^a or Le^X were shown to be also efficient ligands for E-selectin (for general review see Gahmberg et al. 1992; Feizi 1993). Sialyl-Le^X (the sialylated derivative of $Le^{X)}$ also seems to be the ligand of P-selectin which is expressed selectively in α-granules of platelets and in the Wiebel Palade bodies of endothelial cells.

The Le^X antigen (also referred to as CD15) and sialyl-Le^X (also referred to as CD15s) are synthesized in some tissues (especially leucocytes) by the action of $\alpha(1,3)$fucosyltransferases which are different from the fucosyltransferase encoded by the *Le* (*FUT3*) gene. Two groups have recently described a fifth $\alpha(1,3)$fucosyltransferase gene (*FUT7*) that is thought to encode the enzyme involved in the biosynthesis of sialyl-Le^X on human leucocytes (Sasaki et al. 1994; Natsuka et al. 1994). It was evoked that the synthesis of LeX and sialyl-LeX could also depend on the action of the fucosyltransferase encoded by *FUT4* genes (Kannagi and Magnani 1995) (Table 10) The expression of Le^X and sialyl-Le^X in tissues is monomorphic in humans. Thus, although chemically related to the Lewis blood group substances, Le^X and sialyl-Le^X are not true histo-blood group antigens.

Table 10. Different human fucosyltransferase genes, cloned and registered in the Genome Data Base

GDB name	General designation	Transfer	Acceptor chain type	Cloning name	EMBL/ GenBank	Chromosome location
FUT1	H	α-2-fucose	2	H	M35531	19q13.3
FUT2	Secretor	α-2-fucose	1 more than 2	Se	U17894	19q13.3
FUT3	Lewis	α-3/4-fucose	1 more than 2	Fuc-TIII	X53578	19p13.3
FUT4	Myeloid	α-3-fucose	2	Fuc-TIV	M58596	11q21
FUT5	?.	α-3-fucose	2	Fuc-TV	M81485	19p13.3
FUT6	Plasma	α-3-fucose	2	Fuc-TVI	L01698	19p13.3
FUT7	Leukocyte	α-3-fucose	2	Fuc-TVII	U08112	9

The coupling sialyl-Le^X (or Le^X) – E-selectin is involved in the adhesion of leucocytes to endothelial cells which, in turn, preconditions the migration of leucocytes from blood stream to extravascular tissues (for a general review, see Carlos and Harlan 1994).

Leucocyte adhesion is a complex process: at first, leucocytes roll along the endothelium of postcapillary venules that drain the sites affected by an inflammation. Sialyl-Le^X (or Le^X) – E-selectin is involved in the rolling phenomenon, followed by firm adhesion of leucocytes through the interaction of integrins with ICAM. Leucocyte integrins are heterodimeric membrane proteins composed of the CD18 polypeptide (β chain of three leucocytes integrins) associated with various α chains, CD11a, CD11b, CD11c, corresponding to LFA1, Mac1 and p150/95 receptors, respectively. These three integrins and their counter-structures on endothelial cells (ICAM-1 or ICAM-2) play a crucial role in transendothelial migration of leucocytes (see LAD1 syndrome below).

The importance of the expression of sialyl-Le^X at the surface of leucocytes is emphasized by the severity of the immune deficiency related to its absence in a so-called leucocyte adhesion deficiency (LAD 2) syndrome. This syndrome was first described in two Israeli children with Bombay red cells, nonsecretor saliva and Le^a-, Le^b-, and Le^X-negative tissues (Etzioni et al. 1992). The levels of CD18 and CD11 a, b, and c on leucocytes were normal in both cases, and in this respect the syndrome described by Etzioni et al., differed from the earlier described LAD 1 syndrome. The latter syndrome is secondary to a congenital defect of the expression of CD18 (β chain of leucocyte integrins) which results in the absence of three leucocyte integrins (CD18-CD11a; CD18-CD11b; CD18-CD11c) and abolition of the capacity of leucocytes to migrate to inflammatory sites. Since the defects in Etzioni's LAD 2 syndrome affect all the fucosylated glycoconjugates, the primary defect probably concerns fucose metabolism (such as an incapacity to synthesize UDP-fucose).

The homozygosity for silent alleles at the ABO, Hh, Se, Le loci is not known to result in any pathologic condition; the respective genes, therefore, do not seem indispensable for any vital functions. In contrast, some fucosyltransferases (FUT4 and FUT7) are certainly indispensable for life and no silent mutations of these genes were ever observed.

Characterization of the A/B and H Glycosyltransferases

The glycosyltransferases of the ABO, H-h systems can be detected in human plasma and tissue extracts. Purification of the soluble form of the A galactosyltransferase has allowed determination of partial amino acid sequences. Oligonucleotide primers deduced from these amino acid sequences constituted the starting point for the synthesis (by the PCR method) of probes which, in turn, made possible the identification of a cDNA encoding the human group A-specific galactosyltransferase.

Characterization of soluble glycosyltransferases can also be carried out in nonhuman primates. The tests, based on demonstration of the enzymatic activities in the fluid to be studied, use as substrates either low molecular weight synthetic substrates or the red cells expressing the precursor oligosaccharide chain. Depending on the enzyme to be studied, the donor of the sugar is a UDP or GDP sugar. The activity of the enzyme is measured as the quantity of sugar transferred per unit time. With synthetic substrates, the sugar to be transferred is radiolabelled and the products of the reaction are fractionated by chromatography. The radioactivity of the fraction corresponding to the products of the reaction reflects the enzymatic activity. When O red blood cells are employed as the substrate, transfer of the sugar is revealed by agglutination of treated red cells by means of reagents specific for A or B substances.

Results obtained with synthetic sugars (Table 11) show that the H fucosyltransferase activity is detectable in the plasma of all primates studied and independent of their ABO type (Thome 1983). The enzymatic activities vary among species (in decreasing order): humans > chimpanzee >baboon > gorilla > macaque. The activity of chimpanzee plasma is comparable to that of human plasma, while baboons and gorilla show significantly lower activities than chimpanzee and humans. The activity in macaque is the lowest observed but is still easily detectable. A-specific or B-specific plasma enzymatic activity correlates with the ABO blood group of the animal (see Table 11).

The B-specific transferase was found in the plasma of gorillas and group B or AB cynomolgus macaques while the A-specific transferase was found in the plasma of group A chimpanzees as well as in group A and AB cynomolgus macaques. The plasma of baboons contained very low levels of the A-specific transferase.

The nonhuman primate enzymes add sugars to the terminal position of H substrates, as do their human counterparts. The specificities of the nonhuman primates' plasma enzymes, in terms of the sugar they bind to a given precursor substrate, are similar to those of their human homologues. However the B-transferase of gorilla has a parasite A-tranferase activity showing an intermediate specificity for the type of sugar to be transferred (this is reminiscent of the specificity of human transferase of the $B^{(A)}$ type or that of the enzyme encoded by the human *cis AB* gene). The very low expression of the B antigen at the surface of gorilla red cells is not related to a low B-specific transferase activity but to the very low expression of substrate H at the surface of the red cells of this species. The same conclusion can be drawn for cynomolgus macaques, which express sig-

Table 11. A-, B- and H-transferase enzymatic activities in the serum of nonhuman primates (Thome 1983)

Genus	Species	Individual designation	Blood group	Activity A^a cpm	Activity A^a pmoles	Activity B^b cpm	Activity B^b pmoles	Activity H^c cpm	Activity H^c pmoles
Homo	*Homo*	Ruf.	B	0	0	38984	73	46942	169
	sapiens	Hue.	A_1	25274	304	0	0	29193	105
		Bad.	A_1A_2	nt	nt	364	0	83409	300
		Rah.	A_2	12878	155	0	0	22023	79
Anthropoid apes									
Pan	*Pan tro-*	Vaillant	0	0	0	0	0	31725	114
	glodytes	Endoum	0	0	0	0	0	27555	99
		Judo	A_1	6123	74	0	0	69925	220
		Mopia	A_1	2121	25	0	0	35407	128
		Nestor	A_1	5484	65	0	0	30794	111
Gorilla	*Gorilla*	Typhen	B	0	0	67927	127	11202	40
	gorilla	Tèbe	B	645	7	69979	114	24800	89
		Djoutou	B	925	11	87891	164	13509	48
Old World monkeys									
Papio	*Papio*	Mandrill 7	A	390	3	0	0	20017	72
	sphinx	Mandrill 10	A	317	3	0	0	23526	85
		Mandrill 12	A	nt	nt	0	0	19739	71
		Mandrill 7	A	37	0	nt	nt	nt	nt
		Mandrill 10	A	337	4	nt	nt	nt	nt
		Mandrill 14	A	156	1	nt	nt	nt	nt
Macaca	*Macaca*	Cynomolgus 21	B	0	0	14025	26	2196	7
	fasciu-	Cynomolgus 24	A	4251	51	0	0	2981	10
	laris	Cynomolgus 25	B	0	0	6514	12	2658	9
		Cynomolgus 26	AB	3436	41	3354	6	4486	16
		Cynomolgus 27	A	4485	53	0	0	2903	10
		Cynomolgus 28	A	5245	63	0	0	3525	12
		Cynomolgus 63	B	130	1	14723	27	4567	16

nt, not tested; cpm, count per minute of radioactivity transferred to RBCs.

[a] Activity A: 18 h at 37 °C on 2'fucosyllactose.

[b] Activity B: 72 h at 37 °C on 2'fucosyllactose.

[c] Activity H: 72 h at 37 °C on phenylβDgalactoside.

nificant A- or B-transferase activities but only a low level of H substrate and display a very low H-fucosyltransferase activity in their plasma.

The results obtained with synthetic substrates were confirmed by tests performed with human O red cells as substrates (Thome 1983; Palatnik 1986; Nakajima et al. 1993). The in vitro conversion of human blood group O red cells into blood group A or B red cells was achieved by the action of chimpanzee and gorilla transferases, respectively (Thome 1983). The activities of both enzymes were comparable to those of their human counterparts. The white handed gibbon, agile gibbon and chimpanzee showed high A- or/and B-transferase activities (Nakajima et al. 1993). Significantly, the plasma of an orangutan, typed AB by means of hemagglutination tests, failed to convert human O red cells to either A or B (Nakajima et al. 1993).

Old World monkeys have A- and B-transferase activities in plasma that correspond to their A and B phenotypes, as established by saliva testing or by absorption-elution of anti-A and anti-B on the red cells (Thome 1983; Nakajima et al. 1993). The transferase activities in Old World monkey plasmas are significantly lower than those observed in humans and African great apes (chimpanzees and gorillas). B-transferase activity was evidenced in plasma of rhesus monkeys, Formosan and Japanese monkeys. Plasma of cynomolgus macaques contains A-and/or B-transferase activity corresponding to the ABO type of the individual. Neither A- nor B-transferase activities were detected in plasma of type O pig-tailed macaque and O patas monkey. The baboon plasmas failed to convert human O red cells, as judged by the very low A-transferase activity observed (Thome 1983).

Plasmas of the New World monkeys displayed A-or B-transferase activities, correlated with the ABO red cell phenotypes of the animals, as determined by absorption elution of anti-A and anti-B antibodies (Nakajima et al. 1993).

A-transferase activity was detected in the plasma of a Grant galago (*Galago senegalensis granti*) and B-transferase activity was identified in the plasma of a ring-tailed lemur. However, despite the presence of a B-like antigen on the red cells of Grant galago and common tree shrew (Tupaia), no B-transferase activity was found in their plasma (Nakajima et al. 1993).

Experiments using nonhuman primate O red cells as substrates showed that only chimpanzee red cells can be converted into A or B red cells by exposure to glycosyltransferases of human or nonhuman primate plasma (Thome 1983).

The optimal pHof A-transferases of Old World monkeys was comparable to that of human group A_2 (7.4 and 7.6, respectively), while the optimal pH of A-transferases of anthropoid apes was similar to that of human group A_1 (6.4) (Nakajima et al. 1993). The optimal pHs of B-transferases from human and nonhuman primates were between 6.2 and 6.5 (Nakajima et al. 1993).

Molecular Genetics of the ABO Histo-Blood Groups

cDNA Cloning of the A-Transferase

Glycosyl tranferases encoded by the *A* and *B* alleles of the ABO system are present in soluble form in plasma and in tissue extracts. In the late 1980s, a protein which seemed to correspond to the soluble form of human A-transferase was purified from the lungs of blood group A individuals (Clausen et al. 1990). This protein fraction was absent in the lungs of type O individuals. Three mouse monoclonal antibodies reacting with the protein epitopes rather than with carbohydrate epitopes of the purified protein (White et al. 1990) were found to partially inhibit not only the A- but also the B-transferase activity from plasma and the A-transferase from human lungs. Partial amino acid sequences of the purified protein were determined (Clausen et al. 1990), which subsequently allowed cDNA cloning of the protein (Yamamoto et al. 1990b).

Two degenerate oligonucleotides were deduced from the known protein sequence and used as primers to amplify the cDNA sequence. The PCR-amplified fragment was used as a probe to screen the library from a human gastric cancer cell line, MKN 45, chosen for its high expression of A antigen and high A-transferase activity. Several hybridizing clones were isolated and the insert from one of these was later used to rescreen the same library. Nucleotide sequences of the so established clones were determined. The FY-59-5 insert, which contained the entire coding sequence, was then used as a probe for Southern and northern hybridization experiments. No restriction fragment length polymorphism, following analysis with 12 restriction enzymes, was observed among DNA samples from individuals with different ABO blood groups. The result of northern hybridization experiments showed the presence of hybridizing messages in RNA from the cell lines expressing different ABO phenotypes.

The human ABO gene locus has been mapped to chromosome 9q34 through family studies and somatic cell genetics (Ferguson-Smith et al. 1976). DNA from a panel of human-hamster cell hybrids possessing various human chromosomes was amplified by PCR using three separate sets of ABO gene-specific primers.

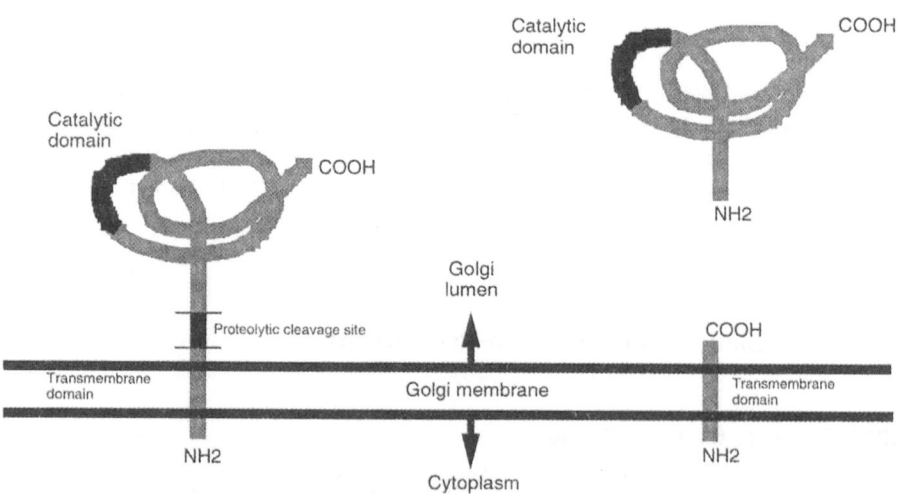

Fig. 4. Type II transmembrane topology of the ABO blood group glycosyltransferases. Mammalian glycosyltransferases are anchored within the membrane of the Golgi apparatus through their short, hydrophobic transmembrane domains. Functional soluble glycosyltransferases derive from membrane-bound proteins through proteolysis that dissociates the catalytic domain from its membrane-spanning domain

Amplification depended on the presence of human chromosome 9 in the hybrids, thus confirming the location of the human ABO locus on this chromosome (Yamamoto, unpublished).

Hydropathy analysis of the amino acid sequence revealed the presence of a hydrophobic region (16–19 amino acid residues) which probably has a double function: it constitutes a noncleavable signal anchor sequence and contains Golgi targeting signals (Joziasse 1992). The soluble forms of the enzymes derive from the anchored proteins by protease cleavage (Fig. 4).

Three Major Alleles

RNA from four human colon adenocarcinoma cell lines (SW 948, SW 48, COLO 205, and SW 1417) were converted into cDNAs which were cloned in a phage (λgt 10). These cell lines were obtained from patients with four different ABO genotypes: O/O, A/B, O/unknown, and B/B. These cDNA libraries were screened with the FY-59-5 probe and the inserts of hybridized clones were sequenced (Yamamoto et al. 1990a).

Comparisons of amino acid sequences deduced from cDNA sequences (see Table 12) lead to the following conclusions:
1. There are four amino acid substitutions (in positions 176, 235, 266, and 268) between A- and B-transferases, namely, arginine, glycine, leucine, and glycine in A-transferase, and glycine, serine, methionine, and alanine in B-transferase.
2. The purified protein from the MKN 45 cell line (as explained earlier, this cell line was used to isolate the initial cDNA probe) actually was an A-transferase.
3. O alleles contain a single nucleotide deletion (G residue at nucleotide 261) which causes frame shifting of codons and results in a protein without enzymatic activity of either A- or B-transferase.
4. ABO genotyping becomes possible based on differences in the nucleotide sequence.

Since this first description, another technique of genotyping was elaborated based on the amplification of ABO genes followed by restriction of the amplified fragments (Clausen et al. 1994; Olsson and Chester 1995).

The genomic organization of human histo-blood group ABO genes was characterized recently by Yamamoto et al. (1995). The entire coding region of the ABO cDNA clones is covered by seven exons. Exons VI and VII encode 77% of the full coding region and 91% of the coding region of the previously established catalytically active, soluble protein.

Subtyping of the ABO histo-blood groups has generally been limited to the coding region covered by exons VI and VII, which were analyzed by PCR amplification of genomic DNA (Grunnet et al. 1994). Alternative splicing of the ABO genes may exclude exons III–VI without changes in the reading frame. By reverse transcription followed by PCR using primers located in exon I and exon VII, it was demonstrated that different exon usages are found in different organs (Clau-

Table 12. Amino acid differences among primate A and B glycosyltransferases (positions 152 to 286)

	Amino acid position[a]																		
	153	156	157	163	169	174	176	195	197	198	211	214	235	240	266	268	276	283	334
Hosa A¹*	Thr	Pro	Ala	Thr	Gln	Glu	Arg	Phe	Glu	Arg	Val	Met	Gly	Ser	Leu	Gly	Gln	Ala	Ala
Hosa B*	–	Leu	–	–	–	–	Gly	–	–	–	–	–	Ser	–	Met	Ala	–	–	–[b]
Hosa A²*	–	Leu	–	–	–	–	–	–	–	–	–	–	–	–	–	Ala	–	–	–
Cis AB*	–	–	–	–	–	–	Gly	–	–	–	–	–	–	–	Met	Ala	–	–	–
B(A)*	–	–	–	–	–	–	Gly	–	–	–	–	–	–	–	Met	Arg	–	–	–
human O²	–	–	–	–	–	–	–	–	–	–	–	–	–	–	–	–	–	–	–
Chimp.1**	–	–	–	–	–	–	–	–	Gln	–	–	–	–	–	–	–	–	–	Ser
Chimp.2**	–	–	–	–	–	–	–	–	Gln	–	–	–	–	–	–	–	–	–	Ser
Patr1***	–	–	Pro	–	–	–	–	–	Gln	–	–	–	–	–	–	–	–	–	–
Patr2***	–	Leu	–	–	–	–	–	–	Gln	–	–	–	–	–	–	–	–	–	–
Patr3***	–	–	–	–	–	–	–	–	Gln	–	–	–	–	–	–	–	–	–	–
Gogo1***	–	–	–	–	–	–	–	–	–	–	–	–	–	–	Met	Ala	–	–	–
Gogo2***	–	–	–	–	–	–	–	–	–	–	–	–	–	–	Met	Ala	Lys	–	–
Gogo5***	–	–	–	–	–	–	–	–	–	–	–	–	–	–	Met	Ala	–	–	–
Gorilla**	–	–	–	–	–	–	–	–	–	–	–	–	–	–	Met	Ala	–	–	Thr
Popy***	Ser	–	–	–	Leu	Gly	Lu	Leu	–	–	–	Ile	–	Thr	–	–	–	–	–
Orang.1**	–	–	–	–	–	Gly	–	–	–	–	–	–	–	Thr	–	–	–	Thr	–
Orang.2**	–	–	–	–	–	Gly	–	–	–	–	–	–	–	Thr	–	–	–	–	–
Macaque**	–	–	–	Ala	–	Gly	–	–	–	Gln	–	–	Ala	–	–	–	–	–	–
Baboon(A)**	–	–	–	Ala	–	Gly	–	–	–	–	Ala	–	Ala	–	–	–	–	–	–
Baboon(B)**	–	–	–	Ala	–	Gly	–	–	–	–	Ala	–	Ala or Gly	–	Met	Ala	–	–	–

The amino acid sequences were deduced from the cDNA sequences. The numbers following the sequences (e.g. Chimp.1) designate the different types of sequences encountered by the authors: * Yamamoto and coworkers; ** Kominato et al. 1992; *** Martinko et al. 1993. For details regarding animals and their respective phenotypes and genotypes, see legend to Fig. 6. Note that in the study of Kominato et al. (1992), the phenotypes of chimpanzees, gorilla and orangutans were not provided.

[a] The numbering is given according to Yamamoto (1995) and Kominato et al. (1993). The underlined positions differentiate the human alleles A² and B.

[b] When compared to human A¹ allele, the human A² allele displays a deletion (on of the three cytosine positions 1059–1061). This deletion induces a frame shift resulting in an additional domain of 21 amino acids at hte COOH-terminal.

sen, unpublished data). Alternative splicing has already been observed with other glycosyltransferases (Joziasse 1992).

A^1 and A^2 Alleles

Transferase A activities in sera from A_1 and A_2 individuals were analyzed and compared: activity in A_1 subjects was found to be five to ten times higher than in A_2 subjects (Schachter et al. 1971). In addition, analyses of kinetic properties have shown qualitative differences in cation requirements, pH optima, and K_m values between A-transferases coded by A^1 and A^2 alleles (Schachter et al. 1973).

Only two differences have been identified between human A^1 and A^2 alleles (Yamamoto et al. 1992). Taking as reference the sequence of A^1, the A^2 allele is characterized by:

- a nucleotide substitution (C->T at nucleotide 467) that results in an amino acid substitution, i.e. proline (A_1-transferase) to leucine (A_2-transferase) at codon 156 (Table 12);
- a single base deletion (one of three C residues at nucleotides 1059–1061) that shifts the reading frame of codons resulting in an additional domain of 21 amino acids at the COOH-terminal of A_2-transferase (Yamamoto et al. 1992).

A and B Subtypes: A^3, B^3, and A^x Alleles

PCR allowed identification of molecular alterations associated with mutations leading to the subtypic alleles A^3, B^3 and A^x (Yamamoto et al. 1993c,d). The following amino acid substitutions were recognized:

- In two A^3 alleles: codon 291 aspartic acid to asparagine
- In one B^3 allele: codon 352 arginine to tryptophan
- In one A^x allele: codon 216 phenylalanine to isoleucine.

Introduction into the expression construction (pAAAA expression construct) of a single mutation (codon 291: aspartic acid to asparagine), which seemed to be specific for some A^3 alleles, diminished the A-transferase activity (Yamamoto 1995).

Cis AB

When compared with A^1 alleles, both cases of *cis-AB* alleles thus far investigated were found to possess two nucleotide substitutions resulting in two amino acid substitutions (proline to leucine at codon 156 and glycine to alanine at codon 268) (Yamamoto et al. 1993a). The $B^{(A)}$ allele had two nucleotide substitutions resulting in one amino acid substitution at codon 235 (serine to glycine) as compared with normal B alleles (Yamamoto et al. 1993c). Codon 156 of the human *cis*

AB allele encodes for a leucine, which was previously identified in all the *A²* alleles analyzed but only in a single *A¹* allele (Yamamoto et al. 1990b, 1992). It is noteworthy that codons 235 and 268 are the second and the fourth positions of the four amino acid substitutions which discriminate between A- and B-transferases (Yamamoto et al. 1990a). The alanine residue at codon 268 is found in B-transferase while the glycine residue at codon 235 is in A-transferase. Thus, the enzymes encoded by *cis-AB* and *$B^{(A)}$* alleles are chimeric A/B-transferases (Table 12).

Another Type of *O* Allele

The first to be identified and the most common *O* allele (now referred to as *O^1*) is characterized by a single nucleotide deletion in the 5′ end of the coding sequence (codon 261). This deletion induces a frameshift and the resulting protein (if it is translated from the mRNA) is a truncated 114 amino acid polypeptide without a functional catalytic site.

A variant, *$O^{1'}$*, allele displays additional substitutions (Yamamoto et al. 1993d). In the course of the structural analysis of subtypes, another type of *O* allele was found to be devoid of the single nucleotide deletion characteristic of *O^1* and *$O^{1'}$* (Yamamoto et al. 1993b). This third *O* allele, now called *O^2*, was originally identified in DNA from an individual with a B_3 phenotype. The genotype of this individual was later found to be *B^3/O*. When compared to the *A¹* allele, *O^2* is characterized by three nucleotide substitutions resulting in two amino acid substitutions (codon 176: arginine to glycine, and codon 268: glycine to arginine). Another case of allele *O^2* concerned an *O/O* individual of Danish origin (Grunnet et al. 1994). The sequence of the Danish gene was identical with that reported by Yamamoto et al. The frequency of the *O^2* allele in the Danish population was estimated at 7.3 %. Another study (Olsson and Chester 1996) demonstrated that the *O^2* allele was ten times less frequent than the *O^1* allele. While the *O^1* allele was found in all human populations studied so far (Caucasians, Blacks and American Indians) (Franco et al. 1994), the *O^2* allele was found in Caucasians and Blacks but not in American Indians (Franco et al. 1995). Further examples of *O* alleles lacking the deletion at nucleotide 261 have been reported recently (David et al. 1993). Experimental introduction of the mutations responsible for the two amino acid substitutions (arginine to glycine and glycine to arginine at codons 176 and 268) in the A-transferase gene, expressed in an in vitro model, resulted in a protein devoid of any transferase activity (Yamamoto et al. 1993b). The presence of an arginine in position 268 in the *O^2*-encoded protein (instead of a noncharged amino acid: glycine for *A* and alanine for *B*) renders the protein inactive. Such a profound effect of this substitution implies that residue 268 plays a crucial role in the catalytic function of the A- and B-transferases.

In early studies, Yoshida and coworkers detected in the serum of O individuals a protein cross-reacting with antibodies raised against the A and B glycosyltranferases (Yoshida et al. 1979). This protein could be the product of the *O^2*

allele because the predicted amino acid sequence of this product must display high cross-reactivity with such antibodies due to its similarity with the A- and B-glucosyltransferases. In contrast, the protein encoded by the allele O^1 most probably remains undetectable in plasma with anti-A-(or B-)glycosyltransferase antibodies.

Thus, the human silent allele O is heterogeneous and mutations which resulted in inactivation of the glycosyltransferase function were most probably not unique during evolution. It is not known at present whether the silent O alleles of chimpanzee, baboon, cynomolgus or vervet monkeys are similar to the O alleles of humans or, rather, are the products of different mutations.

Another Galactosyltransferase Gene

ABO genes show considerable homology with the $\alpha(1,3)$galactosyltransferase genes of cattle (Joziasse et al. 1989), mouse (Larsen et al. 1989), pig (Strahan et al. 1995) and New World monkeys (see the chapter by Galili). The $\alpha(1,3)$galactosyltransferase catalyzes the transfer of a galactose in $\alpha(1,3)$ linkage to Galβ-4GlcNAc. In humans, apes and Old World monkeys this gene is not functional. In humans, two pseudogenes were characterized, one on chromosome 9, close to the ABO locus, and the other on chromosome 12 (Joziasse et al. 1991). This second pseudogene lacks an intron and probably resulted from the insertion of a mature, retrotranscribed messenger (Joziasse et al. 1991). Unlike the situation in humans, only one pseudogene was characterized in Old World monkeys. Interestingly, the functional murine and pig genes are located on chromosomes homologous to human chromosome 9 (Strahan et al. 1995; Joziasse et al. 1991).

The genes of glycosyltransferases A/B and $\alpha(1,3)$galactosyltransferase belong to the same family of genes, which probably resulted from duplication events. As shown by Joziasse (1992), the homology between the two types of genes is limited to the last two coding exons in both genes, namely, exons VI and VII of the ABO genes and exons VIII and IX of the $\alpha(1,3)$galactosyltransferase genes. The sequence homology, as well as the similarity of the exon sizes, suggest an evolutionary relationship based on exon shuffling (Joziasse 1992). The intron/exon structure in the remaining 5' part of the ABO locus shows similarity to that of mouse. Interestingly, a similar structure in the reading frame of the exons was also found.

The ABO gene has a single base excess in the first two small exons (I and II), and the intervening exons III-VI are perfect multimers of three bases, whereas the last coding exon has a single base excess. The bovine $\alpha(1,3)$galactosyltransferase shows the same general pattern but has one coding exon less and thus two excess bases in the first coding exon (exon IV) (Joziasse et al. 1991).

The phylogenetic relationships between all available $\alpha(1,3)$galactosyltransferase sequences (Larsen et al. 1989, 1990; Joziasse et al. 1989, 1991; Galili and Swanson 1991), the sequences of the human A-type transferase (GenBank accession n[ii]. J05175), and a human pseudogene related to both types of gene (Yamamoto et al. 1991) were analyzed using the neighbor-joining method and the Jukes and

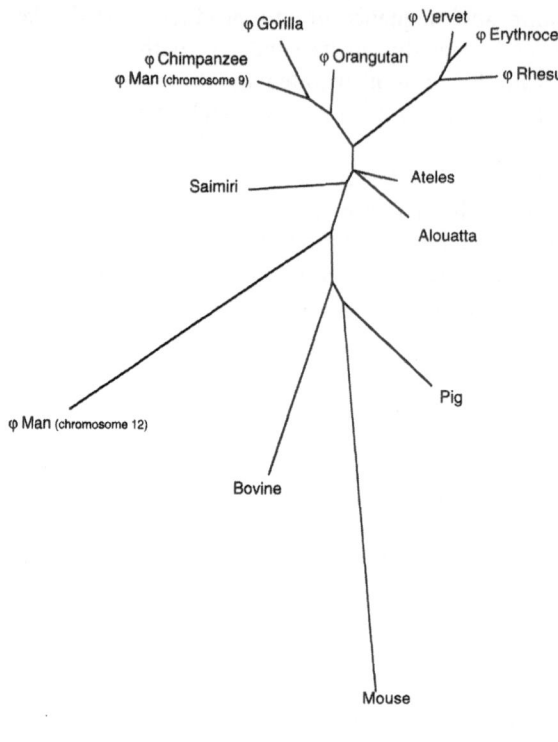

φ Gorilla
φ Vervet
φ Erythrocebus
φ Chimpanzee
φ Orangutan
φ Man (chromosome 9)
φ Rhesus
Saimiri
Ateles
Alouatta
Pig
φ Man (chromosome 12)
Bovine
Mouse

Fig. 5. Phylogenetic relationships between α(1,3)galactosyltransferase genes and corresponding pseudogenes (φ). In humans, pseudogenes are found on chromosomes 9 and 12 (Larsen et al. 1990). In apes, pseudogenes are found in chimpanzee (*Pan panicus*); orangutan (*Pongo pygmaeus*), and gorilla (*Gorilla gorilla*) (Galili and Swanson 1991). In Old World monkeys, pseudogenes are found in rhesus (*Macacca mulatta*), erythrocebus (*Erythrocebus patas*), and vervet (*Ceropithecus aethiops*) monkeys (Galili and Swanson 1991). In New World monkeys, functional genes are found in ateles (*Ateles geoffroyi*), alouatta (*Alouatta caraya*), and saimiri (*Saimiri sciureus*) monkeys. In other species functional genes are found in cows (*Bos taurus*) (Joziasse et al. 1989) and mice (*Mus muscularis*) (Larsen et al. 1989)

Cantor algorithm for distance matrix calculations (Gustafsson et al. 1994; Yamamoto et al. 1991). We carried out a comparable analysis (software courtesy of Dr. Feselstein, University of Washington) (Felsenstein 1995); the results are presented in Fig. 5. The analysis was restricted to the region available for all genes (positions 198 to 290 in the pig amino acid sequence).

The sequences of the pseudogene lying on human chromosome 9 and those of the pseudogenes of the anthropoid apes as well as of Old World monkeys are all situated on a unique arborescence that comprises two master branches: one with the pseudogene of human 9 chromosome and pseudogenes of apes, and the other containing sequences displayed by the Old World monkeys. The New World monkeys, which possess functional α(1,3)galactosyltransferase genes, have their sequences located on a unique arborescence where all functional genes are clustered together with the human pseudogene of chromosome 12. Thus, the two human pseudogenes are widely separated on this tree, which would suggest that the appearance of the human pseudogene of chromosome 12 preceded the inactivation of the human pseudogene located on chromosome 9. The appearance of the pseudogene of human chromosome 12 was estimated to have occurred 30 million years ago (Joziasse et al. 1991), probably before inactivation of the chromosome 9 α1,3-GT gene copy. Gustafsson et al. (1994) suggested that the mutation which resulted in the appearance of a stop codon in position 263 (as compared to the pig α(1,3)galactosyltransferase gene) most probably constituted the initial cause of inactivation of the pseudogenes of human chromosome 9 as well as of the pseudogenes of apes and Old World monkeys. Following that gene inactivation, other silent mutations accumulated which reinforced the nonfunctionality of those pseudogenes.

Evolution of the ABO Genes

It has long been known that ABH antigens are not restricted to humans, but are present in various organisms from bacteria on, up to the highest animal taxons (Kabat 1956; Springer 1970). The presence of A- and B-transferase activities has been documented in various species of animals (Tuppy and Schenkel-Brunner 1969; Baker et al. 1973; Palatnik 1986; Thome 1983). More recently, a human A-transferase cDNA probe was used to search for homologous sequence(s) in other organisms (Kominato et al. 1992). No hybridization was observed with DNA from bacteria (Escherichia coli), yeast (Saccharomyces cerevisiae), nematode (Caernohabditis elegans), fly (Drosophila melanogaster), clam (Mercenaria mercenaria), lobster (Homarus americanus), sea urchin (Strongylocentrotus purpuratus), or toad (Xenopus laevis). A very weak band was discernible in DNA from the chicken (Gallus domesticus). Strong hybridization was observed with DNA from all the mammalian species so far tested.

The nucleotide sequence of PCR-amplified partial fragments of the ABO genes from several primates was determined in order to identify the amino acid residue corresponding to codons at positions 176, 235, 266, and 268 of the human A- and B-transferases (Kominato et al. 1992; Martinko et al. 1993). Of the above four codons, the amino acid residues at positions 266 and 268 always conform with the ABO histo-blood group – as occurs in humans (see above); thus, leucine and glycine are present in A-transferase and methionine and alanine in B-transferase (Table 12). It seems, therefore, that these amino acid substitutions are crucial for the specificity of the transferase for the donor nucleotide sugar. A similar conclusion was drawn from experiments based on the transfection of HeLa cells with chimeric A-B genes (Yamamoto et al. 1990b). In the study of Kominato et al. (1992), only ABO phenotypes of baboons were known while the animals of other species were of unknown ABO types. It is possible to deduce from the residues in positions 266 and 268 that gorilla and chimpanzee transferases are of the B and A types, respectively, which is in accordance with the known high frequency of group A in chimpanzees and with the fact that all gorillas are group B.

It should be recalled that the transferase activity in gorilla plasma showed high B-transferase as well as weak A-transferase activities (see above, Thome 1983). When sequences of gorillas are compared with those of humans, they

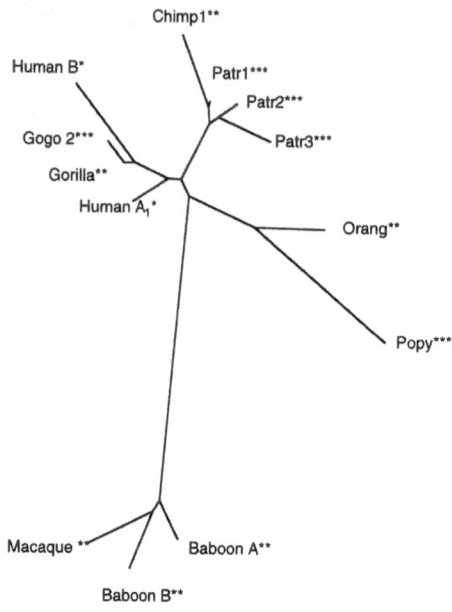

Fig. 6. Phylogenetic relationships between A and B genes in man and nonhuman primates. The sequences are those reported by *Yamamoto and coworkers; ** Kominato et al. 1992; and *** Martinko et al. 1993. In *Pan troglodytes*, (Chimp, Patr), the phenotype is A and the most probable genotype is A/A (or A/O). In *Pongo pygmaeus* (Orang, Popy) the phenotype is A and the most proabable genotype is A/A. In *Gorilla gorilla* (Gorilla, Gogo), the phenotype is B and the most likely genotype is B/B. In *Papio cynocephalus* (Baboon), the phenotype is A and the most likely genotype is A/A (or A/O); or B, with genotype B/B (or B/O); or AB with genotype A/B. In *Macaca fascicularis* (Macaque), both phenotype and genotype are unknown

appear as hybrid: type B in positions 266 and 268 and type A in positions 176 and 235. Thus, it may be that the residues in positions 176 and 235 influence the specificity of the enzyme for the nucleotide donor sugar

The above results provide molecular evidence for the trans-species evolution of ABO polymorphism (Martinko et al. 1993). The critical substitutions which differentiated the *A* and *B* genes occurred before divergence of the lineages leading to humans, chimpanzee, gorilla and orangutan, therefore no less than 13 millions years ago.

The presence of ABH antigens throughout the various taxa of Primates, from prosimians to humans, suggests that the ABO polymorphism predated the divergence of Old and New World monkeys and had been passed on, from species to species, for more than 37 million years (Ruffié 1973). A study of the phylogenetic relationship between A- or B-transferases of apes and those of two Old World monkey species is shown in Fig. 6. The phylogenetic tree was obtained by the neighbor joining method using, for distance calculation, the algorithm of Saitoue and Nei (software kindly provided by Dr J. Feselstein, University of Washington). The three sequences from Old World monkeys are all branched on a unique arborescence, apart from all ape and human sequences. This would indicate that the mutations which led to the appearance of the alleles *A* and *B* in baboon occurred independently from the mutations resulting in the same polymorphism in the common ancestor of humans and apes. This suggests that the polymorphism at the ABO locus was maintained throughout primate evolution because of its advantage for the species. Comparison of sequences of homologues of the ABO genes in species more distantly related to humans, such as New World monkeys and prosimians, could throw light on the emergence of the A/B polymorphism during primate evolution. We have already seen that, at least in humans, the allele O is heterogeneous which suggests that there was more than one event leading to inactivation of the A or B alleles. The silent O alleles were stabilized in the human population and maybe even in human ancestor populations. Comparison of the sequences of alleles O of various nonhuman primate species could be instrumental in tracing the origin of mutations which led to human O alleles.

Molecular Genetics of the Hh Blood Group System and Other Fucosyltransferases

The human locus H is localized on chromosome 19 (19q13.3) (Kelly et al. 1994; Reguine-Arnould et al. 1995). Humans have two alleles, H and h. The H allele encodes the FUT1 $\alpha(1,2)$fucosyltransferase, and the very rare allele h (present in the exceptional Bombay and para-Bombay individuals) encodes for a defective enzyme. The deficient allele h displays punctual mutations responsible for nonsense or missense mutations (Kelly et al. 1994).

The different fucosyltransferase genes that have been cloned are listed in Table 10.

The gene Se was first thought to correspond to a regulatory gene governing the expression of the H gene in cells of exocrine glands. Later, Oriol et al. (1981) hypothesized that the Se gene was rather a structural gene of the FUT2 $\alpha(1,2)$fucosyltransferase. This assumption was subsequently confirmed by biochemical studies (Kumazaki and Yoshida 1984). The gene was recently cloned (Kelly et al. 1995; Rouquier et al. 1995) and shown to encode a protein similar to the FUT1 enzyme. The Se gene is closely linked to the H gene on human chromosome 19 (Reguine-Arnould et al. 1995). Nonsecretor Caucasian individuals (20% of the population) are homozygous for the se allele. This allele was shown to correspond to a mutated Se gene, having a nonsense mutation in the coding region of the Se gene (Kelly et al. 1995). Recently, another se allele was described in Polynesians (Henry et al. 1996).

The FUT3 gene encodes the Lewis $\alpha(1,3/4)$fucosyltransferase. This enzyme catalyzes the synthesis of both the Le^a and Le^b substances, by adding a fucose to the type 1 chains, and the Le^X and Le^Y substances when the fucose is added to type 2 chains (Fig. 1). The Le locus lies on human chromosome 19 (19p13.3) (Reguine-Arnould et al. 1995). The gene is far from the Se and H loci, since no linkage has ever been detected between Se and Le loci. Cloning of the gene (Kukowska-Latallo et al. 1990) elucidated the molecular basis of its inactivated form (the amorphic allele le) (Mollicone et al. 1994b; Elgrem et al. 1993, 1996; Nishihara et al. 1994). Two types of missense mutations were found. The first alters the transmembrane domain of the enzyme and impairs its anchorage in the Golgi membrane; the second affects the catalytic site of the enzyme leading to the synthesis of a completely inactive protein.

As mentioned earlier, the FUT3 (Le) enzyme uses type 2 chains as acceptors, leading to the production of Le^X and Le^Y antigens. However, the latter substances are synthesized in tissues (especially leucocytes and brain) by the action of another fucosyltransferase, $\alpha(1,3)$fucosyltransferase (FUT4), which produces Le^X and Le^Y and a small amount of sialyl-Le^X. The FUT4 structural gene has been cloned (Goeltz et al. 1990; Lowe et al. 1991) and mapped to human chromosome 11 in band 11q21 (Tetteroo et al. 1987; Couillin et al. 1991; Reguigne et al. 1994). This enzyme shows only 60 % similarity with the FUT3 protein.

Three other $\alpha(1,3)$fucosyltransferase genes (FUT5, FUT6, and FUT7) were recently cloned (for a review, see Mollicone et al. 1994a; van den Eijnden and Joziasse 1993). FUT5 and FUT6 genes are located on chromosome 19, on the same band as FUT3. FUT3 lies on the 25 kb segment which separates the FUT5 and FUT6 genes, and the distance between FUT3 and FUT6 is less than 13 kb (Reguine-Arnould et al. 1995). The amino acid sequences of FUT3, FUT5 and FUT6 are very similar. Some 9 % of the population of the island of Java are deficient in FUT6 transferase, which constitutes the main part of fucosyltransferase found in plasma (Mollicone et al. 1994b).

The gene of the FUT7 fucosyltransferase is on human chromosome 9. Its expression seems to be restricted to the myeloid lineage. This enzyme catalyzes the synthesis of the sialyl-Le^X determinant which implies its participation in the synthesis of the lectin ligands on leucocytes (Sasaki et al. 1994; Natsuka et al. 1994).

The high degree of homology between the different human FUT genes suggests that they all derived, by multiple duplication events, from a common ancestral gene. It has been suggested (Weston et al. 1992) that a distant gene duplication event generated two distinct $\alpha(1,3)$fucosyltransferase genes on two different chromosomes (human chromosomes 19 and 11) and that, since then, their sequences have diverged significantly. More recent duplication events could have generated a series of very similar FUT genes (FUT3, FUT5 and FUT6) located on the long chromosome 19. However, other FUT genes (FUT7 on chromosome 9, FUT1 and FUT2 genes on chromosome 19) probably derived from an ancestor gene common to all human fucosyltransferases. By using human FUT1 (H), FUT2 (Se) and FUT3 (Le) cDNA probes, we demonstrated by Southern blot that nonhuman primates possess counterparts of these three genes (Apoil and Blancher unpublished data). Although it is impossible to assert that the sequences detected in the genomic DNA of nonhuman primates correspond to functional fucosyltransferase genes, it was found that each human cDNA probe gave a specific pattern of hybridization in each species. The length of restriction fragments detected by the human cDNA probes varied from species to species.

Thus, nonhuman primate genomic DNA samples were digested by three different restriction enzymes (HindIII or BamHI or EcoRI). After electrophoresis and blotting, restriction fragments were hybridized successively with the human cDNA probes FUT1 and FUT2. Patterns of hybridization revealed by FUT1 human cDNA probe on DNA samples (digested by BamHI or EcoRI or HindIII) of chimpanzee, gorilla and orangutan were identical to those obtained with human DNA samples used as control. The patterns obtained with gibbon and monkey genomic DNA differed from those of humans. Hybridization patterns were compatible in all apes and monkeys studied (chimpanzee, gorilla, orangutan, gibbon, macaques, baboon, marmoset, and squirrel monkey) with the presence of a single *FUT1* and a single *FUT2* gene per haploid genome. The intensity of hybridization was comparable to that observed with human genomic restriction fragments. Significantly, the intensity of hybridization obtained with lemur genomic DNA samples was very low compared to that obtained with human and simian DNA samples. This suggests that the degree of homology between the human *FUT1* and *FUT2* probes and the sequences of fragments they revealed in lemur DNA is probably very low.

Hybridization patterns revealed by the *FUT3* cDNA probe were more difficult to interpret because of the large number of bands revealed by this probe. This result is explained in humans by cross-hybridization of the cDNA *FUT3* probe with *FUT5* and *FUT6*, which are 85%–90% homologous to the *FUT3* gene (Mollicone et al. 1995). Depending on the enzyme used to digest human genomic DNA, the *FUT3* cDNA probe revealed three (*Eco*RI and *Bam*HI) to four (*Hind*III) restriction fragments. The human *FUT3* probe also revealed three to four restriction fragments in ape (chimpanzee, gorilla, orangutan) DNA. This suggests that great apes possess counterparts of the human *FUT3* gene and a number of genes homologous to the *FUT3* gene comparable to those observed in humans. The human cDNA *FUT3* probe hybridized at high stringency with three to four DNA restriction fragments of Old World monkeys. By contrast, the *FUT3* probe did not hybridize at high stringency with restricted DNA from marmoset and showed very faint hybridization with restricted DNA from squirrel monkey. It must be recalled that marmosets and squirrel monkeys do not secrete the Le[a] substance in saliva. One could hypothesize that New World monkeys either do not possess the equivalent of the human *FUT3* gene or that the counterpart of the *FUT3* gene in New World monkeys is not sufficiently homologous to the human probe to allow hybridization. As for the prosimians, no band was revealed by the *FUT3* cDNA probe with DNA from *Lemur macaco* and *Lemur fulvus*, either at low or high stringencies.

Functions of ABH and Lewis Blood Group Structures and Concluding Remarks

Despite some knowledge of the molecular basis of the antigens and of the pathways of their synthesis, and some comprehension of the role of glycosyltransferases and their genes, the physiological functions of the structures depending on the *ABO, Hh, Se* and *Le* loci are still not well understood. Thus, an understanding of the mechanisms that maintained ABO polymorphism in humans and in other primate species also remains inadequate.

None of the alleles at the *ABO, Hh, Se,* and *Le* loci appear to play an essential physiological function. For instance, the fact that group O individuals, who constitute a large proportion of many human populations, completely lack the terminal sugars added by the A- and B-transferases renders difficult the search for any specific function of the human ABO locus. Similarly, the existence of individuals homozygous for the silent alleles *se* or *le* suggests that the functions of the *Le* and *Se* genes are not vital. The discovery of apparently normal individuals carrying the very rare Bombay phenotypes reinforces the notion that the H substance and its derivatives (A, B, Le^a, Le^b, ALe^b, BLe^b, etc.) play no vital role in the development and maintenance of tissue integrity.

However, the presence at the surface of numerous microorganisms of oligosaccharides cross-reacting with the anti-A and anti-B antibodies (see chapter by Eder and Spitalnik) suggests that natural antibodies against the A and B antigens constitute a factor of natural defense. If so, the existence of anti-A and anti-B antibodies, which result from the polymorphic expression of antigens A and B, may be the key element among selective forces which maintained ABO polymorphim in humans and nonhuman primates. Yet, it is difficult to understand why the expression of ABH antigens on red cells varies from one species to another. We have said earlier in this chapter that, while the ABH antigens are expressed at similar levels on human as well as most anthropoid ape red cells, expression of A and B antigens is weak or undetectable on red cells of most Old and New World monkeys.

It seems, therefore, possible that the main function of the ABH substances is related to their presence in tissues and in secretions, while expression of ABH antigens on red cells could be an epiphenomenon limited to humans and anthropoid apes. The role of soluble substances in secretions and body fluids could be to saturate the potential receptors to blood group substances expressed by microorganisms. Saturation of these receptors could prevent the adhesion of microorganisms to the ABH/Le epitopes expressed on endo- and epithelial cells (for a detailed review, see the chapter by Spitalnik). For example, it was found that *Helicobacter pylori*, a bacterium associated with human gastroduodenal ulcers, expresses an adhesion molecule which binds to the Le^b molecules on the surface of epithelial cells (Boren et al. 1993). Adhesion of the bacterium to epithelial cells conditions the pathogenicity of the microorganism. It is probable that the association between blood group O and the risk of gastric ulcer is related to the affinity of the *Helicobacter pylori* receptor to the Le^b oligosaccharide. In fact, in type O individuals the number of Le^b sites available for the bacterial receptors is considerably higher than in individuals of group A or B, because in the latter groups the Le^b substance is modified to form ALe^b or BLe^b, respectively. Similarly, it was demonstrated that binding of uropathogenic *E. coli* to glycolipids extracted from vaginal cells depends on histo-blood group secretor status (Stapelton et al. 1992) and that the Lewis blood group phenotype is associated with recurrent urinary tract infections in women (Sheinfeld et al. 1989). By extension, one could consider that the polymorphisms at *ABO*, *Se*, and *Le* loci were maintained in humans (and in other primates) through pressure exerted by various microorganisms. Variations of allele frequencies among human populations (Bernard and Ruffié 1966, 1972) could result from epidemics which selectively decimated some groups in populations. However, one cannot exclude other mechanisms that could have caused variations in allele frequencies such as genetic drift, founder effect which took place at the time of migrations, or genetic bottlenecks in ancestral populations. Moreover, one has to remember that the numbers of alleles at the *ABO* locus varies among species: gorilla has only one allele (*B*), the chimpanzee has two (*A* and *O*) and humans, rhesus monkeys, crab-eating macaques, some baboon species, etc. possess all three alleles (*A*, *B* and *O*). Comparison of the nucleotide sequences encoding glycosyltransferases supports the notion that ABO polymorphism persisted throughout speciation (Martinko et al. 1993). One must assume, therefore, that the three major alleles at the *ABO* locus were present in the common ancestor of humans, anthropoid apes and Old World monkeys and that changes in allele frequencies took place at the time of, or after, speciation. To explain these interspecies differences, it is necessary to evoke specific causes which led to the disappearance of alleles *A* and *O* in gorilla and allele *B* in chimpanzee. Those factors, if they did exist, remain to be identified. Most probably, the causes which influenced allele fixation and allele frequencies in human and nonhuman primates vary from one species to another. For example, it is highly probable that the pygmy chimpanzee experienced a genetic bottleneck, because this species exhibits a remarkable monomorphism at various loci, such as *ABO*, *RCEF*, *MN*, *VABD* (see elsewhere in this volume). The high frequency of allele *O* in American Indians could be interpreted as a founder effect or as a consequence of the selective disappearance of alleles *A* and *B* in response to the action of some microorganisms which the migrants encountered in the South American continent and which selected the *O* allele over the *A* and *B* alleles.

We have seen that oligosaccharides of blood groups are involved in the adhesion of microorganisms to various cells. These oligosaccharides are also involved in adhesion processes between cells. The interest in the physiological role of the Lewis structures was renewed when it was demonstrated that the sialyl-Le^X structures are ligands for adhesion molecules. However, it must be noted that expression of Le^X on human leucocytes does not depend on the *Le* locus, but on the action of the FUT7 enzyme. Expression of Le^X on leucocytes is monomorphic in humans so Le^X is not a blood group in its true sense. However, we have seen that the gene encoding FUT7 belongs to a family of genes which, most probably, resulted from duplications of a common ancestral fucosyltranferase gene. The redundancy of the fucosyltransferase genes allowed their diversification. Variations between different fucosyltransferase genes are observed at the level of restriction of expression in tissues. All enzymes encoded by these genes transfer fucosyl residues, but the type of linkage between the fucose and the acceptor oligosaccharide varies from one enzyme to another (for a general review, see Oriol et al. 1995). As was certainly the case for other multiple gene families (Ruffié et al. 1982), the gene redundancy was the starting point of gene diversification and of emergence of polymorphisms at certain loci. For example, the silent allele *le* (or *se*) could have been fixed in human populations because the absence of the fucosyltransferases encoded by *FUT3* (or *FUT2*) had no serious deleterious effect. The multiplicity of the fucosyltransferases allowed some loci to undergo mutations, some of which submitted to the pressures of natural selection and to genetic drift, without, however, interference with the initial function of the ancestral gene necessary for maintaining the integrity of the organism. The same could be true for the evolution of the ABO blood group system. The A/B glycosyltransferases belong to a family of genes that encompasses at least the $\alpha(1,3)$galactosyltransferase. Here, again, the multiplicity of loci allowed fixation in populations of numerous mutated alleles at the *ABO* locus, and the study of oligosaccharide blood groups of humans and nonhuman primates illustrates the evolution of genes encoding glycosyltransferase enzymes.

Acknowledgements. We are indebted to Rafael Oriol, Professor Jean Ducos and Doctor Francis Roubinet, who generously accepted to review this text and gave us precious advice. We thank Christel Cherry for technical assitance in the preparation of the manuscript. Experimental work on fucosyltransferase genes in nonhuman primates was performed with funds from the Ministère de la Recherche, France, Université Paul Sabatier, Toulouse, Centre Régional de Transfusion Sanguine de Toulouse.

References

Alexandre GP, Squifflet JP, De Bruyere M, Latinne D, Reding R, Gianello P, Carleir M, Pirson V. Present experiences in a series of 26 ABO-incompatible living donor renal allografts. Transpl Proc 19: 4538–4542, 1987

Alroy J, Kerr D, Gavris V, Heaney JA, Bronson RT, Ucci A. Non-human primates as an animal model for studying the dynamics of A, B and H isoantigens in transitional cell epithelium (i.e., urothelium). In: Levy E. (ed) Advances in pathology, vol 2. Pergamon, Oxford, 1982

André AM, Courtois GM, Lennes G, Ninane G, Osterrieth PM. Mise en évidence d'antigènes de groupe sanguin A, B, O et Rh chez les singes chimpanzés. Ann Inst Pasteur 101:82–95, 1961

Arcilla MB, Sturgeon P. Le^x, the spurned antigen of the Lewis blood-group system. Vox Sang 26:425–438, 1974

Arend P, Nijssen J. A-specific autoantigenic ovarian glycolipid inducing production of "natural" anti-A antibody. Nature 269:255–257, 1977

Baker AP, Griggs LJ, Munro JT, Finkelstein JA. Blood group A active glycoproteins of respiratory mucus and their synthesis by an N-acetyl-galactosaminyltransferase. J Biol Chem 248:880–883, 1973

Ball SP, Tongue N, Gibaud A, Le Pendu J, Mollicone R, Gerard G, Oriol R. The human chromosome 19 linkage group FUT1 (H), FUT2 (SE), LE, LU, PEPD, C3, APC2, D19S7, and D19S9. Ann Hum Genet 55 (pt 3):225–233, 1991

Bernard J, Ruffié J. Hématologie Géographique. Ecologie humaine et caractères héréditaires du sang. Volume 1, Masson, Paris, 1966

Bernard J, Ruffié J. Hématologie géographique. Variations hématologiques acquises. l'hématologies et l'évolution, vol 2. Masson, Paris, 1972

Bernstein F. Ergebnisse einer biostatischen zusammenfassenden Betrachtung über die erblichen Blutstrukturen des Menschen. Klin Wochenschr 3:1495, 1924

Bernstein F. Zusammenfassende Betrachtungen über erbliche Blutstrukturen des Menschen. Z Indukt Abstamm-Vereb Lehre 37:237, 1925

Bevilacqua MP, Nelson RM. Selectins. J Clin Invest 91:379–387, 1993

Bhatia HM. The "Bombay" (Oh) blood group. Vox Sang 52:152–153, 1987

Bhende YM, Deshpande CK, Bhatia HM, Sanger R, Race RR, Morgan WTJ, Watkins WMA. "New" blood group character related to the ABO system. Lancet i:903, 1952

Boren T, Falk P, Roth KA, Larson G, Normack S. Attachment of Helicobacter pylori to human gastric epithelium mediated by blood group antigens. Science 262:1892–1895, 1993

Brett F, Jolly JJ, Socha WW, Wiener AS. Human-like ABO blood groups in wild Ethiopian baboons. Yearbook Phys Anthropol 20:276–280, 1976

Byles RH, Sanders MF. Intertroop variation in the frequency of the ABO alleles in a population of olive baboons. Int J Primatol 2:35–46, 1981

Camerini D, James SP, Stamenkovic I, Seed B. Leu-8/TQ1 is the human equivalent of the Mel-14 lymph node homing receptor. Nature 342:78–82, 1989

Carlos TM, Harlan JM. Leukocyte-endothelial adhesion molecules. Blood 84 (7):2068–2101, 1994

Cavalli-Sforza LL, Menozzi P, Piazza A. The history and geography of human genes. Princeton University Press, Princeton, 1994

Ceppellini R. On the genetics of secretor and Lewis characters; a family study. Proceedings of the 5th Congress of the International Society for Blood Transfusion, Paris. Karger, Basel, pp 207–211, 1956

Chopek MW, Simmons RL Platt JL. ABO incompatible renal transplantation:initial immunopathologic evaluation. Transpl Proc 19:4553–4557, 1987

Clausen H, Hakomori S. ABH and related histo-blood group antigens: Immunochemical differences in carrier isotypes and their distribution. Vox Sang 56:1–20, 1989

Clausen H, Levery SB, Nudelman E, Tsuchiya S, Hakomori S. Repetitive A epitope (type 3 chain A) defined by blood group A_1-specific monoclonal antibody TH-1: chemical basis of qualitative A_1 and A_2 distinction. Proc Natl Acad Sci USA 82:1199–1203, 1985

Clausen H, Levery SB, Nudelman E, Baldwin M, Hakomori S. Further characterization of type 2 and type 3 chain blood group A glycosphingolipids from human erythrocyte membranes. Biochemistry 25:7075–7079, 1986

Clausen H, White T, Takio K, Titani K, Stroud MR, Holmes E, Karkov J, Thim L, Hakomori S. Isolation to homogeneity and partial characterization of a histo-blood group A defined Fuc $\alpha_1 \rightarrow 2$ Gal $\alpha_1 \rightarrow 3$-N-acetylgalactosaminyltransferase from human lung tissue. J Biol Chem 265:1139–1145, 1990

Clausen H, Bennett EP, Grunnet N. Molecular genetics of ABO histo-blood groups. TCB 2:79–89, 1994

Conger JD, Chan MM, De Palma L. Analysis of the repertoire of human B-lymphocytes specific for type A and type B blood group terminal trisaccharide epitopes. Transfusion 33(3):200–207, 1993

Cook DJ, Graver B, Terasaki PI. ABO-incompatibility in cadaver donor kidney allografts. Transplant Proc 19:4549–4552, 1987

Cooper DKC. Clinical survey of heart transplantation between ABO-blood group incompatible recipients and donors. J Heart Transplant 9:376–380, 1990

Cooper DKC, Lexer G, Rose AG, Keraan M, Rees J, DuToit E, Oriol R. Cardiac allotransplantation across major blood group barriers in the baboon. J Med Primatol 17:333–346, 1988

Cooper DKC, Good AH, Koren E, Oriol R, Ippolito RM, Malcolm AJ, Neethling FA, Romano E, Zuhdi N. Specific intravenous carbohydrate therapy – a new approach to the inhibition of antibody-mediated rejection following ABO-incompatible allografting and discordant xenografting. Transplant Proc 25:377–378, 1993a

Cooper DKC, Ye Y, Niekrasz M, Kehoe M, Martin M, Neethling FA, Kosanke S, Debault L, Worsley G, Zuhdi N, Oriol R, Romano E. Specific intravenous carbohydrate therapy –

a new concept in inhibiting antibody-mediated rejection: expereince with ABO-incompatible cardiac allografting in the baboon. Transplant Proc 56:769–777, 1993b

Cooper DKC, Koren E, Oriol R. Oligosaccharides and discordant xenotransplantation. Immunol Rev 141:31–58, 1994

Couillin P, Mollicone R, Grisard MC, Gibaud A, Ravisé N, Feingold J, Oriol R. Chromosome 11q localization of one of the three expected genes for the human α-3-fucosyltransferases, by somatic hybridization. Cytogenet Cell Genet 56:108–111, 1991

David L, Leitao D, Sobrinho-Simoes M, Benett EP, White T, Mandel U, Dabelsteen E, Clausen H. Biosynthetic basis of incompatible histo-blood group A antigen expression: anti-A transferase antibodies reactive with gastric cancer tissue of type O individuals. Cancer Res 53:5495–5500, 1993

Decastello A, von Sturli A. Ueber die Isoagglutine im Serum gesunder und kranker Menschen. München Med Wochenschr 49:1090–1095, 1902

Downing HJ, Benimadho S, Bolstridge MC, Klomfass HL. The ABO blood groups in vervet monkeys (Ceropithecus pygerythrus). J Med Primatol 2:290–295, 1973

Downing HJ, Burgers LE, Getliffe FM. A-B-O blood groups of two subspecies of chacma baboons (Papio ursinus) in South Africa. J Med Primatol 4:103–107, 1975

Dracopoli NC, Jolly CJ. ABH salivary antigens in populations of vervet monkeys (Cercopithecus aethiops) from Kenya. Int J Primatol 4:383–397, 1983

Drickamer K. Two distinct classes of carbohydrate-recognition domains in animal lectins. J Biol Chem 263:9557–9560, 1988

von Dungern E, Hirszfeld L. Ueber Vererbung gruppenspezifische Strukturen des Blutes (III). Z Immunitatsf 8:526–530, 1911

Elmgren A, Rydberg L, Larson G. Genotypic heterogeneity among Lewis negative individuals. Biochem Biophys Res Com 196:515–520, 1993

Elmgren A, Börjeson C, Svensson L, Rydberg L, Larson G. DNA sequencing and screening for point mutations in the human Lewis (FUT3) gene enables molecular genotyping of the human Lewis blood group system. Vox Sang 70:97–103, 1996

Epstein AA, Ottenberg R. Simple method of performing serum reactions. 117–123, 1908

Erskine AG, Socha WW. The principles and practice of blood grouping, 2nd edn. Mosby, St. Louis, 1978

Etzioni A, Frydman M, Pollack S, Avidor I, Phillips L, Paulson JC, Gershoni-Baruch R. Brief report: recurrent severe infections caused by a novel leukocyte adhesion deficiency. N Engl J Med 327:1789–1792, 1992

Feizi T. Demonstration by monoclonal antibodies that carbohydrate structures of glycoproteins and glycolipids are onco-developmental antigens. Nature 314:53–57, 1985

Feizi T. Oligosaccharides that mediate mammalian cell-cell adhesion. Curr Opin Struct Biol 3:701–710, 1993

Felsenstein J. PHYLIP – Phylogeny inference package (version 3.57c), 1995

Ferguson-Smith MA, Aitken DA, Turleau C, Grouchy J. Localization of the human ABO. Np-1: AK-1 linkage group by regional assignment of AK-1 to 9q34. Hum Genet 34:35–43, 1976

Filitti-Wurmser S. Natural antibodies and immune antibodies of human ABO blood group system. Biochimie 58:1345–1353, 1976

Franco RF, Simoes BP, Guerreiro JF, Santos SEB, Zago MA. Molecular bases of the ABO blood groups of Indians from the Brazilian Amazon region. Vox Sang 67:299–301, 1994

Franco RF, Simoes BP, Zago MA. Relative frequencies of the two O alleles of the histo-blood ABH system in different racial groups. Vox Sang 69:50–52, 1995

Froehlich JW, Socha WW, Wiener AS, Moor-Jankowski J. Blood groups of the mantled howler monkeys (Allouatta palliata). J Med Primatol 6:219–231, 1977

Fukuda M. Cell surface glycoconjugates as onco-differentiation markers in hemopoietic cells. Biochimica et Biophysica Acta 780:119–150, 1985

Fukuda M, Fukuda MN. Changes in cell surface glycoproteins and carbohydrate structures during the development and differentiation of human erythroid cells. J Supramol Struct 17:313–324, 1974

Gahmberg CG, Kotovuori P, Tontti E. Cell surface carbohydrate in cell adhesion. Sperm cells and leukocytes bind to their target cells through specific oligosaccharide ligands. APMIS Suppl 27 (100):39–52, 1992

Galili U, Swanson K. Gene sequences suggest inactivation of α-1,3-galactosyltransferase in catarrhines after the divergence of apes from monkeys. Proc Natl Acad Sci USA vol 88:7401–7404, 1991

Gengozian N. Human A- and B-like antigens on red cells of marmosets. Proc Soc Exp Biol (NY) 177:858, 1966

Gerard G, Vitrac D, Le Pendu J, Muller A, Oriol R. H-deficient blood groups (Bombay) of Reunion Island. Am J Hum Genet 34:937–939, 1982

Goeltz SE, Hession C, Goff D, Griffiths B, Tizard R, Newman B, Chi-Rosso G, Lobb R. ELFT: a gene that directs the expression of an ELAM-1 ligand. Cell 63:1349–1356, 1990

Gordon RD, Iwatuki S, Esquivel CO, Todo S, Makowka L, Tzakis A, Marsh JW, Starzl TE. Experience with primary liver transplantation across ABO blood groups. Transplant Proc 19:4575–4579, 1987

Grubb R. Observations on the human blood group system Lewis. Acta Pathol Microbiol Scand 28:61–81, 1951

Grunnet N, Steffensen R, Bennett EP, Clausen E. Evaluation of histo-blood group ABO genotyping in a Danish population: frequency of a novel allele defined as O^2. Vox Sang 67:210–215, 1994

Gustafsson K, Strahan K, Preece A. α1,3galactosyltransferase: a target for in novo genetic manipulation in xenotransplantation. Immunol Rev 141:59–70, 1994

Hakomori S. Aberrant glycosylation in tumours and tumour-associated carbohydrate antigens. Adv Cancer Res 52:257–331, 1989

Hakomori S. Possible functions of tumor-associated carbohydrate antigens. Curr Opin Immunol 3:646–653, 1991

Hakomori S. Aberrrant glycosylation in cancer cell membranes as focused on glycolipids: overview and perpespectives. Cancer Res 45:2405–2414, 1995

Harmening D. Modern blood banking and transfusion practices, 2nd edn. Davis, Philadelphia, 1993

Henry S, Oriol R, Samuelsson B. Lewis histo-blood group system and associated secretory phenotypes. Vox Sang 69:166–182, 1995

Henry S, Mollicone R, Lowe JB, Samuelsson B, Larson G. A second nonsecretor allele of the blood group α(1,2)Fucosyltransferase gene (FUT2). Vox Sang 70:21–25, 1996

Holgersson J, Breimer ME, Samuelsson BE. Basic biochesmistry of cell surface carbohydrates and aspects of the tissue distribution of histo-blood group ABH and related glycosphingolipids. APMIS Suppl 27 (100):18–27, 1992

Inoue H, Hirohashi S, Shimosato Y, Enjoji M, Clausen H, Hakomori SI. Etablishment of an anti-A human monoclonal antibody from a blood group A lung cancer patient: evidence for the occurrence of autoimmune response to difucosylated type-2 chain A. Eur J Immunol 19:2197–2203, 1989

Jolly DJ, Turner TR, Socha WW, Wiener AS. Human-type A-B-O blood antigens in Ethiopian vervet monkeys (Cercopithecus aethiops) in the wild. J Med Primatol 6:54–57, 1977

Joziasse DH. Mammalian glycosyltransferases: genomic organization and protein structure. Glycobiology 2:271–277, 1992

Joziasse DH, Shaper JH, Jabs EW, Shaper NL. Characterization of an α1 → 3-galactosyltransferase homologue on human chromosome 12 that is organized as a processed pseudogene. J Biol Chem 266(11):6991–6998, 1991

Joziasse DH, Shaper JH, Vand den Eijnden DH, Van Tunen AJ, Shaper NL. Bovine α1 → 3-galactosyltransferase: isolation and characterization of a cDNA clone. J Biol Chem 264(24):14290–14297, 1989

Kabat EA. Blood group substances: their chemistry and immunochemistry. Academic, New York, 1956

Kannagi R, Magnani L. CD15s (sLe^x) cluster report. In: Schlossman et al (eds) Leucocyte typing V, white cell differentiation antigens, vol 2. Oxford University Press, Oxford, 1995

Kelly RJ, Ernst LK, Larsen RD, Bryant JG, Robinson JS, Lowe JB. Molecular basis for H blood group deficiency in Bombay (Oh) and Para-Bombay individuals. Proc Natl Acad Sci USA 91:5843–5847, 1994

Kelly RJ, Rouquier S, Giorgi D, Lennon GG, Lowe JB. Sequence and expression of a candidate for the human secretor blood group a1,2-fucosyltransferase gene (FUT2). J Biol Chem 270:4640–4649, 1995

Kominato Y, McNeill PD, Yamamoto M, Russell M, Hakomori S, Yamamoto F. Animal histoblood group ABO genes. Biochem Biophys Res Commun 189:154–164, 1992

Kuijpers TW. Terminal glycosyltransferase activity: a selective role in cell adhesion. Blood 81(4):873–882, 1993

Kukowska-Latallo JF, Larsen RD, Nair RP, Lowe JB. A cloned human cDNA determines expression of a mouse stage-specific embryonic antigen and the Lewis blood group α1,3/1,4-fucosyltransferase. Gen Develop 4:1288–1303, 1990

Kumazaki T, Yoshida A. Biochemical evidence that secretor gene, Se, is a structural gene encoding a specific fucosyltransferase. Proc Natl Acad Sci USA 81:4193–4197, 1984

Laine RA, Rush JS. Chemistry of human erythrocyte polylactosamine glycopeptides (erythroglycans) as related to ABH blood group antigenic determinants. Adv Exp Med Biol 228:331–347, 1988

Landsteiner K. Ueber Agglutinationserscheinungen normalen menschlichen Blutes. Wien Klin Wochenschr 14:1132–1134, 1901

Landsteiner K, Miller CP. Serological studies on the blood of primates, I. The differentiation of human and anthropoid bloods. J Exp Med 42:841–852, 1925a

Landsteiner K, Miller CP Jr. Serological studies on the blood of the primates. II. The blood groups in anthropoid apes. J Exp Med 43:853–862, 1925b

Landsteiner K, Miller CP. Serological studies on the blood of primates, III. Distribution of serological factors related to human isoagglutinogens in the blood of lower monkeys. J Exp Med 42:863–872, 1925c

Larsen RD, Rajan VP, Ruff MM, Kukowska-Latallo J, Cummings RD, Lowe JB. Isolation of a cDNA encoding a murine UDP galactose: β-D-galactosyl-1,4-N-acetyl-D-glucosaminide α-1,3-galactosyltransferase: expression cloning by gene transfer. Proc Natl Acad Sci USA 86:8227–8231, 1989

Larsen RD, Rivera-Marrero A, Ernst LK, Cummings RD, Lowe JB. Frameshift and nonsense mutations in a human genomic sequence homologous to a murine UDP-Gal: β-D-Gal (1,4)-D-GlcNAc α(1,3)-galactosyltransferase cDNA. J Biol Chem 265(12):7055–7061, 1990

Le Pendu J, Clamagirand-Mulet C, Cartron J P, Gerard G, Vitrac D, Oriol R. H-deficient blood groups of Reunion Island. III. α-2-L-fucosyltransferase activity in sera of homozygous and heterozygous individuals. Am J Hum Genet 35:497–507, 1983a

Le Pendu J, Gerard G, Vitrac D, Juszczak G, Liberge G, Rouger P, Salmon C, Lambert F, Dalix AM, Oriol R. H-deficient blood groups of Reunion Island. II. Differences between Indians (Bombay Phenotype) and Whites (Reunion Phenotype). Am J Hum Genet 35:484–496, 1983b

Le Pendu J, Lambert F, Samuelsson BE, Breimer ME, Seitz RC, Urdaniz MP, Suesa N, Ratcliffe M, Francoise A, Poschmann A, Vinas J, Oriol R. Monoclonal antibodies specific for type 3 and type 4 chain-based blood group determinants: relationship to the A1 and A2 subgroups. Glycoconjugate J 3:255–258, 1986

Lowe JB. Biochemistry and biosynthesis of ABH and Lewis antigens. Characterization of blood group-specific glycosyltransferases. In: Cartron J-P, Rouger P (eds) Molecular basis of major human blood group antigens. Plenum, New York, pp 75–115 (Blood cell biochemistry, vol 6), 1995

Lowe JB, Stoolman LM, Nair RP, Larsen RD, Berhend TL, Marks RM. ELAM-1-dependent cell adhesion to vascular endothelium determined by a transfected human fucosyltransferase cDNA. Cell 63:475–484, 1990

Lowe JB, Kukowska-Latallo JF, Nair RP, Larsen RD, Marks RM, Macher BA, Kelly RJ, Ernst LK. Molecular cloning of a human fucosyltransferase gene that determines expression of the Lewisánd VIM-2 epitopes but not ELAM-1-dependent cell adhesion. J Biol Chem 266:17467–17477, 1991

Mandel U. Carbohydrates in oral and secretions: variations with cellular differentiation. APMSI Suppl 27 (100):119–129, 1992

Martinko JM, Vincek V, Klein D, Klein J. Primate ABO glycosyltransferases: evidence for trans-species evolution. Immunogenetics 37:274–278, 1993

Mollicone R, Bara J, Le Pendu J, Oriol R. Immunohistologic patern of type1 (Le^a, Le^b) ad type2 (X, Y, H) blood group-related antigens in the human pylori and duodenal mucosae. Lab Invest 53:219–227, 1985

Mollicone R, Davies DR, Evans B, Dalix AM, Oriol R. Cellular expression and genetic control of ABH antigens in primary sensory neurons of marmoset, baboon and man. Journal of Neuroimmunology 10: 255–269, 1986

Mollicone R, Caillard T, Le Pendu J, François A, Sansonetti N, Villarroya H, Oriol R. Expression of ABH and X (Le^x) antigens on platelets and lymphocytes. Blood 71: 1113–1119, 1988a

Mollicone R, Dalix AM, Jacobsson A, Samuelsson BE, Gerard G, Crainic K, Caillard T, Le Pendu J, Oriol R. Red cell H-deficiency, salivary ABH secretor phenotype of Reunion Island. Genetic control of the expression of H antigen in the skin. Glycoconjugate J 5: 499–512, 1988b

Mollicone R, Candelier JJ, Fletcher A, Reguigne I, Couillin P, Oriol R. Molecular biology of fucosyltransferases (H, Se, Le, FUT4, FUT5 and FUT6). Transfusion Clin Biol 2: 91–97, 1994a

Mollicone R, Reguigne I, Kelly RJ, Fletcher A, Watt J, Chatfield S, Aziz A, Cameron HS, Weston BW, Lowe JB. Molecular basis for Lewis a(1,3/1,4)-fucosyltransferase gene deficiency (FUT3) found in Lewis-negative Indonesian pedigrees. J Biol Chem 269: 20987–20994, 1994b

Mollicone R, Cailleau A, Oriol R. Molecular genetics of H, Se, Lewis and other fucosyltransferase genes. TCB 4: 235–242, 1995

Mollison PL. Blood transfusion in clinical medicine, 6th edn. Blackwell, Oxford, 1979

Mollison PL, Engelfriet CP, Contreras M. Blood transfusion in clinical medicine, 9th edn. Blackwell, London, 1994

Moor-Jankowski J, Wiener AS. Blood groups of apes and monkeys; human-type and simian-type. In: Starck, Schneider and Kuhn (eds) Progress in primatology. Fischer, Stuttgart, pp 384–410, 1967a

Moor-Jankowski J, Wiener AS. Seroprimatology, a new discipline. In: Starck, Schneider and Kuhn (eds) Progress in primatology. Fischer, Stuttgart, pp 378–381, 1967b

Moor-Jankowski J, Wiener AS. Red cell antigens of primates. In: Fiennes RN TT-W (ed) Pathology of simian primates, part I. Karger, Basel, pp 270–317, 1972

Moor-Jankowski J, Wiener AS, Gordon EB. Blood groups of apes and monkeys. The A-B-O blood group in baboons. Transfusions 4: 92–100, 1964

Moor-Jankowski J, Wiener AS, Socha WW, Gordon EB, Mortelmans J. Blood groups of the dwarf chimpanzee (Pan paniscus). J Med Primatol 1: 90, 1972

Moore SJ, Green C., 1987) The identification of Rhesus polypeptide-blood group ABH-active glycoprotein complex in the human red cell membrane. Biochem J 244:735–741

Mourant AE, Kopec AC, Domaniewska-Sobczak K. The ABO blood groups. Blackwell, Oxford, 1958

Nakajima T, Furukawa K, Takenaka O. Blood group A and B glycosyltransferase in non human primate plasma. Exp Clin Immunogenet 10: 21–30, 1993

Natsuka S, Gersten KM, Zenita K, Kannagi R, Lowe JB. Molecular cloning of a cDNA encoding a novel human leukocyte α1,3-fucosyltransferase capable of synthesizing the sialyl Lewis determinant. J Biol Chem 269(24):16789–16794, 1994

Nishihara S, Narimatsu H, Iwasaki H, Yazawa S, Akamatsu S, Ando T, Seno T, Narimatsu I. Molecular genetic analysis of the human Lewis histo-blood group system. J Biol Chem 269: 29271–29278, 1994

Nyholm P, Samuelsson BE, Breimer M, Pascher I. Conformational analysis of blood group A active glycosphingolipids using HSEA calculations. The possible significance of the core oligosaccharide chain for the presentation and recognition of the A-determinant. J Mol Recognition 2:103–113, 1989

Olsson ML, Chester MA. A rapid and simple ABO genotype screening method using a novel B/O^2 versus A/O^2 discriminating nucleotide substitution at the ABO locus. Vox Sang 69:242–247, 1995

Olsson ML, Chester MA. Frequent occurrence of a variant O^1 gene at the blood group ABO locus. Vox Sang 70: 26–30, 1996

Oriol R. ABO, Hh, Lewis, and secretion. Serology, genetics and tissue distribution. In: Cartron J-P, Rouger P (eds) Molecular basis of major human blood group antigens, blood cell biochemistry, vol 6. Plenum, New York, pp 37–73, 1995

Oriol R, Danilovs J, Hawkins BR. A new genetic model proposing that the Se gene is a structural gene closely linked to the H gene. Am J Hum Genet 33: 421–431, 1981

Oriol R, Cooper JE, Davies DR, Keeling PWN. ABH antigens in vascular endothelium and some epithelial tissues of baboons. Lab Invest 50(5): 514–518, 1984

Oriol R, Le Pendu J, Mollicone R. Genetics of ABO, H, Lewis, X and related antigens. Vox Sang 51:161–171, 1986

Oriol R, Samuelsson BE, Messster L. ABO antibodies: serological behaviour and immunochemical characterization. J Immunogenet 17:279–299, 1990

Oriol R, Mollicone R, Couillin P, Dalix A-M, Candelier J-J. Genetic regulation of the expression of ABH and Lewis antigens in tissues. APMIS Suppl 27(100):28–38, 1992

Oriol R, Barthod F, Bergemer A-M, Ye Y, Koren E, Cooper DKC. Monomorphic and polymorphic carbohydrate antigens on pig tissues. Implications for xenotransplantation in the pig-to-human model. Transpl Int 7: 405–413, 1994

Palatnik M. Blood group B gene-specified transferase in rhesus monkey (Macaca mulatta). Intersciencia 11(1): 25–27, 1986

Platt JL. A perspective on xenograft rejection and accommodation. Immunol Rev 141:127–149, 1994

Platt JL, Bach FH. The barrier to xenotransplantation. Transplantation 52:937–947, 1991

Platt JL, Vercellotti GM, Dalmasso AP, Matas AJ, Bolman RM, Najarian JS, Bach FH. Transplantation of discordant xenografts: a review of progress. Immunol Today 11:450–456, 1990

Proceedings of the Second International Workshop on Monoclonal Antibodies Against Red Blood Cells and Related Antigens. Lund, Sweden, 1–4 April 1990). J Immunogenet (Appendix) 17:349–353

Prokop O, Rachwitz A, Schlesinger D. A "new" human blood group receptor A$_{hel}$ tested with saline extracts from Helix hortensis (garden snail). J Forensic Med (South Africa) 12:108–111, 1965

Rearden A, Elmajian DA, Baird SM. Comparison of human and siamang ABH and MN blood groups using monoclonal antibodies. J Med Primatol 13: 315–325, 1984

Reguine I, James MR, Richard III CW, Mollicone R, Seawright A, Lowe JB, Oriol R, Couillin P. The gene of myeloid alpha3 fucosltrnasferase (FUT4) is located between D11S388 and D11S919 on 11q21. Cytogenet Cell Genet 66:104–106, 1994

Reguine-Arnould I, Couillin P, Mollicone R, Faure S, Kelly RJ, Lowe JB, Oriol R. Relative positions of two clusters of human a-L-fucosyltransferases in 19q (FUT1, FUT2) and 19p (FUT3, FUT5, FUT6). Cytogenet Cell Genet 71:158–162, 1995

Rieben R, Tucci A, Nydegger UE, Zubler RH. Self tolerance to human A and B histo-blood group antigens exists at the B cell level and cannot be broken by potent polyclonal B cell activation in vitro. Eur J Immunol 22:2713–2717, 1992

Rouger P, Poupon R, Gane P, Mallissen B, Darnis F, Salmon C. Expression of blood group antigens including HLA markers in human adult liver. Tissue Antigens 27:78–86, 1986

Rouquier S, Lowe JB, Kelly RJ, Fertitta AL, Lennon GG, Giorgi D. Molecular cloning of a human genomic region containing the H blood group α1,2-fucosyltransferase gene and two H locus-related DNA restriction fragments. J Biol Chem 270:4632–4639, 1995

Ruffié J. Les données de l'immunogénétique et le processus de spéciation chez les primates C R. Acad Sci Paris D 276:2101–2104, 1973

Ruffié J, Socha WW. Les groupes sanguins erythrocytaires des primates non-hominiens. Nouv Rev Fr Hematol 22: 147–209, 1980

Ruffié J, Moor-Jankowski J, Socha WW. Immunogenetic evoltion of primates. In: Chiarelli AB, Corruccini RS (eds) Advanced views in primate biology. Springer, Berlin Heidelberg New York, pp 28–34, 1982

Salmon C, Cartron JP, Rouger P. The human blood groups. Masson, New York, 1984

Samuelsson B, Rydberg L, Breimer ME, Bäcker E, Gustavsson M, Holgersson J, Karlsson E, Uyterwa AC I, Cairns T, Welsh K. Natural antibodies and human xenotransplantaticn. Immunol Rev 141: 151–168, 1994

Sasaki K, Kurata K, Funayama K, Nagata M, Watanabe E, Ohta S, Hanai N, Nishi T. Expression cloning of a novel α1,3-fucosyltransferase that is involved in the biosynthesis of the sialyl Lewis'carbohydrate determinants in leukocytes. J Biol Chem 269(20):14730–14737, 1994

Schachter H, Michaels MA, Crookston MC, Tilley CA, Crookston JH. A quantitative difference in the activity of blood group A-specific N-acetyl-D-galactosaminyltransferase in serum from A$_1$ and A$_2$ human subjects. Biochem Biophys Res Commun 45:1011–1018, 1971

Schachter H, Michaels MA, Tilley CA, Crookston MC, Crookston JH. Qualitative differences in the N-acetyl-D-galactosaminyltransferases produced by human A^1 and A^2 genes. Proc Natl Acad Sci USA 70:220–224, 1973

Schmitt J. Immunobiologische Untersuchungen bei Primaten. Bibl Primatol. Karger, Basel, 1968

Seyfried H, Walewska I, Werblinska B. Unusual inheritance of ABO group in a family with weak B antigens. Vox Sang 9:268–277, 1964

Sheinfeld J, Schaffer AJ, Cordon-Cardo C, Rogatko A, Fair WR. Association of the Lewis blood-group phenotype with recurrent urinary tract infections in women. N Engl J Med 320(12):773–777, 1989

Smith HV, Kusel JR, Girdwood RWA. The production of human A and B blood group like substances by in vitro maintained second stage Toxocara canis larvae: their presence on the outer larval surfaces and in their excretions/secretions. Clin Exp Immunol 54:625–633, 1983

Socha WW. Blood groups of pygmy and common chimpanzees: a comparative study. In: Sussman RL (ed) The pygmy chimpanzee evolutionary morphology and behavior. Plenum, New York, pp 13–41, 1984

Socha WW, Ruffié J. Blood groups of primates. Theory, practice, evolutionary meaning. Liss, New York, 1983

Socha WW, Wiener AS. Problem of the C factor of the A-B-O blood group system. A critical historical review. N Y St J Med 73:2144, 1973

Socha WW, Wiener AS, Gordon EB, Moor-Jankowski J. Methodology of primate blood grouping. Transplant Proc 4:107–111, 1972

Socha WW, Wiener AS, Moor-Jankowski J, Mortelmans J. Blood groups of mountain gorillas (Gorilla gorilla beringei-). J Med Primatol 2:364, 1973

Socha WW, Wiener AS, Moor-Jankowski J, Jolly CJ. Blood groups of baboons. Population genetics of feral animals. Amer J Phys Anthropol 47:4453–442, 1977

Socha WW, Moor-Jankowski J, Ruffié J. Blood groups of primates: present status, theoretical implications and practical implications: a review. J Med Primatol 13:11–40, 1984

Socha WW, Marboe CC, Michler RE, Rose EA, Moor-Jankowski J. Primate animal model for the study of ABO incompatibility in organ transplantation. Transplant Proc 6:4448–4455, 1987

Socha WW, Blancher A, Moor-Jankowski J. Red cell polymorphism in nonhuman primates. A review. J Med Primatol 24:282–305, 1995

Solomon JM, Waggoner R, Leyshon WC. A quantitative immunogenic study of gene suppression involving A_1 and H antigens of the erythrocyte without affecting secreted blood group substances. The ABH phenotypes A_m^h O_m^h. Blood 25: 470–485, 1965

Springer GF. Role of human cell surface structures in interactions between man and microbes. Naturwissenschaften 57(4):162–171, 1970

Springer GF, Horton RE. Blood group isoantibody stimulation in man by feeding blood group-active bacteria. J Clin Invest 48(7):1280–1291, 1969

Springer GF, Tegtmeyer H. Absence of B antibody in a blood group A_1 person. Vox Sang 26:247–258, 1974

Stapleton A, Nudelman E, Clausen H, Hakomori SI, Stamm WE. Binding of uropathogenic Escherichia coli R45 to glycolipids extracted from vaginal epithelial cells is dependent on histo-blood group secretor status. J Clin Invest 90:965–972, 1992

Stejskal R, Mlsna J, Delort PJ, Davidsohn I. Localization of human A, B and H isoantigens in Cynomolgus monkey tissues. Experientia 36:1319–1321, 1980

Strahan KM, Feng Gu, Preece AF, Gustavsson I, Andersson L, Gustafsson K. cDNA sequence and chromosome localization of pig α1,3 galactosyltransferase. Immunogenetics 41:101–105, 1995

Szulman AE. The ABH and Lewis antigens of human tissues during prenatal and postnatal life. In: Mohn JF, Plunkett RW, Cunningham RK, Lambert RM (eds) Human blood groups. Karger, Basel, pp 426–436, 1977

Tanner MJA, Martin PG, High S. The complete amino acid sequence of the human erythrocyte membrane anion-transport protein deduced from the cDNA sequence. Biochem J 256:703–712, 1988

Tedder TF, Penta AC, Levine HB, Freedman AS. Expression of the human leukocyte adhesion molecule, LAM1. Identity with the TQ1 and Leu-8 differentiation antigens. J Immunol 144:532–540, 1990

Terao K, Fujimoto K, Cho F, Honjo SH. Inheritance mode and distribution of human-type ABO blood groups in the cynomologus monkeys. J Med Primatol 10:72–80, 1981

Tetteroo PAT, de Heij HT, van den Eijnden DH, Visser FJ, Scoenmaker E, Geurts van Kessel AHM. A GDP-fucose: [Galβ 1 → 4] GlcNAcα 1 → 3-fucosyltransferase activity is correlated with the presence of human chromosome 11 and the expression of the Le^x, Le^y

and sialyl-Le[x] antigens in human-mouse cell hybrids. J Biol Chem 262:15984–15989, 1987

Thome O. Mécanismes de biosynthèse des antigènes de groupes sanguins A, B, O chez les primates non-hominiens. Thesis, University Claude-Bernard, Lyon, 1983

Tuppy H, Schenkel-Brunner H. Formation of blood group A substance from H-substance by an α-N-acetylgalactosaminyltransferase. Eur J Biochem 58:152–157, 1969

Van den Eijnden DH, Joziasse DH. Enzymes associated with glycosylation. Curr Opin Struct Biol 3:711–721, 1993

Watkins WM. Blood-group substances. Sciences 152:172–181, 1966

Watkins WM. Molecular basis of antigenic specificity in the ABO, H and Lewis blood group systems. In: Montreuil J, Schachter H, Vliegenthart J.F.G. (eds) Glycoproteins. Elsevier, Amsterdam, pp 313–390, 1995

Weston BW, Smith PL, Kelly RJ, Lowe JB. Molecular cloning of a fourth member of a human α(1,3) fucosyltransferase gene family. Multiple homologous sequences that determine expression of the Lewis X epitopes. J Biol Chem 267:24575–24584, 1992

White T, Mandel U, Orntoft TF, Dabelsteen E, Karkov J, Kubeja M, Hakomori S, Clausen H. Murine monoclonal antibodies directed to the human histo-blood group A-transferase (UDP-Ga1NAc: Fucα 1 → 2 Gal α 1 → 3-N-acetylgalactosaminyltransferase) and the presence therein of N-linked histo-blood group A determinant. Biochemistry 29:2740–2747, 1990

Wiener AS. Origin of naturally occurring hemagglutinins and hemolysins. J Immunol 66:287–295, 1951

Wiener AS, Gordon EB. The blood groups of chimpanzees: A-B-O and M-N types. Amer J Phys Anthropol 18: 301, 1960

Wiener AS, Moor-Jankowski J. The A-B-O blood groups of baboons. Amer. J phys Anthropol 30:117, 1969

Wiener AS, Moor-Jankowski J. Blood groups of chimpanzees. In: Kratochvil C (ed) Chimpanzee: Immunological specificities of blood. Karger, Basel (Primates in medicine, vol 6), 1972

Wiener AS, Candela PB, Goss LJ. Blood group factors in the blood, organs and secretions of primates. J Immunol 45:229–235, 1942

Wiener AS, Lewis HB, Moores P, Sanger R, Race RR. A gene, y, modifying the blood group antigen A. Vox Sang 2:25–37, 1957

Wiener AS, Moor-Jankowski J, Gordon EB. Blood groups of apes and monkeys II. The A-B-O blood groups, secretor and Lewis types of apes. Amer J Phys Anthropol 21(3):271–281, 1963

Wiener AS, Gordon EB, Moor-Jankowski J. The Lewis blood groups in man. A review with supporting data on non-human primates. J Forensic Med 11:67, 1964a

Wiener AS, Moor-Jankowski J, Gordon EB. Blood groups and antibodies in primates including man. II Studies on the M-N types of orangutans. J Immunol 93:10, 1964b

Wiener AS, Moor-Jankowski J, Gordon EB. Blood groups of apes and monkeys. V. Studies on the human blood factors A, B, H, and Le in Old and New World monkeys. Amer J Phys Anthropol 22:175, 1964c

Wiener AS, Moor-Jankowski J, Gordon EB. Marmosets as laboratory animal. V. Blood groups of marmosets. Lab Anim Care 17:71, 1967

Wiener AS, Moor-Jankowski J, Cadigan FO Jr, Gordon EB. Comparison of the A-B-O blood group specificities and the M-N types in man, gibbons (Hylobates) and siamangs (Symphalangus). Transfusion 8:235, 1968

Wiener AS, Brain P, Gordon EB. Further observations on the hemagglutinins of the snail Achatina granulata. Haematologia (Budapest) 3:9–16, 1969a

Wiener AS, Moor-Jankowski J, Gordon EB. The specificity of hemagglutinating bean and seed extracts (lectins). Implications for the nature of A-B-O agglutinins. Int Arch Allergy Appl Immunol 36:582–586, 1969b

Wiener AS, Moor-Jankowski J, Gordon EB. Blood groups of gorillas. Kriminalistik 6:31, 1971

Wiener AS, Socha WW, Moor-Jankowski J. Homologues of the human A-B-O blood groups in apes and monkeys. Haematologia 8 (1–4):195–216, 1974

Wiener AS, Socha WW, Arons EB, Mortelmans G, Moor-Jankowski J. Blood group of gorillas: further observations. J Med Primatol 5: 317–320, 1976

Yamaguchi H, Okubo Y, Hazama F. An A_2B_3 phenotype blood showing atypical mode of inheritance. Proc Jpn Acad 41:316–320, 1965

Yamamoto FI. Molecular genetics of the ABO histo-blood group system. Vox Sang 69:1–7, 1995

Yamamoto F, Clausen H, White T, Marken J, Hakomori S. Molecular genetic basis of the human histo-blood group ABO system. Nature 345:229–233, 1990a

Yamamoto F, Marken J, Tsuji T, White T, Clausen H, Hakomori S. Cloning and characterization of DNA complementary to human UDP-GalıNAc: Fuc α1,2 Gal α 1,3-GalNAc transferase (histo-blood group A-transferase) mRNA. J Biol Chem 265:1146–1151, 1990b

Yamamoto F, McNeill PD, Hakomori S. Identification in human genomic DNA of the sequence homologous but not identical to either the histo-blood group ABH genes or alpha 1->3galactosyltranferase pseudogene. Biochem Biophys Res Commun 3:986–994, 1991

Yamamoto F, McNeill PD, Hakomori S. Human histo-blood group A^2 transferase coded by A^2 allele, one of the A subtypes, is characterized by a single base deletion in the coding sequence, which results in an additional domain at the carboxyl terminal. Biochem Biophys Res Commun 187:366–374, 1992

Yamamoto F, McNeill PD, Kominato Y, Yamamoto M, Hakomori S, Ishimoto S, Nishida S, Shima M, Fujimura Y. Molecular genetic analysis of the ABO blood group system. 2. cis-AB alleles. Vox Sang 64:120–123, 1993a

Yamamoto F, McNeill PD, Yamamoto M, Hakomori S, Bromilow IM, Duguid JKM. Molecular genetic analysis of the ABO blood group system. 4. Another type of O allele. Vox Sang 64:175–178, 1993b

Yamamoto F, McNeill PD, Yamamoto M, Hakomori S, Harris T. Molecular genetic analysis of the ABO blood group system. 3. A^x and $B^{(A)}$ alleles. Vox Sang 64:171–174, 1993c

Yamamoto F, McNeill PD, Hakomori S. Genomic organization of human histo-blood group ABO genes. Glycobiology 5:51–58, 1995

Yamamoto F, McNeill PD, Yamamoto M, Hakomori S, Harris T, Judd WJ, Davenport RD. Molecular genetic analysis of the ABO blood group system. 1. A^3 and B^3 alleles. Vox Sang 64:116–119, 1993d

Yoshida A, Yamagushi YF, Dave V. Immunologic homology of human blood group glycosyltransferases and genetic background of blood group (ABO) determination. Blood 54:344–346, 1979

3 The MNSs Blood Group System

Serology and Formal Genetics

W.W. Socha and A. Blancher

Antigens M, N, S, and s of Human Blood

The two major human red cell agglutinogens M and N were discovered in 1927 by Landsteiner and Levine by means of reagents obtained from the sera of rabbits immunized with human group O cells (Landsteiner and Levine 1927a,b). The anti-M and anti-N reagents so obtained defined three MN types in human blood, M, N and MN, attributed to the action of two codominant alleles, *M* and *N*, which combined in three possible genotypes: *MM*, *NN* or *MN*. Extensive family studies and the application of gene frequency analysis to studies on the distribution of the MN types in various human populations yielded results consistent with the genetic theory proposed by the discoverers of this blood group system. In 1947, Walsh and Montgomery discovered an alloantibody in the serum of the mother of an erythroblastotic baby. The antibody, called anti-S, defined a new antigen, S, which was found to be associated with M or N. A few years later, Levine et al. (1951) discovered another alloantibody in the serum of the mother of an erythroblastotic infant. The serum gave reactions anthithetical to anti-S and

Molecular Biology and Evolution of Blood Group
and MHC Antigens in Primates
Blancher/Klein/Socha (Eds.)
© Springer-Verlag Berlin Heidelberg 1997

was therefore named anti-s; the corresponding antigen was called s. Antigens S and s were found to depend on two codominant alleles, *S* and *s*. Evidence of recombination between *MN* and *Ss* genes implied the independence of the two loci, but the rarity of the occurrence of such recombination pointed to their genetic closeness.

While the original antibodies that led to the discovery of M and N antigens were xenoantibodies (developed in rabbits), the first anti-S and anti-s antibodies were of human origin. Since M and N antigens are only weakly immunogenic for humans and anti-M and anti-N alloantibodies are rare in humans, the M and N reagents used for routine blood typing are usually derived from rabbit sera. In contrast, antigens S and s are more immunogenic and the resulting antibodies can cause (although rarely) severe erythroblastosis fetalis or transfusion accidents. Although anti-s (but not anti-S) can also be produced in rabbits, human donors remain the principal sources of anti-S and anti-s typing reagents.

It was established that the common antigens for the MNSs system reside in a family of type I transmembrane sialoglycoproteins (GP) (60 % of the glycoprotein is oligosaccharide) (for general review, see Huang et al. 1991a; Huang and Blumenfeld 1995; Blumenfeld and Huang 1995). The first attempts to characterize the molecular basis of the M and N antigens suggested that sialic acid and other sugars (b-D galactopyranosyl) have a critical role in the constitution of the M and N antigens (for general review see Blumenfeld and Huang 1995; Huang and Blumenfeld 1995). Further molecular studies determined that amino acid residues of the GP chains play a determinant role in the configuration of the M and N antigens (Dahr et al. 1977; Wasniowska et al. 1977; Blumenfeld and Adamany 1978; Furthmayr 1978). The two forms of glycophorin A (GPA) that specify the two antigens, M and N, differ from one another at positions 1 and 5 of their polypeptide chains. In the type M molecule there is Ser at position 1 and Gly at position 5, while in the type N molecule there are Leu and Glu at positions 1 and 5, respectively.

Antigens S and s are carried by glycophorin B (GPB) and differ from one another by a single substitution (Met vs Thr) at position 29 (Dahr et al. 1980b). Amino acid residues of glycophorin B at positions 1 and 5 do not vary from person to person and are the same as those in glycophorin A of type N individuals. This provides a molecular explanation for the well known serological observation that human type M red cells cross-react with anti-N antibodies (see later in this chapter) and why, after prolonged immunization with M human red cells, the resulting immune rabbit serum contains not only the expected anti-M but also anti-N antibodies. This also clearly explains why type N red cells do not display any M activity.

Other Antigens of the MN System

Numerous variants of the main antigens M and N (M^c, M^r, M^z N_2, M^a, N^a) were defined not by means of specific antibodies, but by their unusual reactions with standard anti-M or anti-N sera. Among those variants, M^c is particularly relevant for the discussion of MN groups in primates. The M^c antigen was first described by Dunsford et al. (1953), who observed that a human blood sample reacted with 25 out of 27 anti-M antisera and also with five out of 28 anti-N antisera. This particular antigen was detected in several members of one family. Two more examples of antigen M^c have been found in Zurich; one aligned with S and the other, like the first family observed in England, with s. Since the publication of the original paper by Dunsford et al., anti-M and anti-N reagents have been classified as reactors or nonreactors with variant M^c (Metaxas et al. 1968). The latter authors pointed out that M^c antigen, though reactive with rabbit anti-N reagents, did not react with *Vicia graminea* extract. Those observations show some resemblance to the reactions obtained in tests with chimpanzee red cells: although all chimpanzees tested so far reacted with all the anti-M polyclonal reagents tried, those animals which were positive with *Vicia graminea* extract and therefore considered to be N-positive reacted with some, but not all, anti-N polyclonal reagents.

Several other antigens related to the same system were discovered either by experimental immunization of rabbits with human blood of various ethnic origins or defined by antibodies obtained from accidentally immunized persons.

Discussion of numerous human variants of M and N will be limited to those that have been shown to have counterparts in nonhuman primates or are deemed useful to illustrate some human mutations.

Antigens Defined by Human Alloantibodies

Antigen U

In 1953 Wiener et al. (1953, 1954) found an irregular antibody in the serum of a multitransfused African-American woman. The antibody, which was of the IgM class, was tested with hundreds of red cell samples of Caucasians and Blacks and was found to react with all blood samples of Caucasians tested and with most, but not all, samples from Blacks. Because of its prevalent occurrence, the antigen was designated as U (for universal). The initial population study carried out by Wiener et al. showed that the U-negative individuals were either of N or MN types and that all were Ss-negative, which pointed to a relationship between U and the MNSs system (Wiener et al. 1954). Family studies indicated that the gene *U*, responsible for the expression of antigen U, was dominant over its silent allele *u*; therefore the genotype of U-negative persons was designated as *uu*. Similar to antigens S and s, U is immunogenic and anti-U antibodies can cause hemolytic transfusion reactions or fetal erythroblastosis. More recently, biochemical analysis demonstrated that, although the expression of antigen U depends on the presence of glycophorin B (GPB), which carries the antigens S and s (Dahr and Moulds 1987), the U antigen itself is not carried by this glycophorin (Borne et al. 1990). Significantly, it was also demonstrated that the Rh-associated glycoprotein GPRh50 (see chapter on the Rhesus system) is critical for the expression of antigen U, as shown by the fact that, in some Rh$_{null}$ cells, the absence of RhGP50 is correlated with the absence of antigen U (Borne et al. 1990). All these observations indicate that GPB may be part of a larger complex inside the erythrocyte membrane which also includes RhGP50, and an Rh-related molecule (Rh30 polypeptides). Expression of the antigen U depends on this complex. Antigen U is carried by the RhGP50 protein which, in turn, is stabilized in a special configuration by static interference from GPB (Ridgwell et al. 1994).

Other Antigens Associated with the Glycophorins: Wright Antigens

The antigen Wr^a was first detected by means of an alloimmune antibody (Holman et al. 1953). It is a low frequency antigen. In contrast, Wr^b, defined by antibodies in the sera of rare Wr (a+b-) individuals, is a high frequency antigen. Numerous anti-Wr^b monoclonal antibodies (mAbs) were later obtained from lymphocytes of immunized mice and, more recently, by one of us (A.B.), through the immunization of a macaque with human red cells. Antigen Wr^b is a complex epitope which, like antigen U, results from the interaction between two proteins, GPA and protein band 3 (Telen and Schasis 1990). The epitope Wr^b, recognized by human anti-Wr^b and mouse mAbs, seems to reside on GPA because the antigen is present on all human red cells except those lacking GPA, such as En(a-) red cells or those with the rare genotype M^kM^k. The epitope Wr^b is related to the ficin-resistant part of GPA (residues 62–70) and, therefore, one could expect it to be somewhat related to the MNSs system. Surprisingly, however, family studies indicate that expression of Wr^a Wr^b antigens is independent of the MN system. This apparent paradox is a consequence of the fact that two independent genes encoding GPA and band 3 protein are required for the expression of Wr^a and Wr^b antigens. Loss or alteration of either GPA or band 3 proteins may affect the expression of Wr^a or Wr^b antigens.

Antigens M^g, M^v, St^a, SAT

These low-frequency antigens are defined by human immune alloantibodies.

The antigens M^g (Allen et al. 1958) and M^v (Gershowitz and Fried 1966) are the products of two rare alleles of M and N.

The antigen St^a (Clerghorn 1962) is prevalent among Orientals (Madden et al. 1964) but rare in Caucasians. Its occurrence in Blacks is not known. It was found recently that the presence of the St^a antigen depends on a series of rare alleles at the MN locus (see later in this chapter).

The antigen SAT is a private antigen so far found only in two Japanese families.

Antigens of the Miltenberger Series

Low incidence antigens (Vw, Mi^a, Mur, Hil, Gr, Hut, Hop, Nob, MINY, Tsen, DANE) belong to a series called Miltenberger, after the name of a woman whose serum contained antibodies (anti-Mi^a) that defined the first antigen of this series (Cleghorn 1962). These antigens occur in various combinations in rare individuals expressing mutated forms of glycophorins (see later in this chapter). The various combinations of Miltenberger antigens were initially arranged in five classes, I–V (Cleghorn 1966), to which six new classes were later added (see below).

Antigenic Variants Defined by Rabbit Antibodies

The Hunter (Hu) Antigen

This antigen was defined very early by antibodies raised in a rabbit immunized with red cells of an African-American named Hunter (Lansteiner et al. 1934). The antigen was found mostly in the blood of individuals of African origin (22 % of West Africans, 7.3 % of African-Americans) and rarely in Caucasians (0.5 %).

The Henshaw (He) Antigen

 This antigen was first identified by Ikin and Mourant (1951) through antibodies isolated from a rabbit anti-M serum and later by means of an antibody experimentally produced by immunizing a rabbit with the red cells of a Nigerian named Henshaw (Chalmers et al. 1953). The antibody was accordingly designated as anti-He and the antigen as He. The He antigen was observed in 2.7 % of 1390 West Africans tested, but was not found among 1500 Europeans. Most of the He-positive samples were N- and S-positive. Comparative tests with anti-Hu and anti-He proved that the two antigens are not related.

The Antigen M^e

Anti-M^e antibodies seem to define an epitope shared by He and M. Wiener and Rosenfield (1961) postulated the action of a single gene that belongs to the same polyallelic series as M and N. This gene, called M^e, is presumed to be an intermediate between M and He and to simultaneously control the appearence of both antigens M and He.

Null Phenotypes

Three rare variants were found, En(a-), S-s-U-, and M^k, that were characterized by the absence of all or some of the MNSs glycoproteins. The En(a-) phenotype, first described by Darnborough et al. (1965), is identified by the absence of M and N antigens, and the lack of a high frequency antigen called En^a is associated with a particular abnormality of the red cell membrane. The phenotype En(a-) results from the absence of GPA which carries the antigens M or N and En^a (Darnborough et al. 1969). This defect, which causes a sharp decrease in the number of negative charges at the erythrocyte surface, causes increased red cell agglutinability. Recently, the lack of GPA was related to the deletion of its gene.

The S-s-U-phenotype is characterized, in approximately 50 % of subjects with this trait, by the absence of GPB. In some cases, as recently demonstrated, it corresponds to sequences changes in GPB (see later in this chapter).

The rare M^k gene is a silent allele that results in the absence of antigens of the MNSs system and combines with deletion of the GPA and GPB genes (Okubo et al. 1988; Blumenfeld and Huang 1995).

Expression of MNSs Blood Group Antigens in Nonhuman Primates

Shortly after the discovery of the MN system in humans by Landsteiner and Levine (1927a,b), the same authors reported the presence of the M and N factors on red cells of anthropoid apes (Landsteiner and Levine 1928). However, a systematic study of the MN blood groups in nonhuman primates was not undertaken until 10 years later, when tests were carried out on red cells of various apes and monkeys using batches of anti-human M and N reagents (Landsteiner and Wiener 1937). Those extensive experiments showed that, not only are MN factors recognizable in nonhuman primate red cells, but that immunization of rabbits with monkey blood yielded anti-M reagents equal in strength to the most potent anti-human M sera (Landsteiner and Wiener 1937; Wiener 1938).

Tests with Anti-M Antibodies

Tests with Rabbit Anti-M Polyclonal Antibodies

Anti-human M reagents are prepared from the raw sera of rabbits immunized against human M cells. This is done by absorbing the sera with packed, washed human N cells in order to remove rabbit anti-human species-specific antibodies. The resulting reagents are usually of high titer and narrow specificity when tested with human red cells. Immunization of rabbits with the red cells of chimpanzees, baboons or macaques resulted in equally potent anti-M reagents.

When several anti-M reagents were used in parallel tests some differences were observed in their reactivity with ape and Old World monkey blood, although these same reagents uniformly agglutinated all human M or MN red cells tested (Table 1). This proves that the antigens expressed by apes and Old World monkeys are not identical with the human antigen M.

Differences among M antigens of various origins were clearly demonstrated by numerous absorption-fractionation experiments. In one such experiment absorptions with gorilla red cells separated a number of antibody reactions in supposedly monospecific rabbit anti-human M reagents (Table 2). At least three fractions of anti-M antibodies were identified when an anti-M reagent derived from the serum of a rabbit immunized with chimpanzee red cells was fractionated by series of parallel absorptions with red blood cells of chimpanzees and gorillas. This indicated that agglutinogen M molecules of gorilla, chimpanzee, and human red cells share some configurations, while other surface configurations are particularly exclusive to the M agglutinogen of gorilla, chimpanzee, or humans (Wiener et al. 1972). Comparative testing of the red cells of common and dwarf chimpanzees showed that the M antigens of these two species are not identical: absorption of the reagents with red cells of the dwarf chimpanzee left behind a fraction of antibodies still reactive (although weakly) with common chimpanzee red cells as well as human M and MN cells. By contrast, absorption with red cells of the common chimpanzee removed the activity against any chimpanzee and any human M-carrying cells (Wiener 1963). Other experiments, performed with gibbon red cells, demonstrated that some anti-M reagents contained at least three fractions of anti-M: one shared by all gibbons, one absent from all blood samples tested and one shared only by M and MN gibbons (Wiener et al. 1966a,b).

If one disregards subtle differences in the reactivity of various reagents, the results of tests with panels of anti-M sera prepared by immunization of rabbits with human or primate blood can be summarized as follows:

1. *Uniformly M-positive*: chimpanzee (common and pygmy), Old World monkeys (macaques, baboons, vervet monkeys)
2. *Uniformly M-negative:* New World monkeys, lemurs
3. *M-positive or M-negative*: orangutan, gorilla, gibbon, galago (with the latter reactions often weak).

Early research on the MN blood groups of nonhuman primates showed that the M antigens of both humans and nonhuman primates have a complex, mosaic structure (Wiener 1963), which is the source of the difficulties one encounters when trying to establish the M phenotype of apes and monkeys. Accumulated experience has confirmed the need to take special precautions when carrying out typing tests: it is absolutely necessary to perform these tests not with a single reagent but with batteries of antisera of known titers against red cells of several primate species. The specificity of positive reactions must be confirmed by proper absorptions and by control of the resulting reagent with panels of red cells of several species.

Table 1. Results of tests on red cells of gorillas, orangutans and other simians, with anti-M, anti-M^e, anti-He and anti-N^v reagents

Tested against red cells of	Anti-M (anti-human)			Anti-M (anti-chimpanzee)	Anti-M (anti-baboon)	Anti-He (anti-human)	Anti-M^e (anti-human)	Anti-N^v lectin (*Vicia graminea*)
	M1	M2	M6					
Gorillas								
Anka	+++	Traces	-	+++	+++	+++	++	+++
Banga	+++	+±	+	+++	+++	+++	+±	+++
Choomba	+±	-	-	-	+±	++	+	+++
Calabar	+++	-	-	+++	+±	-	-	+++
Chad	Traces	-	+±	-	Traces	+++	+±	+++
Inaki	+++	+±	Traces	+++	++	++±	+±	+++
Jini	+++	+	+	+++	+++	++±	+±	+++
Katoomba	+++	-	-	+++	++±	++±	+	+++
Oban	+++	+±	-	+++	++	-	-	+++
Oko	+++	++	++	-	+++	+++	++	+++
Ozoom	-	-	-	+++	-	+++	++	+++
Paki	+++	++	+++	+++	+++	++	+±	+++
Rann	+±	-	-	-	+±	++	+±	+++
Shamba	+++	+±	++	+++	+++	++	+	+++
Segon	+++	-	Traces	+++	++	++	+±	+++
Human controls								
Type M, He-	+++	+++	++±	+++	+++	-	++	-
Type N, He-	-	-	-	-	-	-	-	+++
Type MN, He-	+++	++±	+±	+++	+++	-	++	+++
Type N, He+	-	-	-	-	-	+++	+++	+++
Orangutans								
Paddi	(+++)	(++)	(++)	(+++)	(+++)	(++)	(++)	-
Tupa	(++)	(+±)	(++)	(++)	(+++±)	(+++±)	(+++±)	-
Chimpanzees								
Five animals	+++	++±	++±	+++	+++	-	++	n.d.

Table 1. (continuet)

Tested against red cells of	Anti-M (anti-human)			Anti-M (anti-chimpanzee)	Anti-M (anti-baboon)	Anti-He (anti-human)	Anti-M^e (anti-human)	Anti-N^v lectin (*Vicia graminea*)
	M1	M2	M6					
Gibbons								
G-12 Abby	+++	n.d.	+±	+++	++±	–	Traces	n.d.
G-3 Blacky	+++	n.d.	–	++	+	–	–	n.d.
Baboons								
Three animals	+++	++	++	+++	+++	–	–	n.d.
Rhesus monkeys								
Two animals	+++	n.d.	+±	+++	+++	–	–	n.d.
Crab-eating macaques								
Two animals	+++	n.d.	±	+++	+++	–	–	n.d.
Geladas								
Two animals	+++	n.d.	++	+++	+++	–	–	n.d.
Celebes black apes								
Two animals	+++	++	++	+++	+++	–	–	n.d.
Capuchin monkey								
C4 Leah	–	–	–	–	±	–	–	n.d.

n.d.: not determined.

Parentheses indicate that the reaction was not removed by absorbing the reagents with the red cells of the type used for immunizing the rabbit

Table 2. Fractionation by absorption of the anti-M reagent M1 prepared from an immune rabbit serum for human red cells

Tested against red cells of	Unabsorbed	Absorbed with red cells of							
		Human M	Gorilla Choomba	Gorilla Katoomba	Chimpanzee	Orangutan Paddi	Baboon	Crab-eating macaquq	Celebes black ape
Gorillas									
Anka	+++	−	+++	−	−	−	−	−	−
Banga	+++	−	+++	−	−	++	−	−	−
Choomba	−	−	−	−	−	−	−	−	−
Calabar	+++	−	++	−	−	−	−	−	−
Chad	−	−	−	−	−	−	−	−	−
Katoomba	+++	−	++±	−	−	−	−	−	−
Oban	+++	−	+++	−	−	+±	−	−	−
Chimpanzee # 335	+++	−	++	−	−	+±	−	−	−
Orangutans									
Paddi	+++	++	++±	++	++	+±	++	++	++
Tupa	++	++	++±	++	+±	+±	+±	++	++
Human controls									
Type M	+++	−	++±	++±	+++	+++	+++	+++	++±
Type N	−	−	−	−	−	−	−	−	−
Type MN	+++	−	++±	++	++	+++	+++	+++	++±
Baboon # 914	+++	−	++±	++	++	++±	+++	+++	−
Crab-eating macaques	+++	−	++±	−	−	±	−	−	−
Rhesus monkey # 55	+++	−	++±	−	−	+±	−	−	−
Celebes black ape # 13	+++	−	++±	−	−	−	−	−	−
Capuchin # 4	−	−	−	−	−	−	−	−	−

Tests with Mouse Monoclonal Antibodies Raised Against Human M Antigen

The complexity of the M antigen is also demonstrated by the results of tests carried out on red cells of various nonhuman primates using batches of human anti-M mAbs. As shown in Fig. 1, all our anti-M mAbs tested gave specific anti-M reactions by saline and antiglobulin methods with human M and MN cells and

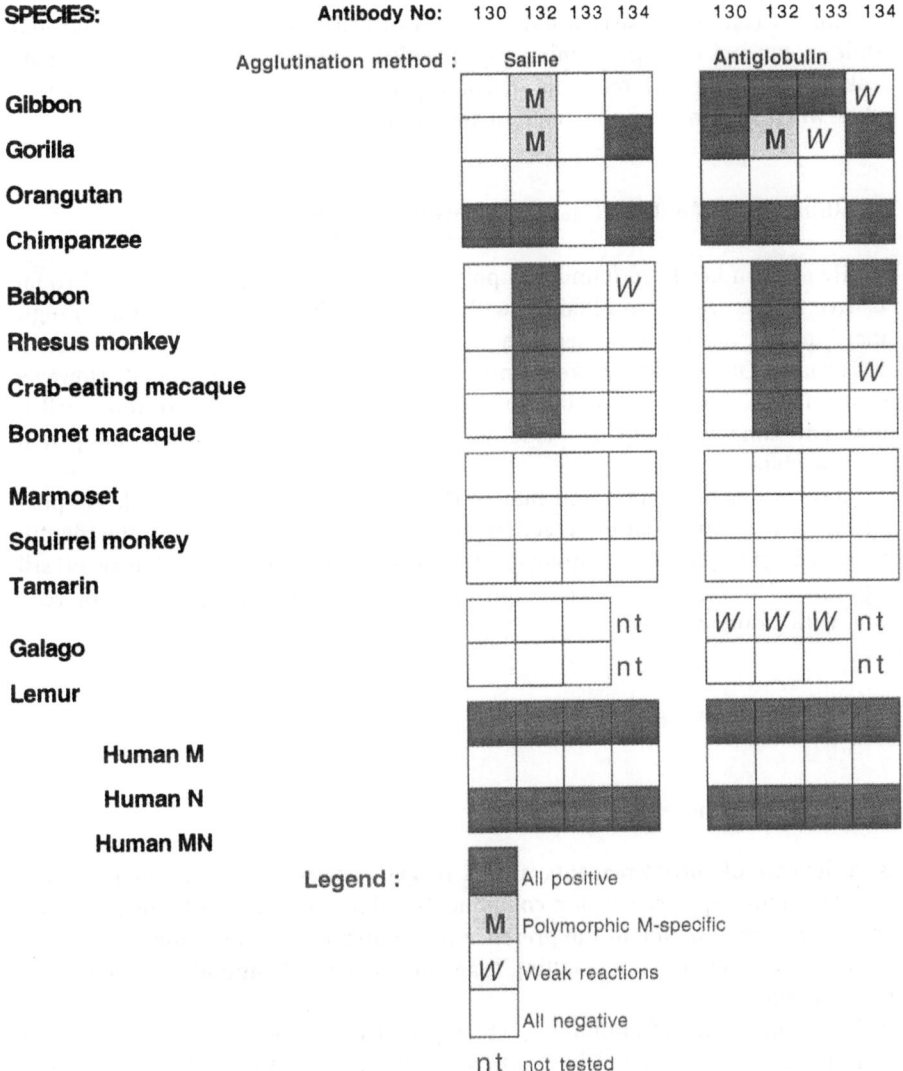

Fig. 1. Anti-human M monoclonal antibodies in tests with nonhuman primate red blood cells. Antibodies were provided by participants of the Second International Workshop on Monoclonal Antibodies Against Human Red Cells and Related Antigens (Lund Sweden, 1–4 April 1990)

uniformly negative reactions with N red cells. But, in tests with ape red cells each of the four mAbs reacted differently. Only one (mAb 132) detected, in tests using the saline method, M epitopes on ape and monkey red cells, as they were defined by reactions with rabbit polyclonal anti-M antibodies. Significantly, however, this reagent, like all the remaining anti-M mAbs, failed to agglutinate M-positive orangutan red cells and, when using the antiglobulin method, it agglutinated all gibbon red cells irrespective of their MN type.

Limited tests were carried out on blood of prosimians (lemurs and galagos). While lemur red cells gave uniformly negative results with all anti-M reagents, weak agglutinations of red cells of some galagos were produced by all anti-M mAbs when the antiglobulin method was used.

Tests Using Rabbit Antibodies Raised Against He Antigen

The He antigen has been found in a polymorphic form in 15 gorillas tested so far. Positive results were also obtained with anti-He antibodies in tests with orangutans, but the results were negative with chimpanzees, gibbons (Wiener et al. 1966a,b), all Old World monkeys and with the single capuchin tested. However, as shown in Table 1, the numbers of animals of some species were too small to allow any conclusions to be made as to the existence of Henshaw antigen in these primates.

Absorptions of a rabbit anti-human He with either human or gorilla He-positive red cells removed all the reactivity from the reagent, confirming the identity of human and gorilla He antigens. By contrast, the so absorbed reagent still agglutinated orangutan red cells, thus putting in doubt the specificity of reactions with orangutan red cells.

Tests with Anti-N Reagents

Anti-N Reagents from Rabbits

As in the case of anti-M reagents, anti-N reagents always require arduous absorptions to remove species-specific components, which is achieved by using human red cells of type M. Due to the presence of N antigen on type M human red cells (see below for details), the resulting type-specific anti-N antibodies are considerably weakened.

Unlike most anti-M antibodies, which produced strong clumping of chimpanzee red cells, only selected rabbit anti-N sera agglutinated red cells of these apes, and the reactions were often weak and sometimes erratic. In their original study, Landsteiner and Levine (1928) found that most anti-N sera did not agglutinate chimpanzee red cells but those rare anti-N sera that gave positive reactions agglutinated the blood of all chimpanzees. Thus, they concluded that all chimpanzees tested by them had an N-like antigen. Wiener (1938) confirmed the expression of an N-like factor by some, but not all, chimpanzee red cells. The polymorphic expression of N in chimpanzees was confirmed by tests performed with *Vicia graminea* seed extracts (see below). Similarly, some but not all gibbons were shown to express an N-like factor (Moor-Jankowski et al. 1964). Limited tests with gorilla red cells produced weakly positive results with some of the animals and negative results with others (as in the case of chimpanzees the N typing of gorilla was done by means of lectins, see below). Relatively strong agglutination of orangutan red cells with rabbit anti-N sera was probably due to the presence of strong species-specific anti-orangutan antibodies which were impossible to remove from rabbit sera.

None of the Old and New World monkeys tested by Landsteiner and Wiener (1937) showed any N activity in the blood.

Extracts from *Vicia graminea* and *V. unijuga* Seeds

A clear distinction between N-positive and N-negative chimpanzee red cells was possible using anti-N^v plant extracts (*Vicia graminea* and *V. unijuga*) that gave distinct reactions with red cells of some chimpanzees (Moor-Jankowski et al. 1964). As in the case of chimpanzees, the N typing of gorillas is routinely carried out with lectins instead of rabbit anti-N sera. All gorillas tested thus far with *Vicia* extracts proved positive.

It must be stressed that combining sites for anti-N antibodies and anti-N^v plant extracts are different because most of the former depend on the presence of sialic acid while the latter are sialic acid-independent (Springer and Hupprikar 1972). Biochemical analysis demonstrated that the immunodominant complementary structure of the combining site for the *Vicia* reagent is a βD-galactopyranosyl residue (Springer and Hupprikar 1972).

Limited tests were carried out on blood of lemurs and galagos using anti-N^v plant extracts (Socha, unpublished). *Vicia unijuga* lectin produced uniformly medium-to-strong agglutination reactions of all lemur red cells (13 samples studied) and weak reactions with red cells of all galagos (17 samples studied). Specificity of the reactions observed with lemur red cells was confirmed by absorption with human N red blood cells, which rendered the *Vicia* extracts inactive against prosimian blood.

Tests with Mouse Anti-N Monoclonal Antibodies

Recently, mouse monoclonal anti-N antibodies were produced by the fusion of splenocytes from mice immunized against human red cells. These reagents, perfectly homogeneous in terms of antibody specificity, provide new tools for pinpointing the antigenic differences between N blood groups in humans and nonhuman primates. In a series of experiments with a panel of anti-human N mAbs, variations among reagents were observed in tests with various nonhuman primate red cells. As shown in Fig. 2, the six anti-N mAbs, tested by the saline

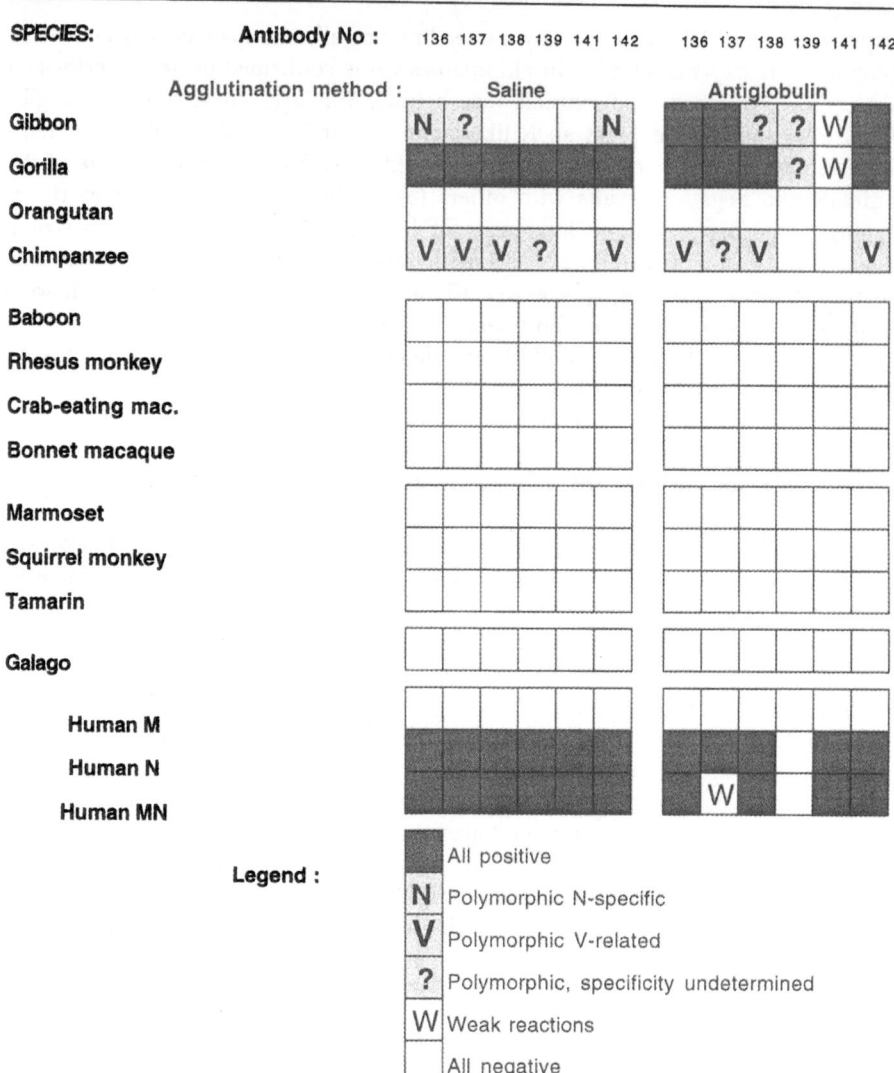

Fig. 2. Anti-human N monoclonal antibodies in tests with nonhuman primate red blood cells. Antibodies were provided by participants of the Second International Workshop on Monoclonal Antibodies Against Human Red Cells and Related Antigens (Lund Sweden, 1–4 April 1990)

method, confirmed the findings obtained with polyclonal reagents, namely, that all gorillas possess N antigen on their red cells while red cells of orangutans and of all lower monkeys (Old and New World) lack N specificity. Two of the antibodies, when tested by the saline method, confirmed the presence of N in polymorphic form in gibbons. In indirect antiglobulin tests these two reagents produced non-specific reactions with gibbon red cells.

Significantly, four out of six of those anti-human-N mouse mAbs detected a chimpanzee alloantigen, V^c, which is closely related, but not identical to, human N and is a component of the chimpanzee blood-group system, the VABD system (see below).

In a few instances the mouse anti-human N mAbs detected (by saline or antiglobulin methods) individual differences in gibbons, gorillas, and in chimpanzees, which were unrelated to any known blood group polymorphisms in those species.

Antigens of the Miltenberger Series

The only evidence for the presence of Miltenberger-related antigens in nonhuman primates was obtained by means of an alloimmune chimpanzee serum which was found to contain antibodies of two specificities: anti-N-related antibody (anti-V^c) and an antibody which was named anti-W^c. The latter antibody, in tests with human red cells, gave results parallel to those obtained with human antibody anti-Mi^a (Socha and Moor-Jankowski 1979). Antigen W^c appears only among common chimpanzees (*Pan troglodytes*) and at low frequencies (Socha 1984). All gorillas tested so far were found to be W^c-negative. Results obtained in tests with red cells of gibbon, orangutan and rhesus monkey were impossible to interpret because of interference by a species-specific component in chimpanzee anti-W^c reagent.

S, s, and U Antigens

The presence of antigens S and s in nonhuman primates is difficult to assess because the reagents are of human origin, and to be suitable for testing nonhuman primate red cells the sera have to be absorbed to remove species-specific antibodies. The so-absorbed reagents were usually too weak to produce unequivocal results in tests with nonhuman primate blood. The few monoclonal anti-S reagents tested so far gave uniformly negative results with all nonhuman primate species red cells (unpublished).

As for antigen U, the limited tests carried out by Wiener and Gordon (1960) produced uniformly negative results with all chimpanzees tested.

Wr^b and En^a Antigens

Out of 29 anti-human red blood cell macaque mAbs (produced by A.B.), only one was found to detect the Wr^b antigen on human red cells (Marion Reid, personal communication). This same reagent gave positive reactions with all 14 chimpanzees, all four gorillas and three out of four gibbons tested (unpublished). The titers of reactions (by the antiglobulin method) with the red cells of those apes were all comparable to those observed with human positive controls. Uniformly negative results were obtained with four orangutans, six crab-eating macaques, six rhesus monkeys, six baboons and four marmosets thus far tested. Similar results were obtained with one murine anti-Wr^b mAb (Socha, unpublished).

Among the same 29 macaque mAbs, three were found to produce, with human red cells, reactions parallel to those obtained with polyclonal anti-En^a reagents (Marion Reid, personal communication). All three mAbs were tested with various primate species red cells: two of them agglutinated only V^c-positive chimpanzee red cells, the third agglutinated all chimpanzee red cells (7 tested), two out of four gorilla samples, and one out of four from orangutans. All three reagents were negative with red cells of baboons (4 samples), rhesus monkey (4), crab-eating macaques (4), vervets (2), tamarins (3), marmosets (4) and lemur (2).

Bearing in mind the narrow specificity of mAbs and the fact that the presence of En^a on ape red cells did not seem to influence expression of their MN-VABD antigens, the question may be raised as to whether a similar relationship exists between En^a and the *MN-VABD* locus in chimpanzees as between En^a and the *MN* locus in humans.

Formal Genetics of MN Blood Groups as Inferred from Serological Reactions

Table 3 summarizes, in most general terms, the results of studies on the homologues of the human MN blood types in various species of apes (Wiener et al.

Table 3. The M-N blood group of anthropoid apes

Species	Blood types observed based on reactions with anti-M sera and anti-N^v lectin)	Corresponding genotypes according to Wiener's theory (Wiener, et al. 1972)
Chimpanzees	M	*MMnn*
Common type	MN	*MMNn*, or *MMNN*
Dwarf chimpanzees	M	*MMnn*
Gorillas	N	*mmNN*,
Lowland and mountain	MN	*MMNN*, or *MmNN*
Gibbons	M	*MMNN*
Various species	N	*mmNN*
	MN	*MmNN*
Orangutans	M	*MMnn*, or *Mmnn*,
	m	*mmnn*

According to Wiener's theory:

The *n* gene is absent from gorilla and gibbon, while gene *m* is missing in chimpanzee and N is not present in orangutan.

In gibbon it is assumed that an interaction between products of the *M* and *N* genes modifies the phenotypic expression of N (see text for details).

1972). As can be seen, among anthropoid apes only gibbons have all three MN types similar to those found in humans. In contrast, all chimpanzees were found to be M-positive, while about 40 % of them were found to be also N-positive and 60 % to be N-negative. Thus, chimpanzees display only two MN types, M and MN. In gorillas, by contrast, all animals proved to be N-positive, while 76 % of them were also M-positive. Thus, gorillas are of two types only: N or MN. Finally, in orangutans, half of the animals have red cells that react with anti-M only, while the red cells of the other half fail to react with anti-M or anti-N sera. In this way, only two MN types are defined in orangutans: M or m. The distribution of the MN types in humans and gibbons is in accordance with the classical theory of Landsteiner and Levine regarding inheritance of the expression of M and N antigens by a pair of allelic genes. This theory can hardly explain the occurrence of the MN types in the other species of anthropoid apes. Available facts are in obvious conflict with the Hardy Weinberg law, if M and N types in chimpanzees and in gorillas were also transmitted by the corresponding allelic genes. Those serological facts on the MN types of apes prompted Wiener and collaborators to propose a new theory that could explain the genetics and serology of the MN system in anthropoid apes. This theory postulated the existence of two closely linked genes, each with two alleles resulting in four theoretically possible haplotypes: *MN, Mn, mN, mn*. Table 3 gives the genotypes of gorillas, chimpanzees and orangutans according to this theory. In this concept, the two MN types observed in chimpanzees can readily be accomodated by postulating that type MN corresponds to the genotypes *MMNN* or *MMNn* and type M to the genotype *MMnn*, with the gene *m* evidently lacking. In gorillas, type N would correspond to genotype *mmNN*, while type MN would have two corresponding genotypes: *MMNN* or *MmNN* with the *n* gene evidently lacking. In orangutans the theory assumes the existence of three genotypes: *MMnn* or *Mmnn* defining type M and the genotype *mmnn* defining the M-negative type. The *N* gene is postulated to be absent in orangutans. This theory is interesting to recall because it was the first time that the existence of two closely linked genes was evoked to explain the transmission of the MN types. As discussed in the second part of this chapter, the molecular biology of the MN antigens did, in fact, confirm that these two antigens are carried by two glycophorins (glycophorin A and B) which are encoded by two closely linked genes.

Wiener extended this theory to the model of heredity of MN types in humans and gibbon. To explain the known serological facts in the context of this theory, he assumed that a heterostasis occurs between M and N: the quantity of N carried by the red blood cells diminishes when M is present in the genotype. For instance, all human beings (and all gibbons) were assumed to be homozygous for gene *N*, but the expressivity of the antigen N was postulated to be affected by the *M/m* genotype. N antigen was detectable only in the absence of gene *M*, as in genotype *mmNN*, or when this gene was present at a single dose, as in genotype *MmNN*. Remarkably, this interpretation intuitively assumed the constant presence of the *N* gene in humans, which is confirmed, as discussed later in this chapter, by the presence of an N-bearing GPB in all human subjects. However, the recent data of the molecular genetics of humans and great apes revealed a much greater complexity of the MN system than assumed by its earlier model, which was based exclusively on serological observations (see the second part of this chapter).

The V-A-B-D Blood Group System

This system, closely related to the human MN groups, was the first to be defined in *P. troglodytes* by means of antisera produced by allo or cross-immunization of chimpanzees (Wiener et al. 1974). The V-A-B-D system is built around a central antigen, the V^c antigen, which has a unique position among specificities of this system. This is indicated by the fact that V^c is demonstrable on the red cells of chimpanzees not only by alloimmune anti-V^c serum, but also by an anti-V^c reagent obtained from the serum of chimpanzees immunized with human red cells. Reagents of the two kinds give reactions paralleling those obtained with the anti-N^v lectin from the seeds of *Vicia graminea* which, as was explained earlier, reacts with human-type antigen N. Significantly, among the various reagents that recognize the V^c epitope are also mouse mAbs directed against human M as well as N epitopes, and an anti-human MN mAb produced by macaque-mouse heterohybridoma (Blancher and Socha 1991). All remaining specificities of this system are defined exclusively by chimpanzee alloimmune sera, namely, anti-A^c, anti-B^c and anti-D^c. A hypothesis postulating the existence of five multiple alleles (v, v^A, v^B, v^D, and V) to account for the 11 regular phenotypes of the system (Table 4) gave a satisfactory fit and was confirmed by family analysis in earlier population studies (Socha and Moor-Jankowski 1979). The heritable nature of the rare irregular types shown in Table 4 is supported by observations of a few families of *P. troglodytes* in which the variants V^{pq} and V^q were transmitted to the offspring from at least one parent. While all the 17 V-A-B-D phenotypes shown in Table 5 are observed in common chimpanzees (with frequencies ranging from under 1% to over 20%), all rare pygmy chimpanzees *(P. paniscus)* tested so far were of type v. D, a blood group occurring in only 1% of *P. troglodytes* (Socha 1984).

The red cells of all lowland gorillas as well as those of the rare mountain gorillas tested so far were uniformly agglutinated by the anti-V reagents but failed to react with any of the sera of anti-A^c, anti-B^c or anti-D^c specificities (Socha et al. 1973; Wiener et al. 1976). Thus, all gorillas were of type V.O. At least five V-A-B-D types, including four irregular forms, have been observed in orangutans. None of the V-A-B-D reagents reacted with the red cells of gibbons.

Table 5 summarizes the occurrence of V-A-B-D blood groups in nonhuman primates.

One of the alloimmune anti-V^c sera was found to contain a second antibody, originally designated as anti-W^c, which specifically reacted not only with red cells of some chimpanzees, but also with the human red blood cells of Mi^a type of the Miltenberger series. The antibodies anti-W^c produced by a chimpanzee (Ch. 639 "Dina") strongly agglutinated the human red cells of Miltenberger

Table 4. Serology and genetics of the V-A-B-D blood group system of apes

Designation	Reaction with serum of specificity				Possible genotypes
	Anti-V^c	Anti-A^c	Anti-B^c	Anti-D^c	
v.O	−	−	−	−	vv
v.A	−	+	−	−	$v^A v^A$, or $v^A v$
v.B	−	−	+	−	$v^B v^B$, or $v^B v$
v.D	−	−	−	+	$v^D v^D$, or $v^D v$
v.AB	−	+	+	−	$v^A v^B$
v.AD	−	+	−	+	$v^A v^D$
v.BD	−	−	+	+	$v^B v^D$
V.O	+	−	−	−	VV, or Vv
V.A	+	+	−	−	Vv^A
V.B	+	−	+	−	Vv^B
V.D	+	−	−	+	Vv^D
Rare types					
V^q.A	+(a)	+	−	−	
V^q.B	+(a)	−	+	−	
V^{pq}.A	+(b)	+	−	−	
V^{pq}.B	+(b)	−	+	−	
V^{pq}.D	+(b)	−	−	+	
V^{pq}.O	+(b)	−	−	−	

(a), positive reactions with two out of three anti-V^c reagents;
(b), positive reactions with one out of three anti-V^c reagents.

Table 5. Summary of tests with V-A-B-D reagents

Species		V-A-B-D polymorphism
Positive		
Common chimpanzee	(Pan troglodytes)	17 types, including variants
Dwarf chimpanzee	(Pan paniscus)	All v.D
Lowland gorilla	(Gorilla g.gorilla)	All V.O
Mountain gorilla	(Gorilla g.beringei)	All V.O
Orangutan	(Pongo pygmaeus)	v.O., V^q.O, V^{pq}.O, V^q.B, V^{pq}.B
Gibbon	(Hylobates lar lar)	v.O, V^p.O
Negative		
Old World monkeys	(Various species)	All negative
New World monkeys	(Various species)	All negative

I, II, III, and IV classes and therefore was identified as an anti-Mi^a serum. Nonetheless, the serum in question did not react with human red cells of the following uncommon antigens: He+, Hu+, M^g, M^k, Mt(a+), Ri(a+), Sul +, Sta+, To(a+), Far +, Di(a+), Bu(a+), V+ (L. Marsh, personal communication). Therefore, the anti-Mi^a specificity of the chimpanzee serum was definitely confirmed. In chimpanzees, W^c-positive and W^c-negative phenotypes are inherited independently of the VABD types, as products of the dominant W^c gene and its recessive, probably silent, allele w^c. Similar to the situation in the Miltenberger blood groups, known to be part of the MN system of humans, the W^c and w^c types are thought to be part of the chimpanzee MN-VABD blood group system (Socha and Moor-Jankowski 1979).

Bearing in mind the homology between N^v and V^c and the allelic relationship between V^c and v^A, v^B, v^D one can conclude that the series of alleles V, v^A, v^B, v^D, and v are the neosequences that resulted from repeated mutations at the N^v locus in the process of speciation of the chimpanzee. As will be shown in the following part of the chapter, these alleles most probably belong to the GPA gene family. Since the W^c of chimpanzee is identical with its human counterpart Mi^a of the Miltenberger series, both these antigens are probably the paleosequences that appeared early during evolution (Socha and Ruffié 1983). The specificities V^c, A^c, B^c, and D^c together with M and W^c are carried by a chimpanzee major membrane glycoprotein that is similar but not identical to human GPA. The NH_2-terminal fragment of the glycoprotein was found to be an intermediate form of the human M and N glyco-octapeptides (Blumenfeld et al. 1983). Using immunoblotting techniques, at least three glycophorin patterns were identified in chimpanzees which could be correlated with particular V-A-B-D serological types (Lu et al. 1990; Rearden 1986). It is noteworthy, however, that, with only one exception, no correlation was found betwen the V-A-B-D groups and restriction site polymorphisms of chimpanzee glycophorin genes.

Molecular Biology of Glycophorins of Human and Nonhuman Primates

O.O. BLUMENFELD, C.-H. HUANG, S.S. XIE, AND A. BLANCHER

Introduction

The achievements in serology and immunogenetics described in the first part of this chapter provided evidence for the existence, incidence and polymorphism of MNSs blood group antigens in humans and nonhuman primates. Those antigens in all higher primates are products of the glycophorin A (GPA) gene family. The chemical structure of the human M, N, S and s genes is known, and the molecular basis for the common antigens and of a large number of variant alleles are understood (for reviews see: Cartron et al. 1990; Huang et al. 1991b; Fukuda 1993; Huang and Blumenfeld 1995; Blumenfeld and Huang 1995). Currently, information is accumulating on the structure and expression of homologous genes in

Molecular Biology and Evolution of Blood Group
and MHC Antigens in Primates
Blancher/Klein/Socha (Eds.)
© Springer-Verlag Berlin Heidelberg 1997

nonhuman primates. That knowledge has already clarified and substantiated a number of serological observations, and its ramifications in the fields of human genetics, population diversity and evolution are to be anticipated.

The following section deals with the chemical and gene structures of glyco-phorins, the major sialoglycoproteins of the erythrocyte membrane. Current knowledge of the human GPA gene family will be summarized briefly and the focus will be on recently acquired information on the homologues of this gene family in anthropoid apes and Old World monkeys, the only nonhuman primates in which the expression of MN blood groups has been documented so far. Results of phylogenetic analysis are included and the molecular mechanisms for allelic diversification are related to the evolutionary divergence processes. Whenever possible, correlation of serological observations with information thus far available on gene or transcript structure will be attempted.

Human Glycophorins – Protein and Gene Structures

GPA

Glycophorin A carries the antigens for the M and N blood groups. Human GPA has a molecular weight of ~31 000 and contains approximately 50 % carbohydrate in the form of a single Asn-linked bisecting biantennary unit and approximately 15 O-linked glycans (Fig. 3). Numerous studies have shown that two forms of GPA, differing by two amino acid residues (at positions 1 and 5 of the polypep-tide), specify the M and N antigens (Fig. 3) (Huang et al. 1991b; Huang and Blumenfeld 1995).

GPA is one of the most fully characterized type I integral membrane proteins, and for many years it has served as a general model for such proteins. The gen-

---

➤

Fig. 3. *Top:* Gene organization, splicing pattern and protein structure of primate glycophorins. Protein domains are drawn in correspondence with individual exons. *EC*, extracellular; *TM*, transmembrane; and *CP*, cytoplasmic denote the corresponding domains; φ marks the exons that are not expressed. The *vertical black box* at the 5′ end of exon V of GPB and GPB/E illustrates a 9 bp insertion relative to GPA or human GPE; the *crossed vertical box* in exon V of human GPE illustrates the 24 bp insertion. *Diagonal lines* in exons VI of GPB and GPE and GPB/E indicate lack of homology to GPA exon VI (see Fig. 10). GPE protein is not shown in humans and chimpanzee because, most probably, it is not expressed. In the gorilla, both GPA lacking the sequence of exon III and GPB/E are expressed in addition to GPA and GPB. Orangutan, gibbon and rhesus monkey, most probably, have only one gene of the GPA gene family. Sites of attachment of O-glycans one indicated by horizontal lines in extracelluar domain of the proteins; the N-linked glycan in GPA shown by a longer line ending with a circle. Gene, protein and glycan structures and attachment sites are known in man, only; in other primates they are deduced. *Bottom:* A representation of the human glycophorin gene cluster at the MNSs locus, on the long arm of chromosome 4 (Cook et al. 1980; Vignal et al. 1990; Onda et al. 1994); the *small box* downstream from GPA represents the precursor gene

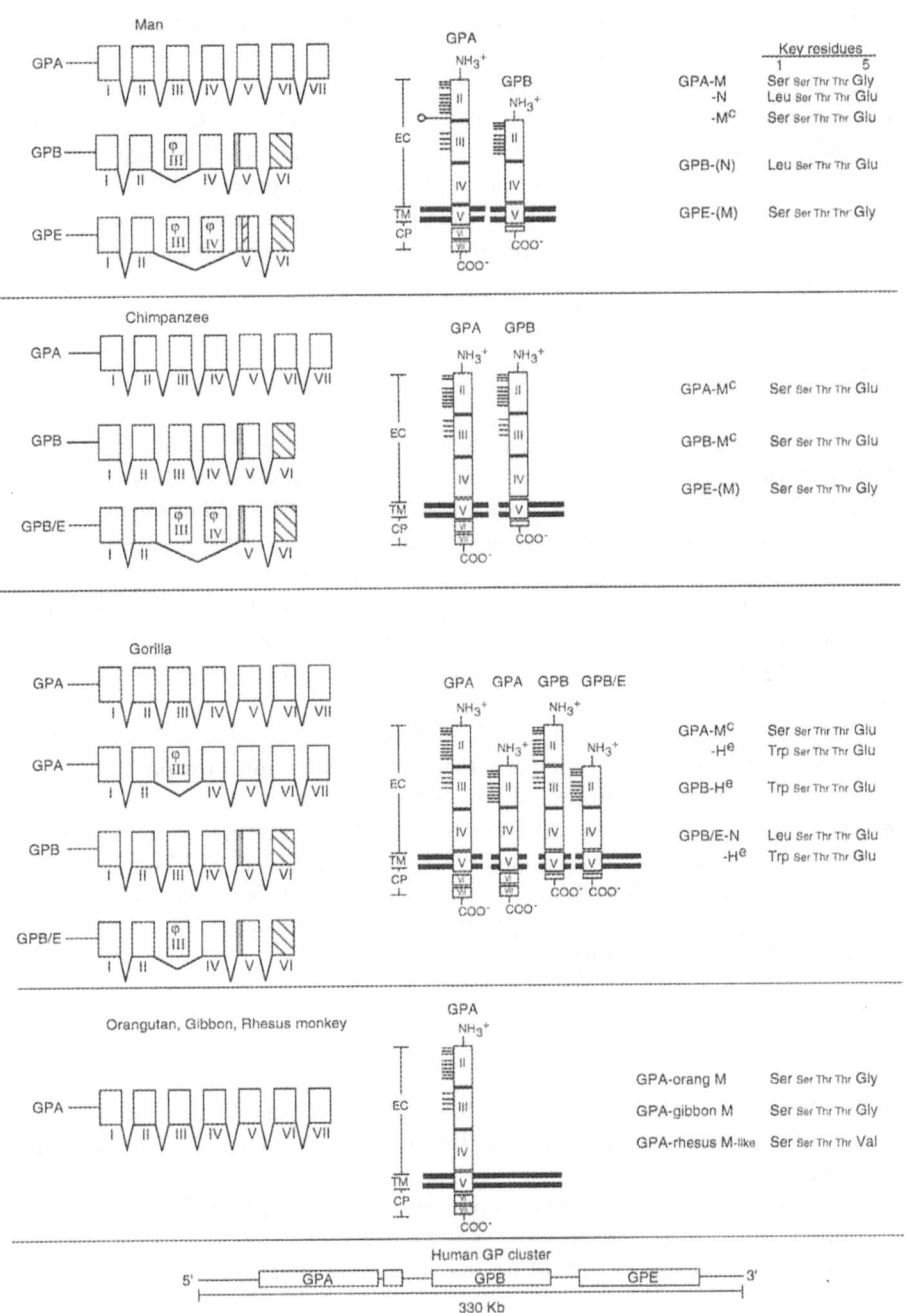

eral organization of the protein domains is superimposable on the organization of the exons. The extracellular domain, encoded by exons II–IV, carries the sites for all the common and variant antigenic epitopes and contains all the glycosylation sites (Fig. 3). Exon V encodes the transmembrane domain, and exons VI and VII the cytoplasmic tail. The sequence for the signal peptide, which is eliminated during protein maturation, is shared by exons I and II. All the exons and introns are of relatively small size, except intron 1 which spans approximately 40 kb; the total *GPA* gene spans about 45 kb (Fukuda 1993; Onda et al. 1994).

GPB

The S and s antigens reside in GPB, encoded by another member of the GPA gene family. Formal genetic studies have shown that the *MN* and *Ss* loci are closely linked (Race and Sanger 1975). Two allelic forms of GPB, differing by a single M $\rightarrow$ T change at residue 29, specify the S and s epitopes (Dahr et al. 1980a,b). Both forms exhibit the N epitope at the NH_2-terminal (Fig. 3). GPB is similar to GPA in sequence content and the pattern of O-glycosylation, but it lacks the N-linked glycan; the major differences include a shorter extracellular domain and the absence of a cytoplasmic tail. Thus, the *GPB* gene differs from *GPA* by:

- A G $\rightarrow$ T transversion at the first nucleotide of the consensus signal of the 5′ splice site in intron 3, which inactivates the expression of exon III.
- A 9 bp in-frame insertion at the 5′ end of exon V.
- The exon VI sequence is different from *GPA* and essentially encodes the 3′ untranslated region; that sequence is homologous to a sequence located downstream from the *GPA* gene (Onda et al. 1994). It is proposed that this sequence was present in the precursor gene of human *GPB* and *GPE* (see the section on evolution).

GPE

GPE is the third member of the gene family, but this gene is most probably not expressed. In humans, *GPE* is very similar to *GPB* but exhibits the following differences:

1. If it were expressed, the gene would encode a glycophorin with the M epitope.
2. Both exons III and IV would be inactivated by alteration of 5′ splicing signals.
3. Compared to *GPB* a 9 bp deletion is present at the 5′ end of its exon V.
4. Compared to *GPA* and *GPB* its exon V contains an in-frame 24 bp insertion (Fig. 3).
5. Compared to *GPB* a premature stop codon, caused by a C $\rightarrow$ T transition, shortened the reading frame by five codons.

6. Exon VI of *GPE* is homologous to exon VI of *GPB* and encodes the 3′ untranslated region (Fukuda 1993).

The three human glycophorin genes are closely linked and reside on the long arm of chromosome 4, somewhere in the q 28–31 region. The cluster is approximately 330 kb in length and the gene order is 5′*GPA-GPB-GPE* 3′ (Fig. 3, bottom) (Cook et al. 1980; Rahuel et al. 1988; Vignal et al. 1990).

The expression of *GPA* and *GPB* genes is confined to erythroid tissues and the proteins are detected in mature forms at the surface of cells at the proerythroblast stage (Ekblom et al. 1985). GPA and GPB account for approximately 2 % of membrane proteins. Their precise role is still not known and the evidence available suggests that they may fulfill one or more of the following functions:

- Due to their abundant O-glycan units, which terminate in sialic acid residues, they provide a negatively charged coat to the erythrocyte and thus maintain the cell in a nonadhesive state in the circulation.
- GPA may contribute to the integrity of the cytoskeleton through interactions with its cytoplasmic tail (Chasis and Mohandas 1992).
- As receptors for malarial merezoites and for certain viruses, they may play a role in the invasion of erythrocytes by those pathogens (Dolan et al. 1994).

Nevertheless, the apparently normal state manifested by the rare variant individuals whose erythrocytes are devoid of both GPA and GPB (the M^kM^k phenotype) (Okubo et al. 1988) suggests that glycophorins may be dispensable under normal physiological circumstances (Blumenfeld and Huang 1995). This requires further study.

Glycophorins in Nonhuman Primates

As described in the first part of this chapter, in the animal kingdom, only the anthropoid apes and most, if not all, species of Old World monkeys express the antigens for the MN blood groups (Socha and Moor-Jankowski 1979). Only humans expresses S and /or s antigens.

Protein Products

Gel electrophoresis and immunoblotting, using antisera that distinguish human GPA from GPB, first indicated that homologues of GPA and/or GPB are present in erythrocytes of chimpanzees, gorillas, orangutans, gibbons and rhesus monkeys (Fig. 4) (Rearden 1986; Lu et al. 1987a). In addition, the amino acid sequence of the Japanese monkey (*Macaca fuscata*) glycophorin demonstrated homology with human GPA (Murayama et al. 1989).

Despite an electrophoretic pattern generally similar to that seen in humans, and similar reactions of monomer and dimer bands with a number of specific

A.

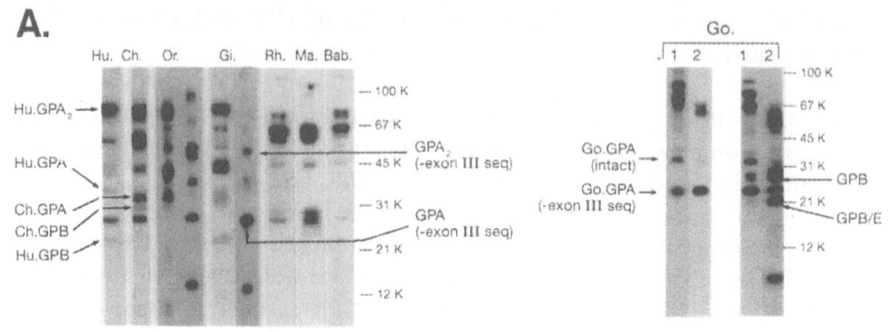

B.   IMMUNOBLOT ERYTHROCYTE LYSATES

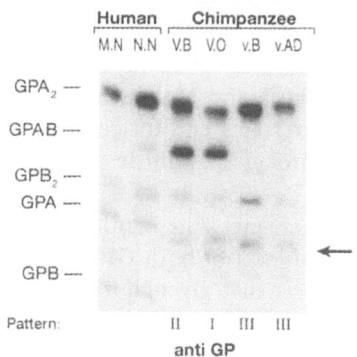

Fig. 4 A,B. A A composite of immunoblots matched from different experiments illustrates the expression and gel electrophoretic patterns of primate glycophorins. Erythrocyte lysates were used for *Hu,* human; *Ch,* chimpanzee; *Rh,* rhesus monkey; *Ma,* macaque; *Bab,* baboon; and isolated membranes were used for *Gi,* gibbon; *Go,* gorilla; *Or,* orangutan. Gels were developed with either glycophorin antiserum (*lanes Hu, Ch, Or, Gi* and right panel: lanes Go.1 and Go.2) or an antiserum to the COOH-terminal portion of human GPA (anti-peptide C serum) (Lu et al. 1987a,b) (lanes Rh, Ma, Bab; and *left panel: lanes Go.1* and *Go.2*). The former antiserum reacts with all glycophorins but the latter reacts only with GPA and not with GPB, GPE or GPB/E. In the case of the gorilla, peptide C antiserum allows clearer visualization of GPA bands; in the case of Rh, Ma, and Bab, both antisera gave an identical band pattern, but the reaction with the peptide C antiserum was stronger. The erythrocyte lysates show a nonspecific band (Lu et al. 1987a,b) at a position slightly ahead of the band marked GPA (exon III sequence). Thus this shortened GPA is present in one orangutan, one gibbon and both gorillas and possibly in Rh and Ma, but is absent in Hu, Ch, and Bab. **B** Correlation of glycophorin band patterns with the VABD blood group phenotypes of representative chimpanzees (the patterns were consistent among over 20 individuals examined (Lu et al. 1987a). Aliquots of erythrocyte lysates were subjected to electrophoresis on 10%–15% gradient gels, transferred to membranes and immunoblotted with glycophorin antiserum. The designation of bands is indicated for human glyco-phorins; note the higher mobility of chimpanzee GPA in relation to human, its slight variation in mobility, and the varying intensities of GPA, GPB and the heterodimer of GPA and GPB bands (GPAB); the *arrow* points to the monomer of GPB in the chimpanzee

antisera, differences were noted between human glycophorins and those of the apes and monkeys. Thus, in all nonhuman primates examined the molecular size of GPA was smaller than that of humans, reflecting the absence in the nonhuman species of the Asn-linked carbohydrate unit (Yoshima et al. 1980). Indeed, as shown later by nucleotide sequence analysis, all members of the glycophorin gene family in all nonhuman primates examined so far lack the consensus sequence for the attachment of this unit. In some gorillas two species of GPA monomers are expressed, one full-sized and one lacking the sequence of exon III (see below) (Xie et al., 1997). The latter form of GPA may also be present in some orangutans and gibbons (Fig. 4) (Xie et al., unpublished).

The size of GPB in the chimpanzee and the gorilla appears larger than that of human GPB and, in contrast to humans, contains the sequence of exon III (Huang et al. 1995). However, no GPB band was observed in the orangutan, gibbon or the rhesus monkey.

Unlike in humans and chimpanzee, the gorilla expressed another glycophorin, tentatively assigned as a product of the *GPB/E* "archaic" gene. Transcripts of this gene lack the sequence of exon III but contain exon IV. These transcripts are most probably translated, leading to the expression of the GPB/E product at the membrane surface (Fig. 4) (see below) (Xie et al., 1997).

Among randomly chosen chimpanzees, small differences in electrophoretic patterns could be correlated with blood groups of the simian-type VABD blood group system (Fig. 4B) (see below) (Lu et al. 1987a).

Glycophorin Genes

So far studies of glycophorin genes in nonhuman primates have been confined to restriction analysis. Patterns of digests with a number of restriction enzymes, probed with human cDNA GPA and GPB fragments, led to the conclusion that a homologue of the human *GPA* gene was present in common and pygmy chimpanzees, gorilla, orangutan and gibbon; in addition, a homologue of the *GPB* gene was detected in chimpanzees and gorillas but not in orangutans and gibbons (Rearden et al. 1990a). In a further study in the chimpanzee, mapping of genomic DNA was performed using as probes synthetic oligonucleotides characteristic of human *GPA*, *GPB* or *GPE* sequences. The results indicated that chimpanzee *GPA* and *GPB* genes were highly homologous to their human counterparts. In addition, it was concluded that a homologue of the human *GPE* gene was also present in the chimpanzee and that organization of the three genes within the cluster may be similar to that in humans (Lu et al. 1990).

Analysis of chimpanzee glycophorin genes with a number of restriction enzymes revealed a relatively large array of nonrandomly distributed restriction fragment length polymorphisms (RFLPs). Rearden et al. (1990b) noted a correlation of certain RFLPs with the expression of the V^c determinant of the simian VABD blood group system; however, such a correlation was not noted by Lu et al. (1990), perhaps due to a difference in the probes used. Since, in a previous, study a clear correlation was observed on immunoblots between VABD determi-

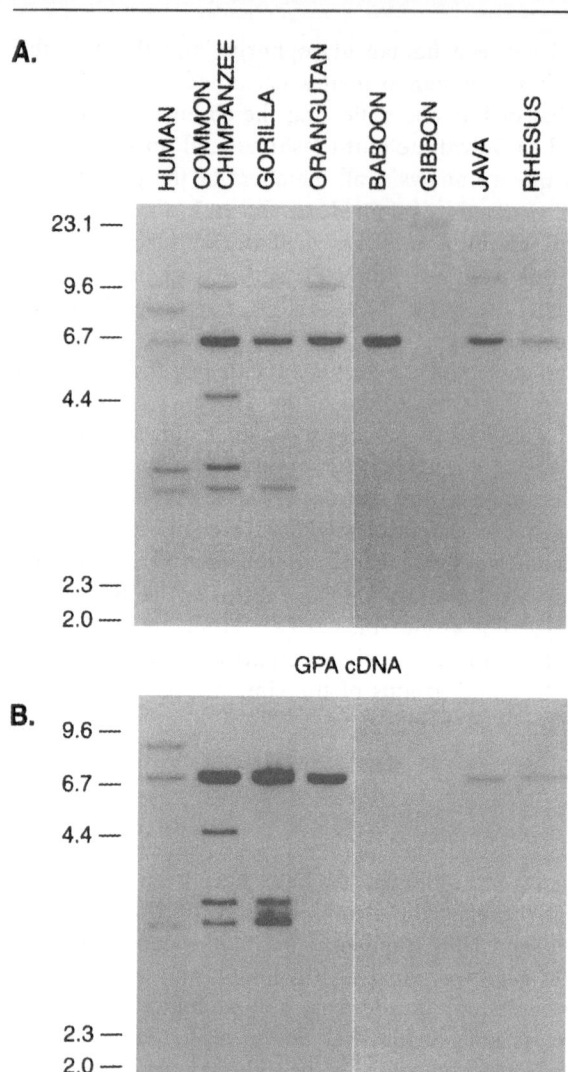

A.

HUMAN
COMMON CHIMPANZEE
GORILLA
ORANGUTAN
BABOON
GIBBON
JAVA
RHESUS

23.1 —
9.6 —
6.7 —
4.4 —
2.3 —
2.0 —

GPA cDNA

B.

9.6 —
6.7 —
4.4 —
2.3 —
2.0 —

IVS 2/3

Fig. 5 A,B. Southern blot of *Sac*I digests of primate genomic DNA probed with **A** human GPA cDNA and **B** a combined intron 2- and intron 3-specific probe (Huang et al. 1991a). The latter probe allows visualization of the portion of the gene that encodes the extracellular domain. The gene origin of fragments is as follows: 9.6 kb, 3′ GPA in all species; 8.0 kb, GPB, in humans only; 6.4 kb, GPE (and /or GPB/E) in humans, chimpanzee and gorilla; 5′ GPA, in all others; 4.4 kb 5′ GPB in chimpanzee; 3.1 kb middle part of GPA in humans, chimpanzee and gorilla; 3′ GPB in chimpanzee; a 2.9 kb fragment of GPB in gorilla; 2.7 kb 5′GPA in humans, chimpanzee and gorilla (Lu et al. 1990; Huang et al. 1991; Huang and Blumenfeld 1995; Xie et al., 1997). In the gibbon the 23 kb band most probably encodes GPA

nants and the protein banding patterns (Lu et al. 1987a), the latter authors concluded that the RFLPs occurred more frequently in the noncoding regions than in coding sequences.

Despite the RFLPs observed with a large number of restriction enzymes, the patterns obtained with *Sac*I digestion in chimpanzees and gorillas are relatively uniform among different individuals and are similar to those of humans (Fig. 5) (Lu et al. 1990). In particular, this digestion has been most informative in revealing that fragments of *GPA* and *GPE* genes of chimpanzees and gorillas are identical to the corresponding human gene fragments; however, the distribution of

*Sac*I sites in GPB differs among the above three species (Huang et al. 1995; Xie et al., 1997). The orangutan, gibbon and rhesus monkey lack both *GPB* and *GPE* and only *GPA* gene fragments are apparent (Fig. 5) (Xie et al., unpublished). Thus, the three-gene framework persists only in humans, chimpanzee and gorilla (Fig. 3).

Transcripts of the GPA Gene Family

Recently, more extensive studies have focused on analysis of transcripts and deduced amino acid sequences of glycophorins. The approach used was based on a reverse transcriptase – polymerase chain reaction (RT-PCR). Total RNA isolated from reticulocytes and primers that were specific for human glycophorins were used to generate cDNAs from individuals in each species (chimpanzee, gorilla, orangutan, gibbon, rhesus monkey). The assignment of such cDNAs as originating from *GPA*, *GPB* or *GPE* was achieved by sequence homology to the respective features of the human counterparts; features examined included sequence content and other specific characteristics of each glycophorin (Huang et al. 1995; Xie et al., 1997).

GPA

All nonhuman primate species investigated thus far, namely, the chimpanzee, gorilla, orangutan, gibbon and rhesus monkey, contained transcripts of GPA whose nucleotide sequences showed about 90 % identity to human GPA (Huang et al. 1995; Xie et al., 1997). The deduced amino acid sequences are shown in Fig. 6 (the relevant nucleotide sequences are deposited in the GenBank). Exon size and organization also appeared identical to those in humans, although in some gorillas exon III was not expressed (see below). The nonhuman GPAs uniformly lacked the sequence for N-glycosylation (residue 26). The O-glycosylation pattern is most likely similar to that in humans, judging from the size of the proteins on polyacrylamide gels and the occurrence of Ser and Thr at the corresponding O-glycosylation sites. (Figs. 4, 6).

The sequence of the MN blood group epitopes has not been the same in all nonhuman primates examined to date. While the epitope at the NH_2-terminal of orangutan and gibbon GPA was identical to the human GPA of the M type, the epitope of the chimpanzee and the gorilla GPAs showed a hybrid sequence of M and N types, with a Ser at position 1 and a Glu at position 5 (hybrid epitope) (Figs. 3, 6). This confirmed previous protein structural studies, which had shown the presence of a hybrid M, N epitope in glycophorins isolated from four chimpanzees (Blumenfeld et al. 1983). In addition, these studies indicated that the epitope region contained only two, instead of three, O-glycans, commonly found in the same region of human GPA (Blumenfeld et al. 1983). The amino acid sequence of Japanese monkey GPA showed an M-like epitope (Ser-1; Val-5), identical to that observed in the Rhesus monkey. That region contained three O-gly-

```
                     II                         III                      IV                   V                            VI                          VII
Human GPA   1  SSTTGVAMHTSTSSSVPKSYISSQTN DTHKRDTYAATFPRAHEV.SEISVRTVYPPEEET GER.VQLAHHFSEP EITLIIFGVMAGVIGTILLISYGIRRLIK KSPSDVKPLPSPDTDVPLSSVEIENPE TSDQ
Human GPA      L---E-----------------------------N-----------------Y-T------------------------Y-S---------------------I-----------------------------------------------
Chimp GPA      ---E-S------E-S------K--N---P------------------------Y-T----------------Y-S-----------------F-Y-S------------R-T------------------------------------I---
Goril GPA      ---E--------E-S------K------P---------------TA-----YN---------V-R--A----------I--------------F-Y-S------------R-T----------Q---E------------------------I-
Goril GPA      W---E-V--------------------N---------------W..I-VPR--A------------I-----------------------------S---------------E----T---G--
Orang GPA      ----EV------TP----K--Q--P--S---------------N----------G--V-R--V----V-------------C----R--Q-----R------------------------------
Gibbo GPA      ----LV--A-I---T----KN--WY--P-.RSVN---D-WQQG-V-R--------I-THS-------------S---CL-R--------Q-----E--------------------------I--
Rhes GPA       ---VP-T---LGPE--V--S---K-TS--HPT--S--F-G--H--DN-R----V-E--L--V-A-------------R---C--R-------Q---P-AE-----ET-S
Monk GPA       ---VP-T--S--LGPEQTV---S---K-TS-SHPTPTS--TT-F-G--H--D------V-E--L--V-A---------F---S-----E---Q---P-AE----D---ET-ELNSFTLRK,BFRNES
Mouse GP                                                                    -D-PAL VMI--L---I----------I-I--LA-VS---R- RP-
Horse GP                                                                    -D--Q- V-TV--L---I----------I-I--F-A-L--M-- -S
Porc GP                                                                    QD--HA --G--A--LLLI-F-A-L--M-- -PL

                     II                        III                      IV                   V
Human GPB   1  LSTTEVAMHTSTSSSVPKSYISSQTN dkhkrdtypa.htanevseisvttvsppekkn GETGQLVHRVTVP APVVLILILCVMAGIIGTILLISYTIRRLIK A
Human GPB      ------------------------------------------------------M---
Chimp GPB      S-----------------------------G------TLG-H--------Y--ED---V---G--L---
Goril GPB      W------------V-----------------------PPR--H--T--Y--EDA--V--M--HS--
Goril GPB      W------------V----------------Q--E--PPR--H--L--T--Y-L-EDG--V--Vg----VFSD*

                     II                        III                      IV                   V                    IV
Human GPE   1  SSTTGVAMHTSTSSSVPKSYISSQTN dkhkrdtypa.hsvnqvseensvttvsppeeen gergqlvhrfpea ..GIVILINWANARVIFEVMLVVGNILLISYCIR....
Chimp GPB/E    ----------I----------------------p ALVEKI---LCP-AG-I--L--Y-S-RLIK A
Goril GPB/E    L---E--V-----------------d---T-----VP ALVEKI...VLCP-AG-I-ML-D-S-RLIK A
```

Fig. 6. Polypeptide sequence of glycophorins in primates. Human sequences were from various sources (see Cartron et al. 1990 for references). Nonhuman primate sequences were deduced from the sequence of cDNAs (see text), except the Japanese monkey GPA sequence which was determined by amino acid sequence analysis (Murayama et al. 1989). For simplification, only one sequence of each animal species was shown, but for gorilla GPA two identified forms are presented, namely, full size and the form missing exon III. The transmembrane regions of porcine, horse and mouse glycophorins are shown for comparison (Honma et al. 1980; Murayama et al. 1982; Matsui et al. 1989). *Capital letters* indicate the expressed sequence; *lower case letters* indicate the non-expressed sequence of pseudoexons, deduced from the corresponding genomic sequence. *Roman numerals* above the sequence indicate correponding exons; *dashes* indicate identical residues; *dots* indicate missing residues; *blanks* in exons VI and/or VII of gorilla and orangutan GPA indicate residues not determined due to location of primers during RT-PCR. Note the residues corresponding to the 9bp insertion and 24bp deletion in exon V of GPB/E in chimpanzee and gorilla, when compared to the sequence in humans

cans (Fig. 6) (Murayama et al. 1989). A sequence analogous to the epitope of human Henshaw (He) antigen was also found on GPA or GPB of some gorillas (see below).

Serological analyses indicate that the MN hybrid sequence, as well as the M-like sequence seen in the rhesus and Japanese monkeys, mainly defines the M specificity; however, the occurrence of such epitopes also may have led to non-uniform serological reactions depending on the difference in carbohydrate side chains and the specificity of the antibodies used as reagents (see Table 1 and below).

GPB

In agreement with restriction analyses, GPB transcripts were obtained only from the chimpanzee and the gorilla (Huang et al. 1995; Xie et al., 1997). Their nucleotide sequences were almost 90 % identical to human GPB; however, the sequence that specifies the human S or s antigen was altered. Accordingly, such reactivites should not be observed in those species. In fact, limited tests with anti-S and anti-s reagents were uniformly negative with the ape and Old World monkey red cells thus far tested (Socha, unpublished).

The important difference from humans was the expression of the GPB exon III sequence in all chimpanzees and gorillas examined so far. As in GPA, GPB transcripts showed a hybrid sequence of the MN epitope in the chimpanzee. However, a He-like sequence was observed in GPBs of the two gorillas examined (Figs. 3, 6).

GPE or GPB/E

Extensive studies of *GPE* or its *GPB/E* "archaic" gene were carried out in the chimpanzee (Huang et al. 1995). Studies of Rearden et al. (1993) of genomic DNAs from nonhuman primates first suggested that a human-like *GPE* gene, containing a 24 bp insertion in exon V, was present in chimpanzees and some gorillas, but not in orangutans or gibbons. However, so far, no transcript equivalent to that gene has been isolated (Huang et al. 1995). Instead, the GPE-like transcript isolated from the chimpanzee most resembled human GPE in sequence content, in the presence of the true M epitope and in the absence of exons III and IV. Nonetheless, the transcript also bore much similarity to human GPB, namely a 9 bp insertion at the 5′ end of exon V and a lack of the 24 bp insertion sequence in exon V. On the basis of these findings it was proposed that the chimpanzee GPB/E-like transcript is a product of the *GPB/E* "archaic" composite gene (see below). Genomic studies have indicated that, in addition to the archaic gene, a GPE-like gene that contains a 24 bp insertion sequence is present in the chimpanzee; however, so far, no transcript of that gene has been isolated and it remains unclear whether the gene is expressed or just occurs as a pseudogene (Huang et al. 1995).

A transcript, tentatively assigned as originating from a GPB/E gene, was also isolated from the gorilla (Xie et al., 1997). Unlike the situation in chimpanzees, expression of a translated product from gorilla at the erythrocyte membrane was demonstrated by western blot (Fig. 4). The structure of the gorilla GPB/E transcript most resembles that of the chimpanzee but it is also similar to human GPB. Thus, the gorilla transcript lacks exon III due to the inactivation of the 5' splice site in intron 3 by a G → A transition; it contains a 9 bp insertion at the 5' end of exon V but lacks the 24 bp insertion. In contrast to human and chimpanzee homologues, the gorilla GPB/E transcript retains a functional exon IV whose sequence is identical to the genomic sequence amplified from the SacI 6.4 kb fragment. The same fragment in the human and the chimpanzee also specifies the GPE gene. Whereas human GPE and chimpanzee GPB/E bear the sequence for the M epitope, the gorilla GPB/E sequence was N in one individual. Perhaps the occurrence of this epitope preceded that of the M epitope which, subsequently, became associated with the chimpanzee GPB/E or human GPE by gene conversion (Kudo and Fukuda 1994). Further genomic studies should address the evolutionary history of the MN epitopes on the three glycophorins; moreover, it is important to establish whether or not the gorilla GPB/E archaic gene occurs in addition to a possibly silent GPE-like gene.

Intraspecies Variation and Allelic Diversity

In Humans

The existence of variants of MNSs blood groups in the human population has been thoroughly documented by serological analyses (Race and Sanger 1975). Thus, in addition to, or instead of, the usual glycophorins, the common MNSs antigens expressed in most humans, erythrocytes of some individuals bear structurally altered glycophorins that exhibit novel epitopes, absent from the common forms. These uncommon epitopes either replace the MNSs epitopes or result in additional ones. Interestingly, the occurrence of variant phenotypes in populations is not random but usually shows a well-defined ethnic or geographic distribution: the incidence of variant forms in Western populations is relatively low, but it can be relatively high (7 %–33 %) in certain regions of the world, particularly among Asians and among Blacks in Central Africa (see Table 1, Blumenfeld and Huang 1995).

Recently, the molecular basis was elucidated for many allelic variations which result in the human variant antigens. Homologous recombinations among alleles of GPA and GPB were found to be the predominant mechanism for diversification of human glycophorins. The molecular basis for close to 25 variant phenotypes has now been established (summarized in: Huang and Blumenfeld 1995; Blumenfeld and Huang 1995). Briefly, most of these variants resulted from rearrangements consisting of unequal homologous recombinations and/or gene conversions. In both GPA and GPB genes, recombinations occurred predominantly in a region of 4 kb, confined to the three exons coding for the extracellular domain. At least two hot-spots of recombination mark this region; a key hot-spot is located in sequences within and surrounding exon III, which is expressed in GPA but is silenced in GPB (therefore, it is called a pseudoexon). An important feature unique to this gene family was the coupling of recombination to splicing events. For example, expression of silent patches of pseudoexon III of GPB occurred by activation of a splicing signal. Such activation was mediated by recombinations, and indeed several Mi variant genes arose by this type of mechanism.

Figure 7 shows a schematic representation of products of the variant genes, as deduced from gene and/or transcript analyses and in some cases confirmed by protein sequencing. The epitope sites are also indicated in the figure.

Concerning the variants of the Miltenberger (Mi^a) series, all but MiV and XI arose by gene conversion. In MiI, II, VII, VIII and IX the GPA gene was the recipient and the GPB gene was the donor. In all cases recombination occurred at different sites in exon III whereby small patches of pseudoexon III of GPB replaced the corresponding region of GPA exon III. The variant epitopes (Vw, Hut, Nob, Hop, Mur) were created by these patches of previously silent, but now expressed sequences from pseudoexon III and by the intra-exon junctions. In the cases of MiIII, IV, VI and X, the reverse occurred and the recipient gene was GPB and the donor was GPA. Again, all recombinations occurred at different sites within and around exon III; here too, the epitopes were created by activation of expression of pseudoexon sequences, albeit by a mechanism involving transfer of an active splicing signal along with the GPA donor segment. Accordingly, the Hil epitope was created by GPA-GPB inter-exon junction.

MiV and MiXI arose by the unequal recombination between GPA and GPB that occurred within intron 3 (Huang and Blumenfeld 1991a). For MiV this resulted in the joining of GPA and GPB (the s allele) as above and hence creation of the Hil epitope. In MiXI, the S allele of GPB participated in a junction with GPA and because of the sequence difference gave rise to the Tsen epitope (Reid et al. 1992).

It should be noted that several variants bear the same epitope. This is because, although the expressed patches of pseudoexon III in different variants vary in length, they may all contain a common sequence determining a common epitope (Fig. 7).

The case of variants that express the St^a epitope is noteworthy. That epitope results from the junction of GPB or GPA exon II (these two exons have an identical sequence) with GPA exon IV. As shown in Fig. 7 the St^a epitope can result from different events or modes of recombination. The most common is unequal homologous recombination between GPB and GPA, occurring at different sites in intron 3. The less common GPMz St^a and ERIK (TFF) St^a occurred by gene conversion, wherein the recipient was GPA and the donor pseudoexon III of GPB (for M^z) or of GPE (for ERIK (TFF); GPERIK St^a is the only exception and has arisen by a splice site mutation at the 3' end of exon III of GPA. Clearly, as noted above, the exon III-intron 3 region is most prone to recombination. Close to 20 variant alleles arose by recombination in this 1 kb region. This condensation of recombi-

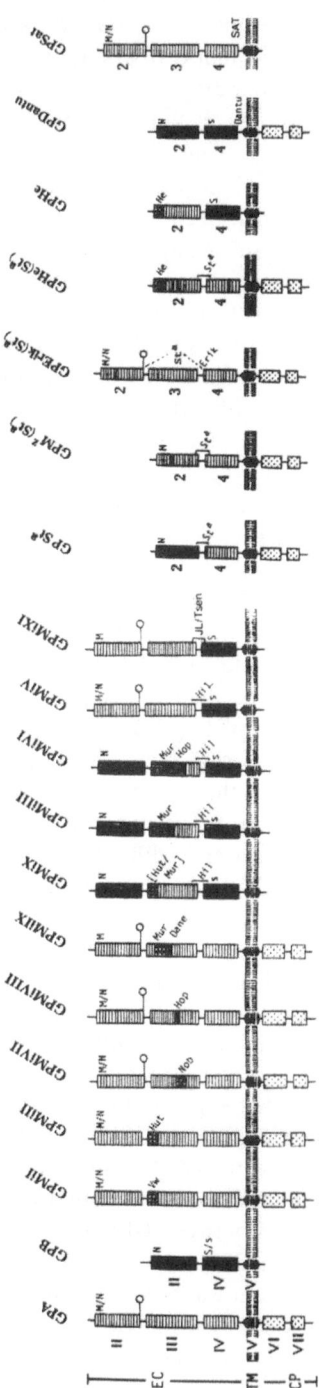

Fig. 7. The protein strucure of glycophorin variants of the Miltenberger complex, the St^a isoforms, He, Dantu and SAT (adapted from Huang and Blumenfeld 1995). Structures of common GPA and GPB are included for comparison. Location of variant MNSs epitopes is shown. *M/N* indicates that the variant may occur in two allelic forms. In most cases hybrid inter-or intra-exon junctions specify the variant epitopes. The latter usually result from expression of patches of previously silent sequence from GPB pseudoexon III. —○: Asn-linked glycan; O-glycans are not shown but they are present in all variant glycophorins and are located at sites similar, but not necessarily identical to, those that occur in common forms of GPA or GPB. *Cross-hatched boxes:* extracellular domain exons of GPA; *stippled boxes:* cytoplasmic domain exons of GPA; *filled-in* boxes: extracellular domain exons of GPB; *black stippled patches:* patches of pseudoexon III sequence of GPB introduced by gene conversion, or in the case of He, the region of untemplated mutations

nation events may be partly attributed to the presence of unusual sequence motifs, characteristic of a hot-spot of recombination (Huang and Blumenfeld 1991b). The two reciprocal variants Dantu and Sat arose from unequal homologous recombination within intron 4, which may have been driven by yet-to-be-defined recombination sequences (Fig. 7).

The M^g, M^c and He variants all occurred by alteration of the sequence that specifies the M or the N epitope, at the NH_2-terminal of GPA (Fig. 3). Thus, the sequence for M^g is Leu-Ser-Thr-Asn-Glu; for M^c it is Ser-Ser-Thr-Thr-Glu; and for He it is Trp-Ser-Thr-Ser-Gly (Blumenfeld et al. 1981; Furthmayr et al. 1981; Dahr et al. 1984; Huang et al. 1994a,b). Those alterations could also have occurred by gene conversion (Huang and Blumenfeld 1995).

The variant phenotypes En(a-), S-s-U- and M^k are characterized by the absence from the erythrocyte membrane of either GPA or GPB or both glycophorins and are caused by gene deletions. Usually a large portion of each gene(s) (except exon I) is deleted through recombination between two genes within the cluster (Huang and Blumenfeld 1995). For example, recombination between GPB and GPE within intron 1 leaves an intact GPA and produces a hybrid of GPB (exon I) and GPE (exons II-VI). The latter event resulted in the absence of GPB-borne S, s, and U antigens giving rise to the S-s-U-phenotype (Huang et al. 1987) (see below). The He, nondeletion S-s-U- and Wr^b variants will be discussed below.

In evaluating the allelic diversification of human glycophorins, it is important to note that, in all cases, a starting point for study was a variant serological phenotype. Perhaps that is why the gene rearrangements documented here are a selected set of alterations, each of which led to a glycophorin bearing a segment of a novel uncommon sequence; such sequences would be immunogenic in those individuals not normally expressing them. However, point mutations can also give rise to variant epitopes, for example in MiI, MiII or M^g, which are possibly due to changes in glycosylation pattern (Blumenfeld et al. 1981) and/or conformation of the polypeptide.

Such "new" sequences are all expressed and tolerated by the organism, indicating that the physiological functions of glycophorins are compatible with considerable sequence alterations. It remains to be established whether the presence of O-glycosylation on the extracellular domain of the protein is sufficient to allow a glycophorin to fulfill its key functions.

In Nonhuman Primates

As indicated previously, chimpanzee erythrocytes bear diverse antigens of the simian-type V-A-B-D blood group system (Socha and Moor-Jankowski 1993). The antigens for those blood groups most probably reside on glycophorins, a conclusion based on the following observations:

1. Isolated chimpanzee glycophorin preparations inhibited antibodies specifically directed against the V-A-B-D antigens (Blumenfeld et al. 1983).

2. Differences in electrophoretic patterns of glycophorins could be correlated with the V-A-B-D blood groups of the subjects on immunoblots developed with specific glycophorin antisera (Lu et al. 1987a); reaction with a specific monoclonal anti-N antibody (NN5) stained GPB bands only in animals that expressed the V determinant (Rearden 1986).

3. GPA- and/or GPB-specific bands were visualized on immunoblots of chimpanzee erythrocyte lysates, developed with chimpanzee anti-V^c, anti-A^c+B^c+C^c or anti-V^c+W^c antisera. The former two reacted exclusively with GPA bands whereas the latter, in addition, made the GPB band apparent (Socha and Blumenfeld, unpublished).

4. Specific RFLPs detected with human glycophorin-specific cDNA probes were correlated, in the chimpanzee, with expression of the V^c determinant (Rearden et al. 1990b).

The fact that chimpanzee erythrocytes express various V-A-B-D antigens but still all react with human M antisera points to intraspecies differences in their glycophorin structures at sites other than the hybrid MN epitope. So far, the molecular basis for that simian blood group system or any variant blood group phenotype has not been established with regard to the chimpanzee or other nonhuman primates. It is possible that gene recombinations akin to those documented in humans also play a role in allelic diversification of chimpanzee glycophorins. That possibility is based on the following:

1. The three glycophorin genes are closely homologous and probably reside in close proximity within the cluster.

2. In all three genes, exon III has a "hot-spot-like" sequence resembling the one found in human glycophorins (Fig. 6) (Huang and Blumenfeld 1991b). In fact, a variant phenotype was defined on chimpanzee red cells by means of a chimpanzee alloimmune anti-W^c serum, which in tests with human red cells produced results identical to those obtained with human anti-Mi^a (Socha and Moor-Jankowski 1993).

3. Subtle differences observed in electrophoretic patterns of chimpanzee GPA could be correlated with the V-A-B-D blood group phenotypes (Lu et al. 1987a), suggesting that alterations involved whole blocks of the gene sequence.

4. Antigens bearing the V-A-B-D epitopes appear to be products of variant alleles (see above).

The first indication that the structures of glycophorins varied among gorillas was obtained by immunoblotting analyses. Three different electrophoretic patterns of GPA were observed among three randomly selected gorillas (Lu et al. 1987a). More recently, comparison of GPA gene transcripts from seven unrelated gorillas revealed unexpected differences in the expression of exon III that could partially explain the electrophoretic differences observed among the protein products (Xie et al., 1997). Thus GPA cDNAs, either containing or lacking exon III, were documented at the sequence level. As shown by immunoblotting, GPAs with or without exon III were expressed; in some individuals both forms were observed, whereas in others a truncated form was found (Fig. 4). Since minor but significant differences in sequence were observed among GPAs from different gorillas, it appears that the two forms of GPA are products of different alleles. This needs to be established at the genomic level. In view of the key role that exon III, in active or inactive form, played in diversification of human glycophorins, it is intriguing that expression of this exon in GPA is variable in the gorilla.

Among the gorillas examined, no single common nucleotide sequence was found among transcripts of GPA. The number of amino acid substitutions among the seven gorillas studied varied from less than 1% to 6% of GPA total amino acid residues. That number may be larger than would be encountered among human glycophorins from randomly chosen individuals. Unfortunately, as yet, such data have not been compiled for humans and still need to be thoroughly documented in nonhuman primates. It is of interest that among the seven gorillas examined, five carried, on GPA, GPB, or GPB/E the epitope for the human variant He antigen (Xie et al., 1997). This was demonstrated on immunoblots by reaction with He mAbs and/or by sequence content (see below).

Clearly, the question of diversification of glycophorins among nonhuman primates has not been thoroughly addressed as yet. As noted earlier, serological evidence suggests that polymorphism does occur within each species, but structural and molecular studies have lagged behind. Information of that kind will be essential to the understanding of the molecular evolution of this gene family.

Variants Common to Human and Nonhuman Primates

Henshaw (He) and S-s Nondeletion Variants

The S-s blood group phenotype in humans results from the absence of the corresponding epitopes that usually reside on GPB. Two different molecular mechanisms are responsible for the occurrence of this phenotype. The S-s- deletion variants show the absence of GPB from the membrane surface due to deletion of the GPB gene (Huang et al. 1987) (see above); in contrast in S-s nondeletion variants, altered GPB is present but, because of its sequence changes, does not express these antigens (Huang et al. 1989). Recently, approximately 50% of S-s specimens were shown to be nondeletion variants. Of those, a relatively large number were He-positive as revealed by serological testing and/or by RT-PCR (Reid et al. 1996).

In humans, He is a low-incidence antigen of the MNSs blood group system encountered predominantly among Blacks, who show a relatively high incidence of this phenotype (Race and Sanger 1975). The He epitope resides on GPB (Dahr et al. 1984). Molecular studies have indicated that GPHe is a product of a variant GPB gene that arose by gene conversion accompanied by several nontemplated nucleotide substitutions. Those alterations occurred at the 5' end of the coding region of the gene; as a result, in the expressed GPB the He epitope replaced the M or N epitope (Huang et al. 1994a,b) In some He-positive individuals, alterations at the transmembrane region (exonV) also occurred resulting in the different disposition of altered GPB at the erythrocyte surface; that probably accounts for the observed differences in serological reactivities (Reid et al. 1996).

Serological studies first demonstrated that human-like He variants occurred with a high incidence among gorillas (Wiener et al. 1972). Those studies showed that of 15 animals tested with human anti-He reagent, 13 gave a positive reaction. With the exception of the orangutan, no other species of anthropoid apes that commonly carry the MN antigens, nor any other tested species of nonhuman primates, reacted with anti-He.

Sequence analysis of cDNAs from several gorillas indicated that an He-like epitope, Trp-Ser-Thr-Thr-Glu, could occur on GPA or GPB (Fig. 6) (Xie et al., 1997). This sequence differs from that of the human epitope Trp-Ser-Thr-Ser-Gly, in the last two residues (Dahr et al. 1984). Nevertheless, immunoblots developed with a human He-specific mAb confirmed that the He antigen was expressed at the membrane surface (Xie et al., inpress). The antigens were detected on either GPA, GPB or GPB/E or on both GPA and GPB/E. It thus appears that the first three residues Trp-Ser-Thr, most probably in O-glycosylated form, are sufficient to specify the He epitope. It should be noted that the original observations of Wiener et al. (1972) clearly showed that all He-positive gorillas also were M-positive. At this time, the coincidence of those two epitopes can best be explained by coexpression at the membrane surface of both He- or the M-reactive glycophorins. Furthermore, in gorillas, this phenotype may be associated with deletions in exon V of GPB or GPB/E identical to those present in GPHe from some human He-positive individuals (Xie et al., 1997; Huang et al. 1994a,b).

The observation that GPHe occurs in the gorilla and can be a variant of GPA, GPB or GPB/E points to an evolutionary relationship with human GPHe. Yet, those gorilla genes contain none of the untemplated nucleotide changes outside of the epitope-defining region, as observed in human GPHe (Huang et al. 1993). Whether this points to a more recent and independent origin of GPHe in humans requires further study.

Wr[b] Antigens

In humans, the Wright antigen (Wr) consists of the low-incidence antigen Wr[a] and the high-incidence Wr[b] (Wren and Issitt 1988). Recent studies have shown that this polymorphism correlates with an allelic substitution of Lys-658-Glu of band 3 (Bruce et al. 1995). Nevertheless, the antigenic site for Wr[b] is a nonlinear structure arising from specific conformationally dependent interactions within the erythocyte membrane between GPA and the band 3 protein (Huang et al. 1996). In particular, contacts between a region with a key Glu residue at position 658 of band 3 and a number of residues within a small stretch (residues 62–71) immediately adjacent to the transmembrane junction of GPA have been implicated.

The Wr[b] antigen was also detected in chimpanzees, gorillas and gibbons tested with a macaque mAb directed against human erythrocytes (see first part of this chapter). The orangutans, macaques, rhesus monkeys and baboons thus far tested were all negative. Human antiserum (MF) also reacted positively with red cells of two chimpanzees but negatively with two orangutans, a gibbon and a Rhesus monkey. Comparison with the human sequence of the critical regions of the cDNA sequence of GPA and band 3 of the latter animals helped to define the structural requirements for expression of the Wr[b] antigen (Huang et al. 1996).

En[a] and Mi[a]

A number of nonhuman primates have shown reactions with anti-Mi[a] reagents (Socha and Moor-Jankowski 1993) or anti-En[a] reagents (Socha and Blancher, first part of this chapter) that paralled those found in humans. Although erythrocytes of most humans react with anti-En[a], and several human Miltenberger variants, namely, MiI, II, III, IV and VI, react with anti-Mi[a] (Dahr 1992), the location of epitopes that define these reactions still need to be clarified (see below).

So far, the above are the only common, human-like variant antigens of the MN blood group system found in nonhuman primates by either serological or molecular analyses.

Correlation of Serological and Molecular Observations

Table 6 gives the peptide sequences from positions 1–5 of glycophorins of humans and nonhuman primates and the antigens carried by the glycophorins in these species. As is well known, the amino acids at positions 1 and 5 of GPA are determinant in humans for the expression of antigens M and N. The M antigen depends on the presence of a serine and a glycine at positions 1 and 5, respectively, while the expression of the N antigen correlates with a leucine and a glutamic acid in those positions. In these same positions, GPB is invariable and carries an N specificity resistant to proteolysis and therefore easily revealed on all human red cells after destruction of GPA following treatment by proteolytic enzymes.

The amino acid sequences of some human glycophorin variants are given in Fig. 6 for comparison with the glycophorins of nonhuman primates. Expression of the He antigen corresponds in humans to a mutated GPB with a tryptophan at position 1, a serine at position 4 and a glycine at position 5 (Dahr et al. 1984). The human antigen M[c] was defined as a variant of antigens M and N because it was recognized by most rabbit anti-M and a minority of rabbit anti-N reagents (see the first part of this chapter). The M[c] phenotype corresponds to a GPA which carries a serine in the first position (like M-type GPA) and a glutamate in the fifth position (like N-type GPA) (Furthmayr et al. 1981).

Table 6. Amino acid residues at the first five positions of the glycophorins of humans and nonhuman primates: correlation with the expression of the antigens carried by the proteins

Species and phenotypes	Antigen	Type of glycophorin	Residues in positions[a]				
			1	2	3	4	5
Humans	M	A^M	Ser	Ser	Thr	Thr	Gly
M or N or MN	N	A^N	Leu	–	–	–	Glu
	N	B	Leu	–	–	–	Glu
	M	(E)[b]	Ser	–	–	–	Gly
	M^c	A	Ser	–	–	–	Glu
	M^g	A	leu	–	–	Asn	Glu
	H^e	B	Trp	–	–	Ser	Gly
Chimpanzee	M^c	A	Ser	–	–	–	Glu
(Pierre)	M^c	B	Ser	–	–	–	Glu
M, N weak N^v, V^c.A^c	M	(B/E)[b]	Ser	–	–	–	Gly
Gorilla	M^c	A	Ser	–	–	–	Glu
N or MN	H^e-like	B	Trp	–	–	–	Glu
H^e 13/15	N	B/E^N	Leu	–	–	–	Glu
	H^e-like	B/E^{He}	Trp	–	–	–	Glu
Orangutan M+ or M–	M	A	Ser	–	–	–	Gly
Gibbon M or N or MN	M	A	Ser	–	–	–	Gly
Rhesus monkey All M+ All N–	M-like	A	Ser	–	–	–	Val
Macaca fuscata All M+	M-like	A	Ser	–	–	–	Val

[a] Deduced from sequence of cDNAs or amino acid analysis of glycophorin peptides.
[b] Probably not expressed.
Numbers of animals examined: chimpanzee: GPA (4), GPB and GPB/E (1 each); gorilla: GPA (7), GPB (2), GPB/E (2), orangutan: GPA (2), all others: GPA (a single animal).

A mutation at position four (threonine to asparagine) is responsible for the human low frequency antigen M^g (Furthmayr et al. 1981). This mutation completely abolishes O-glycosylation of the three residues at positions 2, 3 and 4. This leads to the conclusion that the amino acid residues present at these positions and /or the cluster of three O-linked glycans may be involved in epitopes M and N (Blumenfeld et al. 1981).

One can infer the type of antigen carried by the few nonhuman primates analyzed so far from an examination of residues present at positions 1 and 5 of the polypeptide (Table 6). The deduced sequences of orangutan and gibbon glycophorins A are similar in the five first positions to the human glycophorin of type M. As orangutans are M-positive or M-negative (and always N-negative), it would be important to clarify the molecular basis for the M-negative orangutan type. In the gibbon, as in humans, three groups (M, MN, N) were described. The sequence which is given in the table corresponds to a type M; thus, the substitutions determining type N in gibbon remain to be defined. Serologically, all rhesus monkeys were found to be M, but the sequence of the rhesus GPA differs from the human type M glycophorin at position 5 where a valine is found instead of a glycine. Although this substitution can be considered as conservative, the M epitopes of rhesus monkey and humans are similar but not identical. This is supported by serological observations, namely, absorption experiments of rabbit anti-M reagents on human and nonhuman primate M red cells. The glycophorin of the Japanese monkey (*Macaca fuscata*) was shown to be identical to the glycophorin of the rhesus monkey in the 11 NH_2-terminal positions (Fig. 6). Serologically, the GPA purified from Japanese monkey red cells inhibited both the anti-M and the anti-N reagents (Murayama et al. 1989).

The situation in chimpanzees and gorillas is more complex than in orangutans, gibbons, rhesus and Japanese monkeys because the former species not only express GPA but also GPB, while gorillas, in addition, also express a third glycophorin encoded by a gene called *GPB/E*. The present study has not established the molecular basis of the MN polymorphism of gorillas and chimpanzees because too few chimpanzees were investigated and the phenotypes of the seven gorillas were not known. However, one could conclude that the GPAs of gorilla and chimpanzee are identical in the first five positions to the human glycophorin variant of type M^c. The four chimpanzees so far studied were all M-positive. The polymorphic expression of antigen N was demonstrated in the chimpanzee by means of tests carried out with rabbit anti-N reagents. The results obtained with rabbit anti-N correlated with those obtained with *Vicia graminea* and later with chimpanzee alloimmune or heteroimmune (anti-human) antibodies (Moor-Jankowski and Wiener 1965). The determinant evidenced by the *Vicia graminea* reagent was called N^v and that detected by chimpanzee antibodies V^c. Recently, we produced anti-human macaque mAbs which gave results paralleling those obtained with chimpanzee anti-V^c (Blancher et al. 1993). Studies of chimpanzee glycophorins by western blot evidenced correlations between the polymorphism of GP polypeptides and the animals' VABD phenotypes; the GPA bands reacted with anti-V^c antibody suggesting that V^c epitopes reside on GPA (Socha and Blumenfeld, unpublished). However, Rearden (1986) observed that a monoclonal anti-N antibody reacted with the GPB band in V^c-positive animals only. It is impossible to define the molecular basis of V^c and N^v antigen expression because these phenotypes were known in only one animal, Pierre, who was V^c- and N^v-positive. One can only suppose that the presence of a glutamic acid residue at position 5 in the sequence of the GPA of this chimpanzee rendered his red cells agglutinable by anti-V^c and anti-N^v reagents. One must, however, remember that human M^c red cells, which have also a glutamic acid residue in position 5, are not agglutinated by *Vicia graminea* reagent. Thus, the molecular basis of N^v and V^c antigen expression in the chimpanzee remains unresolved.

Serologically, gorillas are either MN or N. Among the seven gorillas which were studied, five possessed a GPA with a hybrid MN epitope, whereas one animal showed a He epitope and one (most probably a heterozygote) had both forms of GPA, type MN and type He. Since the (M-, N+) phenotype is rare among gorillas (4/15), it is probable that all gorillas included in the present study were M-positive. The presence of a glutamic acid residue at position five of all gorilla glycophorins (A, B, B/E) could explain the positivity with *Vicia graminea* reagent as well as the positive reactions obtained with some anti-N mAbs in tests with all gorillas thus far investigated. However, the presence of a tryptophan in the first position of gorilla GPB or GPA, observed in most animals under study, is concordant with the prevalent (13/15) expression of antigen He in this species. The M-positive phenotype of gorillas can best be explained by coexpression of the M-reactive and He-reactive glycophorins. It is interesting to note that the gorilla sequences are not identical with the GPB variant which carries the He antigen in humans. Comparison of the human and gorilla sequences indicates that the He epitope depends on the first three amino acid residues, Trp-Ser-Ser. Positions 4 and 5 can by occupied either by Ser-Gly (humans) or Thr-Glu (gorilla).

In humans, expression of the low frequency antigen Mi^a correlates with the presence of a methionine (MiI) or a lysine (MiII, MiIII, MiIV, MiVI, MiX) in position 28 of these glycophorin variants (Dahr 1992). In the serological section of this chapter, we have seen that the chimpanzee alloantibodies anti-W^c in tests with human red cells gave results identical to those obtained with anti-Mi^a: only the rare MiI, MiII, MiIII, MiIV red cells were agglutinated while all other human red cells tested were negative (Marsh, personal communication). Interestingly, a lysine is found in position 28 of GPA from chimpanzee Pierre, gorillas, gibbon, orangutan and rhesus monkey (Table 7). However, Pierre and all gorillas tested so far were found to be W^c-negative. This suggests that the presence of a lysine at position 28 by itself is not sufficient to produce positive reactions with anti-W^c antibodies. Reactions obtained with chimpanzee anti-W^c reagent in tests with gibbon, orangutans and rhesus monkeys were difficult to interpret because of interference by species-specific antibodies. We will, therefore, focus our comparisons on human variants, chimpanzee and gorilla.

It is noteworthy that the chimpanzee and gorilla GPB, unlike their human counterpart, displays the sequences encoded by exon III. Comparison of the sequences of GPBs from the chimpanzee Pierre and of some gorillas with those of the human Mi^a+ variants suggests that a single substitution at position 31 of chimpanzee GPB would be sufficient to restore expression of the Mi^a (W^c) epitope. It would therefore be very interesting to study the sequence of a W^c+ chimpanzee to verify whether this point mutation is really involved in expression of the W^c antigen in the chimpanzee. The gorilla 5 GPB is identical to human MiIII, MiIV and MiVI variants at positions 22–36, including a lysine in position 28. Although gorilla 5 was not tested for W^c (Mi^a), one has to recall that all gorillas tested so far were W^c-negative. It is probable, therefore, that the presence of the sequence PAPP (positions 35–38 of gorilla GPB) instead of the sequence PAHT (human GPA) completely modifies the conformation of the region impli-

Table 7. Amino acid residues at positions 22–40 of the glycophorins of humans and non-human primates: correlation with the expression of the antigen Mi[a] (W[c])

		Exon II	Exon III
Decades		22222	22233333333333
Units		23456	78901234567890
Human (M or N)	GPA	SSQTN	DTHKRDTYAATPRA
Human MiI		-----	-M------------
Human MiII		-----	-K------------
Human MiIII		-----	-K------P-HTAN
Human MiIV		-----	-K------P-HTAN
Human MiVI		-----	-K------P-HTAN
Human MiX		-----	-K------------
Chimpanzee	GPA	--E-S	-K--W---P-----
Gorilla	GPA	--E-S	-K------P----
Orangutan	GPA	TP---	-K--Q---P---S-
Gibbon	GPA	-----	-KN-WY--P.RSV
Rhesus monkey	GPA	---S-	-K-TS--HPT--S-
Chimpanzee	GPB	-----	-K--G---P--LG-
Gorilla 5	GPB	-----	-K------P-PPR-
Gorilla 1a	GPB	-----	-K--Q--EP-PPR-
Gorilla 1a'	GPB	-----	-E--Q---P-PPR-

The position of each amino acid is indicated by the number of the decade and the number of units.

The amino acid residues are indicated by uniletter code.

The sequences of the human glycophorin variants which carry the Mi[a] antigen are presented. Note that in the most common form of human GPA, position 28 of the protein is occupied by a threonine (T) while in the Mi[a]+ Miltenberger variants, a lysine (K) is found at this position. The sequences of nonhuman primate GPA and GPB are shown for comparison. All the animals from which the sequences were obtained are W[c]-negative and therefore most probably Mi[a]-negative (for more details, see text).

cated in the epitope Mi[a] (Table 7). This reinforces the notion of epitope Mi[a] being most probably not linear, but conformational.

The expression of antigens S and s in humans depends on position 29 of GPB: a methionine corresponds to the expression of S while a threonine in this position results in s (Dahr 1980a). In nonhuman primates, equivalents of GPB are found only in gorillas and chimpanzees. In both species, exon III, unlike in human GPB, is functional, and GPB contains 104 amino acids. This leads to the expression at the surface of chimpanzee and gorillas red cells of a GPB protein with an extended extracellular domain totally different from that of human GPB. In the position equivalent to position 29 of human GPB, the chimpanzee GPB displays a tryptophan and some gorillas GPB a threonine (like human type s GPB). The presence of a tryptophan in chimpanzee GPB is certainly incompatible with the expression of S or s by chimpanzee GPB. As a matter of fact, all chimpanzees that we have studied so far are S- and s-negative (see first part of this chapter). The presence of a threonine in gorilla GPB could be

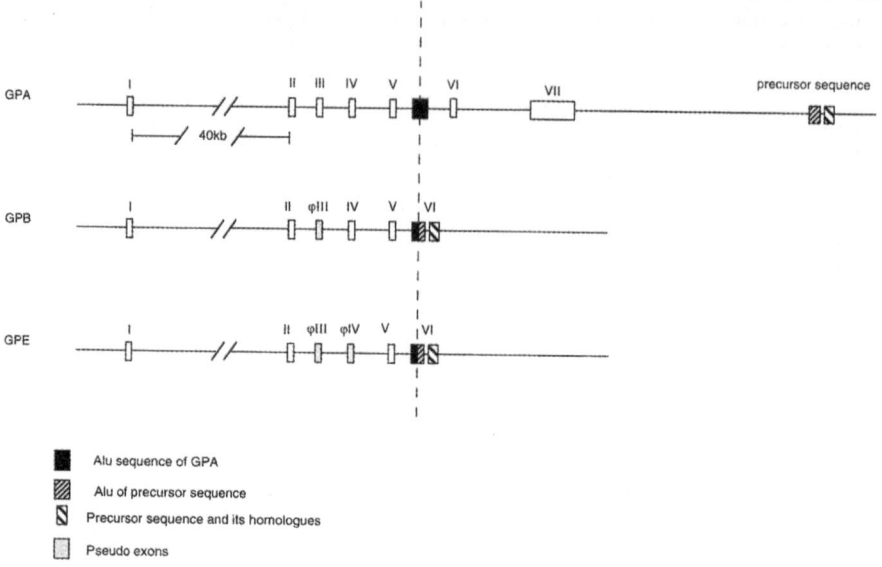

Fig. 8. Genomic organization of human glycophorins (adapted from Onda et al. 1994, Fig. 1). Exons are shown as *open boxes*; *stippled boxes* are silent exons or pseudoexons; precursor gene sequence is indicated and its homologous sequence in intron 5 of GPB and GPE is shown adjoining the *Alu* sequence

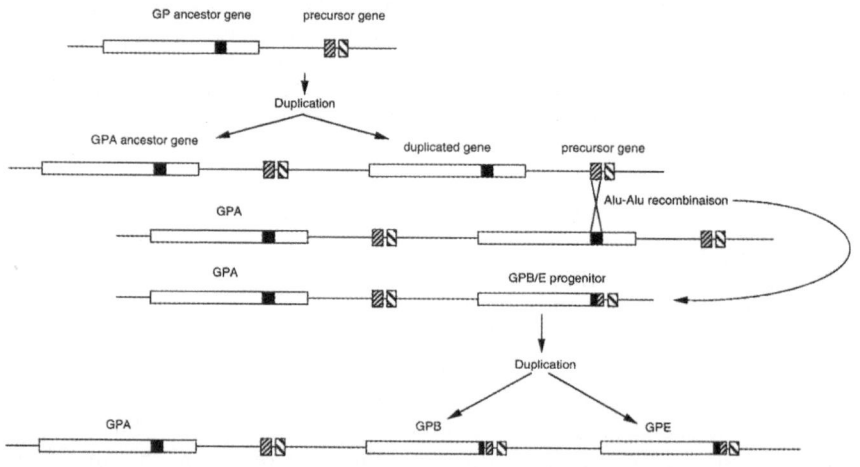

Fig. 9. Evolutionary pathway of the three glycophorin genes (adapted from Onda et al. 1994; Table 6). The designation of the genes is indicated above the *open rectangular boxes*. The *filled box* in GPA illustrates the *Alu* sequence in intron 5; the *Alu* sequence in the precursor gene is the *stippled box* in between the large stretch of the genomic sequence 5' and the 3' end of the gene (*crossed box*). Following *Alu-Alu* recombination, the latter fragment then appears first in the GPB/E progenitor gene and then in GPB and GPE

compatible with the expression of the antigen s. However, the amino acid residues which precede the threonine are not encoded, as in human GPB, by exon II but by exon III, which is functional in gorilla GPB. This certainly inactivates the potential epitope s and could explain why it was impossible to demonstrate the presence of antigens s (or S) in gorillas (see first part of this chapter).

Evolution of Glycophorin Genes

Model Based on Genomic Analyses

Based on comparative analysis of the nucleotide sequence encompassing exon V-intron 5 (transmembrane region) in genomes of humans, common and pygmy chimpanzees, gorilla, orangutan and gibbon, the following scheme was proposed for the evolution of the GPA gene family (Rearden et al. 1993; Onda et al. 1993, 1994; Kudo et al. 1993; Kudo and Fukuda 1994) (Figs. 8, 9). First, an ancestral GPA gene was duplicated. One of the duplicated genes diversified to form GPA, while the other gave rise to a GPB/E progenitor gene through unequal homologous recombination between the Alu repetitive sequence in intron 5 and the Alu sequence of the precursor gene, located 9 kb downstream from the last exon of GPA. The GPB/E progenitor gene finally duplicated by unequal crossing-over with the precursor gene and, following diversification, gave rise to GPB and GPE. This proposed scheme is based largely on the homology of the 3' end (exon V and downstream) in GPB and GPE and the 3' end of the progenitor sequence. More recently, studies of the human gene locus in YAC clones showed the absence of the precursor gene in sequences downstream from GPB and GPE, providing further evidence for the two proposed recombination events. Nevertheless, the precursor gene, or its counterpart, is most probably found downstream from GPA in all anthropoid apes. It was concluded that duplication of the ancestral glycophorin gene did not occur in lower primates such as the orangutan or gibbon (Rearden et al. 1990a; Onda et al. 1993, 1994). This may explain why GPB and GPE-like genes are absent in those species. The model also predicted that the first duplication facilitated recombination with the precursor gene and that those events occurred prior to gorilla divergence approximately 8–10 million years ago. Duplication of the progenitor gene to generate GPB and GPE was estimated to have taken place prior to the divergence of the chimpanzee from the human-gorilla-chimpanzee clade, i.e., 6.3–7.7 million years ago (Onda et al. 1994).

Model vs cDNA Sequence Data

The proposed model was based on early restriction analysis of genomic DNAs, which showed the presence of the GPA gene in all primates and of GPB in

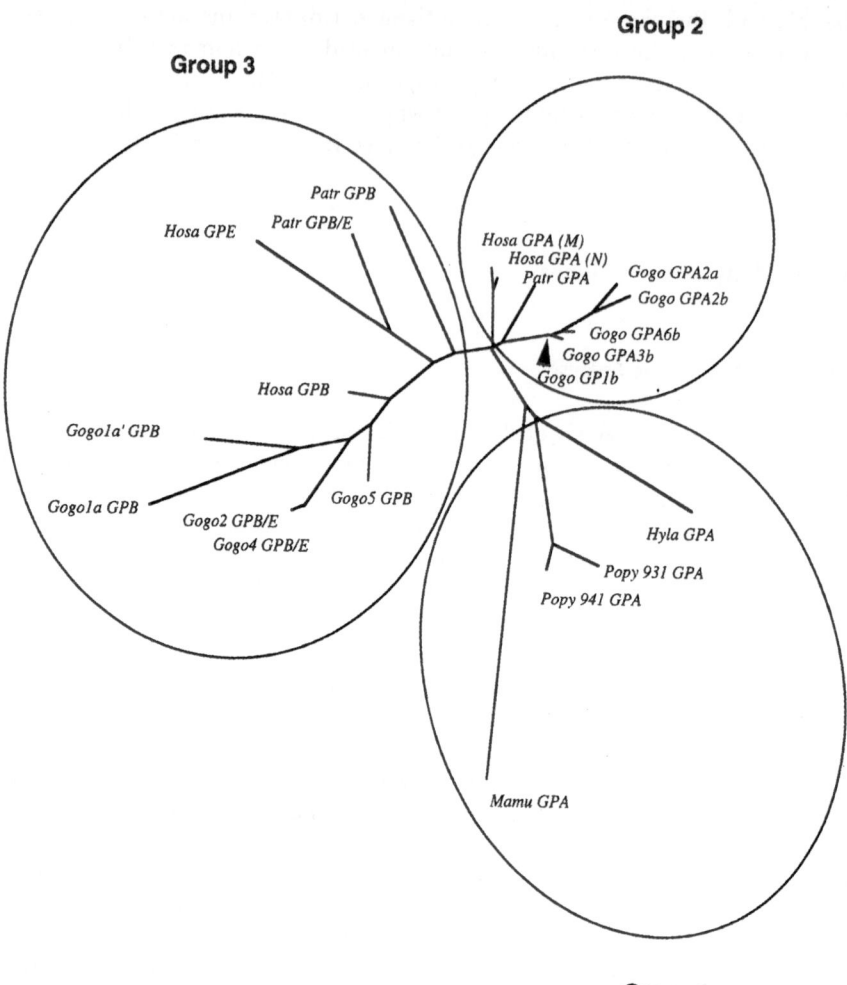

Fig. 10. Sequences of cDNAs encompassing exons II–V were aligned. Distances were calculated using the Kimura two-parameter method (Kimura 1980). The phylogenetic tree was reconstructed by using the neighbor joining method (Saitou and Nei 1987). All algorithms were from Felsenstein (1995). *Hosa* (*Homo sapiens*: humans); *Patr* (*Pan troglodytes*: chimpanzee); *Gogo* (*Gorilla gorilla*: gorilla); *Popy* (*Pongo pygmaeus*: orangutan); *Hyla* (*Hylobates lar*: gibbon); *Mamu* (*Macaca mulatta*: rhesus monkey). Data was compiled for cDNA sequences obtained from seven individual gorillas and two orangutans; for others, single animals were used. In the case of GPA, *a* denotes full sequence; *b* the sequence lacking exon III

humans, chimpanzee and gorilla only (Rearden et al. 1990a). The early immunoblot studies showed, in addition, that GPA was expressed in the red blood cell membranes of all anthropoid apes, the rhesus monkey and the baboon, but that GPB was present only in the chimpanzee and the gorilla (Rearden 1986;

Lu et al. 1987a; Xie et al., 1997). More recent analyses of genes and transcripts of nonhuman primates, summarized above, essentially agree with the proposed model. The data document the homology of GPA and point to its relative conservation of sequence across the species. They confirm the presence of a full transcript of GPB in the chimpanzee and gorilla. As noted, the GPE-like transcript isolated from the chimpanzee and the gorilla bears similarity to both GPB and GPE and it may have kept some archaic traits inherited from the *GPB/E* progenitor gene. Its presence in the gorilla was predicted by the model (Onda et al. 1994). However, a revision may be required concerning the cluster in the chimpanzee and possibly in the gorilla as well. Although both human-like *GPE* (with an insertion sequence in exon V) and *GPB/E* were detected in chimpanzee genomic DNA, only the latter was transcribed (Huang et al. 1995). Thus, with respect to the chimpanzee, the model needs to incorporate the persistence of the *GPB/E* gene. As to whether the same situation prevails in the gorilla remains unclear and more studies are needed to determine the evolutionary time for the branching-out and disappearance of the archaic *GPB/E* gene.

Phylogenetic Reconstruction Based on cDNA Sequence

Figure 10 shows the phylogenetic tree constructed from cDNA sequences, described in the previous sections of this chapter. The unrooted tree shows that the sequences group into three clusters. The first encompasses the sequences of species possessing only one glycophorin gene, namely, the macaque, gibbon and orangutan. Those sequences probably evolved by mutations from the ancestor gene common to all glycophorin genes. The second group consists of all sequences of GPA from species possessing multiple members of the *GPA* gene family, namely humans, chimpanzee and gorilla. The sequences of gorilla GPAs constitute a subgroup of GPA. The third group of sequences encompasses human GPB and GPE and their counterparts in gorilla and chimpanzee. It is important to note that human GPE and chimpanzee GPB/E are on the same branch in this group, whereas gorilla GPB and GPB/E are on a distinct branch together with human GPB.

Results of this analysis concern sequences which are not orthologous. Thus, these results cannot be used for reconstruction of the phylogenesis of glycophorin genes. However, these data are consistent with duplication of an ancestral gene in a common ancestor of humans and African apes; one of the duplicated genes gave rise to the group of genes encompassing all *GPB*, *GPE* and *GPB/E* sequences while the other diversified to give the *GPA* genes currently found in African anthropoid apes and Old World monkeys.

Perspectives

It is important to note again that, in the animal kingdom, the occurrence of antigens of the MN blood group system and the presence of products of the glycophorin A gene family have been observed only in anthropoid apes and some species of Old World monkeys, but not in New World monkeys (Socha and Moor-Jankowski 1979; Lu et al. 1987a). Nevertheless, New World monkeys may contain homologues that belong to the GPA family. This would be analogous to a number of mammalian species (horse, pig, mouse) whose red cells bear glycoproteins closely related to higher primate glycophorins (Honma et al. 1980; Murayama et al. 1982; Matsui et al. 1989; Krotiewski 1988). All these glycoproteins are type I membrane proteins. They differ in the length of the polypeptide chain and show little sequence homology to primate glycophorins except in the transmembrane segment (Fig. 6). Yet their extracellular domains are rich in serine and threonine residues and are heavily glycosylated with sialylated O-linked glycans whose structures are similar to those seen in humans (Krotiewski 1988). It appears that in evolution the major structural requirement for glycophorins was to be anchored in the membrane and accommodate a large number of sites in the extracellular domain for sialylated O-linked glycans. These features alone may satisfy the key functional role for glycophorins, namely, to provide a nonadhesive surface to the erythrocyte. Yet, conservation of the glycophorin sequence in anthropoid apes and Old World monkeys is intriguing and points to a more subtle role for glycophorins. Further studies will be needed to understand the gap between the high level of conservation of the glycophorin sequence among humans, anthropoid apes and Old World monkeys and the striking divergence of its sequences in the lower mammalian species.

In addition to their importance for understanding the molecular aspects of the MNSs blood group system, glycophorins became the first model of membrane proteins based on studies of their disposition across the membrane, the nature of their transmembrane region, the structures and arrangement of carbohydrates on the extracellular domain, and their gene structure and organization (Cartron et al. 1990). Also, they were the first protein blood group antigens whose epitopes were elucidated at the sequence level and in which differences were defined in the products of the two common alleles (Blumenfeld and Adamany 1978). Subsequently, because of the occurrence of a large number of variants well-documented at the serological level, this system served as an ideal model for studies of the molecular mechanisms that led to this polymorphism in the population.

As shown in this chapter, homologous recombination, including gene conversion and unequal crossing-over occurring predominantly at two hot-spots found at sequences encoding the extracellular domain, was the major mechanism for diversification. The presence of recombination motifs within the two hot-spots, homologous in all three members of the gene family, appear to have been partly responsible for this process.

Two important questions concerning the evolution of glycophorin remain to be addressed: (1) how and when such structural motifs were acquired, and (2) whether the modulation of splicing coupled to recombination was unique to humans or was predated in nonhuman primates. It appears from studies summarized here that activation or inactivation of exon III in all three members of the gene family is a key characteristic in humans and in chimpanzee and gorilla, the nonhuman primates in which the three-gene (or four-gene) cluster prevails. The high sequence homology among the three genes and, most probably, an arrangement in tandem within the cluster, similar to that seen in humans allows the inference that allelic diversification has occurred in these primates and that the pattern of recombination observed in humans was predated in the chimpanzee and the gorilla. This can be tested and verified by identification of serological variants in nonhuman primates and by elucidation of the molecular basis for their occurrence. Extended studies in this direction will provide clues to the more general question: Was allelic diversity of glycophorin genes maintained as an advantge for survival of the species or was this diversity fixed by pure chance during evolution?

Acknowledgements. We thank the Wildlife Conservation Society New York N.Y. and the Yerkes Research Center, Atlanta, Georgia, (grant no. RR00165) for providing us with gorilla, orangutan and gibbon blood. We thank our colleague Dr. Sam Seifter for critical reading of the manuscript. This work was supported by NIH grants GM 16389 and 1P50HL54459.

References

Allen FHJr, Cordoran PA, Kenton HB. Breare N M^g a new blood group antigen in the MNSs system. Vox Sang 3:81–91, 1958

Blancher A, Socha WW. Anti-human red cell monoclonal antibodies produced by macaque-mouse heterohybridomas. J Med Primatol 20:352–356, 1991

Blancher A, Clayton J, Socha WW, Ruffié J. Application of maximum likelihood statistics for the population and family studies of inheritance of the chimpanzee R-C-E-F and V-A-B-D blood groups. J Med Primatol 22:44–49, 1993

Blumenfeld OO, Adamany AM. Structural polymorphism within the amino-terminal region of MM, NN, and MN glycoproteins (glycophorins) of the human erythrocyte membrane. Proc Natl Acad Sci USA 75:2727–2730, 1978

Blumenfeld OO, Huang C-H. Molecular genetics of the glycophorin gene family the antigens for MNSs blood groups: multiple gene rearrangements and modulation of splice site usage result in extensive diversification. Hum Mutation 6:199–209, 1995

Blumenfeld OO, Adamany AM, Puglia KV. Amino acid and carbohydrate structural variants of glycoprotein products (M-N glycoproteins) of the M-N allelic locus. Proc Natl Acad Sci USA 78:747–751, 1981

Blumenfeld OO, Adamany AM, Puglia KV, Socha WW. The chimpanzee M blood-group antigen is a variant of the human M-N glycoproteins. Biochem Genet 21:333–348, 1983

Borne AE, Bos ME, Lomas C, Tipett P, Bloy C, Hermand P, Cartron J-P, Admiraal LG, van dem Graaf J, Overbeeke AM. Murine monocloanl antibodies against a unique determinant of erythrocytes related to Rh and U antigens: expression on normal and malignant erythrocyte precursosrs and Rh$_{null}$ red cells. Br J Hamatol 75:254–261, 1990

Bruce LJ, Ring SM, Anstee DJ, Reid ME, Wilkinson S, Tanner MJA. Changes in the blood group Wright antigens are associated with a mutation at amino acid 658 in human erythrocyte band 3: a site of interaction between band 3 and glycophorin A under certain conditions. Blood 85:541, 1995

Cartron J-P, Colin Y, Kudo S, Fukuda M. Molecular genetics of human erythrocyte sialoglycoproteins glycophorins A, B, C, and D. In: Haris JR (ed) Blood cell biochemistry. Plenum, New York, pp 299–335, 1990

Cartron JP, Rahuel C. Human erythrocyte glycophorins: protein and gene structure analyses. Transfusion Med Rev 2:63–92, 1992

Chalmers JNM, Ikin EW, Mourant AE. A study of two unusual blood-goup antigens in West Africans Rare blood group antigen occurring in negroes. Br Med J ii:175–177, 1953

Chasis JA, Mohandas N. Red blood cell glycophorins. Blood 80:1869–1879, 1992

Clerghorn TE. Two human blood group antigens St^a (Stones) and Ri^a (Ridley) closely related to the MNSs system. Nature 195:297–298, 1962

Clerghorn TE. A memorandum on Miltenberger blood groups. Vox Sang 11:219–222, 1966

Cook PJL, Noades JE, Lomas CG, Buckton KE, Robson EB. Exclusion mapping illustrated by the MNSs blood group. Ann Hum Genet 44:61–72, 1980

Dahr W. Miltenberger subsystem of the MNSs blood group system. Vox Sang 62:129–135, 1992

Dahr W, Moulds JJ. High-frequency antigens of human sialoglycoproteins. IV. Molecular properties of the U antigen. Hoppe Seylers Z Physiol Chem 368:659–667, 1987

Dahr W, Uhlenbruck G, Janssen E et al. Different N-terminal aminoacids in the M, N glycoprotein form MM and NN erythrocytes. Hum Genet 35:335–343, 1977

Dahr W, Beyreuther K, Steinbach H et al. Structure of the Ss blood group antigens. II. A methionine/threonine polymorphism within the N-terminal sequence of the Ss glycoprotein. Hoppe Seylers Z Physiol Chem 361:895–906, 1980a

Dahr W, Gielen W, Beyreuther K, Krüger J. Structure of the Ss blood group antigens Isolation of Ss glycopeptides and differentiation of the antigens by modification of methionine. Hoppe Seylers Z Physiol Chem 361:145–152, 1980b

Dahr W, Kordowicz M, Judd WJ, Moulds J, Beyreuther K, Kruger J. Structural analysis of the Ss sialoglycoprotein specific for Henshaw blood group from human erythrocyte membranes. Eur J Biochem 141:51–55, 1984

Darnborough J, Dunford I, Wallace J. The En factor. A genetical modification of human red cells affecting their blood grouping reactions. Programme, British Society of Haematology, meeting, vol 28, 1965

Darnborough J, Dunford I, Wallace J. The En^a antigen and antibody. A genetical modification of human red cells affecting their blood grouping reactions. Vox Sang 17:241–255, 1969

Dolan SA, Proctor JL, Alling DW, Okubo Y, Wellems TE, Miller LH. Glycophorin B as an EBA-175 independent *Plasmodium falciparum* receptor of human erythrocytes. Mol Biochem Parsitol 64: 55–63, 1994

Dunsford E, Ikin EW, Mourant AE. A human blood group gene intermediate between M and N. Nature 172:688–689, 1953

Ekblom M, Gahmberg CG, Andersson LC. Late expression of M and N antigens on glycophorin A during erythroid differentiation. Blood 66:233–236, 1985

Erskine AG, Socha WW. The principles and practice of blood grouping, 2nd edn. Mosby, St Louis, 1978

Felsenstein J. PHYLIP: phylogenesis inference package (version 357c), 1995

Fukuda M. Molecular genetics of the glycophorin A gene cluster. Semin Hematol 30:138–151, 1993

Furthmayr H. Structural comparison of glycophorins and immunochemical analysis of genetic variants. Nature 271:519–521, 1978

Furthmayr H, Metaxas MN, Metaxas-Buhler M. Mg and Mc: mutations within the amino-terminal region of glycophorin A. Proc Natl Acad Sci USA 78:631–635, 1981

Gershowitz H, Fried K. Anti-M^v a new antibody of the MNSs system. I. M^v an inherited variant of the M^v gene. Am J Hum Genet 18:264–281, 1966

Holman CA. A new rare human blood group antigen (Wra). Lancet ii 119, 1953

Honma K, Tomita M, Hamada A. Amino acid sequence and attachment sites of oligosaccharide units of porcine erythrocyte glycophorin. J Biochem 88:1679–1691, 1980

Huang C-H, Blumenfeld OO. Identification of recombination events resulting in three hybrid genes encoding human MiV, MiV(J.L.) and St^a glycophorins. Blood 77:1813–1820, 1991a

Huang C-H, Blumenfeld OO. Multiple origins of the human glycophorin St^a gene. J Biol Chem 266:23306–23314, 1991b

Huang C-H, Blumenfeld OO. MNSs blood groups and major glycophorins. Molecular basis for allelic variation, chapter 5. In: Cartron J-P, Rouger P (ed) Blood cell biochemistry 6: Molecular basis of human blood group antigens. Plenum, New York, 1995

Huang C-H, Johe K, Moulds JJ, Siebert PD, Fukuda M, Blumenfeld OO. Glycophorin (glycophorin B) gene deletion in two individuals homozygous for the S-s-U-blood group phenotype. Blood 70:1830–1835, 1987

Huang C-H, Lu W-M, Boots ME, Blumenfeld OO. Two types of delta glycophorin gene alterations in S-s-U- individuals. Transfusion 29:35S, 1989

Huang C-H, Guizzo ML, McCreary J, Leigh EM, Blumenfeld OO. Typing of MNSs blood group specific sequences in the human genome and characterization of a restriction fragment tightly-linked to S-s- alleles. Blood 77:381–386, 1991a

Huang C-H, Johe KK, Seifter S, Blumenfeld OO. Biochemistry and molecular biology of MNSs blood group antigens. Baillieres Clin Haematol 4:821–848, 1991b

Huang C-H, Blumenfeld OO, Reid M. The S-s-U- and S-s-U+ blood group phenotypes. Molecular heterogeneity and relationship to Henshaw allele. Transfusion 33:46S, 1993

Huang C-H, Reid ME. Blumenfeld OO Remodeling of the transmembrane segment in human glycophorin by aberrant RNA splicing. J Biol Chem 269:10804–10812, 1994a

Huang C-H, Lomas C, Daniels G, Blumenfeld OO. Glycophorin He(St^a) of the human red blood cell membrane is encoded by a complex hybrid gene resulting from two recombinational events. Blood 83:3369–3376, 1994b

Huang C-H, Xie SS, Socha W, Blumenfeld OO. Sequence diversification and exon inactivation in the Glycophrin A gene family from chimpanzee to human. J Mol Evol 41:478–486, 1995

Huang C-H, Reid ME, Xie SS, Blumenfeld OO. Human red cell Wright antigens: a genetic and evolutionary perspective on Glycophorin A-band 3 interaction. Blood 87:3942–3947, 1996

Ikin EW, Mourant AE. A rare blood group antigen occurring in negroes. Br Med J i:456–457, 1951

Kimura M. A simple method for estimating evolutionary rates of base substitutions through comparative studies of nucleotide sequences. J Mol Evol 16:111–120, 1980

Krotiewski H. The structure of glycophorins of animal erythrocytes. Glycoconjugate J 5:35–48, 1988

Kudo S, Fukuda M. Contribution of gene conversion to the retention of the sequence for the M blood group type determinant in Glycophrin E gene. J Biol Chem 269:22969–22974, 1994

Kudo S, Onda M, Rearden A, Fukuda M. Primate genes for glycophorins carrying MN blood group antigens. J Med Primatol 22:7–12, 1993

Landsteiner K, Levine P. A new agglutinable factor differentiating individual human bloods. Proc Soc Exp Biol Med 24:600–602, 1927a

Landsteiner K, Levine P. Further observations on individual differences of human blood. Proc Soc Exp Biol Med 24:941–942, 1927b

Landsteiner K, Levine P. On individual differences in human blood. J Exp Med 47:757–775, 1928

Landsteiner K, Strutton WR, Chase MW. An agglutination reaction observed with some human bloods, chiefly among negroes. J Immunol 27:469–472, 1934

Landsteiner K, Wiener AS. On the presence of M agglutinogens in the blood of monkeys. J Immunol 33:19–25, 1937

Levine P, Kuhmichel AB, Wigod M, Koch E. A new blood factor, s, allelic to S. Proc Soc Exp Biol 78:218–220, 1951

Lu Y-Q, Liu J-F, Socha WW, Nagel RL, Blumenfeld OO. Polymorphism of glycophorins in nonhuman primate erythrocytes. Biochem Genet 25:477–491, 1987a

Lu Y-Q, Nichols ME, Bigbee WL, Nagel RL, Blumenfeld OO. Structural polymorphism of glycophorins demonstrated by immunoblotting techniques. Blood 69:618–624, 1987b

Lu WM, Huang CH, Socha WW, Blumenfeld OO. Polymorphisms and gross structure of glycophorin genes in common chimpanzee. Biochem Genet 28:399–413, 1990

Madden HJ, Cleghorn TE, Allen FH, Rosenfield RE, Mackeprang M. A note on the relatively high frequency of St^a on the red cells of orientals, and report on a thrid example of anti-St^a. Vox Sang 9:502–504, 1964

Matsui Y, Natori S, Obinata M. Isolation of the cDNA clone for mouse glycophorin erythroid-specific membrane protein. Gene 77:325–332, 1989

Metaxas MN, Metaxas-Buhler M, Ikin EW. Complexities of the MN locus. Vox Sang 15:102–117, 1968

Moor-Jankowski J, Wiener AS. Simian type blood factors in nonhuman primates. Nature 205:369–371, 1965

Moor-Jankowski J, Wiener AS. Red cell antigens of primates. In: TT-W-Fiennes RN (ed) Pathology of simian primates, part 1. Karger, Basel, pp 270–317, 1972

Moor-Jankowski J, Wiener AS, Gordon EB. Blood groups of apes and monkeys. III. The M-N blood factors of apes. Folia Primat 2:129–148, 1964

Murayama J-I, Tomita M, Hamada A. Primary structure of horse erythrocyte glycophorin HA. Its amino acid sequence has a unique homology with those of human and porcine erythrocyte glycophorins. J Membr Biol 64:205–215, 1982

Murayama J-I, Utsumi H, Hamada A. Amino acid sequence of monkey erythrocyte glycophorin MK. Its amino acid sequence has a striking homology with that of human glycophorin A. Biochim Biophysica Acta 999:273–280, 1989

Okubo Y, Daniels GL, Parsons SF, Anstee DJ, Yamaguchi H, Tomita T, Seno T. A Japanese family with two sisters apparently homozygous for M^k. Vox Sang 54:107–111, 1988

Onda M, Kudo S, Rearden A, Mattei M-G, Fukuda M. Identification of a precursor genomic segment that provided a sequence unique to glycophorin B and E genes. Proc Natl Acad Sci USA 90:7220–7224, 1993

Onda M, Kudo S, Fukuda M. Genomic organization of glycophorin A gene family revealed by yeast artificial chromosomes containing human genomic DNA. J Biol Chem 269:13013–13020, 1994

Race RR, Sanger R. The MNSs blood groups in Blood groups in man, 6th edn. Blackwell, Oxford, pp 92–138, 1975

Rahuel C, London, d'Auriol L, Mattei MG, Tournamille C, Skrzynia C, Lebouc Y, Galibert F, Cartron JP. Characterization of cDNA clones for human glycophorin A. Eur J Biochem 172:147–153, 1988

Rearden A. Evolution of glycophorin A in the hominoid primates studied with monoclonal antibodies, and description of a sialoglycoprotein analogous to human glycophorin B in chimpanzee. J Immunol 136:2504–2509, 1986

Rearden A, Phan H, Kudo S, Fukuda M. Evolution of the glycophorin gene family in the hominoid primates. Biochem Genet 28:209–222, 1990a

Rearden A, Phan H, Fukuda M. Multiple restriction fragment length polymorphisms associated with the Vc determinant of the MN blood group-related chimpanzee V-A-B-D systems. Biochem Genet 28:223–231, 1990b

Rearden A, Magnet A, Kudo S, Fukuda M. Glycophorin B and glycophorin E genes arose from glycophorin A ancestral gene via two duplications during primate evolution. J Biol Chem 268:2260–2267, 1993

Reid ME, Moore BPL, Poole J, Parker NJ, Asenbryl E, Vengelen-Tyler V, Lubenko A, Galligan B. TSEN: a novel MNS-related blood group antigen. Vox Sang 63:122–128, 1992

Reid ME, Storry JR, Ralph H, Blumenfeld OO, Huang C-H. Expression and quantitative variation of the low incidence blood group antigen He on some S-s- RBCs. Transfusion (36:719–724), 1996

Ridgwell K, Eyers S, Mayby W, Anstee D, Tanner J. Studies on the glycoprotein associated with Rh (Rhesus) blood group antigen expression in the human red blood cell membrane. J Biol Chem 269:6410–6416, 1994

Saitou N, Nei M. The neighbor joining method: a new method for reconstructing phylogenetic trees. Mol Biol Evol 4:406–425, 1987

Socha WW. Blood groups of pygmy and common chimpanzees: a comparative study. In: Sussman RL (ed) The pygmy chimpanzee: evolutionary morphology and behavior. Plenum, New York, pp 13–41, 1984

Socha WW, Moor-Jankowski J. Blood groups of anthropoid apes and their relationship to human blood groups. J Hum Evol 8:453–465, 1979

Socha WW, Moor-Jankowski J. The M-N-V-A-B-D- blood group system of chimpanzee and other apes: serology and genetics. J Med Primatol 22:3–6, 1993

Socha WW, Ruffié J. Blood groups of primates. Theory, practice, evolutionary meaning. Liss, New York, 1983

Socha WW, Wiener AS, Moor-Jankowski J, Mortelmans J. Blood groups of mountain gorillas (Gorilla gorilla beringei). J Med Primatol 2:364–369, 1973

Springer GF, Hupprikar SV. On the biochemical and genetic basis of the human blood-group MN specificities. Haematologia 6:81–92, 1972

Telen MJ, Chasis JA. Relationship of the human erythrocyte Wr[b] antigen to an interaction between glycophorin A and band 3. Blood 76:842–848, 1990

Vignal A, London J, Rahuel C, Cartron JP. Promoter sequence and chromosomal organization of the genes encoding glycophorins A, B, and E. Gene 95:289–293, 1990

Walsh RJ, Montgomery C. A new human isoagglutinin subdividing the MN blood groups. Nature 160:504, 1947

Wasniowska K, Drzeniek Z, Lisowska E. The aminoacids of M and N blood group glycopeptides are diferent. Biochem Biophys Res Commun 76:385–390, 1977

Wiener AS. Blood groups of man and lower primates. A review. Transfusion 3:173–184, 1963

Wiener AS. The agglutinogens M and N in anthropoid apes. J Immunol 3:11–18, 1938

Wiener AS, Gordon EB. The blood groups of chimpanzees. A-B-O groups and M-N types. Am J Phys Anthrop 18:301–311, 1960

Wiener AS, Rosenfield RE. M[e] a factor common to the antigenic properties M and He. J Immunol 87:376, 1961

Wiener AS, Unger LJ, Gordon EB. Fatal Hemolytic transfusion reaction caused by sensitization to a new blood factor U. J Amer Med Assoc 153:1444–1446, 1953

Wiener AS, Unger LJ, Cohen L. Distribution and heredity of blood factor U. Science 119:734–735, 1954

Wiener AS, Moor-Jankowski J, Gordon EB. Blood group antigens and cross-reacting antibodies in primates including man. I. Production of antisera for agglutinogen M by immunization with blood other than human type blood. J Immunol 92:391–396, 1964

Wiener AS, Moor-Jankowski J, Gordon EB, Daumy OM, Davis J. Blood groups of gibbons; further observations. Int Arch Allery Applied Immunol 30:466–470, 1966a

Wiener AS, Moor-Jankowski J, Riopelle AJ, Shell NF. Human-type blood factors in gibbons, with special reference to the multiplicity of serological specificities of human type M blood. Transfusion 6:311–318, 1966b

Wiener AS, Moor-Jankowski J, Cadigan J, Gordon EB. Comparison of the ABH blood group specificities and M-N types in man gibbons (*hylobates*) and siamangs (*symphalangus*). Transfusion 8:235–243, 1968

Wiener AS, Moor-Jankowski J, Gordon EB. Blood groups of gorillas. Kriminal Forens Wissensch 6:31–50, 1971

Wiener AS, Gordon EB, Moor-Jankowski J, Socha WW. Homologues of the human M-N blood types in Gorillas and other nonhuman primates. Heamatologia (Budapest) 6:419–432, 1972

Wiener AS, Moor-Jankowski J, Socha WW, Gordon EB. The chimpanzee V-A-B blood group system. Am J Hum Genet 26:35–44, 1974

Wiener AS, Socha WW, Gordon EB, Mortelmans J, Moor-Jankowski J. Blood groups of gorillas: further observations. J Med Primatol 5:317–320, 1976

Wren MR, Issitt PD. Evidence that Wr[a] and Wr[b] are antithetical. Transfusion 28:113, 1988

Xie SS, Huang C-H, Blancher A, Reid ME, Blumenfeld OO. Glycophorin A gene family in Gorillas: Structure, expression and comparison with the human and chimpanzee homologues Biochem Genet. in press, 1997

Yoshima H, Furthmayr H, Kobata A. Structures of the asparagine-linked sugar chains of glycophorin A. J Biol Chem 255:9713–9718, 1980

4 The Rhesus System

A. Blancher and W.W. Socha

Molecular Biology and Evolution of Blood Group
and MHC Antigens in Primates
Blancher/Klein/Socha (Eds.)
© Springer-Verlag Berlin Heidelberg 1997

Discovery of the Rhesus System

The discovery of the Rhesus system in 1940 by K. Landsteiner and A.S. Wiener (Landsteiner and Wiener 1940) constituted the most important event in the science of human blood groups since the description of the ABO blood groups, by Landsteiner, at the beginning of the century. The discovery of the Rh factor shed light on the process of fetomaternal alloimmunization and explained transfusion reactions, of which the causes had, until then, remained unknown.

As described in the next chapter, by immunizing rabbits with human red blood cells Landsteiner and Levine produced antibodies that led to the discovery of the MN factors. Later, in the search for new variants of the M and N antigens, Landsteiner and Wiener employed rabbit immune sera directed not against human red blood cells, but raised against red cells from various species of apes and monkeys. In particular, by injecting rabbits with rhesus monkey (*Macaca mulatta,* also called *Macacus rhesus)* red cells, the two authors obtained a specific anti-M along with another antibody which, after a series of absorptions, was found to agglutinate the red cells of 85% of U.S. whites (who were called Rhesus-positive: Rh+). The 15% whose red cells were not agglutinated were called Rhesus-negative: Rh- (Landsteiner and Wiener 1940).

Already before the discovery of the Rhesus factor it was suspected that some cases of grave icterus in newborn babies resulted from unspecified immunological incompatibility between the mother and the fetus (Levine and Stetson 1939). The original observation of Levine and Stetson concerned a case of severe transfusion reaction with oliguria and bloody urine in a woman who lost a considerable amount of blood after the delivery of her second baby, a stillborn fetus, and received three blood transfusions from her husband. As both recipient and donor were of the group O blood type, it was obvious that the transfusion reaction was due to a blood group incompatibility other than ABO. The serum of the woman agglutinated not only the red cells of her husband but also 80% of 104 other group O blood samples. The authors suggested that the "agglutinins" responsible for the severe transfusion reactions were induced by an "immunizing property" of the blood of the fetus, probably inherited from the father. Later it became obvious that the "immunizing property of the fetus" was somewhat related to the Rh antigen described by Landsteiner and Wiener. Several observations made following the discovery of the Rh factor firmly established that: (1) the Rh factor was at the root of unexplained transfusion reactions that occurred during transfusions of blood which was of the same ABO, MNSSs and P types as the recipient, and (2) it was the cause of hemolytic disease of the newborn (Wiener and Peters 1940; Levine et al. 1941). In the case reported by Levine and Stetson, in the Rh-negative woman, who was immunized in the course of her previous pregnancy, the intra-uterine death during the following pregnancy was caused by the maternal anti-Rh antibodies. The post-transfusion accident observed in this woman after transfusion of her husband's blood was obviously induced by Rh incompatibility.

The relationship between the antigens defined by the human alloantibodies and the heteroantibody raised in rabbits and guinea pigs injected with red cells from rhesus monkeys was a source of intense debate and controversy (Schmidt 1994). Early on, comparisons of specificities of both types of antibodies led to the conclusion that they were not identical but that their similarity was sufficient to name both kinds of antibodies "anti-Rhesus" antibodies (Landsteiner and Wiener 1941; Levine et al. 1941). Even now it is not clear whether both human and rabbit antibodies define the same antigenic target. The difficulty of defining the strict specificity of the heteroantibodies was amplified by the introduction of the so-called LW antigen. It was defined by some antibodies produced by guinea pigs immunized with human red cells (Levine et al. 1961, 1963) and named LW by Levine et al. (1963), in honor of Landsteiner and Wiener. The LW antigen seemed to be phenotypically related to Rh since it was more strongly expressed on red cells of Rh-positive individuals than on those of Rh-negative individuals (Levine and Cellano 1967). The rare Rh_{null} red cells lacked the LW antigens, thus confirming the association between Rh and LW molecules. However, the LW locus is on human chromosome 19 (Sistonen 1984), and, therefore, is genetically independent of the RH locus which is on human chromosome 1 (Marsh et al. 1974; Chérif-Zahar et al. 1991). The cloning of LW cDNA was recently performed with the aid of peptide sequences of the LW protein immunopurified from human erythrocytes (Bailly et al. 1994; Hermand et al. 1995). The predicted encoded protein (271 amino acids) clearly demonstrated that LW has no structural relationship with Rh proteins, which is in agreement with earlier studies carried out on proteins by two-dimensional fingerprinting analysis (Bloy et al. 1990). The *LW* gene was assigned to chromosome 19p13.3 by in situ hybridization performed with the cDNA probe (Hermand et al. 1995). Both Rh and LW molecules are part of a large complex of molecules that also encompasses CD47, Rh50 glycoprotein, Duffy and glycophorin B (for a review see Cartron 1994; Cartron and Agre 1995).

Although several questions remained unsolved until recently concerning the true nature of rhesus monkey Rh-like antigen, the names "Rhesus system" and "Rhesus antigen" have survived as a part of a universally accepted terminology.

The Rhesus System in Humans

Irregular antibodies in polytransfused subjects or in alloimmunized pregnant women were the source of reagents that allowed definition of a series of blood group antigens linked to the initial Rhesus factor. In fact, the RH system was recognized as one of the most complex polymorphic systems in humans (Issitt 1989; Cartron and Agre 1995).

Table 1. Wiener's designations for the eight common *Rh* gene complexes

Gene	Agglutinogen	Blood factors
r	rh	hr', hr''
r'	rh'	rh', hr''
r''	rh''	rh'', hr'
r^y	rh^y	rh', rh''
R^o	Rh$_0$	RH$_0$, hr', hr''
R^1	Rh$_1$	RH$_0$, rh', hr''
R^2	Rh$_2$	RH$_0$, rh'', hr'
R^Z	Rh$_Z$	RH$_0$, rh', rh''

Wiener postulated that a single gene was responsible for the production of Rh antigens. He designated the two genes that corresponded to the alleles responsible for both expression of antigen Rh and its absence as *Rh* and *rh*, respectively (Landsteiner and Wiener 1941). With the discovery of additional blood groups of the RH system, the gene designations were modified accordingly (Wiener 1943). Wiener's unilocular theory postulated the existence of a single polyallelic series with each allele ensuring the synthesis of a corresponding agglutinogen, itself composed of a number of specificities, or, as we call them now, epitopes. Wiener's concept of the eight common *RH* genes is shown in Table 1.

An alternative theory of inheritance of the Rh blood groups was proposed by British workers (Fisher and Race), who, by 1943, had four sera that appeared to detect distinct antigens belonging to the RH system. Sir Ronald Fisher noted that the antibodies called "anti-C" and "anti-c" gave antithetical reactions, and postulated accordingly the existence of a pair of alleles *C* and *c*. The remaining two sera were called anti-D and anti-E, and Fisher named the genes responsible of their expression *D* and *E*, respectively (Race et al. 1944). The theory proposed by Race, and based on Fisher's model, assumed the existence of three sets of alleles *D/d*, *E/e*, *C/c* whose loci were so closely linked that crossing over between them rarely occurred (Race 1948). The theory was confirmed in part by the discovery of the anti-e antibodies, which gave antithetical reactions with the anti-E antibodies (Mourant 1945). Nonetheless, despite numerous efforts, the anti-d antibodies were never found and, therefore, the existence of the *d* allele was never serologically confirmed. Table 2 summarizes the relationships between Wiener's and Fisher-Race's nomenclatures.

Little by little, Fisher and Race's nomenclature was generally adopted because it was considered more convenient to handle and easier to remember. Most workers favor the Fisher-Race terminology over Wiener's Rh-Hr nomenclature when describing antigens and antibodies (such as D and anti-D, or C and anti-C). However, the nomenclature of Wiener is still used because it is more practical for descriptions of red cell phenotype (as, for instance: R$_1$R$_1$ by Wiener notation, instead of D+C+E-c-e+ or *DCe/DCe* according to Fisher and Race's nomenclature).

Table 2. Common *Rh* gene complexes in Caucasians

Haplotypes		Frequencies (France)	Antigens expressed
Fisher and Race	Wiener		
DCe	R^1	0.410	D, C, e
dce	*r*	0.394	c, e
DcE	R^2	0.116	D, c, E
Dce	R^0	0.067	D, c, e
dcE	*r″*	0.0073	c, E
dCe	*r′*	0.0026	C, e
DCE	R^Z	0.0016	D, C, E
dCE	r^y	Very rare	C, E

A third terminology was proposed by Rosenfield et al. (1962). It was not based on genetic considerations but was conceived as an easy, numerical way to designate all new Rh specificities, which by now number over 50 (Table 3).

More recently, Tippett proposed another genetic model consisting of two closely linked structural loci: *D* and *CcEe* (Tippett 1986). This model was based on the biochemical findings of Moore et al. (1982), which suggested that the Rh antigens are associated with two distinct polypeptides: one carrying the D antigen and the other bearing antigens of the Cc and Ee series. The existence of compound antigens resulting from the combined action of alleles *C*, or *c* and *E* or *e* in *cis* position supports the idea that the Cc and Ee antigens are one and the same polypeptide encoded by a single gene (for detailed discussion of compound antigens, see below).

To follow the current literature we will use the symbol RH for description of the blood group system and corresponding locus, *RH* for the genes, Rh for the proteins.

Without going into minute details of the human RH system the following facts are to be quoted for the comprehension of this system in humans as well as in nonhuman primates:

1. The original Rh antigen, designated Rh_0 in Wiener's nomenclature, D in the Fisher-Race notation, and Rh1 in the numerical notation of Rosenfield, is the most immunogenic alloantigen in humans and occupies a central, privileged position, which is also true, as we shall see, of its equivalent in nonhuman primates. The expression of the antigen D varies in strength, and the phenotype with the weak expression of the antigen is designated by the symbol D^u (Stratton 1946). D^u does not correspond to a single type but includes several graded variants, from the highest, which are agglutinated by almost all anti-D (Rh_0) reagents, down to the lowest degree, revealed only by the antiglobulin technique. Family studies demonstrated that some D^u are inherited, presumably encoded by a gene D^u at the *RHD* locus (Stratton 1946; Renton and Stratton 1950). D^u occurs most frequently as CD^ue or cD^uE in Whites and as cD^ue in Blacks. The number of D^u mutations must be very high, judging from the number of D^u degrees thus far described. Certain D^u phenotypes, espe-

Table 3. RH antigen nomenclature

Rosenfeld (numerical)	Fisher-Race	Wiener	Rosenfeld (numerical)	Fisher-Race	Wiener
Rh1	D	Rho	Rh28[a]		hr^H
Rh2	C	rh'	Rh29[b]	Total Rh	RH
Rh3	E	rh''	Rh30[a]	Go^a (D^{Cor})	
Rh4	c	hr'	Rh31	e-like	hr^B
Rh5	e	hr''	Rh32[a]		product of $\bar{\bar{R}}^N$
Rh6	cis ce(f)	hr	Rh33[a]		product of R^{oHar}
Rh7	cis Ce	rh$_i$	Rh34[b]	Bas	Hr^B
Rh8[a]	C^w	rh^{w1}	Rh35[a]	1114	
Rh9[a]	C^x	rh^x	Rh36[a]	Be^a	
Rh10[a]	V (ce^S)	hr^v	Rh37[a]	Evans	
Rh11[a]	E^w	rh^{w2}	Rh39[b]	C-like	
Rh12	G	rh^G	Rh40[a]	Tar	
Rh17[b]	(Cc)	Hr$_0$	Rh41	Ce-like	
Rh18[b]	(Ee)	Hr^S	Rh42[a]	Ce^S	
Rh19		hr^S	Rh43[a]	Crawford	
Rh20[a]	e^S(VS)		Rh44[b]	Nou.	
Rh21	C^G	rh^G	Rh45[a]	Riv.	
Rh22	cis CE	rh	Rh46[b]	Sec.	
Rh23[a]	D^W	Rh^{Wi}	Rh47[b]	Dav.	
Rh24	E^T	rh^T	Rh48[a]	JAL	
Rh26	c-like	hr^A	Rh49[a]	STEM	
Rh27	cis cE	rh$_{ii}$	Rh50[a]	FPTT	
			Rh51[b]	MAR	

[a] Low-incidence antigens.
[b] High-incidence antigens.

Nota bene: LW(Rh25) and Duclos (Rh38) are not part of the RH system. Rh13, 14, 15 and 16 correspond to Rh^A, Rh^B, Rh^C, and Rh^D, respectively. These four antigens were defined by Wiener as "cognates" of Rh$_o$. These names were used to describe partial Rh$_o$ (D) antigens which are now classified in categories (see text for details).

cially those of high grade, may be due to the effect of a *Cde* or *CdE* complex, which in the *trans* position depresses the expression of a normal *D* on the other chromosome (Ceppellini 1952; Ceppellini et al. 1955). More recent studies indicate that most D^u erythrocytes should carry a normal RhD polypeptide which is expressed at a lower level than in classical phenotypes (Rouillac et al. 1996). Only some of the D^u antigens lack part of the epitopes that constitute the Rh$_o$(D) mosaic (Leader et al. 1990) (see also the following paragraph).

2. The D (Rh$_o$) antigen is not homogeneous but is composed of a mosaic of factors, the first of which (Rh^A, Rh^B, Rh^C, Rh^D) were discovered by A.S. Wiener and his coworkers and called "cognates" of Rh$_o$ (Sacks et al. 1959; Unger et al. 1959a,b; Wiener et al. 1957; Wiener and Unger 1962). The mosaic

aspect of the Rh_o antigen was deduced from puzzling observations of D positive individuals who produced alloanti-D antibodies. These alloantibodies recognized most of the D positive cells and were negative in test with all D negative red cells as well as with the blood of the individual who produced these antibodies. It was assumed that these people had a variant D antigen which lacked some "cognate factors" that are parts of the normal Rh mosaic and against which they can be immunized by incompatible transfusion or pregnancy. The variant D antigens defined the partial Rh_o antigens that can, according to Tippett (1987), be classified into at least seven categories, which became more recently nine (Daniels et al. 1995). Tippett's classification accomodates almost all observations, including the very important discovery that some partial D individuals express low frequency antigen which are specific to these phenotypes (category IV phenotypes have Rh30 (Go^a), while persons in category V have Rh23 (D^W), individuals in category VII express the low frequency antigen Tar (Rh40), those with DFR express FPTT and the DBT phenotype is associated with the expression of Rh32 (see Table 4). The $Rh_o{}^{Har}$ phenotype was recently assimilated to partial D phenotypes (Daniels et al. 1995). This phenotype associates the expression of a low incidence antigen, Rh33, and very low expression of antigen D. In contrast to D^u, the very weak antigen D of $Rh_o{}^{Har}$ is most frequently not detected in the antiglobulin test but more efficiently by tests on papain-treated cells.

Table 4. Epitopes characterized by human monoclonal anti-D in test with red cells of various categories of partial D

Category[a]	II	IIIa	IVa	IVb	Va	VI	VII	DFR	DBT	$R_o{}^{Har}$
Low frequency antigens by name/number			Go^a/ Rh30	Evans[b]/ Rh37	D^W/ Rh23	BARC[c]	Tar/ Rh40	FPTT/ Rh50	Rh32	Rh33[d]
antibody anti-Monoclonal										
epD1	+	+	−	−	−	−	+	+/−[e]	−	−
epD2	+	+	−	−	+	−	+	+/−	−	−
epD3	+	+	−	−	+	+	+	+	−	−
epD4	−	+	+	−	+	+	+	+	−	−
epD5	+	+	+	+	−	−	+	+/−[e]	−	+/−[e]
epD6/7	+	+	+	+	+	−	+	+/−[f]	+/−[f]	+/−[f]
epD8	+	+	+	+	+	−	−	−	−	−
epD9	−	+	−	−	+	+	+	+	−	−

[a] AfterTippett (1988, 1990b), Lomas et al. (1993a) and Daniels et al (1995).
[b] Only some DIVb express the antigen Evans.
[c] BARC is a low frequency antigen associated with $D^{VI}Ce$ (Lomas and Mougey 1989).
[d] $R_o{}^{Har}$ red cells also express the low frequency antigen FPTT.
[e] The serological pattern of reactions of monoclonal anti-D varied with the agglutination method employed.
[f] Tests with DFR, DBT and $R_o{}^{Har}$ red blood cell samples resulted in subdivision of the specificity epD6/7 into eight subspecificities (Daniels et al. 1995).

When large panels of human anti-D monoclonal antibodies were tested against D variants, seven, and then at least nine, classes of antibodies were recognized by their type of reactivity with those variants (Lomas et al. 1989b, 1993a,b; Tippett 1990a,b; Daniels et al. 1995). These nine types of reactivity are considered to correspond to nine epitopes. Most of the mutations responsible for the partial defect of the D antigenic mosaic have been identified (Mouro et al. 1994b; Cartron 1994; Rouillac et al. 1995a–c; Beckers et al. 1995, 1996a,b, Tippett et al. 1996). However, the particular epitopes detected by various classes of monoclonal reagents are still waiting for a tri-dimensional molecular definition. More recently, new variants of the D antigen have been added so that the number of epitopes, inferred from the variety of patterns of reactivity observed in tests with monoclonal reagents, could be as high as 30 (Jones et al. 1995). Yet, as we will see later on, the size of the extracellular part of the molecule deduced from the assumed conformation of the D-related polypeptide is incompatible with such a large number of postulated epitopes.

3. The expression of the G antigen is transmitted by all rhesus haplotypes with the exception of the *cde (r) and cdE (r'')* (Allen and Tippett 1958). Thus, the expression of the G antigen seems to follow the expression of D or C antigen, behaving as a common epitope shared by D and C antigens (Rouillac et al. 1995c; Mouro et al. 1996).

4. Numerous antigenic variants, such as C^w and C^x, are assumed to be transmitted by alleles of *C/c*. Recently, the antigens C^w and C^x were elucidated at the molecular level (Mouro et al. 1995) (see later in this chapter). As for the *E/e* series, only a few variants of E (such as E^W, E^T), but many variants of e, were described (Issitt 1991). The human variants of the Cc and Ee series have no counterparts in nonhuman primates and, therefore, will not be discussed here.

5. There are functional relationships between the two series postulated by Race and Fisher (*C/c* and *E/e*) that give rise to the compound antigens, products of the activity of two adjacent loci situated in *cis* position, e.g., Ce (rh_i), CE (rh), ce (hr), cE (rh_{ii}) (Rosenfield et al. 1973).

An alternative explanation for these compound antigens is that each one is carried by a single molecule which is a product of one of four different alleles (*ce, Ce, CE, cE*), in accordance with the two-locus theory of the RH system (Cartron and Agre 1995; Huang et al. 1996).

6. Only one crossing-over has thus far been suspected within the RH system despite the tens of thousands of families studied throughout the world (whereas it has been observed more frequently for MNSS). In this family, investigated by Steinberg, crossing-over seems to be the most probable explanation, as one of the seven children of *CDe/cde×cde/cde* parents was *Cde/cde*. Since illegitimacy seemed very unlikely in that case, the inherited *Cde* haplotype must had resulted from a crossing-over between *DCe* and *dce* haplotypes during meiosis of the gametogenesis of the father (Steinberg 1965).

7. Rare variants were discovered which fail to express any Rh antigens. These Rh_{null} individuals have clinical symptoms of anemia. Red cells deficient for all Rh antigens (Rh_{null} phenotype) arise from two distinct genetic backgrounds. In both cases the syndrome is transmitted as a recessive trait. The "amorph type" is caused by homozygosity for a silent allele $\overline{\overline{r}}$ at the RH locus, whereas the more common "regulator type" (38 cases out of 42 in the series analyzed by Nash and Shojania 1987) is apparently caused by homozygosity for an autosomal suppressor gene (X^o_r), which is genetically independent of the RH locus. Another gene X^Q (which is not an allele of X^o_r) is responsible for the Rh_{mod} phenotype, which is characterized by residual quantities of Rh antigenic substance (Race and Sanger 1975). Recently, the Rh_{mod} and Rh_{null} of regulator type were shown to correspond to the expression of mutated forms of Rh50 glycoprotein (Chérif-Zahar et al. 1996).

All individuals with Rh deficiency exhibit a unique hematological syndrome which associates: a chronic hemolytic anemia (often well compensated), stomatospherocytosis, an increased osmotic fragility and increased cation permeability. Splenectomy in severe forms of hemolysis leads to a marked clinical improvement with normal red cell survival, although the persistence of a mild residual hemolysis has been noted.

The Rh_{null} cells are not only deprived of Rh-related molecules (proteins of about 32 kDa and called 32 kDa Rh-related proteins), but also of other proteins which are part of the Rh complex (for a general review see Cartron 1994) (Fig. 1):

- Rh50 glycoprotein: a highly glycosylated protein of an apparent molecular weight of 50 kDa composed of 409 amino acids (Ridgweel et al. 1992). The amino acid sequence of Rh50 exhibits a strong similarity to the Rh30 related polypeptides. Although possessing three potential N-glycosylation sites, only the Asp-37 is N-glycosylated. The large oligosaccharide chain was shown to carry ABO blood group specificities (Moore and Green 1987). The general conformation of Rh50 glycoprotein is very similar to that of the Rh30 polypeptides. It was shown that the Rh50 and Rh30 polypeptides are associated as heterodimers in the red cell membrane (Eyers et al. 1994). The precise function of the Rh50 polypeptide, if it has one, remains obscure, as is the case for the functional role of Rh30 polypeptides. However, it was recently demonstrated that the Rh_{null} of regulator type and the Rh_{mod} phenotypes are direct consequences of the expression of mutated forms of Rh50 glycoprotein (Chérif-Zahar et al. 1996). Thus, Rh50 glycoprotein plays an essential role in the expression of Rh antigens at the erythrocyte membrane.
- LW: structurally related to the ICAM proteins and exhibiting adhesive properties to integrins (Hermand et al. 1995; Bailly et al. 1995).
- CD47: defined by mouse monoclonal antibodies. The precise function of the CD47 glycoprotein at the red cell membrane is unknown. This protein has a large tissue distribution. Its expression is severely reduced on red cells of Rh_{null} individuals while outside red cells it is normally expressed. Recent studies on the CD47 protein showed that it is identical to OA3 (an ovarian tumour marker) and to integrin associated protein (IAP) of platelets and

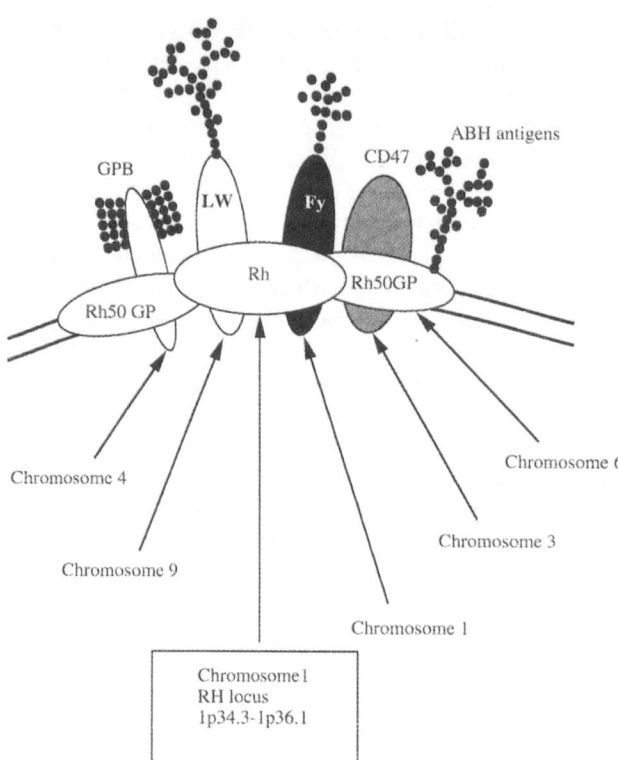

Fig. 1. The Rh protein complex on the red cell membrane. This representation does not take into account the relative mass of each component. It also ignores the fact that some of the proteins are found outside the complex. The interaction between Rh50 glycoprotein (*RhGP50*) and GPB is essential for the expression of antigen U. Duffy molecules interact with Rh proteins since Rh$_{null}$ types lack antigen Fy5. The glycosaccharides of RhGP carry ABH antigens. *Black circles* indicate glycosaccharides

endothelial cells (Schwartz et al. 1993). Whether CD47 has the same role on red cells remains to be demonstrated.

- Duffy protein is a receptor for interleukin-8 (IL8) (see chapter by Pogo and Chaudhuri). Expression of the Fy5 antigen probably requires an interaction with some protein of the Rh complex as this epitope is not detected at the surface of Rh$_{null}$ red cells.
- Glycophorin B: precise function still unknown. It interacts with the Rh50 glycoprotein in expression of the U antigen (see p. 95). The interaction seems to take place early in the maturation of both proteins, since glycosylation of Rh50 is influenced by the presence of glycophorin B. The degree of glycosylation of Rh50 glycoproteins in Rh$_{null}$ (U-negative and U-weak) and in S-s-erythrocytes (U-neg or U-weak), respectively, suggests that the transit time of Rh50 into the Golgi apparatus is modified according to the presence or the absence of Rh proteins or glycophorin B (Ridgwell et al. 1994). Thus, it was proposed that glycophorin B facilitates the transport of Rh50 to the cell surface, whereas Rh proteins slow down this process.

Rh$_{null}$ membranes are characterized by the following abnormalities: hyperactive membrane ATPases, an abnormal distribution of certain phospholipids in the two layers of the membrane, and a relative deficiency of membrane cholesterol.

The serum of Rh$_{null}$ individuals contains antibodies which react with all human red cells expressing Rh molecules (including red cells of D-- individuals). These antibodies, called Rh29, are assumed to define an epitope shared by all Rh molecules.

8. Other rare variants result in partially silent phenotypes, such as: D--, DC^W-, Dc-, which are characterized by an enhanced expression of D, an undectable level of Ee antigens and variable presence of Cc antigens. D·· is very similar to D-- and is characterized by expression of the private antigen Evans (Rh37) and by normal level of expression of antigen D. The phenotype Dc- was described by Tate et al. (1960).

D-- individuals produce anti-Rh17 antibodies which react with all human red cells except those from Rh$_{null}$ and D-- individuals (Race et al. 1951; Wiener 1958).

Individuals who are D^{IV}(C)- produce anti-Rh18 antibodies which are assumed to react with an epitope shared by E and e antigens (Salmon et al. 1969).

Some of the recent discoveries in molecular genetics offer an explanation of the mechanisms of some of these phenotypes but information about regulation of the expression of the *RH* genes remains scant.

The Human Anti-Rh Alloantibodies

The antigens of the RH system are defined by human antibodies which are usually of immune origin. Indeed, naturally occurring antibodies are rare, except anti-E. Antigen D is the most immunogenic blood group antigen. If incompatible D-positive blood is transfused to a D-negative recipient, the probability of the appearance of an immune response is estimated to be between 50 % and 70 %.

Anti-D alloantibodies are either complete agglutinins of the IgM or incomplete agglutinins of the IgG isotypes. The existence of incomplete anti-D was inferred from the presence of antibodies capable of inactivating serum containing an anti-Rh complete agglutinin of high titer. This phenomenon was first elucidated independently by Race (1944) and Wiener (1944). Wiener assumed that the blocking antibodies were monovalent and that they were of greater clinical significance in the pathophysiology of erythroblastosis than the complete agglutinating antibodies (Wiener 1944). Methods for detection of the incomplete antibodies that were more efficient than the original blocking tests were soon developed. The introduction of proteins, such as bovine albumin, as the medium of the reaction enhanced red cell agglutinability and rendered incomplete antibodies capable of agglutinating Rh-positive red cells (Diamond and Denton 1945). The invention of the indirect antiglobulin method (Coombs et al. 1945) offered a powerful tool for use in immunohematology. In this method, incomplete antibodies specifically coat red cells that express the given antigen, without agglutinating them. Agglutination is brought about by addition of rabbit anti-human immunoglobulin antibodies that bind to the coating antibodies. The direct agglutination (direct Coomb's) test, by contrast, detects in vivo sensitization of red cells of babies with hemolytic disease (Coombs et al. 1946). The third technique which allows easy detection of incomplete antibodies was introduced by Morton and Pickles (1947), who found that enzymatic treatment of red cells rendered them agglutinable by incomplete anti-D. The tests using enzyme-treated cells are the most sensitive followed by the antiglobulin test and tests using macromolecules.

Most anti-D antibodies are of the IgG class (incomplete antibodies); only some sera have a significant proportion of IgM and even of IgA. Among the IgG anti-D almost all are of the IgG1 and IgG3 subclasses. We have encountered a unique serum containing anti-D of the IgG2 subclass (Dugoujon et al. 1989; Kumpel et al. 1996). Anti-D antibodies of the IgG4 subclass have never been described. The IgM antibodies are short-lived and occur at the beginning of the first immunizing episode or at the initial phase of the later immune responses. The rare IgM anti-D agglutinins, previously known as "complete" antibodies, are active in saline medium. In view of the high immunogenicity of the D antigen and the relatively high frequency of D-negative people, anti-D is far the most commonly encountered alloantibody. Anti-D antibodies of the IgG isotype, by and large, do not bind complement, probably because the distance between two adjacent antigenic sites at the erythrocyte membrane is too wide, so that two anti-D antibody molecules, bound to their antigenic targets, cannot cooperate in activation of the first fraction of complement. The IgG anti-D alloantibodies cross the placenta barrier and may cause immunologic conflicts, a source of fetomaternal incompatibility (Levine et al. 1941).

Anti-c and anti-E antibodies, less frequent, are encountered mainly in polytransfused patients. Anti-e are even less frequent. There are some known cases of immune hemolytic anemia caused by anti-e auto-antibodies. Anti-C antibodies only rarely occur;, they are usually associated with anti-D, and many sera with anti-D+C also contain anti-G and, frequently, anti-C^W antibodies. The frequencies of occurrence of Rh antibodies of various specificities are (in decreasing order): anti-D, anti-E, anti-c, anti-e and, finally, anti-C. Other specificities are much less frequent, e.g. antibodies against "compound antigens" (for example, anti-Ce) or against antigens of low frequency (anti-Evans, anti-Go[a]). Rare individuals with partially or totally silent phenotypes can produce antibodies against antigens of very high frequency, the so-called public antigens (Rh29, Rh17, Rh18, see above).

Until recently, all anti-Rh typing reagents were polyclonal and derived from the serum of individuals immunized by pregnancy, transfusion or deliberate injection of incompatible red cells. In 1980, the possibility of producing in vitro human anti-D antibodies by transformation of human B lymphocytes by the Epstein Barr virus was demonstrated (Koskimies 1980; Boylston et al. 1980). In 1983, the first human lymphoblastoid cell line cloned and producing monoclonal anti-D antibody was reported (Crawford et al. 1983). Independent efforts by many laboratories led to the development of numerous monoclonal antibodies (for a general review see Rouger and Goossens 1990). Their activities and specificities were compared during the First, Second, and Third International Workshops on Monoclonal Antibodies Against Red Cells and Related Antigens. (Paris, France, September 21–24 1987; Lund, Sweden, April 1–4 1990; Nantes, France, September 25–27 1996). Among monoclonal antibodies, anti-D are the most numerous; those of other specificities are rarer (in decreasing amounts: anti-c, anti-E, anti-e, anti-C).

The Rhesus System in Nonhuman Primates: Rh Factors Recognized by Polyclonal Reagents of Human Origin

The fact that the RH system was discovered by means of an antiserum raised against the red cells of a rhesus monkey inspired immunologists to search for structures on nonhuman primate erythrocytes that might be linked to the RH system of humans. Such studies revealed the existence of immunological structures of extraordinary complexity which, on the phylogenic level, have not yet divulged all their secrets.

The anthropoid apes, the animals closest to humans, are the only nonhuman primates whose red cells cross-react strongly with some polyclonal antibodies specific for the human alloantigens of the Rhesus system (Wiener et al. 1966). Initial study of the blood groups of the great anthropoid apes was, therefore, carried out using antisera of human origin. However, tests performed with such reagents presented some difficult technical problems: human sera always contain anti-primate heteroagglutinins that need to be eliminated in order to obtain sufficiently potent type-specific reagents. For this purpose, the sera must be properly absorbed or diluted, and the residual specific antibody is not always strong enough to be suitable for typing primate red cells. The immunologist is, therefore, forced to maneuver between two thresholds, beneath which the antibody is not specific and above which it is too weak to give interpretable results.

Other factors, defined by antibodies produced by alloimmunization of nonhuman primates, especially chimpanzees, were found to be related to the human antigens of the Rhesus system. The latter blood groups, restricted to some anthropoid apes, will be discussed later (see sections on R-C-E-F and V-A-B-D blood groups and the chapter by Socha).

Studies of the Rh blood groups in nonhuman primates were initiated in 1953 by Wiener et al. and have been continued since then. The results of tests carried out with polyclonal human reagents are summarized in Table 5.

Table 5. Homologues of the human Rh-Hr blood factors in the blood of anthropoid apes, as defined by human reagents

Red cell source	Reactions with human antisera for specificities					Remarks
	Rh_o	rh′	rh″	hr′	hr″	
Chimpanzees	Positive	Negative	Negative	Positive	Negative	Absorptions of human anti-Rh_o sera with chimpanzee c^c-negative (type CC) or R^c-negative (type r^c) red cells yield reagents specific for chimpanzee c^c or R^c specificities, respectively.
Gorillas	Positive	Negative	Negative	Positive	Negative	Weak variants of Rh_o observed.
Gibbons	Negative	Negative	Negative	Posivite	Negative	
Orangutans	Negative	Negative	Negative	Negative	Negative	Absorption of human anti-Rh_o serum with human Rh-positive red cells does not eliminate weak non specific reactions for orangutan red cells.
Old World monkeys	Negative	Negative	Negative	Negative	Negative	Rh-like structures obviously present since immunization of rabbits and guinea pigs with rhesus or baboon red cells yields reagents detecting human Rh_o specificity.

Human Polyclonal Anti-D

Chimpanzees

The first experiments carried out with unabsorbed human polyclonal anti-D indicated that all chimpanzees are Rh_o positive. By tests with reagents specific for the Rh_o cognate antigens, Wiener et al. demonstrated that the antigen Rh^C was fully expressed by the chimpanzee, while the related factors Rh^B, Rh^D appeared as low grade variants of their human counterparts (Wiener and Gordon 1961). Interestingly, absorption of human anti-Rh_o sera with chimpanzee red cells removed the reactivity for chimpanzee blood, but left behind a fraction of antibodies still specifically reactive for human Rh-positive blood (Wiener et al. 1964). That proved that red cells of all chimpanzees had an agglutinogen closely related to, but not identical with, the Rh agglutinogen of human blood.

The homogeneous expression of Rh_o in chimpanzees was contradicted by observations by Masouredis et al. who, using as reagent ^{125}I-labeled anti-Rh_o purified by absorption-elution from D-positive human red blood cells, defined a polymorphism in those apes (Masouredis et al. 1967). These results were later confirmed by Owens et al. (1982). In order to reconcile the seeming contradiction between the two studies, Wiener and coworkers absorbed anti-Rh_o sera with ficinated chimpanzee red cells which were classified as Rh_o-negative by Masouredis (Wiener et al. 1966, 1972). The tests carried out with such absorbed sera revealed individual differences among chimpanzee red cells paralleling those shown by the eluates prepared by Masouredis. The same study also demonstrated an association between the expression of Rh_o-like antigen and a chimpanzee simian-type blood group c^c. This was the first argument suggesting that the chimpanzee blood group system, to which the antigen c^c belongs, is a chimpanzee counterpart of the human RH system. (see discussion of the R-C-E-F blood group system below).

Gorillas

The reactions of a nonabsorbed anti-D serum are always positive with gorilla red blood cells. Tests with anti-Rh_o sera on blood specimens of 11 lowland gorillas *(Gorilla gorilla gorilla)* were all positive, both by saline and by ficin-treated red cell methods (Wiener et al. 1971). It was noticed, however, that three out of 11 gorilla blood samples reacted to somewhat lower titers. Absorption of the anti-Rh_o sera with weakly reacting gorilla red cells produced a reagent still highly reactive with human Rh-positive red cells as well as with red cells of the more strongly reactive gorillas, but negative with red cells of the weakly reacting gorilla red cells. Similar observations were made with blood of six mountain gorillas *(Gorilla gorilla beringei)* (Wiener et al. 1976). By contrast, absorptions of the antisera with human Rh-positive cells removed all activity for gorilla red cells as well as for human Rh-positive red cells. Based on these results, two types of gorilla blood were defined: that expressing an antigen Rh_o-like (called Rh_o^{Go}-positive) and that which is Rh_o-negative (or Rh_o^{Go} variants, if one considers that the weak agglutination observed with some anti-D reagents were type-specific).

Gibbons and Orangutans

Tests on the red cells of gibbons gave uniformly negative reactions with human anti-Rh_o (Wiener et al. 1964; Moor-Jankowski et al. 1973), while results of tests on orangutan red cells were uninterpretable because of very strong anti-orangutan heteroantibodies regularly found in human serum; absorption attempts rendered the sera inactive. However, when the red cells from a female orangutan, immunized by feto-maternal incompatibility, were used for absorption of the human anti-Rh_o serum, the so absorbed reagent agglutinated some but not all orangutan red cells, thus probably differentiating between Rh_o-positive and Rh_o-negative individuals. Since the Rh_o-negative type appears to be quite rare among orangutans, it is possible that the red cells previously used for absorptions of human anti-Rh_o reagents were all of the Rh_o-positive type, and they thus removed from the antisera not only the non-specific agglutinins but the type-specific anti-Rh_o antibodies as well (Socha and van Foreest 1981; Socha 1993a,b).

Old World and New World Monkeys

Red cells of all species of Old and New World monkeys tested with Rh-Hr reagents by Wiener and his collaborators gave uniformly negative results (Moor-Jankowski and Wiener 1972). However, one has to remember that discovery of the human Rhesus factor was made by means of rabbit anti-rhesus antibodies, which would indicate that the rhesus monkey red cells share some Rh-related antigen with humans. Yet, surprisingly, the anti-Rh antibodies produced by Wiener were unable to agglutinate rhesus red cells. Levine and Cellano (1962) demonstrated that rhesus monkeys immunized with human red cells were able to produce anti-D-like antibodies, which suggested that the D-like antigenic determinants could vary from species to species. The presence of an Rh-like antigen in rhesus monkey red cells was confirmed by biochemical characterization of Rh-like polypeptides extracted from monkey red cell membranes (Saboori et al. 1989).

Only recently was the existence of an Rh-like factor on rhesus monkey red cells definitely established by detection of an Rh-like cDNA in the bone marrow of *Macaca mulata* (Mouro et al. 1994a) (see later in this chapter).

Anti-C (Anti-rh′) and Anti-c (Anti-hr′) Antibodies

The red blood cells of all the African anthropoid apes as well as those of the gibbon are agglutinated by anti-c (anti-hr'). The presence or absence of the antigen hr' (c) on red blood cells of orangutans is difficult to ascertain because of interference by very strong heterospecific antibodies in the reagents, making interpretation of the reactions very doubtful. Nonetheless, no ape blood is agglutinated by anti-C (anti-rh').

No agglutination by anti-rh' and anti-hr' was observed in earlier tests with red cells from New and Old World monkeys. However, recently, Westhoff and Wylie (1994a) found that some monkey red cells, namely those of langur and capucin, were agglutinated by polyclonal anti-c reagents.

Anti-E (Anti-rh″) and Anti-e (Anti-hr″) Antibodies

These antibodies do not react with the red blood cells of any of the nonhuman primates tested so far.

Rh-Related Factors Recognized by Reagents of Simian Origin: The R-C-E-F Blood Group System

The simian-type blood groups are defined by allo- or heteroimmune sera against red cell antigens of primate animals. Since among anthropoid apes, only chimpanzees are available for large scale immunizations and for extensive blood group studies, the bulk of information on simian-type blood groups of apes concerns chimpanzees. To date, over 20 simian-type antigens have been defined on the red blood cells of chimpanzees; some of them were also found on the red cells of gorillas and other apes (for more details, see chapter by Socha). One of these simian-type blood groups is the R-C-E-F chimpanzee blood group, which is the equivalent of the human RH system.

The R-C-E-F Blood Group System in the Chimpanzee

Wiener et al. discovered factors of chimpanzee red cells that are linked to the MN system (V^c, A^c, B^c, D^c) by means of antibodies obtained by alloimmunizations among randomly selected chimpanzees (see the next chapter). During the same series of experiments, antibodies were obtained that recognized factors independent of the MN system. Some of those factors were assumed, based on the type of their reactivity, to be homologues of the human Rhesus system antigens. These were: $C^c/c^c/E^c/F^c$ (the exponent "c" was meant to identify the antigen detected by the chimpanzee immune sera and to differentiate it from the similar notation used in the human blood group system). It must be stressed that, despite similarity of notations, there is no serological identity or cross-reactivity between chimpanzee C^c, c^c, E^c and human C, c, E, antigens.

Four specific isoimmune reagents, anti-C^c, anti-c^c, anti-E^c and anti-F^c, defined four corresponding antigens which combined into nine blood types: cef, Ccef, CcEf, CceF, CcEF, CCef, CCEf, CCeF, CCEF. These constituted a blood group system initially called the CEF chimpanzee blood group system. Population analysis indicated that the CEF blood groups were inherited by a series of five codominant alleles, c, C, C^E, C^F, C^{EF} (Wiener et al. 1965). This model was later confirmed by family studies (Wiener and Socha 1974).

Since the C-E-F reagents reacted best by proteolytic enzyme techniques, as do the antibodies for Rh-Hr specificities, the C-E-F system has been considered, since its discovery, to be the chimpanzee counterpart of the Rh-Hr blood groups of humans. A more direct proof of the close relationship between the two blood group systems was later obtained when it was found that human anti-Rh_0 sera, appropriately absorbed, would give reactions with chimpanzee red cells that parallelled those produced by chimpanzee anti-c^c serum (Wiener et al. 1965).

Subsequently, another specificity of chimpanzee red cells, L^c, was described as possibly associated with the C-E-F blood group system (Wiener et al. 1967), but direct proof of such an association was not obtained until 13 years later, when it was found that an alloimmune chimpanzee anti-L^c antibody gave reactions with human red cells paralleling those obtained with anti-Rh_0 reagents

(Socha and Moor-Jankowski 1979). At the same time it was established that properly absorbed, human anti-Rh_o sera detected L^c specificity on chimpanzee red cells (Socha et al. 1989). The c^c factor is linked to R^c as in humans C is linked to D: i.e., it is frequently (but not always) inherited together with it. Furthermore, the human anti-D (anti-Rh_o) reagents often possess, in addition to the anti-R^c specificity, an anti-c^c specificity (just as there are human anti-C+D sera). This has been confirmed by fractionation absorptions. An anti-Rh_o antibody absorbed by chimpanzee R^c-positive, c^c-negative (i.e. RC type) red blood cells agglutinated c^c-positive chimpanzee cells (i.e. of R^c or of r^c type), but did not react with C^c-positive red cells (i.e. type RC or rC). Such absorption removed the anti-R^c antibodies while leaving behind the anti-c^c antibodies.

To emphasize the immunological closeness between the human antigen Rh_o (D) and the chimpanzee antigen L^c, the previously assigned symbol L^c was changed into R^c. Consequently, the letter R was added to the generic name of the blood group system which, since 1980, has been referred to as the RCEF system (Socha 1993a).

Several properties of the chimpanzee R^c antigen reinforce the notion of it being the counterpart of the human D antigen (Socha and Ruffié 1983):

- R^c is one of the most immunogenic red blood cell alloantigens in chimpanzees, as is the case for the D antigen of humans.
- Anti-R^c antibodies proved to be the cause of erythroblastosis fetalis in newborn chimpanzees.
- Like anti-D antibodies, the reactivity of anti-R^c antibodies is enhanced by enzymatic pretreatment of red blood cells with ficin or papain.
- Expression of the R^c factor is transmitted as a dominant character.
- Quantitative and qualitative variations of the antigen R^c exist that are similar to variants of the human D antigen.

It was demonstrated that anti-R^c antibodies found in the serum of a chimpanzee named Sean (an R^c-negative negative animal immunized with R^c-positive cells) can be fractionated by absorption (Socha and Moor-Jankowski 1978). A fraction of the anti-R^c antibodies specifically agglutinated human Rh_o-positive red cells. The reagent absorbed with human Rh_o-positive cells still contained a fraction of anti-R^c antibodies. This result demonstrated that R^c antigen is not identical to human Rh_o antigen but shares several epitopes with it. Parellel tests with human polyclonal anti-Rh_o antibodies and chimpanzee polyclonal anti-R^c antibodies confirmed that the antigens Rh_o and R^c share some epitopes (Socha et al. 1989).

The complexity of the R-C-E-F blood group system was further expanded by addition of another specificity, namely, one defined by a chimpanzee alloimmune antibody, anti-P^c. Anti-P^c reagent agglutinated some, but not all, c^c-positive chimpanzee red blood cells but always gave negative reactions with c^c-negative cells, thus subdividing the c^c type into two subgroups: the c^c_1 subgroup, recognized by positive reactions with both anti-c^c and anti-P^c reagents; and c^c_2, defined by positive reactions with anti-c^c and negative reactions with anti-P^c. The relationship between P^c and c^c was, therefore, a graded one, comparable

Table 6. Standard reagents used for R-C-E-F blood grouping in chimpanzees and other anthropoid apes

Reagent number	Specificity	Origin
1	R^c	Chimpanzee isoimmune (CH-643 Sean), ETC
2	R^c	Chimpanzee isoimmune (CH-177 Karen), ETC
3	R^c	Chimpanzee isoimmune (CH-11 Tom), absorbed, ETC
4	$R^c + G^c$	Chimpanzee isoimmune (CH-490 Chica, 1966), ETC
5	G^c_1	Chimpanzee isoimmune (CH-490 Chica, 1982), ETC
6	C^c	Chimpanzee isoimmune (CH-491 Bonnie), ETC
7	E^c	Chimpanzee isoimmune (CH-38 Doug), ETC
8	E^c	Chimpanzee isoimmune (CH-11 Tom), absorbed, ETC
9	F^c	Chimpanzee isoimmune (CH-34 Jack), ETC
10	F^c	Chimpanzee isoimmune (CH-136 Hope), ETC
11	$C^c + CF^c$	Chimpanzee isoimmune (CH-225 Andy), ETC
12	c^c	Chimpanzee isoimmune (CH-114 Rufe), ETC
13	c^c_1	Chimpanzee isoimmune (CH-355 Leo), absorbed, ETC
14	$O^c + c_1{}^c$	Chimpanzee isoimmune (CH-355 Leo), ETC

ETC, Enzyme treated red cell technique.

to the relationship between A_2 and A_1. Accordingly, the anti-P^c was renamed anti-c^c_1.

The reagents used to define the various R-C-E-F antigens are listed in Table 6. The list of 19 regular R-C-E-F blood types actually observed at the time of this writing is given in Table 7.

Genetic Model of the Chimpanzee R-C-E-F System

Taking into account the results of population analysis, a genetic model of the RCEF system was proposed (Socha and Ruffié 1983) (Tables 7, 8) based on the following observations:

- Anti-c^c gives antithetic reactions to anti-C^c (i.e. when c^c is absent C^c is always present and vice versa, but c^c may occur together with C^c). This implies that c^c and C^c are characters depending on a pair of codominant alleles. Expression of the antigens E^c and F^c depends on the simultaneous presence of antigen C^c. Therefore, it was assumed that the following alleles exist at the single locus: c, C, C^E, C^F, C^{EF}. The subsequent discovery of two subgroups of c^c implied the presence of an additional allele, c_1.
- Expression of R^c seemed to be under the control of a dominant allele, R, at a locus (R/r) closely linked to the locus C/c.

Table 8 shows the nine alleles required by the genetic model of transmission of the R-C-E-F blood groups. This model is compatible with the hypothesis of a single locus or two closely linked loci (C/c and R/r). In the latter case the term *allele* has to be replaced by *haplotype*.

Table 7. The R-C-E-F blood group system of the chimpanzee

Desig-nation	Reaction with serum of specificity						Phenotype frequencies ($n = 599$)	Possible genotype(s)
	Anti-R^c	Anti-C^c	Anti-E^c	Anti-F^c	Anti-c^c	Anti-$c_1^{\,c}$		
rc_1	−	−	−	−	+	+	28.76	r^1/r^1; $r^1 r^2$
rc_2	−	−	−	−	+	−	0.17	r^2/r^2
rCF	−	+	−	+	−	−	1.51	r^{CF}/r^{CF}
$rCFc_1$	−	+	−	+	+	+	8.7	r^{CF}/r^1
$rCFc_2$	−	+	−	+	+	−	0.5	r^{CF}/r^2
Rc_1	+	−	−	−	+	+	23.24	R^1/R^1, R^1/R^2, R^1/r^1, R^1, r^2, R^2/r^1
Rc_2	+	−	−	−	+	−	6.86	R^2/R^2, $R^2 r^2$
RC	+	+	−	−	−	−	0.67	R^C/R^C
RCc_1	+	+	−	−	+	+	0.67	R^C/R^1, R^C/r^1
RCc_2	+	+	−	−	+	−	0.50	R^C/R^2, R^C/r^2
RCF	+	+	−	+	−	−	4.85	R^{CF}/R^{CF}, R^C/R^{CF}, R^{CF}/r^{CF}, R^C/r^{CF}
$RCFc_1$	+	+	−	+	+	+	6.69	R^{CF}/R^1, R^{CF}/r^1, R^1/r^{CF}
$RCFc_2$	+	+	−	+	+	−	6.86	R^{CF}/R^2, R^{CF}/r^2, R^2/r^{CF}
RCE	+	+	+	−	−	−	2.01	R^{CE}/R^{CE}, R^{CE}/R^C
$RCEc_1$	+	+	+	−	+	+	3.18	R^{CE}/R^1, R^{CE}/r^1
$RCEc_2$	+	+	+	−	+	−	1.51	R^{CE}/R^2, R^{CE}/r^2
$RCEF$	+	+	+	+	−	−	3.01	R^{CEF}/R^{CEF}, R^{CEF}/R^{CE}, R^{CEF}/R^{CF}, R^{CEF}/r^{CF}, R^{CEF}/R^C, R^{CE}/R^{CF}, R^{CE}/r^{CF}
$RCEFc_1$	+	+	+	+	+	+	0.17	R^{CEF}/R^1, R^{CEF}/r^1
$RCEFc_2$	+	+	+	+	+	−	0.17	R^{CEF}/R^2, R^{CEF}/r^2
Irregular forms thus far encountered								
$R_{var}Cc_1$	+[a]	+	−	−	+	+		
$R_{var}CF$	+[a]	+	−	+	−	−		
$R_{var}CFc_1$	+[a]	+	−	+	+	+		
$R_{ac}CE$	+[b]	+	+	−	−	−		
$R_{var}CEF$	+[a]	+	+	+	−	−		
rCc_1	−[c]	+	−	−	+	+		

[a] Weak reactions with all three anti-R^c reagents.
[b] Positive reaction with only one out of three anti-R^c, reagent types encountered by pygmy chimpanzees only.
[c] The presence of a very weak R variant could not be excluded.

As we will see later, the avalaible molecular genetics data are consistent with the presence of three Rh-related genes per chimpanzee haploid genome, although it is still impossible to assess whether all these genes are functional and if different loci are responsible for the R/r and for the C/c series.

The nine initially postulated alleles could combine to give $(n + 1)n/2 = 45$ (where n is the number of alleles) possible genotypes, which could theoretically

Table 8. Postulated alleles of the R-C-E-F system and their products

Alleles	Factors expessed
r^1	c^c_1
r^2	c^c_2
r^{CF}	$C^c + F^c$
R^1	$R^c + c^c_1$
R^2	$R^c + c^c_2$
R^C	$R^c + C^c$
R^{CF}	$R^c + C^c + F^c$
R^{CE}	$R^c + C^c + E^c$
R^{CEF}	$R^c + C^c + E^c + F^c$

Table 9. Estimates of the R-C-E-F haplotype frequencies by two methods

Haplo-types	Resulting antigens RCEFc c$_1$	Square root method	Likelihood method				
			F	support limits of F		χ^2	P
r^1	- - - - + +	0.39272	0.47	0.44	0.51	Infinity	0
R^2	+ - - - + -	0.17983	0.181	0.15	0.21	104.1	2×10^{-24}
r^{CF}	- + - + - -	0.10357	0.1	0.08	0.12	Infinity	0
R^{CF}	+ + - + - -	0.11225	0.073	0.05	0.09	Infinity	0
r^2	- - - - + -	0.11828	0.026	0.01	0.05	Infinity	0
R^1	+ - - - + +	0.02254	0.001	0	0.03	< 0.001	> 0.975
R^{CE}	+ + + - - -	0.05369	0.049	0.04	0.06	546.1	10^{-120}
R^C	+ + - - - -	0.00974	0.0139	0.01	0.02	Infinity	0
R^{CEF}	+ + + + - -	0.00783	0.0028	0	0.008	0.01	0.92
"R- - -"	+ - - - - -	0	0.05	0.02	0.09	4.83	0.028
"r- - -"	- - - - - -	0	0.019	0	0.06	0.05	0.823
"r^{Ccl}"	- + - - + +	0	0.019	0	0.005	< 0.001	> 0.975

For details see text.

give rise to 19 phenotypes. Of these, all but two were actually observed in a sample of 203 unrelated adult chimpanzees (*P. troglodytes*). The observed distribution of the R-C-E-F blood types gave a satisfactory fit with the expected frequencies, based on the nine allele model of inheritance. The gene frequencies determined by the square root method are given in Table 9 and were calculated using the formulas published in 1980 (Socha and Moor-Jankowski 1980).

The genetic model of transmission of the R-C-E-F blood groups was confirmed by the likelihood method (Blancher et al. 1993a). This method gave results which, by and large, are compatible with the originally proposed model of transmission of the R-C-E-F blood groups based on nine codominant genes (or haplotypes). Table 9 compares the frequencies of haplotypes calculated by the square root and maximum likelihood methods, and only those haplotypes are shown for which there is a non-zero estimate of frequency.

The frequencies of six haplotypes, R^2, R^C, R^{CE}, R^{CF}, R^{CEF}, r^{CF}, obtained by either method, are comparable, while the frequencies of two haplotypes, r^2 and R^1, are lower when calculated by likelihood method. In fact, the frequency of R^1 is so low that the existence of this allele does not appear to be necessary to support the genetic model. The frequency of another haplotype, r', which allows expression of c_1, is higher than the one obtained by the square root method.

The maximum likelihood method implied the existence of three additional haplotypes: R---, r---, rCc', but only the existence of the R--- haplotype is supported by the χ^2 value. The ultimate confirmation of its existence would depend on observation of homozygous individuals who expressed the antigen R^c alone. To date, no such phenotype has been encountered in the more than 1000 chimpanzees thus far tested. The expected frequency of such a phenotype would be approximately 2 per 1000 (4×10^{-4} to 8.1×10^{-3}) so that the size of the analyzed population is insufficient to confirm or refute the existence of this haplotype. Interestingly, however, this phenotype has been observed in gorillas (Socha 1986). Except for the haplotype rCc', for which there is only weak evidence for its existence, the hypothesis of C/c allelism is confirmed.

As the number of chimpanzees tested has more than doubled in the last 3 years, not only the two missing, but expected, phenotypes (RC and RCEFc$_2$) were encountered, but so were some additional types not originally postulated. Among the latter were type rCc$_1$, and a series of irregular forms, some of which proved to be inheritable (see Table 7).

The R-C-E-F System in Other Nonhuman Primate Species

Pigmy Chimpanzee

The ten pigmy chimpanzees *(Pan paniscus)* so far tested with the battery of R-C-E-F typing reagents were all found to be type RCE. As the data in Table 9 indicate, the RCE type is relatively rare among common chimpanzees. It may be of interest that, unlike chimpanzees of the common type, which are highly polymorphic for all known chimpanzee blood groups, the dwarf chimpanzees are surprisingly uniform with respect to their blood groups. This observation applies not only to animals born in captivity but also to wild animals (Socha 1984).

Gorillas

All gorillas, both lowland *(Gorilla gorilla)* and mountain *(Gorilla beringei)*, are R^c-positive (Socha et al. 1973). This monomorphic factor of gorilla red cells appears either alone, which could be considered the equivalent of the D-- type in humans, or accompanied by the C or CF factors. The c^c allele, either in its c^1 or c^2 form, has never been observed in these apes, nor has the E^c specificity. Therefore, if the observations on relatively small numbers of animals are valid, the existence of three alleles, R, R^C and R^{CF}, at the R-C-E-F locus of gorillas may be assumed. If, instead of the polyallelic unilocus model of inheritance of the R-C-E-F groups, one considers an alternative model consisting of a series of closely linked genes, two loci can be distinguished in the gorilla: (1) R or r and (2) C or C^F or c (where r and c are silent alleles).

Understandably, since the number of gorillas thus far investigated does not exceed 40, there is a good chance that further R-C-E-F alleles will be identified, or that a phenotypic expression of genes so far considered as silent alleles will become known.

Leaving aside differences in the numbers of gorillas and chimpanzees tested, there are striking peculiarities in the distribution of the R-C-E-F blood groups in both species. First, the solitary type R has never been observed among chimpanzees, while two thirds of gorillas tested are of this type (and the rest are types RC or RCF). Second, type RC is extremely rare among common chimpanzees (the first two *Pan troglodytes* to have RC type red cells have been only recently encountered) but quite frequent among gorillas.

Moreover, expression of R-C-E-F antigens on gorilla red cells is always weaker than on those of chimpanzees (Socha and Moor-Jankowski 1979). In fact, absorptions of anti-R^c, anti-C^c, and anti-F^c reagents with positively reacting chimpanzee red cells rendered the sera inactive against chimpanzee as well as gorilla erythrocytes. The reverse, however, is not true: the same sera absorbed with gorilla red cells of type RCF were still able to agglutinate the chimpanzee red cells of the corresponding types. From a serological point of view, these differences seem of secondary significance, but molecular biology data demonstrated that the Rh-like proteins from chimpanzee and gorilla are substantially different (for details, see below).

Gibbons

The red blood cells of 12 gibbons gave uniformly positive reactions with anti-c^c antibodies. In tests with anti-R^c, ten animals were found to be positive and two negative. The tests with the other R-C-E-F reagents (anti-C^c, anti-c^c, anti-E^c, and anti-F^c) were all negative (Table 10).

Orangutans

The results of tests on blood specimens of orangutans *(Pongo pygmaeus)* were difficult to interpret because of interference from potent anti-orangutan heteroagglutinins normally present in chimpanzee sera. These are impossible to absorb without causing the titer of type-specific antibodies to drop. Thus, no readable reactions were obtained with red cells of the six orangutans previously tested.

Table 10. Comparison of R-C-E-F blood group of common and pygmy chimpanzees, gorilla, orangutan and gibbon: a summary of findings

Common chimpanzee	Pygmy chimpanzee	Gorilla	Gibbon	Orangutan
19 regular phenotypes (see Table 7)	RCE	R RC RCF	R-positive R-negative	RCc$_1$ rc$_1$

When, however, large quantities of blood became available from several members of an orangutan family, multiple absorptions, fractionations, and elution experiments, carried out in parallel with orangutan and chimpanzee red cells, led to the conclusion that at least two R-C-E-F types occur in orangutans, namely, RCc$_1$, and rc$_1$ (Socha and Moor-Jankowski 1980).

Old World Monkeys

None of the factors of the R-C-E-F system were detected on red cells of primates that are situated lower on the phylogenic ladder and which separated earlier from the common trunk. All tests carried out with blood specimens of macaques, baboons, vervet monkeys and *Cercopithecus* remained negative.

Rh Antigens Recognized by Human Monoclonal Reagents

Human Monoclonal Anti-D antibodies

Study of Chimpanzee Red Cells

The use of human monoclonal antibodies (mAbs) provided additional confirmation of the antigenic similarity between the human antigen D and its chimpanzee counterpart, R^c. Reactivities of anti-D mAbs in tests with chimpanzee red blood cells of various RCEF phenotypes can be classified into four general types:
- Type 1: positive with most R^c-positive RBCs
- Type 2: negative with all chimpanzee RBCs
- Type 3: positive with all chimpanzee RBCs
- Type 4: positive with some, but not all, chimpanzee RBCs, irrespective of their R^c phenotype

		Hominoidea								Cercopithecidea					
		Pongidae						Hylobatidae		Cercopithecidae					
		Chimp.		Gorilla		Orang		Gibbon		Rhesus		Cynomol		Baboon	
Origin	mab	IAT	ETC	IAT	ETC	IAT	ETC	IAT	ETC	IAT	ETC	IAT	ETC	IAT	ETC
113**	BS 221	R^c	R^c	+	+	0	0	+	0	0	0	0	0	0	0
95**	IA3,3	R^c	+	D^{gor}	D^{gor}	0	0	+	0/+	0	0	0	0	0	0
10w1*	CB6	R^c	+			0	0			0	0	0	0	0	0
10W2*	6D10	R^c	+	D^{gor}	D^{gor}	0	0			0	0	0	0	0	0
26W6*	H2D5D2F5	R^c	+			0	0			0	0	0	0	0	0
10W5*	2A10	0/+	R^c	D^{gor}	D^{gor}										
16W6*	186;5;D2	0/+	R^c	D^{gor}	D^{gor}	0/+	0			0	0	0	0	0/+	0/+
26W7*	QA37C3G6	0/+	R^c	D^{gor}	D^{gor}	0	+			0	0	0	0	0	0
6W3*	FOG3	0/+	R^c	D^{gor}	D^{gor}	0	0			0	0	0	0	0	0
91**	P3x83 Mil	0	R^c	+	+	0	0	+	0/+	0	0	0	0	0	0
211**	BS-228	0	R^c	+	+	0	0	+	0/+	0	0	0	0	0	0
Toulouse	LORI	0	R^c	0/+	0/+	0	0	+	0	0	0	0	0	0	0
105**	AR	0/+	+	D^{gor}	D^{gor}	0	0	+	0	0	0	0	0	0	0
Lille	229-14G4	0/+	+	D^{gor}	D^{gor}	0	0	+	0	0	0	0	0	0	0
Lille	D 190-31	0/+	+	D^{gor}	D^{gor}	0/+	0	+	0	0	0	0	0	0	0
26W8*	V2B114E2	0/+	+	D^{gor}	D^{gor}	0	0			0	0	0	0	0	0
90**	HM16	+	+	+	+	+	+	+	0	0	0	0	0	0	0
103**	ESD4-(D19/1)	+	+	0/+	0/+	0	+	+	0	0	0	0	0	0	0
98**	AB5	+	+	+	+	0	0	+	0	0	0	0	0	0	0
210**	F64	+	+	+	+	0	0	+	0	0	0	0	0	0	0
10W4*	FC3	+	+	D^{gor}	D^{gor}	0	0			0	0	0	0	0	0
6W8*	FOG1	+	+			0	0			0	0	0	0	0	0
10W3*	H27	0/+	0	D^{gor}	D^{gor}	0	0			0	0	0	0	0	0
23W3*	GRL-04	0/+	0			0	0			0	0	0	0	0/+	0
6W6*	PAG1	0/+	0			0	0			0	0	0	0	0	0
6W7*	GAD2	0/+	0			0	0			0	0	0	0	0	0
Toulouse	GAN 4B5	0	0/+	D^{gor}	0/+	+	+	+	0	0	0	0	0	0	0
Toulouse	NOI	0	0/+	D^{gor}	D^{gor}	+	+	+	0	0	0	0	0	0	0
97**	1A3-1	0	0/+	D^{gor}	D^{gor}	0	0	+	0	0	0	0	0	0	0
Toulouse	SAL4E8	0	+	0/+	+	+	0	+	0	0	0	0	0	0	0
Toulouse	LORA	0	+	D^{gor}	0/+					0	0	0	0	0	0
Toulouse	Lor 12E2	0	0	D^{gor}	D^{gor}	+	0	+	0	0	0	0	0	0	0
Toulouse	Lor 15C9	0	0	D^{gor}	D^{gor}	+	0	+	0	0	0	0	0	0	0
Toulouse	SAL12D5	0	0	D^{gor}	+	+	0	+	0	0	0	0	0	0	0
83**	LF.97.3.39.1	0	0	+	+	0	0	+	0	0	0	0	0	0	0
94**	Ch.24	0	0	D^{gor}	D^{gor}	0	0	+	0	0	0	0	0	0	0
96**	H26	0	0	D^{gor}	D^{gor}	0	0	+	0	0	0	0	0	0	0
100**	JAC10	0	0	D^{gor}	D^{gor}	0	0	+	0	0	0	0	0	0	0
101**	ESD-1	0	0	D^{gor}	D^{gor}	0	0	+	0	0	0	0	0	0	0
104**	Co-88	0	0	D^{gor}	D^{gor}	0	0	+	0	0	0	0	0	0	0
107**	21-G6	0	0	+	+	0	0	+	0	0	0	0	0	0	0
112**	BS 227	0	0	D^{gor}	D^{gor}	0	0	+	0	0	0	0	0	0	0
Toulouse	LORE	0	0	D^{gor}	D^{gor}					0	0	0	0	0	0
Toulouse	NOE 2	0	0	D^{gor}	D^{gor}					0	0	0	0	0	0

Fig. 2. Reactivity of anti-D monoclonal antibodies of the IgG isotype. *IAT,* indirect anti-globulin test; *ETC,* enzyme treated cells; *antibody included in the First International Workshop; **antibody included in the Second Internatinal Workshop; R^c, positive with most R^c-positive chimpanzee RBCs; D^{gor}, positive with D^{gor}-positive gorilla RBCs; o, negative with red blood cells of all animals of the species; +, positive with red blood cells of all animals of the species; o/+, positive with red blood cells of some animals of the species (defining polymorphism unrelated to know blood group systems)

The first three types of reactions produced by mAbs were identical to those obtained with polyclonal human anti-D, while the fourth category of reactions was observed only with anti-D mAbs.

Figure 2 summarizes the results of tests with 44 anti-D mAbs of the IgG class carried out on chimpanzee (and other primate) red blood cells by the indirect antiglobulin method (IAT) and on enzyme-treated red cells (ETC). Twelve monoclonal antibodies agglutinated selectively R^c-positive RBCs (type 1 reactions) in at least one of the two techniques used. Among those there was one that produced such reactions by both agglutination methods; four by IAT only; and seven exclusively by the ETC technique.

There were minor differences among human anti-D mAbs that preferentially agglutinated (by appropriate techniques and at proper dilution) chimpanzee R^c-positive RBCs: some reagents failed to react with a few R^c-positive blood specimens, most frequently with those of the phenotypes RCEF, $RCEc_1$, RCF and RCE. It appears that, in these phenotypes, steric interference of secondary epitopes rendered the R^c epitopes inaccessible to some anti-D monoclonal antibodies. In very rare instances, however, these mAbs agglutinated a few R^c-negative RBCs, but the reactions were always weaker and the titers lower than observed with R^c-positive RBCs.

It is noteworthy that, among the seven reagents that produced R^c-specific reactions with enzyme-treated RBCs, four defined a polymorphism unrelated to R^c (type 4 reaction) in a test by antiglobulin; the remaining three failed to clump any chimpanzee RBCs by IAT. One of the antibodies of the latter type, namely, LORI (produced and characterized by one of us, A.B.), detected an R^c-related polymorphism only when used diluted 1:10. As a crude supernatant it agglutinated all ficinated chimpanzee RBCs. The specificity of reactions of antibody LORI with the R^c-bearing molecule was demonstrated in radioimmunoassay: the binding of LORI was inhibited by anti-R^c antibodies while anti-R^c binding was inhibited by LORI (Roubinet et al. 1993). The dissociation constant of antibody LORI in reaction with R^c antigen was significantly higher than that observed in the reaction of this antibody with human antigen D. Thus the functional affinity of LORI for R^c antigen is significantly lower than its affinity for D antigen. Similar results were obtained with another anti-D mAb (D190-31) (Roubinet et al. 1993).

Of the remaining 32 mAbs, 13 failed to agglutinate chimpanzee RBCs by either method (type 2 reactions), while 19 either clumped all chimpanzee RBCs (type 3 reaction) or produced positive reactions with some, but not all chimpanzee RBCs, irrespective of their R-C-E-F phenotype (type 4 reaction). The mAbs that agglutinated all chimpanzee RBCs (type 3 reactions) reacted with R^c-positive and R^c-negative cells with the same avidity and with comparable titers.

Indirect immunofluorescence assay of one mAb of the latter type showed that the quantities of the antibody bound to R^c-positive and R^c-negative RBCs did not differ significantly (Roubinet et al. 1993).

In order to clarify the nature of the reactions observed with anti-D mAbs, competitive inhibition experiments were carried out between a *radiolabeled* type 1 anti-D (selectively agglutinating R^c-positive RBCs) and an *unlabeled* anti-D antibody of type 3 (agglutinating all chimpanzee red blood cells) (Roubinet et al. 1993). The latter antibody totally inhibited binding to chimpanze RBCs of the type 1 anti-D mAb. It is likely that the two epitopes recognized on chimpanzee red cells by these two antibodies are on a single protein, common to all chimpanzees. Subtle allotypic variations of this molecule are the source of the variability of R^c expression.

Each antibody, among those mAbs that produced type 4 reactions, detected a somewhat different polymorphism, unrelated to the R-C-E-F system or to any of the 14 chimpanzee blood group systems thus far known (Socha 1984). As shown in Fig. 2, this type of reaction was more often produced by the antiglobulin technique. Of the six monoclonal reagents that detected new polymorphisms by IAT, two agglutinated ficinated RBCs of all chimpanzees while four reacted exclusively with enzyme-treated R^c-positive RBCs. The latter observation reinforces earlier findings (Socha and Moor-Jankowski 1980), that enzymatic treatment of the cells renders the R^c antigen more accessible to antibodies, probably by destroying adjacent structures that interfere with expression of R^c. Comparison of reactivities of human anti-D mAbs with chimpanzee and human partial D of various categories demonstrated that chimpanzee R^c lacks some D epitopes and thus resembles the human partial D of category IVb (Blancher et al. 1992b). One must recall that Wiener and Gordon (1961) demonstrated that chimpanzee expressed low-grade variants of Rh^B and Rh^D cognates.

The reactivities of 25 monoclonal anti-D antibodies of the IgM class are depicted in Fig. 3. The same four types of reaction were found; however, it is interesting to note that the frequency of R^c-specific cross-reacting antibodies is lower than among the IgG anti-D. By the same token, the frequency of antibodies that agglutinated all samples of chimpanzee blood was higher (Socha et al. 1993).

Study of Gorilla Red Cells

Some monoclonal anti-D antibodies have been found to recognize a D-like structure (D^{gor}) on gorilla red cells, and the specificity of these reactions was confirmed by competitive radioimmune experiments (Roubinet et al. 1993). Similar to humans and chimpanzee, which are polymorphic for the expresion of D and R^c, respectively, gorillas may be either D^{gor}-positive or D^{gor}-negative. The designation D^{gor} was adopted in reference to the original description, by Wiener and collaborators, of the Rh_o-like antigen Rh_o^{Go}. It was impossible to decide whether the specificity Rh_o^{Go}, defined by polyclonal reagents, and the specificity D^{gor}, defined by monoclonal anti-D, were identical because the reagents used by Wiener are no longer available. Most of the IgG anti-D are able to define D^{gor} polymorphism (Fig. 2), while only rare IgM anti-D are D^{gor}-specific (Fig. 3). It is significant that parallel testing of panels of chimpanzee and gorilla red blood cells clearly demonstrated the absence of any correlation between Rc and D^{gor} specificities. Experiments with monoclonal anti-D mAbs emphasize the great complexity and diversity of D-related molecules in various species of anthropoid apes.

		Hominoidea								Cercopithecidea					
		Pongidae						Hylobatidae		Cercopithecidae					
		IAT	ETC	IAT	ETC	IAT	ETC	A	ETC	IAT	ETC	IAT	ETC	IAT	ETC
Origin	mab	Chimp.		Gorilla		Orang.		Gibbon		Rhesus		Cynomol		Baboon	
Lille	86-A11G5	R^c	R^c	0/+	+	0/+	+	+	0	0	0	0	0	0	+
16W5*	171;6C11	R^c	R^c			0/+	0/+			0	0	0	0	0/+	0/+
110**	BS 224	0	R^c	+	+	0	0	+	0	0	0	0	0	0	0
6W1*	MAD2	0/+	+	+	+	0/+	+	+	+	0	0/+	0	0/+	0	0/+
6W5*	HAMA	0/+	+			0	0			0	0	0	0	0	0
93**	D7 F2 F4	+	+	+	+	+	+	+	0	0	0	0	0	0	+
85**	j 187	+	+	+	+	0	0	+	0	0	0	0	0	0	0
Lille	208-33A10	+	+	+	+	+	+	+	0	0	0	0	0	0	+
Lille	208-37D8	+	+	+	+	+	+	0	0	0/+	0	0/+	0	0/+	0
Lille	212-25D10	+	+	+	+	0	+	0/+	0	0	0	0	0	0	·0
Lille	212-19H8	+	+	+	+	+	+	0	0	+	+	0/+	0/+	+	+
Lille	213-8B12	+	+	+	+	+	+	0	0	0	0	0	0	0	0
Lille	225-14D9	+	+	+	+	+	+	0	0	0	0	0/+	0	0	0
Lille	HM-10	+	+	+	0/+	0	0	0	0	0	0	0	0	0	0
Lille	P3-X61	+	+	+	+	+	+	0	0	0	0	0	0	0	+
24W4*	D61	+	+			+	0/+			0/+	0/+	0/+	0/+	0	+
26W6*	BAC9	+	+			0/+	+			0	0/+	0	0/+	0	0
6W4*	FOM1	+	+			0	+			0	0/+	0	0/+	0	0/+
Bordeaux	186-2-D1	0	0/+	D^{gor}	D^{gor}	+	0	0/+	0	0	0	0	0	0	0
Bordeaux	214-4-E3	0	0/+	D^{gor}	0/+	+	0	+	0	0	0	0	0	0	0
Lille	212-11F1	0	0/+	0/+	0/+	0/+	+	0	0	0	0/+	0	0	0	+
111**	BS 226	0	+	+	+	0	0	+	0	0	0	0	0	0	+
Lille	212-18D5	0	+	0/+	+	0	0	0	0	0	0	0	0	0	0
Lille	212-23B10	0	0	D^{gor}	D^{gor}	+	0	+	0	0	0	+	0	0/+	0
Lille	212-24E10	0		D^{gor}	0/+	+		+	0	0	0	0	0	0	0

Fig. 3. Reactivity of the anti-D monoclonal antibodies of the IgM isotype. *IAT,* indirect antiglobulin test; *ETC,* enzyme treated cells; *antibody included in the First International Workshop; **antibody included in the Second Internatinal Workshop; *R^c,* positive with most R^c-positive chimpanzee RBCs; *D^{gor},* positive with *D^{gor}*-positive gorilla RBCs; o, negative with red blood cells of all animals of the species; +, positive with red blood cells of all animals of the species; o/+, positive with red blood cells of some animals of the species (defining polymorphism unrelated to know blood group systems)

Other Anthropoid Apes

A number of anti-D mAbs of the IgM and IgG classes agglutinated orangutan red cells, and some detected individual differences among animals of this species; however, the specificities of the reactions varied from one antibody to another and could not be assessed with confidence. A high proportion of IgM and all IgG mAbs weakly agglutinated gibbon red cells. But the specificity of the latter reactions was not related to human D or chimpanzee R^c, as was evidenced by results of parallel tests with human and chimpanzee polyclonal reagents.

Other Nonhuman Primates

All tests carried out with IgG anti-D on the red cells of Old and New World monkeys were negative, but when human IgM anti-D were used some mAbs were found to agglutinate monkey red cells and a few mAbs detected polymorphisms in macaques and baboons (Socha et al. 1993). The difference in reactivity of IgG and IgM mAbs could be interpreted in two ways: either the tests with IgM mAbs are more sensitive than with IgG mAbs, or IgM antibodies cross-react with membrane structures unrelated to the D antigen. The latter possibility is supported by comparative tests in which we demonstrated that IgM anti-D antibodies frequently cross-react with autoantigens, while this type of cross-reactivity is exceptional among IgG anti-D (Blancher et al. 1991).

As for prosimians, only one IgM and two IgG mAbs reacted weakly with red cells of black lemurs (*Lemur macaco*), while uniformly negative results were obtained with other lemurs and galagos.

Human Anti-c Monoclonal Antibodies

Tests with human monoclonal antibodies specific for other antigens of the RH system confirmed that all chimpanzees express an antigen very close to the human antigen c (hr') (Socha and Ruffie 1990; Westhoff and Wylie 1994a) Anti-c (hr') antibodies partially inhibit the binding of a radiolabeled anti-D mAb cross-reacting with R^c (LORI) (Roubinet et al. 1993). This result suggests that the R^c molecule contains a c-like structure.

As in the chimpanzee, the human monoclonal or polyclonal anti-c reagents gave uniformly positive reactions with all gorillas and gibbons tested so far. Orangutans are negative with anti-c reagents. Most of the Old World monkeys tested are negative; however, positive reactions were reported with sooty mangabey (*Cercocebus torquatus*), mandrill (*Papio sphinx*) and capucin (*Cebus apella*) erythrocytes in tests with anti-c mAbs (Westhoff and Wylie 1994a). This reactivity was confirmed by tests carried out with human polyclonal anti-c purified by absorption-elution on human R_2R_2 red cells. All other New World monkey red cells analyzed so far gave negative results with monoclonal anti-c reagents.

Human Monoclonal Antibodies Other Than Anti-D and Anti-c

The specificities of the weak reaction obtained with human anti-E (rh'') anti-e (hr'') tested against ape and monkey red cells could not be assessed using polyclonal reagents (Socha and Ruffié 1990). The anti-G antibodies did not react with red cells of any of the chimpanzees or other nonhuman primates tested so far (Socha and Ruffié 1990).

A human monoclonal anti-Ce (FOR) was studied on panels of nonhuman primate red cells (Blancher et al. 1996). It agglutinated cells from two out of 16 gorillas and all gibbons and orangutans tested, while it was negative with chimpanzee, and Old World and New World monkey red cells. It is important to note that all the nonhuman primates studied lack C and e antigens when tested with monospecific reagents. Thus, the cross-reactivity observed on the cells of some gorillas, orangutans and gibbons was probably due to an accidental similarity of the structure detected by that antibody to the compound antigen Ce. Another possibility is that expression of the Ce epitope is independent of that of the antigens C and e; thus, the name "compound" assigned to this antigen is invalid.

Tests with Mouse Monoclonal Reagents Against Rh-Associated Molecules

Some hybridomas derived from lymphocytes of mice immunized with human red blood cells produced monoclonal antibodies which agglutinated all human red blood cells, except those rare cells of a few individuals who do not express Rh molecules (Rh_{null}, Rh_{mod}) (Avent et al. 1988b; Rouger and Edelman 1988). As described above, Rh_{null} red cells lack not only Rh-related molecules but also numerous molecules referred to under the collective name of "Rh-associated molecules". Among the mouse mAbs that did not react with human Rh_{null} erythrocytes one can distinguish antibodies specific for one of the following polypeptides missing from the Rh_{null} erythrocytes:

- Rh-related polypeptides which are encoded by *RH* genes
- LW polypeptides
- CD47 polypeptide
- Rh50 glycoprotein.

Mouse monoclonal antibodies which are specific for Rh-related polypeptides are most frequently directed against epitopes common to all human Rh proteins. The epitopes defined by these antibodies do not correspond to blood groups detected by the human alloantibodies. Only some rare mouse monoclonal reagents are able to detect a polymorphism in humans which mimics the polymorphism defined by human alloantibodies. These exceptional mouse mAbs define an antigen similar to e (Bourel et al. 1987) or a blood group somewhat related to e (Fraser et al. 1990). No other mouse mAbs against human Rh blood groups have ever been produced.

The reactivities of most of the murine monoclonal reagents against Rh-related and Rh-associated molecules described here were investigated during the First International Workshop on Monoclonal Antibodies Against Human Red Blood Cell and Related Antigens (Paris , 21–24 Septembre 1987). Description of the results were in part published (Socha and Ruffié 1990). The results of the Second Workshop (Lund, April 1–4 1990) were summarized by Tippett and Moore (1990b). Designations of the mouse reagents followed by their reference number of the First or Second Workshops will be used in the following paragraphs.

Anti-Rh29-Like Monoclonal Antibodies

The antibodies LM14/15 (Paris # 5W2) and 2.23 (Paris # 20W7) agglutinated ficin-treated red blood cells of all primates studied (chimpanzee, orangutan, gorilla, rhesus monkey, crab-eating macaque, baboon). Tests performed by the antiglobulin method gave uniformly negative results which were difficult to interpret because of the use of human or rhesus monkey antiglobulin reagents that may have limited affinity to mouse proteins.

The antibody BRIC 69 (Lund # 115) (Avent et al. 1988b) reacted with gorilla and chimpanzee cells while monoclonal BRIC 207 (Lund # 116) (Avent et al. 1988b) reacted only with gorilla cells. In this respect BRIC 69 was similar to R6A (Rigwell et al. 1983). The mAb R6A was found to react with red cells from gorillas and chimpanzees but was negative in tests with Old and New World monkeys (Shaw 1986). The three mouse antibodies (BRIC 69, BRIC 207, and R6A) react with human Rh-related molecules (Eyers et al. 1994). Their cross-reaction with red cells of African apes (gorillas and chimpanzees) confirms that these species express Rh-like molecules.

Antibodies Against *RHCE* Gene-Encoded Proteins

The antibody BS58 (Paris #19W6) was found to react with human red blood cells as an anti-C or anti-e antibody in antiglobulin testing (Tippett 1988; Sonneborn et al. 1990) but as an anti-Rh17 in tests on enzyme-treated cells. It reacted only with some chimpanzee red cells and was negative with cells of gorilla, orangutan, baboon, macaques and marmosets.

The reactivity of mAb LM131/191 (Lund # 121) suggested that it recognized an epitope on e-related proteins (the antibody reacted more strongly with untreated ee cells than with untreated EE red cells). This antibody reacted only with gorilla bromelin-treated red cells.

The antibody V2 (Lund # 126) reacted as anti-e in tests with human red cells. It reacted with all gibbon red cells (six samples) in the antiglobulin test and with one of two gorilla samples in the enzyme method. It agglutinated one of two samples of rhesus monkey. It was uniformly negative with all chimpanzee, orangutan, baboon, and cynomolgus red cells.

Monoclonal Antibodies Against LW Molecules

Mice immunized against human red blood cells produced mAbs reacting with human LW molecules. These antibodies allowed identification and partial characterization of LW-associated proteins (Mallison et al. 1986). Some of these anti-LW antibodies were tested against the red cells of various primates (Shaw 1986; Sonneborn et al. 1984):

- mAb BS46 reacted with red cells of gorillas and rhesus monkey but not with those of orangutans, baboons or marmosets.

- NIM-M8 reacted with the cells of all nonhuman primates studied (chimpanzee, gorillas, orangutans, baboons, rhesus monkeys and marmosets), while BS56 reacted with all primate red cells except chimpanzee RBCs.

Antibodies BS46 (Paris # 19W3) and BS56 (Paris # 19W4) were also analyzed during the First International Workshop. The two anti-LW antibodies reacted with red cells of Old World monkeys but not with those of chimpanzees, orangutans or gorillas, confirming the observations made using polyclonal reagents. These results were obtained by the antiglobulin method using rabbit anti-human, or anti-rhesus monkey or anti-baboon immunoglobulins (Socha and Ruffié 1990) but not with anti-mouse immunoglobulin, as in tests by Shaw, which could explain the discrepancies between the two series of experiments.

Antibodies Against CD47

The antibodies BRIC125 (Lund # 114) and 1/7B5 (Lund # 120) reacted with all primate cells tested. Immunoblotting experiments showed similar reactions. It was therefore concluded that both antibodies recognized epitopes expressed on the polypeptide CD47. The expression of CD47 is very weak in human Rh_{null} red cells but is normal on other peripheral blood cells of Rh_{null} individuals. This suggests that the gene of CD47 is functional in Rh_{null} individuals and that there is a deficiency in the transport of CD47 proteins to the erythrocyte membrane. The reactivity of the anti-CD47 antibodies with all nonhuman primate red cell samples suggests a high degree of conservation of the structure of this protein during primate evolution.

Antibodies Against Rh50 Glycoprotein

Three antibodies, 2D10 (Lund # 117), LA 18.18 (Lund # 118) and LA 23.40 (Lund # 119), were found to stain a diffuse band with a leading edge of 39 kDa. All of these antibodies were found to react with the same human antigenic target, called 2D10 polypeptide (Mallison et al. 1990), and to agglutinate gorilla, chimpanzee and patas red cells. Antibody 2D10 reacts with an epitope carried by the Rh50 glycoprotein, which is found in Rh_{null} U+ but not on Rh_{null} U- cells. This suggests that the U antigen may depend on the association of glycophorin B (GPB) with Rh50 glycoprotein (von dem Borne et al. 1990) (see discussion of antigen U in the next chapter).

Studies of Rh variants suggest that the Rh proteins Rh50 glycoprotein and glycophorin B interact during their biosynthesis (Ridgwell et al. 1994). In keeping with this hypothesis, it is postulated that the D polypeptide also facilitates the cell surface expression of LW, since there are more LW antigens (4400 vs 2800 molecules/cell) on RhD-positive than RhD-negative adult red cells. Because of these protein interactions it is reasonable to assume that some (if not all) Rh antigenic determinants might be shared by various components of the Rh protein complex.

The positive reactions of anti-Rh50 in tests with gorilla, chimpanzee and patas red cells suggest that these species most probably possess an Rh50 glyco-protein very close to that of humans.

Molecular Biology of Rh Antigens and Rh-Related Polypeptides

Initial Characterization and Purification of the Human Rh-Related Polypeptides

The molecular basis of the RH antigens is difficult to elucidate because the antigenic properties of the molecules are lost after their extraction from the membrane or their transfer to artificial membranes, such as nitrocellullose sheets. Thus, for many years, it was impossible to characterize Rh polypeptide by classical methods, such as western blot or purification on immunosorbents. Numerous studies reported that anti-D antibodies failed to specifically detect the Rh D-related molecule in a western blot. Yet, 21 anti-D mAbs produced in our laboratory were screened by this method. These antibodies displayed various patterns of reactivity when tested with panels of human partial D red cells. Only one out of the 21 antibodies (LOR15-C9), an IgGK produced by a human-mouse hybridoma, reacted in the western blot with a peptide present only in RhD-positive human red cell samples. The molecular weight of the revealed polypeptide was near 30 kDa (see later in this chapter) (Apoil, unpublished).

Immunoprecipitation of D-related molecules by anti-D-specific polyclonal antibodies was the first method that allowed characterization of Rh-related polypeptides (Moore et al. 1982). The surface membrane proteins of human D-positive erythrocytes were radiolabeled and anti-D antibodies were incubated with membrane preparations before extraction of proteins. The immune complexes (D-related molecules + anti-D antibodies) were purified and submitted to SDS PAGE. Under these conditions autoradiography of the gel revealed a diffuse but intense radiolabeled band corresponding to a protein with an apparent molecular weight between 28 and 33 kDa. These results implied that RhD-related polypeptide possessed an exofacial tyrosine suitable for surface ^{125}I-labeling. Polyclonal anti-c or anti-E antibodies immunoprecipitated radiolabeled protein related to c and E antigens, respectively. Careful analysis of one-dimensional SDS-PAGE gradient gels of immunoprecipitated Rh polypeptides showed a small but reproducible difference in the mobilities of the different polypeptides. The RhD-related protein migrated with an apparent molecular weight of 31.9 kDa, whereas c and E migrated with an apparent molecular weight of 33.1 kDa (Moore and Green 1987).

In other studies, the 32 kDa bands were cut from SDS-PAGE gels of membranes or immunoprecipitated with Rh-specific mAbs from surface ^{125}I-labeled red blood cells of various Rh phenotypes. The isolated Rh polypeptides were proteolytically digested and analyzed by one-dimensional SDS-PAGE autoradiography. Variations in the degradation patterns indicated that Rh D was distinct from C/c and E/e polypeptides (Bloy et al. 1988b; Krahmer and Prohaska 1987).

RhD-related polypeptide shares some properties with other proteolipids: it binds dicyclohexylcarbodiimide (DCCD), which inactivates its antigenic properties, and it is soluble in chloroform/methanol. This particular solubility allowed purification of the polypeptide. The apparent molecular weight of the purified molecule was 35 kDa, as determined by SDS-PAGE.

Meanwhile, the Rh polypeptides were isolated from surface ^{125}I-labeled RhD-positive human red cells by nonimmune methods, such as hydroxyapatite chromatography and preparative electrophoresis of SDS-solubilized membrane skeleton S51 or membrane vesicles (Agre et al. 1987; Saboori et al. 1988). The nonimmune method resulted in a nearly 200-fold purification and permitted calculation of the total number of Rh polypeptides per native red blood cell, namely, approximately 60 000. Immunoprecipitation of Rh proteins by human mAbs allowed purification of the RhD polypeptide to homogeneity, as judged by SDS-PAGE (Bloy et al. 1987; Avent et al. 1988b; Suyama and Goldstein 1988). The amino acid composition of the Rh polypeptides obtained by immunopurification (Bloy et al. 1988a) and nonimmune purifications (Saboori et al. 1988) were very similar, with approximately 37% hydrophobic residues.

The Rh polypeptides isolated by the nonimmune methods showed distinct polymorphisms when RhD-negative and RhD-positive preparations were compared by two-dimensional iodopeptide maps (Saboori et al. 1988) using Elder's method (Elder et al. 1977). Moreover, Rh polypeptides immunoprecipitated from a single unit of blood from an individual (with presumed Rh genotype CDE/cDE) by means of mAbs specific for c, or D, or E were also analyzed by two-dimensional iodopeptide maps. The Rh c, D, and E polypeptides were found to be closely related but distinct proteins (Blanchard et al. 1988). While c and E polypeptides were found to be nearly identical, polypeptide D was significantly different (although related to the former). A single iodopeptide shared by c, D, and E was identified by two-dimensional electrophoresis as the one that contained the exofacial ^{125}I-labeled tyrosine residue on intact red cells.

In all the studies reported above, the apparent molecular weight of the Rh-related proteins was between 30 and 35 kDa. However, other methods produced different results:

- The molecular size of the RhD antigen estimated by the radiation inactivation method was 60 kDa (Green et al. 1983b). This is significantly higher than that found by SDS-PAGE and could reflect association of the D-polypeptide with other membrane proteins. The association of Rh-related polypeptide in a large complex at the red cell membrane has been confirmed by considerable evidence (see above discussion on Rh-associated proteins defined by mouse mAbs).
- Hartel-Shenk and Agre (1992) used palmitic acid associated with the Rh-related peptides as a probe to determine the size of the complexes containing the Rh-associated proteins in nonionic detergent-solubilized membranes. The estimated weight of the proteins in the complex was 170 kDa, after correction for bound detergent.

- The molecular weight of the RhD polypeptide predicted from the cDNA sequence was calculated as 45 kDa. It is assumed that the difference between this calculated value and that found by SDS-PAGE is related to an abnormal migration resulting from the abnormally high level of SDS that binds to these very hydrophobic proteins.

No detectable carbohydrates were found on the Rh polypeptides when examined with methods for labeling terminal sugars and no endogenous phosphorylation was detected (Ghamberg et al. 1983).

Chemical Properties of Human Rh Polypeptides in Relation to Their Antigenicity

Investigations of the chemistry of Rh antigens have established some important properties of these molecules, such as the thiol and lipid dependence, the high hydrophobicity of the 30–32 kDa nonglycosylated membrane carriers and possible association with the membrane skeleton (for review, see Cartron and Agre 1995).

The Rh Antigens on Intact Red Cells Resist Proteolytic Degradation

Routinely, agglutination tests by anti-Rh reagents are performed on protease-treated cells (trypsin, chymotrypsin, papain, bromelain, ficin). The protein is probably embedded deeply in the membrane lipid bilayer and is therefore almost inaccessible to external proteases. Enzymatic treatment of the red cells facilitates their agglutination by incomplete antibodies (antibodies of the IgG class) which do not agglutinate in saline medium. The same phenomenon is observed with reagents used for RCEF typing (see above). Digestion of intact RBCs with phospholipase A2, followed by digestion with papain, resulted in partial degradation of the D polypeptide, but not of the Rh C/c or E/e polypeptides, confirming an inherent difference between these Rh polypeptides (Suyama and Goldstein 1990). Avent et al. (1992) showed that extracellular treatment with bromelain cleaves Rh-related protein at a site similar to that cleaved by papain.

Thiol Groups Are Essential for Expression of Rh Antigens

Using a variety of sulfhydryl-reactive probes, it was established that oxidation of a single sulfhydryl was responsible for loss of the D antigen (Green 1967, 1983). This loss was reversible by reduction of disulfides and could be prevented by prior incubation with anti-D antibodies (Green et al. 1983).

The sequences of the Rh polypeptides led to the conclusion that only one cysteinyl residue (Cys-285) is extracellular. This exofacial-free sulfhydryl radical is probably involved in destruction of the Rh antigens by an impermeant maleimide reagent. Radiolabeled N-maleoylmethionine [^{35}S]sulfone binds to two membrane proteins of normal red cells, but not of Rh$_{null}$ cells. The respective weights of the so labeled proteins were 32 and 34 kDa (Ridgwell et al. 1983).

Histidine Involvement in Rh Antigens

This was suggested by the sharp profile of the plot representing the binding of labeled antibody to RBC vs the pH of the medium (Masouredis 1959). Between pH 6.0 and pH 6.5, the binding rises almost 50-fold, a pH range which corresponds closely to the pK exhibited by the imidazolium ion in proteins. Photooxidation of red blood cells in the presence of rose bengal inactivates all Rh antigens. It remains to be determined whether histidine(s) contribute(s) to the Rh epitopes or play(s) an indirect role in their expression or interactions with lipids (or with other proteins).

Relationship with Lipids

Early investigation demonstrated a relationship between membrane lipids and Rh antigenic reactivity. Indeed, RhD reactivity, inactivated by treatment of red cell membranes with butanol, can be restored with exogenous phosphatidylcholine (Green 1968, 1972). Digestion of intact red cells with phospholipase A2 also produced partial loss of D antigenic reactivity. This could be prevented by prior addition of anti-D antibodies (Hughes-Jones et al. 1975). Alterations in the membrane cholesterol/phospholipid ratio by cholesterol enrichment or depletion produced enhancement or reduction of Rh antigenic reactivity, respectively (Shinitzky and Souroujon 1979; Basu et al. 1980). More recently, the Rh polypeptides were found to be a major fatty acylated membrane protein in human red cells (de Vetten and Agre 1988). Palmitic acid appears to be covalently attached to the Rh polypeptides by thioester linkages onto free sulfhydryls on cysteine residues located on the cytoplasmic side of the membrane (Anstee and Tanner 1993). The fatty acylation was found to be reversible (de Vetten and Agre 1988), but it remains to be established whether the palmitylation of Rh proteins is important for antigenic reactivity or for a putative, and as yet unrecognized, function of these proteins. Rh polypeptides share with proteolipids solubility in chloroform/methanol (Brown et al. 1983; Sinor et al. 1984).

Linkage of the Rh Polypeptides with the Cytoskeleton

Under certain conditions, Rh polypeptides precipitate with the membrane skeleton, which was shown to preserve Rh antigenic activity when membranes were solubilized in nonionic detergents (Ghamberg et al. 1984; Ridgwell et al. 1984). Although there may be a relatively weak interaction between Rh polypeptides and the membrane skeleton, this may only reflect the relative insolubility of the polypeptides in Triton X-100 (Agre and Cartron 1992).

Primary Sequence and Membrane Topology of Human Rh Proteins

Amino acid sequences of human Rh polypeptides were independently reported by three research groups, each employing different isolation (Bloy et al. 1988a; Saboori et al. 1988; Avent et al. 1988b). Identical amino acid sequences were found for the first NH$_2$-terminal amino acid residues from Rh polypeptides isolated by hydroxyapatite chromatography from red cells of several RhD-negative or -positive phenotypes, thus suggesting that the D, Cc and Ee antigens have identical NH$_2$-terminals. The NH$_2$-terminal sequence was used to design oligonucleotide primers, which allowed amplification of cDNA templates prepared from β-thalassemic spleen erythroblasts (Chérif-Zahar et al. 1990) and peripheral reticulocytes (Avent et al. 1990). This technology led to the cloning and sequencing of two identical cDNA fragments from a human bone marrow library. By in situ hybridization of human metaphase chromosomes it was shown, using a Rh cDNA probe, that *RH* genes are located on human chromosome 1p34.3-p36.1 (Chérif-Zahar et al. 1991), a location corresponding to that of the *RH* locus, as demonstrated by cytogenetic analysis (Marsh et al. 1974). Northern blot experiments indicated the presence of a major 1.7 kb and a minor 3.5 kb transcript in preparations from erythroblasts. The 1.7 kb transcripts were found in RNA preparations from fetal liver and cells of HEL and K562 erythroleukemia cell lines, but not in preparations from tissues which do not express the Rh antigens (adult human liver, kidney). Studies with polyclonal antibodies, raised against synthetic peptides designed from the cDNA coding sequences (clone RhIXb), demonstrated that the 1.7 kb cDNA clone encoded for proteins carrying E or e antigens (Hermand et al. 1993). More recently, other cDNAs were cloned which were shown to correspond to *RHD* transcripts (Le Van Kim et al. 1992b; Arce et al. 1993). The RhD cDNAs differ from the previous one (clone RhIXb) by 41 codons, resulting in 36 amino acid substitutions. Further experimental evidence was obtained by PCR analysis of genomic DNA and cloning and sequencing of *RHD* and *RHCE*-specific gene fragments. The data unambiguously demonstrated that the new mRNA corresponded to the *RHD* gene transcript. Direct confirmation was obtained by retroviral-mediated gene transfer of RhD and RhcE cDNA transcripts in cells of the K562 erythroleukemia cell line (Smythe et al. 1996). Transduction of K562 cells with RhD cDNA induced expression of RhD and RhG antigen, while transduction of K562 cells with RhcE cDNA induced expression of Rhc and RhE antigens (Smythe et al. 1996).

Rh Proteins (D and Cc/Ee)

The D and non-D cDNA clones encode for polypeptides with similar sequences, as was expected from previous investigations indicating that D, c and E antigens are carried by homologous, but not identical, proteins. The predicted translation products of both cDNAs are polypeptide chains of 417 amino acids with NH$_2$-terminal methionine removed from the mature protein. The sequences of RhcE and RhD proteins exhibit 35 amino acid (8.4 %) substitutions spread throughout

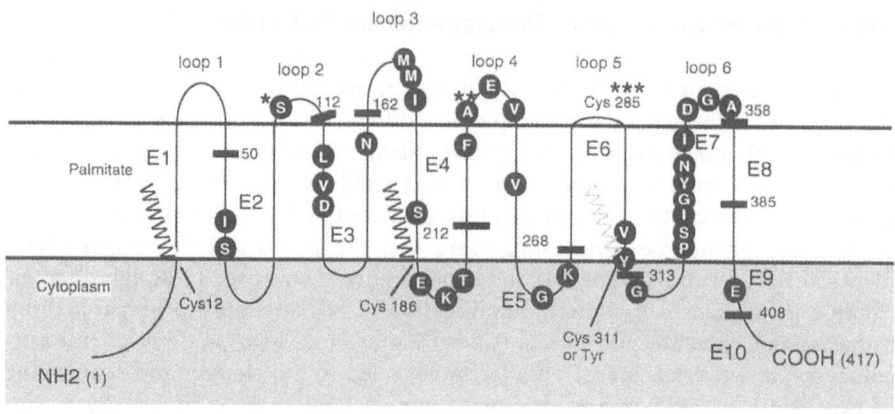

Fig. 4. The membrane topology of the Rh and Rh-like proteins. The model of insertion of Rh-related polypeptides was deduced from hydropathy plots (Cartron and Agre 1993). The 35 positions in *black circles* correspond to substitutions differentiating polypeptide RhD from RhcE (letters in black circles correspond to amino acids observed in RhD protein). Position 1 corresponds to a methionine which is removed in the course of maturation. The approximate limits of externally exposed loops are: loop 1, amino acids (aa) 31–44; loop 2, aa 102–113; loop 3, aa 154–175; loop 4, aa 226–240; loop 5, aa 283–287; loop 6, aa 349–358; *black bars*, limits of the exons (numbers correspond to composite codons or to the last codon of exon 3 (162), exon 6 (313), exon 9 (409); *black zig zag line*, palmitate linked to the cysteine (CLP motive); *gray zig zag line*, palmitate linked only to *RHCE*-encoded proteins; *position 103 is determinant for the expression of antigens C, c and G; **position 226 is determinant for the expression of antigens E and e (see text); ***only one cysteinyl residue (Cys-185) is externally exposed

the polypeptide chain, while the NH$_2$- and COOH-terminal domains of the Rh proteins are relatively well conserved (Fig. 4). The calculated molecular weight of the proteins was 45.5 kDa, higher than the molecular weight measured by mobility on SDS-PAGE (approximately 30–32 kDa). As mentioned before, this difference was probably due to abnormal binding of SDS, rather than to posttranslational proteolytic cleavage (Helenius and Simons 1975; Suyama et al. 1991). Although the D protein is non-glycosylated (Gahmberg 1983) there is a potential N-glycosylation site at Asn-331, but this residue is located within a putative hydrophobic transmembrane domain. Of the six cysteine residues of the RhCcEe polypeptide, five are conserved at the same positions in RhD, including the single exofacial cysteine at position 285 (Fig. 4). Cys-311 of the RhCcEe protein is replaced by a Tyr residue in RhD.

Membrane Topology

Secondary structure predictions suggest that all Rh proteins are organized similarly: the molecules reside almost entirely within the phospholipid bilayer with 12 transmembrane domains. Only short hydrophilic loops protrude beyond the membrane surface between adjacent transmembrane domains (Fig. 4).

Recent findings based on immune precipitation, ELISA and western blot analysis with intact cells, ghosts and membrane vesicles using antipeptide antibodies indicated that both the NH_2- and COOH-terminals of the Rh polypeptides are located intracellularly (for a review, see Cartron and Agre 1995).

The two human Rh proteins carry a negatively charged residue in transmembrane helices 1, 2, 3 and 5; there is another negative charge in the putative helix 12 of the D protein. Such charged groups are often present in hydrophobic domains of ion transporters and channels.

Cysteine-285 is probably located in the outer leaflet of the lipid bilayer in the sequence: Cys-His-Leu-Ile-Pro (CHLIP). This is in agreement with the observation that an exofacial sulfhydryl is involved in the immunoreactivity of RhD and that, in addition, all Rh antigens are sensitive to histidine-reactive compounds (Victoria et al. 1986). Two other cysteines (Cys-12 and Cys-186) are part of the motif Cys-Leu-Pro (CLP) and expected to be located near the inner leaflet of the bilayer. The CLP motifs are likely to be the sites of fatty acylation by palmitoyl residues (Anstee and Tanner 1993). The first cysteine (Cys-311) of another motif, Cys-Leu-Pro-Val-Cys-Cys (CLPVCC), is probably also fatty-acylated by a palmitoyl residue; it is found in CcEe polypeptides but is replaced by tyrosine in the RhD protein (YLPVCC).

Nonhuman Homologues of the Rh Polypeptide

Using the hydroxyapatite method for isolating Rh polypeptides, proteins of similar molecular weight (32 kDa) were isolated from the membranes of rhesus monkey, cow, cat, and rat red blood cells and analyzed by two-dimensional iodopeptide maps (Saboori et al. 1989). Of approximately one dozen iodopeptides contained in each preparation, only two were found to be shared by human and all animal preparations. Thus, the animal counterparts of the human Rh polypeptides are markedly different from the latter in their primary structures. In addition the iodopeptide corresponding to the surface ^{125}I-labeled exofacial tyrosine was found exclusively in the human preparations.

Western blot analysis, using a rabbit antiserum to purified and denatured RhD polypeptide (Suyama and Goldstein 1988), allowed characterization of Rh-like proteins on red cells from various nonhuman primates (Salvignol et al. 1995). The same method of analysis, using rabbit polyclonal antibodies (MPC8 reagent) raised against human Rh peptides encoded by the tenth exon (amino acid residues 408–416) (Hermand et al. 1993), confirmed that the red cell membranes of anthropoid apes (chimpanzee, gorilla, orangutan) and macaques express proteins equivalent to Rh polypeptides (Salvignol et al. 1995). The molecular weights of these Rh-like polypeptides appear similar to those of their human homologues. Red cell preparations from other monkey species (*Papio, Aotus, Cebus, Saimiri*) reacted similarly in western blot with anti-Rh polypeptide antibodies, although with variable intensity. Interestingly, the MPC8 reagent did not react with proteins extracted from gibbon red cells. This is in accordance with the finding of a transcript lacking exon 7 in this species (Salvignol et al.

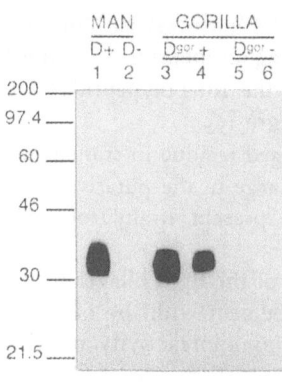

Fig. 5. Western blot proteins extracted from human and gorilla red cells. Proteins extracted from membrane red cells were separated by 10 % SDS-PAGE and transferred to a nitrocellulose sheet. This was incubated with monoclonal antibody (*mAb*) LOR 15-C9 reagent for 120 min at room temperature. After extensive washings, the nitrocellulose sheet was incubated with peroxidase conjugated goat anti-human IgG. The reaction was revealed by a chemiluminescent substrate (exposure time: 30 s). LOR 15-C9 mAb revealed a single band of 33 kDa in protein extracted from human D-positive red cells (*lane 1*), and from two D^{gor}-positive animals (*lane 3*, "Mabeke"; *lane 4*, "Yasmina"). LOR 15-C9 was unreactive with proteins extracted from erythrocytes of a human D-negative individual and from red cells of two D^{gor}-negative gorillas (*lane 5*, "Kessala"; *lane 6*, "Dian")

1995). The lack of exon 7 induces a frame shift which results in the nonexpression of the exon 10-encoded region, which, in turn, renders impossible recognition of the gibbon polypeptide by the MPC8 reagent.

Antibodies raised in rabbit against a peptide encoded by the first exon of the human *RH* gene (MPC1 reagent) were found to give results similar to those obtained with the MPC8 reagent in tests with red cells from anthropoid apes and Old World monkeys; however, they failed to react with the red cells of New World monkeys. We will see later in this chapter that the first exon is probably not translated in South American monkeys.

Using a human mAb, LOR15-C9, which specifically recognizes the human RhD-related polypeptide by western blot, we demonstrated the presence of an RhD-like polypeptide in gorilla red cells (Fig. 5). The LOR15-C9 antibody failed to react with proteins extracted from blood samples from all other primates. The molecular weight of the gorilla protein is similar to that of its human counterpart.

Structure of the Human *RH* Locus

Marsh et al. (1974) demonstrated by cytogenetic analysis that the human *RH* locus is on the short arm of chromosome 1. In situ hybridization with a Rh cDNA probe confirmed the precise location of that locus, on chromosome 1p34.3-p36.1 (Chérif-Zahar et al. 1991).

The Two-Gene Model

Southern blot analysis and genomic studies have shown that the haploid genome of unrelated RhD-positive individuals of Caucasian origin is composed of two

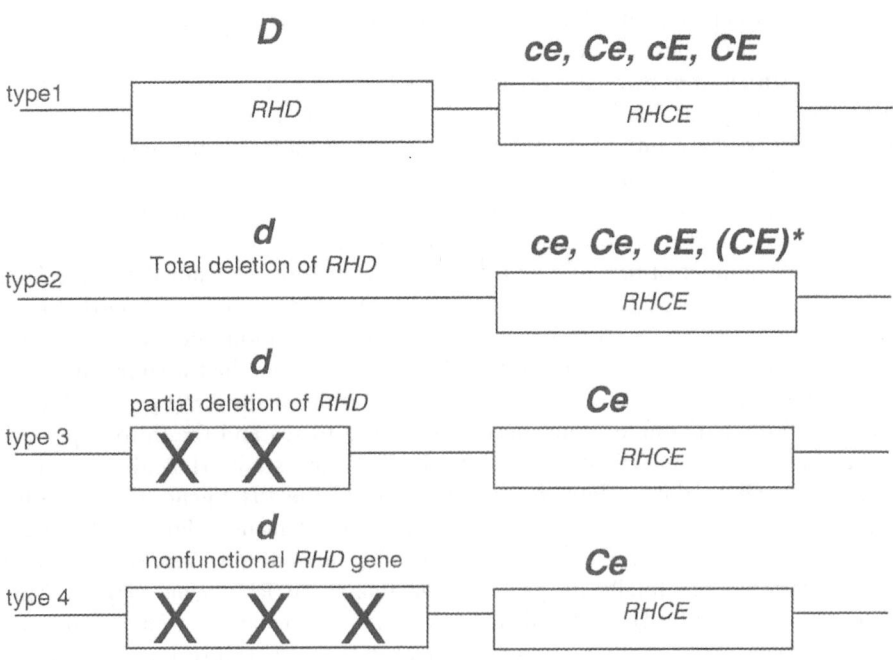

Fig. 6. The *RH* locus in humans. There are two *RH* genes: *RHD* and *RHCE*. The poly-morphism at the *RHD* locus consists of the presence or absence of a functional *RHD* gene. The *RHD* gene may be present and functional and encode the RhD protein (type 1); totally deleted (type 2) (Colin et al. 1991; Hyland et al. 1994a); or nonfunctional (type 3 and 4) (Hyland *et al*.1994b). In the latter case it was found to be either partially deleted (type 3) or apparently intact but nonfunctional (type 4) (Hyland et al. 1994b). Variants of the *RHD* gene are of low frequency and result in partial D phenotypes of various categories (see text). The *RHCE* gene encodes for the RhCcEe proteins. Four regular alleles are identified: *ce, Ce, cE, CE*. *The rare haplotype *dCE* has not been analyzed for the presence or absence of the *RHD* gene. *X*, nonfunctional gene

homologous genes *(RHD* and *RHCE)* which are both similarly organized (Colin et al. 1991) (Fig. 6). The D-positive/D-negative polymorphism corresponds to the presence or absence of a functional *RHD* gene, respectively. The *RHD* gene is most frequently totally deleted, although in some cases the deletion is partial or undetectable. In the two latter cases, the gene is not functional and does not encode for the RhD protein. Allele *d* of the Fisher and Race nomenclature corresponds to a silent allele, itself corresponding to a total or partial deletion or punctual silencing mutations. This explains why anti-d antibodies were never observed. RhC/c and RhE/e polymorphisms are correlated with allotypic variations on a single *RHCE* gene (Mouro et al. 1993). The *RHCE* gene encodes for a single protein carrying the antigens of the C/c and E/e series (Smythe et al. 1996).

The *RHD* gene encodes for the D protein and the *RHCE* gene encodes for the CcEe protein(s). These results are consistent with a two-gene model of inheritance in which the two closely linked *RHD* and *RHCE* genes are inherited together as a single haplotype (Colin et al. 1991). It is not known how these structural genes are physically organized on chromosome 1 with respect to the direction of transcription.

Although the *RHD* gene has been found absent in all D-negative individuals of Caucasian origin, the meaning of that observation is uncertain due to the relatively small number of *dce, dCe and dcE* haplotypes thus far studied and the fact that the very rare *dCE* haplotype was not investigated (Colin et al. 1991). Upon examination of CCee blood samples of three Australian blood donors (of unspecified ethnic origin), Hyland et al. (1994b) found that one had a complete deletion of the *RHD* gene, another had a partial deletion of the *RHD* gene at the 3' region (presumably due to a deletion from exon 7 to 10) and the third appeared to have a normal (but silent) *RHD* gene. In the same study, Hyland et al. confirmed that all 25 *dce* haplotypes examined lacked the *RHD* gene. Unusual Rh-negative gene complexes have also been found in Africans (Blunt et al. 1994). These findings indicate that the "RhD-negative phenotype" may result from rare mutations that totally or partially inactivate the *RHD* gene (Fig. 6). The observations made by Hyland and Blunt provided an explanation for the discovery of a cDNA clone (RhPII), almost identical to the RhD transcript, in Japanese individuals typed as RhD-negative (Kajii et al. 1993)

RH Genes and Rh Transcripts

The *RHCE* gene is composed of 10 exons spread over 75 kb of DNA (Chérif-Zahar et al. 1994a). The structure of the gene is not yet completely clarified, but preliminary investigations indicate that it is organized like the *RHCE* gene (Huang et al. 1996). However, a variation of the length of intron 4–5 has been identified: the intron 4–5 of the *RHCE* gene was reported to be 1200 bp and that of *RHD* was 600 bp (Arce et al. 1993).

In erythroid tissues and cell lines, the main transcripts of the two genes are 1.5 and 3.0–3.5 kb. Three mRNA isoforms of the *RHCE* gene lack exons 4–6, exons 4, 5, 8, and exon 7, respectively (Le Van Kim et al. 1992a). The two major D mRNA isoforms are missing from exon structures 7 and 7–9, respectively (Suyama et al. 1994; Westhoff and Wylie 1994b; Huang et al. 1995). The D isoform, lacking exons 7–9, has been found in the erythroleukemia cell line K562 (Westhoff and Wylie 1994b). Other isoforms were characterized by Kajii et al. (1995). The Rh mRNA isoforms are theoretically capable of encoding Rh protein isoforms that may exhibit new structural and antigenic properties. The translated proteins, however, have not yet been observed.

Molecular Basis of Human Rh Blood Group Specificities

Molecular Analysis of D Variants and Blood Group D Epitopes

The analysis of variant antigens associated with the lack of D epitopes, discussed earlier in this chapter, was instrumental in the recognition of the molecular basis of antigen D. Twelve categories of partial D have been defined, each characterized by the absence of some epitopes of the D antigen (for general review, see Daniels et al. 1995). Important rearrangements of the *RHD* gene have been found in these variants, as schematically shown in Fig. 7.

The following section summarizes the data existing on the molecular biology of the partial D variants (Rouillac et al. 1995a,c; Mouro et al. 1994b; Beckers et al. 1995). Studies of D category cells indicated that different types of exchanges between *RHD* and *RHCE* genes (most probably by conversion or unequal crossing-over) are responsible for each phenotype: D^{IIIb}, D^{IIIc}, D^{IVa}, D^{IVb}, D^{Va}, DFR, DBT. The regions involved were shown to encompass exon 2, exon 3, exons 2 and 7, exons 7–9, exon 5 and exon 4, and exons 5–9 respectively (Fig. 7). These variants are assumed to encode the 417 residue hybrid D-CE-D proteins. Due to the replacement of certain exons of *RHD* by their homologues encoded by *RHCE* exons, these hybrid molecules express only some of the D epitopes present on common RhD-positive cells. Correlations between the lack of expression of some D epitopes and the type of exon replacement at the genomic level allowed determination of the amino acid residues of the RhD protein involved in the expression of each epitope. In another rare phenotypes R_o^{Har}, D- - DC^W-, the mutated genes were found to be hybrids of type *CE-D-CE* in which the exon 5, 4–9 and 3–9 , respectively, came from *RHD* gene (Beckers et al. 1995; Chérif-Zahar et al. 1994b) (Fig. 7).

The red cells of category D^{VI} are characterized by loss of most of the D epitopes (lacking epitopes epD1, epD2 and epD5 to epD8). Two types of D^{VI} were identified at the genomic level (Mouro et al. 1994b). Some variants exhibited a deletion of the region encoding exons 4–6 whereas in others a gene conversion event had replaced exons 4–6 of the *RHD* gene by exons 4–6 of the *RHCE* gene, therefore generating a *D-CE-D* hybrid gene. Both rearranged genes are transcribed and are predicted to encode, respectively, for a short protein form (267 residues) and for a protein of normal size (417 residues), both carrying only epitopes D3, D4 and D9. Recently we characterized by western blot the proteins produced by the D^{VI} gene of the hybrid type using an exceptional human mAb (LOR15-C9), which is capable of recognizing protein D by western blot (Apoil, unpublished). The antibody revealed two proteins in the extract from D^{VI} individuals: one with a migration identical with that of the regular D protein, and a shorter one with a molecular weight of 21 kDa. All the D^{VI} variants tested so far (24 samples) displayed this pattern by western blot. We concluded that this protein resulted from the translation of an alternatively spliced RhD mRNA. The most likely hypothesis is that the RHD^{VI} hybrid gene encodes for mRNA which

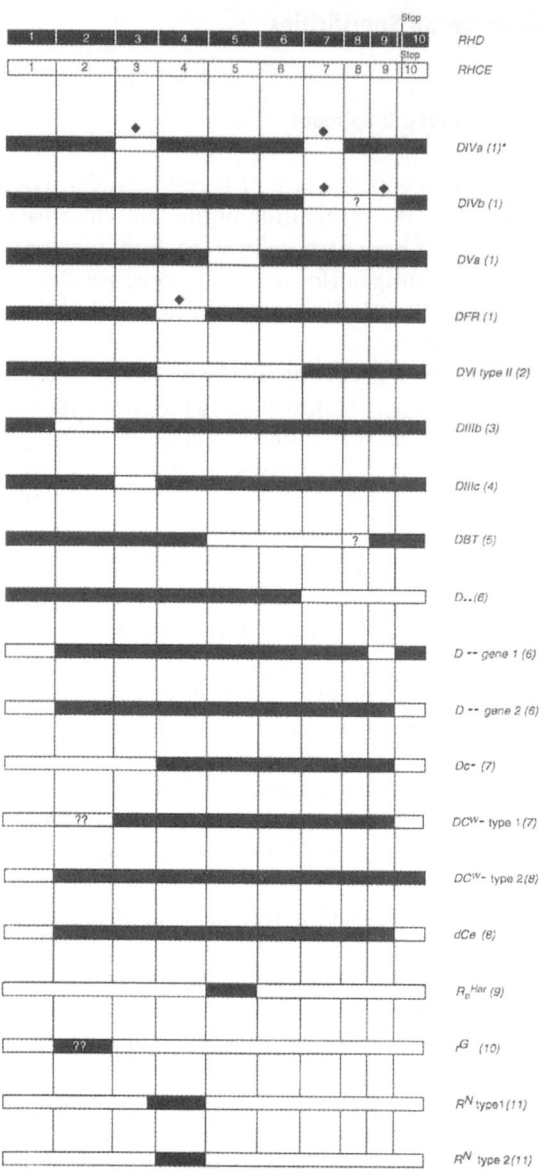

Fig. 7. Variants of human *RH* genes resulting from exchanges between *RHD* and *RHCE* genes. *?*, Exon 8 of uncertain origin because the *RHCE* and *RHD* genes are identical in this region; *??*, exon 2 from *RHC(E or e)* or from *RHD*. References are shown in *parentheses:* (1) Rouillac et al. 1995a; (2) Mouro et al. 1994; (3) Rouillac et al. 1995c; (4) Beckers et al. 1995; (5) Beckers et al. 1996 a; (6) Chérif-Zahar et al. 1996; (7) Chérif-Zahar et al. 1994; (8) Huang et al. 1996; (9) Beckers et al. 1996b; (10) Mouro et al. 1996; (11) Rouillac et al. 1996. ◆, Only part of the relevant exons could be involved in the segmental DNA exchanges. The intergenic exchange between *RHD* and *RHCE* genes was demonstrated in both the genomic and the cDNA for D^{VI} (typeII), D^{IIIc}, D^{IIIb}, *DBT*, R_o^{Har}, *Dc-*, *DC^w-*, r^G, R^N type I, R^N type II. As for D^{IVa}, D^{IVb}, D^{Va} and *DFR*, intergenic exchanges were suspected only from cDNA sequences. Although RFLPs were revealed by hybridization with the full length RhcDNA probe in case of D^{IVa}, D^{IVb}, D^{Va}, hybridization with exon-specific probes demonstrated that the presence of polymorphic fragments in the restriction patterns was caused by nucleotide polymorphisms located in that part of the genomic region not involved in the expression of the D^{IVa}, D^{IVb}, D^{Va} category phenotypes. The hybrid r^G gene was derived either from interallelic or intergenic conversion between *RHce* and *RHC(E or e)* or *RHD* genes in a region encompassing part of exon 2, or from a crossing-over between positions 159 and 178 of *RHce* and *RHCe* alleles. In R^N type I, the genomic fragment encompassing the 3' end of exon 3 and the entire exon 4 of the *RHCE* gene (*Ce* allele) has been replaced by the equivalent of the *RHD* gene. In R^N type II the region of intergenic exchange was limited to exon 4. On the D- - chromosome two converted genes were found

can by spliced differently than the messengers deriving from the regular *RHD* gene (Apoil, in preparation).

From analysis of the D variants, a tentative model has been proposed that correlates D polymorphic positions on external loops of the D protein with the individual D epitopes (Cartron 1994). However, this tentative model should not be accepted without some caution: Rh epitopes are conformational. This could explain how amino acid substitutions at intramembrane or intracytoplasmic locations may modulate the D protein conformation thus greatly influencing the expression of some RhD epitopes. Moreover, several other factors may be critical for D antigen expression, such as the lipid composition of the membrane and the fatty acylation of the Rh proteins or protein interactions between D and other components (Rh50, CD47, LW, etc.) of the Rh protein complex (see below). Thus, substitutions of some amino acids essential for these complex interactions could indirectly alter the expression of D epitopes.

In only one case was it possible to locate some epitopes of the Rh molecule: the reactivity of LOR15-C9 mAb in the western blots indicated that the epitopes recognized by this special antibody were resistant to purification and denaturation. It was therefore obvious that those epitopes most probably depended exclusively on RhD protein. Studies with partial D further specified that the sequence involved was encoded by exon 7 of the RhD gene. The arguments supporting this hypothesis are:

- The antibody specifically immunoprecipitated the RhD protein translated in vitro but failed to react with an in vitro translated RhD protein lacking exons 7–10 (Colin and Mouro, personal communication).
- The human D^{IVa} and D^{IVb} variants are not agglutinated by the LOR15-C9 antibody. Both variants show the replacement of exon 7 of the *RHD* gene by its homologue from the *RHCE* gene.
- Most probably only a short portion of the protein encoded by exon 7 protrudes outside of the membrane, and in this portion all chimpanzee proteins resemble the *RHCE*-encoded protein,while some of the gorilla Rh-like proteins display here a motif specific to *RHD*. Accordingly, some gorilla red cell samples were agglutinated by LOR15-C9, while all chimpanzees so far tested have been negative.

Most of the human variants so far detailed correspond to hybrid *RH D-CE-D* or *RHCE-D-CE* genes; other rare variants were shown to be related with punctual mutations. The partial D of category VII, which expresses the low prevalence antigen Tar, was related to a single mutation in codon 110 (Pro to Leu) (Rouillac et al. 1995b). A single mutation in codon 103 (Ser to Pro) is able to suppress expression of antigen G (Faas et al. 1995). Two rare variants of antigen C, namely, C^W and C^X, were shown to correspond to amino acid substitutions in positions 41 (Glu to Arg) and 36 (Ala to Thr), respectively (Mouro et al. 1995).

Positions Responsible for C/c and E/e Polymorphisms

The genetic basis of the human C/c and E/e polymorphisms has been established by sequencing *RHCE* transcripts from reticulocytes of donors homozygous for the *dce, dcE and dCe* haplotypes (lacking the structural *RHD* gene) and by sequencing genomic DNA fragments from unrelated individuals of different Rh phenotypes (Mouro et al. 1993). The C and c alleles differed by one nucleotide change in exon 1 and five nucleotide changes in exon 2 of the *RHCE* gene, thus resulting, at the protein level, in four amino acid substitutions: Cys-16-Trp, Ile-60-Leu, Ser-68-Asn and Ser-103-Pro.

Comparison of the above human sequences with Rh-like proteins of nonhuman primates showed that prolines 102 and 103 are necessary for the expression of the c or c-like antigens by human and nonhuman primate Rh proteins (see later in this chapter and Fig. 3).

The substitutions which give rise to the rare variants C^W and C^X confirmed that the first external loop of the Rh protein is the determinant in the polymorphism detected by anti- C, -c, -C^X, -C^W, -MAR (RH51) (Mouro et al. 1995).

The E/e polymorphism corresponds, at the genomic level, to a single nucleotide substitution in exon 5 of the *RHCE* gene, resulting, at the protein level, in a Pro (for E) to Ala (for e) substitution in position 226. These results confirmed, as previously suggested from Southern blot analysis, that the C/c and E/e antigens are encoded by a single gene, since both these polymorphisms are found on the same transcripts.

Since human RhD and Rh-like proteins of all nonhuman primates have an alanine in position 226, and, at the same time, lack antigen e (and E), one has to assume that position 226, although indispensible, is not sufficient for the expression of antigen e. It seems plausible that the amino acid substitutions adjacent to position 226 in RhD and in all nonhuman primate Rh-like proteins are incompatible with the expression of e (see later in this chapter).

Study of *RH*-Like Genes in Nonhuman Primates

Southern Blot Analysis

Southern blot experiments using human Rh cDNA probe demonstrated that nonhuman primates carry *RH*-like genes and indicated that the number of *RH*-like genes varies from one nonhuman primate species to another (Fig. 8). Gorilla and chimpanzee most probaly carry two and three *RH*-like genes, respectively, whereas other nonhuman primates (orangutans, gibbons and all Old and New World monkeys) carry a single *RH*-like gene (Blancher et al. 1992a; Salvignol 1993; Salvignol et al. 1993; Westhoff and Wylie 1994a). However, most rhesus monkeys, and one among the ten gorillas so far investigated, were found to also carry small Rh-related 5'-fragments, which correspond, most probably, to

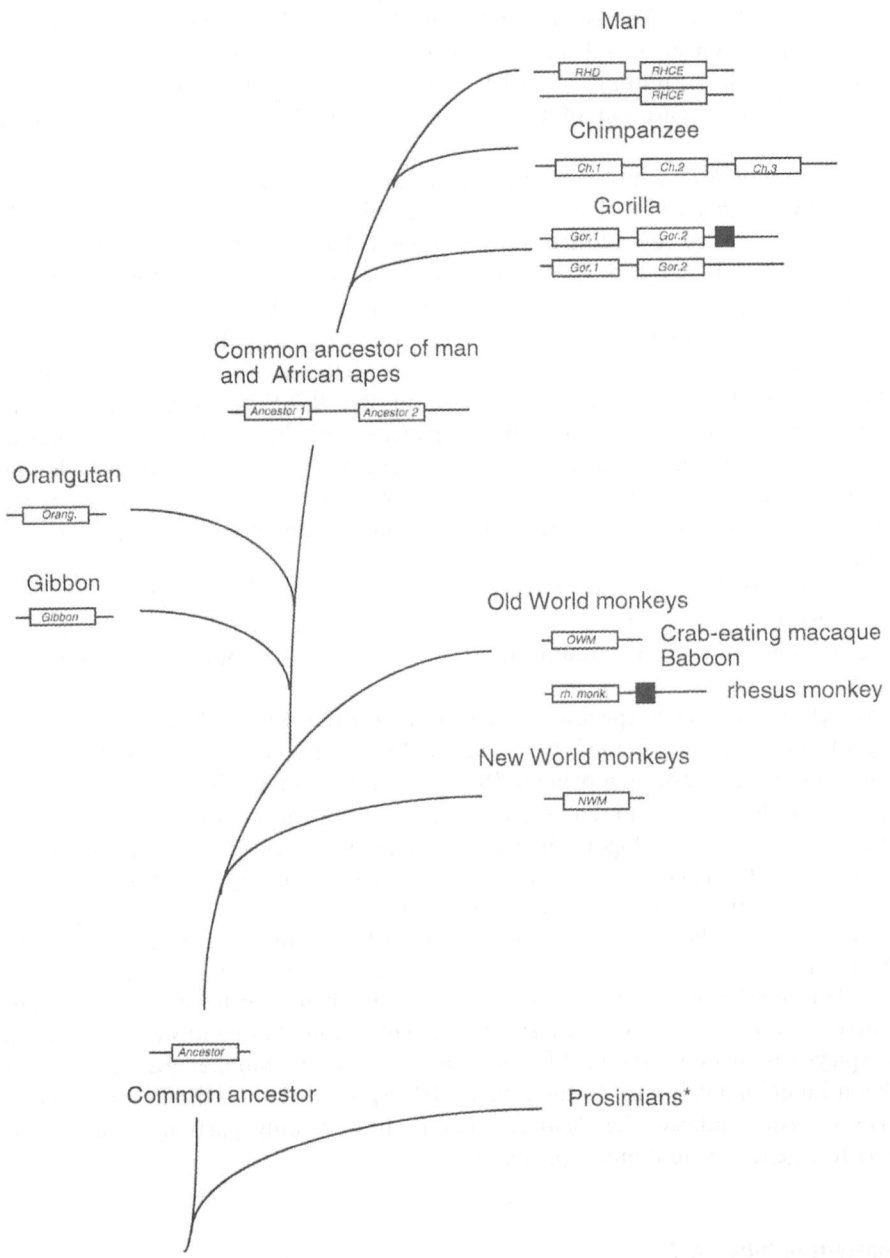

Fig. 8. Proposed evolutionary pathway of the *RH* genes in nonhuman primates (for explanation see text). *In prosimians (lemurs) the presence of *RH*-like genes was not demonstrated by Southern blot experiments. *Black rectangles* correspond to truncated *RH*-like genes

truncated *RH* genes. The latter may be the result of either a partial duplication or the vestige of an ancestral, partially deleted gene (Salvignol 1993; Mouro et al. 1994a). All these findings suggest that stable duplication of the *RH* genes occurred after separation of the African and Asian apes, namely 8–11 million years ago.

Studies of DNA samples from a large panel of chimpanzees of various R-C-E-F types by Southern blot demonstrated a complex polymorphism of the *RH*-like genes, as defined by human cDNA probes. Furthermore, associations were demonstrated between R-C-E-F types and polymorphic restriction fragments hybridized by human Rh cDNA probes (Salvignol et al. 1993, 1994). Family studies of heredity of chimpanzees of R-C-E-F types and of polymorphic restriction fragments demonstrated that *RH*-like genes, detected by Southern blot using human Rh cDNA probes, corresponded to the *RCEF* genes (Salvignol et al. 1994). The in situ hybridization of chimpanzee metaphase chromosomes showed that the *RH*-like genes are on chimpanzee chromosome 1, in a location homologous to that of the *RH* genes on human chromosome 1 (Calvas et al. 1994). This confirms that the R-C-E-F locus is the chimpanzee counterpart of the human *RH* locus.

Southern blot analysis of nonhuman primate DNA digested by the restriction enzyme *SphI* confirmed that the number of *RH*-like genes varies among species (Huang, personal communication; see Fig. 9). The use of a human probe corresponding to exons 4–7 of the RhcE cDNA allowed definition of the *SphI* restriction fragments specific for *RHCE* and *RHD* genes (Huang et al. 1995, 1996). The results obtained by Southern blot, using this probe in tests with *SphI*-restricted nonhuman primate DNA, corroborate that chimpanzee and gorilla possess at least two different *RH*-like genes. Chimpanzee and gorilla restriction fragments were homologous in size to fragments derived either from human *RHD* or *RHCE* genes. Orangutan displays a pattern of hybridization with two bands corresponding in size to human RHCe 6–7 and RHCe 4–5 fragments. This reinforces the notion that the orangutan has a single *RH*-like gene, similar to the human *RHCe* allele at least with respect to *SphI* restriction sites.

The restriction profiles of gibbon, rhesus monkey, and baboon share a band corresponding in size to the RHD6–7 fragment. While that band was the only one displayed by baboon, the profiles of gibbon and rhesus monkey display a second band homologous in size to the human RHCe4–5 band or to the RHCe6–7 band. These results indicate that *SphI* restriction sites are only partially conserved in *RH*-like genes of nonhuman primates.

Analysis of Intron 4–5

The length of this intron was assessed recently in nonhuman primates (Westhoff 1993; Westhoff and Wylie 1996). The intron was amplified by PCR using, as templates, genomic DNA from chimpanzee and gorilla. DNA samples of seven chimpanzees allowed PCR amplification of fragments of the size of those derived from the human *RHD* intron. Individual variations of the length of gorilla introns were

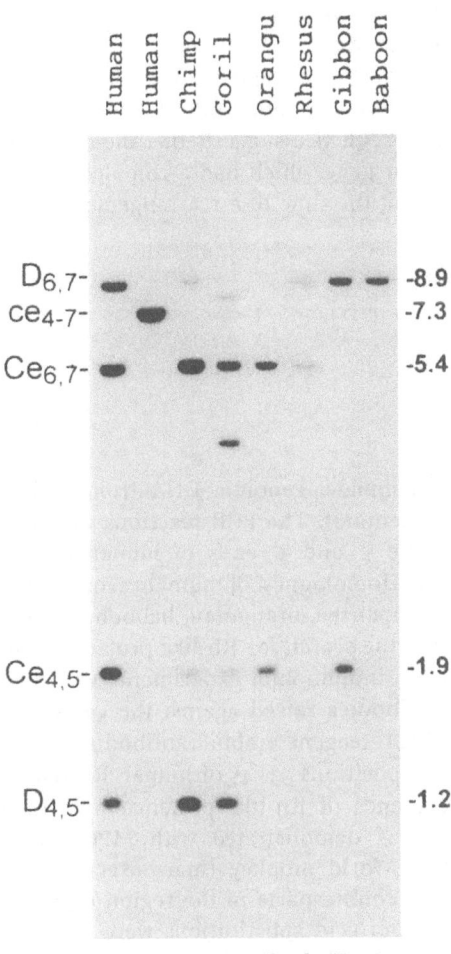

Fig. 9. Southern blot analysis of *RH* locus equivalents in nonhuman primates (courtesy of C.H. Huang, New York Blood Center). Genomic DNA was digested by *SphI* and hybridized on Southern blot with a human cDNA probe encompassing exons 4–7

observed: two gorillas had introns of 1200 and 600 bp, two had only the shorter form, and one only the larger. These results suggest that all *RH*-like genes of chimpanzee share with the human *RHD* gene an intron 4–5 of 600 bp. In gorillas, however, two types of genes (or alleles) are encountered. The one carrying the smaller intron resembles the human *RHD* gene, while the other, which carries the larger intron, corresponds to the human *RHCE* gene. The absence in chimpanzees and in some gorillas of *RH*-like genes with the 1200 bp intron does not result in nonexpression of the c-like antigen. As mentioned earlier, all chimpanzees and gorillas express a c-like antigen on their erythrocytes.

Amplified introns from chimpanzee and the smaller form of gorilla were sequenced and compared to their human counterparts. The gorilla intron was cloned and found to be exactly the size of the human *RHD* (427 bp), while the chimpanzee intron was 439 bp. The human *RHCE* intron lacks a stretch of 654 bp, which contains an *Alu* Sx element that is present in the human *RHCE* intron. Interestingly, the chimpanzee intron contains an *Alu* Sx flanking sequence. The presence of such an insertion sequence suggests that the chimpanzee *RH*-like gene is derived from an ancestor gene which had a complete *Alu* Sx element. This element was probably deleted at the time of a rearrangement of the chimpanzee *RH*-like locus.

The Analysis of Genomic Sequences

Exon 1

Exon 1 was amplified by PCR using, as template, genomic DNA from various nonhuman primates including prosimians (lemurs). The PCR reactions were carried out with primers corresponding to the 5′ and 3′ ends of human exon 1. Amplified fragments were and sequenced. Homologues of human exon 1 were obtained from all nonhuman primates (chimpanzee, orangutan, baboon, squirrel monkey, marmoset) but not from lemurs. In the search for Rh-like proteins in the red cell membranes of New World monkeys, immunoblot experiments were performed with the MPC8 reagent (rabbit antibodies raised against the carboxylic end of the human Rh proteins) and MPCI reagent (rabbit antibodies raised against a synthetic peptide coresponding to positions 33–45 of human Rh-related proteins) (Hermand et al., 1993). The presence of Rh-like proteins in extracts from red cells of New World monkeys was demonstrated with MPC8 while MPC1 was totally unreactive. In fact, New World monkey (marmoset, squirrel monkey) exons 1 diverge from their human counterparts in the region encoding amino acid positions 33–45 so that five amino acid substitutions were observed (Apoil, unpublished). One can hypothesize that these substitutions are responsible for the absence of the MPC1 epitope in the New World monkeys Rh-proteins, while the conservation of the epitope recognized by the MPC8 reagent leads to suppose that the carboxylic end of the Rh-like proteins of the New World monkeys is homologous to its human counterpart.

Exon 7

Exon 7 was amplified, cloned and sequenced from genomic DNA of various nonhuman primates (Westhoff and Wylie 1996). Exon 7 sequences were obtained from seven chimpanzees and six gorillas. Failure to amplify exon 7 from DNA of gibbons and orangutan indicates that the nucleotide sequences at the primer regions of exon 7 are too different from those of humans for correct annealing. However, the use of human exon 7 as a probe of a Southern blot revealed the presence of homologues of exon 7 in genomic DNA of gibbons and orangutans

(Westhoof and Wylie 1996). Thus, the absence of exon 7 in the cDNA (Salvignol et al. 1995) is most likely related to the elimination of exon 7 during mRNA splicing (Westhoof and Wylie 1996).

Exon 7 is the region where the density of substitutions is highest between human *RHCE* and *RHD* genes. This is also true when comparison with human sequences is extended to all nonhuman primates. Interestingly, as stressed earlier, the part of the Rh protein encoded by exon 7 is essential for expression of the epitope recognized by the human monoclonal anti-D LOR15-C9. This antibody failed to agglutinate chimpanzee red cells but it defined in gorilla a polymorphic antigen called D^{gor} (Socha et al. 1993; Roubinet et al. 1993). In humans, as in chimpanzees and gorillas, the presence of the motif $D_{350}G_{353}A_{354}$ in the exon 7-encoded region is correlated with reactivity with LOR15-C9. Interestingly, the D^{gor} polymorphism was correlated with restriction fragment length polymorphisms (RFLPs) evidenced by Southern blots, using human exon 4 as probe. The RFLPs in gorilla were observed by means of three different restriction enzymes (*Hind*III, *Eco*RI, *Bam*HI). One could speculate that the gorilla alleles, which express the human RhD motif $D_{350}G_{353}A_{354}$, also contain modifications of restriction sites which result in these three RFLPs.

Study of Rh-Like Transcripts of Nonhuman Primates

Isolation of Rh-Like Transcripts

To characterize the structure of the *RH*-like genes of monkeys, the transcripts from bone marrow cells of each animal (three chimpanzees, one gorilla, one gibbon and one cynomolgus) were amplified by RT-PCR with oligonucleotide primers covering the untranslated 5' and 3' regions, which are common to the human RhD and RhCE cDNAs. Amplified products were obtained in all animals but varied slightly in size from one species to another. Transcripts from all chimpanzees and a gorilla were 1.3 kb, similar to the major human transcripts. Two amplified fragments of 1.4 and 1.2 kb were identified in the crab-eating macaque (*Macaca fascicularis*), whereas transcripts from the gibbon (*Hylobates pileatus*) were approximately 1.2 kb. Upon hybridization under stringent conditions with human Rh cDNA probes, the PCR fragments obtained from monkeys gave strongly positive signals, suggesting a significant sequence similarity with the human cDNAs.

Nucleotide Sequence Analysis of Rh-Like Transcripts

Transcripts from chimpanzee, gorilla, gibbon and crab-eating macaque were cloned and sequenced. The complete nucleotide sequences of these transcripts are available through the GeneBank under the access numbers L37048 (Ch. 211-

IIF), L37049 (Ch. 317-IA), L37048 (Ch. 317-IIR), L37048 (Gib.-IH), L37048 (Gor.-IC), L37048 (Gor.-ID), L37048 (Mafa-IIIE). All nonhuman primate sequences exhibit significant homology with the human Rh transcripts. Nucleotide positions that diverge from the human RhcE and RhD cDNA clones (RhIXb and RhXIII, respectively) were investigated by Salvignol et al. (1995).

Chimpanzee

Nine recombinant clones were sequenced from the three different chimpanzees. Five transcripts exhibited the same open reading frame of 1251 base pairs as found in the human Rh cDNAs (Fig. 10) and two represented isoforms lacking

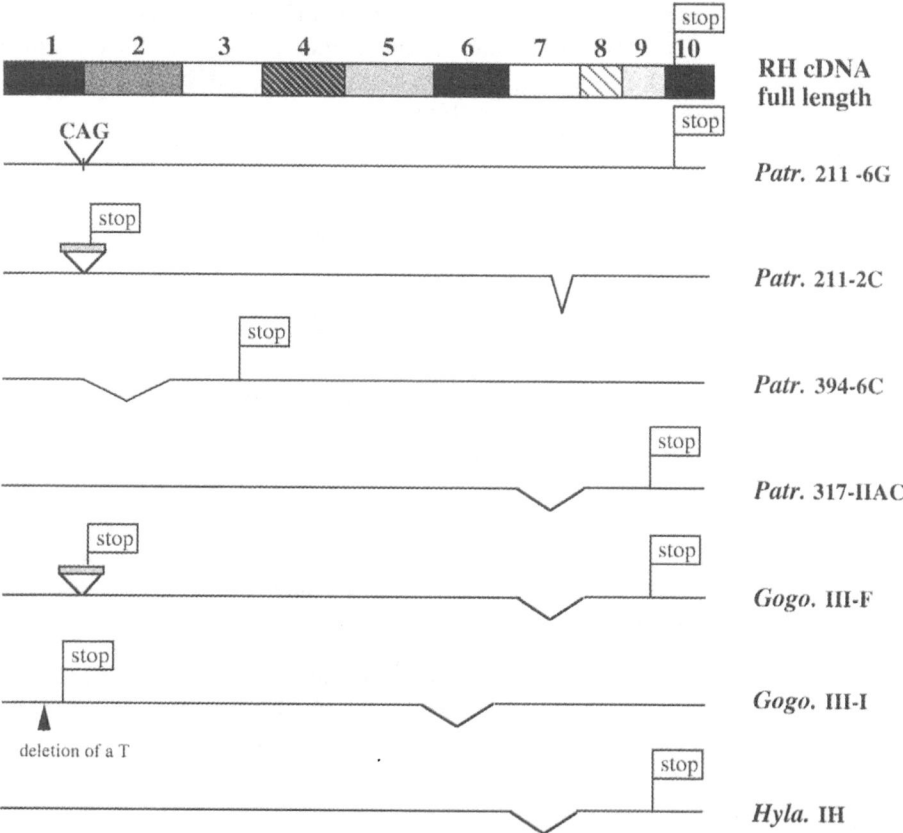

Fig. 10. Nonhuman primate Rh-like cDNA clones which exhibit addition or deletion of nucleotides when compared to the full length transcripts. *Patr., Pan troglodytes* (common chimpanzee); *Gogo, Gorilla gorilla* (gorilla); *Hyla, Hylobates lar* (gibbon); ▽, addition of 44 base pairs (see text)

sequences encoded by exon 2 (Ch.394–6C) and exon 7 (Ch.317-IIAC). These iso-forms may have resulted from alternative splicing. Another clone (Ch.211–6G) has an insert of a CAG triplet at the boundary of exons 1 and 2. Clone (Ch.211–2C) lacks 3' nucleotides from exon 7 (data not shown) and carries a 44 bp insert (GGGCTGGGAAGTCTGCATGCTGTCTGTAAATCCAGAACCAGAAG) at the boundary of exons 1 and 2, most probably the result of aberrant splicing of intron 1. PCR experiments demonstrated the presence of these 44 nucleotides in the genomic DNA of all the animals tested (ten different samples) as well as in humans (one sample). A highly homologous insert of 44 bp was found in humans in a RhcE cDNA clone (Blumenfeld, personal communication) and in a RhD cDNA clone (Chérif-Zahar et al. 1994a). The human and chimpanzee inserts dif-fer only by a base transition in the position immediately preceding the stop codon underlined in the sequence given above (G in chimpanzee, A in humans).

The 44 bp insertion and the deletion of exon 2 drastically alter the reading frame of the transcripts and generate premature stop codons which interrupt the translation, probably leading to unstable truncated proteins. The absence of exon 7 induces a frame shift which leads to a putative truncated protein of 354 residues with a COOH-terminal comparable to that encoded by the human Rh4 clone (Table 8). The Rh4 clone is a human cDNA corresponding to human transcripts with deletion of exon 7, which have been found in human reticulo-cytes (Le Van Kim et al. 1992a). The transcript with the CAG insertion potentially encodes for a protein with an additional residue (Gln) as compared to normal Rh proteins.

Comparison of all the full length transcripts led to the identification of a first group of five homologous sequences (referred to as type 1) among which the high-est level of divergence between two sequences was 13 substitutions (Ch.317-IIR vs Ch.317-IA), eight being in exon 7. Sequences that belong to type 1 were recovered from the three chimpanzees under study. The most similar sequences (Ch.211–6E and Ch.317-IIR) were derived from two animals. When all type 1 sequences were examined, 19 nucleotide positions appeared polymorphic and considered as result-ing from allelism at an RH-like locus. One sequence (Ch.211-IIF), which differs from the type 1 consensus sequences by at least 35 nucleotide positions, was con-sidered as the unique member of a second group referred to as type 2. The type 1 sequences are slightly closer to human RhD (divergence from 3.2 % to 3.6 %) than to human RhcE (divergence from 4.1 % to 4.7 %), while the type 2 sequence (Ch.211-IIF) exhibits equal divergences (3.6 %).

The number of substitutions between type 1 and type 2 sequences is compar-able to that observed between RHD and RHCE transcripts (Le Van Kim et al. 1992b), suggesting that they could correspond to transcripts of two distinct chimpanzee RH-like genes. The large number of type 1 sequences found in our experiments may reflect either a relative abundance of these transcripts in the bone marrow or a bias introduced by the PCR technique and the primers used in this study. Amplification and sequence analysis of genomic DNA fragments extending from exons 1 to 4 from these three chimpanzees (Salvignol et al. 1994) confirmed the transcript sequencing data and indicate that the genome of the chimpanzee carries at least two distinct homologous RH-like genes. The type 1 and type 2 transcripts identified here might be derived from two of these genes. However, the minimal number of *RH*-like genes, deduced from Southern blot analysis, was three (Salvignol et al. 1993, 1994). Thus, a third type of transcript might be expected, although it is not known whether all the postulated *RH*-like genes are transcribed or whether one of the aberrant RNA isoforms identified above may represent the product of a third gene.

Gorilla

Sequence analysis of four recombinant clones revealed the presence of transcripts exhibiting the same 1251 nucleotide open reading frame as in the human Rh cDNAs. In addition, two isoforms, lacking exon 6 (Gor.-III-I) and exon 7 (Gor.-III-F), were found (Fig. 10). These clones correspond to messengers that are most probably nonfunctional since the former has a nucleotide deletion between position 87 and 93 (deletion of a T in a stretch of Ts) and the latter carries a 44 bp insertion at the junction of exons 1 and 2. In both cases, the reading frames are shortened by the generation of a premature stop codon. Note that the sequence of the 44 bp insert was identical to that in the chimpanzee clone (Ch. 211–2C). Like the chimpanzee transcripts, the gorilla sequences have been classified into two groups. Type 1 sequences (Gor.-IC and Gor.-IIA2b) differ by two substitutions only, whereas type 2 sequences (Gor.-ID) differ from type 1 sequences by 31 substitutions. Both types might represent alleles at a given locus and are thought to represent the products of two distinct genes. This is compatible with the identification of two *RH*-like genes in gorilla, as deduced from genomic analysis by Southern blot analysis (Blancher et al. 1992a).

The level of substitutions in type 1 and human RhcE and RhD sequences is 4.6 % and 4.1 %, respectively, and for type 2 sequences, 4.4 % and 3 %. The highest degree of similarity is observed between chimpanzee type 2 and gorilla type 1 sequences.

Gibbon

The two transcripts isolated from a single animal (Gib.-IH and Gib.-IIE3) were shorter than those found in humans and both lacked nucleotides corresponding to exon 7. These transcripts diverge from one another by seven substitutions. In addition, one clone (Gib.-IIE3) has a nucleotide deletion between position 87 and 93 (stretch of Ts). Transcripts lacking exon 7 potentially encode for truncated proteins of 354 residues (see above). Excluding exon 7, the nucleotide sequence of the gibbon transcripts exhibited equivalent divergence (5.3 %) when compared to RhD or RhCcEe. Attempts to amplify exon 7 from gibbon DNA with human primers were unsuccessful. However Southern blot experiments demonstrated that the genomic DNA of gibbon contains a sequence homologous to human exon 7 (Westhoff and Wylie 1996). This suggests that exon 7 was eliminated from the transcripts by splicing.

Macaques

Two independent recombinant clones from a crab-eating macaque (*Macaca fascicularis*) were sequenced. Both exhibited the same open reading frame of 1251 nucleotides, but diverged by nine substitutions (7.2 %). The level of substitutions in these sequences vs human RhcE and RhD cDNAs was 8.3 % and 9.5 %, respectively. When compared to the sequence Mac.1, obtained from a rhesus monkey (*Macaca mulatta*) (Mouro et al. 1994a), the rates of divergence were extremely low (from 1.1 % to 1.3 %).

Membrane Topology of Rh-Like Proteins and Tentative Correlations Between Their Antigenic Properties and Their Amino Acid Sequences

Comparison of the nonhuman primate sequences with the known Rh sequences indicates a high degree of similarity. As expected, chimpanzee and gorilla sequences were closer to human sequences (95 % identity) than to macaque sequences (82 % identity).

Hydropathy profiles of the amino acid sequences of the Rh-like proteins of chimpanzee, gorilla, and macaque are similar to those of human Rh proteins. Several cysteine residues (Cys-12, -186, -285 and -315) were found in all human and nonhuman proteins. Importantly, Cys-285 is supposed to be externally exposed (loop 5), and in humans essential for expression of RhD and RhC antigens (Green 1965, 1967, 1983). The three cysteinyl residues, Cys-12, -186 and -311, are assumed to be positioned near the inner leaflet of the phospholipid bilayer and to belong to the motif Cys-Leu-Pro, which is most probably involved in linkage of fatty acids to Rh proteins by thioester bond formation (de Vetten and Agre 1988; Hartel-Schenk and Agre 1992). It is not known if these fatty acids play any role in the membrane anchorage or antigenic properties of Rh proteins, but they do display a high degree of conservation, as do the Cys-Leu-Pro motifs. Significantly, Cys-311 is substituted by a tyrosine in the human RhD polypeptide and the same substitution is observed in chimpanzee polypeptides.

Other cysteinyl residues (Cys-16, -311, -316), present in human proteins, are observed irregularly in nonhuman primates. The remaining cysteinyl residues (Cys-11, -51 and -178) although occasionally found in nonhuman primates are never present in human proteins.

The conformational model proposed for the human Rh proteins probably also applies to their nonhuman primate counterparts. Typically, the NH_2- and COOH-terminals of these proteins are intracellular and the proteins are predicted to traverse the membrane 12 times (Avent et al. 1992; Hermand et al. 1993; Eyers et al. 1994). There are six external and five internal loops between each pair of transmembrane domains (see Fig. 11). As the Rh antigens most probably mainly depend on these external loops, comparison of the sequences will be focused on these parts of the molecule. The predicted primary sequence of the external loops of Rh-like proteins from each animal species is presented in Fig. 11 in align-

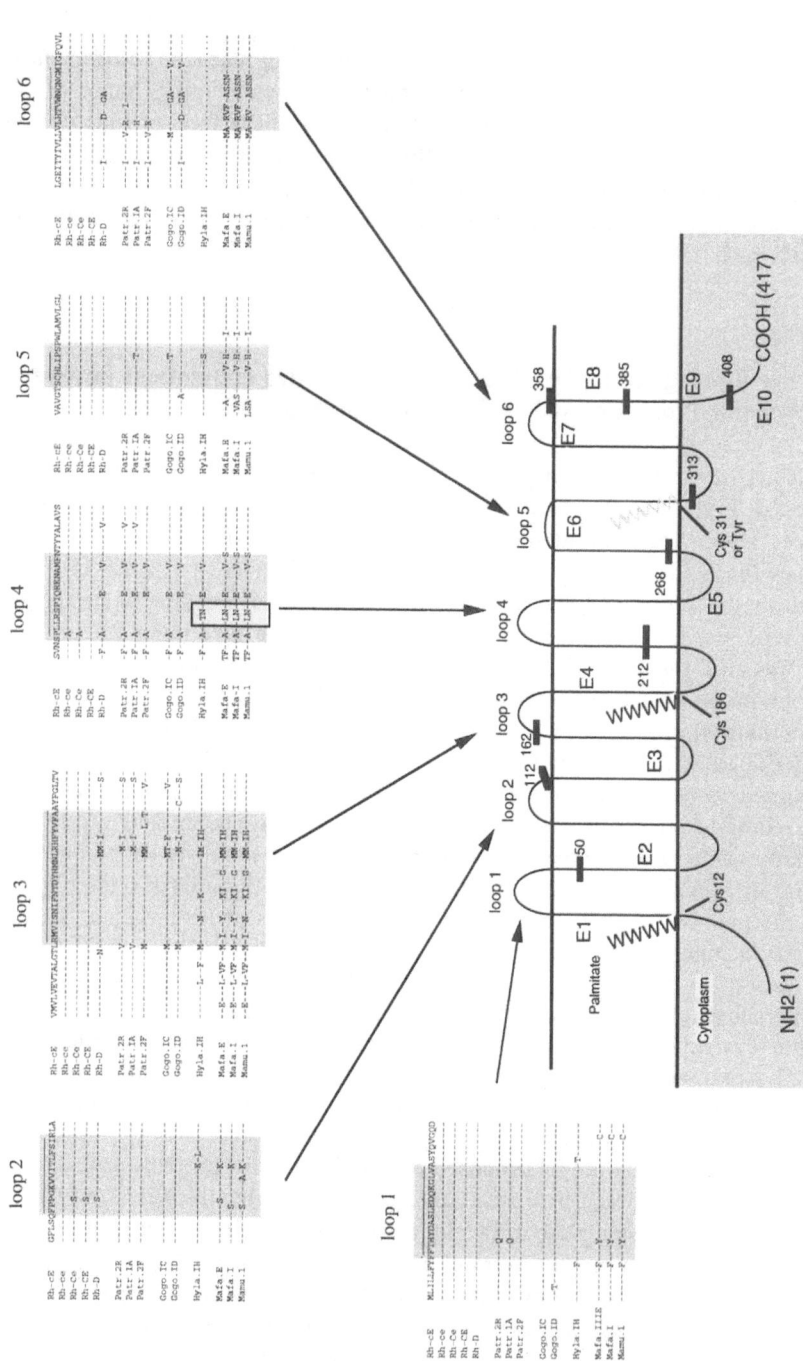

Fig. 11. Amino acid sequences of external loops of Rh polypeptides in humans and nonhuman primates. *Black bars,* limits of the exons; *black zig zag line,* palmitate linked to the cysteine (CLP motive); *gray zig zag line,* palmitate linked only to RHCE-encoded proteins

ment with the human RhD protein and the four allotypic proteins (RhcE, RhCe, RhCE, Rhce) encoded by the respective alleles of the *RHCE* gene (Mouro et al. 1993). Several motifs important for antigenic properties of Rh proteins were examined.

Blood Group D Epitopes

Human D differs from non-D proteins in 36 amino acid positions (Le Van Kim et al. 1992b), some corresponding to external hydrophilic loops predicted by the model (Fig. 4). Residues in these positions may be directly correlated with antigenic properties because they are accessible to antibodies. However, residues located outside the external loops may influence protein conformation over long distances and, therefore, may be essential for the expression of antigen D. Within those 35 positions that differentiate D from non-D, chimpanzee and gorilla proteins exhibit a higher degree of homology with the human D sequence (16–21 positions for chimpanzee; 16–18 for gorilla) than all other nonhuman primate species (10 for gibbon and 12–13 for macaques). Thus, it is not surprising that only chimpanzee and gorilla red cells cross-react type-specifically with human anti-D reagents.

Rh-like proteins of nonhuman primates share some residues which are, in humans, specific to RhD (Fig. 11). These residues can be divided into three categories:

(1) those that are observed exclusively in species (chimpanzee, gorilla) which may express D-like antigens (examples are positions L-121, D-128, V-245, Y-311, G-314, N-331, I-342, D-350, G-353); (2) those that are conserved in all nonhuman primates (examples are K -198, F-223, E-233, V238, and S-325); and (3) those that are inconsistently observed in species irrespective of D-like antigen expression (examples are V-127, M-159, M-160, I-162, S-182, T-201, G-353, and A-354).

In positions 152 and 267, all nonhuman primates share residues common to all human RHCE encoded proteins.

These observations confirm that, in most instances, D epitopes cannot be reduced to individual D polymorphic positions by themselves, but most likely represent conformation-dependent structures resulting from protein folding.

As mentioned before in this chapter, the D-specific motif D-350, G-353, A-354 (external loop 6) is absent from the RhD protein expressed by partial D of category IVb (Rouillac et al. 1995a). Red cells from chimpanzee also lack this motif and resemble, in serological studies, human D^{IVb} variants (Blancher et al. 1992b). In gorilla protein the motive D-350, G-353, A-354 is well conserved and gorilla red cells are agglutinated by antibodies unreactive with DIVb or chimpanzee red cells (Socha et al. 1993). This motif is certainly crucial in the constitution of the epitope recognized by the human monoclonal anti-D, LOR15-C9.

In the fourth external loop there are three residues specific to the RhD polypeptide: F-223, E-233, V-238. These are conserved in all nonhuman primates. However, in proteins from RhD-negative species (gibbon and both species of macaque) residues R-229 and S-230 are substituted by T-229 and N-230, or

L-229 and N-230. These substitutions, which are predicted to be in the middle of the fourth external loop, may restrict the capacity of gibbon and macaque Rh-like polypeptides to express those D epitopes related to this loop.

Blood Group C/c Epitopes

The C/c polymorphism in humans has been correlated with six nucleotide changes (nucleotides 48, 150, 178, 201, 203 and 307) resulting in only four amino acid substitutions encoded by exon 1 (residues 16) and exon 2 (residues 60, 68 and 103) of the *RHCE* gene (Mouro et al. 1993). When these positions were examined in the transcripts and predicted Rh-like proteins of monkeys, interesting variations were observed regarding the correlation with the blood group C/c specificities (Fig. 3).

The C to c polymorphism in humans is associated with amino acid substitutions C16W, I60L, S68N and S103P (Mouro et al. 1993), but, in c-positive monkey species, the protein substitutions are either CISP (chimpanzee, gorilla) or WLNP (gibbon). In addition, all macaque proteins that are typed as c-negative and C-negative have the combination WLNP at the respective positions. Examination of the human and nonhuman protein sequences reveals that all C-positive and c-positive proteins carry the dipeptide PS or PP, respectively, at positions 102–103. Proteins from individuals unreactive with anti-c or anti-C sera exhibit either the PS (human RhD protein) or SP (macaque proteins) motifs. This demonstrates that the genetic basis of the C/c polymorphism is strictly correlated with amino acid substitutions at position 103 exclusively and that the C or c epitopes are correctly presented to anti-C/c reagents only when a proline residue is present at position 102. Our results support the recent findings that nucleotide changes at position 48 (encoding residue 16) were not strictly correlated to C/c polymorphism (Hyland et al. 1994a).

Blood Group E/e Epitopes

The E/e polymorphism in humans has been clearly associated with a polymorphism at position 226 (see the above description of the fourth external loop and also the description of human Rh proteins). A proline or an alanine at this position corresponds to the expression of the antigens E or e, respectively (Mouro et al. 1993). In all nonhuman Rh-like proteins as well as in the human RhD protein, position 226 is occupied by an alanine. Since all nonhuman primates are e-negative, as are human red cells which express only the RhD protein (D– phenotypes), an alanine residue in position 226 is not sufficient for expression of antigen e. Most probably, the expression of antigen e is not compatible with the human RhD-specific motif (F-223, E-233, V-238) which includes position 226 and which is conserved in all nonhuman primate proteins.

The definitive proof of these findings will be obtained when recombinant Rh proteins are expressed at the surface of eukaryotic cells. However, all attempts have been unsuccessful to date (Hermand et al. 1993; Suyama et al. 1993), most likely because Rh proteins are present at the cell surface as a complex with other membrane proteins (see Cartron and Agre 1993; Anstee and Tanner 1993). However, key deductions were obtained from the present studies both on the sequence similarity of Rh proteins and on the structural basis of Rh blood group specificities.

Evolutionary Pathway of the *RH* Locus

Reconstruction of evolutionary pathways of the *RH* genes was based on comparison of cDNA sequences and the results obtained by Southern blot analysis. Phylogenetic reconstructions were performed using software kindly provided by Felsenstein (1995). A sample of sequences was analyzed that included human RhcE (as a representative of the set of sequences of *RHCE* alleles) and RhD together with examples of all the types of sequences observed in ape and monkey species. The genetic distances were calculated by means of the two-parameter method of Kimura (1980), and the matrix of genetic distances was used for phylogenetic reconstruction by the neighbor joining method (Saitou and Nei 1987). Consistency of the resulting tree was tested by the bootstrap method (1000 random sets of data were generated), followed by the reconstruction of trees by neighbor joining, and the results were published (Salvignol et al. 1995). A similar tree was obtained by the maximum likelihood method (see Fig. 12), and one of the two trees obtained by the parsimony method (314 mutations) was found to be identical with that produced by maximum likelihood, and the second one very similar.

Whichever the method used for reconstruction of the phylogenetic trees, all chimpanzee and gorilla sequences cluster together with the human RhD sequence. It is noteworthy that one chimpanzee (animal # 211) showed two RhcDNA sequences: one closer to gorilla than to human RhD, the other closer to RhD. This is compatible with the presence in chimpanzee of two types of genes (or alleles), one resembling RhD, the other being closer to gorilla sequences. This could be interpreted as preservation of some genes (or alleles) throughout the speciation of apes. The human RhcE sequence lies on a branch distinctly different from the master branch which carries human RhD together with chimpanzee and gorilla sequences. This suggests the ancientness of the separation of *RHCE* from the ancestor of chimpanzee and gorilla *RHD* sequences. The distances calculated for gibbon and macaque sequences are in accordance with current knowledge of the evolution of primates.

Genomic DNA analysis by Southern blot with human Rh probes indicated that the number of *RH*-like genes varied from one monkey species to another (Fig. 8). Gorilla and chimpanzee, which express analogues of blood factor D, carry two and three *RH*-like genes, respectively. Other monkey species, such as orangutan, gibbon as well as all Old and New World monkeys have a single *RH*-like gene (Blancher et al. 1992a; Salvignol 1993). However, most rhesus mon-

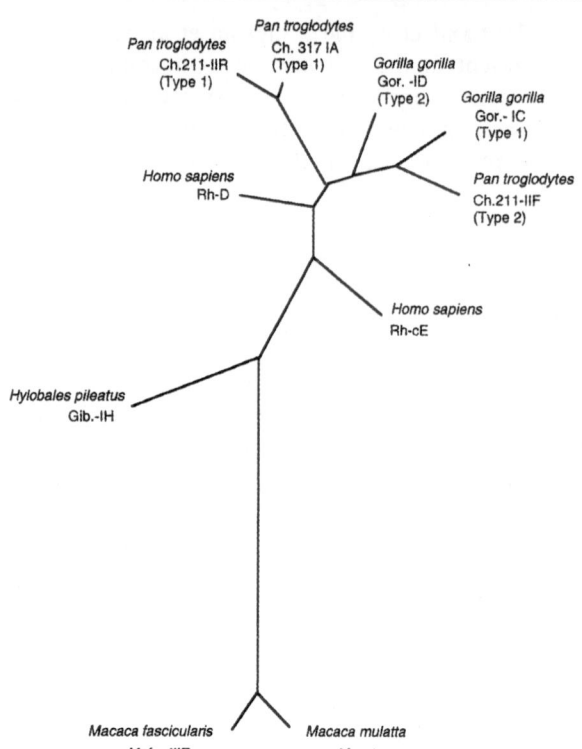

Fig. 12. Unrooted phylogenetic tree obtained by the maximum likelihood method as described by Felsenstein (1981)

keys and one gorilla also carry a small Rh-related 5'-fragment, which corresponds to a truncated *RH* gene that may be a remnant of a second ancestral gene (Fig. 8). Allowing for the small numbers of gibbons and orangutans examined, these findings suggest that duplication of the *RH* genes occurred relatively recently, between 8 and 11 million years ago (Fig. 8). The Southern blot technique recognized two types of *RH* locus: one with a single *RH* gene (*RHCE*), and the other with two tandem-linked genes, *RHD* and *RHCE*. Taking into account the probable ancientness of the *RHCE* gene, suggested by sequence comparison (see above), it appears more likely that the locus with the single *RH* gene resulted from deletion of the *RHD*.

Genomic DNA analysis confirmed close relationships among the human RH, chimpanzee R-C-E-F, and gorilla D^{gor} systems. However, unlike the situation in RhD-negative humans, none of the chimpanzees, irrespective of the R-C-E-F type, showed deletion of one of the *RH*-like genes. Thus, the R^c-negative phenotype of chimpanzees does not result from the absence of the R^c gene, but depends on another, as yet uncharacterized, polymorphism.

All the above observations support the following hypothesis: the common ancestor of humans, chimpanzees and gorillas had two *RH*-related genes. One gene encoded for an early type of D polypeptide, and derivatives of that gene encoded for the D in humans and the R^c in chimpanzees, and D^{gor} in gorillas. The second gene persisted in humans only as the *RHCE* gene. In chimpanzees and gorillas, duplication of one of the two genes of the anthropoid common ancestor led to C-E-F polymorphism. Whichever mechanism led to the diversification of the *RH* genes, it resulted in polymorphisms (D±, R^c ± and D^{gor} ±) in humans, chimpanzees and gorillas which were stabilized during evolution. At present, it is unknown whether the same mechanisms maintained these three polymorphisms; however, it seems plausible that a common factor of selection played a role.

Finally, it is important to recall here that exchange of gene fragments (most probably by gene conversion) between the two human *RH* genes led to the appearance of most of the rare Rh variants. The frequency of such events in humans suggests that they could have also played an essential role during the evolution of *RH* genes in the common ancestor of African anthropoid apes. This latter hypothesis is supported, as we mentioned before, by the probable existence of two *RH* genes in the common ancestor of humans and anthropoid apes. If multiple exchanges took place in the course of evolution, conclusions drawn from the comparison of sequences must be taken with caution, because each sequence obtained from living animals could be a mosaic of genes.

Conclusion

Despite the considerable amount of information available the functions of Rh-related proteins remain elusive. The membrane anomalies observed in the rare individuals lacking Rh proteins (Rh_{null}) suggest that these proteins are involved in the maintenance of red cell membrane integrity. However, Rh_{null} anomalies may result from multiple defects, as indicated by observations that Rh_{null} red cells lack not only Rh proteins, but also express very weakly, or not at all, LW, CD47, GPB, and Rh 50 glycoprotein. Moreover, an interaction with Duffy proteins is suggested by the absence of the Fy5 antigen in Rh_{null} phenotypes (see above and, for more details, Cartron and Agre 1995). Thus, dysfunction of the Rh_{null} red cell membrane seems to be related to the defect in the Rh protein complex rather than to the absence of Rh proteins.

Without knowing the exact functions of the Rh complex and of the Rh molecules, it is impossible to identify the mechanisms of selection that ensured the maintenance of *RH* genes throughout primate evolution. The same is true of the mechanisms that maintained Rh polymorphism in humans, chimpanzees, gorillas and orangutans. It was statistically documented that fetomaternal incompatibility could exert some selective pressure, the immunization of Rh-negative mothers favoring the survival of Rh-negative fetuses (Socha 1966). Comparable effects were observed in chimpanzees and orangutans (Socha 1993b). However, the selective pressure due to fetomaternal incompatibility is very low, and alone could not explain the maintenance of the Rh D+/D- polymorphism in humans and other polymorphisms observed in nonhuman primates.

The variations of the *RH* gene frequencies observed among human populations were probably the result of genetic drift or other, local environmental factors or a combination of both. Extreme examples of such variations are the Basques (with frequency of *d* gene surpassing 50 %) and the Japaneese (frequency of *d* varying between 0 % and 20 %) (Mourant et al. 1976; Bernard and Ruffié 1966, 1972).

In the absence of any clearly defined vital function, one can assume that the *RH* locus has accumulated a number of alleles that encode for highly variable proteins. The amino acid substitutions characteristic of these variants are neutral (without any detectable influence on the survival of the individuals bearing the resulting molecules). This may explain why the RH and RCEF systems are the most polymorphic erythrocytic blood group systems in humans and chimpanzee, respectively. The presence of two linked homologous *RH* genes (*RHD* and *RHCE*) favored recombination processes which were the main source of most of the variants so far characterized. However, other mechanisms of gene diversification (point mutations, deletion, silencing mutations, inequal crossing-over, mutations of the splicing consensus, etc.) have certainly been implicated in diversification of the *RH* genes in humans and nonhuman primates. Other examples of similar mechanisms will be found in the chapters on MNSS and MHC genes.

Acknowledgements. We thank Professor Jacques Ruffié, Director of the Laboratory of Physical Anthropology, Collège de France, Paris, for his continuous encouragement and advice. We are indebted to Professor Jean Ducos, Dr. Yves Colin, Dr. Francis Roubinet and Dr. Pol-André Apoil who generously agreed to read this text and offered precious advice. We thank Christel Cherry for technical assistance in preparation of the manuscript.

Table 11. Amino acid residues at positions 102 and 103 of the Rh polypeptides, relationship with the expression of C, c and G

Type of sequence	Amino acid positions[a]					Reactions with		
	16	60	68	102	103[b]	Anti-c	Anti-C	Anti-G
Human c	W[c]	L	N	P	P	+	−	−
Human C	Cys	I	S	P	S	−	+	+
Human D	W	I	S	P	S	−	−	+
Chimpanzee	Cys	I	S	P	P	+	−	−
Gorilla	Cys	I	S	P	P	+	−	−
Gibbon	W	L	N	S	P	−	−	−
Crab-eating macaque	W	L	N	S	P	−	−	−
Rhesus monkey	W	L	N	S	P	−	−	−

[a] Amino acid residues at positions 16, 60, 68, and 103 differ between human porteins carrying antigens C or c.

[b] Amino acid residues at positions 102 and 103 are determinants for the expression of antigens C and G.

[c] Amino acid residues are given by single-letter code. To avoid confusion with antigen RhC, cysteine is encoded by Cys.

This work was carried out with funds from the Collège de France (Paris, France), Agence Française du Sang (contrat de recherche 9500173123), Ministère de la Recherche (contrat J. E. 1966) and private funds from the Centre Régional de Transfusion Sanguine de Toulouse. We thank Johnson and Johnson, Eurobio, and Diamed for their financial support.

References

Agre P, Cartron J-P. Molecular biology of the Rh antigens. Blood 78(3):551–563, 1991

Agre P, Cartron J-P. The molecular biology of the Rh antigens. In: Agre P.C, Cartron J-P (eds) Protein blood group antigens of the red cells. John Hopkins University Press, Baltimore, pp 20–52, 1992

Agre P, Saboori AM, Asimos A, Smith BL. Purification and partial characterization of the M_r 30,000 integral membrane protein associated with the erythrocyte Rh(D) antigen. J Biol Chem 262:17497, 1987

Allen FH, Tippett PA. A new Rh blood type which reveals the Rh antigen G. Vox Sang 3:321–330, 1958

Anstee DJ, Tanner MJA. Biochemical aspects of the blood group Rh (Rhesus) antigens. In: Tanner MJA, Anstee DJ (eds) Red cell membrane and red cell antigens, vol 6, no 2. pp 402–422 (Clinical haematology) , 1993

Arce MA, Thompson ES, Wagner S, Coyne KE, Ferdman BA, Lublin DM. Molecular cloning of RhD cDNA derived from a gene present in RhD-positive, but not RhD-negative individuals. Blood 82:651–655, 1993

Avent ND, Ridgwell K, Mawby WJ, Tanner MJA, Anstee DJ, Kumpel B. Protein-sequence studies on Rh-related polypeptides suggest the presence of at least two groups of proteins which are associated in the human red-cell membrane. Biochem J 256:1043–1046, 1988a

Avent ND, Judson PA, Parsons SF, Mallinson G, Anstee DJ, Tanner MJA, Evans PR, Hodges E, Macivers AG, Holmes C. Monoclonal antibodies that recognize different membrane proteins that are deficient in Rh_{null} human erythrocytes. Biochem J 251:499–505, 1988b

Avent ND, Ridgwell K, Tanner MJA, Anstee DJ. cDNA cloning of a 30 kDa erythrocyte membrane protein associated with Rh (Rhesus)-blood-group-antigen expression. Biochem J 271:821–825, 1990

Avent ND, Butcher SK, Liu W, Mawby WJ, Mallinson G, Parsons SF, Anstee DJ, Tanner MJA. Localization of the C termini of the Rh (Rhesus) polypeptides to the cytoplasmic face of the human erythrocyte membrane. J Biol Chem 267:15134–15139, 1992

Bailly P, Hermand P, Callebaut I, Sonneborn HH, Khamlichi S, Mornon JP, Cartron J-P. The LW blood group glycoprotein is homologous to intercellular adhesion molecules. Proc Natl Acad Sci USA 91:5306, 1994

Bailly P, Tontti E, Hermand P, Cartron J-P, Gahmberg CG. The red cell LW blood group protein is an intercellular adhesion molecule which binds to CD11/CD18 leukocyte integrins. Eur J Immunol 25:3316–3320, 1995

Basu MK, Flamm M, Schacter D, Bertles JF, Maniatis T. Effects of modulating erythrocyte membrane cholesterol. Biochem Biophys Res Commun 95:887, 1980

Beckers EAM, Faas BHW, Simsek S, Overbeeke MAM, Van Rhenen DJ, Wallace M, Von Dem Borne AEGK, Van Der Schoot CE. The genetic basis of a new partial D antigen: D^{DBT}. Br J Haematol 93:720–727, 1996a

Beckers EAM, Faas BHW, Von Dem Borne AEGK, Overbeeke MAM, Van Rhenen DJ, Van Der Schoot CE. The R_o^{Har} Rh:33 phenotype results from substitution of exon 5 of the

RHCE gene by the corresponding exon of the RHD gene. Br J Haematol 92:751–757, 1996b

Beckers E, Faas B, von dem Borne A, Overbeeke M, van Rhenen D, van der Schoot C. Rh rearrangements leading to D^{IIIc}, DBT, R_o^{Har}. Transfusion 35 [Suppl] (abstr S201), 1995

Bernard J, Ruffié J. Ecologie humaine, caractères héréditaires du sang. Masson, Paris (Hématologie géographique, vol 1), 1966

Bernard J, Ruffié J. Variations hématologiques acquises: l'hématologie et l'évolution. Masson, Paris (Hématologie géographique, vol 2) , 1972

Blanchard D, Bloy C, Hermand P, Cartron J-P, Saboori A, Smith SL, Agre P. Two-dimensional iodopeptide mapping demonstrates erythrocyte Rh D, c, and E polypeptides are structurally homologous but nonidentical. Blood 72:1424, 1988

Blancher A, Roubinet F, Oksman F, Ternynck T, Broly H, Chevaleyre J, Vezon G, Ducos J. Polyreactivity of human monoclonal antibodies: human anti-Rh monoclonal antibodies of IgM isotype are frequently polyreactive. Vox Sang 61:196–204, 1991

Blancher A, Calvas P, Ruffié J. Etude des équivalents des antigènes Rhésus chez les primates non hominiens. C R Soc Biol 186:682–695, 1992a

Blancher A, Socha W. W, Ruffié J. Diversity of human anti-D monoclonal antibodies revealed by reactions with chimpanzee red blood cells. Vox Sang 63:112–118, 1992b

Blancher A, Clayton J, Socha WW, Ruffié J. Application of maximum likelihood statistics for the population and family studies of inheritance of the chimpanzee R-C-E-F and V-A-B-D blood groups. J Med Primatol 22:44–49, 1993a

Blancher A, Ruffié J, Socha WW., 1993b) The R-C-E-F blood group system of chimpanzee: serology and genetics. J Med Primatol 22:13–18

Blancher A, Reid M, Alié-Daram S. J, Dugoujon J M, Roubinet F.., 1996) Characterization of a human anti-hr^B-like monoclonal antibody. Immunohematology 12:119–121, 1996

Bloy C, Blanchard D, Lambin P, Goossens D, Rouger P, Salmon C, Cartron J-P., 1987) Human monoclonal antibody against Rh(D) antigen: partial characterization of rh Rh(D) polypeptide from human erythrocytes. Blood 69:1491

Bloy C, Blanchard D, Dahr W, Beyreuther K, Salmon C, Cartron J-P. Determination of the N-terminal sequence of human red cell Rh(D) polypeptide and demonstration that the Rh(D), (c) and (E) antigens are carried by distinct polypeptide chains. Blood 72:661, 1988a

Bloy C, Blanchard D, Lambin P, Goossens D, Rouger P, Salmon C, Masouredis SP, Cartron J-P. Characterization of the C, c, E, and G antigens of the Rh blood group system with human monoclonal antibodies. Mol Immunol 25:925, 1988b

Bloy C, Blanchard D, Hermand P, Kordowicz M, Sonneborn HH, Cartron J-P. Properties of the blood group LW glycoprotein and preliminary comparison with Rh proteins. Mol Immunol 26:1013–1019, 1989

Bloy C, Hermand P, Cherif-Zahar B, Sonneborn H, Cartron J-P. Comparative analysis by two-dimensional iodopeptide mapping of the RhD protein and LW glycoprotein. Blood 75:2245, 1990

Blunt T, Daniels G, Carritt B. Serotype switching in a partially deleted gene. Vox Sang 67:397–401, 1994

Bourel D, Lecointre M, Genetet N, Guiguen-Duchesne M, Genetet B. Murine monoclonal antibody suitable for use as an Rh reagent anti-e. Vox Sang 52:85–88, 1987

Boylston AN, Gardner B, Anderson RL, Hughes-Jones NC. Production of human IgM anti-D in tissue culture by EB-virus transformed lymphocytes. Scand J Immunol 12:355–358, 1980

Brown PJ, Evans JP, Sinor LT, Tilzer LT, Plapp FV. The Rhesus D antigen: a dicyclohexylcarbodiimide-binding proteolipid. Am J Pathol 110(2):127–134, 1983

Calvas P, Blancher A, Depétris D, Salvignol I, Chérif-Zahar B, Mattéi MG. Chimpanzee Rh-like blood group genes map to chromosome region (1p34.2–1p36.1) by in situ hybridization. Cytogenet Cell Genet 65:247–249, 1994

Cartron J-P. Defining the Rh blood group antigens. Blood Rev 8:199–212, 1994

Cartron J-P, Agre P. Rh blood group antigens: protein and gene structure. Semin Hematol 30(3):193–208, 1993

Cartron J-P, Agre P. Rh blood groups and Rh-deficiency syndrome. In: Cartron J-P, Rouger P (eds) Molecular basis of human blood group antigens. Plenum, NewYork (Blood cell biochemistry, vol 6) , 1995

Ceppellini R, Nasso S, Tecilazich F. (eds) La malattia emolitica del Milano, 1952

Ceppellini R, Dunn LC, Turri M. An interaction between alleles at the Rh locus in man which weakens the reactivity of the Rh_o factor (D^u). Proc Natl Acad Sci USA 41:283–288, 1955

Chérif-Zahar B, Bloy C, Le Van Kim C, Blanchard D, Bailly P, Hermand P, Salmon C, Cartron J-P, Colin Y. Molecular cloning and protein structure of a human blood group Rh polypeptide. Proc Natl Acad Sci USA 87:6243–6247, 1990

Chérif-Zahar B, Mattéi MG, Le Van Kim C, Bailly P, Cartron JP, Colin Y. Localisation of the human Rh blood group gene structure to chromosome 1p34.3–1p36.1 region by in situ hybridization. Hum Genet 86:398–400, 1991

Chérif-Zahar B, Le Van Kim C, Rouillac C, Raynal V, Cartron JP, Colin Y. Organization of the gene encoding the human blood group RhCcEe antigens and characterization of the promoter region. Genomics 19:68–74, 1994a

Chérif-Zahar B, Raynal VC, D'Ambrosio AA, Cartron JP, Colin Y. Molecular analysis of the structure and expression of the RH locus in individuals with D– – Dc– DC^W– gene complexes. Blood 12:4354–4360, 1994b

Chérif-Zahar B, Raynal V, Gane P, Mattei MG, Bailly P, Gibbs B, Colin Y, Cartron J-P. Candidate gene acting as a suppressor of the RH locus in most cases of Rh-deficiency. Nature Genet 12:168–173, 1996a

Chérif-Zahar B, Raynal VC, Cartron JP. Lack of RHCE encoded proteins in the D– phenotype may result from homolgous recmbination between the two RH genes. Blood (in press), 1996b

Colin Y, Chérif-Zahar B, Le Van Kim C, Raynal V, Van Huffel V, Cartron J-P. Genetic basis of the RhD-positive and RhD-negative blood group polymorphism as determined by Southern analysis. Blood 78:2747, 1991

Colin Y, Cherif-Zahar B, Le Van Kim C, Mouro I, Cartron J-P. Recent advances in molecular and genetic analysis of Rh blood group structures. J Med Primatol 22:36–43, 1993

Coombs RRA, Mourant AE, Race RR. Detection of weak and incomplete Rh agglutinins: a new test. Lancet ii:15, 1945)

Coombs RRA, Mourant AE, Race RR. In vivo isosensitization of red cells in babies with haemolytic disease. Lancet i:264–266, 1946

Crawford DH, Barlow MJ, Harrisson JF, Winger L, Huehns ER. Production of human monoclonal antibody to Rhesus D antigen. Lancet i:386–388, 1983

Daniels G, Lomas C, Wallace M, Tippett P. Epitopes of RhD: serology and molecular genetics. In: Sylberstein LE (ed) Molecular and functional aspects of blood group antigens. American Association of Blood Banks, Bethesda, pp 193–228, 1995

de Vetten MP, Agre P. The Rh polypeptide is a major fatty acid acylated erythrocyte membrane protein. J Biol Chem 263:18193–18196, 1988

Diamond LK, Denton RL. Rh agglutination in various media with particular reference to the value of albumin. J Lab Clin Med 30:821–830, 1945

Dugoujon JM, de Lange GG, Blancher A, Alie-Daram S, Marty Y. Characterization of an IgG2 G2m(23) anti-Rh-D antibody. Vox Sang 57:133–136, 1989

Elder JH, Pickett RA, Hampton H, Lerner RA. Radioiodination of proteins in single polyacrylamide gel slices. J Biol Chem 252:6510, 1977

Eyers SAC, Ridgwell K, Mawby WJ, Tanner MJA. Topology and organization of human Rh (Rhesus) blood group-related polypeptides. J Biol Chem 269 9:6417–6423, 1994

Faas B, Beckers E, Simek S, Overbeeke M, Cuijpers H, Pepper R, van Rhenen D, von dem Borne A, van der Schoot C. Involvement of Ser 103 in the RH G epitope. Transfusion 35 [Suppl] (abstr S200), 1995

Felsenstein J. Evolutionary trees from DNA sequences: a maximum likelihood approach. J Mol Evol 17:368–376, 1981

Felsenstein J. PHYLIP: Phylogenesis inference package (version 3.57c) , 1995

Fraser RH, Inglis G, Allan JC, Murphy MT, Allan EK, Mackie A, Mitchell R. Murine monoclonal antibody with anti-e-like specificity: suitability for screening for e-negative cells. Transfusion 30(3):226–229, 1990

Gahmberg CG. Molecular characterization of the human red cell Rh_o (D) antigen. EMBO J 2:223, 1983

Ghamberg CG, Karhi KK. Association of Rh_o (D) polypeptides with the membrane skeleton in Rh_o (D)-positive human red cells. J Immunol 133:334, 1984

Green FA. Studies on the Rh(D) antigen. Vox Sang 10:32–53, 1965

Green FA. Erythrocyte membrane sulfhydryl groups and Rh antigen activity. Immunochemistry 4:247, 1967

Green FA. Phospholipid requirement for Rh antigen activity. J Biol Chem 243:5519, 1968

Green FA. Erythrocyte membrane lipids and Rh antigen activity. J Biol Chem 247:881, 1972

Green FA. The mode of attenuation of erythrocyte membrane Rh_o (D) antigen activity by 5,5'-dithiobis-(2-nitrobenzoic acid) and protection against loss of activity by bound anti-Rh_o (D) antibody. Mol Immunol 20:769–774, 1983a

Green FA, Owens NA, Hui HL, Jung CY, Cuppoletti J. Molecular size of the Rh_o(D) antigen of the human erythrocyte in situ by radiation inactivation. Mol Immunol 20(4):361–365, 1983b

Hartel-Schenk S, Agre P. Mammalian red cell membrane Rh polypeptides are selectively palmitoylated subunits of a macromolecular complex. J Biol Chem 267:5569–5574, 1992

Helenius A, Simons K. Solubilization of membranes by detergents. Biochim Biophys Acta 415:29, 1975

Hermand P, Mouro I, Huet M, Bloy C, Suyama K, Goldstein J, Cartron J-P, Bailly P. Immunochemical characterization of Rhesus proteins with antibodies raised against synthetic peptides. Blood 82:669–676, 1993

Hermand P, Gane P, Mattei MG, Sistonen P, Cartron J-P, Bailly P. Molecular basis and expression of the LW^a/LW^b blood group polymorphism. Blood 84:1590–1594, 1995

Hermand P, Le Pennec PY, Rouger P, Cartron J-P, Bailly P. Characterization of the gene encoding the human LW blood group protein in LW^+ and LW^- phenotypes. Blood 87(7):2962–2967, 1996

Huang CH. Alteration of RH gene structure and expression in human dCCee and DC^w-red cells: phenotypic homosygosity versus genotypic heterozygosity. Blood (in press), 1996

Huang CH, Reid ME, Chen Y. Identification of a partial internal deletion in the RH locus causing the human erythrocyte D- phenotype. Blood 86:784–790, 1995

Huang CH, Reid ME, Chen Y, Coghlan G, Okubo Y. Molecular definition of red cell Rh haplotypes by tightly linked SphI RFLPs. Am J Hum Genet 58 (in press) , 1996

Hughes-Jones NC, Green EJ, Hunt VA. Loss of Rh antigen activity following the action of phospholipase A_2 on red cell stroma. Vox Sang 29:184, 1975

Hyland CA, Wolter LC, Liew YW, Saul AA. Southern analysis of Rh blood group genes: association between restriction fragment length polymorphism patterns and Rh serotypes. Blood 83:566–572, 1994a

Hyland CA Wolter LC, Saul A. Three unrelated RhD gene polymorphisms identified among blood donors with Rhesus CCee (r'r') phenotypes. Blood 84:321–324, 1994b

Issitt PD. The Rh blood group system 1988: eight new antigens in nine years and some observations on the biochemistry and genetics of the system. Transfus Med Rev 3(1):1–12, 1989

Issitt PD. An invited review: the Rh antigen e its variants and some closely related serological observations. Immunohematologyume 7 2:29–36, 1991

Issitt PD. The Rh blood groups. In: Garratty G (ed) Immunobiology of transfusion medicine. Marcel Dekker, New York, pp 111–141, 1994

Jones J, Scott ML, Voak D. Monoclonal anti-D specificity and Rh D structure: criteria for
selection of monoclonal anti-D reagents for routine typing of patients and donors.
Transfus Med 5:171–184, 1995

Kajii E, Umeneshi F, Iwamoto S, Ikemoto S. Isolation of a new cDNA clone encoding an Rh
polypeptide associated with Rh blood group system. Hum Genet 91:157–162, 1993

Kajii E, Umenishi F, Omi T, Ikemoto S. Intricate combinatorial patterns of exon splicing
generate multiple Rh-related isoforms in human erythroid cells. Hum Genet 95:657–
665, 1995

Kimura M. A simple method for estimating evolutionary rates of base substitutions
through comparative studies of nucleotide sequences. J Mol Evol 16:111–120, 1980

Koskimies S. Human lymphoblastoid cell line producing specific antibody against Rh-anti-
gen D. Scand J Immunol 11:73–77, 1980

Krahmer M, Prohaska R. Characterization of human red cell Rhesus specific polypeptides
by limited proteolysis. FEBS Lett 226:105, 1987

Kumpel BM, Van de Winkel JGJ, Westerdaal NAC, Hadley AG, Dugoujon JM, Blancher A.
Antigen topography is critical for interaction of IgG2 anti-red cell antibodies with
Fcγ receptors. Br J Haematol (in press) , 1996

Lansteiner K, Wiener AS. An agglutinable factor in human blood recognized by immune
sera for blood. Proc Soc Exp Biol Med 43:223, 1940

Lansteiner K, Wiener AS. Studies of an agglutinogen (Rh) in human blood reacting with
anti-rheusus antisera and human isoantibodies J Exp Med 74:309–320, 1941

Le Van Kim C, Chérif-Zahar B, Raynal V, Mouro I Lopez M, Cartron J-P. Multiple Rh messenger RNA isoforms are produced by alternative splicing. Blood 80:1074–1078, 1992a

Le Van Kim C, Mouro I, Chérif-Zahar B, Raynal V, Cherier RC, Cartron JP, Colin Y. Molecular cloning and primary structure of the human blood group RhD polypeptide. Proc
Natl Acad Sci USA 89:10925–10929, 1992b

Leader KA, Kumpel BM, Poole GD, Kirkwood JT, Merry AH, Bradley BA. Human monoclonal anti-D with reactivity against category D^{VI} cells used in blood grouping and determination of the incidence of the category D^{VI} phenotype in the D^U population. Vox
Sang 58:106–111, 1990

Levine P, Celano MJ. Presence of "D-like" antigens on various monkey red blood cells.
Nature 193 4811:184–185, 1962

Levine P, Celano MJ. Agglutinating specificity of LW factor in Guinea pig and rabbit anti-
Rh serums. Science 156 3783:1744–1746, 1967

Levine P, Stetson RE. An unusual case of intragroup agglutination. JAMA 113:126–127, 1939

Levine P, Burnham L, Katzin EM, Vogel P. The role of isoimmunization in the pathogenesis
of erythroblastosis fetalis. Am J Obstet Gynecol 42:925–937, 1941

Levine P, Celano MJ, Fenichel R, Pollack W, Singher HA. "D-like" antigen in rhesus monkey human Rh positive and human Rh negative red blood cells. J Immunol 87:747–
752, 1961

Levine P, Cellano MJ, Wallace J, Sanger R. A human "D-like" antibody. Nature 198:596–597,
1963

Lomas C, Mougey R. Rh antigen D: variable expression in DVI phenotypes; a possible subdivision of category VI by a low frequency antigen (abstr). Transfusion 29
[Suppl]:14S, 1989

Lomas C, Tippett P, Thompson KM, Melamed MD, Hughes-Jones N. Demonstration of
seven epitopes on the Rh antigen D using monoclonal anti-D antibodies and red
cells from D categories. Vox Sang 57:261–264, 1989

Lomas C, McColl K, Tippett P. Further complexities of the Rh antigen D disclosed by testing category D^{II} cells with monoclonal anti-D. Transfus Med 3:67–69, 1993a

Lomas C, Tippett P, Mannessier L. Abolition of the D^{Vc} subcategory. Transfusion 33 6:535, 1993b

Mallinson G, Anstee DJ, Avent ND, Ridgwell K, Tanner MJA, Daniels GL, Tippett P, Von Dem Borne AEG. Murine monoclonal antibody MB-2D10 recognizes Rh-related glycoproteins in the human red cell membrane. Transfusion 30(3):222–225, 1990

Mallinson G, Martin PG, Anstee DJ, Tanner MJA, Merry AH, Tills D, Sonneborn HH. Identification and partial characterization of the human erythrocyte membrane component(s) that express the antigens of the LW blood-group system. Biochem J 234:649–652, 1986

Marsh WL, Chaganti RSK, Gardner FG, Mayer K, Nowell PC, German J. Mapping human autosomes: evidence supporting assignment of Rhesus to the short arm of chromosome n°1. Science 184:966–968, 1974

Martel-Shenk S, Agree P. Mammalian red cell membrane Rh polypeptides are selectively palmitoylated subunits of a macromolecular complex. J Biol Chem 267:5569–5574, 1992

Masouredis SP. Reaction of I-131 trace labeled human anti-Rh_0(D) with red cells. J Clin Invest 38:279–290, 1959

Masouredis SP, Dupuy ME, Elliot M. Distribution of the Rh (D) antigen in the red cells of non-human primates. J Immunol 98:8–16, 1967

Moor-Jankowski J, Wiener AS. Red cells of primates. In: Fiennes RNTW (ed) Pathology of simian primates. Part I: general pathology. Karger, Basel, pp 270–317, 1972

Moor-Jankowski J, Wiener AS, Socha WW, Gordon EB, Kaczera Z. Blood group homologues in orang-utans and gorillas of the human Rh-Hr and chimpanzee C-E-F systems. Folia Primatol (Basel) 19:360–367, 1973

Moore S, Green C. The identification of specific Rhesus polypeptide blood group ABH active glycoprotein complexes in the human red cell membrane. Biochem J 244:735–741, 1987

Moore S, Woodrow CF, McClelland DBL. Isolation of membrane components associated with human red cell antigens Rh_0(D) (c) (E) and Fy^a. Nature 295:529–531, 1982

Morton JA, Pickles M. Use of trypsin in the detection of incomplete anti-Rh antibodies. Nature 158:880, 1947

Mourant AE. A new rhesus antibody. Nature 155:542, 1945

Mourant AE, Kopec ADA, Domaniewska-Sobczak K. The distribution of the human blood groups and other polymorphisms, 2nd edn. Oxford University Press, Oxford, 1976

Mouro I, Colin Y, Chérif-Zahar B, Cartron JP, Le Van Kim C. Molecular genetic basis of the human rhesus blood group system. Nature Genet 5:62–65, 1993

Mouro I, Le Van Kim C, Chérif-Zahar B, Salvignol I, Blancher A, Cartron J-P, Colin Y. Molecular characterization of the Rh-like locus and gene transcripts from the rhesus monkey (Macaca mulatta). J Mol Evol 38:169–176, 1994a

Mouro I, Le Van Kim C, Rouillac C, Van Rhenen DJ, Le Pennec PY, Cartron J-P, Colin Y. Rearrangements of the blood group RhD gene associated with the D^{VI} category phenotype. Blood 83:1129–1135, 1994b

Mouro I, Colin Y, Petti Sistonen, Le Pennec PY, Cartron JP, Le Van Kim C. Molecular basis of the RhC^W (Rh8) and RhC^X (Rh9) blood group specificities. Blood 83:1196–1201, 1995

Mouro I, Colin Y, Gane P, Collec E, Zelinski T, Cartron JP. Le Van Kim C. Molecular analysis of blood group Rh transcripts from a $r^G r$ variant. Br J Haematol 93:472–474, 1996

Nash R, Shojania AM. Hematological aspect of Rh deficiency syndrome: a case report and review of the literature. Am J Hematol 24:267–275, 1987

Owens NA, Hui HL, Green FA. The binding characteristics of anti-Rh_0(D)-positive human and chimpanzee erythrocytes measured with ^{14}C protein A. J Immunol 129(4):1471–1473, 1982

Race RR. An incomplete antibody in human serum. Nature 153:171, 1944

Race RR. The Rh genotypes and Fisher's theory. Blood 3 [Suppl 2]:27–42, 1948

Race RR, Sanger R. Blood groups in man, 6th edn. Blackwell, Oxford, 1975

Race RR, Taylor GL, Ikin EW, Prior AM. The inheritance of the allelomorphs of the Rh gene in fifty-six families. Ann Eugen Lond 12:206–210, 1944

Race RR, Sanger R, Selwyn JG. A possible deletion in a human Rh chromosome: a serological and genetical study. Br J Exp Pathol 32:124–135, 1951

Renton PH, Stratton F. Rhesus type D^u. Ann Eugen Lond 15:189–209, 1950

Ridgwell K, Roberts SJ, Tanner MJA, Anstee DJ. Absence of two membrane proteins containing extracellular thiol groups in Rh_{null} human erythrocytes. Biochem J 213:267, 1983

Ridgwell K, Tanner MJA, Anstee DJ. The Rhesus (D) polypeptide is linked to the human erythrocyte cytoskeleton. FEBS Lett 174:7, 1984

Ridgwell K, Spurr NK, Laguda B, MacGeoch C, Avent ND, Tanner MJA. Isolation of cDNA clones for a 50 kDa erythrocyte membrane glycoprotein of the human erythrocyte membrane associated with Rh (Rhesus)-blood-group-antigen expression. Biochem J 287:223–228, 1992

Ridgwell K, Eyers SAC, Mawby WJ, Anstee DJ, Tanner MJA. Study of the glycoprotein associated with Rh(Rhesus) blood group antigen expression in the human red blood cell membrane. J Biol Chem 269(2):6410–6416, 1994

Rosenfield RE, Allen FH, Swisher SN, Kochwa S. A review of Rh serology and presentation of a new terminology. Transfusion Phild 2:287–312, 1962

Rosenfield RE, Allen FH, Rubinstein P. Genetic model for the Rh blood-group system. Proc Natl Acad Sci USA 70(5):1303–1307, 1973

Roubinet F, Blancher A, Socha WW, Ruffié J. Quantitative study of chimpanzee and gorilla counterparts of the human D antigen. J Med Primatol 22:29–35, 1993

Rouger P, Edelman L. Murine monoclonal antibodies associated with Rh17 Rh29 and Rh46 antigens. Transfusion 28:52–55, 1988

Rouger P, Goossens D. Human monoclonal antibodies against human red blood cells. In: Borrebaeck C, Larrick J (eds) Therapeutic monoclonal antibodies. Stockton, Basingstoke, pp 263–286, 1990

Rouillac C, Colin Y, Hughes-Jones NC, Beolet M, D'Ambrosio AM, Cartron J-P, Le Van Kim C. Transcript analysis of D category phenotypes predicts hybrid Rh D-CE-D proteins associated with alteration of D epitopes. Blood 85(10):2937–2944, 1995a

Rouillac C, Le Van Kim C, Beolet M, Cartron J-P, Colin Y. Leu110Pro substitution in the RhD polypeptide is responsible for the D^{VII} category blood group phenotype. Am J Hematol 49:87–88, 1995b

Rouillac C, Le Van Kim C, Blancher A, Roubinet F, Cartron J-P, Colin Y. Lack of G blood group antigen in D^{IIIb} erythrocytes is associated with segmental DNA exchange between RH genes. Br J Haematol 89:424–426, 1995c

Rouillac C, Gane P, Cartron J, Le Pennec PY, Cartron J-P, Colin Y. Molecular basis of the altered antigenic expression of RhD in weak D (D^u) and RhC/e in R^N phenotypes. Blood 87(11) (in press) , 1996

Saboori AM, Smith BL, Agre P. Polymorphism in the M_r 32,000 Rh protein purified from Rh(D) positive and negative erythrocytes. Proc Natl Acad Sci USA 85:4042, 1988

Saboori A, Denker BM, Agre P. Isolation of proteins related to the Rh polypeptides from non-human erythrocytes. J Clin Invest 83:187–191, 1989

Sacks MS, Wiener AS, Jahn EF, Spurling CL, Unger LJ. Isosensitization to a new blood factor Rh^D with special reference to its clinical importance. Ann Int Med 51:740, 1959

Saiki RK, Gelfand DH, Stoffel S, Scharf SJ, Higuchi R, Horn GT, Mullis KB, Erlich HA. Primer-directed enzymatic amplification of DNA with a thermostable DNA polymerase. Science 239:487–491, 1988

Saitou N, Nei M. The neighbor joining method: a new method for reconstructing phylogenetic trees. Mol Biol Evol 4:406–425, 1987

Salmon C, Gerbal A, Liberge G, Sy B, Tippett P, Sanger R. Le complexe génique D^{IV} (C)-. Rev Fr Transfus 12:239–247, 1969

Salvignol I. Les gènes Rh-like chez les primates nonhumains. Doctoral thesis, Université Paul Sabatier Toulouse, 1993a

Salvignol I, Blancher A, Calvas P, Socha WW, Colin Y, Cartron J-P, Ruffié J. Relationship between chimpanzee Rh-like genes and the R-C-E-F blood group system. J Med Primatol 22:19–28, 1993b

Salvignol I, Blancher A, Calvas P, Clayton J, Socha WW, Colin Y, Ruffié J. Molecular genetics of chimpanzee Rh-related genes: their relationship with R-C-E-F blood group system the chimpanzee counterpart of human RH system. Biochem Genet 32:201–221, 1994

Salvignol I, Calvas P, Socha WW, Colin Y, Le Van Kim C, Bailly P, Ruffié J, Cartron J-P, Blancher A. Structural analysis of the RH-like blood group gene products in nonhuman primates. Immunogenetics 41:271–281, 1995

Schmidt PJ. Rh-Hr: Alexander Wiener's last campaign. Transfusion 34(2):180–182, 1994

Schwartz MA, Brown E, Fazeli B. A 50 kDa integrin-associated protein is required for integrin-regulated calcium entry in endothelail cells. J Biol Chem 268:19931–19934, 1993

Shaw MA. Monoclonal anti-LW^{ab} and anti-D reagents recognize a number of different epitopes. J Immunogenet 13:377–386, 1986

Shinitzky M, Souroujon M. Passive modulation of blood group antigens. Proc Natl Acad Sci USA 76:4438, 1979

Sinor LT, Brown PJ, Evans JP, Plapp FV. The Rh antigen specificity of erythrocyte proteolipid. Transfusion 24(2):179–180, 1984

Sistonen P. Linkage of the LW blood group locus with the complement C3 and Lutheran blood group loci. Ann Hum Genet 48:239, 1984

Smythe JS, Avent ND, Judson PA, Parsons SF, Martin P, Anstee DJ. Expression of the RHD and RHCE gene products in K562 establishes the molecular basis of the RH blood group antigens. Blood:86:473a (abstr 1878), 1995)

Smythe JS, Avent ND, Judson PA, Parsons SF, Martin PG, Anstee DJ. Expression of RHD and RHCE gene products using retroviral transduction of K562 cells establishes the molecular basis of Rh blood group antigens. Blood 87(7):2968–2973, 1996

Socha WW. Serological differentiation of human populations. Polish State Medical Publishers, Warsaw, 1966

Socha WW. Blood groups of pygmy and common chimpanzees: a comparative study. In: Susman RL (ed) The pygmy chimpanzee: evolutionary morphology and behavior. Plenum, New York, pp 13–41, 1984

Socha WW. Blood groups of nonhuman primates. In: Swindler DR, Erwin J (eds) Comparative primate biology systematics evolution and anatomy, vol 1. Liss, New York, 1986

Socha WW. Blood groups of apes and monkeys. In: Jones TC, Mohr U, Hunt RD (eds) Nonhuman primates II. Springer, Berlin Heidelberg New York, pp 208–215, 1993a

Socha WW. Erythroblastosis fetalis. In: Jones TC, Mohr U, Hunt RD (eds) Nonhuman primates II. Springer, Berlin Heidelberg New York, pp 215–220, 1993b

Socha WW, Moor-Jankowski J. Rh antibodies produced by an isoimmunized chimpanzee: reciprocal relationship between chimpanzee simian-type isoimmune sera and human anti-Rh$_o$ reagents. Int Arch Allergy Appl Immunol 56:30–38, 1978

Socha WW, Moor-Jankowski J. Blood groups of anthropoid apes and their relationship to human blood groups. J Hum Evol 8:453–465, 1979

Socha WW, Moor-Jankowski J. Chimpanzee R-C-E-F blood group system: a counterpart of the Rh-Hr blood groups. Folia Primatol (Basel) 33:172–188, 1980

Socha WW, Ruffié J. Blood groups in primates. Theory practice evolutionary meaning. Liss, New York, pp 75–90 (Monographs in primatology, vol 3,7), 1983

Socha WW, Ruffié J. Monoclonal antibodies directed against human red cell antigens in tests with the red cells of non human primates. Rev Fr Transfus Hemobiol 33:39–48, 1990

Socha WW, van Foreest AW. Erythroblastosis fetalis in a family of captive orangutans. Am J Primatol 3:326 (abstract) , 1981

Socha WW, Wiener AS, Moor-Jankowski J, Mortelmans J. Blood groups of mountain gorillas (Gorilla gorilla beringei). J Med Primatol 2:364–369, 1973

Socha WW, Rouger P, Ruffié J, Moor-Jankowski J. Complexity of the Rh antigen demonstrated by comparative tests using antisera of human and primate origins. Exp Clin Immunogenet 6:150–155, 1989

Socha WW, Blancher A, Ruffié J. Comparative study of human monoclonal anti-D antibodies of IgG and IgM classes in tests with red cells of nonhuman primates. Rev Fr Transfus Hemobiol 36:485–497, 1993

Sonneborn HH, Uthemann H, Tills D, Lomast CG, Shawt MA, Tippett P. Monoclonal anti-LW^{ab}. Biotest Bull 2:145–149, 1984

Sonneborn HH, Ernst M, Tills D, Lomas CG, Gorick BD, Hughes-Jones NC. Comparison of the reactions of the Rh-related murine monoclonal antibodies BS58 and R6A. Vox Sang 58:219–223, 1990

Steinberg AG. Evidence for a mutation or crossing-over at the Rh locus. Vox Sang 10:721–724, 1965

Stratton F. A new Rh allelomorph. Nature 158:25, 1946

Suyama K, Goldstein J. Antibody produced against isolated Rh(D) polypeptide reacts with other Rh-related antigens. Blood 72:1622–1626, 1988

Suyama K, Goldstein J. Enzymatic evidence for differences in the placement of Rh antigens within the red cell membrane. Blood 75:255, 1990

Suyama K, Goldstein J, Aebersold R, Kent S. Regarding the size of Rh proteins. Blood 77:411, 1991

Suyama K, Roy S, Lunn R, Goldstein J. Expression of the 32-kd polypeptide of the Rh antigen. Blood 82:1006–1009, 1993

Suyama K, Lunn R, Haller S, Goldstein J. Rh(D) antigen expression and isolation of a new Rh(D) cDNA isofrom in human erythroleukemic K562 cells. Blood 84:1975–1977, 1994

Tate H, Cunningham C, McDade MG, Tippett PA, Sanger R. An Rh gene complex Dc-. Vox Sang 5:398–402, 1960

Tippett P. A speculative model for the Rh blood groups. Ann Hum Genet 50:241–247, 1986

Tippett P. Rh blood group system: the D antigen and high- and low-frequency Rh antigens. In: Vengelen-Tyler, Pierce (eds) Blood group systems: Rh. American Association of Blood Banks, Arlington, 1987

Tippett P. Co-ordinator's report on group 3: monoclonal Rh antibodies: serological and biological studies. Rev Fr Transfus Immunohematol 31(2):249–258, 1988

Tippett P. Serologically defined Rh determinants. J Immunogenet 17: 247–257, 1990

Tippett P, Moore S. Monoclonal antibodies against Rh and Rh-related antigens. J. Immunogenetics 17:309–319, 1990

Tippett P, Lomas-Francis C, Wallace M. The Rh antigen D: partial D antigens and associated low incidence antigens. Vox Sang 70:123–131, 1996

Umenishi F, Kajii E, Ikemoto S. Molecular analysis of Rh polypeptides in a family with RhD-positive and RhD-negative phenotypes. Biochem J 299:207–211, 1994

Unger LJ, Wiener AS. A "new" antibody anti-Rh^C resulting from isosensitization by pregnancy with special reference to the heredity of a new Rh-Hr agglutinogen Rh^c. J Lab Clin Med 32:499, 1959a

Unger LJ, Wiener AS, Wiener L. A new antibody (anti-Rh^B) in an Rh-positive patient resulting from blood transfusion. JAMA 170:1380, 1959b

Victoria EJ, Branks MJ, Masouredis SP. Rh antigen immunoreactivity after histidine modification. Mol Immunol 23(10):1039–1044, 1986

Von Dem Borne AEGK, Bos MJE, Lomas C, Tippett P, Bloy C, Hermand P, Cartron J-P, Admiraal LG, Van de Graaf J, Overbeeke MAM. Murine monoclonal antibodies against a unique determinant of erythrocytes related to Rh and U antigens: expression on normal and malignant erythrocyte precursors and Rh$_{null}$ red cells. Br J Haematol 75:254–261, 1990

Westhoff CM. Investigation of the Rh blood group system in humans and non-human primates. The University of Nebraska, Lincoln, 1993

Westhoff CM, Wylie DE. Investigation of the human Rh blood groups system in nonhuman primates and other species with serologic and southern blot analysis. J Mol Evol 39:87–92, 1994a

Westhoff CM, Wylie DE. Identification of a RhD -specific mRNA from K562 cell. Blood 83:3098, 1994b

Westhoff CM, Wylie DE. Investigation of the Rh locus in gorillas and chimpanzees. J Mol Evol 43:658–668, 1996

Wiener AS. Genetic theory of the Rh blood types. Proc Soc Exp Biol NY 54:316–319, 1943

Wiener AS. A new test (blocking test) for Rh sensitization. Proc Soc Exp Biol NY 56:173–176, 1944

Wiener AS. Blood group nomenclature. Science 128:849–852, 1958

Wiener AS, Gordon EB. The blood groups of chimpanzees: the Rh-Hr (CDE/cde) blood types. Amer J Phys Anthropol 19 1:35–43, 1961

Wiener AS, Peters HR. Hemolytic reactions following transfusions of blood of the homologous group with three cases in which the same agglutinogen was responsible. Ann Int Med 13:2306–2322, 1940

Wiener AS, Socha WW. Macro and micro differences in blood group antigens and antibodies. Int. Arch. Allergy 44:547–561, 1974

Wiener AS, Unger LJ. Blood factors Rh^A Rh^B Rh^C and Rh^D. Transfusion 2(4):230–233, 1962

Wiener AS, Gavan JA, Gordon EB. Blood group factors in anthropoid apes and monkeys. 11. Further studies on the Rh-Hr factors. Amer J Phys Anthropol 11:39–45, 1953

Wiener AS, Geiger J, Gordon EB. Mosaic nature of the Rh$_o$ blood factor. Exp Med Surg 15:75, 1957

Wiener AS, Moor-Jankowski J, Gordon EB. Blood groups of apes and monkeys IV. The Rh-Hr blood types of anthropoid apes. Amer J Hum Genet 16(2):246–253, 1964

Wiener AS, Moor-Jankowski J, Riopelle AJ, Shell WF. Simian blood groups. Another blood group system C-E-F in chimpanzee. Transfusion 5:508–515, 1965

Wiener AS, Moor-Jankowski J, Gordon EB, Kratochvil CL. Individual differences in chimpanzee blood demonstrable with absorbed human anti-Rho sera. Proc Natl Acad Sci USA 56:458–462, 1966

Wiener AS, Wisecup W, Moor-Jankowski J. A new simian-type blood factor L^c associated with the C-E-F blood group system of chimpanzee. Transfusion 7:351–354, 1967

Wiener AS, Moor-Jankowski J, Gordon EB. Blood groups of gorillas. Kriminalistik Forensische Wissenschaft 6:31–43, 1971

Wiener AS, Kratochvil CH, Moor-Jankowski J. Implications of studies in chimpanzees for the serology of the human Rh-Hr blood types. In: Golsmidt EI, Moor-Jankowski J (eds) Primates in medicine, vol 1. Karger, Basel pp 95–99, 1972

Wiener AS, Socha WW, Arons EB, Mortelmans G, Moor-Jankowski J. Blood group of gorillas: further observations. J Med Primatol (Basel) 5:317–320, 1976

5 The Duffy Blood Group System and Its Extensions in Nonhuman Primates

A.O. Pogo and A. Chaudhuri

Discovery and Background

The Duffy blood group system was named for a polytransfused hemophiliac man in whose serum a new antibody was found (Cutbush et al. 1950). "Duffy" was the patient's last name and the authors used this name for the new blood group system (Cutbush and Molisson 1950). The system consists of two principal antigens, Fy^a and Fy^b, produced by $FY*A$ and $FY*B$ codominant alleles. Antisera, anti-Fy^a and anti-Fy^b define four phenotypes: Fy(a+b-), Fy(a-b+), Fy(a+b+) and Fy(a-b-). Subsequently, several Duffy antigens were found and named Fy3 (Albrey et al.

Molecular Biology and Evolution of Blood Group
and MHC Antigens in Primates
Blancher/Klein/Socha (Eds.)
© Springer-Verlag Berlin Heidelberg 1997

1971), Fy4 (Behzad et al. 1973) and Fy5 (Colledge et al. 1973). Recently, the murine monoclonal antibody, anti-Fy6, identified the Fy6 antigen (Nichols et al. 1987; Riwom et al. 1994).

A significant observation related to malaria was made by Sanger et al. (1955). They found that red cells in the majority of American blacks did not react with anti-Fy^a and anti-Fy^b sera; these individuals had Fy(a-b-) erythrocytes. Twenty years later, Miller et al. (1975) found that the erythrocytes of these individuals were not invaded by certain species of malarial parasites.

The Duffy locus was the first to be assigned to a specific autosome in humans, when it was shown to be strongly linked to an element, *Un*, postulated as controlling the uncoiling of the paracentric region on the long arm of chromosome 1 (Donahue et al. 1968). Subsequently, the *FY* locus was assigned to the 1q21Dq25 region by linkage analysis (Dracopoli et al. 1991). Collins et al. (1992) narrowed the assignment to the 1q22 → q23 region using multiple pairwise analysis of sex-specific LOD scores. By using fluorescence in situ hybridization (FISH) with a cDNA clone that encodes the major subunit of Duffy protein, this region was confirmed (Mathew et al. 1994). *FY* and *Rh* loci are synthenic, e.g., both reside on chromosome 1; however, *Rh* is near the tip of the short arm and *FY* is on the long arm.

Duffy Antigenic Determinants

Fy^a and Fy^b Antigens

The three major Duffy phenotypes are Fy(a+b-), Fy(a-b+), and Fy(a+b+) and family studies show that Fy^a or Fy^b antigens are expressed in a single dose (Marsh 1975; Salmon et al. 1984; Beattie 1988). The fourth major phenotype, Fy(a-b-), predominant in blacks, is rarely found outside the black population (Table 1). The erythrocytes of 88 % of South Africans blacks (Hitzerroth et al.

Table 1. Duffy blood group system

Antisera	Antigens	Phenotypes	Alleles
Anti-Fy^a	Fy^a	Fy(a+b-)	*FY*A/FY*A*
Anti-Fy^b	Fy^b	Fy(a-b+)	*FY*B/FY*B*
Anti-Fy^a/anti-Fy^b	Fy^a/Fy^b	Fy(a+b+)	*FY*A/FY*B*
–	–	Fy(a-b-) blacks	*FY*B^{bms}/FY*B^{bms}* [a]
–	–	Fy(a-b-) non-blacks	?
Anti-Fy3	Fy3	Fy3	*FY*A or FY*B*
Anti-Fy5	Fy5 (+Rh)[b]	Fy5	*FY*A or FY*B*
Anti-Fy6	Fy6	Fy6	*FY*A or FY*B*

[a] bms, bone marrow silent gene.
[b] Fy5, requires the presence of Rh polypeptide.

1982) and 68 % of Americans blacks (Sanger et al. 1955) do not react with anti-Fy^a and Fy^b antibodies. This condition led to the hypothesis of a null phenotype through the inheritance of two silent *FY* genes in double doses. However, *FY* is not silent in Fy(a-b-) blacks; it produces the Duffy protein but in nonerythroid cells (see below).

Fy3 Antigen

The anti-Fy3 antibody defines Fy3 antigen. The antibody has been detected in three non-black women with the rare Fy(a-b-) phenotype. However, among many transfused Fy(a-b-) blacks, anti-Fy3 is an exceptional finding (Albrey et al. 1971; Oberdorfer et al. 1974; Buchanan et al. 1976). Fy3 antigen is always present when Fy^a and/or Fy^b are present, but it is missing in the Fy(a-b-) erythrocytes of black and non-black individuals (Salmon et al. 1984; Beattie 1988). This indicates that the antigen is linked to the presence of Duffy protein in the red cells (see below). Recently, the antigenic domain of Fy3 has been mapped at the third loop of the Duffy protein (Zhao-hai et al. 1996).

Fy4 Antigen

A weak Duffy antibody, anti-Fy4, reacts with all Fy(a-b-) and most Fy(a+b-) and Fy(a-b+) cells of blacks (Behzad et al. 1973). It does not recognize Fy(a+b+) red cells in blacks or non-blacks. It also reacts with one, and possibly two, of the non-black women with rare Fy(a-b-) phenotype. The Fy4 antigenic determinant is not related to the Duffy system since anti-Fy4 antibody reacts with erythrocytes that lack the Duffy protein (see below).

Fy5 Antigen

Anti-Fy5 antibody, which defines the Fy5 antigen, agglutinates all Duffy-positive cells but not Fy(a-b-) red cells from blacks (Colledge et al. 1973; DiNapoli et al. 1976). Thus, the Fy5 antigen is present in all Duffy-positive red cells but absent in all Duffy-negative red cells. Anti-Fy5 does not react with Duffy-positive red cells of the Rh$_{null}$ phenotype and reacts weakly with D- - red cells. The Fy5 antigen is somehow related to the presence of the Rh protein. The antigenic determinant of Fy5 must include both Duffy and Rh protein domains (Bloy et al. 1987).

Fy6 Antigen

A mAb obtained by Nichols et al. (1987) and Riwom et al. (1994) interacts with all Duffy-positive cells except Fy(a-b-) red cells and it defines a Duffy antigen, Fy6. Anti-Fy6 has been a valuable reagent for the purification and characterization of the Duffy complex (Chaudhuri et al. 1989). Anti-Fy6, unlike anti-Fy5, reacts with Rh_{null} erythrocytes of both the regulator and the amorphic types (Nichols et al. 1987).

Fy^x, a Variant of Fy^b

Some Fy(a-b+) red cells react weakly with anti-Fy^b and this property is genetically determined (Chown et al. 1965). There is no specific anti-Fy^x antibody and the Fy^x antigen is detected as a weak reaction of anti-Fy^b. The red cell of an $FY*A/FY*X$ individual can be mistaken for Fy(a+b-) and a homozygous $FY*X/FY*X$ individual can be mistaken for Fy(a-b+) if the anti-Fy^b used recognizes the Fy^x antigen. Individuals with the Fy^x antigen also possess the weak Fy3 antigen. Detection of the Fy ^{x} antigen is difficult and many cases are not recognized.

Duffy-Related Human Antigens in Nonhuman Primates

The study of the Duffy gene in nonhuman primates provides clues to the origin of the Duffy gene. Fy(a-b+) erythrocytes are prevalent in chimpanzee and rhesus monkeys; therefore $FY*B$ is their predominant allele (Palatnik and Rowe 1984; McGinnis and Miller 1977). Although there are few cases of simian Fy(a-b-) erythrocytes, the phenotype is common in new world monkeys (Palatnik and Rowe 1984). $FY*A$ is exceedingly scarce in nonhuman primates and may not exist. The first Duffy allele was probably $FY*B$, then gene duplication and a point mutation (at nucleotide 306) generated $FY*A$ in humans (see below).

Biochemistry

Isolation and Characterization of the Duffy Complex

The isolation of human red cell membrane components associated with some blood group antigens has been difficult, largely due to their small quantities and the instability of antigen-antibody complexes upon detergent solubilization of the lipid bilayer. Moore et al. (1982) demonstrated the feasibility of obtaining a relatively stable antigen-antibody complex between Duffy antigens and anti-Fy^a and anti-Fy^b sera. The sera immunoprecipitated proteins of 39, 64 and 88 kDa. The wide range of molecular sizes and the absence of sharp bands were interpreted, like in glycophorins, as being a result of variable glycosylation. Hadley et al. (1984) identified protein(s) with an apparent molecular mass of 35–43 kDa in Fy(a+b+), but not in Fy(a-b+), red cell ghosts immunoblotted with a potent anti-Fy^a serum. They showed also the unusual property of aggregation when red cell ghosts were boiled in 5% sodium dodecyl sulfate (SDS) with

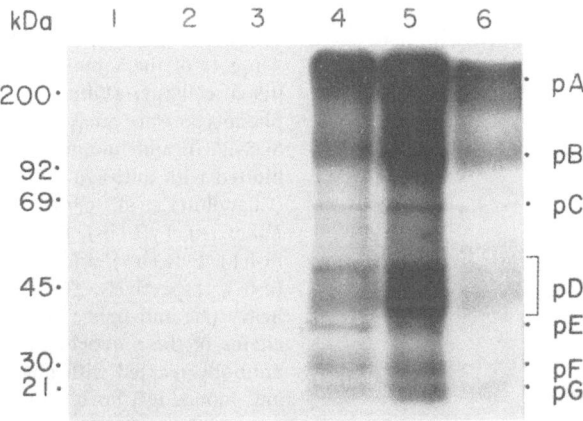

Fig. 1. Protein composition of purified Duffy complex. ^{125}I surface-labeled red cells were incubated with and without anti-Fy6, hemolyzed, and membrane proteins solubilized by Triton X-100. The antigen-antibody complex was purified by affinity chromatography using anti-mouse IgG Sepharose beads (Chaudhuri et al. 1989). Equivalent amounts of cells per Duffy phenotype were run on SDS-PAGE. *Lanes 1, 2,* Fy(a-b-) cells incubated without and with anti-Fy6, respectively. *Lanes 3, 4,* Fy(a+b-) incubated without and with anti-Fy6, respectively. *Lane 5,* Fy(a-b+) cells; *Lane 6,* Fy(a+b+) incubated with anti-Fy 6. The gel was vacuum-dried and autoradiographed

5% β-mercaptoethanol. Since all red cell ghost proteins are treated, it is not clear whether the aggregates are due to interactions of Duffy protein(s) with itself or with other proteins.

The availability of the murine mAb anti-Fy6 played a crucial role in the purification and characterization of the Duffy blood group system. Thus, Chaudhuri et al. (1989), using anti-Fy6, developed a procedure for the purification of Duffy antigens in human red cells. Anti-Fy6 reacts efficiently when the antigen is in the cell membrane but not when it is in solution; nonetheless, the antibody-antigen complex is quite stable upon dissolution of the lipid bilayer (Chaudhuri et al. 1989).

The affinity purified Triton X-100-soluble fraction, obtained from Fy(a+b-), Fy(a-b+) and Fy(a+b+) red cells and surface-labeled with ^{125}I, yielded a complex pattern of proteins when separated by sodium dodecyl sulfate-polyacrylamide gel electrophoresis (SDS-PAGE) (Fig. 1). The absence of protein bands from Fy(a-b-) erythrocytes indicates the specificity of the antibody. An identical set of proteins was observed when Fy(a+b-) erythrocytes were challenged with anti-Fy[a] serum and the affinity-purified material was immunoblotted with anti-Fy6 antibody (Chaudhuri et al. 1989). Both antibodies reacted with the same set of proteins.

The bands were assigned alphabetical labels starting from the top of the gel, and the range of molecular mass was greater than 200–18 kDa. The presence of many bands indicated that there were several proteins or that there was a single protein with various degrees of glycosylation or aggregation. However, not all the

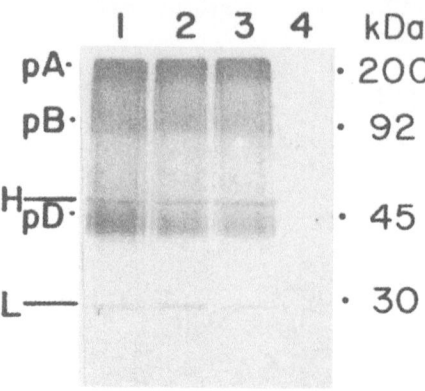

Fig. 2. Immunoblots of affinity-purified Duffy complex. Aliquots of the same number of cells per Duffy phenotype were analyzed on SDS-PAGE and immunoblotted with anti-Fy6 (Chaudhuri et al. 1989). *Lanes 1–4*, Fy(a+b−), Fy(a−b+), Fy(a+b+), and Fy(a−b−), respectively. The heavy (*H*) and light (*L*) chains of the monoclonal antibody reacted with goat anti-mouse IgG horse radish peroxidase conjugate

isolated proteins reacted with the anti-Fy6 antibody (Fig. 2). Proteins that reacted were in the pA, pB and pD regions and were classified as showing Duffy blood group activity. pD protein was the most rapidly migrating protein with Duffy activity. Obviously, this protein (the monomer) forms polymeric aggregates readily. For example, gel-purified pD protein incubated at either −20°C or 4°C yielded proteins with mobility of pA and pB bands (Chaudhuri et al. 1989). Conversely, gel purified pA and pB proteins, when re-ran, yielded pA and pB bands as well as the pD band (Chaudhuri et al. 1989). However, a quantitative conversion of pA and pB into pD was never observed although the polymers were treated with strong dissociating reagents, such as urea and SDS in the presence of reducing reagents (Chaudhuri and Pogo 1995). The monomeric form of Duffy protein was originally named pD, but it was renamed gp-Fy, to avoid confusion with Rh "D" (Chaudhuri et al. 1995).

Carbohydrate Composition

The polymeric and monomeric form of the major subunit of the Duffy antigen have small quantities of mannose, galactose, *N*-acetylglucosamine, sialic acid, and fucose (Tanner et al. 1988; Chaudhuri and Pogo 1995). Glycosidase digestion of the monomeric form of gp-Fy, showed that N-glycosidase F increased the migration but O-glycosidase did not (Chaudhuri and Pogo 1995). Neither endo-β-galactosidase nor endoglycosidase F increased the electrophoretic mobility, even in the presence of β-mercaptoethanol and detergent. The deglycosylation by N-glycosidase F is consistent with the chemical detection of *N*-acetylglucosamine and the absence of *N*-acetylgalactosamine is consistent with the insensitivity to O-glycosidase digestion. The slight increase in mobility after neuraminidase digestion (Hadley et al. 1984) is consistent with the chemical detection of a small quantity of sialic acid.

Molecular Biology

The Primary Structure of gp-Fy

A major advance in the study of the Duffy blood group system was the cloning of the Duffy gene (Chaudhuri et al. 1993). *FY* codes for an acidic glycoprotein (gp-Fy) of 337 amino acid residues with a molecular weight (Mr) of 35 733, the same as the deglycosylated major subunit of the Duffy protein complex (Chaudhuri et al. 1989). (Most proteins are initially synthesized with methionine at their NH_2-terminal amino acid, which is removed after translation by a specific aminopeptidase. It is assumed that in gp-Fy, like most stable proteins, the methionine is removed. Thus, the number of amino acids is 337 in the finished molecule but is 338 when newly synthesized.) The hydropathy map explicitly developed for membrane proteins by Engelman et al. (1986) predicts an exocellular NH_2-terminal domain of 64 residues, nine transmembrane α-helices, short protruding hydrophilic loops and an endocellular COOH-terminal domain of 23 residues (Chaudhuri et al. 1993). By using another hydropathy plot, seven is the predicted number of transmembrane helices, in agreement with the interleukin-8 chemokine receptor (Darbonne et al. 1991; Horuk et al. 1993).

Glycoprotein Duffy (gp-Fy)

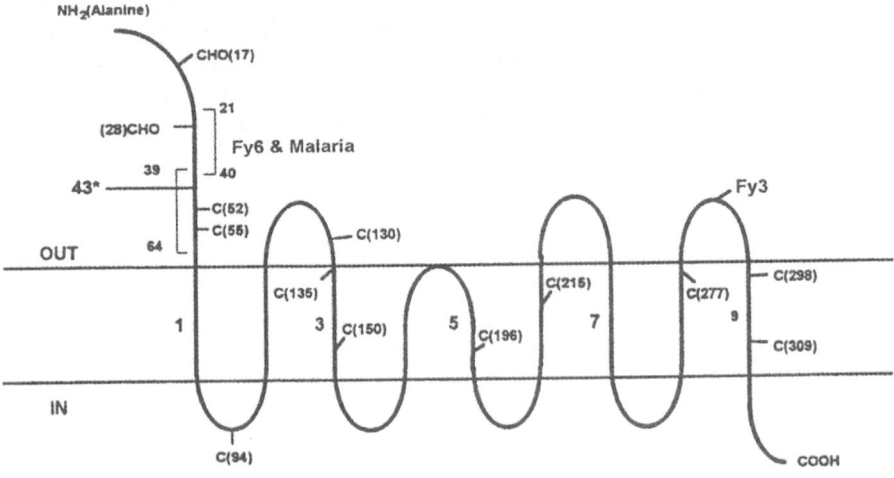

Fig. 3. This model for membrane orientation of gp-Fy was based on the prediction of the hydropathy profile of Engelman et al. (1986) and on the exocellular localization of the NH_2-terminal. Alanine is the first residue at the NH_2-terminal. Glycosidated (CHO) asparagines at residues 17 and 28 are indicated. Residues 21–40 delineate a region of specific binding for Fy6 and parasites (region I, see Fig. 6A). Residues 39–64 delineate a region for binding of anti-Fy^a and anti-Fy^b (region II, see Fig. 6A). The location of 11 cysteines are indicated. 43* is the residue that determines Fy^a (glycine) or Fy^b (aspartic acid) specificity. Fy3 binding specificity at the third exocellular loop is indicated. (From Zhao-hai et al. 1996)

The topology of the NH_2-terminal exocellular domain is validated by the finding of the two N-glycosylation sites at residues 17 and 28 (Chaudhuri et al. 1993) (Fig. 3). There is a putative third N-glycosylation site at residue 34, but the aspartic acid at residue 35 restricts glycosylation of the asparagine at residue 34 (Marshall et al. 1972). A further confirmation of the NH_2-terminal topology is that anti-Fy6 reacts with a chemically synthesized 35-mer polypeptide encompassing residues 9–44 of this domain (to be published). Finally, a rabbit polyclonal antibody against this synthetic polypeptide agglutinates all Duffy-positive red cells (unpublished results). The number of transmembrane α-helices, the number and length of the exocellular and endocellular loops, and the topology of the COOH-terminal domain remain to be experimentally determined.

Duffy mRNAs

On RNA blot analysis, the Duffy cDNA clone detects an approximately 1.3 kb mRNA in the bone marrow of Duffy-positive, but not Duffy-negative, individuals (Chaudhuri et al. 1993). The same size mRNA has been detected in the kidney, spleen, heart, lung, muscle, duodenum, pancreas, fetal liver (but not adult liver), and placenta of Duffy-positive and Duffy-negative individuals (Chaudhuri et al. 1993, 1994; Neote et al. 1994). The human brain synthesizes a Duffy-related mRNA. We and others have detected a prominent 8.5 kb and a minor 2.2 kb mRNA in the brain of Duffy-positive individuals (Chaudhuri et al. 1993; Neote et al. 1994). These large Duffy-related mRNAs encode larger proteins which have a domain with extensive homology with gp-Fy (unpublished results). It appears that *FY* is part of a larger gene. This novel human gene is fully transcribed in the brain, but in other organs only the region codifying for gp-Fy is transcribed.

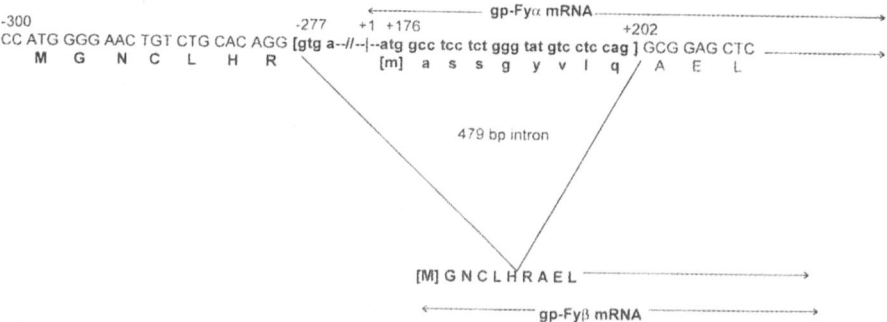

Fig. 4. The origin of gp-Fyα and gp-Fyβ mRNAs. The 5′ end (cap site) of gp-Fyα mRNA is denoted as +1 and upstream sequences are in negative numbers. The open reading fram (ORF) of gp-Fyα mRNA starts at +176 nucleotides. The intron of gp-Fyβ mRNA is within *brackets* and in *bold lowercase letters*. A single-letter code below each codon indicates the ORF. The NH_2-terminals of gp-Fyα and gp-Fyβ are in *bold lowercase* and *bold uppercase letters*, respectively

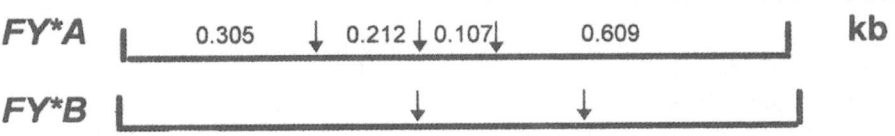

Fig. 5. Approximate fragment sizes generated by BanI digestion of FY*A and FY*B alleles

Two variants of Duffy mRNAs and their differential expression in bone marrow and several nonerythroid organs have been reported recently (Iwamoto et al. 1996; Chaudhuri et al., to be published). Nonsplicing and splicing mechanisms generate the two variants. The spliced RNA is the most abundant and is generated by the removal of an intron of 479 nucleotides (nts) and the splicing of a short 5' exon of 59 nts to a longer 3' exon of 1040 nts (Fig. 4). The structural difference between the two encoded proteins is a sequence of six residues for the spliced mRNA and eight residues for the nonspliced mRNA at the NH_2-terminal domain. However, both proteins have identical immunological and physiological properties (Iwamoto et al. 1996; Chaudhuri et al., to be published).

Molecular Basis for Duffy Polymorphism

The difference between the FY*A and FY*B alleles is a single nucleotide change at position 306: guanine is in FY*A and adenine is in FY*B. The mutation in this nucleotide produces a codon change that modifies the amino acid at position 43: glycine is in Fy(a+b-) and aspartic acid is in Fy(a-b+) red cells (Chaudhuri et al. 1995; Iwamoto et al. 1995; Mallinson et al. 1995; Tournamille et al. 1995). FY*B has two and FY*A has three BanI restriction sites (Fig. 5). The guanine at position 306 creates an additional BanI restriction site in FY*A. BanI digestion products of polymerase chain reaction (PCR) of genomic DNA of FY*B and FY*A yields

Table 2: Duffy alleles and restriction fragment length polymorphisms

	Phenotypes	Alleles	Number of fragments[c]
Non-blacks	Fy(a+b-)	FY*A/FY*A	4
	Fy(a-b+)	FY*B/FY*B	3
	Fy(a+b+)	FY*A/FY*B	5
	Fy(a-b-)	?	?
Blacks	Fy(a+b-)	FY*A/FY*B[a]	5
	Fy(a-b+)	FY*B/FY*B	3
	Fy(a+b+)	FY*A/FY*B	5
	Fy(a-b-)	FY*B/FY*B[b]	3

[a] It remains to be determined if FY*B allele is expressed in nonerythroid tissues.
[b] The allele is silent in the bone marrow only.
[c] BanI digestion fragments.

three and four fragments, respectively. Restriction fragment length polymorph-ism (RFLP) studies demonstrate that: (1) Fy(a+b-) and Fy(a-b+) non-blacks are homozygous *FY*A* and *FY*B*, respectively; (2) most Fy(a-b-) blacks are homozygous *FY*B*; (3) Fy(a+b-) blacks are heterozygous *FY*A /FY*B*; and Fy(a-b+) blacks are homozygous *FY*B*. In the black population, a bone marrow silent *FY*B* is very common, but a silent *FY*A* has not been found yet (Table 2).

The molecular basis for the erythroid-specific repression of *FY* in Duffy-negative blacks is the disruption of a GATA 1 erythroid motif in the promoter region of the gene (Tournamille et al. 1995). At the 5' flanking coding region of *FY*, there are two GATA elements, one located at -46 nts and the other at about -396 nts upstream of the transcription initiation site (Tournamille et al. 1995). In the closer GATA element, a T→C substitution has been found in most of the Duffy-negative blacks.

Expression in Nonerythroid Tissues

The Duffy antigen belongs to the class of red cell proteins that are found in other cell types. Erythroid proteins such as spectrin, ankyrin and band 4.1 (called synapsin in the brain) are expressed in other cells (Bennett 1985). A truncated form of band 3, the anion transporter, is in kidney (Brosius et al. 1989). As a rule, these proteins in nonerythroid cells are products of genes that differ from the homologous erythroid gene.

The Duffy antigen is produced in brain, breast, colon, fetal liver (but not adult liver), kidney, lung, spleen, thymus and thyroid. With the exception of brain, nonerythroid gp-Fy is the product of the same gene (Neote et al. 1994; Peiper et al. 1995; Chaudhuri et al. 1995; Chaudhuri et al., to be published). In lung, thyroid and spleen, gp-Fy is produced by the plasma membrane of fene-strated and nonfenestrated endothelial cells of capillary and postcapillary venules (Chaudhuri et al., to be published). In kidney, the plasma membrane of fene-strated endothelial cells of the glomerulus and the nonfenestrated endothelial cells of vasa recta express gp-Fy. The epithelial cells of the collecting ducts of kidney and the epithelial cells (type I) of pulmonary alveolus also produce gp-Fy (Chaudhuri et al., to be published).

Gp-Fy is in caveolae or non-clathrin-coated invaginated pits or vesicles pre-sent in several cell types (Chaudhuri et al. 1995). This membrane specialization is capable of sealing off the extracellular environment and creating a unique mem-brane-bound compartment of the cell surface (Anderson et al. 1992). It is tenta-tively proposed that gp-Fy, via binding to a specific ligand, uses caveolae to incorporate the ligand into the cell interior.

The Primary Structure of gp-Fy in Nonhuman Primates

The study of the primary structure of gp-Fy in nonhuman primates has produced interesting findings. As mentioned, the erythrocytes of chimpanzee (*Pan troglo-*

dytes), rhesus (*Macaca mulatta*), squirrel (*Saimiri sciureus*), and aotus monkeys (*Aotus trivirgatus*) are Fy(a-b+) and *FY*B* is their allele. The amino acid diversions between human and nonhuman primates are: 1, 32, 37 and 40 residues in chimpanzee, rhesus, squirrel, and aotus monkeys, respectively (Chaudhuri et al. 1995). These diversions correlate with the genetic distances. Thus, aotus and squirrel monkeys are more distant from humans than chimpanzee monkeys (Fleagle 1988). In addition, the changes at the NH$_2$-terminal domain highlight several key residues for antibody and malaria binding specificity (Chaudhuri et al. 1995). The residues are clustered in two overlapping regions (Fig. 6A). In region I, residues 22–40 show the changes that affect anti-Fy6 and *Plasmodium vivax* binding specificity. Amino acid substitutions and deletions that correlate with modifications in ligand binding were seen in region I of rhesus monkey. As previously shown, this simian's erythrocytes are not recognized by anti-Fy6 and are not invaded by *P. vivax* (Nichols et al. 1987; Wertheimer and Barnwell 1989; Barnwell et al. 1989) (Fig. 6B). In rhesus, the phenylalaline at position 23 is deleted and substitutions occur at the residues 22 (aspartic acid to asparagine),

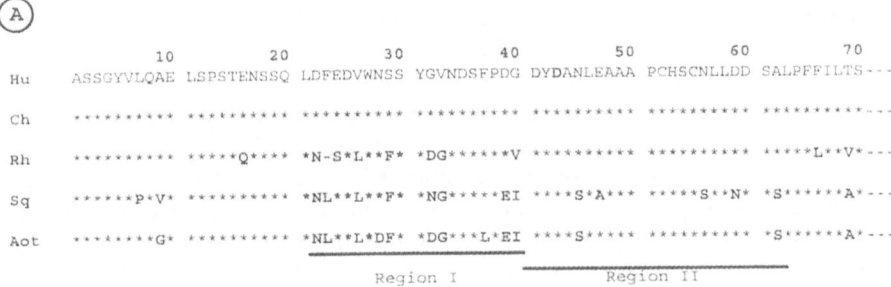

Fig. 6 A,B. **A** The deduced amino acid sequences from genomic DNAs of humans and simians. NH$_2$-terminal exocellular domains are shown. *Hu*, human *FY*B*; *Ch*, chimpanzee; *Rh*, rhesus; *Sq*, squirrel; *Aot*, aotus monkey. Amino acid sequences are numbered at the top. *Asterisks* denote the same residues while *bold letters* indicate different residues.
Bold underlining indicates two regions of binding specificity for antibody and merozoite.
B Antibody reactivity and erythrocyte invasion by parasites. The data were obtained from Nichols et al. (1987), Wertheimer and Barnwell (1989), and Barnwell et al. (1989).
P. vivax invasion in squirrel monkey erythrocytes may not be gp-Fy related (Wertheimer and Barnwell 1989)

24 (glutamic acid to serine), 26 (valine to leucine), 29 (serine to phenylalanine), 32 (glycine to aspartic acid), 33 (valine to glycine) and 40 (glycine to valine). The remaining NH_2-terminal of gp-Fy in rhesus is identical to that of human and chimpanzee, with the exception of residue 16, where glutamic acid is substituted by the homologous glutamine.

Region I of the squirrel monkey, when compared with rhesus, shows other modifications that may affect ligand binding specificity in a different way. Thus, residue 23 is not deleted and residue 24, glutamic acid, is not replaced by serine (Fig. 6A). Residue 32 is asparagine instead of aspartic acid as seen in rhesus, but they are homologous amino acids; residue 39 is glutamic acid instead of aspartic acid and 40 is isoleucine instead of valine. These amino acid substitutions may be significant in preserving some affinity to anti-Fy6 (Fig. 6B). It has been shown that squirrel monkey erythrocytes are recognized by anti-Fy6, albeit with less affinity (Wertheimer and Barnwell 1989). Although squirrel monkey erythrocytes are invaded by P. vivax merozoites, they failed, like rhesus monkey, to bind P. vivax 135–140 kDa Duffy-specific protein (Wertheimer and Barnwell 1989).

Aotus monkey erythrocytes are recognized by anti-Fy6 and by P. vivax 135–140 kDa Duffy-specific protein (Wertheimer and Barnwell 1989). They are invaded by P. vivax merozoites but at a lower rate (Barnwell et al. 1989). Aotus monkey gp-Fy region I shows modifications similar to those in squirrel, except at residue 28, where asparagine is substituted by the homologous aspartic acid, and at residue 37, where phenylalanine is substituted by leucine (Fig 6A).

In humans, chimpanzee and rhesus monkeys, the first 15 residues are the same (Fig. 6A). In aotus monkeys, residue 9 is glycine instead of alanine and in squirrel monkey, residues 7 and 9 are proline and valine instead of leucine and alanine, respectively. We do not know if these changes modify anti-Fy6 and/or merozoite binding specificity; however, the first 10 residues may not be necessary for anti-Fy6 and merozoite binding. The first 10 residues of gp-Fy were not essential for anti-Fy6 reactivity with the 35-mer peptide in an ELISA assay (unpublished results).

Region II, residues 39–62, delineate human antisera binding. Thus, residue 43 is critical for anti-Fy^a and anti-Fy^b recognition. However, the study of aotus and squirrel monkeys revealed the role of other residues in determining whether erythrocytes can react with anti-Fy^b. Both simians had FY*B alleles and aspartic acid at position 43, but only aotus monkey erythrocytes reacted weakly with anti-Fy^b serum (McGinnis et al. 1977). Substitutions seen at residue 39 (aspartic acid to glutamic acid), residue 40 (glycine to isoleucine), residue 47 (glutamic acid to alanine), residue 56 (asparagine to serine) and residue 59 (aspartic acid to asaparagine) may be critical for anti-Fy^b binding. Other substitutions, at residue 45 (asparagine to serine) and at residue 62 (alanine to serine), may be less critical for anti-Fy^b binding, since the same substitutions were found in the in red cells of aotus monkeys that reacted (weakly) with anti-Fy^b (Fig. 6B).

Duffy Protein and Malaria

A remarkable aspect of the Duffy antigen is its role as a receptor for malarial parasites. Duffy-negative human red cells are resistant to invasion by *Plasmodium knowlesi*, a monkey malarial parasite which is known to invade human red cells of Duffy-positive individuals and, occasionally, to infect humans (Miller et al. 1975). The studies on *P. knowlesi* were extended to *Plasmodium vivax*, a human malaria parasite that invades the red cells of Duffy-positive individuals and yet is incapable of infecting the red cells of Duffy-negative individuals (Miller et al. 1976). Thus, blacks with Fy(a-b-) erythrocytes cannot be infected by *P. vivax*, and their red cells are also resistant to in vitro invasion by *P. knowlesi* (Miller et al. 1975, 1976; Galinski et al. 1992). The parasites rely on a Duffy antigen-parasite ligand interaction for invasion. It should be noted that *P. vivax* merozoites primarily invade reticulocytes, although both reticulocytes and mature erythrocytes express gp-Fy. Evidently, the reticulocyte membrane has additional receptor(s) essential for *P. vivax* invasion (Galinski et al. 1992). Anti-Fy^a, anti-Fy^b and anti-Fy6 block invasion, which indicates that parasite and antibody determinants are located close to each other at the NH$_2$-terminal exocellular domain of gp-Fy (Nichols et al. 1987). This observation is consistent with our study of *FY* in nonhuman primates and indicates that the specificity of binding malarial parasites and Duffy antibodies are clustered in two overlapping regions at the exocellular NH$_2$-terminal domain of gp-Fy (see below).

Chemokine Receptor

The binding of chemokines to Duffy-positive but not to Duffy-negative erythrocytes suggests that the antigen may be another proinflammatory peptide receptor (Darbonne et al. 1991; Horuk et al. 1993; Neote et al. 1994). Furthermore, the binding of these peptides to erythroleukemia K562 cells transfected with Fy^b cDNA demonstrates that gp-Fy is the human erythrocyte chemokine receptor (Chaudhuri et al. 1994). The same result is obtained with human embryonic kidney 293 cells transfected with cDNA containing the coding region of gp-Fy (Neote et al. 1994). The role of this protein as a red cell chemokine receptor is unclear, since the function of chemokines is to mediate migration of leukocytes. The receptor is postulated to function as a "sink" or a scavenger for chemokines (Darbonne et al. 1991; Horuk et al. 1993). It should be noted that the protein is not essential for the normal structure and function of erythrocytes. Erythrocytes that lack gp-Fy do not have any manifested abnormalities and individuals with a Duffy-negative phenotype are healthy. Does the expression of this protein in other cell types compensate for the lack of erythroid expression in Duffy-negative individuals? What is the role of gp-Fy in epithelial cells of the collecting ducts of human kidney and type I squamous epithelial cells of human pulmonary alveoli? Further studies of the molecular biology of the Duffy gene will answer these basic questions.

Summary

A major advance in understanding the Duffy blood group system has been achieved with the cloning of *FY*, a single copy gene located in the 1q22 → q23 region of chromosome 1. The product of the Duffy gene is an acidic glycoprotein of 337 residues which spans the plasma membrane nine times, has an exocellular NH$_2$-terminal domain of 64 residues, an endocellular COOH-terminal domain of 23 residues, and whichcontains all the Duffy antigenic determinants. Several cell types, including the endothelial cells of capillary and postcapillary venules of some organs and the epithelial cells of kidney collecting ducts and lung alveoli, produce Duffy protein. It is also made in the brain.

Duffy polymorphism is simple; there are two proteins, which are the products of two codominant alleles *FY*A* and *FY*B*. A single nucleotide change, guanine in *FY*A* and adenine in *FY*B*, is the difference between the alleles. The mutation produces a codon change that modifies residue 43, hence, glycine is in the protein of Fy(a+b-) erythrocytes and aspartic acid is in the protein of Fy(a-b+) erythrocytes. Blacks of Fy(a-b-) phenotype do not produce the protein in their erythrocytes; they have a homozygous *FY*B* allele which is silent only in the bone marrow. The erythroid-specific repression in Duffy-negative blacks is produced by a disruption of a GATA motif in the promoter region of the gene. In nonhuman primates, *FY*B* is the predominant allele, whereas the *FY*A* allele is exceedingly scarce and indeed may not exist. Hence, the first Duffy allele in nonhuman primates was probably *FY*B*, with gene duplication and a point mutation subsequently generating *FY*A* in humans.

Although lack of Duffy protein rarely causes a severe transfusion reaction, the protein plays a central role in malaria infection and in the disposal of deleterious proinflammatory peptides. Thus, the Duffy protein is a newly discovered member of the family of chemokine receptors as well as the membrane protein that malarial parasites require for invasion of red cells. Interestingly, the malarial parasite-specific binding, the binding of chemokines, and the major antigenic domains are located in two overlapping regions at the exocellular NH$_2$-terminal of the Duffy protein.

Acknowledgements. The authors thank T. Huima-Byron for photographic/printing assistance and V.M. Sarnicole for secretarial assistance. This work was supported by Grant HL 53297 of the National Heart, Lung, and Blood Institute of the National Institutes of Health.

References

Albrey JA, Vincent EER, Hutchinson J, Marsh WL, Allen FH, Gavin J, Sanger R. A new antibody, anti-Fy3, in the Duffy blood group system. Vox Sang 20:29–35, 1971

Anderson RGW, Kamen BA, Rothberg KG, Lacey SW. Potocytosis: sequestration and transport of small molecules by caveolae. Science 255:410–411, 1992

Barnwell JW, Nichols ME, Rubinstein P. In vitro evaluation of the role of the Duffy blood group in erythrocyte invasion by Plasmodium vivax. J Exp Med 169:1795–1802, 1989

Beattie KM: The Duffy Blood Group System. In: Pierce SR, Macpherson CR (eds) Blood group systems: Duffy, Kidd, and Lutheran. American Association of Blood Banks, Arlington VA, pp 1–25, 1988

Behzad O, Lee CL, Gavin J, Marsh WL. A new anti-erythrocyte antibody in the Duffy system: anti-Fy4. Vox Sang 24:337–342, 1973

Bennett V. The membrane skeleton of human erythrocytes and its implications for more complex cells. Annu Rev Biochem 54:273–304, 1985

Bloy C, Blanchard D, Lambin P, Goossens D, Rouger P, Salmon C, Cartron J-P. Human monoclonal antibody against Rh(D) antigen: partial characterization of the Rh(D) polypeptide from human erythrocytes. Blood 69:1491–1497, 1987

Brosius FC, Alper SL, Garcia AM, Lodish HF. The major kidney band 3 gene transcript predicts an amino-terminal truncated, band 3: polypeptide. J Biol Chem 264:7784–7787, 1989

Buchanan DI, Sinclair M, Sanger R, Gavin J, Teesdale P. An Alberta Cree Indian with a rare Duffy antibody anti-Fy3. Vox Sang 27:619–624, 1976

Chaudhuri A, Pogo AO. The Duffy blood group system and malaria. In: Cartron J-P, Rouger P (eds) Blood cell biochemistry, vol 6. Plenum, New York, pp 243–265, 1995

Chaudhuri A, Zbrzezna V, Johnson C, Nichols ME, Rubinstein P, Marsh WL, Pogo AO. Purification and characterization of an erythrocyte protein complex carrying Duffy blood group antigenicity. J Biol Chem 264:13770–13774, 1989

Chaudhuri A, Polyakova J, Zbrzezna V, Williams K, Gulati S, Pogo AO. Cloning of glycoprotein D cDNA, which encodes the major subunit of the Duffy blood group system and the receptor for the Plasmodium vivax malaria parasite. Proc Natl Acad Sci USA 90:10793–10797, 1993

Chaudhuri A, Zbrzezna V, Polyakova J, Pogo AO, Hesselgesser J, Horuk R. Expression of the Duffy antigen in K562 cells. J Biol Chem 269:7835–7838, 1994

Chaudhuri A, Polyakova J, Zbrzezna V, Pogo AO. The coding sequence of Duffy blood group gene in humans and simians: restriction fragment length polymorphism antibody and malarial specificities and expression in nonerythroid tissues in Duffy-negative individuals. Blood 85:615–621, 1995

Chitnis et al., 1996 (see Note added) in Proof.

Chown B, Lewis M, Kaita H. The Duffy blood group system in Caucasians: evidence for a new allele. Am J Hum Genet 17:384–389, 1965

Colledge KI, Pezzulich M, Marsh WL. Anti-Fy5 an antibody disclosing a probable association between the Rhesus and Duffy blood group genes. Vox Sang 24:193–199, 1973

Collins A, Keats BJ, Dracopoli N, Shields DC, Morton NE. Integration of gene maps: chromosome 1. Proc Natl Acad Sci USA 89:4598–4602, 1992

Cutbush M, Mollison PL. The Duffy blood system. Heredity 4:383–389, 1950

Note Added in Proof

Since the submission of this article, it was found that carbohydrates block the receptor on rhesus erythrocytes for binding P. vivax ligand. This finding suggests that the P. vivax ligand can bind the peptide backbones of the human and rhesus Duffy antigen. The same observation was obtained with erythrocytes of the squirrel monkey. The amino acid modifications in the NH_2-terminal domain in rhesus may be not so critical for parasite recognition. It remains to demonstrate whether P. vivax invades rhesus deglycosylated-erythrocytes. (Chitnis CE, Chaudhuri A, Horuk R, Pogo AO, and Miller LH. The domain on the Dufy Blood Group antigen for binding Plasmodium vivax and P. knowlesi malarial parasite to erythrocytes. J Exp Med 184, 1531–1536, 1996).

Cutbush M, Mollison PL, Parkin DM. A new human blood group. Nature 165:188–189, 1950

Darbonne WC, Rice GC, Mohler MA, Apple T, Hebert CA, Valente AJ, Baker JB. Red blood cells are a sink for interleukin 8, a leukocyte chemotaxin. J Clin Invest 88:1362–1369, 1991

DiNapoli J, Garcia A, Marsh WL, Dreizin D. A second example of anti-Fy5. Vox Sang 30:308–311, 1976

Donahue RP, Bias WB, Renwick JH, McKusick VA. Probable assignment of the Duffy blood group locus to chromosome 1 in man. Proc Natl Acad Sci USA 61:949–955, 1968

Dracopoli NC, O'Connell P, Elsner TI, Lalouel JM, White RL, Beutow KH, Nishimura DY, Murray JC, Helms C, Mishra SK, Donis-Keller H, Hall JM, Lee MK, King MC, Attwood J, Morton NE, Robson EB, Mahtani M, Willard HF, Royle NJ, Patel I, Jeffreys AJ, Verga V, Jenkins T, Weber JL, Mitchell AL, Bale AE. The CEPH consortium linkage map of human chromosome 1. Genomics 9:686–700, 1991

Engelman DM, Steitz TA, Goldman A. Identifying nonpolar transbilayer helices in amino acid sequences of membrane proteins. Annu Rev Biophys Chem 15:321–353, 1986

Fleagle JG. Primate adaptation and evolution. Academic, London, p 259, 1988

Galinski MR, Corredor Medina C, Ingravallo P, Barnwell JW. A reticulocyte-binding protein complex of plasmodium vivax merozoites. Cell 69:1213–1226, 1992

Hadley TJ, David PH, McGinnis MH, Miller LH. Identification of an erythrocyte component carrying the Duffy blood group Fy^a antigen. Science 223:597–599, 1984

Hadley TJ, Lu Z, Wasniowska K, Martin AW, Peiper SC, Hesselgesser J, Horuk R. Postcapillary venule endothelial cells in kidney express a multispecific chemokine receptor that is structurally and functionally identical to the erythroid isoform, which is the Duffy blood group antigen. J Clin Invest 94:985–991, 1994

Hitzeroth HW, Bender K, Burckhardt K. South African negroes: serogenetic polymorphisms (ABO, Rhesus, MNS, Duffy and Kell) and inter-ethnic genetic distances. Acta Anthropogenet 6(3):171–193, 1982

Horuk R, Chitnis CE, Darbonne WC, Colby TJ, Rybicki A, Hadley TJ, Miller LH. A receptor for the malarial parasite Plasmodium vivax: the erythrocyte chemokine receptor. Science 261:1182–1184, 1993

Iwamoto S, Omi T, Kajii E, Ikemoto S. Genomic organization of the glycoprotein D gene: duffy blood group Fy^a/Fy^b alloantigen system is associated with a polymorphism at the 44-amino acid residue. Blood 85:622–626, 1995

Iwamoto S, Li J, Omi T, Ikemoto S, Kajii E. Identification of a novel exon and spliced form of Duffy mRNA that is the predominant transcript in both erythroid and postcapillary venule endothelium. Blood 87:378–385, 1996

Mallinson G, Soo KS, Schall TJ, Pisacka M, Anstee DJ. Mutations in the erythrocyte chemokine receptor (Duffy) gene: the molecular basis of the Fy^a/Fy^b antigens and identification of a deletion in the Duffy gene of an apparently healthy individual with the Fy(a-b-) phenotype. Br J Haematol 90:823–829, 1995

Marsh WL. Present status of the Duffy blood group system. Crit Rev Clin Lab Sci 5:387–412, 1975

Marshall RD. Glycoproteins. Ann Rev Biochem, pp 673–702, 1972

Mathew S, Chaudhuri A, Murty VVVS, Pogo AO. Confirmation of Duffy blood group antigen locus (FY) at 1q22 → q23 by fluorescence in situ hybridization. Cytogenet Cell Genet 67:68, 1994

McGinniss MH, Miller LH. Malaria, erythrocyte receptors and the Duffy blood group system. In: Steane EA (ed) Cellular antigens and disease. American Association of Blood Banks, Washington DC, pp 67–77, 1977

Miller LH, Mason SJ, Dvorak JA, McGinniss MH, Rothman IK. Erythrocyte receptors for (Plasmodium knowlesi) malaria: Duffy blood group determinants. Science 189:561–563, 1975

Miller LH, Mason SJ, Clyde DF, McGinniss MH. The resistance factor to Plasmodium vivax in blacks. The Duffy blood group genotype FyFy. N Engl J Med 295:303–304, 1976

Moore S, Woodrow CF, McClelland DBL. Isolation of membrane components associated with human red cell antigens Rh(D),(c), (E) and Fy^a. Nature 295:529–531, 1982

Neote K, Mak JY, Kolakowski LF, Schall TJ. Functional and biochemical analysis of the cloned Duffy antigen: identity with the red blood cell chemokine receptor. Blood 84:44–52, 1994

Nichols ME, Rubinstein P, Barnwell JD, de Cordoba SR, Rosenfield RE. A new human Duffy blood group specificity defined by a murine monoclonal antibody. J Exp Med 166:776–785, 1987

Oberdorfer CE, Kahn B, Moore V, Zelenski K, Øyen R, Marsh WL. A second example of anti-Fy3 in the Duffy blood group system. Transfusion 14:608–611, 1974

Palatnik M, Rowe AW. Duffy and Duffy-related human antigens in primates. J Hum Evol 13:173–179, 1984

Peiper SC, Wang Z-X, Neote K, Martin AW, Showell HJ, Conklyn MJ, Ogborne K, Hadley TJ, Zhao-hai L, Hesselgesser J, Horuk R. The Duffy antigen/receptor for chemokines (DARC) is expressed in endothelial cells of Duffy negative individuals, who lack the erythrocyte receptor. J Exp Med 181:1311–1317, 1995

Riwom S, Janvier D, Navenot JM, Benbunan M. Muller JY, Blanchard D. Prooduction of a new murine monoclonal antibody with Fy6 specificity and characterization of the immunopurified N-glycosylated Duffy-active molecule. Vox Sang 66:61–67, 1994

Salmon C, Cartron J-P, Rouger P. The human blood groups. Masson, New York, pp 249–255, 1984

Sanger R, Race RR, Jack JA. The Duffy blood groups of New York Negroes The phenotype Fy(a-b-). Br J Haematol 1:370–374, 1955

Tanner MJA, Anstee DJ, Mallinson G, Ridgwell K, Martin PG, Aventi ND, Parsons SF. Effect of endoglycosidase F-peptidyl N-glycosidase F preparations on the surface components of the human erythrocyte. Carbohydr Res 178:203–212, 1988

Tournamille C, Colin Y, Cartron J-P, Le Van Kim C. Disruption of a GATA motif in the Duffy gene promoter abolishes erythroid gene expression in Duffy-negative individuals. Nat Genet 10:224–228, 1995

Waters AP, Higgins DG, McCutchan TF. Plasmodium falciparum appears to have arisen as a result of lateral transfer between avian and human. Proc Natl Acad Sci USA 88:3140–3144, 1991

Wertheimer SP, Barnwell JW. Plasmodium vivax interaction with human Duffy blood group glycoprotein: identification of a parasite receptor-like protein. Exp Parasitol 69:340–350, 1989

Zhao-hai L, Zi-xuan W, Horuk R, Hesselgesser J, Yan-chun L, Hadley TJ, Peiper SC. The promiscuous chemokine binding profile of the antigen/receptor for chemokines (DARC) is primarily localized to sequences in the amino terminal domain. J Biol Chem (in press)

6 The α-Galactosyl Epitope (Galα1–3Galβ1–4GlcNAc-R) and the Natural Anti-Gal Antibody

U. Galili

Introduction

The carbohydrate structure Galα1–3Galβ1–4GlcNAc-R (termed the α-galactosyl epitope) and the natural antibody which interacts specifically with this epitope (termed anti-Gal) display a unique pattern of distribution in primates. The α-galactosyl epitope is abundantly expressed on red cells and nucleated cells of prosimians and New World monkeys as well as on cells of nonprimate mammals. It is absent, however, from Old World monkeys, apes, and humans. In contrast, anti-Gal is produced in large amounts in Old World monkeys, apes, and humans but is absent from New World monkeys, prosimians, and nonprimate mammals. The studies reported on in this chapter: (1) identify the α-galactosyl epitope as the "B-like" antigen reported by Landsteiner 70 years ago to be present on New World monkey red cells; (2) describe the reciprocity in the distribution of the α-galactosyl epitope and of anti-Gal in primates; (3) discuss the molecular aspects of the evolutionary event that led to suppression of α-galactosyl epitope expression and the appearance of anti-Gal in ancestral Old World primates (i.e., Old World monkeys and apes); and (4) discuss the pathophysiologic outcomes of the interaction between the α-galactosyl epitope and anti-Gal in the areas of red cell aging, xenotransplantation and autoimmunity. Finally, this chapter describes

Molecular Biology and Evolution of Blood Group
and MHC Antigens in Primates
Blancher/Klein/Socha (Eds.)
© Springer-Verlag Berlin Heidelberg 1997

the potential exploitation of anti-Gal and the α-galactosyl epitope for the augmentation of viral and tumor vaccine immunogenicity.

Distribution of Anti-Gal and *a*-Galactosyl Epitopes in Primates

Affinity chromatography of human, ape, or Old World monkey sera on columns of α-galactosyl epitope, Galα1–6Glc-R (melibiose), or α-methyl galactoside coupled to beads results in the binding of as much as 1% of serum IgG to these adsorbents (Galili et al. 1984; Davin et al. 1987; Galili 1993a). The eluted antibody (termed anti-Gal or anti-α-galactosyl IgG) was found to be polyclonal and to interact specifically with α-galactosyl epitopes and with no other carbohydrate epitopes produced by mammalian cells. This highly restricted interaction of anti-Gal with α-galactosyl epitopes could be demonstrated by immunostaining of mammalian glycolipids with anti-Gal on thin layer chromatography (TLC) plates (Galili et al. 1985, 1987a; Suzuki and Naiki 1984) and by binding to glycoproteins with α-galactosyl epitopes (Towbin et al. 1987; Gabrielli et al. 1991; Galili 1993a) or to synthetic oligosaccharides in their free form (Galili et al. 1985; Towbin et al. 1987), either linked to proteins (i.e., neoglycoproteins) (Weislander et al. 1990) or to solid matrix (e.g., silica beads) (Avila et al. 1989; Tsuji et al. 1990; Galili 1993a). All of these studies have also shown that anti-Gal does not interact with other carbohydrates commonly found on mammalian glycoconjugates such as sialyl, fucosyl, mannosyl, β-galactosyl, N-acetyllactosamine, or Galα1–4Gal residues.

Studies on the expression of α-galactosyl epitopes on primate red cells and on the presence of anti-Gal in serum of various primates demonstrated a unique reciprocal pattern of distribution (Table 1). The α-galactosyl epitope is abundantly expressed on red cells of all New World monkeys studied (Galili et al. 1987b). These epitopes can readily be detected by agglutination of the red cells with anti-Gal purified from normal human serum or with the lectin *Bandeiraea (Griffonia) simplicifolia* IB$_4$, which also interacts specifically with α-galactosyl epitopes (Wood et al. 1979). The α-galactosyl epitope is also expressed in abun-

Table 1. Distribution of anti-Gal and α-galactosyl epitopes in mammals

Species	α-Galactosyl epitope expression	Anti-Gal production
Nonprimate mammals	+	−
Prosimians	+	−
New World monkeys	+	−
Old World monkeys	−	+
Apes	−	+
Humans	−	+

dance on red cells of nonprimate mammals such as rat, rabbit, cow, pig, and dog (Galili et al. 1987b).

Early studies on α-galactosyl epitope expression on primate red cells have demonstrated the presence of this epitope on lemur red cells (Galili 1988b). We recently analyzed this epitope expression on red cells from a variety of other prosimians in blood samples received from Duke University Primate Center. Red cells from loris, galago and potto are readily agglutinated by human anti-Gal and by the *Bandeiraea simplicifolia* IB$_4$ lectin, indicating that prosimians residing in Asia and Africa, like those of Madagascar (i.e., lemurs), produce the α-galactosyl epitope on red cells.

Analysis of α-galactosyl epitopes on glycolipid molecules of red cells from various species demonstrated the presence of ceramide pentahexoside with the structure Galα1-3Galβ1-4GlcNAcβ1-3Galβ1-4Glc-Cer as a major glycolipid molecule in red cells of rabbit, cow, and squirrel monkey (a New World monkey) (Galili et al. 1985, 1987b). The evolutionary conservation, in red cells as well as nucleated cells, of this glycolipid and of glycolipids with longer carbohydrate chains which have terminal α-galactosyl epitopes could be demonstrated in nonprimate mammals and New World monkeys by immunostaining of glycolipids separated on TLC plates using anti-Gal and the *Bandeiraea simplicifolia* IB$_4$ lectin (Galili et al. 1987b; Hendriks et al. 1990). The presence of α-galactosyl epitopes on glycolipids from red cells of New World monkeys, but not of Old World monkeys or humans, was further demonstrated with a mouse monoclonal antibody specific for α-galactosyl epitopes (Galili et al. 1987c). The α-galactosyl epitope is also found on glycoproteins produced in nonprimate mammals (Spiro and Bohyroo 1984; Galili et al. 1988b; Thall and Galili 1990; Thall et al. 1991; Santer et al. 1989; Eckhardt and Goldstein 1983). The same epitopes also were found in New World monkeys (Galili et al. 1988b; Thall and Galili 1990; Thall et al. 1991). Analysis of α-galactosyl epitope expression on nucleated cells have indicated that it is produced as 1×10^6 to 35×10^6 epitopes per cell on various cells of New World monkeys, lemurs, and nonprimate mammals (marsupial as well as placental mammals), but is absent from Old World monkey, apes, and human cells (Galili et al. 1988b). The α-galactosyl epitope is not detected on cells of nonmammalian vertebrates, such as fish, amphibians, reptiles, or birds. It was found, however, on the cobra venom factor glycoprotein as the unique structure Galα1-3Galβ1-4(Fucα1-3)GlcNAc-R (Gowda et al. 1992), which is capable of interaction with anti-Gal (Gowda et al. 1994).

Since the α-galactosyl epitope is the major carbohydrate structure on rabbit red cell glycolipids (Eto et al. 1968; Stellner et al. 1973; Egge et al. 1985), production of anti-Gal in primates could be studied by a simple hemagglutination assay with rabbit red cells, using rabbit anti-human Ig antibodies (Coombs' reagent) as a secondary antibody. No other anti-rabbit red cell antibodies were detected in human serum (Galili 1993a). Anti-Gal activity in human serum results in agglutination of rabbit red cells in titers of 640–2560 (Avila et al. 1989; Galili et al. 1995b). Similar titers of anti-Gal were observed in apes and Old World monkeys; however, no reactivity was detected in the sera of New World monkeys, prosimians, or nonprimate mammals (Galili 1988b; Galili et al. 1987b, 1995a). This result was expected since all of these species are immunotolerant to the α-galactosyl epitope.

Studies in humans have shown that anti-Gal is likely to be produced throughout life as a result of antigenic stimulation by gastrointestinal bacteria that have carbohydrate epitopes on their lipopolysaccharides with structures similar to that of the α-galactosyl epitope (Galili et al. 1988a; Hamadeh et al. 1992). The continuous stimulation of the immune system by gastrointestinal bacteria is effective to the extent that approximately 1% of circulating B lymphocytes in humans are capable of producing this antibody (Galili et al. 1993).

By isolating Epstein-Barr virus (EBV) transformed B lymphocytes which produce anti-Gal, it was possible to clone and identify some of the immunoglobulin genes that encode the variable region of the heavy chain (V_H) of anti-Gal. In eight of the nine clones studied, the V_H genes belonged to the V_H3 family, and most of these genes clustered in a specific region of this family, i.e., they displayed a high level of sequence identity in the complementarily determining regions (CDRs) (Wang et al. 1995b). These findings suggest that many of the genes producing anti-Gal are under structural constraints to produce immunoglobulins which can interact with the highly defined structure of the α-galactosyl epitope. In New World monkeys, prosimians, and nonprimate mammals, although genes potentially capable of producing anti-Gal may be present in the genome, anti-Gal is not produced because of tolerance against autologous α-galactosyl epitopes. Recent studies on mice lacking α-galactosyl epitopes (i.e., "knock-out" mice for the α1,3galactosyltransferase gene) have indicated that, in the absence of autologous α-galactosyl epitopes, the immune system is capable of producing anti-Gal (Thall et al. 1995).

The level of anti-Gal production and the affinity of the antibody may greatly differ in various individuals, and thus may reflect the relative participation of different anti-Gal producing lymphoid clones in the synthesis of this polyclonal antibody. Upon exposure of the human immune system to α-galactosyl epitopes on protozoa, anti-Gal titers may increase by as much as 20-fold. Such an increase was observed in patients with Chagas' disease (Almeida et al. 1991; Avila et al. 1989; Gazinelli et al. 1988; Towbin et al. 1987), in which the immune system reacts against α-galactosyl epitopes expressed on the cell surface glycoproteins of the infecting *Trypanosoma cruzi* parasite (Couto et al. 1990; Almeida et al. 1994). An intensive production of high affinity anti-Gal was also observed in diabetic patients transplanted with fetal porcine islet cells (Groth et al. 1994). In these patients the response of the immune system to α-galactosyl epitopes on the transplanted porcine cells resulted in an increase of up to 50-fold in the titer of anti-Gal, within 7–8 weeks post-transplantation (Galili et al. 1995b; Satake et al. 1994). The observations on the increased titer and affinity of anti-Gal in these patients and in Chagas' disease patients suggest that exposure of the immune system to α-galactosyl epitopes on cells other than the normal gastrointestinal flora results in the preferential proliferation of B lymphoid clones which are capable of producing high affinity anti-Gal molecules.

It is of interest to note that, in a large proportion of elderly individuals, the affinity of anti-Gal is significantly lower than that observed in most young individuals (Wang et al. 1995a). It is not clear, as yet, whether this decrease in anti-Gal activity is related to the well-documented overall age-associated decrease in the "quality" of antibodies, which is thought to cause the higher susceptibility of elderly individuals to opportunistic infections (Kohn 1982), and the inability of many in this population to produce effective antibodies in response to vaccination (Howell et al. 1975).

Anti-Gal is not limited only to the IgG class. Human serum was also found to contain the IgM isotype (Sandrin et al. 1993; Parker et al. 1994) and the IgA isotype (Davin et al. 1987; Hamadeh et al. 1995); however, the IgG isotype of anti-Gal is the most prevalent in the blood (Galili et al. 1995b). This IgG anti-Gal isotype includes all four subclasses of IgG_{1-4} (Ravindran et al. 1988; Gabrielli et al. 1991). In secretory fluids, such as milk, colostrum and saliva, anti-Gal is primarily of the IgA class (Hamadeh et al. 1995). Upon stimulation of the immune system by α-galactosyl epitopes on transplanted porcine cells, anti-Gal was found to increase in activity in all three immunoglobulin classes (Galili et al. 1995b).

Biosynthesis of a-Galactosyl Epitopes in Primates

The abundant expression of the α-galactosyl epitope on nonprimate, prosimian, and New World monkey cells vs its absence from Old World monkeys, apes, and humans is the result of the differential activity of the glycosyltransferase, α1,3galactosyltransferase (α1,3GT). α1,3GT synthesizes the α-galactosyl epitope on N-acetyllactosamine core structures of carbohydrate chains in glycolipids and glycoproteins (Basu and Basu 1973; Betteridge and Watkins 1983; Blake and Goldstein 1981; Blanken and van den Eijnden 1985) by the following reaction:

$$\text{Gal}\beta 1\text{-}4\text{GlcNAc-R} + \text{UDP-Gal} \xrightarrow{\alpha 1,3GT} \text{Gal}\alpha 1\text{-}3\text{Gal}\beta 1\text{-}4\text{GlcNAc-R} + \text{UDP}$$
$$N\text{-acetyllactosamine} \qquad\qquad\qquad \alpha\text{-galactosyl epitope}$$

Catalytic activity of α1,3GT could be demonstrated in microsomal fractions from nonprimate mammal and New World monkey cells but not in Old World monkey or human cells (Galili et al. 1988b; Thall et al. 1991). The gene for α1,3GT has been cloned from mouse (Larsen et al. 1989), cow (Joziasse et al. 1989), pig (Sandrin et al. 1995; Strahan et al. 1995), and New World monkey (Henion et al. 1994) cDNA libraries. Comparison of sequences revealed more than 78% sequence identity in the various species. The α1,3GT gene was found to be conserved in a nonexpressed form (i.e., as a pseudogene) in Old World monkeys (Joziasse et al. 1989; Galili and Swanson 1991) and in humans (Joziasse et al. 1989, 1991; Larsen et al. 1990). No mRNA of this gene could be detected in cells from these species (Joziasse et al. 1989).

Evolution of the *a*-Galactosyl Epitope and *a*1,3-Galactosyltranserase in Primates

The absence of α-galactosyl epitopes from nonmammalian vertebrates vs its abundant expression on cells of marsupials (e.g., opossum and kangaroo) and placental mammals (Galili et al. 1988b) suggests that the α1,3GT gene appeared early in mammalian evolution, before the divergence between placental and marsupial mammals. This enzyme has been conserved in an active form in the various lineages of mammals (including primates), subsequent to the great mammalian radiation, estimated to have occurred some 80 million years ago (Eisenberg 1981). The lack of α1,3GT activity in Old World monkeys, apes, and humans vs its expression in New World monkeys implies that the activity of this enzyme was suppressed in ancestral Old World higher primates after their divergence from New World monkeys, estimated to have occurred about 35–40 million years ago (Pilbeam 1984). It is possible that an infectious agent detrimental to primates, which was endemic to the Old World and which expressed α-galactosyl epitopes, exerted a powerful selective pressure for the survival of small groups of ancestral Old World primates that could protect themselves against such a pathogen by producing anti-Gal. The production of the antibody required the suppression of autologous α-galactosyl epitope expression in order to lose immune tolerance to this epitope and to prevent autoimmune phenomena. Presently, there are many pathogens known to express α-galactosyl epitopes. These include enveloped viruses propagated in nonprimate mammals and in New World monkeys, in which α-galactosyl epitopes on the envelope glycoproteins are synthesized by the host α1,3GT (Geyer et al. 1984; Repik et al. 1994; Galili et al. 1996), bacteria (Galili et al. 1988a), and protozoa such as *Trypanosoma cruzi* (Avila et al. 1989; Couto et al. 1990; Almeida et al. 1994). According to this scenario, lemurs, which diverged from primates approximately 60 million years ago and which reside in Madagascar, have been geographically separated from Old World primates and thus were not subjected to the selective pressure that led to the evolutionary inactivation of the α1,3GT gene. New World monkeys, which also have been geographically separated from Old World primates, like lemurs, did not have to cope with the selective pressure for α1,3GT inactivation and thus continue presently to synthesize α-galactosyl epitopes. The expression of α-galactosyl epitopes on cells of loris, potto, and galago suggests that the ancestors of these prosimians were not affected by the putative pathogen that was detrimental to higher primates in Asia and Africa.

Based on distinct differences in nucleotide sequences of the α1,3GT pseudogene in Old World monkeys, apes, and humans, we suggested that the α1,3GT gene was inactivated in ancestral Old World primates less than 28 million years ago, after the divergence between ancestral Old World monkeys (cercopithecoids) and apes (hominoids) (Galili and Swanson 1991). The selection of ancestral Old World primates which could suppress autologous α1,3GT gene expression and produce anti-Gal was likely to have been associated with a major extinction of primates during the Miocene. Although no proof can be provided, it may be possible that the observations on the absence of apes from the fossil record in the period of 5–10 million years ago, vs their success in number and species diversity in the early Miocene, and the late beginning of Old World monkey radiation (estimated to have occurred 8 million years ago) are related to this evolutionary event (Galili and Andrews 1995).

The exact molecular mechanism which caused the inactivation of α1,3GT in ancestral Old World primates is not clear at present. The observation on the absence of α1,3GT mRNA in Old World monkey and human cells (Joziasse et al. 1989) implies that this inactivation was primarily the result of regulatory mutations which affected the promoter of this gene and thus prevented transcription of the α1,3GT gene. In addition, studies on the human α1,3GT pseudogene have demonstrated two deletions which cause shifts in the reading frame and the appearance of premature stop codons (Larsen et al. 1990). These studies imply that, in case of aberrant expression of the α1,3GT gene in human cells, these mutations would ensure the inactivation of the enzyme product and thus would prevent the de novo synthesis of α-galactosyl epitopes. Aberrant synthesis of α-galactosyl epitopes in humans is of pathologic consequence since it is likely to result in an autoimmune reaction mediated by anti-Gal binding to these epitopes (Galili 1989). Such a scenario seems to be part of the pathogenesis of Graves' disease. When anti-Gal binds to α-galactosyl epitopes on the TSH (thyroid stimulating hormone) receptor on porcine thyrocytes, it exerts a stimulatory effect similar to that of TSH (Winand et al. 1993). We have further reported that anti-Gal binds in vitro to Graves' disease thyrocytes but not to normal human thyrocytes and that binding to the pathologic thyrocytes results in stimulation of iodine uptake, increased cAMP synthesis, and increased cell proliferation (Winand et al. 1994). Furthermore, specific removal of anti-Gal from the serum of Graves' disease patients resulted in a 50%–80% decrease in the in vitro stimulatory effect of these sera on autologous thyrocytes (Winand et al. 1994). All of these observations suggest the aberrant expression of α-galactosyl epitopes on Graves' disease thyrocytes by a mechanism which has not been elucidated as yet.

Studies of the α1,3GT pseudogene in primates demonstrated that the two deletions also occur in the chimpanzee pseudogene, which displays only one base difference compared to the human pseudogene (Galili and Swanson 1991). Gorilla and orangutan were found to have only one of the two deletions, and their α1,3GT sequences were found to display 97% identity with the human sequence (Galili and Swanson 1991). Introduction of the one deletion common to apes into the New World monkey cDNA indeed resulted in complete elimination of the catalytic activity of α1,3GT when expressed in COS cells (Henion et al. 1994). It is of interest to note that Old World monkeys were found to lack these deletions in their α1,3GT pseudogenes, suggesting that only apes developed an additional mechanism of structural mutations (i.e., deletions) to ensure prevention of α1,3GT expression (Galili and Swanson 1991).

Anti-Gal, Anti-Blood Group B Antibodies, and Landsteiner's B-Like Antigen

The blood group B antigen is very similar in its molecular structure to the α-galactosyl epitope. The only difference between the two structures is the fucose linked to the penultimate galactose in blood group B antigen (Fig. 1). This structural difference result in subtle differences in the specificity of anti-Gal in individuals of various blood types. As expected, anti-Gal antibodies from blood group B or AB individuals interact with α-galactosyl epitopes but not with

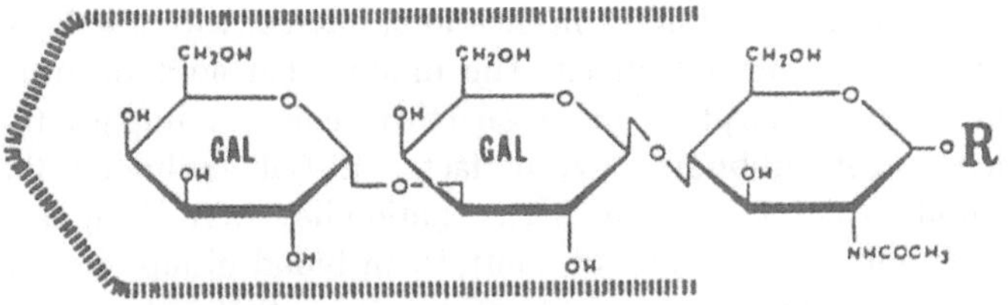

α-galactosyl epitope
Galα1-3Galβ1-4GlcNAc-R

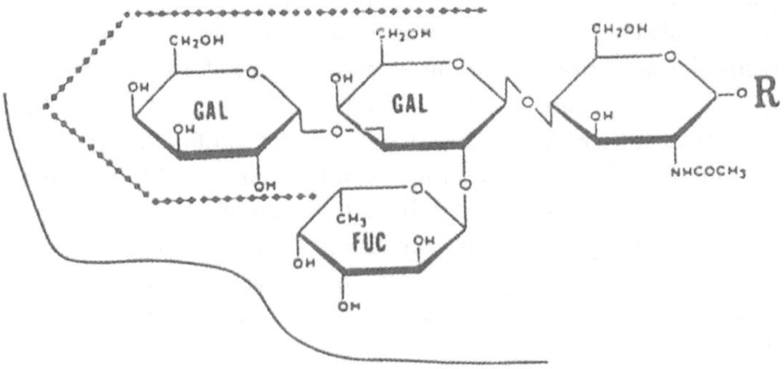

blood group B antigen
Galα1-3(Fucα1-2)Galβ1-4GlcNAc-R

Fig. 1. The interaction between anti-Gal and the α-galactosyl epitope or the blood group B antigen. The *hatched line* represents the combining site of anti-Gal from AB or B sera which only interacts with the α-galactosyl epitope. The *dotted line* represents anti-Gal from A or O sera which may bind to the α-galactosyl epitope with or without the branching fucose residue on the penultimate galactosyl (i.e., anti-Gal B). The *solid line* represents anti-B antibody interacting exclusively with the blood group B antigen. In humans, approximately 90 % of anti-B antibodies in the serum have the specificity displayed by the *dotted line*, and 10 % have the specificity displayed by the *solid line*. Rabbits immunized with blood group B red cells produce anti-B antibodies with a specificity displayed by the *solid line*. (From Galili et al. 1987a)

blood group B antigen, unless it is defucosylated (Galili et al. 1987a). However, anti-Gal from the serum of individuals of blood group A or O interacts also with blood group B antigen (Galili et al. 1987a). Moreover, approximately 90 % of anti-blood group B antibodies in A or O individuals can be adsorbed on α-galactosyl epitopes and only 10 % of anti-B antibodies interact exclusively with the fucosylated form of the α-galactosyl epitope (i.e., blood group B antigen) (Galili et al. 1987a; Galili 1988b). These observations suggest that anti-Gal, being a polyclonal antibody, can recognize multiple "facets" of the α-galactosyl epitope. In blood group A and O individuals, many of the anti-Gal clones can interact with α-galactosyl epitopes regardless of the presence of a fucosyl linked to the penultimate galactose. These clones can, therefore, interact with α-galactosyl epitopes and with blood group B antigen. The finding that 90 % of anti-B activity can be removed by adsorption on α-galactosyl epitopes implies that most of the so-called anti-B antibodies are, in fact, anti-Gal antibodies that also can bind to blood group B antigen. These antibodies were designated anti-Gal B antibodies (Galili et al. 1987a). In contrast, in blood group B or AB individuals, in whom blood B antigen is a self-antigen, no anti-Gal clones capable of binding to blood group B are produced because of immune tolerance. These individuals produce anti-Gal antibodies only against the nonfucosylated α-galactosyl epitopes, i.e., the Galα1–3Galβ1–4GlcNAc-R epitope (see Fig. 1).

These observations explain the original report of Landsteiner and Philip-Miller (1925) on the occurrence of the so-called B-like antigen on red cells of New World monkeys. In their seminal studies on blood group antigens in primates, these investigators observed that anti-B antibodies are, in fact, heterogeneous antibodies recognizing more than one antigen. They found that New World monkey red cells are agglutinated by human anti-B antibodies, but these red cells do not bind anti-B antibodies from sera of rabbits immunized with human blood group B red cells. Therefore, Landsteiner and Philip-Miller (1925) suggested that New World monkey red cells express an antigenic structure that closely resembles blood group B antigen, but differs from it since it does not bind rabbit anti-B antibodies. They designated this antigen as the B-like antigen and stated that it is present on New World monkey red cells but absent from human, ape, and Old World monkey red cells.

The studies on anti-Gal B described above (Galili et al. 1987a) indicate that the B-like antigen is the α-galactosyl epitope, which is abundant on New World monkey red cells and on rabbit red cells (Galili et al. 1987b). Since most of anti-B antibodies are anti-Gal B, these antibodies can agglutinate both human red cells and New World monkey red cells. However, anti-B antibodies produced in rabbit by immunization with human blood group B red cells can interact exclusively with the fucosylated form of the α-galactosyl epitope (i.e., blood group B antigen). Rabbit anti-B antibodies do not bind to the α-galactosyl epitope, since this antigen is an autologous structure on rabbit red cells; thus rabbits are immunotolerant to it. Therefore, the rabbit anti-B antibodies can interact with blood group B red cells but not with New World monkey red cells, which lack blood group B antigen but express the α-galactosyl epitope (Fig. 1).

Anti-Gal Binding to Human Senescent Red Cells

Removal of human senescent red cells from the circulation was found to be mediated by the binding of several hundred IgG molecules to cryptic antigens exposed de novo in course of red cell aging (Kay 1975). Studies aimed at identifying the specificity of these antibody molecules have indicated that many of them display anti-Gal specificity. A large proportion of IgG molecules on senescent red cells could be eluted by galactose, α-methyl-galactoside or melibiose (Galα1–6Glc) but not by other carbohydrates (Galili et al. 1984, 1986b). Furthermore, in these studies anti-Gal isolated from normal sera was found to bind in vitro to senescent red cells depleted of autologous IgG but not to young red cells. Anti-Gal was also found to bind in vivo to a large proportion of pathologic red cells with intrinsic deformability defects, such as red cells from patients with β-thalassemia or sickle cell anemia (Galili et al. 1983, 1984, 1986a).

All these findings led to the hypothesis that human red cells have on their surface cryptic α-galactosyl epitopes that are exposed as the cell ages (Galili et al. 1986a,b; Galili 1988a). Alternatively, there could be a cryptic epitope, uncharacterized as yet, which may bind anti-Gal upon exposure. It was further suggested that senescent red cells, being denser and thus less deformable than young red cells, are retained for longer periods in the small passages of the reticuloendothelial system in the spleen. At these sites, the red cells are likely to be subjected to the activity of proteases produced by macrophages lining the passages. These proteases may expose the cryptic epitopes that are capable of binding anti-Gal (Galili 1988a). In sickle cell anemia and in β-thalassemia the intrinsic defects in the deformability capacity of the red cells result in the retention of young pathologic red cells within the small sinuses of the reticuloendothelial system and thus cause the subsequent premature exposure of cryptic anti-Gal binding epitopes. The observed binding of anti-Gal to these red cells seems to greatly contribute to their early removal from the circulation (Galili et al. 1983, 1984, 1986a).

It was estimated that there are approximately 2000 anti-Gal binding cryptic epitopes per red cell (Galili et al. 1986a). Because of their small number per cell, it has been difficult to isolate and characterize these epitopes on human red cells. Thus, the mechanism leading to their biosynthesis is not yet clear. It should be stressed that the anti-Gal-mediated destruction of normal senescent red cells and of some pathologic red cells is one of several mechanisms that contribute to the removal of senescent red cells. Evidently, this mechanism does not exist in nonprimate mammals and New World monkeys, all of which lack anti-Gal. In humans, however, anti-Gal together with other autoantibodies (Kay et al. 1983; Sorette et al. 1991) mediates removal of senescent red cells.

Interaction of Anti-Gal and the *a*-Galactosyl Epitope as an Obstacle for Xenotransplantation in Humans

The evolutionary event which led to the suppression of α-galactosyl epitope expression and the appearance of anti-Gal in ancestral Old World primates is of clinical significance, since it generated an immunological barrier for the transplantation of organs and tissues from nonprimate mammalian donors into human recipients. In recent years, the severe shortage in human organs for transplantation has increased interest in the possible use of xenogeneic organs for this purpose (Auchincloss 1988; Platt and Bach 1991). The use of primate donors is restricted by their number in captivity and by concerns for animal rights. These limitations prompted intensive research into the possible use of nonprimate mammals (in particular pigs) as xenograft donors. Attempts to transplant porcine organs into humans or into Old World monkeys have shown that the transplanted organs undergo rapid rejection (termed hyperacute rejection), which is mediated by preexisting natural antibodies that bind to the endothelium of the graft, fix complement, and lead to cytolytic destruction of the endothelial cells and collapse of the vascular bed (Leventhal et al. 1993; Platt and Bach 1991). Since the α-galactosyl epitope is expressed as millions of epitopes per cell on porcine endothelial cells as well as on other porcine cells (Galili et al. 1988b), it was reasonable to assume that binding of the natural primate or human anti-Gal to these epitopes mediates the observed hyperacute rejection (Galili 1993b). Indeed, specific removal of anti-Gal from human serum, by adsorption on α-galactosyl epitopes linked to silica beads, resulted in the elimination of the complement-mediated cytolysis induced by human sera on various porcine cells (Good et al. 1992). Furthermore, incubation of porcine endothelial cells, epithelial cells, or fibroblasts with anti-Gal and with human mononuclear cells resulted in lysis of the porcine cells by antibody-dependent cell cytotoxicity (ADCC) (Galili 1993b). It is likely that, in vivo, anti-Gal IgM molecules mediate complement cytolysis whereas anti-Gal IgG molecules induce the adhesion of granulocytes, monocytes, macrophages, and NK cells to the xenograft cells via the interaction between their Fc receptors and the Fc portion of anti-Gal. This adhesion may result in cell mediated destruction of the graft. Indeed, transplantation of New World monkey, and porcine organs into Old World monkeys was shown to result in hyperacute rejection mediated by complement activated by anti-Gal (Collins et al. 1995).

The immune system in humans also was found to respond vigorously to α-galactosyl epitopes on porcine grafts, resulting in a 50-fold increase in the anti-Gal titer of diabetic patients transplanted with fetal porcine islet cells (Galili et al. 1995b). These findings imply that abolishing complement activity may be insufficient for preventing xenograft rejection since the rejection can be induced by anti-Gal-mediated ADCC. A theoretical approach for avoiding this immunological barrier of anti-Gal is the generation of pig strains which lack α1,3GT activity. Disrupting this gene by homologous recombination may be achieved in the future (i.e., knockout of the α1,3GT gene). Currently, there are no porcine embryonic stem cells which may facilitate such studies. Furthermore, it is not yet clear whether pigs devoid of α-galactosyl epitopes will undergo normal development. A recent study on the knock-out of the $\alpha 1,3$GT gene in mice suggests that such a gene disruption may theoretically be feasible also in pigs (Thall et al. 1995). In addition, the fact that ancestral Old World primates succeeded in eliminating this epitope, as a result of evolutionary constraints, further suggests that elimination of this epitope in species that may serve as organ donors is a reasonable objective for the foreseeable future.

Opsonization of Tumor and Viral Vaccines by Anti-Gal

The studies discussed above describe the detrimental effect of anti-Gal. This antibody, however, may theoretically be exploited in the future for improving immunogenicity of human autologous and allogeneic tumor vaccines as well as the immunogenicity of various inactivated virus vaccines. This is because the antibody is present in all humans, and the α-galactosyl epitope can be readily synthesized in vitro, on human tumor cells and on viruses by recombinant (rec.) $\alpha 1,3$GT, as described below.

The objective of cancer immunotherapy is the induction of an effective immune response against tumor-associated antigens (TAA) and thus stimulation of the immune-mediated destruction of tumor cells expressing these antigens (Livingston 1991; Pardoll 1993). An effective anti-TAA response requires the effective uptake, processing, and presentation of these antigens by antigen presenting cells (APCs), so that the receptors on T helper cells can recognize the TAA peptides in association with class II MHC molecules on the membrane of APCs (Unanue and Allen 1987; Lanzavecchia 1993).

A rate limiting step in the antigenic processing of TAA is the inadequate uptake of the vaccinating tumor cells or cell membranes by APC, since these membranes usually lack specific signals directing their uptake by macrophages or dendritic cells. The need for increased uptake of tumor vaccines was summarized by Nossal (1993) as: "To have the best chance of creating an immunogenic anti-tumor vaccine ... the antigen in question must be prepared in such a way as to be palatable to professional antigen-presenting cells, particularly dendritic cells and macrophages." A number of studies with protein antigens, such as tetanus toxoid, β-galactosidase and hepatitis B virus envelope protein, have shown that their processing and presentation are much more effective when they are presented to human monocytes/macrophages in the form of an antigen-antibody complex with a specific IgG antibody, rather than as the pure antigen (Gosselin et al. 1992; Manca et al. 1991; Celis and Chang 1984). This augmentation of antigen processing was found to be mediated by the increased uptake of the antigen-antibody complexes via Fc receptors on the APCs that bind the Fc portion of the complexed antibody. Since anti-Gal is present in all humans and since this antibody readily binds in vivo to α-galactosyl epitopes, as shown in xenotransplantation studies (Collins et al. 1995), it is probable that this antibody would opsonize autologous tumor vaccines if the cell membranes in these vaccines expressed α-galactosyl epitopes.

As indicated above, human cells do not express α-galactosyl epitopes. However, such epitopes can be readily synthesized on human cells in vitro by using rec. α1,3GT. As we have recently shown (Galili and Anaraki 1995), this synthesis could be demonstrated on human red cells by a two-step reaction using the recombinant New World monkey α1,3GT produced in bacterias in a soluble form.

I. SA-Galβ1–4GlcNAc-R   *neuraminidase*   Galβ1–4GlcNAc-R + SA
$$\xrightarrow{\hspace{2cm}}$$
N-acetyllactosamine   sialic acid

II. Galβ1–4GlcNAc-R + UDP-Gal   *rec. α1,3GT*   Galα1–3Galβ1–4GlcNAc-R + UDP
$$\xrightarrow{\hspace{2cm}}$$
α-galactosyl epitope

In step I, the *N*-acetyllactosamine residues, which serve as acceptors for α1,3GT are exposed on the human cells by removal of sialic acid residues with neuraminidase. In step II, the exposed *N*-acetyllactosamine residues are "capped" by terminal α1,3galactosyl with rec. α1,3GT. As many as 40 000 α-galactosyl epitopes could be synthesized de novo per cell. Furthermore, these epitopes readily bind anti-Gal, implying that a similar interaction may occur in vivo upon immunization with tumor membranes manipulated by these reactions (Galili and Anaraki 1995).

A similar rationale may apply to inactivated virus vaccines. Inactivated virus vaccines or subviral vaccines usually display low immunogenicity because of low level uptake by APCs. By using rec. α1,3GT, as described above, α-galactosyl epitopes can be synthesized on the asparagine (N)-linked carbohydrate chains of viral envelope glycoproteins, thereby facilitating the in vivo binding of anti-Gal IgG to vaccinating virions or viral glycoproteins. We have recently demonstrated such augmentation of immunogenicity in an in vitro system containing helper T cell clones specific for influenza virus hemagglutinin and APCs (Galili et al. 1996). Presentation of this viral antigen by APCs to the T cells was tenfold more effective when influenza virus expressing α-galactosyl epitopes was incubated in the presence of anti-Gal than in the absence of anti-Gal. This antibody did not augment immunogenicity of influenza virus lacking α-galactosyl epitopes. These are preliminary studies which require the support of in vivo vaccination data. Nevertheless, the occurrence of anti-Gal in all humans as a result of a peculiar evolutionary event 10–20 million years ago and our ability to synthesize the epitope recognized by this antibody is likely to provide an effective tool for manipulation of the immune system for an improved immune response to various antigens.

References

Almeida IC, Milani SR, Gorin AJ, Travassos LR. Complement mediated lysis of Trypanosoma cruzi trypomastigotes by human anti-α-galactosyl antibodies. J Immunol 146:2394–2401, 1991

Almedia IC, Ferguson MAJ, Schenkman S, Travassos LR. Lytic anti-α-galactosyl antibodied from patients with chronic Chagas' disease recognize novel O-linked oligosaccharides on mucine-like glycosyl-phosphatidylinositol-anchored glycoproteins of Trypanosoma cruzi. Biochem J 304:793–802, 1994

Auchincloss H. Xenogeneic transplantation. Transplantation 45:1–20, 1988

Avila JL, Rojas M, Galili U. Immunogenic Galα1→3Gal carbohydrate epitopes are present on pathogenic American Trypanosoma and Leishmania. J Immunol 142:2828–2834, 1989

Basu M, Basu S. Enzymatic synthesis of blood group related pentaglycosyl ceramide by an α-galactosyltransferase. J Biol Chem 248:1700–1706, 1973

Betteridge A, Watkins WM. Two α-3-D galactosyltransferases in rabbit stomach mucosa with different acceptor substrate specificities. Eur J Biochem 132:29–35, 1983

Blake DD, Goldstein IJ. An α-D-galactosyltransferase in Ehrlich ascites tumor cells: biosynthesis and characterization of a trisaccharide (α-D-galacto(1-3)-N-acetyllactosamine). J Biol Chem 256:5387–5393, 1981

Blanken WM, van den Eijnden DH. Biosynthesis of terminal Galα1-3Galβ1-4GlcNAc-R oligosaccharide sequence on glycoconjugates: purification and acceptor specificity of a UDP-Gal: N-acetyllactosamine α1,3galactosyltransferase. J Biol Chem 260:12972–12934, 1985

Celis E, Chang TW. Antibodies to hepatitis B surface antigen potentiate the response of human T lymphocyte clones to the same antigen. Science 224:297–299, 1984

Collins BH, Cotterell AH, McCurry KR, Alvarado CG, Magee JC, Parker W, Platt JL. Cardiac xenografts between primate species provide evidence for the importance of the α-galactosyl determinant in hyperacute rejection. J Immunol 154:5500–5510, 1995

Couto AS, Conclaves MF, Colli W, deLederkremer RM. The N-linked carbohydrate chain of the 85-kilodalton glycoprotein from Trypanosoma cruzi trypomastigotes contains sialyl, fucosyl and galactosyl (α1-3) galactose units. Mol Biochem Parasitol 39:101–109, 1990

Davin JC, Malaise M, Foidart JM, Mahieu P. Anti-α-galactosyl antibodies and immune complexes in children with Henoch-Schönlein purpura or IgA nephropathy. Kidney Int 31:1132–1139, 1987

Eckhardt AE, Goldstein IJ. Isolation and characterization of α-galactosyl containing glycopeptides from Ehrlich ascites tumor cells. Biochemistry 22:5290–5303, 1983

Egge H, Kordowicz M, Peter-Katalinic J, Hanfland P. Immunochemistry of I/i-active oligo- and polyglycosylceramides from rabbit erythrocyte membranes. J Biol Chem 260:4927–4935, 1985

Eisenberg JF. The mammalian radiation: an analysis of trends in evolution adaptation and behavior. University of Chicago Press, Chicago, 1981

Eto T, Iichikawa Y, Nishimura K, Ando S, Yamakawa T. Chemistry of lipids of the posthemolytic residue or stroma of erythrocytes. XVI. Occurance of ceramide pentasaccharide in the membrane of erythrocytes and reticulocytes in rabbit. J Biochem (Tokyo) 64:205–213, 1968

Gabrielli A, Candel M, Ricciatti AM, Caniglia ML, Wieslander J. Antibodies to mouse laminin in patients with systemic sclerosis (scleroderma) recognize galactose (α1,3)-galactose epitopes. Clin Exp Immunol 86:367–373, 1991

Galili U. Abnormal expression of α-galactosyl epitopes in man: a trigger for autoimmune processes? Lancet ii:358–361, 1989

Galili U. The natural anti-Gal antibody the B-like antigen and human red cell aging. Blood Cells 14:205–220, 1988a

Galili U. The two antibody specificities within human anti-blood group B antibodies. Transfusion Med Rev 2:112–121, 1988b

Galili U. Evolution and pathophysiology of the human natural anti-Gal antibody. Springer Semin Immunopathol 15:155–171, 1993a

Galili U. Interaction of the natural anti-Gal antibody with α-galactosyl epitopes: a major obstacle for xenotransplantation in humans. Immunol Today 14:480–482, 1993b

Galili U, Anaraki F. α-Galactosyl (Galα1–3Galβ1–4GlcNAc-R) epitopes on human cells: synthesis of the epitope on human red cells by recombinant primate α1,3-galactosyltransferase expressed in E. coli. Glycobiology 5:783–789, 1995

Galili U, Andrews P. Suppression of α-galactosyl epitopes synthesis and production of the natural anti-Gal antibody: a major evolutionary event in ancestral Old World primates. J Hum Evol 29:433–442, 1995

Galili U, Swanson K. Gene sequences suggest inactivation of α1,3 galactosyltransferase in catarrhines after the divergence of apes from monkeys. Proc Natl Acad Sci USA 88:7401–7404, 1991

Galili U, Korkesh A, Kahane I, Rachmilewitz EA. Demonstration of a natural anti-galactosyl IgG antibody on thalassemic red blood cells. Blood 61:1258–1264, 1983

Galili U, Rachmilewitz EA, Peleg A, Flechner I. A unique natural human IgG antibody with anti-α-galactosyl specificity. J Exp Med 160:1519–1531, 1984

Galili U, Macher BA, Buehler J, Shohet SB. Human natural anti-α-galactosyl IgG. II. The specific recognition of α(1→3)-linked galactose residues. J Exp Med 162:573–582, 1985

Galili U, Clark MR, Shohet SB. Excessive binding of the natural anti-α-galactosyl IgG to sickle red cells may contribute to extravascular cell destruction. J Clin Invest 77:27–33, 1986a

Galili U, Flechner I, Kniszinski A, Danon D, Rachmilewitz EA. The natural anti-α-galactosyl IgG on human normal senescent red blood cells. Br J Haematol 62:317–324, 1986b

Galili U, Buehler J, Shohet SB, Macher BA. The human natural anti-Gal IgG. III. The subtlety of immune tolerance in man as demonstrated by crossreactivity between natural anti-Gal and anti-B antibodies. J Exp Med 165:693–704, 1987a

Galili U, Clark MR, Shohet SB, Buehler J, Macher BA. Evolutionary relationship between the anti-Gal antibody and the Galα1→3Gal epitope in primates. Proc Natl Acad Sci USA 84:1369–1373, 1987b

Galili U, Basbaum CB, Shohet SB, Buehler J, Macher BA. Identification of erythrocyte Galα1–3Gal glycosphingolipids with a mouse monoclonal antibody. J Biol Chem 262:4683–4687, 1987c

Galili U, Mandrell RE, Hamadeh RM, Shohet SB, Griffis JM. Interaction between human natural anti-α-galactosyl immunoglobulin G and bacteria of the human flora. Infect Immun 56:1730–1737, 1988a

Galili U, Shohet SB, Kobrin E, Stults CLM, Macher BA. Man apes and Old World monkeys differ from other mammals in the expression of α-galactosyl epitopes on nucleated cells. J Biol Chem 263:17755–17762, 1988b

Galili U, Anaraki F, Thall A, Hill-Black C, Radic M. One percent of circulating B lymphocytes are capable of producing the natural anti-Gal antibody. Blood 82:2485–2493, 1993

Galili U, Gregory CR, Morris RE. Contribution of anti-Gal to primate and human IgG binding to porcine endothelial cells. Transplantation 60:210–213, 1995a

Galili U, Tibell A, Samuelsson B, Rydberg L, Groth CG. Increased anti-Gal activity in diabetic patients transplanted with fetal porcine islet cell clusters. Transplantation 59:1549–1556, 1995b

Galili U, Repik PM, Anaraki F, Mozdzanowska K, Washko G, Gerhard W. Enchancement of antigen presentation of influenza virus hemagglutinin by the natural anti-Gal antibody. Vaccine 14:321–328, 1996

Gazzinelli R, Galvao LMC, Dias JCP, Gazzinelli G, Brener Z. Anti-laminin specific antibodies in acute Chagas disease. Transact R Soc Trop Med Hyg 82:574–576, 1988

Geyer R, Geyer H, Strim S, Hunsmann G, Schneider J, Dabrowski U, Dabrowski J. Major oligosaccharides in the glycoprotein of Friend murine leukemia virus: structure elucidation by one and two dimensional proton nuclear magnetic resonance and methylation analysis. Biochemistry 23:5628–5634, 1984

Good AH, Cooper DCK, Malcolm AJ, Ippolito RM, Koren E, Neethling FA, Ye Y, Zuhdi N, Lamontage LR. Identification of carbohydrate structures which bind human anti-porcine antibodies: implication for discordant xenografting in man. Transplant Proc 24:559–562, 1992

Gosselin EJ, Wardwell K, Gosselin DR, Alter N, Fisher JL, Guyre PM. Enhanced antigen presentation using human Fcγ receptor (monocyte/macrophage) specific immunogens. J Immunol 149:3477–3481, 1992

Gowda DC, Schultz M, Bredehorst R, Vogel CW. Structure of the major oligosaccharide of cobra venom factor. Mol Immunol 29:335–344, 1992

Gowda DC, Petrella EC, Raj TT, Bredehorst R, Vogel CW. Immunoreactivity and function of oligosaccharides in cobra venom factor. J Immunol 152:2977–2986, 1994

Groth CG, Korgsen O, Tibell A, Tollerman J, Möller E, Bolinder J, Ostman J, Reinholt FP, Hellerstrom C, Andersson A. Transplantation of fetal porcine pancreas to diabetic patients: biochemical and histological evidence for graft survival. Lancet 344:1402–1404, 1994

Hamadeh RM, Jarvis GA, Galili U, Mandrell RE, Zhou P, Griffis JM. Human natural anti-Gal IgG regulates alternative complement pathway activation on bacterial surfaces. J Clin Invest 89:1223, 1992

Hamadeh RM, Galili U, Zhou P, Griffis JM. Anti-α-galactosyl immunoglobulin A (IgA), IgG, and IgM in human secretions. Clin Diagn Lab Immunol 2:125–131, 1995

Hendriks SP He P Stults CLM and Macher BA Regulation of the expression of Galα1-3Galβ1-4GlcNAc glycosphingolipids in kidney. J Biol Chem 256:17621–17626, 1990

Henion TR, Macher BA, Anaraki F, Galili U. Defining the minimal size of catalytically active primate α1,3-galactosyltransferase: structure function studies on the recombinant truncated enzyme. Glycobiology 4:193–201, 1994

Howell CHL, Vesselinova-Jenkins CK, Evans JD, James J. Influenza vaccination and mortality from bronchopneumonia in the elderly. Lancet i:381–383, 1975

Joziasse DH, Shaper JH, Van den Eijnden DH, Van Tunen AH, Shaper NL. Bovine α1,3-galactosyltransferase: isolation and characterization of a cDNA clone Identification of homologous sequences in human genomic DNA. J Biol Chem 264:14290–14297, 1989

Joziasse DH, Shaper JH, Jabs EW, Shaper NL. Characterization of an α1→3-galactosyltransferase homologue on human chromosome 12 that is organized as a processed pseudogene. J Biol Chem 266:6991–6998, 1991

Kay MMB. Mechanism of removal of red cells by macrophages in situ. Proc Natl Acad Sci USA 72:3521–3525, 1975

Kay MMB, Goodman SR, Sorensen K, Whitfield CF, Wong P, Zaki L, Rudolff V. Senescent red cell antigen is immunologically related to band 3. Proc Natl Acad Sci USA 80:1631–1636, 1983

Kohn RP. Cause of death in very old people. JAMA 247:2793–2796, 1982

Landsteiner K, Philip-Miller C. Serological studies on the blood of the primates. III. Distribution of serological factors related to human isoagglutinogens in the blood of lower monkeys. J Exp Med 42:863–875, 1925

Lanzavecchia A. Identifying strategies for immune intervention. Science 260:937–944, 1993

Larsen RD, Rajan VP, Ruff M, Kukowska-Latallo J, Cummings RD, Lowe JB. Isolation of a cDNA encoding murine UDP galactose: βD-galactosyl-14-N-acetyl-D-glucosaminide α1,3-galactosyltransferase: expression cloning by gene transfer. Proc Natl Acad Sci USA 86:8227–8231, 1989

Larsen RD, Rivera-Marrero CA, Ernst LK, Cummings RD, Lowe JB. Frameshift and nonsense mutations in a human genomic sequence homologous to a murine UDP-Galβ-D-Gal(1,4)-D-GlcNAcα(1,3) galactosyltransferase cDNA. J Biol Chem 265:7055–7062, 1990

Leventhal JR, Dalmaso AP, Cromwell JW, Platt JL, Manivel CJ, Bolman RM, Matas AJ. Prolongation of cardiac xenograft survival by depletion of complement. Transplantation 55:857–866, 1993

Livingston P. Active specific immunotherapy in the treatment of patients with cancer. Immunol Allerg Clin North Am 11:401–423, 1991

Manca F, Fenoglio D, Li-Pira G, Kunkel A, Celada F. Effect of antigen/antibody ratio on macrophage uptake processing and presentation to T cells of antigen complexed with polyclonal antibodies. J Exp Med 173:37–48, 1991

Nossal GJV. Tolerance and ways to break it. NY Acad Sci 690:34–41, 1993

Pardoll DM. Cancer vaccines. Immunol Today 14:310–316, 1993

Parker W, Bruno O, Holzkecht ZE, Platt JE. Characterization and affinity isolation of xenoreactive human natural antibodies. J Immunol 153:3791–3803, 1994

Pilbeam D. The decent of hominoids and hominids. Sci Am March 84–95, 1984

Platt JL, Bach FH. The barrier to xenotransplantation. Transplantation 52:937–947, 1991

Ravindran S, Satapathy AK, Das MK. Naturally occurring anti-α-galactosyl antibodies in human plasmodium falciparum infection: a possible role for autoantibodies in malaria. Immunol Lett 19:137–142, 1988

Repik PM, Strizki JM, Galili U. Differential host dependent expression of α-galactosyl epitopes on viral glycoproteins: a study of eastern equine encephalitis virus as a model. J Gen Virol 75:1177–1181, 1994

Sandrin M, Vaughan HA, Dabkowski PL, McKenzie IFC. Anti-pig IgM antibodies in human serum react predominantly with Galα1-3Gal epitopes. Proc Natl Acad Sci USA 90:11391–11395, 1993

Sandrin MS, Dabkowski PL, Henning MM, Mouthouris E, McKenzie IFC. Characterization of cDNA clones for porcine α1,3galactosyltransferase The enzyme generating the Galα(1,3)Gal epitope. Xenotransplantation 41:101–105, 1995

Santer UV, DeSantis R, Hard KJ, van Kuik JA, Vliegenthart JFG, Won B, Glick MC. N-linked oligosaccharide changes with oncogenic transformation require sialylation of multiantennae. Eur J Biochem 181:249–260, 1989

Satake M, Kawagishi N, Rydberg L, Samuelsson BE, Tibell A, Groth CG, Möller E. Limited specificity of xenoantibodies in diabetic patients transplanted with fetal porcine islet cell clusters. Main antibody reactivity against α-linked galactose-containing epitopes. Xenotransplantation 1:89–101, 1994

Sorette MP, Galili U, Clark MR. Comparison of serum anti-band 3 and anti-Gal antibody binding to density separated human red blood cells. Blood 77:628–636, 1991

Spiro RG, Bhoyroo VD. Occurance of α-D-galactosyl residues in the thyroglobulin from several species Localization in the saccharide chains of the complex carbohydrate units. J Biol Chem 259:9858–9866, 1984

Stellner K, Saito H, Hakomori S. Determination of aminosugar linkage in glycolipids by methylation. Aminosugar linkage of ceramide pentasaccharides of rabbit erythrocytes and of Forssman antigen. Arch Biochem Biophys 133:464–472, 1973

Strahan KM, Gu F, Preece AF, Gustavsson I, Andersson L, Gustafsson K. DNA sequence and chromosome localization of pig α1,3galactosyltransferase. Immunogenetics 41:101–105, 1995

Suzuki E, Naiki M. Heterophile antibodies to rabbit erythrocytes in human sera and identification of the antigen as a glycolipid. J Biochem (Tokyo) 83:103–108, 1984

Thall A, Galili U. Distribution of Galα1-3Galβ1-4GlcNAc residues on secreted mammalian glycoproteins (thyroglobulin fibrinogen and immunoglobulin G) as assayed by a sensitive solid-phase radioimmunoassay. Biochemistry 29:3959–3968, 1990

Thall A, Etienne-Decerf J, Winand R, Galili U. The α-galactosyl epitope on mammalian thyroid cells. Acta Endocrinol (Copenh) 124:692–699, 1991

Thall AD, Maly P, Lowe JB. Oocyte Galα1-3Gal epitopes implicated in sperm adhesion to the zona pellucida glycoprotein ZP3 are not required for fertilization in the mouse. J Biol Chem 270:21437–21442, 1995

Towbin H, Rosenfelder G, Weislander J, Avila JL, Rojas M, Szarfman A, Esser K, Nowack H, Timple R. Circulating antibodies to mouse laminin in Chagas disease American cutaneous Leishmaniasis and normal individuals recognize terminal galactosyl (α1-3) galactose epitopes. J Exp Med 166:419–432, 1987

Tsuji J, Noma S, Suzuki J, Okumura K, Shimizu N. Specificity of human natural antibody to recombinant tissue-type plasminogen activator (t-PA) expressed on mouse C127 cells. Chem Pharm Bull 38:765–768, 1990

Unanue ER, Allen PM. The basis for the immunoregulatory role of macrophages and other accessory cells. Science 236: 551–557, 1987

Wang L, Anaraki F, Henion TR, Galili U. Variations in activity of the human natural anti-Gal antibody in young and elderly populations. J Gerontol (Med Sci) 50A:M227–33, 1995a

Wang L, Radic MZ, Galili U. Human anti-Gal heavy chain genes: preferential use of $V_{H}3$ and the presence of somatic mutations. J Immunol 155:1276–1285, 1995b

Weislander J, Mannson O, Kallin E, Gabrielli A, Nowack H, Timpl R. Specificity of human antibodies against Galα1-3Gal carbohydrate epitope and distinction from natural antibodies reacting with Galα1-2Gal or Galα1-4Gal. Glycoconjugate J 7:85–100, 1990

Winand RJ, Anaraki F, Etienne-Decerf J, Galili U. Xenogeneic thyroid-stimulating hormone-like activity of the human natural anti-Gal antibody. Interaction of anti-Gal with porcine thyrocytes and with recombinant human thyroid stimulating hormone receptors expressed on mouse cells. J Immunol 151:3923–3934, 1993

Winand RJ, Winand-Devigne J, Meurisse M, Galili U. Specific stimulation of Graves' disease thyrocytes by the natural anti-Gal antibody from normal and autologous serum. J Immunol 153:1386–1395, 1994

Wood C, Kabat EA, Murphy LA, Goldstein IJ. Immunochemical studies of the combining sites of two isolectins A4 and B4 isolated from Bandeiraea simplicifolia. Arch Biochem Biophys 198:1–8, 1979

7 Simian-Type Blood Groups of Nonhuman Primates

W. W. SOCHA

Introduction

The study of nonhuman primate blood groups was a logical extension of human serohematology and, understandably, the first attempts to define the specificities of primate red cells were carried out using reagents originally prepared for typing human erythrocytes. The blood groups, such as ABO, MN, RhHr, Lewis, Ii, defined in that way in nonhuman primates, were called *human-type blood groups* and were considered to be homologues of human red cell antigens. The second category of red cell specificities included those detected by antisera specifically produced for typing primate animals and obtained either by immunizing laboratory animals (rabbit, guinea pig, mice, etc.) with the red cells of apes and monkeys or, preferably, by allo-or cross-immunizations of a monkey with red cells of an animal of the same or a closely related species. Although alloimmunizations are often time-consuming and not always productive, the resulting reagents are highly specific and free from interfering heteroagglutinins. The allo- or cross-immune antisera are, therefore, reagents of choice, as they provide the best tool for distinguishing subtle, minute differences among red cell specificities. The sera of this latter category detect the so-called *simian-type blood groups,* which were believed to be primate-restricted specificities of which some could be analogues of the human red cell antigens. Certain simian-type specificities,

Molecular Biology and Evolution of Blood Group
and MHC Antigens in Primates
Blancher/Klein/Socha (Eds.)
© Springer-Verlag Berlin Heidelberg 1997

Table 1. Chimpanzee immune antisera specific for simian-type antigens on the red cells of chimpanzees and other anthropoid apes

Sample number	Specificity	Origin	Blood group system
1	V^c	Chimpanzee crossimmune (man) (CH-169 Possum)	VABD
2	$V^c + W^c$	Chimpanzee isoimmune (CH-639 Dina)	VABD
3	V^{c1}	Chimpanzee isoimmune (CH-19 Gabriel)	VABD
4	A^c	Chimpanzee isoimmune (CH-194 Herbie)	VABD
5	B^c	Chimpanzee isoimmune (CH-355 Leo) (absorbed)	VABD
6	D^c	Chimpanzee isoimmune (CH-192 Lindsay)	VABD
7	$B^c + D^c$	Chimpanzee isoimmune (CH-355 Leo) (absorbed)	VABD
8	$A^c + B^c D^c$	Chimpanzee isoimmune (CH-336 Stu)	VABD
9	N^c	Chimpanzee crossimmune (gibbon) (CH-208 Mandy)	Nn (VABD-related?)
10	R^c	Chimpanzee isoimmune (CH-643 Sean), ETC	RCEF
11	R^c	Chimpanzee isoimmune (CH-177 Karen), ETC	RCEF
12	R^c	Chimpanzee isoimmune (CH-11 Tom), absorbed, ETC	RCEF
13	C^c	Chimpanzee isoimmune (CH-491 Bonnie), ETC	RCEF
14	E^c	Chimpanzee isoimmune (CH-38 Doug), ETC	RCEF
15	E^c	Chimpanzee isoimmune (CH-11 Tom), absorbed, ETC	RCEF
16	F^c	Chimpanzee isoimmune (CH-34 Jack), ETC	RCEF
17	F^c	Chimpanzee isoimmune (CH-136 Hope), ETC	RCEF
18	$C^c + CF^c$	Chimpanzee isoimmune (CH-225 Andy), ETC	RCEF
19	c^c	Chimpanzee isoimmune (CH-114 Rufe), ETC	RCEF
20	c^c_1	Chimpanzee isoimmune (CH-355 Leo), absorbed, ETC	RCEF
21	G^c	Chimpanzee isoimmune (CH-490 Chica), ETC	GG_1g (RCEF-related?)
22	G^c_1	Chimpanzee isoimmune (CH-490 Chica, absorbed), ETC	GG_1g (RCEF-related?)
23	H^c	Chimpanzee isoimmune (CH-488 Ginger)	Hh
24	K^c	Chimpanzee crossimmune (man) (CH-85 Billy), ETC	Kk
25	O^c	Chimpanzee isoimmune (CH-355 Leo), absorbed, ETC	Oo
26	S^c	Chimpanzee isoimmune (CH-490 Chica, 1982)	Ss
27	T^c	Chimpanzee isoimmune (CH-17 Mack), ETC	Tt

ETC, Ezyme-treated red cell technique.

originally defined on the red cells of apes, were later detected, in polymorphic form, on human red cells as well (Socha and Moor-Jankowski 1978, 1979a).

Tables 1–3 list the simian-type blood grouping reagents produced and used by us for defining simian-type blood groups of anthropoid apes and Old World monkeys. Primate immune antisera of various specificities were also developed in the past in a few other laboratories, by and large for typing macaques and rhesus monkeys. Most of these reagents were produced in small quantities and their specificities could not be compared with those of our antisera; a few antisera that were made available to us and tested side-by-side with our standard

Table 2. Rhesus simian-type blood grouping reagents

Sample number	Immunized animal	Specificity of the reagent
1	Rh-1 Caesar	A^{rh}
2	Rh-2 Eve	B^{rh}
3	Rh-216 Path	B^{rh}
4	Rh-4 Eleanor	C^{rh}
5	Rh-6 Antigone	D^{rh}_1
6	Rh-942 Kate	D^{rh}_2
7	Rh-212 Mino	D^{rh}_3
8	Rh-924 Fanny	D^{rh}_3
9	Rh-218 Minim	D^{rh}_4
10	Rh-224 Remple	D^{rh}_4
11	Rh-228 Harper	D^{rh}_4
12	Rh-930 Lois	D^{rh}_4
13	Rh-3 Barney	F^{rh}
14	Rh-178 Rachel	G^{rh}
15	Rh-926 Sally[1]	$J^{rh}_1 + V$
16	Rh-926 Sally[2]	J^{rh}_2
17	Rh-1062 Avery	J^{rh}_3
18	Rh-796 Osmia[3]	$J^{rh}_4 + Z$
19	Rh-796 Osmia[4]	J^{rh}_4
20	Rh-938 Camille[5]	J^{rh}
21	Rh-946 Laura	L^{rh}
22	Rh-936 Dana	M^{rh}
23	Rh-321 Harry	N^{rh}
24	Rh-932 April	O^{rh}
25	Rh-922 Jane	P^{rh}
26	Rh-928 Cindy	R^{rh}

1, unabsorbed; 2, absorbed; 3, unabsorbed; 4 absorbed; 5, absorbed.

reagents often displayed very low titers and avidity and gave, at least in our hands, irreproducible or doubtful results (Duggleby et al. 1971; Duggleby and Stone 1971: Edwards 1971a,b,c; for general reviews see: Socha and Ruffié 1983; Socha et al. 1984). Several alloimmune typing sera produced in crab-eating maca-ques (*Macaca fascicularis*) by Terao et al. (1986) resulted in definition of three simian-type blood group systems in that monkey species. We had the opportunity to tests some the Terao's reagents. Most of the sera showed impressive titers and specificities of some of the antibodies parallelled our anti-B^{rh} and anti-O^{rh} reagents while others were of new type unmatched by our rhesus reagents. The only successfull attempt to produce alloimmune antisera for typing New World monkeys (marmosets) was reported by Gengozian et al. (1966).

Table 3. Baboon simian-type blood grouping reagents

Sample number	Immunized animal	Donor	Specifity
1	B-2 Paula Olive baboon	B-11 Chubby Olive baboon	A^P
2	B-13 Jack Yellow baboon	B-9 Eugene Yellow baboon	$A^P + Y^P$
3	B-127 Blue Hybrid	B-15 Irwin Yellow baboon	$A^P + X^P$
4	B-9 Eugene Yellow haboon	B-13 Jack Yellow baboon	$B^P + Z^P$
5	B-9 Eugene[1] Yellow baboon	B-13 Jack Yellow baboon	B^P_1
6	B-15 Irwin Yellow baboon	B-127 Blue Hybrid	$B^P + O^P$
7	B-15 Irwin[2] Yellow baboon	B-127 Blue Hybrid	B^P_2
8	B-808 Agnes[3] Hybrid	Human O, Rh positive	B^P_2
9	Rh-212 Mino[4] Rhesus monkey	Rh-214 Iodine Rhesus monkey	$B^P_3 (D^{rh}_3)$
10	B-77 Chatta Olive baboon	G-5 Benton Gelada	$B^P + G^P$
11	B-11 Chubby Olive baboon	B-2 Paula Olive baboon	$C^P + Q^P$
12	B-11 Chubby[5] Hamadryas baboon	B-2 Paula Olive baboon	C^P
13	B-32 Lynn Hybrid	B-60012 Janna Hybrid	G^P
14	B-268 Grayma Olive baboon	B-85 Nosmo Chacma baboon	hu
15	B-176 Camilla Hamadryas baboon	B-252 Eberle Yellow baboon	ca
16	B-87 Souds Olive baboon	B-553 Benita Olive baboon	$ca + E^P$
17	B-60012 Janna Olive baboon	B-32 Lynn Olive baboon	O^P_1
18	B-17 Janny Yellow baboon	B-133 Roman Hybrid	P^P
19	B-131 Ralph Hybrid	B-776 Ilex Hamadryas baboon	$S^P + T^P$
20	B-131 Ralph[6] Hybrid	B-776 Ilex Hamadryas baboon	S^P
21	B-133 Roman Hybrid	B-17 Janny Yellow haboon	U^P
22	B-143 Dave Hybrid	B-230 Vivian Yellow baboon	$V^P + V^P_1$
23	B-143 Dave[7] Hybrid	B-230 Vivian Yellow baboon	V^P
24	B-145 Thorb Hybrid	B-774 Yucca Hamadryas baboon	L^P
25	B-776 Ilex[8] Olive baboon	B-131 Ralph Olive baboon	M^P

1, absorbed with pooled blood of B-2 and B-11; 2, absorbed with blood of B-2; 3, transfused with human blood; 4, cross-reacting rhesus isoimmune serum; 5, absorbed with pooled blood of B-9 and B-13; 6, absorbed with blood of B-9; 7, absorbed with pooled blood of V_2 type; 8, absorbed with blood of B-790.

Simian-Type Blood Grouping Methodology

Some tests for simian-type blood groups can be carried out by saline agglutination technique, in which the red cells to be blood grouped, suspended in normal saline (or a low-ionic strength medium), are mixed with typing sera to produce direct agglutination. However, the majority of reagents produced by immunization of primate animals are small molecular antibodies of IgG class and, therefore, give the best results using the antiglobulin technique or the enzyme-treated red cell method. While anti-human globulin sera are used in tests on ape red blood cells, anti-baboon or anti-rhesus globulin reagents are preferable for testing red cells of Old and New Worlds monkeys. Proteolytic enzymes (ficin, bromelin, papain, etc.) modify the erythrocyte membrane in such a way as to produce clumping of the red cells by antibodies that normally only attach themselves to red cells without causing their visible agglutination.

For details of blood grouping techniques in humans and nonhuman primates, readers are referred to specialized articles and textbooks (Erskine and Socha 1978; Socha and Ruffié 1983; Socha et al. 1972).

Simian-Type Blood Groups of Anthropoid Apes

Due to limited availability of blood specimens and specific typing reagents, not all ape species were investigated to the same extent for their simian-type blood groups. Understandably, the most abundant data relate to chimpanzees, a species that is often employed in biomedical research and, therefore, available for immunization experiments and testing. In the course of large-scale immunization attempts started in 1964 (Moor-Jankowski et al. 1965) and continued for the next 20 years, a number of allo- and hetero-immune antisera were developed specific for various epitopes on chimpanzee red cells (see Table 1).

Some of those reagents cross-reacted, often type-specifically, with red cells of other anthropoid apes, such as gorilla, orangutan, siamang and gibbon. The so defined specificities were initially assigned consecutive letters of the alphabet: A^c, B^c, C^c, D^c, E^c, F^c, G^c, H^c, etc.) (superscript "c" for chimpanzee). Some of those specificities were found to be parts of two complex blood group systems, V-A-B-D and R-C-E-F, related to MN and Rh blood group systems, respectively, and discussed in other parts of this book.

Several other alloimmune antisera were produced in chimpanzees that define a number of simian-type specificities beyond those belonging to the two above-mentioned blood group systems. These include specificities tentatively designated as G^c, G_1^c, H^c, K^c, N^c, O^c, S^c, T^c, and W^c, detected on the red cells of chimpanzees and, in a few instances, also in the blood of other apes.

Parallel tests with anti-G^c and anti-G_1^c reagents defined three phenotypes, G_1, G_2 and g. All three types are encountered in chimpanzees and in gorillas, but not on the red cells of Asiatic apes. Reactions of anti-G^c reagents are enhanced by enzyme treatment, thus resembling the Rh (R-C-E-F) system-related antigens.

Antigens N^c and W^c, by contrast, show some characteristics of the MN (V-A-B-D) related antigens in that they are destroyed by treatment with proteolytic enzymes. Moreover, it has been shown that chimpanzee anti-W^c serum reacted with human red blood cells in a manner paralleling the reaction obtained with human antisera of anti-Mi(a) specificity. Mi(a) is one of the Miltenberger specificities at the MNSs locus (Socha and Ruffié 1983) (for details, see Chap. 4). The W^c antigen was detected only on the red cells of some chimpanzees, but not in the blood of any other anthropoid ape or monkey species.

Chimpanzee alloimmune anti-N^c reagent agglutinates blood of some, but not all, chimpanzees and gorillas, the frequencies of N^c-positive and N^c-negative types being comparable in those two species . Nonspecific reactions or no reactivity were observed with red cells of other apes.

As is the case with anti-G^c, anti-G_1^c and anti-N^c sera, chimpanzee alloimmune reagents of anti- H^c, anti-K^c, anti-O^c , anti-S^c and anti-T^c specificities react only with chimpanzee and gorillas red cells, but not with those of Asiatic apes or lower monkeys. The so-defined specificities do not seem to be associated with any other, thus far defined blood group systems and are, therefore considered part of the independent chimpanzee blood group systems, each composed of a dominant gene and its silent recessive allele. Gorillas were found to be polymorphic for O^c, S^c and T^c antigens, but all animals of this species tested so far were H^c- and K^c-positive.

Limited isoimmunization experiments carried out in gibbons resulted in the production of blood grouping reagents designated anti-A^g, anti-B^g and anti-C^g. Of these, A^g appeared to be a species-specific trait of *Hylobates lar lar*, while B^g was detected only in *Hylobates lar pileatus* and C^g defined individual differences among white-handed gibbons (Moor-Jankowski et al. 1965; Wiener et al. 1966).

The first orangutan alloimmune antibody, tentatively called anti-Sj^{or}, was identified in the serum of a pregnant female bearing an erythroblastotic fetus (Socha and Van Foreest 1981). The serum was tested against only a few orangutans and the population frequency of Sj^{or} could not be established, but the serologic properties of the serum and the patterns of its reactivity showed some resemblance with human anti-Rh_o and chimpanzee anti-R^c.

Simian-Type Blood Groups of Monkeys

Red Cell Antigens Specific for Old World Monkeys

Specificities of this category are defined by means of antisera raised against monkey red cells either in animals of the same or closely related species (allo- or crossimmune antisera) or, as was practiced in the past, in small laboratory animals such as rabbits and guinea pigs (heteroimmune antisera). For obvious reasons, the most common sera are those developed in species most readily available for immunization experiments and which are the easiest to handle and maintain in captivity.

Blood Groups Defined by Rhesus Alloimmune Antisera

Rhesus monkeys have been and still remain the monkey species most frequently used in biomedical research and therefore are a frequent source of blood typing reagents. Some of the 25 red cell specificities identified so far on the red cells of various macaque species and baboons by means of our rhesus isoimmune antisera (Table 2) were found to be part of complex, multiallelic blood group systems. Among those first to be described was the so-called graded D^{rh} blood group system, detectable, in more or less complete form, on the red cells of several macaque species (Socha et al. 1976b). The five regular types of the D^{rh} system are

Table 4. Serology and genetics of the graded D^{rh} blood group system of macaques

Blood group designation	Specificities present on the red cells	Rections of red cells with reagents				Possible genotypes
		Anti-D_1	Anti-D_2	Anti-D_3	Anti-D_4	
D_4 (D standard)	D_1, D_2, D_3, D_4	+	+	+	+	$D^4D^4, D^4D^3, D^4D^2, D^4D^1, D^4d$
D_3	D_1, D_2, D_3	+	+	+	−	$D^3D^3, D^3D^2, D^3D^1, D^3d$
D_2	D_1, D_2	+	+	−	−	D^2D^2, D^2D^1, D^2d
D_1	D_1	+	−	−	−	D^1D^1, D^1d
d	None	−	−	−	−	dd
Irregular types						
D_3'		+	−	+	−	
D_3''		−	+	+	−	

defined by the set of four reagents anti-D_1, anti-D_2, anti-D_3, and anti-D_4 (Table 4), each detecting a subspecificity of D^{rh} antigen.

Each D^{rh} blood type corresponds to an autonomous mutation. Each new factor implies the presence of preceding factors to which new sequences are added. The direction of dominance goes from the most complex to the most simple factors; finally, a recessive d^{rh} mutation exists that is considered a silent allelle because it produces no detectable factor. Thus, $D^{rh4} > D^{rh3} > D^{rh2} > D^{rh1} > d^{rh}$.

Distributions of the D^{rh} phenotypes observed in large samples of unrelated rhesus monkeys agreed well with theoretically expected frequencies based on an inheritance model involving a series of multiple graded alleles.

Table 5. Distribution of blood groups of the D^{rh} graded system in various species of macaques and in baboons

Blood group	Percentage of positive reactions with red cells of:						
	Macaca mulatta	Macaca nemestrina	Macaca fascicularis	Macaca arctoides	Macaca radiata	Macaca sylvanus	Papio hamadryas
D_4	34.2	34.9	8.7	0.0	0.0	0.0	0.0
D_3	6.3	2.2	4.0	0.0	0.0	78.1	0.0
D_2	23.0	45.0	21.4	3.3	0.0	21.9	0.0
D_1	1.6	5.6	23.0	0.0	0.0	0.0	41.0
d	34.9	12.3	42.9	66.7	100.0	0.0	0.0
Irregular types							
D_3'	0.0	0.0	0.0	0.0	0.0	0.0	59.0
D_3''	0.0	0.0	0.0	30.0	0.0	0.0	0.0
Number of animals tested	126	89	126	30	52	33	52

As shown in Table 5, all the five D^{rh} types occur in rhesus monkeys (*Macaca mulatta*), pig-tailed macaques (*M. nemestrina*) (Socha et al. 1978) and in crab-eating macaques (*M. fascicularis*) (Socha and Ruffié 1983). The D^{rh} system appears in somewhat truncated form in stump-tailed macaques (*M. arctoides*) (Moor-Jankowski and Socha 1978), barbary macaques (*M. sylvanus*) (Socha et al. 1981) and sacred baboons (*Papio hamadryas*) (Moor-Jankowski and Socha 1978). Tests for the specificities of the graded D^{rh} system were negative with the red cells of all bonnet macaques (*M. radiata*) (Socha et al. 1976a) and Japanese macaques (*M. fuscata*) investigated to date (unpublished).

Although the appearance of identical or similar antigens in the blood of various, even not so closely related, species is not infrequent, propagation of sets of related specificities belonging to one and the same system probably reflects an intimate relationship among species that share these specificities. It is believed that the degree of cross-reactivity of isoimmune reagents (as measured by the level of polymorphism they detect) reflects the taxonomic proximity of the spe-

Table 6. Distribution of the simian type red cell specificities in various species of macaques

Antiserum of specificity	Percentage of positive reactions with red cells of:					
	Rhesus monkeys (N = 30)	Pig-tailed macaques (N = 89)	Crab-eating macaques (N = 62)	Barbary macaques (N = 32)	Stump-tailed macaques (N = 30)	Bonnet macaques (N = 62)
Anti-A^{rh}	94.7	71.9	91.8	nd	100.0	100.0
Anti-$B^{rh}_{(1)}$	90.3	4.5	70.2	100.0	100.0	nd
Anti-$B^{rh}_{(1)}$	90.3	96.6	53.3	100.0	100.0	100.0
Anti-C^{rh}	100.0	98.9	100.0	100.0	100.0	100.0
Anti-D^{rh}	67.1	86.5	89.6	100.0	0.0	0.0
Anti-D^{rh}_3	43.5	80.9	82.6	100.0	33.3	nd
Anti-D^{rh}_2	34.4	39.3	0.0	0.0	0.0	0.0
Anti-D^{rh}_2+Anti-Q	43.7	39.3	0.0	87.5	30.0	nd
Anti-D^{rh}_1+Anti-X	33.6	34.8	0.0	0.0	0.0	nd
Anti-D^{rh}_1+Anti-Y	34.4	34.8	0.0	0.0	0.0	0.0
Anti-D^{rh}_1+Anti-Z	53.5	79.8	0.0	0.0	0.0	0.0
Anti-D^{rh}_1	61.8	94.4	0.0	0.0	0.0	nd
Anti-F^{rh}	39.4	96.6	83.3	100.0	100.0	0.0
Anti-G^{rh}	84.7	76.4	100.0	0.0	100.0	100.0
Anti-J^{rh}_3	29.1	68.5	100.0	0.0	0.0	nd
Anti-L^{rh}	29.1	100.0	0.0	0.0	0.0	nd
Anti-M^{rh}	47.6	68.5	28.6	100.0	0.0	nd
Anti-$N^{rh}_{(a)}$	96.8	100.0	100.0	100.0	100.0	nd
Anti-$N^{rh}_{(a)}$	96.8	100.0	21.4	100.0	100.0	nd
Anti-J^{rh}_1	29.4	14.6	100.0	100.0	0.0	nd
Anti-J^{rh}_2	43.2	9.0	100.0	0.0	0.0	nd

[a] Two anti-B^{rh} and two anti-N^{rh} reagents were produced in different animals. Although of identical specificity in test with rhesus monkeys, the antibodies gave disparate results in parallel test with other species of macaques.

cies tested with these reagents. In this way, for example, as shown in Table 5, macaque species which share with rhesus monkeys all of the D^{rh} specificities, namely pig-tailed and crab-eating macaques, are judged as being very close taxonomically to *M. mulatta* , while stump-tailed and bonnet macaques are considered more distant and Barbary macaques most remote from rhesus monkeys (Moor-Jankowski and Socha 1979; Socha and Ruffié 1983).

In addition to the specificities of the D^{rh} series, our rhesus isoimmune sera detected a number of unrelated specificities of rhesus red cells, some of which are also found, not infrequently in polymorphic form, on the red cells of several other macaque species. Comparative distributions of these specificities in the various macaque species (Table 6) reveal significant differences among the frequencies with which some of the specificities appear in various monkey species. This points to the usefulness of simian-type blood groups as additional parameters in the taxonomic classification of these animals (Moor-Jankowski and Socha 1978, 1979). Allowing for the relatively small number of animals of each species tested so far, the fact that some of the specificities do not occur in certain species of macaques but are present in all animals of another species suggests that such blood types may become valuable tools when morphological parameters alone are insufficient to definitely assign the animal to one or another species.

Blood Groups Defined by Baboon Alloimmune Antisera

Prolonged alloimmunizations of panels of baboons resulted in development of several antisera that defined at least 25 distinct red cell specificities in various species of baboon (Table 3). Five of these specifcities were found to be parts of a complex multi-allelic blood group system, the B^P graded blood group system of baboon (Socha et al. 1983). This system, in many respects comparable to the D^{rh} system of macaques, was found first in sacred baboons (*Papio hamadryas*), but its existence was also confirmed in other species of baboons, namely, olive (*P. anubis*) and yellow (*P. cynocephalus*) baboons.

As in the D^{rh} system of macaques, the B^P system also displays the presence of several graded types:

- B^P_4 (or B^P standard), which contains four factors: B_1, B_2, B_3, and B_4
- B^P_3, which contains three factors, B_1, B_2, and B_3
- B^P_2, which contains two factors, B_1 and B_2
- B^P_1, which contains only the factor B_1
- B^P-negative type (b^P), probably a product of a recessive silent mutation.

It is assumed that the graded B^P system is (as is the D^{rh} system of macaques) the result of several progressive "step-like" mutations in which each new mutation contains more antigen sites than the previous one and dominates over the previous mutation.

The graded phenotypes of the B^P system are defined by a set of anti-B^P reagents: anti-B^P_1 (standard), anti-B^P_2, anti-B^P_3, and anti-B^P_4, each containing antibodies for one or more specificities characterizing different grades of the B^P antigen. All B^P reagents were obtained by intentional allo-immunization, but several "normal" sera of apparently nonimmunized baboons were discovered that contained spontaneous antibodies of various anti-B^P specificities. Some of these "natural agglutinins" were probably the result of transplacental immunization in the course of incompatible pregnancies, but the origin of some others, particularly those found in the sera of feral male baboons (unpublished observation), remains unknown.

The D^{rh} and B^P graded blood group systems are not only structurally similar but also display some kind of serologic overlapping: it has been established that the anti-D_3^{rh} antibody recognizes the B_1^P factor on baboon red cells. The reverse, however, is not true. It appears, therefore, that there is a certain phylogenetic connection between the D^{rh} macaque system and the B^P blood group system of baboons in the form of a common antigenic zone situated at the level of the D_3^{rh} subroup of the first system and at the B_1^P subgroup for the second. Anti-B_3^P serum is inactive against the red cells of macaques. It is noteworthy, however, that anti-B_4^P was found to detect polymorphic specificity (independent from the D^{rh} system) on the red cells of *Macaca mulatta* and *M. nemestrina* and in the blood of geladas (*Theropithecus gelada*). As was mentioned earlier, some of the alloimmune sera developed in crab-eating macaques were found to have specificities identical to those of some of the rhesus alloimmune reagents. This further strengthens the idea of an immunological kinship among species of Old World monkeys.

A number of specificities unrelated to the B^P system are detected on baboon red cells by means of the baboon alloimune sera currently available. These specificities (listed in Table 3), together with the products of the respective recessive silent alleles, probably constitute independent blood group systems characteristic of baboon red cell membranes and not displayed on the red cells of any other species except for, perhaps, geladas.

Tests for Simian-Type Blood Groups in New World Monkeys and Prosimians

For lack of specific typing reagents, little is known about the simian-type red cell antigens of New World monkeys. A notable exception are marmosets, for which alloimmune antisera were developed by Gengozian et al. (1966). The sera detected individual differences among marmosets and confirmed the existence of blood chimerism.

Attempts to detect simian-type epitopes on the red cells of prosimians using typing reagents of chimpanzee, rhesus or baboon origin were unsuccessful because of intereference by potent heterospecific components in those reagents that could not be completely eliminated by multiple absorptions.

Some Practical Aspects of Simian-Type Antigen Polymorphisms

Blood Transfusion and Organ Transplantation

Transfusion reactions due to simian-type blood group incompatibilities usually come in to play in the course of multiple transfusions, when agglutinating antibodies are formed as the result of previous exposure to mismatched, highly immunogenic, simian-type red cell antigens. As we have seen in a series of cross-transfusions among intentionally mismatched baboons and rhesus monkeys (Socha et al. 1982), the half-life of transfused incompatible red cells was significantly shortened in the presence of corresponding antibodies in the recipient's serum. In some cases, the survival of transfused incompatible red cells was so drastically shortened as to make the transfusion therapeutically worthless.

Rare occurrences of acute reactions to the first mismatched transfusion are ascribed to the action of preexisting "natural" antibodies (i.e, antibodies not resulting from deliberate immunization) in the recipient's blood. Such antibodies, directed against red cell simian-type epitopes, are detected by cross-matching tests (performed, in parallel, by various agglutination techniques and at various temperatures) among the red cells and sera of recipients and prospective donor animals. The tests for spontaneous antibodies were positive in 12 % of 195 cynomolgus macaques and 13 % of 330 baboons investigated (Socha and Moor-Jankowski 1991). Observations that such antibodies are usually found in the sera of multiparous females would point to maternofetal incompatibility as the cause of immunization (Socha et al. 1976c; Socha 1993).

An assessment of the role of simian-type blood groups in the survival of primate allo- and xenografts was attempted in a retrospective analysis of the results of a series of macaque-to-baboon and baboon-to-baboon cardiac transplantations (Michler et al. 1987). Although no correlation could be found between a positive pretransplantation cross-match and graft survival, there were indications that fewer siman-type mismatches between donor and recipient may have resulted in longer cardiac xenograft survival. In a few cases titrations of simian-type antibodies in recipients' sera were carried out before and after multiple pretransplantation donor-specific transfusions and in the post-transplantation period. The post-transfusion build-up of antibodies was followed by a transient decrease in antibody titer immediately after transplant. This suggested that adsorption of antibodies onto the graft occurred either because of the presence of donor blood cells in the transplant or the presence of simian-type epitopes on grafted tissues (Michler et al. 1987; Socha and Moor-Jankowski 1991).

Maternofetal Incompatibility

Exposure to red blood cell immunogens occurs naturally during incompatible pregnancies in humans and various other species of mammals. The resulting erythroblastosis fetalis, or hemolytic disease of the newborn, may lead to perinatal death or to serious postnatal developmental disturbances. As for nonhuman primates, cases of erythroblastosis fetalis caused by feto-maternal blood group incompatibility were observed first by Genozian and coworkers in marmosets (Gengozian et al. 1966).

Transplacental immunization due to feto-maternal blood group incompatibility was reported to be responsible for the significant frequency of "pathologic outcome" of pregnancies in the Sukhumi hamadryas baboon colony (Verbickij 1972; Verbitsky et al. 1969). The same study also described successful experimental induction of erythroblastosis fetalis in the offspring of female baboons immunized during pregnancy with the red cells of their (incompatible) breeding mates. It is noteworthy that neither Sullivan and coworkers (Sullivan et al. 1972) nor we (Socha and Moor-Jankowski 1979b; Wiener et al. 1975) were able to reproduce these results in baboons or macaques. Even though the newborn of the hyperimmunized macaque or baboon female had their red cells maximally coated with maternal antibodies, the infants did not show any clinical symptoms of erythroblastosis fetalis. There seems to be a protective mechanism which prevents the intravascular destruction even of heavily coated red cells. The nature of this mechanism is still unclear (Stong et al. 1977).

In anthropoid apes, by contrast, the situation seems to resemble that in humans, as shown by the cases of fetal erythroblastosis we observed in captive chimpanzees (Wiener et al. 1977) and, more recently, orangutans (Socha and Van Foreest 1981). This latter case is of practical importance since the transplacental immunization was diagnosed early in pregnancy, allowing life-saving measures to be undertaken in time to save the erythroblastotic infant. Incidentally, the serum of the pregnant orangutan became a rare source of allo-immune antibodies that detected Rh-related polymorphisms in this species of anthropoid ape.

References

Duggleby CR, Blystad C, Stone WH. Immunogenetic studies of rhesus monkeys. II. The HIJK and L blood group systems. Vox Sang 20:1124–1136, 1971

Duggleby CR, Stone WH. Immunogenetic studies of rhesus monkeys. I. The blood group system. Vox Sang 20:109–123, 1971

Edwards RH. Rhesus monkey blood group systems, J, K, L, M, N and O. J Hered 62:149–156, 1971a

Edwards RH. The G and H rhesus monkey blood group system. J Hered 62:79–86, 1971b

Edwards RH. The I blood group system of rhesus monkeys. J Hered 62:142–148, 1971

Erskine AG, Socha WW. The principles and practice of blood grouping, 2nd edn. Mosby, St Louis, 1978

Gengozian N, Lushbaugh CC, Humason GL, Knisely RM. Erythroblastosis fetalis in the primate Tamarinus nigricollis. Nature 209:731, 1966

Michler RE, Marboe CC, Socha WW, Moor-Jankowski J, Reemtsma K, Rose EA. Simian-type blood group antigens in nonhuman primate cardiac xenotransplantation. Transplant Proc 19:4456–4462, 1987

Moor-Jankowski J, Wiener AS, Rogers C. Blood groups of chimpanzee; demonstrated with iso-immune sera. Science 145:1441, 1964

Moor-Jankowski J, Wiener AS, Gordon EB. Simian blood groups. The new blood factors of gibbon blood, A^g and B^g. Transfusion 5:235–262, 1965

Moor-Jankowski J, Socha WW. Blood groups of macaques, a comparative study. J Med Primatol 7:136–145, 1978

Moor-Jankowski J, Socha WW. Blood groups of Old World monkeys. Evolutionary and taxonomic implications. J Hum Evol 8:445–451, 1979

Socha WW. Erythroblastosis fetalis. In: Jones TC, Mohr U, Hunt RD (eds) Nonhuman primates II. Springer, Berlin Heidelberg New York, pp 215–220 (Monographs on pathology of primate animals), 1993

Socha WW, Moor-Jankowski J. Rh antibodies produced by an isoimmune chimpanzee: reciprocal relationship between chimpanzee isoimmune sera and human anti-Rh_o reagents. Int Arch Allergy Appl Immunol 56:30–38, 1978

Socha WW, Moor-Jankowski J. Blood groups of anthropoid apes and their relationship to human blood groups. J Hum Evol 8:453–465, 1979a

Socha WW, Moor-Jankowski J. Serological maternofetal incompatibility in nonhuman primates. In: Ruppenthal GC (ed) Nursery care of nonhuman primates. Premium, New York, pp 35–42, 1979b

Socha WW, Moor-Jankowski J. Nonhuman primate blood group serology: some implications for xenotransplantation. In: Cooper DKC, Kemp E, Reemtsma K, White DJG (eds) Xeno-transplantation. The transplantation of organs and tissues between species. Springer, Berlin Heidelberg New York, 1991

Socha WW, Ruffié J. Blood groups of primates. Theory, practice, evolutionary meaning. Liss, New York, 1983

Socha WW, Van Foreest AW. Erythroblastosis fetalis in a family of captive orangutans. Am J Primatol 1:326, 1981

Socha WW, Wiener AS, Gordon EB, Moor-Jankowski J. Methodology of primate blood grouping. Transplant Proc 4:107–111, 1972

Socha WW, Moor-Jankowski J, Scheffrahn W, Wolfson SK. Spontaneously occurring agglutinins in primate sera. Int Arch Allergy Appl Immunol 51:656–670, 1976a

Socha WW, Moor-Jankowski J, Wiener AS, Risser DR, Plonski H. Blood groups of bonnet macaques (*Macaca radiata*) with a brief introduction to seroprimatology Am. J Phys Anthrop 45:485–491, 1976b

Socha WW, Wiener AS, Moor-Jankowski J, Valerio D. The first isoimmune blood group system of rhesus monkeys (*Macaca mulatta*): the graded D^{rh} system. Int Arch Allergy Appl Immunol 52:355–363, 1976c

Socha WW, Moor-Jankowski J, Sackett GP. Blood groups of pig-tailed macaques (*Macaca nemestrina*) A. J Phys Anthrop 48:321–330, 1978

Socha WW, Mertz E, Moor-Jankowski J. Blood groups of Barbary apes (Macaca sylvanus). Folia Primatol 36:212–225, 1981

Socha WW, Rowe AW, Lenny LL, Lasano SG, Moor-Jankowski J. Transfusion of incompatible blood in rhesus monkeys and baboons. Lab Anim Sci 32:48–56, 1982

Socha WW, Moor-Jankowski J, Ruffié J. The B^p graded blood group system of the baboon: its relationship with macaque red cell antigens. Folia Primatol 40:205–216, 1983

Socha WW, Moor-Jankowski J, Ruffié J. Blood groups of primates: present status, theoretical implications and practical applications: a review. J Med Primatol 13:11–40, 1984

Stong RC, Houser WD, Stone WH. The absence of hemolytic disease in the newborn rhesus moneky (*Macaca mulatta*). Vet Immunol Immunopathol 3:611–627, 1982; 27:348–352, 1977

Sullivan PT, Duggleby PC, Clystad B, Stone WH. Transplacental immunization in rhesus monkeys. Fed Proc 31:792, 1972

Terao K, Fujimoto K, Cho F, Honjo S. The inheritance mode of simian-type EFG and H blood groups in cynomolgus monkeys. Am J Primatol 11:245–251, 1986

Verbitsky M, Volkova L, Kuksova M, Lapin B, Andreyev A, Gvaszava I. A study of haemolytic disease of the foetus and the newborn occuring in hamadryas baboons under
natural conditions. Z Versuchstierkd 2:136–142, 1969

Verbickij MS. The use of hamadryas baboons for the study of the immunological aspects of
human reproduction. In: Diczfalusy E, Standley CC (eds) The use of nonhuman primates in research in human reproduction. WHO Research and Training Centre on
Human Reproduction, Karolinska Institute, Stockholm, pp 492–505, 1972

Wiener AS, Moor-Jankowski J, Gordon EB, Daumy OM, Davis J. Blood groups of gibbons;
further observations. Int Arch Allery Appl Immunol 30:466–470, 1966

Wiener AS, Socha WW, Niemann W, Moor-Jankowski J. Erythroblastosis models. A review
and new experimental data in monkeys. J Med Primat 4:179–187, 1975

Wiener AS, Socha WW, Moor-Jankowski J. Erythroblastosis models. II. Maternofetal
incompatability in chimpanzee. Folia Primat 27:68, 1977

8 Blood Group Antigens as Receptors for Pathogens

A. Frattali Eder and S.L. Spitalnik

Introduction

Hundreds of blood group antigens are represented by epitopes on erythrocyte membrane proteins, glycoproteins, or glycolipids. Eight blood group systems comprising 22 antigens have paramount clinical significance. Exposure to incompatible blood within these systems through transfusion or parturition may result in hemolytic transfusion reactions or hemolytic disease of the newborn, respec-

Molecular Biology and Evolution of Blood Group
and MHC Antigens in Primates
Blancher/Klein/Socha (Eds.)
© Springer-Verlag Berlin Heidelberg 1997

tively. Although blood group antigens are critically important in transfusion medicine, questions regarding their physiologic functions in the erythrocyte membrane and on epithelial surfaces remain unanswered (Lublin 1995). Nevertheless, pathogenic roles for blood group antigens in infectious diseases have been defined (Lublin 1995; Garratty 1995; Mourant 1989; Blackwell 1989). Moreover, several human blood group systems, such as ABH, Lewis, Rh, Ii, and the glycophorin antigens MNSs, have counterparts in nonhuman primates (see elsewhere in this volume). Simian models for bacterial and parasitic infections initiated by interaction with erythrocyte receptors have provided insight into the pathology and immune response in human disease (Waters et al. 1993; Roberts et al. 1984b). In addition, the evolution of blood group diversity within populations may have been influenced by selective pressure for resistance to various endemic infectious diseases (Glass et al. 1985; Miller 1994).

Epidemiological studies have revealed statistical associations between several human infectious diseases and carbohydrate blood group antigens (Tables 1, 2; Garratty 1995; Mourant 1989; Blackwell 1989; Reid and Bird 1990). Overrepresentation of certain blood groups in human populations occurs with respect not only to infectious diseases but also to coagulation disorders, autoimmune diseases and cancer (Garraty 1995; Reid and Bird 1990). Conflicting data regarding several of these disease associations have been presented in different populations. As one example, several groups examined the relationship of *Neisseria gonorrhea* to ABO blood groups in Caucasian and Black populations and reached different conclusions: the occurrence of gonorrhea was significantly higher in Black patients with blood group B (Foster and Labrum 1976); the frequency of blood group B was significantly higher in Caucasian, but not Black, patients with gonorrhea (Miler et al. 1977); and the distribution of blood types in Caucasian gonorrhea patients was not significantly different from a control population (Johnson et al. 1983). Because of the inherent difficulties in designing and interpreting epidemiological studies, such as sample population size and heterogeneity, the significance of associations between diseases and blood groups has been controversial (Manuila 1958; Weiner 1970; Evans et al. 1972). If there is an association between a blood group antigen and a disease, the blood group polymorphisms may play a pathogenic role in the relationship of the human population with pathogens, and selective pressure may influence the prevalence of specific blood group alleles (Mourant 1980; Glass et al. 1985; Miller 1994). However, linkage disequilibrium in a population may also be attributed to random genetic drift (Mourant 1980).

Nevertheless, several hypotheses have been proposed to account for the association between specific blood group antigens and susceptibility to certain infectious diseases. First, isohemagglutinins to carbohydrate blood group antigens may cross-react with bacterial or viral antigens to confer "natural" humoral immunity or decreased infectivity (Blackwell 1989). The role of isohemagglutinins in opsonization and phagocytosis of bacteria has been suggested for both *Streptococcus pneumoniae* and *Escherichia coli* strain 086, but has not been relevant for eight other serotypes of *Escherichia coli* or *Neisseria gonorrhea* (Reed et al. 1974; Check et al. 1972; Blackwell et al. 1983, 1984). Moreover, anti-A and

Table 1. ABO Blood group phenotype and infectious disease associations

Increased susceptibility associated with ABO blood group phenotype:

A	Adenovirus	McDonald and Zuckerman 1962
	Giardiasis	Barnes and Kay 1977; Zisman 1977; Roberts-Thomas 1993
	Plasmodium sp.	Gupta and Chawdhuri 1980
	Schistosoma mansoni	Periera et al. 1979
B	*Coccidiodes immitis*	Deresinski et al. 1979
	Escherichia coli, uropathogenic	Cruz Coke et al. 1965
	Escherichia coli, enteric	Socha et al. 1969
	Escherichia coli, enterotoxigenic	Robinson et al. 1971
	Influenza virus	Mackenzie and Fimmel 1978; Cuadrado and Davenport 1970
	Neisseria gonorrhea	Miler et al. 1977; Foster and Labrum 1976; Kinane et al. 1983
	Salmonella sp.	Robinson et al. 1971
	Schistosoma mansoni	Trangle et al. 1979
AB	*Escherichia coli,* enteropathogenic	Robinson et al. 1971
	Salmonella sp.	Robinson et al. 1971
	Periodontal disease	Pradhan et al. 1971
O	Bacillary tuberculosis	Viskum 1975
	Candida albicans, carriage	Burford-Mason et al. 1988, 1993
	Escherichia coli, enterotoxigenic	Black et al. 1987
	Influenza virus	McDonald and Zuckerman 1962; Potter and Schild, 1967; Tyrell et al. 1968; Curadrado and Davenport 1970
	Periodontal disease	Pradhan et al. 1971
	Vibrio cholera	Barua and Paguio 1977; Chadhuri and De 1977; Levine et al. 1979; Swerdlow et al. 1994; Sircar et al. 1981

Decreased susceptibility associated with ABO blood group phenotype:

A	Bacillary tuberculosis	Viskum 1975
	Influenza virus	McDonald and Zuckerman 1962
	Vibrio cholera	Barua and Paguio 1977
B	*Streptococcus pneumonia*	Reed et al. 1974
O	*Streptococcus pyogenes,* group A	Haverkorn and Goslings 1969
	Adenovirus	McDonald and Zuckerman 1962
	Plasmodium sp.	Gupta and Chawdhuri 1980
	Schistosoma mansoni	Pereira et al. 1979

No association with ABO blood group phenotype:

Escherichia coli, enterotoxigenic	van Loon et al. 1991; Glass et al. 1985
Neisseria gonorrhea	Johnson et al. 1983
Influenza virus	Potter 1969; Evans et al. 1972
Parainfluenza viruses; adenoviruses; *Mycoplasma pneumonia*	Evans et al. 1972
Urinary tract infection	Rosenstein et al. 1984; Sheinfeld et al. 1989; Jantausch et al. 1994; Navas et al. 1994
Helicobacter pylori	Mentis et al. 1991
Filarial and hemoparasitic disease	Higgins et al. 1985; Emeribe and Ejezie 1989; Srividya and Pani 1993
Leishmania sp.	Esterre and Dedet 1989
Schistosoma sp.	Tangle et al. 1979; Kassim and Ejezie 1982
Plasmodia sp.	Kassim and Ejezie 1982

Table 2. Secretor status and infectious disease associations

Increased susceptibility associated with nonsecretor phenotype:

Candida albicans, carriage	Burford-Mason et al. 1988, 1993
	Aly et al. 1992; Blackwell et al. 1989a
Candida albicans, genital	Thom et al. 1989
Candida albicans, oral	Aly et al. 1991; Thom et al. 1989
Caries	Arneberg et al. 1976; Holbrok and Blackwell 1980
Escherichia coli, urinary tract infection, pyelonephritis, renal scarring, recurrent urinary tract infection	Jantausch et al. 1994, Lomberg et al. 1989; May et al. 1989a; Kinane et al. 1982; Sheinfeld et al. 1989
Hemophilus influenza	Blackwell et al. 1986b, 1989b
Neisseria meningitidis	Blackwell et al. 1986a, 1989b
Nesseria meningitidis, carriage	Blackwell et al. 1990; Zorgani et al. 1994
Salmonella typhi, carriage	Bothamley et al. 1993
Streptococcus pyogenes, group A	Haverkorn and Goslins 1969
Streptococcus pneumonia	Blackwell et al. 1986a, 1989b
Vibrio cholera	Chadhuri and Das Adhikary 1978

Increased susceptibility associated with secretor phenotype:

Influenza viruses A and B, respiratory syncytial virus, echovirus	Raza et al. 1991
Schistosoma mansoni	Trangle et al. 1979

No association with secretor/nonsecretor phenotype:

Helicobacter pylori	Höök-Nikanne et al. 1990; Dickey et al. 1993; Mentis et al 1991
Neisseria gonorrhea	Johnson et al. 1983; Kinane et al. 1983; Foster and Labrum 1976
Neisseria meningitidis, carriage	Blackwell et al. 1989b
Periodontal disease	Pradham et al. 1971
Pseudomonas aeruginosa	Mulherin et al. 1990
Recurrent urinary tract infection	Rosenstein et al. 1984; Navas et al. 1994
Vibrio cholera	Glass et al. 1985

anti-B antibodies had no effect on the infectivity of influenza virus in vitro (Potter and Schild 1967). Second, blood group antigens may occur, albeit coincidentally, on complement components or other important molecules involved in the immune response to infectious organisms (Garratty 1995). As examples, the Duffy antigens are found on the chemokine receptor which may be involved in regulating the concentration of cytokines in the circulation; the Dr antigen is on a portion of decay accelerating factor (DAF, CD55) that regulates the complement cascade and protects host erythrocytes from being lysed; and the Rogers and Chido antigens are on complement component C4 (Lublin 1995). Third, blood group antigens may be indirectly involved in host defenses by affecting levels of inflammatory proteins. For example, differences in serum levels of complement components or in serum and salivary levels of immunoglobulin A and M between secretors and nonsecretors have been reported (Blackwell 1989; Black-

Table 3. Blood group antigens as receptors for pathogens

Blood group antigen	Pathogen	Reference
A, B	*Escherichia coli* heat labile toxin	Barra et al. 1992
AnWj	*Hemophilus influenza*	van Alphen et al. 1986
Dr^a (Cromer, DAF)	*Escherichia coli*, uropathogenic	Nowicki et al. 1988b
	Echoviruses	Bergelson et al. 1994; Ward et al. 1994
	Coxsackie viruses	Bergelson et al. 1995
Duffy (Fy^a or Fy^b)	*Plasmodium vivax/knowlesi*	Adams et al. 1992
Glycophorins (MNSs)		
A-C	*Plasmodium falciparum*	Sim et al. 1994; Hermentin and Enders 1984
A-D	Influenza viruses A and B	Ohyama et al. 1993
A	Encephalomyocarditis virus	Allaway and Burness 1986
A^M	*Escherichia coli* (serotype O$_2$)	Väisänen et al. 1982
A	*Escherichia coli* (serotype O18:K1:H7)	Parkkinen et al. 1986
Sialylated I and i	*Mycoplasma pneumonia*	Loomes et al. 1984
	Sendai virus	Suzuki et al. 1985
	Influenza A virus	Suzuki et al. 1986
Lewis a	*Candida* sp.	May et al. 1989b
	Pertussis toxin	van't Wout et al. 1992
	Staphylococcus aureus	Saadi et al. 1993, 1994
	Staphylococcus aureus, enterotoxin B	Essery et al. 1994a
Lewis b	*Helicobacter pylori*	Borén et al. 1993
P blood group	Parvovirus	Brown et al. 1993
	Uropathogenic *Escherichia coli*	Leffler and Svanborg-Edén 1980
	Streptococcus suis	Haataja et al. 1993
	Shiga toxin	Lindberg et al. 1987
	Shiga-like toxins (verotoxins)	Lingwood et al. 1987; Bitzan et al. 1994

well et al. 1989b; Zorgani et al. 1992, 1994). However, these quantitative differences were not observed in all studies, and functional differences in the biological activities of inflammatory proteins from secretors as compared to nonsecretors have not been well defined (Blackwell et al. 1989b; Zorgani et al. 1992, 1994). Fourth, genes encoding blood group antigens may be linked to genes predisposing individuals to various diseases. The chromosomal locations of the 23 established blood group systems have been determined (Zelinski 1995). If the gene encoding a blood group antigen is located on the same chromosome in close proximity to a gene affecting development of a disease, inheritance of the blood type will be associated with that disease. Examples include linkage between the Duffy blood group and Charcot-Marie-Tooth neuropathy and the Rh locus and noninsulin-dependent diabetes (Reid and Bird 1990).

Fifth, and finally, blood group antigens on hematopoietic cells, epithelial surfaces, and extracellular matrix components may serve as receptors for infectious agents. Conversely, secreted forms of blood group antigens, such as the ABH and Lewis carbohydrate antigens in the body fluids of secretors, may saturate the adhesins on the surfaces of microorganisms, thus preventing attachment of the pathogen to host cells. Alternatively, the product of the secretor gene may shield cryptic receptor sites on the surface of the epithelium, thus interfering with pathogen attachment to host cells. These hypotheses have been supported experimentally, and several blood group antigens have been implicated as receptors for various infectious agents (Table 3). In addition, membrane glycolipids and glycoproteins not belonging to human blood group systems also serve as host cell receptors for pathogens (Karlsson 1989; Weir 1989).

Criteria to Unequivocally Assign Blood Group Antigens as Receptors for Microorganisms

Similar to Koch's postulates, which provided the basis to unequivocally link microorganisms with specific diseases, criteria to unequivocally assign a blood group antigen as a receptor for a microorganism may be formulated (Table 4). The initial hypothesis may stem from an epidemiologic association between an infectious disease and a blood group antigen. That is, the disease will occur in individuals possessing the blood group antigen; conversely, the disease will not occur in individuals lacking the blood group antigen. Furthermore, evidence of prior disease, such as the presence of antibodies against a microorganism, will be present only in individuals with the specific blood group phenotype.

Direct evidence supporting the hypothesis involves a demonstration of the ability of the microorganism or, more specifically, a protein expressed on the surface of a microorganism, to bind to the relevant blood group antigen. A preliminary demonstration of the ability of a microorganism to bind to epitopes expressed on the erythrocyte membrane involves adding the microorganism to a suspension of erythrocytes and observing for agglutination. If the microorganism causes direct hemagglutination, the biochemical nature of the receptor may be a specific component of the erythrocyte membrane or nonspecific attachment sites such as negatively charged sialic acid residues. Sialic acid is abundant on membrane glycoproteins of erythrocytes as well as other somatic cells and often mediates either specific or nonspecific, charge-induced cellular attachment of microorganisms. To demonstrate specific binding of a microorganism to a component of the erythrocyte membrane, the glycolipids or glycoproteins are extracted and immobilized on a solid support, and analyzed by thin layer chromatography or western blotting, respectively. Incubation of a radiolabeled microorganism to these immobilized glycolipids or glycoproteins may demonstrate direct binding (Goldhar 1994; Deal and Krivan 1994; Falk et al. 1994). Antibodies against the blood group antigen or against the binding protein of the microorganism, as well as soluble forms and synthetic analogues of the host cell receptor or microbial binding protein, define the specificity of the interaction.

Table 4. Criteria for assigning a blood group antigen as a receptor for a microorganism

Epidemiology

  Statistical association between an infectious disease and blood group:

    Disease in the presence of the blood group phenotype

    Serum antibodies against a microorganism (evidence of previous infection) only in individuals with the blood group phenotype

    No disease in the absence of the blood group phenotype

Demonstrate physical interaction

  Serology – direct hemagglutination

  Biochemistry/molecular biology – purification, cloning, western blotting or thin layer chromatography

    Purify/clone the host receptor (blood group antigen) and microbial attachment protein.

    Immobilize the blood group receptor (glycolipid or glycoprotein) on a solid support and demonstrate interaction with the microorganism.

    Immobilize the microbial attachment protein on a solid support and demonstrate interaction with the appropriate RBCs.

  Cellular biology – transfection/complementation experiments

    Transfect cultured cells that are resistant to infection with cDNA encoding the blood group antigen receptor and demonstrate acquiered susceptibility to infection.

    Transfect cultured cells with the cDNA for the microbial binding protein and demonstrate acquired interaction with red cells of specific phenotype.

  Transgenic animal model

    Establish a transgenic animal line in which susceptibility to a microorganism is conferred by introduction of the gene encoding the blood group antigen receptor.

Block physical interaction

  Antibodies against blood group antigen or against microbial attachment protein (e.g. fimbrial proteins) block interaction.

  Soluble forms of blood group antigen receptors or microbial attachment protein block interaction.

    Complementation experiments provide additional support for the hypothesis. Cells resistant to infection because they lack the blood group antigen receptor become sensitive to infection when they are manipulated to express the receptor by transfection of the cDNA encoding the putative antigenic determinant. The converse experiment to identify the erythrocyte antigen receptor may be performed if the erythrocyte binding protein of the microorganism has been cloned. In this review, the hypothesis that an individual blood group antigen serves as a receptor for a microorganism or a microbial toxin is considered in this framework.

Blood Group Antigens as Receptors for Bacteria

Lewis Antigens and *Helicobacter pylori*

One of the first and most firmly substantiated statistical correlations between a human disease and a blood group was the increased frequency of gastroduodenal ulcers in blood type O individuals as compared to type A and type B individuals. This association was initially described in 1954 and has been supported by numerous additional reports in many human populations (Aird et al. 1954; Mourant et al. 1978; Mentis et al. 1991). Moreover, nonsecretors also demonstrate increased susceptibility to gastric and duodenal ulcers in most but not all epidemiological studies (Mourant et al. 1978; Mentis et al. 1991). Gastroduodenal disease has recently been strongly associated with the bacterium *Helicobacter pylori*, formerly *Campylobacter pylori* (Buck 1990). The mechanism by which these bacteria bind and effectively colonize the epithelium in the acidic milieu of the stomach is an area of active investigation. Binding mechanisms for pathogens are often redundant; thus, not surprisingly, more than one target for *H. pylori* attachment to gastric epithelium has been identified. At least three carbohydrate moieties on the surface of human gastric mucosa mediate bacterial attachment: sialic acid residues, sulfatide (SO_3-Galβ1–1'cer) and the Lewis b (Le^b) antigen (Evans et al. 1993; Saitoh et al.1991; Born et al. 1993). The *H. pylori* gene that codes for a protein that binds specifically to sialic acid has been cloned (Evans et al. 1993).

The identification of the Le^b antigen binding activity of *H. pylori* stemmed from the observation that human colostrum secretory immunoglobulin A (IgA), but not the less glycosylated serum IgA, inhibited *H. pylori* attachment to human gastric surface mucous cells (Falk et al. 1993). Fucose was implicated as the relevant carbohydrate moiety since fucosidase treatment of the secretory IgA eliminated its ability to inhibit bacterial binding. A panel of fucosylated carbohydrate chains chemically linked to albumin were then evaluated for their ability to inhibit *H. pylori* binding to human gastric epithelium in situ (Born et al. 1993). Le^b (20 µg/ml) inhibited binding by 93 % and H type 1 (20 µg/ml) inhibited binding by 48 %. The inhibitory action of the Le^b and H type 1, but not of the H type 2, Le^a, Le^x, and Le^y neoglycoconjugates, demonstrates that the receptor epitope for *H. pylori* is confined to the terminally fucosylated lacto series type 1 chains. This interaction was confirmed by demonstrating strong bacterial binding to Le^b, weak binding to H type 1, and no binding to Le^a glycoconjugates immobilized on protein immunoblots. Additional data supporting the conclusion that *H. pylori* binds to Le^b include the demonstration that anti-Le^b-monoclonal antibodies but not anti-H type 2 antibodies inhibited bacterial binding. Moreover, histologic sections of gastric mucosa from Le^b individuals, but not from Le^a individuals, bound *H. pylori*. Finally, a transgenic mouse model has recently been developed. The human α1,3/4-fucosyltransferase was expressed in gastric cell lineages in mice. The fucosyltransferase resulted in the production of the Le^b epitope on gastric epithelium, and clinical isolates of *H. pylori* bound to these cells (Falk et al. 1995).

Born and colleagues may have provided part of the explanation for the original epidemiological observations that blood type O individuals are more susceptible than type A and type B individuals to peptic ulcers (Born et al. 1993). In individuals with blood type A, the Le^b antigen is substituted with a terminal GalNAc(α1–3) residue. This structure, known as ALe^b, did not bind to *H. pylori* in solution nor did it inhibit bacterial adherence in situ. The interpretation extrapolated from this result is that there are fewer available *H. pylori* receptors in type A and type B individuals than in type O individuals since type A and type B individuals express modified forms of the Lewis b antigen, ALe^b and Ble^b, respectively.

The hypothesis that Le^b, formed by Lewis-positive secretors, is a receptor for *H. pylori* is also supported by some of the available epidemiological data. In one study of patients with gastric ulcers, the prevalence of *H. pylori* in gastric biopsies or cultures obtained during gastroscopy was higher among secretors (67%) than non-secretors (12.5%) (Mentis et al. 1991). Other studies of patients with gastroduodenal disease, however, have not supported a direct correlation between *H. pylori* infection and the secretor phenotype. In a series of 101 patients with symptoms of dyspepsia, there was no significant association between secretor status and *H. pylori* infection (Dickey et al. 1993). Another study failed to show a significant correlation between increased levels of *H. pylori* antibodies and secretor status (Höök-Nikanne et al. 1990). These groups have concluded that *H. pylori* and secretor status are independent risk factors for gastroduodenal disease.

In most of the epidemiological studies, nonsecretors actually have an increased risk of gastroduodenal ulcers (Mourant et al. 1978). In patients with duodenal ulcer, the prevalence of *H. pylori* was higher among nonsecretors (100%) than secretors (84%) (Mentis et al. 1991). Nonsecretors cannot be Le^b positive and only express Le^a. Presumably, nonsecretors are still infected by *H. pylori* because, even though these individuals lack the Le^b target antigen, *H. pylori* can bind to other receptors on gastric epithelium. Some investigators propose that Le^a, in fact, is a receptor for *H. pylori* (Mentis et al. 1991; Essery et al. 1994a,b). Alternatively, some secretors may be protected from ulcers because the soluble form of Le^b occupies some of the receptor sites on the surface of *H. pylori*, blocking attachment of the bacteria to epithelial cells.

Blood Group Antigens and Uropathogenic *Escherichia coli*

P Blood Group Antigens and Uropathogenic *Escherichia coli*

The vast majority of recurrent urinary tract infections are caused by *Escherichia coli* (Neu 1992). A prerequisite for infection involves the initial adherence of the organism to uroepithelial host cells despite the host defenses which ordinarily maintain urinary tract sterility. Uropathogenic *E. coli* serotypes express P fimbrial adhesins on their surface which initiate the attachment process by recognizing mammalian cell surface glycosphingolipids expressed on epithelial surfaces of the human urinary tract (Leffler and Svanborg-Eden 1980, 1981; Vuokko et al. 1981; Källenius et al. 1981a). P fimbriae bind specifically to carbohydrate antigens containing the sequence Galα1-4Gal, whether this sequence appears as an internal or terminal determinant (Leffler and Svanborg-Eden 1981; Bock et al. 1985; Källenius et al. 1980, 1981b, 1982; Karr et al. 1990; Lindstedt et al. 1991). This sequence is found on the P blood group system antigens as an internal determinant in the P, LKE and globo-A antigens and as a terminal determinant on the P_1 and P^k (CD77) antigens. Glycosphingolipids related to the P blood group system have been used to block fimbrial adhesion in vitro and to characterize the specificity of this binding (Leffler and Svanborg-Eden 1980, 1981; Källenius et al. 1980, 1981a, 1982; Lindstedt et al. 1991; Johnson et al. 1992; Johnson and Ross 1993). In mice, receptor analogues have been used to inhibit experimentally induced ascending urinary tract infection (Svanborg-Eden et al. 1982; Johnson and Berggren 1994).

Murine glycoconjugate receptors in the kidney for P-fimbriated *E. coli*, however, are not identical to human or simian isoforms (Lanne et al. 1995). Primates other than humans, including *Macaca mulatta*, *M. fascicularis* and *Papio anubis*, possess receptors containing the Galα1-4Gal moiety on erythrocytes and epithelial cells in the urinary tract (Roberts et al. 1984b). The close structural similarity in the human and siman receptor isoforms for P-fimbriated *E. coli* is among the reasons that nonhuman primates provide an appropriate animal model for studying human pyelonephritis. In baboons, the synthetic P fimbriae-specific receptor analogue Galα1-4Galβ1-Me, when incubated with bacteria before ureteral inoculation, delayed the onset of acute and chronic pyelonephritis (Roberts et al. 1984b). Immunization of nonhuman primates with purified P fimbriae, synthetic oligosaccharide-protein conjugates, or multiple strains of heat-killed uropathogenic bacteria have protected against cystitis and pyelonephritis (Roberts et al. 1984a, 1993; Uehling et al. 1994). These reagents may be useful in preventing renal scarring from recurrent urinary tract infection in humans caused by P-fimbriated strains of *E. coli*.

Consequently, predicting which individuals are at risk of developing recurrent urinary tract infections is a central clinical goal so that appropriate and pathogen-specific intervention can be undertaken. Laboratory tests are available both for typing clinically isolated strains of *E. coli* for the presence of P fimbriae and for determining an individual's P blood group phenotype (Rydberg and Helin 1991). The clinical utility of this endeavor, however, has not clearly been defined because of conflicting epidemiological data (Lomberg et al. 1989). Population studies examining the relationship between P blood group antigen expression in individuals with and without urinary tract infections have yielded discordant results. Some studies show an increase in individuals of the P_1 phenotype in populations with recurrent urinary tract infections without vesicoureteral reflux and in populations with recurrent pyelonephritis (Leffler et al. 1982; Lomberg et al. 1984; Tomisawa et al. 1989). Other studies in similar populations show either a decrease in the number of individuals of the P_1 phenotype or no effect (Mulholland et al. 1984; Rosenstein et al. 1984; Jacobson et al. 1985).

These discrepancies presumably reflect additional complexity in the relationship between uropathogenic *E. coli* and its human host. Multiple bacterial attachment proteins and numerous host attachment sites are likely involved in human bacterial infections. The gene cluster *pap*, encoding P fimbriae in *E. coli*, has been found in 40 % of bacterial strains isolated from children with pyelonephritis (Plos et al. 1990). A second homologous copy of this sequence, *pap-2*, encodes fimbriae recognizing the LKE antigen on human kidney cells and erythrocytes (Karr et al. 1990). The availability of cloned fimbrial adhesins will lead to an increased understanding of the fine specificity of the interactions between the bacterial proteins and mammalian glycosphingolipid receptors (Johanson et al. 1992). In addition, *E. coli* expresses other adhesins that recognize host cell glycoconjugates not related to the P blood group system (Labigne-Roussel et al. 1984; Payne et al. 1993). With the increased knowledge of the specificity of different bacterial adhesins, the improved ability to subclassify strains of *E. coli* according to various virulence characteristics, and the results of additional carefully planned clinical population studies which will further delineate the host distribution, it will be possible to gain greater understanding of the pathogenicity and relationship to the P blood group antigen system of uropathogenic *E. coli* (Lindstedt et al. 1991; Johanson et al. 1992).

Secretor Phenotype and Uropathogenic *Escherichia coli*

In addition to the P blood group, the nonsecretor phenotype has also been associated with a predisposition to recurrent urinary tract infection (Kinane et al. 1982; Jantausch et al. 1994; Sheinfeld et al. 1989, 1990). Nonsecretors of ABH antigens are more likely to have increased binding of uropathogenic *E. coli* to their uroepithelial cells and therefore may be predisposed to pyelonephritis (Lomberg et al. 1986; Jacobson and Lomberg 1990). However, these observations have not been supported in every population study (Navas et al. 1994; Rosenstein et al. 1984). This discrepancy may be partially explained by variation of blood group antigen expression which influences bacterial adherence to host cells and modifies susceptibility to urinary tract infection. The expression of blood group antigens is a dynamic process. The level of ABH and Lewis antigen expression by cultured, normal, ureteral epithelial cells, collected vaginal cells and vaginal mucus varies over time and is affected by, as yet, unidentified systemic factors (Gaffney et al. 1992; Navas et al. 1993; Schaeffer et al. 1994). If samples are collected for analysis at a time when patients are not infected, the differences in antigen expression between control women and women with recurrent urinary tract infection may not be documented. Serial sampling from the same individual during asymptomatic and symptomatic periods may document differences in pattern and/or quantity of blood group antigen expression compared to the control population and may reveal periods of increased susceptibility to urinary tract infection (Navas et al. 1994).

The association between urinary tract infection and the nonsecretor phenotype may exist because nonsecretors may express unique receptors for the relevant bacteria or because secretors may synthesize molecules that hinder bacterial adherence by masking receptors or saturating adhesins. Gaffney and colleagues examined whether expression of the difucosylated Lewis determinants Le[b] and Le[y] conferred relative resistance to *E. coli* by masking adherence receptors (Gaffney et al. 1994). COS-1 cells, cultured fibroblast-like cells which do not produce Lewis antigens, were transfected with plasmids encoding α1,2- fucosyltransferase and α1,3/4- fucosyltransferse. Expression of Le[b] and Le[y] antigens was confirmed by flow cytometry. *E. coli* strains expressing various adhesins bound equally well to nontransfected cells and to transfected cells expressing these antigens. This group concluded that the presence of Le[b] and Le[y] antigens on cells of secretors is not sufficient to mask receptors for *E. coli* pili or hinder bacterial adherence.

Relationship Between the Se Locus and the P Blood Group: Effect on Host Interaction with Uropathogenic *Escherichia coli*

Stapleton and colleagues, however, provided evidence that nonsecretors express unique receptors for uropathogenic *E. coli* (Stapleton et al. 1992). The mechanism underlying the predisposition to urinary tract infections among individuals with the nonsecretor phenotype, whether they are Le(a+b-) or Le(a-b-), may be linked to the inability to modify a P blood group antigen that serves as a receptor site in secretors. The LKE glycosphingolipid antigen and a disialyated form of this antigen are present in vaginal epithelial cells obtained from nonsecretors but not those obtained from secretors. The R45 uropathogenic strain of *E. coli* expressing P and F adhesins bound to these particular glycosphingolipids. One interpretation of these data is that the LKE antigen is fucosylated by secretors of ABH antigens and this modification prevents attachment of certain strains of *E. coli*. Although other glycosphingolipids containing the Galα1-4Gal sequence in a terminal or internal position are receptors for uropathogenic *E. coli*, the authors speculate that the sialylated, but not the fucosylated, globo series glycosphingolipids may be the most accessible for bacterial adhesion to the surface of vaginal epithelial cells (Stapleton et al. 1992).

Other Blood Group Antigens and Uropathogenic *Escherichia coli*

Blood group antigens other than those in the P and Lewis blood groups have been implicated as attachment sites for *E. coli* on uroepithelial cells. Nowicki and colleagues used a systematic approach with erythrocytes representing 30 different phenotypes to identify the Dr[a] antigen, a component of the Cromer-related blood group complex and a portion of decay accelerating factor, as the receptor for the *E. coli* O75X adhesin in a direct hemagglutination assay (Nowicki et al. 1988a). The "Dr-positive" strains are associated with symptomatic urinary tract infections in 28 % of 700 *E. coli* strains tested (Nowicki et al. 1989). The relevance of the observed hemagglutination to urinary tract infection was further supported by the demonstration that the nature of the bacterial receptors expressed in human kidney histologic sections was similar to that of the receptor on erythrocytes involved in hemagglutination. In particular, attachment of bacteria expressing the O75X adhesin to kidney and erythrocytes was inhibited by chloramphenicol, suggesting that both receptors are similar. The Dr hemagglutinin purified from *E. coli* was also used as a reagent to demonstrate the presence of the Dr receptor in the digestive, urinary, genital, integumentary and respiratory systems (Nowicki et al. 1988b). Specifically, the Dr antigen was abundant on the transitional epithelium lining the urethra, bladder, ureter and renal pelvis as well as in the tubular basement membrane and Bowman's capsule of the human kidney. The Dr hemagglutinin belongs to a family of *E. coli* adhesins that recognizes different epitopes of the Cromer blood group (Nowicki et al. 1990). Binding of the Dr hemagglutinin to cells is abolished by a single amino acid substitution in the DAF membrane protein that converts the Dr^a antigen to the Dr^b antigen (Nowicki et al. 1993). The molecular nature of the Dr hemagglutinin was elucidated, and four genes, draA, draC, draD and draE, encoding polypeptides of 15.5, 18, 90 and 32 kDa, respectively, were shown to be required for full mannose-resistant hemagglutinin expression (Nowicki et al. 1989). To investigate the role of the Dr hemagglutinin in the pathogenesis of urinary tract infection, 658 bacterial strains isolated from patients with urinary tract infections or from fecal samples were screened for the presence of draD gene sequences. A significantly higher frequency of draD-related sequences were found among *E. coli* strains from patients with lower urinary tract infection (cystitis) than other clinical forms of urinary tract infection such as pyelonephritis. This correlation, as well as the tissue distribution of the Dr receptor, is consistent with the hypothesis that the Dr hemagglutin functions as a virulence factor in *E. coli* strains associated with lower urinary tract infection (Nowicki et al. 1988b, 1989; Virkola et al. 1988).

P Blood Group Antigens and Other Bacterial Infections

The Galα1–4Gal disaccharide present in the P_1 and P^k blood group antigen structures is recognized by adhesins expressed not only by *E. coli* but also by *Streptococcus suis* (Haataja et al. 1993), a zoonotic pathogen causing meningitis, septicemia and pneumonia in pigs and meningitis in humans. Major determinants of the binding specificity and strength of protein-oligosaccharide interactions are hydrogen bonds between the oligosaccharide and the protein. Using deoxy and other synthetic derivatives of the receptor disaccharide, hydrogen bonding patterns were determined for *S. suis* strains and for the *E. coli* PapG396 adhesin (Haataja et al. 1994). Comparing the models of the interactions for *S. suis* and *E. coli* adhesins revealed that these two unrelated bacterial species recognize the same oligosaccharide by two different mechanisms.

AnWj and *Hemophilus influenza*

Hemagglutination experiments are often used to gain insight into the binding of bacteria to other somatic cells that may be more relevant in the natural history of an infection. However, the erythrocyte receptors that mediate hemagglutination are not always identical to the receptors expressed on other cell types. In hemagglutination experiments, the erythrocyte receptor for fimbriated *Hemophilus influenzae* was identified as the blood group antigen AnWj, formerly the Anton or Wj blood group antigen (van Alphen et al. 1986). Although the genes involved in its expression have not been identified, the AnWj antigen is found on all erythrocytes except Lu(a–b–) erythrocytes of the dominant In(Lu) type, cord erythrocytes, and erythrocytes from a small number of individuals with anti-AnWj in their sera (Marsh et al. 1983; Poole and van Alphen 1988; Poole and Giles 1982). Several lines of evidence suggest that the AnWj blood group antigen is similar, but not identical, to the receptor for fimbriated *H. influenza* on oropharyngeal epithelial cells. Anti-AnWj antibodies inhibit agglutination of erythrocytes by bacteria but not adherence of bacteria to epithelial cells from AnWj-positive adults (van Alphen et al. 1987). Moreover, some individuals possess erythrocytes that are not agglutinated by fimbriated *H. influenza* but whose buccal epithelial cells are bound by bacteria (van Alphen et al. 1987). Nevertheless, expression of the erythrocyte antigens and the epithelial receptor may have a common genetic basis because AnWj-negative individuals have been described who possess erythrocytes that are not hemagglutinated by *H. influenza* and buccal epithelial cells that do not bind fimbriated *H. influenza* (van Alphen et al. 1990).

Despite evidence that the receptors on epithelium and erythrocytes are different and may be expressed independently, a similar profile of saccharides and glycoconjugates inhibited both hemagglutination and epithelial adherence due to fimbriated *H. influenza* (van Alphen et al. 1991). These experiments identified the ganglioside GM3, which contains the same oligosaccharide as the Pr_2 blood group antigen, as the minimal structure for fimbrial-dependent binding of *H.influenza* to its receptor on oropharyngeal epithelial cells and erythrocytes. Interestingly, the free oligosaccharide and sialylglycoproteins were poor inhibitors of bacterial recognition. Ganglioside receptor analogs may be useful pharmaceutical agents to interfere with the initial attachment and colonization by *H. influenza*.

Glycophorin A and *Escherichia coli* (O18:K1:H7) in Neonatal Infections

Similar carbohydrate sequences which serve as glycoprotein and/or glycolipid receptors for pathogens may occur on different integral membrane components on erythrocytes and epithelial cells, or on secreted components such as mucins and extracellular matrix molecules. For example, hemagglutination experiments identified the O-linked sialyloligosaccharides of glycophorin A as a receptor utilized by a strain of *E. coli* (O18:K1:H7) associated with sepsis and meningitis in neonates (Parkkinen et al. 1986). The bacterial components mediating attachment to sialyl galactosides were designated as S fimbriae, which are distinct from P fimbriae (Korhonen et al. 1984). S fimbriae bind to the NeuNAcα2–3Galβ1–3Gal-NAc sequence of the O-linked oligosaccharides of glycophorin A. Although bacterial binding to erythrocytes is probably not of particular relevance in the pathogenesis of neonatal meningitis, similar terminal sialyllactose oligosaccharide moieties are also found on the basement membrane glycoprotein laminin (Virkola et al. 1993). The ability of S fimbriae to bind to choroid plexus epithelium, subarachnoid epithelium and basement membrane may be important in localizing *E. coli* O18:K1:H7 infections to the brain (Parkkinen et al. 1988a). Once localized via fimbrial binding, bacteria may immobilize plasminogen and tissue-type plasminogen activator to generate proteolytic plasmin activity on their surface, thus facilitating bacterial penetration through basement membranes (Korhonen et al. 1992).

Although S fimbriae may play a pathogenic role in neonatal meningitis and sepsis, they occur only rarely in uropathogenic *E. coli* strains despite the fact that they can mediate binding to the same epithelial tissues in the human urinary tract as P fimbriae (Virkola et al. 1988). The virulence of P fimbriated *E. coli* in pyelonephritis, therefore, cannot be fully explained by the mere presence of receptor binding sites on the epithelial tissue of the human urinary tract. Normal human urine inhibits hemagglutination by S fimbriae, type 1 fimbriae, and, variably, the O75X adhesin, but not by P fimbriae (Parkkinen et al. 1988b). The major inhibitor of S fimbriae in normal urine was identified as Tamm-Horsfall glycoprotein which contains terminal NeuAcα2–3Galβ1–4GlcNAc sequences, a receptor determinant for S fimbriae similar to that found on glycophorin A (Parkkinen et al. 1986). Consequently, the ability of P fimbriae to function as a virulence factor may be due to binding to tissue receptors at the site of infection without competitive binding to excreted receptor-like oligosaccharides normally present in urine (Parkkinen et al. 1988b; Virkola et al. 1988).

Ii Antigens and *Mycoplasma pneumoniae*

Mycoplasma pneumoniae infects the human respiratory tract leading to pneumonia. The sialylated oligosaccharide antigens of the Ii blood group system serve as the major cellular receptors for *M. pneumonia* (Loomes et al. 1984, 1985). Sialylated I antigen, containing a branched backbone, demonstrates higher affinity for the pathogen than sialylated i antigen, containing a linear backbone. These carbohydrate structures also have been implicated as receptors for Sendai and influenza viruses (Suzuki et al. 1985, 1986). Interestingly, *M. pneumonia* infection often results in the transient production of cold agglutinins, IgM autoantibodies directed against I antigen present on erythrocytes and epithelial cells. These autoantibodies may cause immune-mediated hemolysis in rare cases. Anti-I production may be triggered by an autoimmunogenic complex formed between the host oligosaccharide receptors and the lipid-rich mycoplasma organism (König et al. 1988).

Immunohistochemical studies have shown that the I and i antigens, as well as their sialylated forms, are abundantly expressed at the primary site of mycoplasma infection, the human bronchial epithelium (Loveless and Feizi 1989). In the Syrian hamster, an animal model commonly used to study *M. pneumoniae* infection, these carbohydrate structures are also expressed in abundance on the goblet cells and in the intracellular globules (Loveless et al. 1992). Receptor-bearing secreted mucins may have a protective role in infection by binding to the microorganisms and leading to their clearance by bronchociliary action (Loveless et al. 1992). This observation may explain the relatively large doses of *M. pneumoniae* required to establish experimental infection in this animal model.

Blood Group Antigens as Receptors for Bacterial Toxins

ABH Antigens, Cholera Toxin, and *Escherichia coli* Heat-Labile Enterotoxins

Blood group antigens may predispose an individual to a disease or affect the severity of disease by serving as a receptor, not only for bacterial fimbriae, but also for elaborated bacterial toxins. For example, even though group O individuals show a greater propensity for developing cholera than do individuals with other ABO phenotypes, they are not more likely to be colonized with *Vibrio cholera* (Barua and Paguio 1977; Chadhuri and De 1977; Glass et al. 1985; Levine et al. 1979). A similar relationship exists between ABO blood group status and the occurrence of diarrhea after ingestion of strains of *E. coli* that produce the heat-labile enterotoxin (Black et al. 1987). These epidemiological trends may reflect the structural, functional and immunological properties shared by cholera toxin and heat-labile enterotoxin (Black et al. 1987; Barra et al. 1992).

Both cholera toxin and heat-labile enterotoxin interact with ganglioside GM1 on the plasma membrane of intestinal epithelial cells (Finkelstein et al. 1987; Yamamoto et al. 1984). Differences in the binding specificity of cholera toxin and heat-labile enterotoxin, however, have also been described (Holmgren et al. 1985; Barra et al. 1992). To investigate the relationship between ABO blood group and predisposition to toxin-mediated diarrheal disease, the ability of blood-group active glycolipids to interact with cholera toxin and heat labile enterotoxin was examined. Porcine glycosphingolipids and mucin-derived glycopeptides bearing blood group A antigenic activity interacted with cholera toxin more efficiently than the corresponding compounds lacking blood group A-active substances (Bennun et al. 1989; Monferran et al. 1990). Porcine gastric mucin-derived glycopeptides containing human blood group A antigenic activity inhibited cholera toxin binding to ganglioside GM1 receptors (Monferran et al. 1990). Heat-labile enterotoxin, like cholera toxin, interacted with several blood group-active glycolipids from pig intestinal mucosa preferentially recognizing glycolipids isolated from animals carrying A-blood group antigenic determinants as compared to those isolated from animals lacking these antigens (Barra et al.

1992). Since the internal carbohydrate structure of the ABH glycolipids differs among animal species, the ability of the toxin to interact with human ABH-active glycolipids was also evaluated. While the heat labile toxin preferentially interacted with A- and B-active glycolipids compared to H glycolipid antigens from human erythrocytes, cholera toxin did not bind to any human erythrocyte antigen. These results suggest the carbohydrate structural requirements for interaction of the toxins with blood group-active glycolipids are different. Although binding to human ABH-active glycolipids was not demonstrated for cholera toxin, the internal carbohydrate structure of glycolipids carrying ABH determinants vary greatly depending on the tissue source (McKibbin 1978). Consequently, a substance in the human gastrointestinal tract similar, but not identical, to the blood group A and/or blood group B antigens may interact with the cholera toxin or heat-labile enterotoxin to prevent their binding to GM1 membrane receptors, thus lessening the severity of the disease in individuals of blood group A, B and AB compared to individuals of blood group O (Barra et al. 1992).

P Blood Group Antigens and *Escherichia coli* Verotoxin

Another family of enterotoxins associated with diarrheal disease produced by *E. coli* strains consists of the Shiga-like toxins (verotoxins). The P^k antigen is a cell surface receptor for Shiga-toxin from *Shigella dysenteriae* type 1 and Shiga-like toxins from *E. coli* (Lindberg et al. 1987; Lingwood et al. 1987). Shiga-like toxins recently have been associated with several other clinical syndromes such as hemorrhagic colitis (Pai et al. 1984) and hemolytic-uremic syndrome (Karmali et al. 1985). An association between thrombotic thrombocytopenic purpura and intestinal disease caused by verotoxin producing enterhemorrhagic *E. coli* has also been reported, but is less frequent (Hofmann 1993). The verotoxin gains entry to the bloodstream presumably through breaks in the intestinal epithelial lining and then binds to the P^k antigen on renal vascular endothelial cells. The ensuing endothelial damage results in renal failure, microangiopathic hemolytic anemia and thrombocytopenia. In addition to its abundant expression in renal tissue (Boyd and Lingwood 1989), differential expression of this verotoxin receptor in various tissues is well correlated with the existence of pathological findings in those locations (Zoja et al. 1992; Boyd et al. 1993). Therefore, the tropism of this toxin for its target tissues is directly related to the expression of its blood group antigen receptor.

An inverse relationship was observed between the strength of P_1 antigen expression on erythrocyes, as judged by the hemagglutination strength with anti-P_1 antiserum, and the outcome of verotoxin-associated, classical hemolytic uremic syndrome (Taylor et al. 1990). This observation led to the hypothesis that strong P_1 expression on erythrocytes results in verotoxin adsorption thereby lowering the concentration of circulating toxin and reducing toxin binding to P blood group antigen receptors on vulnerable endothelial cells. Consequently, transfusion of P_1-positive erythrocytes may effectively neutralize circulating verotoxin and diminish the pathological consequences (Milford et al. 1989). In serological and biochemical experiments, verotoxins from *E. coli* bound to P blood group antigens of human erythrocytes in vitro (Bitzan et al. 1994). Evidence that P blood group status is related to the susceptibility to systemic verotoxin-associated disease also includes the report that the ratio of P^k antigen to its precursor glycolipid, lactosylceramide, was decreased in erythrocytes from patients with enteropathic hemolytic uremic syndrome compared with erythrocytes from children with verotoxin-associated diarrhea who did not develop hemolytic uremic syndrome (Newburg et al. 1993). Alternatively, verotoxins, through their interaction with human erythrocytes, may result in the subsequent hemolysis observed in patients with postenteropathic hemolytic uremic syndrome. Verotoxin binding to erythrocytes in verotoxin-associated disease has not yet been demonstrated, and its role in pathogenesis of hemolytic uremic syndrome remains an open area of investigation.

Lewis Antigens and Pertussis Toxin

A final example of a bacterial toxin that recognizes a blood group antigen is Pertussis toxin. Pertussis toxin mediates attachment of *Bordatella pertussis* to macrophages and respiratory epithelial cells, recognizes several glycoproteins and glycolipids, and produces a number of distinct biological effects (Van't Wout et al. 1992). Pertussis toxin receptors expressed on macrophage surface molecules may carry the Lewis determinant Le^a and the Lewis-related determinant Le^x (Van't Wout et al. 1992).

Unraveling the specificity of interactions between bacterial adhesins and toxins and their receptors will likely aid in the production of pharmaceutical compounds designed to neutralize binding activity by competing for host receptors. Ideally, this approach will lessen the severity of the corresponding disease. Conversely, bacterial toxins acting through their interaction with blood group antigens may be used to treat disease. Human tonsillar germinal center lymphocytes and Burkitt's lymphoma cells both express high levels of the P^k antigen (CD77) and both readily undergo apoptosis (Mangeney et al. 1991). Recent studies showed that a subunit of verotoxin induces apoptosis of these cells after binding to the P^k antigen (Mangeney et al. 1993). This is the first report of inducing programmed cell death in a human lymphoma through the interaction of a toxin with a cell surface glycospingolipid.

Blood Group Antigens as Receptors for Parasites

Malaria affects human and nonhuman primates, and simian models of infection have provided insight into the pathology and immune responses in human disease. The simian parasites, *Plasmodium knowlesi* and *P. cynomolgi*, share similarities with the human parasite *P. vivax* (Waters et al. 1993). *P. vivax* and the simian parasites can infect humans and result in relatively mild symptomatic episodes of the disease manifested as fever, anemia and splenomegaly. *P. falciparum,* however, is the most important human pathogen and is responsible for significant mortality and morbidity in endemic areas. The erythrocytic stage of *P. falciparum* accounts for the death of an estimated 2 million children annually (Sim et al. 1994). *P. falciparum* will infect Columbian night monkeys *(Aotus sp.)* and *Pan troglodytes* (Geiman and Meagher 1967; Waters et al. 1993). In contrast to the morbidity caused in the human host, *P. falciparum* infections are asymptomatic in the laboratory host (Waters et al. 1993). Invasion of erythrocytes by the different malaria parasites is mediated through specific interactions between ligands found on unique vesicular structures, micronemes, of the parasite and receptors found on the surface of the human erythrocyte membrane (Gratzer and Dluzewski 1993). Different *Plasmodium* species express related but distinct erythrocyte binding proteins and attach to different erythrocyte receptors: *P. knowlesi* and *P. vivax* require interaction with the Duffy blood group antigen; whereas *P. falciparum* binds to determinants on glycophorins and other membrane components (Miller et al. 1975, 1976; Mitchell et al. 1986; Hermentin and Enders 1984; Adams et al. 1992).

The step-wise invasion sequence of malaria parasites can be summarized as follows (Adams et al. 1990; Gratzer and Dluzewski 1993): The merozoite form first attaches to the erythrocyte membrane surface nonspecifically, then reorients such that its apical end is in contact with the erythrocyte. Attachment and reorientation of the *P. knowlesi* merozoite occur equally well on Duffy-positive and Duffy-negative erythrocytes. An electron dense plaque forms at the point of contact, and cytoplasmic organelles of the parasite, such as the micronemes which contain the erythrocyte binding proteins, may fuse with the apical membrane to release their contents and translocate parasite ligands to the apex. Finally, the parasite enters the erythrocyte in a vacuole formed from the erythrocyte membrane which ultimately surrounds the merozoite in the erythrocyte cytoplasm.

Although merozoites can attach to erythrocytes that do not express their cognate blood group antigen receptors, the Duffy antigens or glycophorins, specific junction formation occurs only between *P. knowlesi* or *P. falciparum* merozoites and Duffy-positive or glycophorin-positive erythrocytes, respectively (Adams et al. 1990). This phenomenon may represent an additional mechanism of immune evasion since the ligand for specific erythrocyte attachment is sequestered in the microneme until after the merozoite attaches to the erythrocyte membrane surface (Adams et al. 1990).

Duffy Antigens and *Plasmodium vivax*

Invasion of human erythrocytes by *P. vivax* and *P. knowlesi* requires Duffy (Fy^a or Fy^b) blood group determinants (Miller et al. 1975, 1976). Erythrocytes expressing Duffy antigens [Fy(a–b+)] were used for affinity purification of the erythrocyte binding proteins expressed by *P. knowlesi* (Adams et al. 1990). The genes encoding the Duffy-binding proteins of *P. knowlesi* and *P. vivax* were recently cloned (Adams et al. 1992). The gene for a homologous erythrocyte binding protein from *P. falciparum* with a molecular mass of 175 kDa (EBA175) has also been cloned (Adams et al. 1992). Because of its binding specificity, EBA175 is also called the *P. falciparum* sialic acid binding protein. These parasite proteins are transmembrane proteins assigned to the same gene family because they share extensive amino acid sequence homology in several regions of the predicted protein sequence (Adams et al. 1992). One homologous region of these proteins, region II, is involved in binding to the different erythrocyte receptors for *P. falciparum* and *P. vivax/P. knowlesi* (Sim et al. 1994; Adams et al. 1992). Other regions of homology may reflect other essential functions shared by the different species of *Plasmodia* despite their evolutionary divergence (Adams et al. 1992; Waters et al. 1993).

In addition to its role as the receptor for the human malarial parasite *P. vivax*, the Duffy antigen is a binding protein for a family of chemotactic polypeptides known as the chemokines (Horuk et al. 1993). Chemokines, for example interleukin (IL)-8, are small secreted proteins involved in chemoattraction of immune cells. The Duffy gene has been cloned and encodes a 35–46 kDa glycoprotein that, as predicted, shares homology with other members of the chemokine receptor family (Chaudhuri et al. 1993; Horuk et al. 1994). Thus, the Duffy glycoprotein on erythrocytes may be involved in regulating intravascular levels of chemokines. Alternatively, this chemokine receptor may initiate second messenger signal transduction pathways in response to IL-8 in erythrocytes or in other cells in which the protein is expressed (Horuk et al. 1994).

Glycophorin Antigens and *Plasmodium falciparum*

Invasion of erythrocytes by *P. falciparum* requires erythrocyte membrane surface sialic acid residues (Hermentin and Enders 1984; Gratzer and Dluzewski 1993). A protein from *P. falciparum* that binds to erythrocytes, EBA175, interacts specifically with sialic acid residues (Adams et al. 1992). In addition, antibodies against EBA175 block merozoite invasion of erythrocytes in vitro (Sim et al. 1990). Glycophorin A, the major sialoglycoprotein in erythrocytes, is required for optimal invasion by all *P. falciparum* clones (Miller et al. 1977; Pasvol et al. 1982). Glycophorins B and C have also been implicated as receptors for *P. falciparum* (Hermentin and Enders 1984). Moreover, some strains of *P. falciparum* possess receptors that bind a sialic acid-independent ligand in the erythrocyte membrane (Mitchell et al. 1986).

The precise receptor and ligand domains mediating invasion of erythrocytes by *P. falciparum* were delineated (Sim et al. 1994). First, the domains of EBA175 involved in erythrocyte binding were defined by expressing different regions of EBA175 on the surface of COS cells as chimeric proteins and evaluating the ability of the transfected cells to bind erythrocytes (Sim et al. 1994). A cysteine-rich region (region II) of EBA175 was implicated in mediating attachment to human erythrocytes. In similar studies, the homologous region of the related molecule from *P. vivax* was involved in binding to Duffy determinants on erythrocytes (Miller 1994).

To determine the precise domain on the erythrocyte membrane with which region II of EBA175 interacts, erythrocytes treated with enzymes and those with glycophorin null phenotypes were applied to the transfected cells expressing region II of EBA175 (Sim et al. 1994). Sialic acid was necessary for optimal binding as demonstrated by the fact that EBA175 did not bind to erythrocytes that were treated with neuraminidase or that lacked both glycophorin A and glycophorin B (M^kM^k) erythrocytes. Sialic acid, however, was not sufficient for binding because En(a–) erythrocytes, which contain sialic acid residues on glycophorin B but do not express glycophorin A, did not form rosettes on COS cells expressing EBA175. In contrast, SsU(–) erythrocytes, which possess glycophorin A but lack glycophorin B, did form rosettes on COS cells expressing EBA175.

Since the copy number on the erythrocyte membrane of glycophorin B is 10 % that of glycophorin B, differences in sialic acid content may have accounted for these observations (Sim et al. 1994). To evaluate the relative binding efficiency of glycophorins A and B, inhibition of erythrocyte binding to region II expressed on COS cells by soluble glycophorin A and glycophorin B was measured. At similar sialic acid concentrations, glycophorin B did not inhibit binding of erythrocytes to COS cells expressing region II; whereas glycophorin A inhibited binding significantly. Since glycophorins A and B are identical for the first 25 amino acids, and since glycophorin A or a glycopeptide corresponding to the first 64 amino acids of glycophorin A, but not purified glycophorin B, blocked the binding of EBA175 to erythrocytes, an amino acid sequence specific for glycophorin A, in addition to sialic acid, is probably necessary for binding of EBA175 to glycophorin A.

In conclusion, the results of many studies indicate that glycophorin A is an erythrocyte receptor for *P. falciparum*, and, more specifically, for the parasitic protein EBA175 (Hermentin and Enders 1984; Sim et al. 1994). Defining the domains on glycophorin A that interact with domains of the parasitic ligand may lead to new targets for vaccines and to the rational design of therapeutic agents that can prevent or ameliorate this disease (Sim et al. 1994).

Blood Group Polymorphisms and Natural Selection

Since erythrocytes with the Fy(a–b–) phenotype are resistant to invasion by *P. vivax* malaria merozoites, this may represent an example of natural selection in human populations, in which malaria is the agent inducing the selective pressure (Gratzer and Dluzewski 1993; Miller 1994). In the West African population, the frequency of Duffy-negative, Fy(a–b–), individuals approaches 100 %, whereas the frequency of this genotype is virtually 0 % among Caucasians. Fy(a–b–) individuals are not infected with *P. vivax* (Miller et al. 1976). Alternatively, the selective pressure ultimately resulting in the predominant Fy(a–b–) population may not have been directly related to selection for individuals lacking the *P. vivax* receptor. Instead, the possible involvement of Duffy antigens in chemokine metabolism may confer a survival advantage onto Fy(a–b–) individuals in terms of the ability to control other infections, such as that caused by *P. falciparum* (Miller 1994). Interestingly, evidence of malaria-induced selective pressure is not apparent in simian populations. Although *P. knowlesi* is related to *P. vivax*, uses the Duffy antigen as an erythrocyte receptor, and causes malaria in Rhesus monkeys, the Fy(a–b–) phenotype has not been described in these nonhuman primates (Adams et al. 1990).

In addition to the Duffy antigens, there are several other interesting blood group differences between African and Caucasian populations (Miller 1994). For example, glycophorin Dantu results from a recombination between the glycophorin A and glycophorin B genes. The resulting glycoprotein is composed of the extracellular domain of glycophorin B and the transmembrane and cytoplasmic domains of glycophorin A. The frequency of the Dantu phenotype is as high as 4 % in some African populations and may have been influenced by selective pressure for erythrocytes lacking glycophorin A, one of the receptors for *P. falciparum*. Similarly, the parasites demonstrate evolutionary flexibility, and mutation in the erythrocyte binding proteins may also be under selective pressure. For example, *P. falciparum*, unlike *P. vivax*, has developed multiple alternative pathways for invasion and may attach to glycophorin B via a different parasite ligand (Miller 1994). In some Pygmy populations, the frequency of glycophorin B-negative individuals approaches 20 %. Studies performed in vitro suggest SsU(–) erythrocytes, which lack glycophorin A but possess glycophorin B, are partially resistant to *P. falciparum* invasion (Pasvol et al. 1982). The susceptibility of the Pygmy population to malaria has not yet been reported.

Blood Group Antigens as Receptors for Viruses

P antigen and Parvovirus

Involvement of a blood group antigen in mediating infection with parvovirus was first suggested by hemagglutination studies, and the relevance of this observation to the pathogenesis of disease has been supported by clinical studies. Infection with parvovirus B19 produces several clinical syndromes in different patient populations (Luban 1994): (1) fifth disease (erythema infectiosum), a self-limited disease in children characterized by a cutaneous rash, (2) transient aplastic crisis in patients with underlying hemolytic diseases such as sickle cell disease or hereditary spherocytosis, (3) chronic anemia and pure red cell aplasia in immunocompromised patients or (4) intrauterine infection resulting in hydrops fetalis and fetal loss or congenital infection. Parvoviruses are small, single-stranded DNA viruses and B19 is the only known pathogenic parvovirus in humans. Parvovirus B19 agglutinates erythrocytes and demonstrates strict tropism for human erythroid progenitors in vitro. The B19 parvovirus binds specifically to human progenitor cells in the stages between burst-forming units-erythroid (BFU-E) and erythroblasts and inhibits hematopoietic colony formation by replicating in these cells (Mortimer et al. 1983; Takahashi et al. 1990). Parvovirus replication in vitro is specifically supported by the presence of erythropoietin. The virus has been propagated in bone marrow, peripheral blood, fetal liver, and erythropoietic cell lines with erythroid characteristics in vitro (Brown et al. 1994).

The molecular basis of the viral tropism for erythroid cells in vitro has recently been elucidated with the identification of a receptor for parvovirus (Brown et al. 1993). The ability of parvovirus B19 viral capsids to agglutinate erythrocytes with different P blood group phenotypes initially defined the viral binding specificity. Erythrocytes of p and P_1^k phenotypes are not agglutinated by B19, implicating the P antigen in viral attachment. The ability of pure lipid extracts to interfere with agglutination in this assay was then examined. Only P antigen (globoside) and a related glycosphingolipid, the Forssman antigen, inhibited hemagglutination. In addition, the B19 capsids bound to P antigen, but not other glycolipids, immobilized on a thin-layer chromatographic plate. P antigen or a monoclonal antibody directed against the P antigen also inhibited B19 infection of erythroid precursors in vitro.

Parvovirus B19 inhibits erythroid colony formation in cultures of bone marrow obtained from individuals of all P blood group phenotypes tested except for that obtained from individuals with the rare p blood group phenotype (Brown et al. 1994). Hybridization assays for viral DNA and RNA and an immunofluorescent assay for B19 proteins revealed no viral replication in cells with the p phenotype in situ. Conversely, double labeling with anti-B19 antibody and anti-P antibody of infected bone marrow cells of other P blood group phenotypes demonstrated that bone marrow cells containing parvovirus B19 proteins also expressed the P antigen (Kerr et al. 1995).

Consistent with these in vitro experimental results, individuals lacking the P antigen on their erythroid cells are protected from parvovirus B19 infection (Brown et al. 1994). Individuals (17 subjects) with the p phenotype showed no evidence of previous infection and were negative for antibodies to parvovirus B19. In striking contrast, the frequency of anti-parvovirus antibodies in the control population expressing the P antigen was 71%, a rate similar to the seroprevalence in other population studies performed in the United States. This evidence supports the conclusion that bone marrow cells with the p phenotype are not infected by parvovirus B19. The only shortcoming of this clinical study, acknowledged by the authors, is that erythrocytes with the P_1^k phenotype, which are very rare, were not tested for their susceptibility to parvovirus B19 infection in vitro (Brown et al. 1994). Erythrocytes from P_1^k individuals lack P antigen (globoside) but do express the P^k (precursor of P) and P_1 antigens. However, it is known that parvovirus B19 does not agglutinate P_1^k erythrocytes in vitro (Brown et al. 1993).

The P antigen is expressed not only on erythrocytes, but also on endothelial cells, fetal liver and heart, and placenta (Spitalnik and Spitalnik 1995). The presence of the P antigen on cells not known to be permissive for viral replication in vitro may have pathophysiological relevance (Brown et al. 1993). For example, parvovirus-infected endothelial cells may mediate transplacental transmission of the virus or contribute to the rash in fifth disease. The myocarditis observed in parvovirus-infected fetuses may reflect direct infection of cardiac myocytes.

Blood Group Antigens and Other Hemagglutinating Viruses

In addition to parvovirus, numerous other viruses such as influenza, reovirus, and encephalomyocarditis virus agglutinate erythrocytes in vitro (Burness and Pardoe 1981; Paul and Lee 1987). The pathophysiologic relevance of the erythrocyte receptors for these viruses is not as clear as for parvovirus B19 and the *Plasmodia* species. Attachment to erythrocytes may be important for viral dissemination in the host or for viral propagation in natural reservoirs (Eaton and Crameri 1989). As for some bacterial pathogens, the erythrocyte receptors may be different from other somatic cell receptors for these viral pathogens. Nonetheless, important insights into the viral attachment process in a well-characterized membrane system have been obtained from this experimental approach.

Glycophorins and Influenza and Encephalomyocarditis Viruses

In 1941, Hirst presented the first evidence that influenza virus recognized sialic acid-containing receptors on erythrocytes (Hirst 1941). In 1961, Kathan et al. demonstrated that glycophorin A serves as the influenza virus receptor on the surface of human erythrocytes (Kathan et al. 1961). Glycophorins A, B, C, and D, biochemically purified from human erythrocyte membranes, inhibited hemagglutination by influenza viruses A and B (Ohyama et al. 1993). Thus, influenza virus may bind to several glycophorins on erythrocytes. Interestingly, virus binding to respiratory mucins and epithelial cell membranes is also mediated by sialic acid containing oligosaccharides similar to those found on the erythrocyte glycophorins (Couceiro et al. 1993). The glycophorin family was similarly shown to serve as the erythrocyte receptor for encephalomyocarditis virus (Allaway and Burness 1986).

Cromer Antigen (DAF, CD55) and Enteroviruses

Unlike the glycophorins and other membrane proteins which are anchored in the cell membrane by a hydrophobic transmembrane domain, decay-accelerating factor (DAF; CD55) is anchored to the cell membrane by a fatty acid tail attached to its COOH-terminal by a glycosylphosphatidylinositol (GPI) linkage. The GPI-anchored DAF glycoprotein carries the Cromer blood group antigens and protects cells from lysis by autologous complement. Several enteroviruses, including at least seven echovirus serotypes and coxsackie virus B3, utilize DAF as a receptor (Ward et al. 1994; Bergelson et al. 1994, 1995). Enteroviruses can cause asymptomatic infections, nonspecific febrile illnesses, aseptic meningitis, encephalitis, herpangina, pericarditis and myocarditis.

Indian Blood Group Antigens (CD44) and Poliovirus

Another enterovirus recently associated with an integral membrane protein which carries a blood group antigen is poliovirus. The hematopoietic isoform of CD44 contains the Indian blood group antigens and serves as a lymphocyte homing receptor (Spring et al. 1988; Stamenkovic et al. 1989). An antibody that blocks poliovirus binding to cells recognizes CD44 (Shepley and Racaniello 1994). CD44 is involved in poliovirus attachment but does not itself act as a poliovirus receptor since expression of CD44 in cultured cells does not confer poliovirus infectivity (Shepley and Racaniello 1994). CD44 may interact with the poliovirus receptor (PVR) which is a member of the immunoglobulin superfamily. Expression of PVR, unlike CD44, is sufficient for poliovirus infection in some, but not all, cell types (Mendelsohn et al. 1989; Shepley and Racaniello 1994). These data suggest that the poliovirus receptor and CD44 may interact to form a functional receptor site for poliovirus attachment. Although PVR is expressed in a wide range of human tissue, poliovirus replication is limited to oropharyngeal and intestinal mucosa, the Peyer's patches of the ileum and motor neurons within the CNS (Mendelsohn et al. 1989). The viral tropism cannot be explained by the distribution of PVR expression. However, the tissue distribution of the hematopoietic isoform of CD44 correlates with poliovirus tropism (Shepley and Racaniello 1994). Consequently, CD44 may play a role in the pathogenesis of poliovirus (Shepley and Racaniello 1994)

Blood Group Related Antigens and Human Immunodeficiency Virus

Viral attachment to eukaryotic cells involves not only interaction with preexisting carbohydrate and protein moieties but also interaction with virally induced neoantigens. Viral infection may result in altered glycosylation in the host cells (Hansen et al. 1990). For example, the Le^y determinant, an isomer of the Le^b antigen, is highly expressed on the surface of human T cell lines after infection with HIV, as well as on T lymphocytes from patients infected with HIV, but not on normal lymphocytes from healthy individuals (Adachi et al. 1988). Expression of this new antigen has been implicated as a gp120-CD4-independent mechanism of HIV attachment to T cells late in infection (Glinsky 1992). The novel glycan structures produced by HIV-infected cells may also be incorporated into the viral envelope. Monoclonal antibodies directed against three different N- and O-linked carbohydrate blood group related-epitopes (Le^y, A_1, and sialyl-Tn) blocked infection by cell-free HIV and inhibited syncytium formation (Hansen et al. 1990). Preincubation of the virus, but not preincubation of the cells, with the monoclonal antibodies inhibited infection, suggesting that the monoclonal bound to the viral envelope. In addition, gp120 was immunoprecipitated by these antibodies. These glycan epitopes represent potential therapeutic targets for viral neutralization (Hansen et al. 1990).

Conclusion

Considerable progress has been made towards understanding the interaction between pathogens and host cellular targets, and this information will provide insight into the initiation and pathogenesis of infection. Virulence determinants, such as bacterial fimbrial proteins, enable pathogens to elude the normally efficient host defense mechanisms. Many of the host cell attachment sites were first identified as blood group antigens. The erythrocyte receptors likely play a key role in pathogenesis of diseases such as parvovirus infections and malaria. Alternatively, the erythrocyte glycoproteins and glycolipids have facilitated biochemical identification of homologous receptors on epithelial cells, extracellular matrix and mucins. Primate models of infections such as malaria have advanced our understanding of human disease. Delineating the molecular basis of the interaction between glycolipid, glycoprotein, and proteoglycan receptors and pathogens will enable novel interventional strategies. Specific host receptor and microbial analogues may be designed to mask attachment sites on infectious organisms or compete for receptors on host cells, respectively.

References

Adachi M, Hayami M, Kashiwagi N et al. Expression of Le^y antigen in human immunodeficiency virus-infected human T cell lines and in peripheral lyphocytes of patients with acquired immune deficiency syndrome (AIDS) and AIDS-related complex (ARC). J Exp Med 167:323-331, 1988

Adams JH, Hudson DE, Torii M, Ward GE et al. The Duffy receptor family of *Plasmodium knowlesi* is located within the micronemes of invasive malaria merozoites. Cell 63:141-153, 1990

Adams JH, Sim BKL, Dolan SA, Fang X, Kaslow DC, Miller LH. A family of erythrocyte binding proteins of malaria parasites. Proc Natl Acad Sci USA 89:7085-7089, 1992

Aird I, Bentall HH, Bentall HH, Mehigan JA, Roberts JAF. The blood groups in relation to peptic ulceration and carcinoma of colon, rectum, breast and bronchus. Br Med J 2:315-332, 1954

Allaway GP, Burness ATH. Site of attachment of encephalomyocarditis virus on human erythrocytes. J Virol 59:768-770, 1986

Aly FZ, Blackwell CC, MacKenzie DA et al. Chronic atrophic oral candidiasis among patients with diabetes mellitus-role of secretor status. Epidemiol Infect 106:355-63, 1991

Aly FZ, Blackwell CC, Mackenzie DA et al. Factors influencing oral carriage of yeasts among individuals with diabetes mellitus. Epidemiol Infect 109:507-518, 1992

Arneberg P, Kornstad L, Nordbo H Gjermo P. Less dental caries among secretors than among nonsecretors of blood group substance. Scand J Dent Res 84:362-366, 1976

Barnes GL, Kay R. Blood groups in giardiasis (letter). Lancet 1:808, 1977

Barra JL, Monferran CG, Balanzino LE, Cumar FA. *Escherichia coli* heat-labile enterotoxin preferentially interacts with blood group A-active glycolipids from pig intestinal mucosa and A- and B-active glycolipids from human red cells compared to H-active glycolipids. Mol Cell Biochem 115:63-70, 1992

Barua D, Paguio AD ABO blood groups and cholera. Ann Hum Biol 4:489, 1977

Bennun FR, Roth GA. Binding of cholera toxin to pig intestinal mucosa glycosphingolipids: relationship with the ABO blood group system. Infect Immun 57:969–74, 1989

Bergelson JM, Chan M, Solomon KR, St. John NF et al. Decay-accelerating factor (CD55), a glycosylphosphatidylinositol-anchored complement regulaory protein, is a receptor for several echoviruses. Proc Natl Acad Sci USA 91:6245–6248, 1994

Bergelson JM, Mohanty JG, Crowell RL, St John NF et al. Coxsackievirus B3 adapted to growth in RD cells binds to decay-accelerating factor (CD55). J Virol 69:1903–1906, 1995

Bitzan M, Richardson S, Huang C, Boyd B, Petric M, Karmali MA. Evidence that verotoxins (shiga-like toxins) from *Escherichia coli* bind to P blood group antigens of human erythrocytes in vitro. Infect Immun 62:3337–3347, 1994

Black RE, Levine MM, Clements ML, Hughes T, O'Donnell S. Association between O blood group and occurence and severity of diarrhoea due to *Escherichia coli*. Trans R Soc Trop Med Hyg 81:120–123, 1987

Blackwell CC. The role of ABO blood groups and secretor status in host defenses. FEMS Microbiol Immunol 47:341–350, 1989

Blackwell CC, Kowolik M, Winstanley FP et al. ABO blood group and susceptibility to gonococcal infection. I. Factors affecting phagocytosis of *Neisseria gonorrhoeae*. J Clin Lab Immunol 10:173–178, 1983

Blackwell CC, Andrew S, May SJ et al. ABO blood group and susceptibility to urinary tract infection: no evidence for involvement of isohaemagglutinins. J Clin Lab Immunol 15:191–194, 1984

Blackwell CC, Jonsdottir K, Hanson M, Todd WTA, Chaudhuri AKR, Mathew B, Brettle RP, Weir DM. Non-secretion of ABO antigens predisposing to infection by *Neisseria meningitidis* and *Streptococcus pneumoniae*. Lancet ii:284–285, 1986a

Blackwell CC, Jonsdottir K, Hanson MF, Weir DM. Non-secretion of ABO blood group antigens predisposing to infection by *Haemophilis influenzae*. Lancet 2:687, 1986b

Blackwell CC, Aly FZ, James VS, Weir DM, Collier A, Patric AW, Cumming CG, Wray D, Clarke BF. Blood group, secretor status and oral carriage of yeasts among patients with diabetes mellitus. Diabetes Res 12:101–104, 1989a

Blackwell CC, Jonsdottir K, Weir DM et al. Blood group, secretor status and susceptibility to bacterial meningitis. FEMS Microbiol Immunol 47:351–356, 1989b.

Blackwell CC, Weir DM, James VS et al. Secretor status, smoking and carriage of *Neisseria meningitidis*. Epidemiology and Infection 104:203–209, 1990

Bock K, Breimer ME, Brignole A et al. Specificity of binding of a strain of uropathogenic *Escherichia coli* to Galα1–4Gal-containing glycosphingolipids. J Biol Chem 260:8545–8551, 1985

Born T, Falk P, Roth KA, Larson G, Normark S. Attachment of Helicobacter pylori to human gastric epithelium mediated by blood group antigens. Science 262:1892–1895, 1993

Bothamley GH, Schreuder GMT, de Vries RRP, Ivanyi J. Blood group antigen secretion and gallstone disease in the *Salmonella typhi* chronic carrier state. J Infect Dis 167:993–994, 1993

Boyd B, Tyrrell G, Maloney M et al. Alteration of the glycolipid binding specificity of the pig edema toxin from globotetrasyl to globotriosyl ceramide alters *in vivo* tissue targeting and results in a verotoxin 1-like disease in pigs. J Exp Med 177:1745–1753, 1993

Boyd G, Lingwood C. Verotoxin receptor glycolipid in human renal tissue. Nephron 51:207–210, 1989

Brown KE, Anderson SM, Young NS. Erythrocyte P antigen: cellular receptor for B19 parvovirus. Science 262:114–117, 1993

Brown KE, Hibbs JR, Gallinella G, Anderson SM et al. Resistance to parvovirus B19 infection due to lack of virus receptor (erythrocyte P antigen). N Engl J Med 330:1192–1196, 1994

Buck GE. *Campylobacter pylori* and gastric duodenal disease. Clin Microbiol Rev 3:1–12, 1990

Burford-Mason AP, Weber JCP, Willoughby JMT. Oral carriage of *Candida albicans*. ABO blood group and secretor status in healthy subjects. J Med Vet Mycol 26:49–56, 1988

Burford-Mason AP, Willoughby JMT, Weber JC. Association between gastrointestinal tract carriage of *Candida*, blood group O, and nonsecretion of blood group antigens in patiens with peptic ulcer. Dig Dis Sci 38:1453–1458, 1993

Burness AT, Pardoe IU. Effect of enzymes on the attachment of influenza and encephalo-myocarditis viruses to erythrocytes. J Gen Virol 55:275–288, 1981

Chaudhuri A, Das Adhikary CR. Possible role of blood-group secretory substances in the aetiology of cholera. Trans R Soc Trop Med Hyg 72:664–665, 1978

Chadhuri A, De S. Cholera and blood groups. Lancet ii:404, 1977

Chaudhuri A, Polyakova J, Abrzezna V, Williams K et al. Cloning of glycoprotein D cDNA, which encodes the major subunit of the Duffy blood group system and the receptor for the Plasmodium vivax malaria parasite. Proc Natl Acad Sci USA 90:10793–10797, 1993

Check JH, O'Neill EA, O'Neill K, Fuscaldo KE. Effect of Anti-B antiserum on the phagocytosis of Escherichia coli. Infect Immun 6:95–96, 1972

Couceiro JN, Paulson JC, Baum LG. Influenza virus strains selectively recognize sialyloligosaccharides on human respiratory epithelium; the role of the host cell in selection of hemagglutinin receptor specificity. Virus Res 29:155–165, 1993

Cruz-Coke R, Paredes L, Montenegro A. Blood groups and urinary microorganisms. J Med Genet 2:185–188, 1965

Cuadrado RR, Davenport FM. Antibodies to Influenza viruses in military recruits from Argentina, Brazil and Colombia: their relation to ABO blood group distribution. Bull World Health Org 42:873–884, 1970

Deal CD, Krivan HC. Solid-phase binding of microorganisms to glycolipids and phospholipids. Methods Enzymol 236:346–353, 1994

Deresinski SC, Pappagianis D, Stevens DA. Association of ABO blood group and outcome of coccidioidal infection. Sabouraudia 17:261–264, 1979

Dickey W, Collins JSA, Watson RGP, Sloan JM, Porter KG. Secretor status and *Helicobacter pylori* infection are independent risk factors for gastroduodenal disease. Gut 34:351–353, 1993

Eaton BT, Crameri GS. The site of Bluetongue virus attachment to glycophorins from a number of animal erythrocytes. J Gen Virol 70:3347–3353, 1989

Emeribe AO, Ejezie CG. Haemoparasites of blood donors in Calabar. Trop Geogr Med 41:61–65, 1989

Essery SD, Saadi AT, Twite SJ et al. Lewis antigen expression on human monocytes and binding of pyrogenic toxins. Agents Actions 41:108–110, 1994a

Essery SD, Weir DM, James VS, Blackwell CC et al. Detection of microbial surface antigens that bind Lewis (a) antigen. FEMS Immunol Med Microbiol 9:15–21, 1994b

Esterre P, Dedet J-P. The relationship of blood-group to American cutaneous leishamiasis. Ann Trop Med Parasitol 83:345–348, 1989

Evans AS, Shepard KA, Richards VA. ABO blood groups and viral diseases. Yale J Biol Med 45:81–92, 1972

Evans DG, Karjalainen TK, Evans DJ Jr et al. Cloning, nucleotide sequence, and expression of a gene encoding an adhesin subunit protein of *Helicobacter pylori*. J Bacteriol 175:674–683, 1993

Falk P, Roth KA, Born T et al. An *in vitro* adherence assay reveals that *Helicobacter pylori* exhibits cell lineage-specific tropism in the human gastric epithelium. Proc Natl Acad Sci USA, 90:2035–2039, 1993

Falk P, Boren T, Normark S. Characterization of microbial host receptors. Methods Enzymol 236:353–374, 1994

Falk PG, Bry L, Holgersson J, Gordon JI. Expression of a human alpha-1,3/4-fucosyltransferase in the pit cell lineage of FVB/N mouse stomach results in production of Le[b]-containing glycoconjugates: a potential transgenic mouse model for studying Helicobacter pylori infection. Proc Natl Acad Aci USA 92:1515–1519, 1995

Finkelstein RA, Burks MF, Zupan A, Dallas WS et al. Epitopes of the cholera family of enterotoxins. Rev Infect Dis 9:544–561, 1987

Foster MT, Labrum AH. Relation of infection with Neisseria gonorrhoea to blood groups. J Infect Dis 133:329–330, 1976

Gaffney RA, Schaeffer AJ, Duncan JL. Lewis blood group antigen expression by cultured normal ureteral epithelial cells. J Urol 148:1341–1346, 1992

Gaffney RA, Schaeffer AJ, Anderson BE, Duncan JL. Effect of Lewis blood group antigen expression on bacterial adherence to COS-1 cells. Infect Immun 62:3022–3026, 1994

Garratty G. Blood group antigens as tumor markers, parasitic/bacterial/viral recptors and their association with immunologically important proteins. Immunol Invest 24:213–232, 1995

Geiman QM, Meagher MJ. Susceptibility of a New World monkey to Plasmodium falciparum from man. Nature 215:437–439, 1967

Glass RI, Holgren J, Haley CE et al. Predispostion for cholera of individuals with O blood group: possible evolutionary significance. Am J Epidemiol 121:791–796, 1985

Glinsky GV. The blood group antigen-related glycoepitopes: key structural determinants in immunogenesis and AIDS pathogenesis. Med Hypotheses 39:212–224, 1992

Goldhar J. Bacterial Lectinlike adhesins: determination and specificity. Methods Enzymol 236:211–231, 1994

Gratzer WB, Dluzewski AR. The red blood cell and malaria parasite invasion. Semin Hematol 30:232–247, 1993

Gupta M, Chowdhuri ANR. Relationship between ABO blood groups and malaria. Bull World Health Organ 58:913–915, 1980

Haataja S, Tikkanen K, Liukkonen J, Grancois-Gerard C, Finne J. Characterization of a novel bacterial adhesion specificity of Streptococcus suis recognizing blood group P receptor oligosaccharides. J Biol Chem 268:4311–4317, 1993

Haataja S, Tikkanen K, Nilsson U, Magnusson G, Karlsson K-A, Finne J. Oligosaccharide-receptor interaction of the Galα1-4Gal binding adhesin of Streptococcus suis. J Biol Chem 269:27466–27472, 1994

Hansen J.-E.S, Clausen H, Nielsen C, Teglbjrg LS et al. Inhibition of human immunodeficiency virus (HIV) infection in vitro by anticarbohydrate monoclonal antibodies: peripheral glycosylation of HIV envelope glycoprotein gp120 may be a target for viral neutralization. J Virol 64:2833–2840, 1990

Haverkorn MJ, Goslings WRO. Streptococci, ABO blood groups and secretor status. Am J Hum Genet 21:360–375, 1969

Hermentin P, Enders B. Erythrocyte invasion by malaria (Plasmodium faciparum) merozoites: recent advances in the evaluation of receptor sites. Behring Inst Mitt 76:121–141, 1984

Higgins DA, Jenkins DJ, Partono F. Timorian filariasis and ABO blood groups. Trans R Soc Trop Med Hyg 79:537–538, 1985

Hirst GK. The agglutination of red cells by allantoic fluid of chick embryos infected with influenza virus. Science 94:22–23, 1941

Hofmann SL. Southwestern Internal Medical Conference: shiga-like toxins in hemolytic-uremic syndrome and thrombotic thrombocytopenic purpura. Am J Med Sci 306:398–406, 1993

Holbrook WP, Blackwell CC. Secretor state and dental caries in Iceland. FEMS Microbiol Immunol 47:397–400, 1980

Holmgren J, Lindblad M, Fredman P et al. Comparison of receptors for cholera toxin and *Escherichia coli* enterotoxins in human intestine. Gastroenterology 89:27–35, 1985

Höök-Nikanne J, Sistonen P, Kosunen TU. Effect of ABO blood group and secretor status on the frequency of *Helicobacter pylori* antibodies. Scand J Gastroenterol 25:815–818, 1990

Horuk R, Chitnis CE, Darbonne WC, Colby TJ et al. A receptor for the malarial parasite *Plasmodium vivax*: the erythrocyte chemokine receptor. Science 261:1182–1184, 1993

Horuk R, Zi-xuan W, Peiper SC, Hesselgesser J. Identification and characterization of a promiscuous chemokine-binding protein in a human erythroleukemic cell line. J Biol Chem 269:17730–17733, 1994

Jacobson SH, Lins LE, Svenson SB et al. Lack of correlation of P blood group phenotype and renal scarring. Kidney Int 28:797–800, 1985

Jacobson SH, Lomberg H. Overrepresentaion of blood group non-secretors in adults with renal scarring. Scand J Urol Nephrol 24:145–150, 1990

Jantausch BA, Criss VR, O'Donnell R et al. Association of Lewis blood group phenotypes with urinary tract infection in children. J Pediatr 124:863–868, 1994

Johanson I, Lindstedt R, Svanborg C. Roles of the *pap*- and *prs*-encoded adhesins in *Escherichia coli* adherence to human uroepithelial cells. Infect Immun 60:3416–3422, 1992

Johnson AP, Osborn MF, Hanna NF, Dawson SG, McManus TJ, Taylor-Robertson, D. A study of the relationship between ABO groups, secretor status and infection with *Neisseria gonorrhoeae*. J Infect 6:171–174, 1983

Johnson JR, Berggren T. Pigeon and dove eggwhite protect mice against renal infection due to P fimbriated *Escherichia coli*. Am J Med Sci 307:335–339, 1994

Johnson JR, Ross AE. P₁-antigen-containing avian egg whites as inhibitors of P adhesins among wild-type *Escherichia coli* strains from patients with urosepsis. Infect Immun 61:4902–4905, 1993

Johnson JR, Swanson JL, Neill MA. Avian P1 antigens inhibit agglutination mediated by P fimbriae of uropathogenic *Escherichia coli*. Infect Immun 60:578–583, 1992

Källenius G, Möllby R, Svenson SB et al. The Pᵏ antigen as receptor for the haemagglutinin of pyelonephritic *Escherichia coli*. FEMS Microbiol Lett 7:297–302, 1980

Källenius G, Möllby R, Hultberg H et al. Structure of carbohydrate part of receptor on human uroepithelial cells for pyelonephritogenic *Escherichia coli*. Lancet 2:604–606, 1981a

Källenius G, Svenson SB, Hultberg H et al. Occurrence of P-fimbriated *Escherichia coli* in urinary tract infections. Lancet 2:1369–1372, 1981b

Källenius G, Svenson SB, Möllby R et al. Carbohydrate receptor structures recognized by uropathogenic *Escherichia coli*. Scand J Infect Dis 33 [Suppl]:52–60, 1982

Karlsson KA. Animal glycosphingolipids as membrane attachment sites for bacteria. Annu Rev Biochem 58:309–350, 1989

Karmali MA, Petric M, Lim C et al. The association between idiopathic hemolytic uremic syndrome and infection by verotoxin-producing *Escherichia coli*. J Infect Dis 151:775–782, 1985

Karr JF, Nowicki BJ, Truong LD et al. *pap*-2-Encoded fimbriae adhere to the P blood group-related glycosphingolipid stage specific embryonic antigen 4 in the human kidney. Infect Immun 58:4055–4062, 1990

Kassim OO, Ejezie GC. ABO blood groups in malaria and schistosomiasis hematobium. Acta Trop (Basel) 39:179–84, 1982

Kathan RH, Winzler RJ, Johnson CA. Preparation of an inhibitor of viral hemagglutination from human erythrocytes. J Exp Med 113:37–45, 1961

Kerr JR, McQuaid S, Coyle PV. Expression of P antigen in parvovirus B19-infected bone marrow (letter). N Eng J Med 332:128, 1995

Kinane DF, Blackwell CC, Brettle RP et al. ABO blood group, secretor state, and suscept-ibility to recurrent urinary tract infection in women. Br Med J 285:7–9, 1982

Kinane DF, Blackwell CC, Winstanley FP, Weir DM. Blood group secretor status and sus-ceptibility to infection by *Neisseria gonnorhoeae*. Br J Vener Dis 59:44–46, 1983

König AL, Dreft H, Hengge U, Braun RW, Roelcke D. Coexisting anti-I and Anti-F₁/Gd Cold agglutinins in infections by *Mycoplasma pneumoniae*. Vox Sang 55:176–180, 1988

Korhonen TK, Väisänen-Rhen V, Rhen M Pere A, Parkkinen J, Finne J. *Escherichia coli* fimbriae recognizing sialyl galactosides. J Bacteriol 159:762–766, 1984

Korhonen TK, Virkola R, Lahteenmaki K, Bjorkman Y, Kukkonen M, Raunio T, Tarkkanen AM, Westerlund B. Penetration of fimbriated enteric bacteria through basement membranes: a hypothesis. FEMS Microbiol Lett 79:307–312, 1992

Labigne-Roussel AF, Lard D, Schoolnik G et al. Cloning and expression of an afimbrial adhesin (AFA-1) responsible for P blood group-independent, mannose-resistant hemagglutination from a pyelonephritic *Escherichia coli* strain. Infect Immun 46:251–259, 1984

Lanne B, Olsson BM, Jovall PA et al. Glycoconjugate receptors for P-fimbriated *Escherichia coli* in the mouse. An animal model of urinary tract infection. J Biol Chem 270:9017–25, 1995

Leffler H, Svanborg-Eden C. Glycolipid receptors for uropathogenic *Escherichia coli* on human erythrocytes and uroepithelial cells. Infect Immun 34:920–929, 1981

Leffler H, Svanborg-Eden C. Chemical identification of a glycosphingolipid receptor for *Escherichia coli* attaching to human urinary tract epithelial cells and agglutinating human erythrocytes. FEMS Microbiol Lett 8:127–134, 1980

Leffler H, Lomberg H, Botschlich E et al. Chemical and clinical studies on the interaction of *Escherichia coli* with host glycolipid receptors in urinary tract infection. Scand J Infect Dis 33:46–51, 1982

Levine MM, Nalin DR, Rennels MB, Hornick RB, Sotman S, van Blerk G, Hughes TP, O'Donnell S. Genetic susceptibility to cholera. Ann Hum Biol 6:369–374, 1979

Lindberg AA, Brown JE, Stromberg N et al. Identification of the carbohydrate receptor for Shiga toxin produced by *Shigella dysenteriae* type I. J Biol Chem 262:1779–1785, 1987

Lindstedt R, Larson G, Falk P et al. The receptor repertoire defines the host range for attaching *Escherichia coli* strains that recognize globo-A. Infect Immun 59:1086–1092, 1991

Lingwood CA, Law H, Richardson S et al. Glycolipid binding of purified and recombinant *Escherichia coli* produced verotoxin in vitro. J Biol Chem 262:8834–8839, 1987

Lomberg, H, Hellström M, Jodal U et al. Virulence-associated traits in *Escherichia coli* causing first and recurrent episodes of urinary tract infection in children with or without vesicoureteral reflux. J Infect Dis 150:561–569, 1984

Lomberg H, Cedergren B, Leffler H et al. Influence of blood group on the availability of receptors for attachment of uropathogenic *Escherichia coli*. Infect Immun 51:919–926, 1986

Lomberg H, de Man P, Svanborg-Eden C. Bacterial and host determinants of renal scarring. APMIS 97:193–199, 1989

Loomes LM, Uemura K, Childs RA, Paulson JC, Rogers GN, Scudder PR, Michalski J-C, Hounsell EF, Taylor-Robinson D, Feizi T. Erythrocyte receptors for *Mycoplasma pneumoniae* are sialylated oligosaccharides of Ii antigen type. Nature 307:560–563, 1984

Loomes LM, Uemura K, Feizi T. Interaction of *Mycoplasma pneumoniae* with erythrocyte glycolipids of I and i antigen types. Infect Immun 47:15–20, 1985

Loveless RW, Feizi T. Sialo-oligosaccharide receptors for *Mycoplasma pneumoniae* and related oligosaccharides of poly-N-acetyllactosamine series are polarized at the cilia and apical-microvillar domains of the ciliated cells in human bronchial epithe-lium. Infect Immun 57:1285–1289, 1989

Loveless RW, Griffiths S, Fryer PR, Blauth C, Feizi T. Immunoelectron microscopic studies reveal differences in distribution of sialo-oligosaccharide receptors for *Mycoplasma pneumoniae* on the epithelium of human and hamster bronchi. Infect Immun 60:4015–4023, 1992

Luban NL. Human parvoviruses: implications for transfusion medicine. Transfusion 34:821–827, 1994

Lublin DM. Functional roles of blood group antigens. In: Silberstein LE (ed) Molecular and functional aspects of blood group antigens. American Association of Blood Banks, Bethesda, MD, pp 163–192, 1995

Mackenzie JS, Fimmel PJ. The effect of ABO blood groups on the incidence of epidemic influenza and on the response to live and attenuated and detergent split influenza vaccines. J Hyg 80:21–30, 1978

Mangeney M, Richard Y, Coulaud D et al. CD77: an antigen of germinal center B cells entering apoptosis. Eur J Immunol 21:1131–1140, 1991

Mangeney M, Lingwood CA, Taga S et al. Apoptosis induced in Burkitt's lymphoma cells via Gb$_3$/CD77, a glycolipid antigen. Cancer Res 53:5314–5319, 1993

Manuila A. Blood groups and disease-hard facts and delusions. J Am Med Assoc 167:2047–2053, 1958

Marsh WL, Brown PJ, DiNapoli J, Beck ML, Wood M, Wojcicki R, de la Camara C. Anti-Wj: an autoantibody that defines a high-incidence antigen modified by the In(Lu) gene. Transfusion 23:128–130, 1983

May SJ, Blackwell CC, Brettle RP, MacCallum CJ, Weir DM. Non-secretion of ABO blood group antigens: a host factor predisposing to recurrent urinary tract infections and renal scarring. FEMS Microbiol Immunol 1:383–387, 1989a

May SJ, Blackwell CC, Weir DM. Lewis^a blood group antigen of non-secretors: a receptor for *Candida* blastospores. FEMS Microbiol Immunol 47:407–410, 1989b

McDonald JC, Zukerman AJ. ABO blood groups and acute respiratory virus disease. Br Med J 2:89–90, 1962

McKibbin JM. Fucolipids. J Lipid Res 19:131–147, 1978

Mendelsohn CL, Wimmer E, Racaniello VR. Cellular receptor for poliovirus: molecular cloning, nucleotide sequence and expression of a new member of the immunoglobulin superfamily. Cell 56:855–865, 1989

Mentis A, Blackwell CC, Weir DM et al. ABO blood group, secretor status and detection of *Helicobacter pylori* among patients with gastric or duodenal ulcers. Epidemiol Infect 106:221–229, 1991

Miller JJ, Novotny P, Walker PD, Harris JRW, MacLennan IPB. *Neisseria gonorrhoeae* and ABO isohaemagglutinins. Infect Immun 15:713–719, 1977

Milford DV, Taylor CM, Rose PE, Roy TCF, Rowe B. Immunologic therapy for hemolytic-uremic syndrome (correspondence). J Pediatr 115:502–503, 1989

Miller LH. Impact of malaria on genetic polymorphism and genetic diseases in Africans and African Americans. Proc Natl Acad Sci USA 91:2415–2419, 1994

Miller LH, Mason SJ, Dvorak JA et al. Erythrocyte receptors for *(Plasmodium knowlesi)* malaria and Duffy blood group determinants. Science 189:561–63, 1975

Miller LH, Mason SJ, Clyde DF, McGinniss MH. The resistance factor to *Plasmodium vivax* in blacks: the Duffy-blood-group genotype, FyFy. N Engl J Med 295:302–304, 1976

Mitchell GH, Hadley TJ, McGinniss, Klotz FW, Miller LH. Invasion of erythrocytes by *Plasmodium falciparum* malaria parasites: evidence for receptor heterogeneity and two receptors. Blood 67:1519–1521, 1986

Monferran CG, Roth GA, Cumar FA. Inhibition of cholera toxin binding to membrane receptors by pig gastric mucin-derived glycopeptides: differential effect depending on the ABO blood group antigenic determinants. Infect Immun 58:3966–3972, 1990

Mortimer PP, Humphries RK, Moore JG et al. A human parvovirus-like virus inhibits hematopoietic colony formation in vitro. Nature 302:426–429, 1983

Mourant AE, Kopec AC, Domaniewska-Sobczak K. Blood groups and disease. Oxford University Press, Oxford, 1978

Mourant AE. Linkage equilibrium and disequilibrium in human population studies. Ann Hum Biol 7:109–114, 1980

Mourant AE. Recent advances in the study of associations between infection and genetic markers. FEMS Microbiol Immunol 47:317–320, 1989

Mulherin D, Coffey MJ, Keogan MJ, O'Brien P, FitzGerald MX. *Pseudomonas* colonization in cystic fibrosis: lack of correlation with secretion of ABO blood group antigens. Ir J Med Sci 159:217–218, 1990

Mulholland SG, Mooreville M, Parsons CL. Urinary tract infections and P blood group antigens. Urology 24:232–235, 1984

Navas EL, Venegas MF, Duncan JL et al. Blood group antigen expression on vaginal and buccal epithelial cells and mucus in secretor and nonsecretor women. J Urol 149:1492–1498, 1993

Navas EL, Venegas MF, Duncan JL et al. Blood group antigen expression on vaginal cells and mucus in women with and without a history of urinary tract infections. J Urol 152:345–349, 1994

Neu HC. Urinary tract infections. Am J Med 92:63S–70S, 1992

Newburg DS, Chaturvedi P, Lopez EL, Devoto S, Fayad A, Cleary TG. Susceptibility to hemolytic-uremic syndrome relates to erythrocyte glycosphingolipid patterns. J Infect Dis 168:476–479, 1993

Nowicki B, Moulds J, Hull R, Hull S. A hemagglutinin of uropathogenic *Escherichia coli* recognizes the Dr blood group antigen. Infect Immun 56:1057–1060, 1988a

Nowicki B, Truong L, Moulds J, Hull R. Presence of the Dr receptor in normal human tissues and its possible role in the pathogenesis of ascending urinary tract infection. Am J Pathol 133:1–4, 1988b

Nowicki B, Svanborg-Eden C, Hull R, Hull S. Molecular analysis of the Dr hemagglutinin of uropathogenic Escherichia coli. Infect Immun 57:446–451, 1989

Nowicki B, Labigne A, Moseley S, Hull R, Hull S. The Dr hemagglutinin, afimbrial adhesins AFA-I and AFA-III, and F1845 fimbriae of uropathogenic and diarrhea-associated *Escherichia coli* belong to a family of hemagglutinins with Dr receptor recognition. Infect Immun 58:279–281, 1990

Nowicki B, Hart A, Coyne KE, Lublin DM, Nowicki S. Short consensus repeat-3 domain of recombinant decay accelerating factor is recognized by *Escherichia coli* recombinant Dr adhesin in a model of a cell-cell interaction. J Exper Med 178:2115–2121, 1993

Ohyama K, Endo T, Ohkuma S, Yamakawa T. Isolation and influenza virus receptor activity of glycophorins B, C and D from human erythrocyte membranes. Biochim Biophys Acta 1148:133–138, 1993

Pai CH, Gordon R, Sims HV et al. Sporadic cases of hemorrhagic colitis associated with *Escherichia coli* O157:H7. Clinical, epidemiological, and bacteriological features. Ann Intern Med 101:738–742, 1984

Parkkinen J, Rogers GN, Korhonen TK, Dahr W, Finne J. Identification of the O-linked sialyloligosaccharides of glycophorin A as the erythrocyte receptors for S-fimbriated *Escherichia coli*. Infect Immun 54:37–42, 1986

Parkkinen J, Korhonen TK, Pere A, Hacker J, Soinila S. Binding sites in the rat brain for *Escherichia coli* S fimbriae associated with neonatal meningitis. J Clin Invest 81:860–865, 1988a

Parkkinen J, Virkola R, Korhonen TK. Identification of factors in human urine that inhibit the binding of *Escherichia coli* adhesins. Infect Immun 56:2623–2630, 1988b

Pasvol G, Jungery M, Weatherall DJ, Parsons SF et al. Glycophorin as a possible receptor for *Plasmodium falciparum*. Lancet ii:947–948, 1982

Paul RW, Lee PW. Glycophorin is the reovirus receptor on human erythrocytes. Virology 159:94–101, 1987

Payne D, O'Reilly M, Williamson D. The K88 fimbrial adhesin of enterotoxigenic *Escherichia coli* binds to b1-linked galactosyl residues in glycosphingolipids. Infect Immun 61:3673–3677, 1993

Pereira FEL, Boroloni EP, Carneiro JLA et al. ABO blood groups and hepatosplenic form of schistosomiasis mansoni (Symmer's fibrosis). Trans R Soc Trop Med Hyg 73:238, 1979

Plos K, Carter T, Hull S et al. Frequency and organization of *pap* homologous DNA in relation to clinical origin of uropathogenic Escherichia coli. J Infect Dis 161:518–524, 1990

Poole J, Giles CM. Observations on the Anton antigen and antibody. Vox Sang 43:220–222, 1982

Poole J, van Alphen L. *Haemophilus influenzae* receptor and the AnWj antigen (letter). Transfusion 28:289, 1988

Potter CW. Haemagglutination inhibition antibody to various influenza viruses and adenoviruses in individuals of blood groups A and O. J Hyg 67:67–74, 1969

Potter CW, Schild GC. The incidence of HI antibody to influenza virus A2/Singapore 1/57 in individuals of blood groups A and O. J Immunol 98:1320–1325, 1967

Pradhan AC, Chawla TN, Samuel KC, Pradhan S. The relationship between periodontal disease and blood groups and secretor status. J Periodontal Res 6:294–300, 1971

Raza MW, Blackwell CC, Molyneaux P, James VS, Ogilvie MM, Inglis JM, Weir DM. Association between secretor status and respiratory viral illness. Br Med J 303:815–818, 1991

Reed WP, Drach GW, Williams RC Jr. Antigens common to human and bacterial cells IV. Studies of human pneumococcal disease. J Lab Clin Med 83:599–610, 1974

Reid ME, Bird GWG. Associations between human red cell blood group antigens and disease. Transfus Med Rev 4:47–55, 1990

Roberts JA, Hardaway K, Kaack B et al. Prevention of pyelonephritis by immunization with P-fimbriae. J Urol 131:602–607, 1984a

Roberts JA, Kaack B, Källenius G, Möllby R et al. Receptors for pyelonephritogenic *Escherichia coli* in primates. J Urol 131:163–168, 1984b

Roberts JA, Kaack MB, Baskin G, Svenson SB. Prevention of renal scarring from pyelonephritis in nonhuman primates by vaccination with a synthetic *Escherichia coli* serotype O8 oligosaccharide-protein conjugate. Infect Immun 61:5214–5218, 1993

Roberts-Thomson IC. Genetic studies of human and murine giardiasis. Clin Infect Dis 16 [Suppl 2]:S98–104, 1993

Robinson MG, Tolchin D, Halpern C. Enteric bacteria and the ABO blood groups. Am J Hum Genet 23:135–145, 1971

Rosenstein IJ, Hazlehurst GR, Burroughs AK et al. Recurrent bacteriuria and primary biliary cirrhosis: ABO blood group, P1 blood group, and secretor status. J Clin Pathol 37:1055–1058, 1984

Rydberg J, Helin I. A simple reliable agglutination test for screening P-fimbriated *Escherichia coli* in children with urinary tract infections gives valuable clinical information. Scand J Infect Dis 23:573–575, 1991

Saadi AT, Blackwell CC, Raza MW et al. Factors enhancing adherence of toxigenic *Staphylococcus aureus* to epithelial cells and their possible role in sudden infant death syndrome. Epidemiol Infect 110:507–17, 1993

Saadi AT, Weir DM, Poston IR et al. Isolation of an adhesin from *Staphylococcus aureus* that binds Lewis a blood group antigen and its relevance to sudden infant death syndrome. FEMS Immunol Med Microbiol 8:315–320, 1994

Saitoh T, Natomi H, Zhao W, Okuzumi K et al. Identification of glycolipid receptors for *Helicobacter pylori* by TLC-immunostaining. FEBS Lett 282:385–387, 1991

Schaeffer AJ, Navas EL, Venegas MF et al. Variation of blood group antigen expression on vaginal cells and mucus in secretor and nonsecretor women. J Urol 152:859–864, 1994

Sheinfeld J, Schaeffer AJ, Cordon-Cardo C et al. Association of the Lewis blood-group phenotype with recurrent urinary tract infections in women. N Engl J Med 320:773–777, 1989

Sheinfeld J, Cordon-Cordo C, Fair WR et al. Association of type 1 blood group antigens with urinary tract infections in children with genitourinary structural abnormalities. J Urol 144:469–473, 1990

Shepley MP, Racaniello VR. A monoclonal antibody that blocks poliovirus attachmant recognizes the lymphocyte homing receptor CD44. J Virol 68:1301–1308, 1994

Sim BK, Orlandi PA, Haynes JD et al. Primary structure of the 175 K *Plasmodium falciparum* erythrocyte binding antigen and identification of a peptide which elicits antibodies that inhibit malaria merozoite invasion. J Cell Biol 111:1877–1884, 1990

Sim BKL, Chitnis CE, Wasniowska K, Hadley TJ, Miller LH. Receptor and ligand domains for invasion of erythrocytes by *Plasmodium falciparum*. Science 264:1941–1944, 1994

Sircar BK, Dutta P, De·SP, Sikdar SN et al. ABO blood group distributions in diarrhoea cases including cholera in Calcutta. Ann Hum Biol 8:289–291, 1981

Socha W, Bininska M, Kaczera Z, Padjak E, Stankiewicz P. *Escherichia coli* and ABO blood groups. Folia Biol (Krakow) 17:259–269, 1969

Spitalnik PF, Spitalnik SL. The P blood group system: biochemical, serological and clinical aspects. Transfus Med Rev 9:110–122, 1995

Spring FA, Dalchau R, Daniels GL et al. The In[a] and In[b] blood group antigens are located on a glycoprotein of 80,000 MW (the Cdw44 glycoprotein) whose expression is influenced by the In(Lu) gene. Immunology 64:37–43, 1988

Srividya A, Pani SP. Fiariasis and blood groups. Nat Med J India 6:207–209, 1993

Stamenkovic I, Amiot M, Pesando JM, Seed B. A lymphocyte molecule implicated in lymph node homing is a member of the cartilage link protein family. Cell 56:1057–1062, 1989

Stapleton A, Nudelman E, Clausen H et al. Binding of uropathogenic *Escherichia coli* R45 to glycolipids extracted from vaginal epithelial cells is dependent on histo-blood group secretor status. J Clin Invest 90:965–972, 1992

Suzuki Y, Hirabayashi Y, Suzuki T, Matsumoto M. Occurrence of O-glycosidically peptide-liked oligosaccharides of poly-N-acetyllactosamine type (erythroglycan II) in the I-antigenically active Sendai virus receptor sialoglycoprotein GP2. J Biochem 98:1653–1659, 1985

Suzuki Y, Nagao Y, Kato H, Matsumoto M, Nerome K, Nakajima K, Nobusawa E. Human influenza A virus hemagglutinin distinguishes sialyloligosaccharides in membrane-associated gangliosides as its receptor which mediates the adsorption and fusion processes of virus infection. J Biol Chem 261:17057–17061, 1986

Svanborg-Eden C, Freter R, Hagberg L et al. Inhibition of experimental ascending urinary tract infection by an epithelial cell-surface receptor analogue. Nature 298:560–562, 1982

Swerdlow DL, Mintz ED, Rodrigues M Tejada E et al. Severe life-threatening cholera associated with blood group O in Peru: implications for the Latin American epidemic. J Infect Dis 170:468–472, 1994

Takahashi T, Ozawa K, Takahashi K et al. Susceptibility of human erythropoietic cells to B19 parvovirus in vitro increases with differentiation. Blood 75:603–610, 1990

Taylor CM, Milford DV, Rose PE, Roy TCF, Rowe B. The expression of blood group P1 in post-enteropathic haemolytic uraemic syndrome. Pediatr Nephrol 4:59–61, 1990

Thom SM, Blackwell CC, MacCallun CJ, Weir DM, Brettle RP, Kinane DF, Wray D. Nonsecretion of ABO blood group antigens and susceptibility to infection by *Candida* species. FEMS Microbiol Immunol 47:401–406, 1989

Tomisawa S, Kogure T, Kuroume T et al. P blood group and proneness to urinary tract infection in Japanese children. Scand J Infect Dis 21:403–408, 1989

Trangle KL, Goluska MJ, O'Leary MJ, Douglas SD. Distribution of blood groups and secretor status in schistosomiasis. Parasite Immunol 1:133–140, 1979

Tyrrell DAJ, Sparrow P, Beare AS. Relation between blood groups and resistance to infection with Influenza and some picornaviruses. Nature 220:819–820, 1968

Uehling DT, Hopkins WJ, James LJ, Balish E. Vaginnal immunization of monkeys against urinary tract infection with a multi-strain vaccine. J Urol 151:214–215, 1994

Väisänen V, Korhonen TK, Jokinen M et al. Blood group M specific haemagglutinin in pyelonephritogenic Escherichia coli. Lancet 1:1192, 1982

van Alphen L, Poole J, Overnbeek M. The Anton blood group antigen is the erythrocyte receptor for Haemophilus influenzae. FEMS Microbiol Lett 37:69–71, 1986

van Alphen L, Poole J, Geelen L, Zanen HC. The erythrocyte and epithelial cell receptors for Haemophilus influenza are expressed independently. Infect Immun 55:2355–2358, 1987

van Alphen L, Lenene C, Geelen-van den Broek L, Poole J, Bennett M, Dankert J. Combined inheritance of epithelial and erythrocyte receptors for Haemophilus influenzae. Infect Immun 58:3807–3809, 1990

van Alphen L, Geelen-van den Broek L, Blass L, van Ham M, Dankert J. Blocking of fimbria-mediated adherence of Haemophilus influenzae by sialyl gangliosides. Infect Immun 59:4473–4477, 1991

van Loon FP, Clemens JD, Sack DA, Rao MR et al. ABO blood groups and the risk of diarrhea due to enterotoxigenic Escherichia coli. J Infect Dis 163:1243–1246, 1991

van't Wout J, Burnette WN, Mar VL, Rozdzinski E et al. Role of carbohydrate recognition domains of Pertussis toxin in adherence of Bordetella pertussis to human macrophages. Infect Immun 60:3303–3308, 1992

Virkola R, Westerlund B, Holthofer H et al. Binding characteristics of Escherichia coli adhesins in human urinary bladder. Infect Immun 56:2615–2622, 1988

Virkola R, Parkkinen J, Hacker J, Korhonen TK. Sialyloligosaccharide chains of laminin as an extracellular matrix target for S fimbriae of Escherichia coli. Infect Immun 61:4480–4484, 1993

Viskum K. The ABO and Rhesus blood groups in patients with pulmonary tuberculosis. Tubercle 56:329–334, 1975

Vuokko V, Tallgren LG, Makela PH et al: Mannose-resistant haemagglutination and P antigen recognition are characteristic of Escherichia coli causing primary pyelonephritis. Lancet 2:1366–1369, 1981

Ward T, Pipkin PA, Clarkson NA, Stone DM et al. Decay-accelerating factor CD55 is identified as the receptor for echovirus 7 using CELICS, a rapid immuno-focal cloning method. EMBO J 13:5070–5074, 1994

Waters AP, Higgins DG, McCutchan TF. Evolutionary relatedness of some primate models of Plasmodium. Mol Biol Evol 10:914–923, 1993

Weir DM. Carbohydrates as recognition molecules in infection and immunity. FEMS Microbiol Immunol 1:331–340, 1989

Weiner AS. Blood groups and disease. Am J Hum Genet 22:476–483, 1970

Yamamoto T, Nakazawa T, Miyata T, Kaji A, Yokota T. Evolution and structure of two ADP-ribosylation enterotoxins, Escherichia coli heat-labile toxin and cholera toxin. FEBS Lett 169:241–246, 1984

Zelinski T. Chromosomal localization of human blood group genes. In: Silberstein LE (ed) Molecular and functional aspects of blood group antigens. American Association of Blood Banks, Bethesda, MD, pp 41–74, 1995

Zisman M. Blood group A and giardiasis (letter). Lancet 2:1285, 1977

Zoja C, Corna D, Farina C et al. Verotoxin glycolipid receptors determine the localization of microangiopathic process in rabbits given verotoxin-1. J Lab Clin Med 120:229–238, 1992

Zorgani AA, Stewart J, Blackwell CC, Elton RA, Weir DM. Secretor status and humoral immune responses to *Neisseria lactamica* and *Neisseria meningitidis*. Epidemiol Infect 109:445–452, 1992

Zorgani AA, Stewart J, Blackwell CC, Elton RA, Weir DM. Inhibitory effect of saliva from secretors and non-secretors on binding of meningococci to epithelial cells. FEMS Immunol Med Microbiol 9:135–142, 1994

9 Evolution of Blood Group Antigen Polymorphism[1]

J. Klein, C. O'hUigin, and A. Blancher

What are Blood Group Antigens?

Three characteristics define a substance as a blood group antigen. First, the substance is present in the blood, most commonly on the blood cells, but often also in the blood plasma. Substances present exclusively in the plasma and other body fluids are usually not included in the category of blood group antigens. Of the blood cells, the erythrocyte is generally the *type locality*, the site at which an antigen was originally detected. Although leukocyte and platelet antigens are not excluded from the blood group antigen category, their coverage is usually relegated, both for historical and practical reasons, to separate compendia. Historically, erythrocyte antigens were the first to be described, the existence of leukocyte and platelet antigens was reported much later. The leukocyte antigens – now largely assigned either to the major histocompatibility complex (Mhc) or the cluster of differentiation (CD) series – have become so numerous and their characterization so extensive that justice can be done to them only by separating them from the erythrocyte antigens. Second, to qualify as a blood group antigen, a substance must be capable of eliciting an immune response, usually of the

---

[1] This concluding chapter of the blood group antigen part of this book draws heavily on the information described in the preceding contributions. In the interest of brevity, references have been restricted to a minimum; for complete documentation of the discussed findings, the reader is referred to the relevant chapters.

Molecular Biology and Evolution of Blood Group
and MHC Antigens in Primates
Blancher/Klein/Socha (Eds.)
© Springer-Verlag Berlin Heidelberg 1997

humoral type. Some blood group antigens (e.g., those of the ABO, H, Lewis systems) can probably induce only humoral responses, whereas others (e.g. those of the Rh, MNS, Luteran, Kell, Duffy, Kidd, Diego, and other systems) have at least the potential of also eliciting T lymphocyte-based cellular responses. The prevailing reliance on the humoral response has dictated the use of techniques based almost exclusively on antibody reactions. Serology remains to this day emblematic of blood group antigen studies, although in the future it will probably be replaced either entirely, or at least to a large extent, by molecular biology techniques. Third, the genes encoding blood group antigens are polymorphic so that some individuals of a species possess one form of the antigen and others another. This characteristic is the necessary condition for elicitation of immune response against the antigen which is commonly of the allogeneic type. The blood group antigens are, with few exceptions, alloantigens.

All three above-cited characteristics are unnatural in the sense that they group together substances that in fact are not related to one another. Any protein or carbohydrate expressed on the red blood cell surface is a potential blood group antigen. Its encoding gene is bound to mutate sooner or later in some individual and thus produce potentially alloantigenic variants which, under the right circumstances, will elicit the formation of antibodies. There is no a priori reason to expect all the erythrocyte-borne substances to be "blood relatives", tied together by a common evolutionary descent. Recent advances in characterization of the known antigens have indeed established the grouping to be artificial. The individual antigens differ in their biochemical nature, in their expression, and in their function.

Biochemically, the antigens are either carbohydrates (antigens ABO, H, Lewis, Ii, P, Sid, and Cad) or proteins (antigens MNS, Rh, Duffy, Kell, Kx, Lutheran, Gerbich, Dombrock, Scianna, Xg, Chido/Rodgers, Diego, Cromer, Cartwright, Indian, Knops, Kidd, and Colton; for references see other chapters in this book and also Schenkel-Brunner 1995, Bailly et al. 1994, Parsons et al. 1995). The carbohydrate-based antigens are borne by either glycoproteins, proteoglycans (mucins), glycolipids, or free carbohydrate moieties, often the same antigen by all these types of molecules. The enzymes involved in the biosynthesis of the carbohydrate-based antigens include N-acetylgalactosaminyl transferases (antigens A, P, Sd, and Cad), galactosyltransferases (antigens B and P), N-acetylglucosaminyl transferases (antigens I and i), and fucosyltransferases (antigens I and i). The protein-based antigens are contributed by a number of protein families: glycophorins (antigens MNS and Gerbich); multispan membrane proteins (antigens of the Rh system); membrane transporter proteins, including anion transporters (Diego), urea transporters (Kidd), water transporters (Colton), and others (Kx); interleukin (IL-8) family of chemokine receptors (Duffy); thioester (C4) proteins (Chido/Rodgers); immunoglobulin superfamily proteins (Lutheran); ICAM family of adhesion molecules (Landsteiner-Wiener antigen); zinc metalloprotein endopeptidases (Kell); acetylcholinesterases (Cartwright); complement regulatory proteins such as the decay accelerating factor (DAF, CD55; Cromer antigen); complement receptor 1, (CR1; Knops antigen); and single-span membrane glycoprotein CD44 (Indian antigen). Some of the antigens are expressed on cells of the erythroid lineage exclusively (e.g., Rh and Kell antigens), others on a variety of cells (in addition to red blood cells); some are also present in the plasma and in tissue fluids and secretions. The function of the carbohydrate-based antigens is not known, except for the Lewis antigens which are ligands of adhesion molecules. As detailed above, the blood group antigen-bearing proteins are involved in a wide range of functions. Human blood group antigen loci have been mapped to chromosomes 1, 2, 4, 6, 7, 9, 17, 18, 19, 22, and X. The genes occupying these loci and belonging to different systems do not show significant sequence similarity and hence are apparently all of independent origin. All these observations indicate that blood group antigens are a highly heterogeneous assembly of compounds lacking any natural ties to one another.

Polymorphism

As pointed out earlier, every gene has the potential of becoming polymorphic; in fact, most (and perhaps all) genes are polymorphic, if not in their coding, then at least in their noncoding regions. There are two principal kinds of polymorphism – transient and balanced (Nei 1987). Transient polymorphism arises when a new mutation is not eliminated immediately but is maintained in a population for a variable period of time. If this time period is relatively short, the mutant allele remains at a low frequency in the population before becoming extinct. Occasionally, however, a mutation persists in the population and rises in frequency until it becomes fixed. The climb toward fixation can be the result of random genetic drift or of directional positive selection favoring the mutant over the wild-type allele. In a balancing polymorphism, the polymorphic genes are maintained in a stable equilibrium by balancing selection favoring two or more alleles simultaneously. Most polymorphisms are probably of the transient type; balanced polymorphisms are relatively rare.

The total number of known human blood group antigens is estimated to be between 400 and 600. The uncertainty about the actual number stems primarily from difficulties in establishing which of the independently described antigens are in fact identical (Daniels et al. 1993). The majority of the antigens are either of the private or the public type; only about 30 antigens belonging to 12 distinct systems are nonprivate, nonpublic. Private antigens occur at a low frequency in a group of related individuals. Public antigens, by contrast, are widely distributed, and are absent in only a few, often related, individuals. Both the private and the public antigens most probably owe their existence to recent mutations: the former to mutations imposing alloantigenicity on the mutant molecule, and the latter to mutations enabling the bearer to respond to the product of the wild-type allele. Although a few of the private/public antigenic systems might represent balanced polymorphisms in the making, the majority are probably transient polymorphisms destined to disappear from the population. The general absence of the human private/public polymorphisms in nonhuman primates supports this conclusion. (The genes at the loci encoding these antigens undoubtedly have their homologues in the nonhuman primates and the homologues very likely display their own private/public antigen variation, but the polymorphisms have arisen independently in each species and overlap very little, if at all, between species. Because human blood group typing is much more extensive than typing of nonhuman primates, the private/public polymorphisms of the nonhuman primates remain largely unidentified.) Most of the known human private/public antigens do not, in fact, represent true polymorphisms because their frequencies fall below the 1% mark, conventionally regarded as the polymorphism's threshold; they can thus best be classified as "rare variants". The group of loci coding for the private/public antigens therefore behaves like the great majority of loci in that they continuously produce rare variants most of which disappear from the population soon after their appearance. Although the antigens may be useful markers for identification of the molecules carrying them and some might be of clinical significance, from the standpoint of evolution they are relatively uninteresting.

The 30 or so antigens that occur in the human population at appreciable frequencies could theoretically represent mutations on the rise toward fixation or balanced polymorphisms. For most of them, not enough information is available to decide between these two possibilities, but for some at least an attempt can be made in this direction. The latter include the ABH, Rh, MNS, Duffy, and P systems. Here, we shall focus on the ABH and Rh systems, the former being a representative of carbohydrate-based polymorphisms, the latter of protein-based polymorphisms.

Carbohydrate-Based Polymorphisms

Carbohydrate moieties are assembled by the sequential addition of monosaccharides to growing oligosaccharide chains (glycans). These additions are catalyzed by sets of glycosyltransferases which are either membrane-bound or exist in soluble form in body fluids (Paulson and Colley 1989). The oligosaccharides can then become chemically linked to either lipids or proteins in either the particulate or fluid phase (some also remain as free glycans in the fluid phase and others can adsorb to cells passively). Polymorphism can therefore be discerned at three levels: (1) at the level of the glycans differing by the presence or absence of specific monosaccharide residues; (2) at the level of the proteins (enzymes) differing by amino acid replacements or other changes; and (3) at the level of the genes (DNA) encoding the glycosyltransferases and differing by nucleotide substitutions or more complex alterations. Analysis at the protein level can reveal polymorphism not detected at the carbohydrate level because some amino acid replacements have no, or very little, effect on the catalytic activity of the glycosyltransferases. Similarly, analysis at the DNA level may reveal variation not visible at the protein level because synonymous nucleotide substitutions do not result in amino acid replacements.

A remarkable feature of the carbohydrate-based polymorphisms is a high frequency of null alleles in the population at some of the loci. Null alleles lack the ability to produce functional enzymes as a consequence of inactivating mutations. The absence of a particular enzymatic activity then leads to a failure to incorporate a specific monosaccharide into the glycan. The high frequencies are puzzling because nonfunctionality of a protein is normally a deleterious trait and hence is subject to negative selection. The high frequency of null alleles can, theoretically, be the result of their being ignored by selection or their being subject to positive selection. Before we attempt to choose between these two possibilities, let us compare the situation with that at other loci.

Until recently, the only time the existence of a null allele at a locus, other than one encoding blood group antigens, became known was when the allele in a homo-, hemi-, or heterozygote state caused a detectable abnormality or disease. Without these effects, the presence of a null allele is not revealed. This state of affairs has changed dramatically with the advent of transgenic and "knockout" animals. Techniques leading to the production of such animals enable investigators to test null alleles of any cloned gene (Brandon et al. 1995). The testing has shown that genes fall roughly into two categories in terms of null allele effects. Some genes are absolutely indispensable, their inactivation being incompatible with survival or reproduction. Others have proven to be dispensable under certain conditions, their inactivation seemingly not having any adverse effect on their homozygous carriers. The standard explanation for this dispensability is that the genes are a part of a redundant series in which the knocked out gene can be substituted functionally by a related but different gene (Hochgeschwender and Brennan 1994).

This explanation is, however, unlikely to be true for two reasons. First, the postulated redundancy would ultimately lead to fixation of the null alleles by random genetic drift and thus total dispensation of the functional gene. Second, the observed population frequency of the null alleles at the "redundant" loci is exceedingly low, indicating that any such alleles found in the population have been generated by recent mutations and are on the way out. A more likely explanation of the gene knockout results is that the laboratory environment, in which the animals with null alleles are maintained, is not a proper test of gene dispensability. To give an extreme example: By raising knockout animals under germ-free conditions, one might want to conclude that the genes of the vertebrate adaptive immune system (e.g., those coding for immunoglobulins, T cell receptors, or Mhc molecules) are dispensable because the animals survive, which of course would be totally false. In this case, the artificiality of the test situation is obvious; in most cases, however, the investigator may not be aware of the biases the laboratory conditions provide. For a gene to be indispensable, an animal does not need to continually encounter conditions in which the gene is required; a single encounter during the reproductive phase of life is enough. Similarly, not all the individuals need to be tested by natural selection all the time to keep a gene in a population; an occasional test of some individuals may suffice.

The Puzzle of the Blood Group Null Alleles

The redundancy argument has also been used to explain the persistence of null alleles at the blood group loci. It is, however, equally unlikely to be valid for them as it is for other loci. Although the null alleles of the blood group loci can reach high frequencies in some populations, they don't appear to have become fixed in any species. Why not?

The two principal possibilities are: spreading through founder effect and positive selection. We find the former explanation unsatisfactory because a founder effect would have influenced gene frequencies at not only the blood group but also at other loci. Most populations with high frequencies of null alleles at blood group loci, however, do not give any indication of a founder effect from studies of other loci. Assuming that positive selection might be responsible, what might be the agent selecting antigen-negative individuals? It has been suggested that the selection might be effected by viruses, bacteria, or other parasites that have evolved means of misappropriating the blood group antigens for their own benefit, for example, by using them as receptors to gain a foothold in the host tissues and cells (see Frattali and Spitalnik 1997). This suggestion seems reasonable in view of the growing list of parasites known or suspected of using specific blood group antigens as receptors. In this regard, it may not be a coincidence that the loci with high frequencies of null alleles are also those responsible blood group antigens with highest density of expression on the cell surfaces.

The fact that the null alleles do not become fixed must indicate that the functional alleles are not dispensable. Apparently, the possession of the functional allele is advantageous relative to the null allele until a parasite is introduced that misuses the substance specified by the former. It then becomes more advantageous to lack the substance and hence to possess the null allele. In this manner, the null allele may spread through the populations threatened by the parasite. Once the parasite has been eliminated, the functional alleles begin to replace the null alleles again. It is interesting to note that the frequency of null alleles is particularly high in human populations and much lower in nonhuman primates (see Blancher and Socha 1997a). This difference may reflect the dramatic recent expansion of the human species into new environments and thus encounters with new parasites.

Functional Alleles

Most carbohydrate-based blood group antigens occur in two polymorphic forms, one encoded by the functional allele and the other by the null allele; if variants in functional genes occur, they are usually rare. The one exception to this rule is the human *AB* system at which three major functional alleles occur in most populations, *A1*, *A2*, and *B* (see Blancher and Socha 1997a). The *A1* and *B* alleles differ by seven nucleotide substitutions in the coding sequence of approximately

1 kilobase (kb). Four of these substitutions result in amino acid replacements, the remaining three are of the synonymous type. This relatively large interallelic difference suggests that the alleles diverged a long time ago (Martinko et al. 1993). The divergence at the nonsynonymous sites may have been accelerated by selection, and not knowing the selection intensity makes estimates of divergence times difficult. The appearance of the three synonymous differences, by contrast, was presumably not influenced by selection. Taking the synonymous substitution rate of 3×10^{-9} substitutions per site per year, we estimate that the $A1$ and B alleles diverged at least 1 million years ago. Because of the small genetic distance between the alleles, however, the estimate has a large standard error. A more reliable estimate could be obtained from the comparison of intron and untranslated region sequences which, unfortunately, are not available. There are, however, other data that support the long divergence time of the human $A1$ and B alleles. They derive from the comparison of human sequences with those of nonhuman primates.

In a phylogenetic tree based on the available primate AB locus sequences (Fig. 1), the human B sequence clusters with the gorilla B sequences, separately from the human A sequence. This pattern of clustering is suggestive of *trans*-species polymorphism, amply documented for the various Mhc loci (Klein et al. 1993). The essence of the *trans*-species hypothesis is that some alleles (allelic lineages) diverged before the divergence of the species carrying them so that the genetic distances between them are often larger than the distances between genes of two different species (Klein 1980, 1987). The tree in Fig. 1 suggests that the human $A1$ and B alleles diverged before the divergence of human and gorilla ancestors some 5 million years ago (Martinko et al. 1993). An alternative interpretation of these data would be that the interspecific similarities arose by parallel evolution – by independent occurrence of identical mutations in humans and the gorilla, followed by selection-driven spreading of the mutations. Since, however, interspecific sharing is observed also at synonymous sites, the convergence argument is not convincing. Nonetheless, the number of interallelic differences is small and until more extensive data become available, parallelism remains a possibility (but see note added in proofs).

Both explanations require that the interallelic differences in the region encoding the catalytic site accumulated by natural selection. For the *trans*-species polymorphism argument, this requirement follows from the population genetics theory which states that the average coalescence (divergence) time of two neutral genes is $2N_e$ generations, where N_e is the effective population size (Nei 1987). Taking N_e equal to 10^4 breeding individuals and the generation time of 20 years for the hominid lineage, we obtain the average coalescence time of neutral genes of 0.4 million years. Hence, it is inconceivable that neutral polymorphism could be shared by humans and gorillas; the shared substitutions in the A gene must therefore be the result of natural selection. We thus come to the conclusion that both the functional and the null alleles at the AB locus evolve under the influence of positive selection.

The functional A and B polymorphism is unique in the sense that the $A1$ and B alleles specify different enzymes (see Blancher and Socha 1997a). Although the

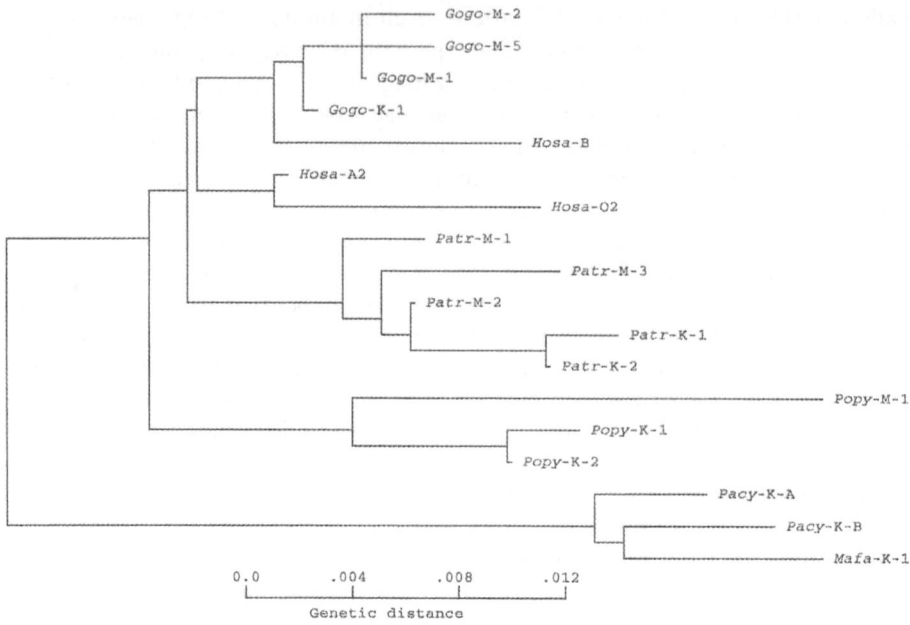

Fig. 1. Phylogenetic tree of primate AB locus sequences. Sequences covering the region from codons 152 to 287 from humans (*Hosa*), gorilla (*Gogo*), chimpanzee (*Patr*), orangutan (*Popy*), baboon (*Pacy*) and macaca (*Mafa*) were aligned. For nonhuman sequences, the single letter following the species designation indicates the published source of the sequences: *M*, Martinko et al. (1993); *K*, Kominato et al. (1992). The final character indicates the clone number given in the original publications. Human allelic variants are designated according to Yamamoto (1995). For these sequences, the final character indicates the ABO type designation, with allele number where relevant. Genetic distances between sequences were measured by Kimura's (1980) two-parameter method and the dendrogram was constructed by the neighbor-joining method (Saitou and Nei 1987)

A and B transferases use the same acceptor substrate (the type 2 H substance), they differ in their use of donor substrate, the A1 transferase being specific for N-acetyl-D-galactosamine and the B transferase for D-galactose. Such a change in enzyme specificity normally occurs during the evolution of separate loci derived from an ancestral gene by duplication and subsequent mutational divergence. In such a case, each individual expresses both enzymes; in the case of the *AB* locus, only the heterozygotes do. Could there be an evolutionary advantage in producing two different enzymes via polymorphism of the enconding locus rather than by duplicated loci? If there were, it would have to be in having individuals with single enzymatic activities because the polymorphic heterozygotes are phenotypically equivalent (except, perhaps, for a dosage difference) to homozygotes with duplicated loci. The advantage could then be of the same nature as in having homozygotes for null alleles – in the absence of certain glycan types in some members of a population, perhaps to avoid their corruption by parasites.

Evolution and Polymorphisms of Fucosyltransferase Genes

The essential precursor substance in the synthesis of carbohydrate-based blood group antigens of the ABH family is the disaccharide lacto-N-biose consisting of a galactose (Gal) linked to N-acetyl-D-glucosamine (GlcNAc). Depending on the linkage between the two monosaccharide residues, several types of precursors are distinguished, of which the type 1 [Galβ(1,3)GlcNAc] and type 2 [Galβ(1,4) GlcNAc] are most pertinent to our discussion (see Blancher and Socha 1997a). These precursors and their various derivatives are the acceptor substrates for a set of fucosyltransferases, each of which catalyzes the addition of a single D-fucose residue to the precursors. Seven fucosyltransferases have thus far been identified in humans and their encoding genes (designated *FUT1–FUT7*) cloned (for references see Figs. 2, 3). They are all related to one another in their sequence, having all presumably originated from a single ancestor (Figs. 2, 3) and they all use GDP-fucose as their donor substrate. They differ in acceptor substrate specificity, requirement for bivalent metal cations, temperature, pH, sensitivity to ethylmaleimide inactivation, and tissue expression patterns. They fall into two groups which diverged early in their evolution: the α(1,2) fucosyltransferases, which catalyze the attachment of a fucose residue to the D-galactose of the disaccharide precursor via the α(1,2) linkage; and the α(1,3/4) fucosyltransferases which attach a fucose residue to the N-acetyl-D-glucosamine of the precursor via either the α(1,3) linkage (most of them) or the α(1,4) linkage. The former group encompasses fucosyltransferases encoded in the *FUT1* and *FUT2* genes (as well as an *FUT2*-derived pseudogene); the latter group includes enzymes encoded in the *FUT3* through *FUT7* genes. The two groups of the *FUT* genes are apparently derived from different ancestral genes (Fig. 3). In the α(1,3/4) group, the three enzymes prominently active in the plasma (encoded by *FUT3*, *FUT5*, and *FUT6* genes) are more closely related to one another than they are to the enzymes encoded in the *FUT4* and *FUT7* genes (Fig. 3). The three genes, again, presumably descended from a single ancestral gene. This putative origin of the *FUT* genes is reflected in their chromosomal distribution, the *FUT1*, *FUT2*, and *FUT2P* genes forming a single cluster on the long arm of human chromosome 9 (19q13.3), the *FUT6*, *FUT3*, and *FUT5* forming another cluster on the short arm of chromosome 19 (19p13.3), and the *FUT4* and *FUT7* genes located on chromosomes 11 and 9, respectively. The distances between the genes in the clusters are <35 kb (the order being as indicated), which is consistent with their origin by duplication. We estimate that the α(1,2) group diverged from the α(1,3/4) group more than 600 million years ago. Within the α(1,2) group, the *FUT1* and *FUT2* genes diverged long before the radiation of the eutherian mammals, more than 180 million years ago. In the α(1,3/4) group, the *FUT4* and *FUT7* genes split from the rest of the group approximately 300 million years ago, whereas the *FUT3*, *FUT5*, and *FUT6* genes separated from one another as recently as 2 million years ago. Evolution introduced important alterations in the promoter region of the gene and the regions coding for the catalytic site. The former resulted in changes in the expression of the genes in the different tissues (*FUT1* being expressed in mesodermally and ectodermally derived tissues, *FUT2* in endodermally derived tissues, *FUT3*, *FUT5*, *FUT6* in body fluids, and *FUT4* in cells of the myeloid lineage and the brain). The latter resulted in the modification of the acceptor substrate specificity (*FUT2*- and *FUT3*-encoded

```
              1          11         21         31         41         51         61         71         81         91
CONSENSUS     MGAPWMDPLG AAKPQWPWR  CLALLLPQVL AAVCFFSYLR VSRDDATGSP RPGLMAVEPV TCAPLGSPCP DSTATPAHPT LLILLMTWPF NGRDTPVALP
Hosa-FUT-1    .......... .......... ...-F--VC- SVIP-LHIHQ D-F******* ********** ********** *-LG-SIL-- PPVAIFCL-G ******-QI-
Hosa-FUT-2    ....MSSL-P MLVV-M-FSF PM-HFILP-F TVSTI-HYQQ R********* ********** **********LA KIQAM-EL-* ****-*QI--
Hosa-FUT-2    ....MSSL-P T-VKGFWAT- PSFSTFYF-F -IFVVSTIFH C********* *********** *********HQ H-A-VPAPW* ******-*A
Hosa-FUT-3    ...-GS-TA -GGRRG---- GRG-PWTVCV L-AAGLTCTA LITYACW-QL P********* **-S--SRQ **T-TR-- H***I---S
Hosa-FUT-4    ...-...... -P------L- -G------L- -V-------- M--------- *PL-WA--T- S*****-*RP VGV---WE-- G---SAPRP-
Hosa-FUT-5    ...-...... -P--S--C- -TT----L- M--------- --Q--P-VY* N****** --N--R-Q- --M------ ****-----
Hosa-FUT-6    ...-...... -P--S--C- ---T----K- -------I- --Q--P-VY* N****** --N--R-Q- --SI P---K-I--
Hosa-FUT-7    ....M NNAGHG-T- LRG-GVLAGV -LLAALWL-W LLG******* *S--R-T-A- Q********* IT--V-H--- T**-Q-PE--

              101        111        121        131        141        151        161        171        181        191
CONSENSUS     RCAEMVPGTA DCNLTADRTV YPQADAVGVH HGEYATLPAL ALPWGIQAH TAEEVDLRVL DYERAAAAAE ALATSSGRPQ GQRWIWASME SPSHAPGLRA
Hosa-FUT-1    PN-SSSCPQH PAS-SGTW-- -NGR*F-NQ M-Q---L-- -QLN****** ********** ********** ********** GQRWIWASME A*AL--VF-I
Hosa-FUT-2    VL-STSKALG PSQ-RGMW-I NAIGR*L-NQ M---Y-- -KMN****** ********** ********** **AF-P-Q-H **TL--IF-I
Hosa-FUT-2    YS-RV-LAPR HLPREDLF-I NSKGR*L-NQ M---Y-- -KMN****** ********** ********** **AF-P-Q-H **RV--IF-I
Hosa-FUT-3    -S---I---K- -HI---K- -T-I-- -WDIMSN-KS R--******* ********** ****-----P *AF-P-Q-H P-PNCQH-E-
Hosa-FUT-4    PDCRLRPNIS G-R-LT--AS -GE-Q--LF- -RDLVKG-PD WP-------- N--******* ****-**PT- --V-MNF- -------S--S
Hosa-FUT-5    --S-----A- -I---SS- I-- -WDIMYN-SA N--*****-- N--******* ****-**PT- F--- ----NCRH-E-
Hosa-FUT-6    --S-----I- I-- -K- -R-VMYN-SA Q--N-*-VF- N--E-*V **** F--- --CWQ-K--
Hosa-FUT-7    SDTCTRY-I- R-H-S-N-SL LAS-----VF- -R-LQ-RRSH LP-****** -V-NFPQ- -P-V-- ----TH--SH

              201        211        221        231        241        251        261        271        281        291
CONSENSUS     LLGNVFNLTM SYRSDSDIFT PYGWLEPWSG QGDPPAHPPL PLSAKTGLPC AWTVSHHLRE QIARVRTYHD LQRHLAVDVL GRLHAGPAGA RPLTFVGVHV
Hosa-FUT-1    T-P*--LAPEV DS-TPWRELQ LHD-MSEEYA D*****#-LRD FLFLS-*-- S--FF----- -R-EF-L- HL-EE-QS- -Q-RL-RT-D R-Y--
Rano-FUT-A    ---------- ---------- ---------- ---------- -------I-G EYVRFV-Y-- S--FY----Q -R-EF-L-- HL-ED-QRL- SG-RI---I R-Y--
Hosa-FUT-2    T-P*-LHSAT AS-IPWQNYH LND-M-EEYR H*****I-G EYVHL--Y-- S--FY----Q E-LQEF-L- HV-EE-QKP- RG-QVN**-S G---
Hosa-FUT-2    T-P*-LHSAT AS-IPWQNYH LND-M-EEYR H*****I-G RCVHL--Y-- -N---A-*WKP E-LQEF-L- HV-EE-QKP- RG-Q-K*W- GQA-
Hosa-FUT-3    -DR*Y- -I- -*** -**** N---E-*V DS---Y-QS --S-----*-F K-PKGTMME
Hosa-FUT-4    -AS-L--W-L -A---V-V -Y-R-H P---SGLAP --R-Q-*V -V---WD- RQ---SQ-VT- -GGP-**** Q-VPEI-LLH
Hosa-FUT-5    -D--Y---- -------K- -*** N---E-*V DS----WKP -A--K--Y -S-****** K--PKGTMME
Hosa-FUT-6    MD--Y---- -------K- -*** N---E-*V DS----WKP -A--K--Y K--PKGTMME
Hosa-FUT-7    -R-*I--WVL R---V R----HWA S*******P -P-SRV*A --V--NFPQ- RQL-A-L-RQ -AP--F- -AN*----F CASCLVP

              291        301        311        321        331        341        351        361        371        381        391
CONSENSUS     TRGDYKFYLP APFNWGVHADR AYLTEALDWF NALEAAWAPV VTGPGMAWCY ENILPSHGDA FFAGDGFQGS PAKIAALLTQ CNHTIATYGT FFRWRAYLAG
Hosa-FUT-1    R---LQVM- QR-K--VG-S --RQ-M- -R-RHEAP-F- -SN--E--K -DT-Q--V T---QEAT -W-DF- -KD--RY-QE LDKDH-R-LS Y--------- GF-A---
Rano-FUT-A    R----LEVM- NR-K--VG-- -QK-M- -R-RHKDPIF- -SN--R--L -DT--R--V V---N-QE-T G-DF- -KD--RY-QE LDRNP-V-RR Y-H-----* M-I-
Hosa-FUT-2    R-...... .T NV-K--V- G--EK-M- -R-RYSSP-F- -SN-----R -NA-R--G V---N-IE* -DF- -KD--RY-QE LDKDH-R-LS Y-------* GI-A----T
Hosa-FUT-2    R-- -VRVM- KV-K--V- G--QQ-- -R-RYSSLIF- -SDD-----R -DT----V V---N-L- -DF- -KD--RY-QE LDKDH-R-LS Y-------* GI-A----T
Hosa-FUT-3    -LSR--- -E-SL-P-* -I--K-*-R -L-G- -L-SRS*N- -RF-P**- -IHV-D-* Y--------- I-V--- GV-A--
Hosa-FUT-4    -VAR----- -E-SQ-L-* -I--K-*-R -L-DR-*N -L-DR-*N -RFV-R**G- -IHV-D-P* ASS-SY-LF LDRNP-V-RR Y-H-----*
Hosa-FUT-5    -LSR--- -E-SL-P-* -I--K-*-R -L-G- -L-SRS*N- -RF-P**- -IHV-D-* -KD--RY-QE LDKDH-R-LS Y-------*
Hosa-FUT-6    -LSR--- -E-SL-P-* -I--K-*-R -L-G- -L-SRS*N- -RF-P**- -IHV-D-* -KD--RY-QE LDKDH-R-LS Y-------*
Hosa-FUT-7    -VAQ-R-- S-E-SQ-R-* -I--KF*-R -V-GT- -L--PR-*T- -AFV-A**-- -VHV-D-G* -ARE--F-G M-ES*-R-QR

              401        411        421        431        441
CONSENSUS     GDTVYLANFT LMDSFPCKAC WFPLAARFLPER VGTAADLAAW FFAATP
Hosa-FUT-1    R----LQVM- *-ALD-- -K--QO*ES* YQ-VRSI--- F-AATP
Rano-FUT-A    R----LEVM- *RSTAVHTS F--E-W-RV- QAVQR*AGD PKSIRN--S- ER-
Hosa-FUT-2    -I----Y-- *E-LRPRS-S *-ALA-- K-QO*ES* YQ-VRSI---
Hosa-FUT-2    R-....... .T *E-LRPRS-S *-ALA-- K-QE*ES* YQ*RGI---
Hosa-FUT-3    -PN---NVVF *E-LRPRS-S *-ALA-- DRYPH---* SQVYE--EG-
Hosa-FUT-4    *-S-RVRL-- D-RER--AI
```

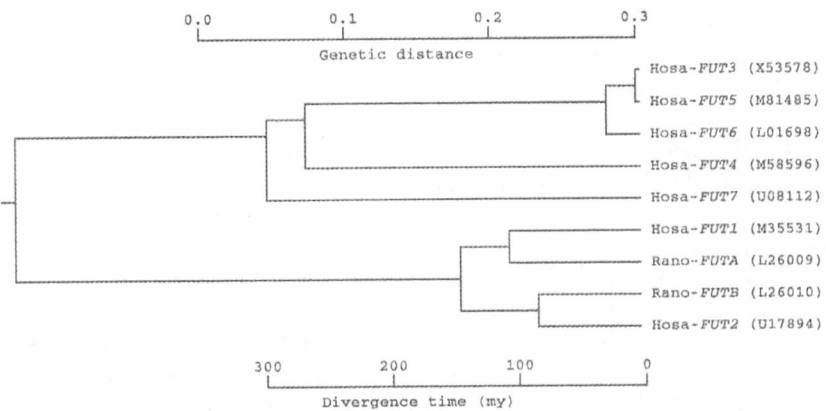

Fig. 3. Phylogenetic tree of FUT sequences. Designations are those used in Fig. 2 upon which alignment the tree is based. The analysis includes only the 3' end region from amino acids 249 to 446, for which most sequence data are available. Genetic distances were estimated between sequences using the proportion of differences in pairwise comparisons. These distances were used to draw a phylogenetic tree according to the UPGMA method using the MEGA program (Kumar et al. 1993). The time scale was placed taking the rat-human divergence at 100 million years (*my*) as a guide

enzymes prefer type 1 over type 2 substrates, while the remaining *FUT*-encoded enzymes use type 2 substrates). Some of the enzymes (e.g., *FUT1*) are predominantly membrane-bound, presumably located in the Golgi system, whereas others (in particular *FUT3*, *FUT5*, and *FUT6*) are secondarily released from the membranes into the fluid phase by proteolytic cleavage. The activity of the *FUT1*-encoded enzyme results in formation of the cell-bound type 2 H substance; activity of the *FUT2(Se)*-encoded enzyme leads to the appearance of H type 1 substance in secretions; the products of the *FUT3*-encoded activity are the Le-a and Le-b antigens; and the products of the activities encoded in the *FUT4* through *FUT7* genes are the various forms of the Le-x and Le-y substances.

Of the seven *FUT* genes only *FUT2(Se)* and *FUT3(Le)* are known to be polymorphic, each having one functional (*Se*, *Le*) and one or more null (*se*, *le*)

◀

Fig. 2. Alignment of fucosyltransferase amino acid sequences. The human *FUT* genes are indicated with the prefix *Hosa*, followed by the gene number and DNA database accession code. The references are: *FUT1*, Larsen et al. 1990; *FUT2*, Kelly et al. 1995; *FUT3*, Kukowska-Latallo et al. 1990; *FUT4*, Goelz et al. 1990; *FUT5*, Weston et al. 1992a; *FUT6*, Weston et al. 1992b; *FUT7*, Natsuka et al. 1994. The rat sequences have the prefix *Rano* followed by the published designation A or B (Piau et al. 1994) and accession codes. The alignment was generated using the Clustal W program (Thompsom et al. 1994). A single letter amino acid code is used and the consensus is based on simple majority. A *dash* indicates agreement with the consensus, a *dot* unavailable data, and an *asterisk* a gap introduced to optimize the alignment

alleles; the remaining genes are monomorphic. The frequencies of the null alleles are remarkably high: approximately 20 % at the *FUT2* locus in Caucasoids and between 10 % and 80 % at the *FUT3* locus, depending on the population. At the remaining *FUT* loci, null alleles arise occasionally by mutations but are apparently eliminated soon after their appearance (the null alleles at the *FUT1* locus result in the Bombay phenotype in *h/h*, *se/se* individuals and in the para-Bombay phenotype in *h/h*, *Se/se* or *Se/Se* individuals). Even though no clinically apparent defects have been observed in the null allele homozygotes at these loci, the inactivating mutations appear nevertheless to be eliminated from the population by negative selection.

There are thus two different situations at the *FUT* loci in terms of their polymorphism. At the *FUT2* and *FUT3* loci, inactivating mutations are not only tolerated, but may actually be positively selected; at the remaining *FUT* loci such mutations are eliminated by purifying selection. Why this difference? First, as noted above, the monomorphic *FUT* genes all encode enzymes that use the type 2 substance as a substrate, whereas the enzymes encoded by the polymorphic *FUT* genes preferentially use the type 1 substrate. Second, the type 2 substance is the precursor of Le-x and related substances which function as ligands for the selectin family of adhesion molecules. There may therefore exist a functional difference between the structures synthesized with the help of enzymes encoded in the monomorphic and polymorphic *FUT* genes. The monomorphism of some of the genes may reflect structural constraints imposed on the enzymes by the substances they help to build. The existence of null alleles at the other *FUT* loci may have the same basis as that proposed earlier for the *O* allele at the *AB* locus: escape from infection by parasites that may have evolved to use as receptors those substances indirectly encoded in the functional alleles.

Protein-Based Polymorphisms

In this category, a blood group antigen is a direct product of a gene transcript, and polymorphism is the result of a difference in amino acid sequence between polypeptides of the donor and the recipient of the combination used to produce the defining immune reagent. The best known representative of this category is the Rhesus system, with its 40 antigens identified in humans and several other antigens described in nonhuman primates (see Blancher and Socha 1997b). Among the 40 human antigens, one (D) occupies a special position because it represents a polymorphism based on the presence or absence of a locus; all the other antigens are the result of a sequence variation in the encoding genes. The human D antigen is the product of the *RHD* locus which is absent or inactivated in up to 40 % of individuals, depending on the population. The high frequency of the null allele at the *RHD* locus has puzzled geneticists for many years (Haldane 1944; Race 1944; Li 1953; Levin 1967; Feldman et al. 1969; Nei et al. 1981, and others). Why does the allele (locus) persist in human populations when it is selectively disadvantageous regarding its deleterious effect in Rh-incompatible fetal-maternal combinations? Why has it not been eliminated by negative selection?

Nei and coworkers (1981) have suggested that the *RHD* polymorphism is transient and reflects the history of the human population in the last 60 000 years. They postulate that for most of the Pleistocene, but particularly in the Wurm-Wisconsin Period, there were many small, isolated human populations in which the occurrence of the *RHD* null allele was generally kept at low frequencies by the balance of mutations, incompatibility selection, and genetic drift. In a few populations, including the ancestors of present-day Europeans, however, by chance genetic drift drove the frequencies to values in excess of 50 %. When the glaciation ended and the populations began to expand and mix, the frequencies of the null allele began to decline. Since, however, the allele frequency change per generation is very small when the frequency is near the 50 % mark, the polymorphism has persisted to present times.

The apparent weakness of this hypothesis at the time of its proposal was that it required a relatively high recurrent mutation rate (10^{-4}). This weakness seems less of a problem now that the nature of the null allele has been elucidated. The high sequence similarity of the *RHD* and *RHCE* loci (~92 %) and their close linkage provide conditions for frequent unequal crossing-over which can repeatedly generate the null allele through the deletion of the *RHD* locus. The existence of *RHD* variants bearing parts of the *RHCE* gene indicates that exchanges of all sorts do indeed occur continuously between the two loci.

In view of the new findings, the *RHD* polymorphism can best be understood in the context of the long-term evolution of the entire *RH* system. This evolution, in turn, probably follows a path similar to that of other recently duplicated loci such as those controlling complement component 4 and the enzyme 21-hydroxylase (Kawaguchi et al. 1991). The available information concerning the *RH*, *C4*, *CYP21*, and other loci leads to the following reconstruction of *RH* evolution.

The original (primigenial) duplication that created the two *RH* loci was a unique event. It may have occurred a long time ago (certainly before the divergence of African and Asian great apes, but possibly much earlier), when short sequences flanking the ancestral single *RH* gene accidentally attained a high degree of similarity. Misalignment of the flanks, either unassisted by or involving a transposon, led to the first unequal crossing-over and the generation of two adjacent, identical *RH* loci. This event started cycles of gene expansions and contractions by subsequent misalignments and unequal recombinations, generating numerical chromosomal polymorphism in the population. In some chromosomes, unequal crossing-over would contract the genes to a single copy, while in others it would expand them to three or more copies. The fate of the numerical variants was decided primarily by random genetic drift, but it depended also on the population structure of the species. Because of the latter factor, the single-copy chromosome may have become fixed in some species, but in others (those on the evolutionary lineage leading to *Homo sapiens*), the cycling has continued, accompanied by homogenization of the sequences at the two loci. The consequence of the homogenization is that the divergence time of the two loci present in a given species is shorter than the time since the primigenial duplication. The

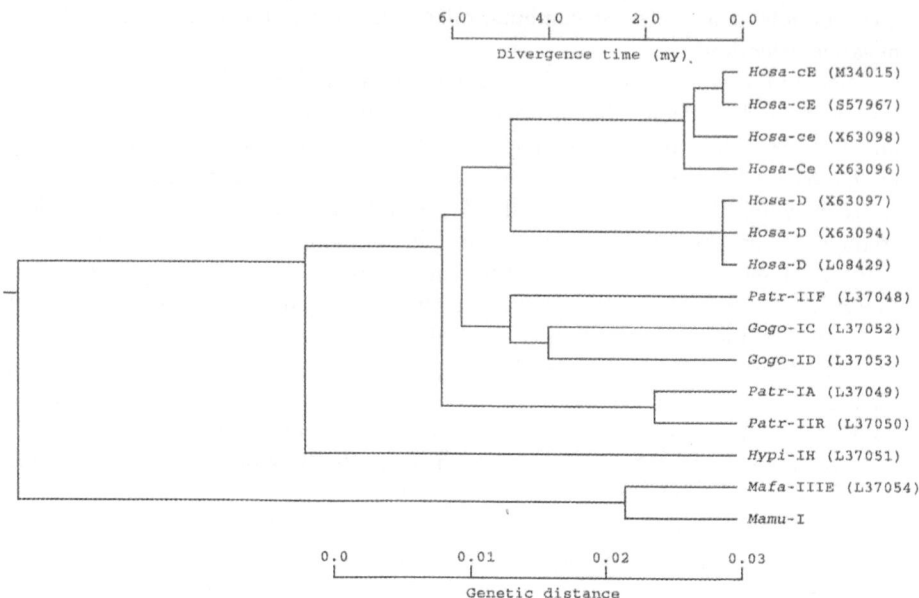

Fig. 4. Phylogenetic tree of Rhesus gene sequences. Designations are those used in Fig. 1 in addition to *Hypi* for the gibbon (*Hylobates pileatus*), followed by the clone number or, in the case of *Hosa* sequences, the rhesus type. Sequences obtained from the DNA database have accession codes given in brackets following sequence designations. The *Hosa* sequences are fom Le Van Kim et al. (1992a,b), apart from the *Hosa*-D (Lo8429) which is from Arce et al. (1993), *Hosa*-cE (S57967) which is from Kajii et al. (1993) and *Hosa*-D (M34015) which is from Cherif-Zahar et al. (1990). The nonhuman sequences are from Salvignol et al. (1994). Genetic distances are estimated using Kimura's two-parameter method and the dendrogram drawn by the UPGMA method using the MEGA program (Kumar et al. 1993). The time scale was placed taking the chimpanzee-human divergence at 6 million years (*my*) as a guide

genetic distance between the human *RHD* and *RHCE* loci (Fig. 4) suggests that they diverged about 4 million years ago, after separation of the lineages leading to humans and chimpanzees (gorillas). This deduction explains why orthology relationships between the human and chimpanzee genes cannot be resolved: the two loci of these two species appear equidistant. Since their divergence 4 million years ago, the *RHD* and *RHCE* loci have continued accumulating polymorphic substitutions.

Crossing-over and possibly other forms of genetic exchange have been the source of the null alleles in the human population. In the past, subdivision of the human population created conditions for random drift to drive the null allele frequencies to high values in some populations, as postulated by Nei and colleagues (1981). This has resulted in a long-lasting, but transient, polymorphism which persists even after the populations have expanded and mixed. Apparently, the presence or absence of the *RHD* locus is of no functional significance to the species, and the persistence of the null alleles is the consequence of the special form of evolution of duplicated genes.

Most, if not all, of the alleles at the *RHCE* locus are probably the result of recent mutations. The *C(c)* and *E(e)* alleles have, however, been in existence long enough to accumulate multiple substitutions. We estimate that the most distant alleles at the *RHCE* locus diverged approximately 1 million years ago (see Fig. 4). Taking into account the large standard error of this estimate, this divergence time can be regarded as being within the range of neutral allele divergence. The apparently random distribution of the allelic substitutions over much of the coding sequence may also be regarded as being indicative of neutrality. The function of the Rh proteins, however, remains unknown and so any deductions about the significance of the *RHCE* locus polymorphism must be considered tentative.

Acknowledgements. We thank Ms. Moira Burghoffer for editorial assistance.

References

Arce MA, Thompson ES, Wagner S, Coyne KE, Ferdman BA, Lublin DM. Molecular cloning of RhD cDNA derived from a gene present in RhD-positive, but not RhD-negative individuals. Blood 82:651–655, 1993

Bailly P, Hermand P, Callebaut I, Sonneborn HH, Khamlichi S, Mornon JP, Cartron JP. The LW blood group glycoprotein is homologous to intercellular adhesion molecules. Proc Natl Acad Sci USA 91:5306–5310, 1994

Blancher A, Socha WW. The ABO Hh and Lewis blood group in man and nonhuman primates. In: Blancher A, Klein J, Socha WW (eds) Molecular biology and evolution of blood group and MHC antigens in primates. Springer, Berlin Heidelberg New York, pp. 30–92, 1997a

Blancher A, Socha WW. The Rhesus system and its counterparts in nonhuman primates. In: Blancher A, Klein J, Socha WW (eds) Molecular biology and evolution of blood group and MHC antigens in primates. Springer, Berlin Heidelberg New York, pp. 147–219, 1997b

Brandon EP, Idzerda RL, McKnight GS. Targetting the mouse genome: a compendium of knockouts. Curr Biol 5: 569–694; 627–634; 873–881, 1995

Cherif-Zahar B, Bloy C, Van Kim C, Blanchard D, Bailly P, Hermand P, Salmon C, Cartron JP, Colin Y. Molecular cloning and protein structure of a human blood group Rh polypeptide. Proc Natl Acad Sci USA 87:6243–6247, 1990

Daniels GL, Moulds JJ, Anstee DJ, Bird GWG, Brodheim E, Cartron JP, Dahr W, Engelfriet CP, Issitt PD, Jorgensen J, Kornstad L, Lewis M, Levene C, Lubenko A, Mallory D, Morel P, Nordhagen R, Okubo Y, Reid M, Rouger P, Salmon C, Seidl S, Sistonen P, Wendel S, Woodfield G, Zelinski T. ISBT working party on terminology for red cell surface antigens. Sao Paulo report. Vox Sang 65:77–80, 1993

Feldman MW, Nabholz M, Bodmer WF. Evolution of the Rh polymorphism: a model for the interaction of incompatibility, reproductive compensation, and heterozygote advantage. Am J Hum Genet 21:171–193, 1969

Frattali AF, Spitalnik SL. Blood group antigens as receptors for pathogens. In: Blancher A, Klein J, Socha WW (eds) Molecular biology and evolution of blood group and MHC antigens in primates. Springer, Berlin Heidelberg New York, pp. 268–304, 1997

Goelz SE, Hession C, Goff D, Griffiths B, Tizard R, Newman B, Chi-Rosso G, Lobb R.: A gene that directs the expression of an ELAM-1 ligand. Cell 63:1349–1356, 1990

Haldane JBS. Mutation and the Rhesus reaction. Nature 153:106, 1944

Hochgeschwender U, Brennan MB. Rendundant genes? Nat Genet 8:219–220, 1994

Kajii E, Umenishi F, Iwamoto S, Ikemoto S. Isolation of a new cDNA clone encoding an Rh polypeptide associated with the Rh blood group system. Hum Genet 91:157–162, 1993

Kawaguchi H, O'hUigin C, Klein J. Evolution of primate C4 and CYP21 genes. In: Klein J, Klein D (eds) Molecular evolution of the major histocompatibility complex. Springer, Berlin Heidelberg New York, pp 357–381, 1991

Kelly RJ, Roquier S, Giorgi D, Lennon GG, Lowe JB. Sequence and expression of a candidate for the human Secretor blood group α(1,2) fucosyltransferase gene (FUT2); homozygosity for an enzyme-inactivating nonsense mutation commonly correlates with the non-secretor phenotype. J Biol Chem 270:4640–4649, 1995

Kimura M. A simple method for estimating evolutionary rates of base substitutions through comparative studies of nucleotide sequences. J Mol Evol 16:111–120, 1980

Klein J. Generation of diversity at MHC loci: Implications for T-cell receptor repertoires. In: Fougereau M, Dausset J (eds) Immunology, vol 80. Academic, London, pp 239–253, 1980

Klein J. Origin of major histocompatibility complex polymorphism: the trans-species hypothesis. Hum Immunol 19:155–162, 1987

Klein J, Satta Y, O'hUigin C, Takahata N. The molecular descent of the major histocompatibility complex. Annu Rev Immunol 11:269–295, 1993

Kominato Y, McNeill PD, Yamomoto M, Russell M, Hakomori S, Yamomoto F. Animal histo-blood group ABO genes. Biochem Biophys Res Comm 189:154–164, 1992

Kukowska-Latallo JF, Larsen RD, Nair RP, Lowe JB. A cloned human cDNA determines expression of a mouse stage-specific embryonic antigen and the Lewis blood group α(1,3/1,4) fucosyltransferase. Genes Dev 4:1288–1303, 1990

Kumar S, Tamura K, Nei M. MEGA: molecular evolutionary genetic analysis version 1.0. Pennsylvania State University, University Park, PA, 1993

Larsen RD, Ernst LK, Nair RP, Lowe JB. Molecular cloning sequence and expression of a human GDP-L-fucose:β-D-galactoside 2-α-L-fucosyltransferase cDNA that can form the H blood group antigen. Proc Natl Acad Sci USA 87:6674–6678, 1990

Le Van Kim C, Cherif-Zahar B, Raynal V, Mouro I, Lopez M, Cartron JP, Colin Y. Multiple Rh mRNAs isoforms are produced by alternative splicing. Blood 80:1074–1078, 1992

Le Van Kim C, Mouro I, Cherif-Zahar B, Raynal V, Cherrier C, Cartron JP, Colin Y. Molecular cloning and primary structure of the human blood group RhD polypeptide. Proc Natl Acad Sci USA 89:10925–10929, 1992

Levin BR. The effect of reproductive compensation on the long term maintenance of the Rh-polymorphism: the Rh crossroad revisited. Am J Hum Genet 19:288–302, 1967

Li CC. Is the Rh facing a crossroad? A critique of the compensation effect. Am Nat 87:257–261, 1953

Martinko JM, Vincek V, Klein D, Klein J. Primate ABO glycosyltransferases: evidence for trans-species evolution. Immunogenetics 37:274–278, 1993

Natsuka S, Gersten KM, Zenita K, Kannagi R, Lowe JB. Molecular cloning of a cDNA encoding a novel human leukocyte α-1,3-fucosyltransferase capable of synthesizing the sialyl Lewis determinant. J Biol Chem 269:16789–16794, 1994

Nei M. Molecular evolutionary genetics. Columbia University Press, New York, 1987

Nei M, Li W-H, Tajima F, Narain P. Polymorphism and evolution of the Rh blood groups. Jpn J Hum Genet 26:263–278, 1981

Parsons SF, Mallinson G, Holmes CH, Houlihan JM, Simpson KL, Mawby WJ, Spurr NK, Warne D, Barclay AN, Anstee DJ. The Lutheran blood group glycoprotein, another member of the immunoglobulin superfamily, is widely expressed in human tissues and is developmentally regulated in human liver. Proc Natl Acad Sci USA 92:5496–5500, 1995

Paulson JC, Colley KJ. Glycosyltransferases. Structure, localization, and control of cell type-specific glycosylation. J Biol Chem 264:17615–17618, 1989

Piau JP, Labarriere N, Dabouis G, Denis MG. Evidence for two distinct α(1,2)-fucosyl-transferase genes differentially expressed throughout the rat colon. Biochem J 300:623–626, 1994

Race RR. Some recent observations on the inheritance of blood groups. Br Med Bull 2:160–165, 1944

Saitou N, Nei M. The neighbor-joining method: a new method for reconstructing phyloge-netic trees. Mol Biol Evol 4:406–425, 1987

Salvignol I, Blancher A, Calvas P, Clayton J, Socha WW, Colin Y, Ruffie J. Molecular genet-ics of chimpanzee Rh-related genes: their relationship with the R-C-E-F blood group system, the chimpanzee counterpart of the human rhesus system. Biochem Genet 32:201–221, 1994

Schenkel-Brunner H. Human blood groups. Chemical and biochemical basis of antigen specificity. Springer, Vienna New York, 1995

Thompson JD, Higgins DG, Gibson TJ. CLUSTAL W: Improving the sensitivity of progres-sive multiple sequence alignment through sequence weighting, positions-specific gap penalties and weight matrix choice. Nucleic Acids Res 22:4673–4680, 1994

Weston BW, Nair RP, Larsen RD, Lowe JB. Isolation of a novel human α(1,3) fuco-syltransferase gene and molecular comparison to the human Lewis blood groups α(1,3/1,4) fucosyltransferase gene. J Biol Chem 267:4152–4160, 1992a

Weston BW, Smith PL, Kelly RJ, Lowe JB. Molecular cloning of a fourth member of a human α(1,3) fucosyltransferase gene family: multiple homologous sequences that determine expression of the Lewis x, sialyl Lewis x, and difucosyl sialyl Lewis x epitopes. J Biol Chem 267:24574–24584, 1992b

Yamamoto FI. Molecular genetics of the ABO histo-blood group system. Vox Sang 69:1–7, 1995

Note added in proofs: The recently obtained intron sequences of human, chimpanzee, and gorilla *AB* genes (C. O'hUigin, A. Sato, and J. Klein, manuscript submitted for publication) indicate that the showing of A and B antigens between humans and non-human primates is the result of convergent evolution. Nevertheless, some of the substitutions must have been retained transspecifically between different species of the genus *Homo*.

III Major Histocompatibility Complex

1 Geography and History of the Genes in the Human MHC: Can We Predict MHC Organization in Nonhuman Primates?

R. Tazi Ahnini, J. Henry, and P. Pontarotti

Quand j'étais enfant,
à l'école communale,
je n'aimais pas la géographie.
PP, Marseille, 5/12/1995

Introduction

We will provide, in this chapter, an overview of genes that are present in the human MHC and compare the organization of the human and mouse MHC. It will be shown that properties common to the human and mouse MHC correspond to a pleisomorphic, i.e., ancestral, organization. That is, these shared properties are found in all of the ancestors that led up to humans, including primate last common ancestors (belonging to the group Catarrhini).

Map of the Human MHC

The question that occurred to us when constructing a map of the human MHC was: What should we include? (1) only true genes? (2) transcribed sequences (including transcribed pseudogenes)? (3) sequences with other features present in the MHC, such as highly and middle repetitive sequences, for example,

Molecular Biology and Evolution of Blood Group and MHC Antigens in Primates
Blancher/Klein/Socha (Eds.)
© Springer-Verlag Berlin Heidelberg 1997

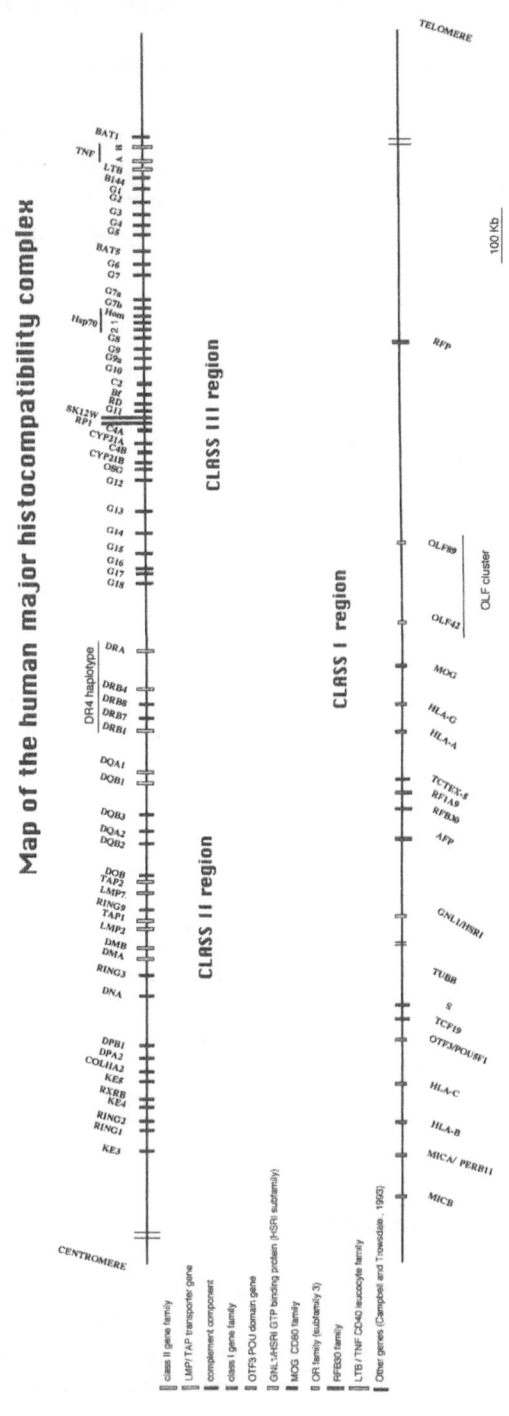

Fig. 1. Map of the "true" genes found in the MHC. Also see Table 1. (From Trowdale and Campbell 1993)

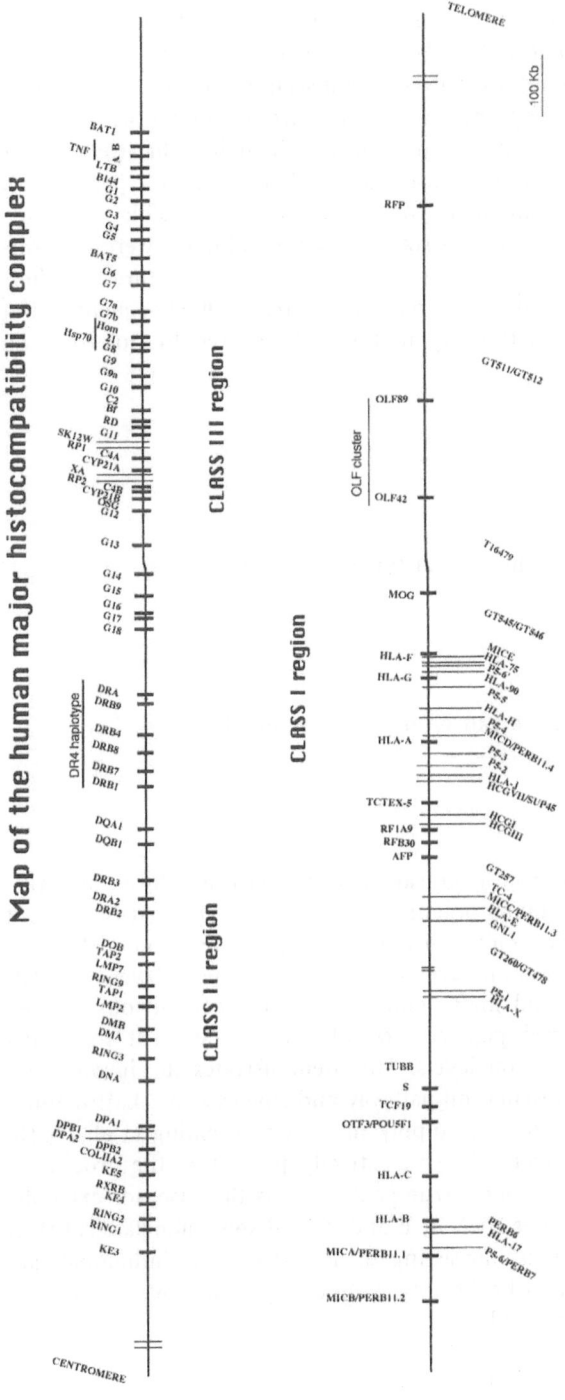

Fig. 2. Map of the MHC complex including genes, pseudogenes, duplicate sequences, genes obtained by cDNA selection (Trowsdale and Campbell 1993; Totaro et al. 1996; Leelayuwat et al. 1995; see also Table 1). Clones obtained by cDNA selection, described by Wei et al. (1993) and Goei et al. (1994), are not included in this map

LINEs (long interspersed repetitive elements) and SINEs (short interspersed repetitive elements)? (4) pseudogenes that are replicates of genes present in the MHC, such as MHC class I pseudogenes, MIC (Perb 11) pseudogenes? (5) sequences of unknown status that correspond to different transcripts isolated by different cDNA selection techniques and by different groups throughout the world?

We therefore decided to draw two maps: one that includes what we believed to be are "true" genes (Fig. 1) and one that includes other sequence types (Fig. 2). The repetitive sequences are so abundant that it would be impossible to include all of them in the map. Sequences whose status remains unknown were very difficult to analyze, as different names were given for the same structure by different groups. For example, Wei et al. (1993) reported 30 transcribed sequences, but we are not sure about their relationship to those discovered by Totaro et al. (1996.)

Sequence Classification

Here we will define each of the several types of sequences.

True, or Functional, Genes

Evidence for a functional gene can be direct or indirect.

Direct Evidence

Proof of the functionality of a given gene can be obtained if a natural, or artificial, mutant (for a cell line or a living organism) has a phenotype that is different from that of the wild type. For example, mice with a targeted deletion of the gene encoding LMP-7, a proteasome subunit (see below), have reduced levels of MHC class I cell surface expression and inefficiently present the endogenous antigen HY. Addition of the HY-derived peptides to splenocytes deficient in LMP-7 restores wild-type class I expression levels. This demonstrates the involvement of LMP-7 in the MHC class I presentation pathway and suggests that LMP-7 functions as an integral part of the peptide supply machinery (Fehling et al. 1994).

Existence of a protein product, while not strictly proof that the product is functional, also provides evidence for a true gene. This is the case, for example, with the classical HLA class I genes HLA-A,-B and -C and some nonclassical HLA class I genes such as HLA-G. It is interesting to note that some individuals are negative for HLA-A and seem to be healthy, although perhaps over time such individuals would be counterselected.

Indirect Evidence

Sequence and comparative (evolutionary) analyses could provide clues as to the function of a DNA sequence. A functional gene should be under either conservative selection (purifying selection) or positive (Darwinian) selection.

The molecular clocks of genes under conservative selection are slower than those of nonfunctional genes (pseudogenes). A direct consequence of this is that such sequences will be conserved at the interorder level, i.e., between humans and mice (see Monaco et al. 1986). Furthermore, the ratio between synonymous replacement and nonsynonymous replacement is equal to one in the case of a pseudogene, while for a gene under purifying selection, the ratio will be greater than one. This is the case for the gene HSR1, which most likely codes for a GTP binding protein (Vernet et al. 1994).

If the gene is under positive selective pressure, then the ratio of synonymous to nonsynonymous replacement will be less than one and the compared sequences should thus be functional. In the case of classical MHC class I and class II genes, the peptide binding domain seems to be under positive selective pressure as the ratio of synononymous to nonsynonymous mutation is less than one. However, the other part of the molecule is under negative selective pressure (Hughes and Nei 1989).

Figure 1 includes only those genes that are likely to be functional, as based on the above criteria. Most of the genes have been described by Campbell and Trowsdale (1993). Genes that have been described since then and their potential function are listed in Table 1.

Transcribed Sequences

This category includes transcribed pseudogenes such as the MHC class I gene HLA-H (Chorney et al. 1990), which is very similar to HLA-A. Although the HLA-H sequence can be transcribed, several lines of evidences show that it is a pseudogene, a phenylalanine replaces a cysteine in position 164, which forms the essential disulfide bond of the $\alpha 2$ domain. Furthermore, a single base deletion produces a translation termination codon in exon 4.

Another example is the P5-1 sequence, which corresponds to a chimeric sequence derived from the 5′ part of a MHC class I gene (promoter region exon 1 and the first part of intron 1 linked to a structurally unrelated sequence, P5). (Avoustin et al. 1994). A "small" (119 amino acids) open reading frame (ORF) is found in the second exon. P5 is not conserved outside primates, meaning that this sequence is not under negative selective pressure. Rather, either this gene is a new product of evolution or it is a sterile transcribed sequence. The mirror example of P5-1 is the HLA-DRB6 gene (Mayer et al. 1993). This class II gene lacks exon 1 and the first four amino acid residues of the mature protein. It also lacks the HLA class II promoter, but is nonetheless transcribed at low levels in a chimpanzee B lymphoma cell line in which the DRB6 homologue is truncated, as in humans. This truncation was due to the insertion of a retrovirus

Table 1. New genes that have been localized to the human MHC since publication of Trowsdale and Campbell (1993)

Gene	Region	Possible function	References
RXRB	Class II	Retinoid X receptor B.	(Nagata et al. 1995)
RP1	Class III	Nuclear protein that interacts with DNA or acidic domains of transcriptional factors.	(Shen et al. 1994)
SKI2W	Class III	Helicase. It contains leucine zipper motif that may be involved in dimerization and a RGD motif that could serve as a ligand for integrin.	(Dangel et al. 1995)
LST1 (B144)	Class III	Leucocyte specific transcript-1 encoding a transmembrane protein. LST1 has possible role in immune response.	(Holzinger et al. 1995)
MICA	Class I	MICA chain may have the capacity to bind peptide or other short ligands.	(Bahram et al. 1994)
MICB	Class I	MICB chain may have the capacity to bind peptide or other short ligands.	(Bahram et al. 1994; Bahram and Spies 1996)
OTF3 (POU5F1)	Class I	Octamer transcription factor 3. It contains a POU specific domain and an homeodomain. It could play role in early development.	(Crouau-Roy et al. 1994; Krishnan et al. 1995)
S	Class I	S may be involved in developmental program in differentiation of keratinocytes.	(Zhou and Chaplin 1993)
TUBB	Class I	Mutation at the TUBB locus (β-tubulin) could be the cause for certain forms of HGLA-linked microtubule dysfunction.	(Volz et al. 1994)
R1	Class I	GTP-binding protein.	(Vernet et al. 1994)
RF1A9	Class I	Ring finger protein, potential binding of nucleic acids.	unpublished data
AFP	Class I	Acid finger protein, potential binding of nucleic acids.	(Chu et al. 1995)
RFB30	Class I	Ring finger protein, potential binding of nucleic acids.	(Vernet et al. 1993)
TCTEX-5	Class I	Unknown.	(Amadou et al. 1995)
MOG	Class I	Myelin oligodendrocyte glycoprotein; candidate target antigen for autoimmune-mediated demyelinisation.	(Amadou et al. 1995; Roth et al. 1995)
OLF42	Class I	Olfactory receptor.	(Amadou et al. 1995 and unpublished data)
OLF89	Class I	Olfactory receptor.	(Amadou et al. 1995 and unpublished data)
FAT11	Class I	Olfactory receptor.	(Fan et al. 1993)
RFP	Class I	Ring finger protein, potential binding of nucleic acids.	(Amadou et al. 1995)

(related to the mouse mammary tumor viruses) into intron 1 of DRB6. This was accompanied or followed by loss of both exon 1 and the promoter region of DRB6. In the 3' long terminal repeat (LTR) of the retrovirus, a new exon arose with a functional donor splice site at its 3' end, which enables it to be spliced in register with DRB6 exon 2. This new exon encodes an ORF comprised mostly of hydrophobic amino acid residues; therefore the sequence could function as a leader for the truncated DRB6 gene. Upstream from the new exon is a promoter enabling transcription of the DRB6 gene. Several other examples, similar to the one described for DRB6, are present in the human genome (see Feutcher et al. 1992).

Repetitive Sequences

A significant percentage of all mammalian genomes consists of interspersed repetitive DNA sequences. These are generally classified as SINEs or LINEs. The SINEs range in size from 90 to 400 bp, while LINEs can be as large as 7000 bp. New copies of both types of elements find their way into the genome via reverse transcription of an RNA intermediate, a process called retroposition or retrotransposition (Weiner et al. 1986). The RNA intermediates involved in the retroposition of SINEs are transcribed by RNA polymerase III, while those of LINEs are thought to be produced by RNA polymerase II. Many of the major SINE and LINE families can be divided into subfamilies, defined in terms of common nucleotide variations at specific locations. In primates the most abundant SINE is the Alu family. The LINEs-1 or L1 family is several orders of magnitude more abundant than any other LINE subfamily. In the primate MHC, several copies of L1 or Alu sequences are found, most likely having occurred either by retroposition or *cis* duplication (Mnukova-Fajdelova et al. 1994).

Replicate Pseudogenes

Duplication events in the MHC class I or class II region not only concern classical class I or class II loci but also surrounding stretches of DNA (Avoustin et al. 1994). In the case of the class I region, one of the ancestral duplication units is formed by an HLA-A-like, P5-like, HLA-80-like, MIC/Perb11-like sequence (Avoustin et al. 1994; Leelayuwat et al. 1995). For example five copies of Perb11/MIC genes are found within the human MHC class I region, always close to class I genes. While at least three of these sequence are pseudogenes, they are nonetheless called MIC-C, -D, -E or perb11-3, -4, -5 (Barham and Spies 1996). Similarly, several MHC class I genes are most likely pseudogenes; however, they are called HLA-J, H, etc. Note that P5 sequences are present in multiple copies in the MHC class I region. One of the sequences is transcribed under the MHC class I promoter, but it is doubtful that the other P5 members are expressed.

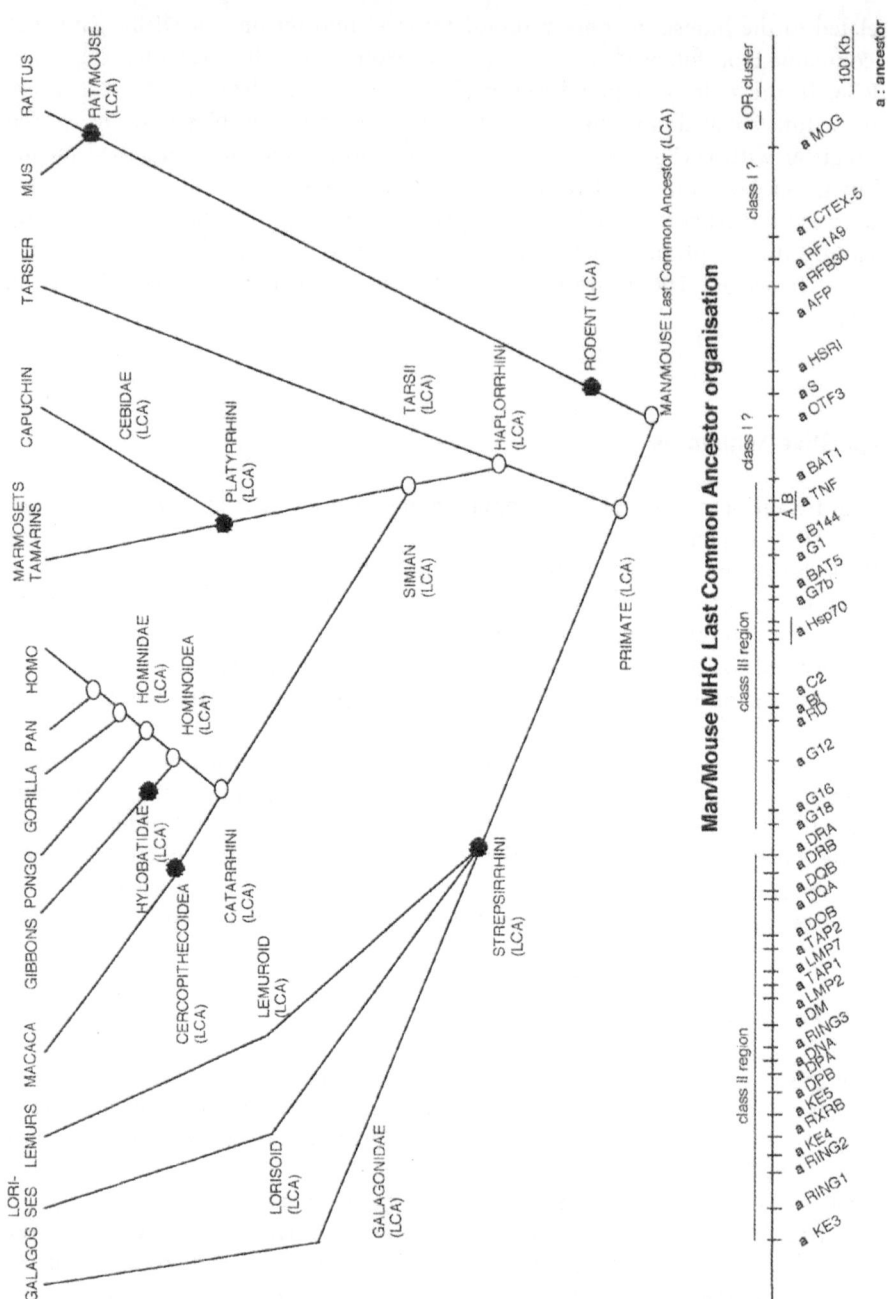

Fig. 3. Ancestral organization of the primate MHC and simplified primates phylogenetic tree

Some of the duplicate pseudogenes, including HLA-H, can be transcribed (Chorney et al. 1990).

Sequences of an Unknown Status

Several group have employed cDNA selection techniques, using the yeast artificial chromosome from the MHC class I region (Wei et al. 1993; Fan et al. 1993; El Kahloun et al. 1993; Goei et al. 1994; Totaro et al. 1996) to isolate transcribed sequence present in the insert. Some of the sequences are still under investigation and their status thus remains unknown.

The map presented in Fig. 2 shows true genes, transcribed sequences, replicate pseudogenes and sequences of an unknown status. Some of the latter sequences are not included in this map (see Wei et al. 1993; Fan et al. 1993; Goei et al. 1994).

Comparative Mapping with the Mouse MHC

The MHC map of another species, mouse, is also well-known (Himmelbauer et al. 1993). Mice belong to the Order Rodentia. Based on phylogenetic analysis and fossil records, rodents and ancestral primates split around 100 million years ago (Novacek 1992). The fossil records also show that the common primate ancestor of the present day primate existed around 65 million years ago. Thus, comparison of the human and mouse MHCs could help in determining the MHC map of their common ancestor – or, more exactly, to decipher the common features (pleisomorphism). It is also likely that this common organization was present in the common primate ancestor (Fig. 3). This organization should be found in the human lineage (Fig. 3, open circles): the primate LCA (last common ancestor), the Haplorrhini LCA, the simian LCA, the Catarrhini LCA, the Hominoidea LCA, the Hominidae LCA, the pan/homo/ gorilla LCA, the homo/pan LCA. It is possible that the organization of the other sublineages (Fig. 3, closed circles) differs from that of the ancestral one.

In order to simplify the discussion we will present comparisons for each subregion: MHC class I, MHC class II, and MHC class III.

The MHC Class I Region

The MHC class I genes (Fig. 4a) underwent several rounds of independent duplication after separation of the mouse/human lineages. The consequence of this is that human MHC class I genes are more similar to one another than to any of the mouse MHC class I loci. However, on the basis of genetic distances measured between the mouse H2-M and the other class I genes, it seems that this class I subfamily is very old but has been deleted in the human lineage (Klein et al. 1991).

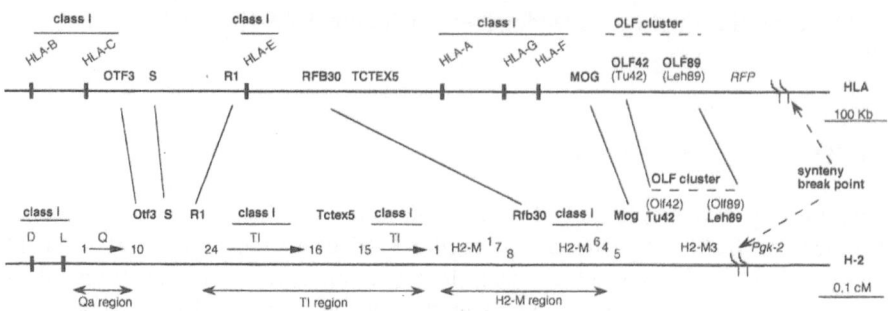

a   Comparative map of mouse and human MHC class I region

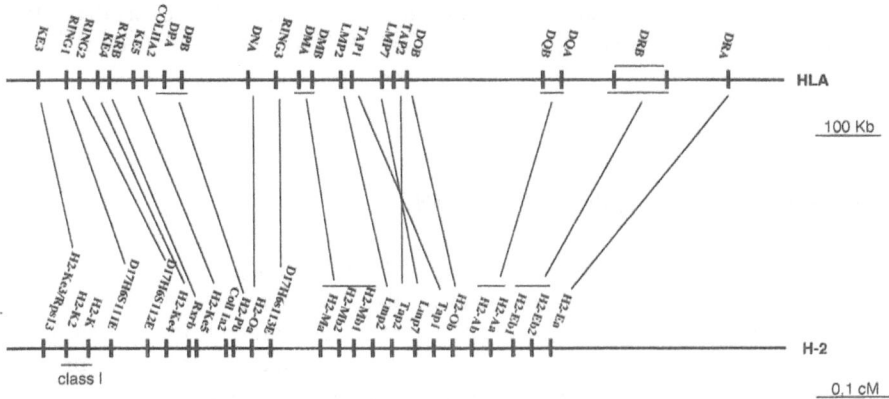

b   Comparative map of mouse and human MHC class II region

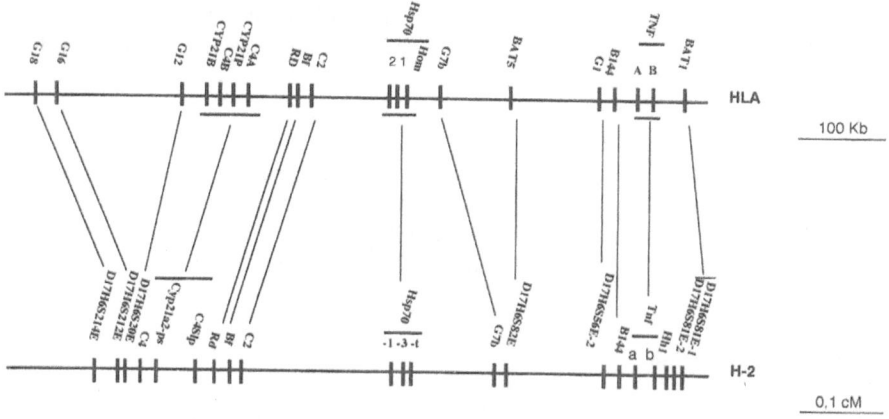

c   Comparative map of mouse and human MHC class III region

Fig. 4A–C. Mouse comparative map. **A** MHC class I; **B** MHC class II; **C** MHC class III

Other genes present in the MHC class I region seem to have an orthologous relationship (Amadou et al. 1995), i.e., they duplicated before separation of the mouse and human lineages and thus arose directly from a gene present in a common ancestor of mice and humans. Moreover, these orthologous genes often correspond to single copy probes (i.e. they hybridize to just one locus), even if they belong to a multigenic family, as deciphered by sequence comparison analysis. For example, RFP (Ret finger protein) and RFB30 (ring finger B30) belong to the same family (Vernet et al. 1993); however, the B30 probe does not cross-react with RFP at the genomic level and vice versa. Finally, it should be noted that mapping of orthologous genes shows that they are arranged in the same order in the human and mouse MHC class I region (Amadou et al. 1995; Jones et al. 1995).

We would also like to comment on the speed of gene duplication. Some genes present in the human and mouse MHCs have undergone duplication after separation of the two lineages, for example, MHC class I genes. However, at the inter-order level, MHC class I genes are orthologously related. For example, HLA-A-like and HLA-B-like genes seem to be found in all Old World monkeys (Catarrhini) (Chen et al. 1992).

The MHC Class II Region

This region consists of class II and non-class II genes (Fig. 4a,b).

Class II Genes

In humans, these can be separated into different subfamilies: DR, DP, DQ, DN/ DO and DM. All of them are also present in the mouse. It should be noted that these subfamilies have undergone independent duplication in the human and mouse lineages. The different subfamilies have been described in several Old and New World monkeys (Watkins 1994). This will be discussed elsewhere in this volume.

Non-class II Genes

Other human and mouse genes also have an orthologous relationship, e.g., LMP2 and LMP7, and TAP1 and TAP2. The former are likely involved in the degradation of intracellular and viral peptides. The latter code for two ATP binding cassette (ABC) transporter products, which transport peptides through the endoplasmic membrane into the lumen of the endoplasmic reticulum. The peptides are then loaded onto MHC class I molecules. These genes may have arisen within the MHC; alternatively, they could have been captured from other parts of the genome. It is possible that the LMP /TAP genes cluster was captured by the MHC because of an evolutionary advantage in having the antigen processing and presenting genes linked. Another view is that the genes for antigen processing and transport to the cell surface were previously associated in an operon-like structure; hence the genes have been kept together.

The MHC Class III Region

The class III region (Fig. 4c) is highly conserved between mice and humans. To the best of our knowledge, all the genes have an orthologous relationship, except for obvious duplications such as C4 p450c21B-Y (Gasser et al. 1994).

Ancestral Organization of the Primate MHC

The human-mouse comparison allows us to draw a map of the ancestral organization (Fig. 3), which should be present in all primate species. However, specific changes (apomorphism) are more likely to be detected in present day primates. We would like to stress the point that all the human-derived probes that cross-react with the mouse genome (and vice versa) must cross-react with primate genomes. These probes can be used as fixed points in mapping the MHC of the different living primates species. The data obtained from such experiments will help in furthering our understanding of primate MHC evolution.

References

Amadou C, Ribouchon MT, Mattei MG, Jenkins NA, Gilbert DJ, Copeland NG, Avoustin P, Pontarotti P. Localization of new genes and markers to the distal part of the human major histocompatibility complex (MHC) region and comparison with the mouse: new insights on the evolution of mammalian genomes. Genomics 26:9–20, 1995

Avoustin P, Ribouchon MT, Vernet C, N'Guyen B, Crouau-Roy B, Pontarotti P. Non-Homologous recombination within the major histocompatibility complex leads to de novo creation of a transcribed hybrid sequence. Mammalian Genome 5:771–776, 1994

Barham S, Spies T. Nucleotide sequence of a human MHC class I MICB cDNA. Immunogenetics 43:230–233, 1996

Barham S, Bresnahan M, Geraghty DE, Spies T. A second lineage of mammalian major histocompatibility complex class I genes. Proc Natl Acad Sci USA 91:6259–6263, 1994

Campbell RD, Trowsdale J. Map of the human MHC. Immunol Today 14:349–352, 1993

Chen IT, Dixit A, Rhoads DD, Roufa DJ. Homologous ribosomal protein in bacteria, yeast and human. Proc Natl Acad Sci USA 83:6907–6911, 1986

Chen ZW, McAdam SN, Hugues AL, Dogon AL, Letvin NL, Watkins DI. Molecular cloning of arangutan and gibbon MHC class I cDNA. The HLA-A and -B loci diverged over 30 million years ago. J Immunol 148(8):2547–2554, 1992

Chorney MJ, Sawada I, Gillepsie GA, Srivastava R, Pan J, Weissman SM. Transcription analysis, physical mapping and molecular characterization of a non-classical human leucocyte antigen class I gene. Mol Cell Biol 10:243–253, 1990

Chu TW, Capossela A, Coleman R, Goei VT, Nallur G, Gruen JR. Cloning of new "finger" protein gene withn the class I region of the Human MHC. Genomics 26:229–239, 1995

Crouau-Roy B, Amadou C, Bouissou C, Clayton J, Vernet C, Ribouchon MT, Pontarotti P. Localization of the OTF3 gene within the human MHC class I region by physical and meiotic mapping. Genomics 21: 241–243, 1994

Dangel AW, Shen L, Mendoza AR, Wu L, Yu K. Human helicase gene SK12 W in the hla class III region exhibits striking structural similarities to the Yeast antiviral gene SK12 and to the human gene K1AAOO52: emergence of a new gene family. Nucleic Acid Res 42:315–322, 1995

El Kahloun A, Chauvel B, Mauvieux V, Dorval A, Jouanolle AM, Gicquel I, Le Gall J-Y, David V. Localization of seven new genes around the HLA-A locus. Hum Mol Genet 21:55–60, 1993

Fan WF, Wei X, Shukla H, Parimoo S, Xu H, Sankhavaram P, Li Z, Weissman SM. Application of cDNA selection technques to region of the human MHC. Genomics 17:575–581, 1993

Fehling HJ, Swat W, Laplace C, Kuhn R, Rajewsky K, Muller U, von Boehmer H. MHC class I expression in mice lacking the proteasome subunit LMP-7. Science 265:1234–1237, 1994

Feutcher AE, Freema JD, Mager DL. Strategy for detecting celllular transcrits promoted by numan endogenous long terminal repeats: identification of a novel gene (CDC4L) with homology to yeast CDC4. Genomics 13:1237–1246, 1992

Gasser DL, Sternberg NL, Pierce JC, Goldner-Sauve A, Feng H, Haq AK, Spies T, Hunt C, Buetow KH, Chaplin DD. P1 and cosmid clones define the organization of 280kb of the mouse H-2 complex containing theCps-1 and Hsp70 loci. Immunogenetics 39:48–55, 1994

Goei VL, Parimoo S, Capossela A, Chu TW, Gruen JR. Isolation of novel non HLA gene fragments from the hemochromatosis region (6p21.3) by cDNA hybridization selection. Am J Hum Genet 54:244–251, 1994

Himmelbauer H, Artzt K, Barlow D, Fisher-Lindhal K, Lyon M, Klein J, Silver L. Encyclopedia of the Mouse genome III. October 1993. Mouse chromosome 17. Mamm Genome 4:S230–252, 1993

Holzinger I, Baey A, Messer G, Kick G, Zwierzinz H, Wess EH. Cloning and genomic characterization of LST1: a new gene in the human TNF region. Immunogenetics 42:315–322, 1995

Hughes AL, Nei M. Nucleotide substitution at major histocompatibility complex class II loci. Evidence for overdominant selection. Proc Natl Acad Sci USA 86:958–962, 1989

Jones EP, Xiao H, Schultz RA, Flaherty L, Trachtulec Z, Vinceck V, Larin Z, Lechrach H, Fisher-Lindhal K. MHC class I gene organization in a >1.5 Mb YAC contigs from the H2-M region. Genomics 27:40–51, 1995

Klein J, Zhu Z, Gutknecht J, Figueroa P, Kasahara M. MHC lessons in evolution. In: Srivastava R, Ram B, Tyle P (eds) Immunogenetics of the major histocompatibilty complex. VCH, New York, pp 65–99, 1991

Krishnan BR, Jamry I, Chaplin DD. Feature mapping of the HLA class I region/Localization of the POU5F1 and TC19 genes. Genomics 30:53–58, 1995

Leelayuwat C, Townsend DC, Degli-Espositi MA, Abraham LA, Dawkins RL. A new polymorphic and multicopy MHC gene family related to nonmammalian classI gene. Immunogenetics 40:339–351, 1994

Leelayuwat C, Pinelli M, Dawkins RL. Clustering of diversed duplicated sequences in the MHC. Evidence for en bloc duplication. J Immunol 155:692–698, 1995

Mayer WE, O'hUigin C, Klein J. Resolution of the HLA-DRB6 puzzle: a case of grafting de novo-generated exon on an existing gene. Proc Natl Acad Sci USA 90:10720–10724, 1993

Mnukova-Fajdelova M, SattaY, O'hUigin C, Maye WE, Figueroa F, Klein J. Alu elements of the primate major histocompatibility complex. Mamm Genome 5:405-415, 1994

Monaco AP, Neve RL, Colletti-Feener C, Bertelson CJ, Kurnit DM, Kunkel LM. Isolation of candidate cDNAs for portions of the Duchesne muscular dystrophy gene. Nature 323:646-650, 1986

Nagata T, Weiss EH, Abe K, Kitagawa K, Ando A, Yara-Kikuti Y, Seldin MF, Ozato K, Inoko H, Taketo M. Physical mapping of the retinoid X receptor B gene in mouse and human. Immunogenetics 41:83-90, 1995

Novacek MJ. Mammalian phylogeny: shaking the tree. Nature 356:121-125, 1992

Pichon L, Carn L, Bouric A, Giffon T, Chauvel B, Chauvel B, Lepourcelet M, Mosser J, Legall JY, David V. Structural analysis of the HLA-A/HLA-F subregion: Precize localization of two new multigene families closely associated with the HLA class I sequence. Genomics (in press), 1996

Roth MP, Malfroy L, Offer C, Sevin J, Enault G, Borot N, Pontarotti P, Coppin H. The human myelin oligodendrocyte glycoprotein (MOG) gene: complete nucleotide sequence and structural characterization. Genomics 28: 241-250, 1995

Shen L, Wu C, Sanglioglu S, Chen R, Mendoza AR, Dangel AW, Caroll MC, Zipf WB, Yu CY. Structure and genetics of the of the partially duplicated gene RP located immediately upstream of the complement C4A and C4B genes in the HLA class III region. J Biol Chem 269:8466-8476, 1994

Totaro A, Rommens JM, Grida A, Lunardi C, Carella M, Huizenga JJ, Roetto A, Camaschella C, De Sandre G, Gasparini P. Heriditary hemochromatosis: generation of transcrption map within a refined and extended map of the HLA class I regon. Genomics 31:319-336, 1996

Vernet C, Boretto J, Mattei MG, Takahashi M, Jack JW, Mather IH, Rouquier S, Pontarotti P. Evolutionary study of multigenic families mapping close to the MHC class I region. J Mol Evol 37:600-612, 1993

Vernet C, Ribouchon MT, Chimini G, Pontarotti P. Structure and evolution studies of a member of a new subfamily of putative GTP-binding proteins mapping to the human MHC class I region. Mamm Genome 5:100-105, 1994

Voltz A, Weiss E, Trowsdale J, Ziegler A. Presence of un expressed b-tubulin gene (TUBB) in the HlA class I region may provide the genetic basis for HLA-linked microtubule dysfunction. Hum Genet 93:42-46, 1994

Watkins DI. MHC of non human primates. In: Letvin NL, Desrosiers RC (eds) Simian immunodeficiency virus. Springer, Berlin Heidelberg New York, pp 145-159 (Current topics in microbiology and immunology, vol 188), 1994

Wei H, Fan WF, Xu H, Parimoo S, Shukla H, Chalin DD, Weissman SM. Genes in one megabase of the HLA class I region. Proc Natl Acad Sci USA 90:9470-9474, 1993

Weiner AM, Deininger PL, Efstratiadis A. Nonviral retroposons: genes, pseudogenes and transposable elements generated by reverse flow of genetic information. Annu Rev Biochem 55:631-661, 1986

Zhou Y, Chaplin DD. Identification in the HLA Class I region of a gene expressed late in the keratinocyte differentiation. Proc Natl Acad Sci USA 90:9470-9474, 1993

2 MHC Class I Genes in Nonhuman Primates

L.F. Cadavid and D.I. Watkins

Introduction

The primate order includes two suborders, the Prosimii and the Anthropoidea (Martin 1970). The Prosimii include tarsiers, lemurs and lorises whereas the Anthropoidea include two infraorders, Catarrhini and Platyrrhini. Catarrhini are Old World primates (for example, rhesus monkeys), apes (chimpanzees, gorillas, orangutans and gibbons) and humans. The taxonomic status of the 16 genera that form the infraorder Platyrrhini (commonly known as the New World primates) has been controversial. However, both morphological studies (Ford 1986; Kay 1990; Rosenberger 1984) and molecular data (Schneider et al. 1993) agree on dividing the 16 extant platyrrhine genera into seven phyletic groups: Callitrichidae (tamarins and marmosets (Fig. 1); genera *Callithrix*, *Cebuella*, *Leontopithecus*, *Saguinus* and *Callimico*), Pithecinae (genera *Pithecia*, *Chiropotes*, and *Cacajao*), Atelinae (genera *Lagothrix*, *Brachyteles*, *Ateles* and *Alouatta*), genus *Cebus*, genus *Saimiri* (squirrel monkeys), genus *Aotus* (owl monkeys) and genus *Callicebus*.

The pioneering studies of the MHC in nonhuman primates were carried out by H. Balner and co-workers (1967, 1978) at the TNO Primate Center in Holland and by Dorf and Metzgar (1970) at Duke University. With classical serological

Molecular Biology and Evolution of Blood Group
and MHC Antigens in Primates
Blancher/Klein/Socha (Eds.)
© Springer-Verlag Berlin Heidelberg 1997

Fig. 1. The cotton-top tamarin in its natural habitat. (Drawing by Louis Agassiz Fuertes)

methodologies, they laid the ground work for the description of MHC class I genes in nonhuman primates. Working with chimpanzees, Balner's group was able to generate alloantisera that described 12 alleles at the A locus and 20 alleles at the B locus, and Dorf and Metzgar extended these studies utilizing the cross-reactivity of human alloantisera with chimpanzee lymphocytes. With the arrival of molecular technologies, the study of the MHC in nonhuman primates turned towards the search for general mechanisms that might explain the origin and maintenance of MHC polymorphism in natural populations. MHC class I sequences from a variety of nonhuman primates have been examined, showing that, in contrast to the instability of MHC class I alleles and allelic lineages, the MHC class I loci themselves have been well conserved during the evolution of Old World primates and apes. In New World primates, however, there is no evidence to date for the *HLA-A*, *-B*, *-C* or *-E* loci. The classical New World primate MHC class I genes appear to have derived from the ancestor of the non-classical *HLA-G* locus.

MHC Class I Genes in Old World Primates and Apes

MHC Class I Genes in the Chimpanzee and Bonobo

A Locus in the Chimpanzee (Pan troglodytes) and the Bonobo (Pan paniscus)

On the bases of sequence similarity and locus-specific motifs, the HLA-A locus alleles fall into six well-defined families derived from at least two ancient lineages (Lawlor et al. 1990a; Wagner et al. 1993; Domena et al. 1993; Starling et al. 1994). One of the ancient lineages (A3) gave rise to the HLA-A9, -A80 and -A1/-A3/-A11 families, whereas the other ancient lineage (A2) gave rise to the HLA-A2/-A28, -A10 and -A19 families.

Earlier serological studies in chimpanzees demonstrated that many of the human alloantisera generated against human A1/A3/A11 molecules cross-reacted with chimpanzee lymphocytes. Molecular analysis of chimpanzee class I alleles later confirmed this HLA-A1, -A3, -A11 cross-reactivity. Fourteen different A locus alleles have been described in chimpanzees and five in bonobos (Lawlor et al. 1988, 1990, 1995; Mayer et al. 1988; Chen et al. 1993; McAdam et al. 1995). These chimpanzee and bonobo A sequences are surprisingly similar to their human A locus counterparts. HLA-A*0101 differs from Patr-A*11 (Pan troglodytes) by five residues in the $\alpha 1$ domain and only four in the $\alpha 2$ domain. These alleles have an overall nucleotide similarity of 98.2 %; similar values have been obtained in chimpanzees for β-globin genes (Chang and Slightom 1984) and for Mhc-DRA genes (Fan et al. 1989).

There is less nucleotide diversity at the A locus in chimpanzees and bonobos than there is at the A locus in humans. The number of nucleotide differences between chimpanzee and bonobo A locus cDNAs range from 3 to 45, whereas in humans they range from 3 to 60. However, the number of alleles at the chimpanzee and bonobo A locus is greater than the number of alleles found in the HLA-A1, -A3, -A11 family. Moreover, the number of nucleotide differences between the chimpanzee and bonobo A alleles is greater than that present in the HLA-A1, -A3, -A11 allelic family (3–45 in chimpanzee and bonobo A locus alleles vs 15–34 differences in HLA-A1, -A3, -A11 alleles).

Since the A locus alleles in bonobos and chimpanzees are related to only one of the six HLA-A allelic lineages, it is possible that the ancestor of these two Pan species may have had a very low effective population size at one time during their evolutionary history. It is unlikely that the reduction in their allelic repertoire is due to a sampling error since out A alleles have been cloned from 12 unrelated chimpanzees and four unrelated bonobos. After this hypothetical low effective population size, A locus alleles in chimpanzees and bonobos may have undergone diversification using a variety of different mechanisms, including point mutations and, in some cases, interlocus segmental exchange.

The condensation of the allelic repertoire of the MHC class I A locus in Pan is consistent with the low average heterozygosity reported for other nuclear genes in this genus (King and Wilson 1975; Lucotte 1983; Bruce and Ayala; 1979). This limited diversity of the *A* locus, however, stands in apparent contradiction to the diversity of mitochondrial DNA (mtDNA) in these species (Ferris et al. 1981; Tamura and Nei 1993). This may be a reflection of differences between the rate of evolution of nuclear and mitochondrial genes. If we postulate that a bottleneck has taken place in chimpanzee and bonobo evolution, this could account for the reduction in the levels of nuclear DNA polymorphism. Since mtDNA evolves more rapidly than nuclear genes do, more variability may have been generated in the mtDNA after the postulated bottleneck.

B Locus in the Chimpanzee and Bonobo

Nineteen different *B* locus alleles have been described in chimpanzees and bonobos (Lawlor et al. 1988; Mayer et al. 1988; McAdam et al. 1994). Comparisons of *B* locus allele sequences among humans, chimpanzees and bonobos reveal an unusual difference between gene trees of exon 2 and exons 3–8. When comparisons are made with exon 2, it is possible to assign each chimpanzee and bonobo *B* locus allele to one of four human allelic families *HLA-B15, -B48, -B57/-58* and *-B27/-7*. However, when exons 3–8 are compared a completely different pattern is observed. Unlike gene trees of exon 2, in gene trees of exons 3–8 the *HLA-B* homologues of chimpanzees and bonobos cluster separately from their human counterparts. This relationship between trees of exon 2 and exons 3–8 is not found in comparisons made with the *A* locus. At the *A* locus the chimpanzee and bonobo alleles group with the *HLA-A1/-A3/-A11* family in gene trees of exon 2 and exons 3–8.

Many of the unique substitutions found in chimpanzees and bonobo *B* locus alleles are also present in the alleles of the *HLA-A1, -A3, -A11* family. This might indicate a role for interlocus segmental exchange in the generation of *B* locus diversity. Although HLA-B*0702 and its bonobo homologues Papa-B*01 and Papa-B*04 (*Pan paniscus*) are very similar in the α1 domain, they differ in the α2 domain by more than nine amino acid substitutions. Thus, it is possible that the ape and human *B7*-related alleles derived from an *HLA-B*07*-like ancestral gene that gave rise to *HLA-B*0702* in humans and *Papa-B*01* and *Papa-B*04* bonobos. Subsequent to their divergence from an ancestral gene, both inter- and intralocus recombination in exon 3, which encodes the α2 domain, served to create new *B* locus alleles in humans, chimpanzees and bonobos. Interestingly, it was sequence changes in exon 3 that were largely responsible for the creation of the new *HLA-B* alleles found in South American indigenous tribes (Belich et al. 1992; Watkins et al. 1992). The conservation of the α1 domain between chimpanzee, bonobo and human B locus alleles implies that the peptide-binding pockets encoded by this domain may be under strong structural and functional constraints.

The observation that an unusually high proportion of alleles at the *B* locus are products of segmental exchange does not necessarily imply that this locus is a "hot spot" for recombination. If selection had maintained more recombinant alleles at the *B* locus than it did at other *Mhc* loci, recombinant alleles would be observed more frequently at the *B* locus than at other loci even if the rate of recombination is similar. A similar phenomenon has been observed in bacteria and other microorganisms, in which recombinants are more likely to be observed at loci where diversity is selectively favored (Hughes 1992; Smith et al. 1990).

C Locus in the Chimpanzee

Since only one allele of the *C* locus has been reported in chimpanzees (Lawlor et al. 1990), it is difficult to draw any conclusion about the evolutionary nature of this locus. *Patr-C*01* has *C*-specific substitutions and is clearly grouped with *HLA-C* alleles in gene trees. *Patr-C*01* differs from HLA-C*0801 by four amino acids in the α1 domain and seven amino acids in the α2 domain.

MHC Class I Genes in the Gorilla

A Locus in the Gorilla (*Gorilla gorilla*)

The gorilla *A* locus alleles *Gogo-A*0101, -A*0201, -A*0401* and *-A*0501* (*Gorilla gorilla*) have many of the specific nucleotide substitutions that characterize the *A2* ancient lineage (Lawlor et al. 1991). Within the *A2* ancient lineage the gorilla locus *A* alleles are more related to the *HLA-A10* and *HLA-A19* families than they are to the *HLA-A2* family, but there is no clear segregation with either of the two families. Thus, it is possible that the gorilla *A* locus alleles are related to the common ancestor of the *HLA-A10* and *HLA-A19* families, suggesting that these families were formed after the divergence of the human and gorilla lineages.

Interestingly, one of the gorilla *A* locus alleles has remarkable nucleotide sequence similarity to the MHC class I pseudogene *HLA-H* (Zemmour et al. 1990) in exon two. Two hundred ten bases of exon 2 (nucleotide positions 3–210, encoding amino acids 2–71 in the α1 domain) of *Gogo-A*0301* (also called *Gogo-OKO*) differ by only two bases from the sequence of *HLA-H*. The remainder of *Gogo-A*0301*, however, was more similar to classical primate *A* locus genes. These data suggest that interlocus segmental exchange occurred between the gorilla *A* locus gene and the ancestor of *HLA-H* (Watkins et al. 1991a). Whether this allele is a product of the *Gogo-A* locus or whether it represents an entirely new gorilla MHC class I locus is uncertain (Lawlor et al. 1991).

It is intriguing that the *Gogo-A* alleles are derived from one ancient lineage (the *A2* ancient lineage) whereas the *Patr-A* and *Papa-A* alleles are derived from another (the *A3* ancient lineage). Although both ancient lineages were present in the common ancestor of gorillas, chimpanzees, bonobos and humans, only one family has survived in each phyletic group (*Gorilla* and *Pan*) whereas both have survived in humans.

B Locus in the Gorilla

Contrary to the high polymorphism and variability observed at the *HLA-B* locus, the gorilla *B* locus seems to have relatively limited polymorphism, characterized by a low number of nucleotide differences between the alleles. In an analysis of four unrelated lowland gorillas, Lawlor and co-workers (1991) isolated the same allele, *Gogo-B*0103*, from two individuals and three of the four distinct alleles (*Gogo-B*0101*, *-B*0102* and *-B*0103*) only differed by one to four nucleotide substitutions. The high sequence similarity among *Gogo-B*0101*, *-B*0102* and *-B*0103* indicates that they may have been derived from the same ancestral allele.

One of the gorilla *B* locus alleles is quite unlike the other *Gogo-B*01* alleles and may have a different evolutionary origin. The *Gogo-B*0201* allele differs from the *Gogo-B*01* group by 58–60 nucleotide substitutions. This allele shares motifs with *Patr-B*03* and with the *HLA-B*40* family. *Gogo-B*0201* also has an additional amino acid in the transmembrane region that is present in *Patr-B*01* and *Patr-B*03* alleles (Lawlor et al. 1988; Mayer et al. 1988). Thus, the origin of *Gogo-B*0201* may predate the divergence of these two ape species.

C Locus in the Gorilla

The five *Gogo-C* alleles described by Lawlor and co-workers (1991) fall into two distinct groups: one group is represented by *Gogo-C*0101* and *Gogo-C*0102* and the other by *Gogo-C*0201*, *Gogo-C*0202* and *Gogo-C*0203* alleles. The two alleles of the first group differ by only one synonymous (silent) substitution. Among the alleles of the second group, *Gogo-C*0203* and *Gogo-C*0201* differ by one nonsynonymous substitution in exon 1. *Gogo-C*0202* and *Gogo-C*0203* differ by a cluster of five nucleotide substitutions, probably the result of a segmental exchange event. The *Gogo-C*02* group is similar to *HLA-C*07* whereas the *Gogo-C*01* group does not correspond to any of the human *HLA-C* alleles. Interestingly, the unique *C* locus allele described in chimpanzees (Lawlor et al. 1990) clusters with the *Gogo-C*01* group.

Differences between the two groups of *Gogo-C* alleles are found in every domain with a bias in the 3' region of the coding region (encoding the α3, transmembrane and cytoplasmic domains). This pattern of substitutions is unlike the usual pattern seen among alleles of the same locus where differences are clustered in exons 2 and 3 (Parham et al. 1989). This might indicate that the two groups of *Gogo-C* sequences are not alleles but products of distinct but closely related genes. However, there is no strong evidence to support this hypothesis.

MHC Class I Genes in the Orangutan and the Gibbon

Orangutans (*Pongo pygmaeus*) and gibbons (*Hylobates lar*) last had a common ancestor with humans at a much earlier time than did the chimpanzee and gorilla (Miyamoto et al. 1987; Miyamoto and Goodman 1990; Goodman 1991). Comparing the orangutan and gibbon MHC class I genes with their human counterparts, Chen and co-workers (1992) found evidence for *HLA-A* and *-B* homologues in these two species indicating that the *A* and *B* loci are at least 30 million year old. Surprisingly, three *B* locus alleles were expressed by the lymphocytes from an orangutan, suggesting that there were two *B* loci in this individual. No evidence for the presence of *HLA-C* homologues was found in either the orangutan and gibbon, indicating that the *C* locus may be absent in these primates.

Several amino acid motifs found in the peptide-binding region (PBR) of human MHC class I molecules were found in the PBR of orangutan and gibbon MHC class I molecules. This suggests that new variants may be maintained by selection, and recombination mechanisms may then assort these new variant residues in several new combinations. New combinations that confer selective advantage may then be maintained en bloc. Thus, in addition to selection for variability at the PBR, there may be an additional selection for conservation of specific amino acid motifs within the PBR.

MHC Class I Genes in Macaques

It has been difficult to generate useful reagents for the serological analysis of MHC class I in macaques. Alloantisera have been generated for MHC characterization in the cynomolgus monkey (*Macaca fascicularis*) and pigtailed macaques (*Macaca nemestrina*) (Keever and Heise 1985a,b; Heise et al. 1987; Gaur et al. 1989). However, these sera do not define all the MHC class I molecules of these species. One-dimensional isoelectrofocusing (1-D IEF) analysis of immunoprecipitated rhesus macaque (*Macaca mulatta*) MHC class I molecules showed that rhesus lymphocytes express at least three MHC class I molecules with considerable polymorphism (Watkins et al. 1988a). Rhesus macaques express the products of MHC class I loci homologous to the *HLA-A* and *HLA-B* loci (Yasutomi et al. 1995; Watanabe et al. 1994; Miller et al. 1991; Boyson et al., 1996a). Analysis of rhesus monkey haplotypes indicates that both the *A* and *B* loci can be duplicated in rhesus monkeys (R. Bontrop, personal communication; Boyson et al., 1996a). Indeed, some haplotypes can have up to three *B* loci and two *A* loci. Additionally, there is no evidence for the *C* locus in the rhesus monkey, suggesting that this locus is of fairly recent origin in humans and apes.

The rhesus monkey classical MHC class I alleles do not appear to be related to any of the classical human MHC class I allelic lineages (Fig. 2). Indeed, in gene trees of exon 2 of primate classical MHC class I alleles, all of the rhesus monkey *A* alleles cluster outside the six different families of human *A* locus alleles. Similarly, the majority of the rhesus monkey *B* locus alleles do not cluster with the other primate *B* locus alleles. The exception to this pattern of clustering is *Mamu-B*01*, which clusters with human, chimpanzee and gorilla *B* locus alleles. However, comparison of this rhesus monkey allele to its most similar human allele *HLA-B*2702* shows that they differ by 48 nucleotides in exon 2 and 3. This number of differences is as large as the number of nucleotide differences separating the most divergent *HLA-B* locus alleles (Lawlor et al. 1988). Interest-

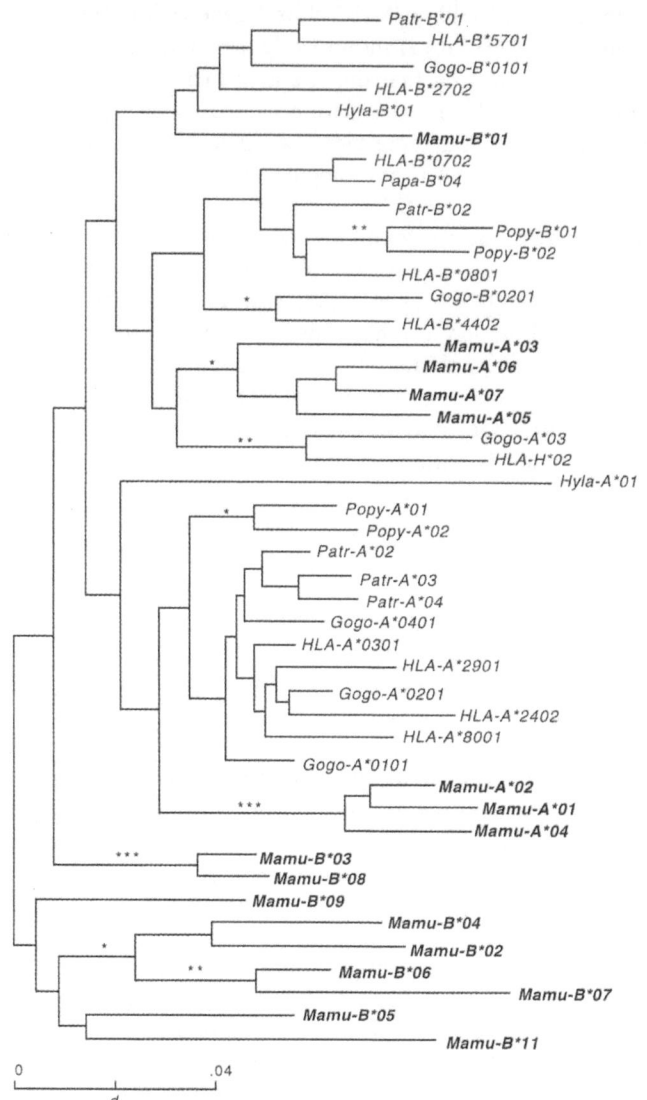

Fig. 2. Gene tree based on exon 2 of the Catarrhini classical MHC class I alleles. Rhesus monkey MHC class I alleles (*bold*) do not evolve in a trans-specific fashion. A subset of *Mamu-A* alleles cluster with *B* locus alleles and are related to *HLA-H* and *Gogo-A*03*. The gene tree was constructed using the neighbor-joining method (Saitou and Nei 1986) based on the number of nucleotide substitutions per site (*d*), estimated by the Jukes and Cantor method (Jukes and Cantor 1969). Test of the hypothesis that the branch length equals zero: *P <0.05; **P <0.01; ***P <0.001 (Rzhetsky and Nei 1992)

ingly, some of the rhesus monkey A locus alleles (*Mamu-A*03, -A*05, -A*06* and *-A*07*) cluster with exon 2 of other primate B locus alleles, suggesting that these rhesus monkey A locus alleles were the product of an ancient recombination event. Additionally, these rhesus monkey A locus alleles are also closely related to *HLA-H* and *Gogo-A*03*.

Nonclassical MHC Class I Genes in the Catarrhini

G Locus

Homologues of *HLA-G* are present in chimpanzees and gorillas (Corell et al. 1994). DNA sequence comparisons of exon 2 shows that *Patr-G* and *Gogo-G* differ from *HLA-G*0101* by two and one nonsynonymous substitutions, respectively.

Attempts to isolate *HLA-G* homologues using RT/PCR with a variety of different primers in the rhesus monkey have met with limited success. Use of primers specific for *HLA-G* at high annealing temperatures has led to the isolation of a family of unusual MHC class I cDNAs from rhesus monkey trophoblast, amniotic membrane and fetal liver tissues (Boyson et al., 1996b). The cDNA encoding this molecule has a stop codon at the beginning of the cytoplasmic domain, one amino acid away from the stop codon found in *HLA-G* (Geraghty et al. 1987). In addition to this unusual cDNA, a pseudogene with numerous premature stop codons and frameshift mutations has been isolated from the rhesus monkey. Interestingly, this cDNA has significant sequence identity to *HLA-G* in exon 2 (Boyson et al. 1996b). This might suggest that the homologue of *HLA-G* has been inactivated in the rhesus monkey and another MHC class I gene has become the functional *HLA-G* analog in this species.

E Locus

The E locus is remarkably well conserved in primates. This is especially evident in the putative PBR of these molecules. An E molecule from the orangutan differs by only nine amino acids substitutions from its human homologue in the α_1 and α_2 domains (Lawlor et al. 1990).

Several *HLA-E* homologues have been isolated from macaques and comparison with their human counterparts sheds some light on the evolution of the E locus (Boyson et al. 1995). Interestingly, the putative PBR of the E molecule is well conserved between humans and macaques. This conservation of the PBR contrasts with regions of the E molecule outside of the PBR that appear to have diverged more rapidly. Furthermore, comparison of the rhesus and human intron B (the intron between exons 2 and 3) revealed that this region of the gene has diverged significantly more rapidly than the exons have. This conservation of the PBR encoded by the E locus in primates suggests that the E molecule may perform a specialized function.

F Locus

An *HLA-F* homologue has been cloned from the chimpanzee (Lawlor et al. 1988). This allele was similar to its human counterpart, with only six amino acids separating the Patr-F molecule from the HLA-F molecule. Furthermore, the chimpanzee F molecule differs from the human F molecule by only one amino acid in the α1 and 2 domains that fold to form the PBR. An *HLA-F* locus homologue has also been cloned from the rhesus monkey (Otting and Bontrop 1993). Again this nonclassical gene seems to be remarkably well conserved between humans and rhesus monkeys.

MHC Class I Genes in New World Primates

Fossil evidence for the New World primates (Platyrrhini) was found in the early Oligocene with no previous record of their presence in South America (Simpson 1980). The earliest known primate found in South America (*Branisella boliviana*) lived at La Salla, Bolivia, 34–35 million years ago (Hoffstetter 1969). New World primates appear to be a monophyletic group with the Catarrhini being their closest sister group (Ford 1986). Although the primate ancestor of the Platyrrhini is yet unknown and there is no agreement about the origin of the Platyrrhini (Conroy 1990), accumulating geological and biogeographical data suggest that Africa is the most likely origin of New World primates. Since the South American continent was isolated by water in the early Oligocene, the founder population of the Platyrrhini may have been rather small.

MHC Class I Genes in Tamarins and Marmosets

The cotton-top tamarin (*Saguinus oedipus*) is unusual among the primates because it expresses MHC class I genes with limited variability and polymorphism (Watkins et al. 1988b, 1990a, 1991). All captive and wild cotton-top tamarins analyzed thus far (more than 120 unrelated individuals) express the same MHC class I allele *Saoe-G*08*. Additionally, two other alleles, *Saoe-G*04* and *Saoe-G*06*, are found at a frequency of more than 90 %. Comparative analysis of the interlocus and intralocus variability in the cotton-top tamarin, human and mouse classical MHC class I genes shows that nucleotide sequences of the cotton-top tamarin are significantly less variable than those of human and mouse. Whether the cotton-top tamarins' restricted MHC class I variability has something to do with their extreme susceptibility to viral infections remains to be formally proven. Another unusual aspect of the functional MHC class I genes in the cotton-top tamarin is that they are more similar to the nonclassical *HLA-G* gene than they are to the classical *HLA-A*, *-B* or *-C* genes (Watkins et al. 1990b) (Fig. 3). Indeed, when gene trees are constructed with the human and the cotton-top tamarin MHC class I genes, all the cotton-top tamarin genes cluster

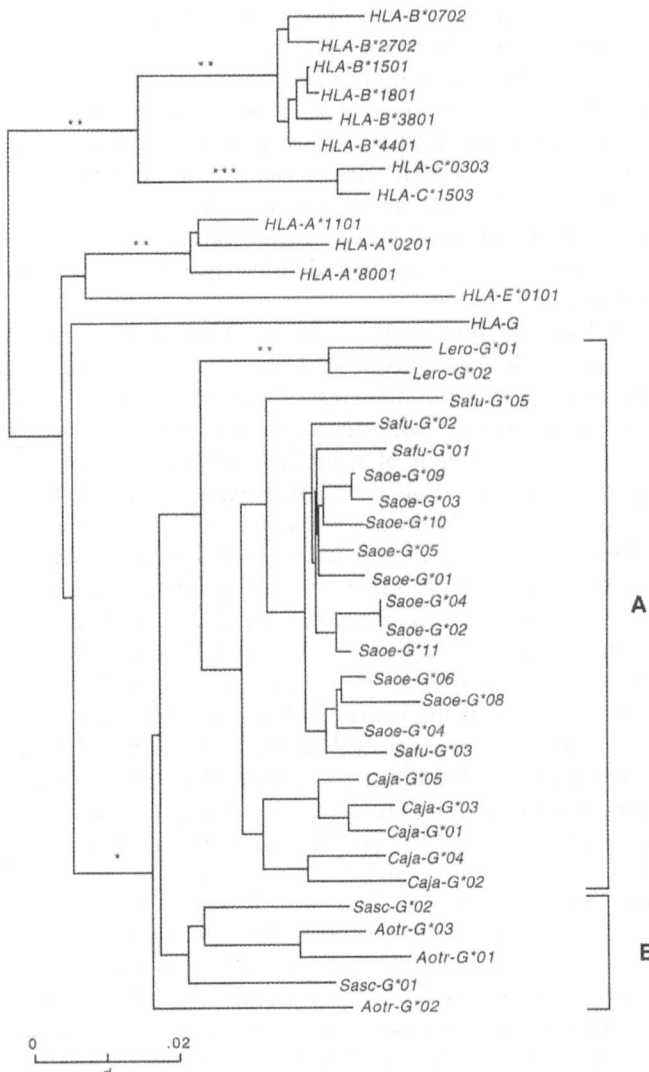

Fig. 3. The MHC class I genes of New World primates are related to *HLA-G*. *A*, Species-specific distribution of MHC class I alleles in Callitrichidae species. *B*, In the non-Callitrichidae New World primate species the MHC class I loci are distributed in a trans-specific fashion. Neighbor-joining tree of exons 4–8 based on number of nucleotide substitutions per site (*d*). *Saoe, Saguinus oedipus* (cotton-top tamarin); *Safu, Saguinus fuscicollis* (saddle-back tamarin); *Lero, Leontopithecus rosalia* (golden-headed tamarin); *Caja, Callitrhix jacchus* (common marmoset); *Sasc, Saimiri sciureus* (squirrel monkey); *Aotr, Aotus trivirgatus* (owl monkey). Sequences are from unpublished data from our laboratory. Test of the hypothesis that the branch length equals zero: *P <0.05; **P <0.01; ***P < 0.001 (Rzhetsky and Nei 1992)

with the nonclassical *HLA-G* gene rather than with the classical human MHC class I genes. Additionally, there are eight HLA-G-specific amino acids substitutions found in the cotton-top tamarin MHC class I sequences. This suggests that the MHC class I genes in the cotton-top tamarin may have been derived from a gene that was the ancestor of *HLA-G*. No evidence for the presence of homologues of the human classical MHC class I genes have been found in the cotton-top tamarin genome. A cDNA from a locus homologous to *HLA-F* has also been isolated from the cotton-top tamarin indicating that all of the cotton-top tamarin MHC class I genes have evolved from ancestral homologues of human nonclassical MHC class I genes.

The unusual placentation of the cotton-top tamarin may have something to do with the tamarin's limited MHC class I diversity. Most tamarins and marmosets are born as bone marrow chimeras (Benirschke et al. 1962). During gestation, placental fusion occurs with anastamosis of the placental blood vessels, resulting in a cross-circulation of genetically distinct hematopoietic elements (Wislocki 1939). It seems possible then that expression of MHC class I genes with limited diversity might help to maintain stable tolerance between two genetically distinct lymphocytes populations. 1-D IEF and RFLP analysis of the MHC class I loci of different tamarin and marmoset species showed that all of them, except the cotton-top tamarin and the red-crested tamarin (*S. geoffroyi*), express polymorphic MHC class I glycoproteins (Watkins and Letvin 1991). Ten different MHC class I molecules were expressed by the lymphocytes of nine unrelated red-crested tamarins. Interestingly, all red-crested tamarins expressed an MHC class I molecule with an isolectric point identical to an MHC class I molecule expressed by all cotton-top tamarins (*Saoe-G*08*). The limited MHC class I polymorphism seen in the cotton-top tamarin and in the red-crested tamarin is, therefore, unique among the Callitrichidae, suggesting that limited polymorphism at these loci may not be necessary for the establishment of stable bone marrow chimerism. This analysis of MHC class I polymorphism, however, does not rule out the possibility that these other species of tamarins and marmosets express MHC class I molecules with limited nucleotide variation.

To determine whether the bone marrow-chimeric callitrichids express MHC class I alleles with limited variation, we cloned, sequenced and expressed MHC class I genes from the saddle-back tamarin (*Saguinus fuscicollis*), the golden-headed lion tamarin (*Leontopithecus rosalia*) and the common marmoset (*Callithrix jacchus*). The functional MHC class I genes in these Callitrichidae species are also similar to the human nonclassical *HLA-G* gene (Cadavid et al., manuscript submitted). Additionally, there was no evidence for the presence of the human classical MHC class I homologues in these Callitrichidae. Surprisingly, the MHC class I genes of all of the marmosets and tamarins exhibited low nucleotide diversity. This is consistent with the hypothesis that bone marrow chimerism may be selecting for limited MHC class I diversity in the Callitrichidae.

Unlike all other primate MHC class I alleles, the alleles of the different species of Callitrichidae clustered together in a species-specific fashion (Fig. 3). Indeed, alleles from the closely related saddle-back tamarins and the cotton-top tamarins clustered together and this cluster was separate from the common marmosets' MHC class I alleles. Since common marmosets and tamarins are probably as closely related as chimpanzees and humans, this pattern of clustering is puzzling. The *A* locus alleles of the human and chimpanzee cluster together as do their *B* locus alleles. There is no species-specific clustering in humans and chimpanzees. Even the *A* locus alleles of the rhesus monkey cluster with the *A* locus alleles of the human. These two species last had a common ancestor at least 35 million years ago. That MHC class I loci in tamarins and in marmosets cluster in separate clades may indicate that these two species express a different set of MHC class I genes. This possibility is supported by the high sequence similarity observed between the cotton-top tamarin MHC class I pseudogene *So-N3* (Watkins et al. 1990b) and the expressed MHC class I genes in the common marmoset (Cadavid et al., manuscript submitted). This similarity might indicate that *So-N3* is the inactive counterpart of the common marmoset functional MHC class I genes. These data suggest that a process of gene duplication and inactivation with differential usage of the duplicated genes has been part of the evolutionary history of Callitrichidae MHC class I genes. Although a high frequency of segmental exchange may be responsible for homogenization of MHC class I loci within each Callitrichidae species (concerted evolution; Dover 1982; Rada et al. 1990), the presence of a homologue of *HLA-F* in the cotton-top tamarin argues against this possibility.

MHC Class I Genes of the Owl Monkey and the Squirrel Monkey

Different karyotypic races of owl monkeys (*Aotus* sp.) have two distinct forms of β2-microglobulin (Smith et al. 1984). It was first noted that the W6/32 antibody generated against human MHC class I molecules reacted with MHC class I molecules of Old world primates and apes, yet in New World primates this antibody only reacted with owl monkey MHC class I molecules. Furthermore, W6/32 only reacted with the lymphocytes from certain defined karyotypic races (karyotypes VI and V). It was later discovered that these different owl monkey karyotypes express β2-microglobulins with different isoelectric points and these differences correlated with W6/32 reactivity and karyotypic group (Parham and Ploegh 1980). 1-D IEF analysis has showed that lymphocytes from owl monkeys and squirrel monkeys (*Saimiri sciureus*) express at least three polymorphic MHC class I molecules (Watkins et al. 1988a). Additionally, sequence analysis of the expressed MHC class I genes of these two species has shown that they are similar to the human nonclassical *HLA-G* gene. Interestingly, phylogenetic analysis of MHC class I alleles in the owl monkey and the squirrel monkey shows a trans-species mode of evolution of their MHC class I loci, indicating that the species-specific clustering of MHC class I alleles seen in tamarins and marmosets is restricted to the Callitrichidae. This analysis also suggests that the owl monkey and the squirrel monkey express similar MHC class I loci which are different from those expressed by the Callitrichidae. The inter- and intra-species variability of MHC class I alleles in the owl monkey and squirrel monkey, although lower than that observed in apes and Old World primates, is higher than the variability displayed by the Callitrichidae MHC class I alleles.

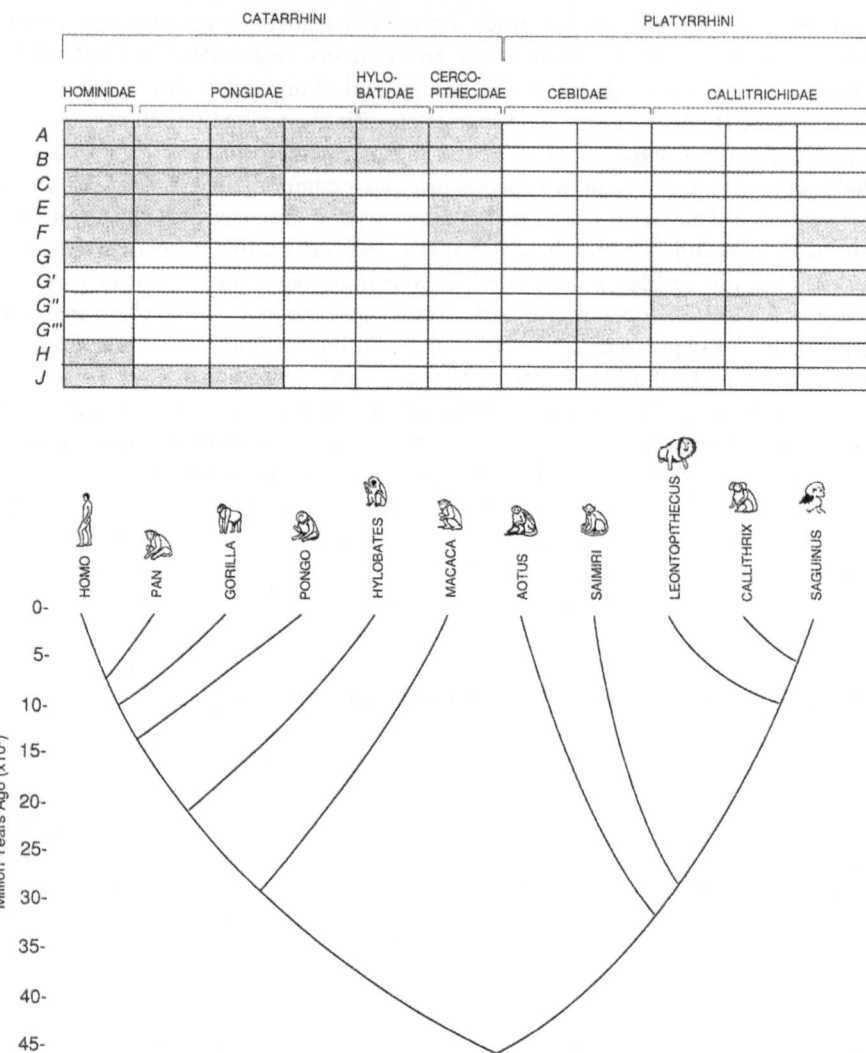

Fig. 4. Contrasting evolutionary histories of the MHC class I loci in Catarrhini and Platyrrhini. The *A* and *B* loci are found in humans, apes and Old World primates, whereas in the New world primates, the MHC class I loci appear to have evolved by a continual process of duplication and subsequent inactivation or expansion of loci. The functional MHC class I loci of the New World primates are more similar to the ancestor of the nonclassical *HLA-G* gene

Conclusions

The study of MHC class I genes in nonhuman primates has contributed to our understanding of the evolution of this multigene family. Since the human MHC class I loci are the most extensively studied among primates, it has served as a point of reference for the analysis of this gene family in other primates. A comparison of the nonhuman primate MHC class I loci shows that the classical MHC class I *A* and *B* loci are conserved in all Catarrhini species (Fig. 4) and, interestingly, they are found in multiple copies in Old World primates and in one of the ape species. Although the *A* and *B* loci are well preserved in the Catarrhini, allelic lineages at these loci do not appear to be shared by humans and rhesus monkeys. The *C* locus is restricted to humans, chimpanzees and gorillas and no evidence exists for the presence of this locus in other primate species. This may indicate that the *C* locus is of relatively recent origin. Given its similarity to the *B* locus, it is likely that a duplication event of the ancestral *B* locus was responsible for the origin of the *C* locus.

The New World primates express MHC class I genes that are related to *HLA-G*, and no evidence for the presence of catarrhine classical MHC class I genes has as yet been found in the Platyrrhini. The small founding population of the New World primates may have played a crucial role in shaping their MHC class I diversity. Additionally, it appears that the MHC class I genes of tamarins and marmosets have been subjected to repeated cycles of duplication and inactivation or expansion, causing the species-specific distribution of alleles and loci in these species. The MHC class I region in New World primates, therefore, appears to have evolved by a continual process of expansion and contraction of loci that coincided with the adaptive radiation of major phyletic groups. The distribution of polymorphism in these species may have been determined principally by the size of the founder population.

Acknowledgements. We thank Dr. Leslie Knapp and Jonathon Boyson for critical review of this manuscript. This work was supported by National Institutes of Health grants (RR00167, HD34215, DK44886 and HD34215) and a Biomedical Science Grant from the Arthritis Foundation.

Not added in proof: An HLA-E orthologue has been identified in the Cotton-top tamarin (Knapp et al., Manuscript Submitted). An HLA-B orthologue has also been identified in the saki monkey and in the spider monkey (Cadavid et al., Manuscript Submitted).

References

Balner H, van Leeuwen A, Dersjant H, van Rood JJ. Defined leukocyte antigens in chimpanzees: use of chimpanzee isoantisera for leukocyte typing in man. Transplantation 5:624–2, 1967

Balner H, van Vreeswijk W, Roger JH, D'amaro J. The major histocompatibility complex in chimpanzees: identification of several new antigens controlled by the the A and B loci of ChLA. Tissue Antigens 12:1–18, 1978

Belich MP, Madrigal JA, Hidebrand WH, Zemmour J, Williams RC, Luz R, Petzl-Erler ML, Parham P. Unusual HLA-B alleles in two tribes of Brazilian Indians. Nature 357:326–329, 1992

Benirschke K, Anderson JM, Brownhill LC. Marrow chimerism in marmosets. Science 138:513–515, 1962

Boyson JE, Shufflebotham C, Cadavid LF, Urvater JA, Knapp LA, Hughes AL, and Watkins DI. The MHC Class I Genes of the Rhesus Monkey: Different Evolutionary Histories of MHC Class I and II Genes in Primates. J. Immunol. 156:4656–4665, 1996a

Boyson JE, Iwanaga KK, Golos TG, and Watkins DI. Identification of the Rhesus Monkey HLA-G Ortholog: Mamu-G is a Pseudogene. J. Immunol. 157:5428–5437, 1996b

Boyson JE, McAdam SN, Gallimore A, Golos TG, Liu X, Gotch FM, Hughes AL, Watkins DI. The MHC E locus in macaques is polymorphic and is conserved between macaques and humans. Immunogenetics 41:59–68, 1995

Bruce EJ, Ayala FJ. Phylogenetic relationships between man and apes:electrophoretic evidence. Evolution 33:1040–1052 1979

Chang L-YE, Slightom JL. Isolation of and nucleotide sequence analysis of the b-type globin pseudogene from human, gorilla and chimpanzee. J Mol Evol 180:767–784, 1984

Chen ZW, McAdam SN, Hughes AL, Dogon AL, Letvin NL, Watkins DI. Molecular cloning of orangutan and gibbon MHC class I cDNA: the HLA-A and -B loci diverged over 30 million year ago. J Immunol 148:2547–2554, 1992

Chen ZW, Hughes AL, Ghim SH, Letvin NL, Watkins DI. Two more chimpanzee Patr-A locus alleles related to the HLA-A1/A3/A11 family. Immunogenetics 38:238–240, 1993

Conroy GC. Primate evolution. Norton, New York, 1990

Corell A, Morales P, Martínez-Laso J, Martín-Villa JM, Varela P, Paz-Artal E, Allende LM, Rodríguez C, Arniz-Villena A. New species-specific alleles at the primate MHC-G locus. Hum Immunol 41:52–55, 1994

Domena JD, Hildebrand WH, Bias WB, Parham P. A sixth family of HLA-A alleles defined by HLA-A*8001. Tissue Antigens 42:156–159, 1993

Dorf ME, Metzgar RS. Medical primatology. Proceedings of the 2nd conference on experimental medical surgery of primates. The distribution of human HLA-A antigens in chimpanzees and gorillas, New York, 1970

Dover GA. Molecular drive: a cohesive mode of species evolution. Nature 299:111–117, 1982

Fan WM, Kasahara M, Gutknecht J, Klein D, Mayer WE, Jonker M, Klein J. Shared class II MHC polymorphism between humans and chimpanzees. Hum Immunol 26:107–121, 1989

Ferris SD, Brown WM, Davidson WS, Wilson AC. Extensive polimorphism in the mitochondrial DNA of apes. Proc Natl Acad Sci USA 78:6319–6322, 1981

Ford SM. Systematics of the New World Monkey. In: Swindler DR, Erwin J (eds) Comparative priamte biology, vol 1. Liss, New York, pp 73–135, 1986

Gaur LK, Bowden DM, Tsai C-C, Davis A, Clark EA. The major histocompatibility complex MnLA of pigtailed macaques: definition of fifteen specificities. Hum Immunol 24:277–294, 1989

Geraghty DE, Koller DH, Orr HT. A human histocompatibility complex class I gene that encodes a protein with a shortened cytoplasmic segment. Proc Natl Acad Sci USA 84:9145–9149, 1987

Goodman M. Molecular evolution in primates. In: Ehara TKA, Takenaka O, Iwamoto M (eds) Primatology today. Elsevier, New York, pp 11–18, 1991

Heise ER, Cook DJ, Shepart BS, Manning CH, McMahan MR, Chedid M, Keever CA. The major histocompatibility complex of primates. Genetica 73:53–68, 1987

Hershkovitz P. Living new world monkeys (platyrrhini) with an introduction to primates, vol 1. University of Chicago Press, Chicago, 1977

Hoffstetter R. Un primate de l'Ologocene inferieour sudamericain:Branisella boliviana gen et sp nov. CR Acad Sci 69:434–437, 1969

Hughes AL. Positive selection and interallelic recombination at the merozoite surface antigen-1 (MSA-1) locus of plasmodium. Mol Biol Evol 9:381–393, 1992

Jukes TH, Cantor RC. Evolution of protein molecules. In: Munro HN (ed) Mammalia protein methabolism. Academic, New York, 1969

Kay RF. The phyletic relationships of exatant and fossil pitheciinae (platyrrhini anthropoidea). J Hum Evol 19:175–208, 1990

Keever CA, Heise ER. The major histocompatibility complex in cynomolgus monkeys: absorption of 24 CyLA antisera. Hum Immunol 12:75–90, 1985a

Keever CA, Heise ER. The major histocompatibility complex in cynomolgus monkey. II. Polymorphism at three serologically defined loci and correlation of haplotypes with stimulation in MLC and skin graft survival. Hum Immunol 12:143–164, 1985b

King MC, Wilson AC. Evolution at two levels in humans and chimpanzees. Science 188:107–116, 1975

Lawlor DA, Ward FE, Ennis PD, Jackson AP, Paharm P. HLA-A and HLA-B polymorphism predate the divergence of humans and chimpanzees. Nature 335:268–271, 1988

Lawlor DA, Warren E, Ward FE, Parham P. Comparision of class I MHC alleles in human and apes. Immunol Rev 113:147–185, 1990

Lawlor DA, Warren E, Taylor P, Parham P. Gorilla class I major histocompatibility complex alleles:comparision to human and chimpanzee class I. J Exp Med 174:1491–1509, 1991

Lawlor DA, Edelson BT, Parham P. MHC-A locus molecules in pygmy chimpanzees: conservation of peptide pockets. Immunogenetics 42:291–295, 1995

Lucotte G. Chimpanzees show less variation than man: electrophoretic evidence. J Hum Evol 12:419–424, 1983

Martin RD. Primate origins and evolution: a phylogenetic reconstruction. Chapman and Hall, London, 1990

Mayer WE, Jonker M, Klein D, Ivanyi P, van Seventer G, Klein J. Nucleotide sequences in chimpanzee MHC class I alleles: evidence for trans-species mode of evolution. EMBO J 7:2765–2774, 1988

McAdam SN, Boyson JE, Liu X, Garber T, Hughes AL, Bontrop RE, Watkins DI. A uniquely high level of recombination at the HLA-B locus. Proc Natl Acad Sci USA 91:5893–4897, 1994

McAdam SN, Boyson JE, Liu X, Garber TL, Hughes AL, Bontrop RE. Watkins DI. Chimpanzee MHC class I A locus alleles are related to only one of the six families of human A locus alleles. J Immunol 154:6421–6429, 1995

Miller MD, Yamamoto H, Hughes AL, Watkins DI, Letvin NL. Definition of an epitope and an MHC class I molecule recognized by gag-specific cytotoxic T-limphocytes in SIVmac-infected rhesus monkeys. J Immunol 147:320–329, 1991

Miyamoto MM, Goodman M. DNA systematics and evolution in primates. Annu Rev Ecol Syst 21:197–214, 1990

Miyamoto MM, Slightom JL, Goodman M. Phylogenetic relations of humans and African apes from DNA sequences in psi eta-globin region. Science 238:369–373, 1987

Otting N, Bontrop RE. Charactirization of the rhesus macaque (Macaca mulatta) equivalent of HLA-F. Immunogenetics 38:141–145, 1993

Parham P, Ploegh HL. Molecular characterization of HLA-A B homologues in owl monkeys and other nonhuman primates. Immunogenetics 11:131–143, 1980

Parham P, Lawlor DA, Lomen CE, Ennis PD. Diversity and diversification of HLA-A B C alleles. J Immunol 142:3937–3950, 1989

Rada C, Lorenzini R, Powis SJ, van den Bogearde J, Parham P, Howard JC. Concerted evolution of class I genes in the major histocompatibility complex of murine rodents. Proc Natl Acad Sci USA 87:2167–2171, 1990

Rosemberger AL. Fossil new world monkeys dispute the molecular clock. J Hum Evol 13:737–742, 1984

Rzhetsky A, Nei M. A simple method for estimating and testing minimun-evolution trees. Mol Biol Evol 9:945–967, 1992

Saitou N, Nei M. The neighbor joining method: a new method for reconstructiong phylogenetic trees. Mol Biol Evol 4:406–425, 1986

Schneider H, Schneider MPC, Sampaio I, Harada ML, Stanhope M, Czelusniak J, Goodman M. Molecular phylogeny of the new world monkeys (platyrrhini primates). Mol Phy Evol 2:225–242, 1993

Simpson GG. Splendid isolation. Yale University Press, New Haven, 1980

Smith KK, Parham P, Ma NSF. Two distinct forms of beta2-microglobulin in different races of owl monkey (Aotus trivirgatus). Immunogenetics 20:459–464, 1984

Smith NH, Beltran P, Selander RK. Recombination of salmonella phase 1 flagellin genes generates new serovars. J Bacteriol 172:2209–2216, 1990

Starling GC, Witkowski JA, Speerbrecher LS, McKinney SK, Hansen JA, Choo SY. A novel HLA-A*8001 allele identified in an African-American population. Hum Immunol 39:163–168, 1994

Tamura K, Nai M. Stimation of the number of nucleotide substitutions in the control region of mitochondrial DNA in humans and chimpanzees. Mol Biol Evol 10:512–526, 1993

Wagner AG, Hughes AL, Iandoli ML, Stewart D, Herbert S, Watkins DI, Hurly CK, Rosen-Brosen S. HLA-A*8001 a member of a newly discovered ancient family of HLA-A alleles. Tissue Antigens 42:522–529, 1993

Watanabe N, McAdam SN, Boyson JE, Piekarczyk MS, Yasutomi Y, Watkins DI, Letvin NL. A simian immunodeficiency virus envelope V3 cytotoxic T-lymphocyte epitope in rhesus monkey and its restricting major histocompatibility camplex class I molecule Mamu-A*02. J Virol 68:6690–6696, 1994

Watkins DI, Kannagi M, Stone ME, Letvin NL. Major histocompatibility complex class I molecules of nonhuman primates. Eur J Immunol 18:1425–1432, 1988a

Watkins DI, Hodi FS, Letvin NL. A primate species with limited major histocompatibility complex class I polymorphism. Proc Natl Acad Sci USA 85:7714–7718, 1988b

Watkins DI, Letvin NI, Hughes AL, Tedder TF. Molecular cloning of cDNA that encode MHC class I molecules from a new world primate (Saguinus oedipus): natural selection acts at points that may affect the peptide presentation to T cells. J Immunol 144:1136–1143, 1990a

Watkins DI, Chen ZW, Hughes AL, Evans MG, Tedder TF, Letvin NL. Evolution of MHC class I genes of a New World primate from ancestral homologues of human nonclassical genes. Nature 346:60–63. 1990b

Watkins DI, Letvin NL. Limited MHC polymorphism is not essential for bone marrow chimerism in New World primates. Immunogenetics 33:194–197, 1991

Watkins DI, Chen ZW, Garber TL, Hughes AL, Letvin NL. Segmental exchange between MHC class I genes in a higher primate:recombination in the gorilla between the ancestor of a human non-functional gene and an A locus gene. Immunogenetics 34:185–191, 1991

Watkins DI, McAdam SN, Liu X, Strang CR, Milford EL, Leviene CG, Garber TL, Dogon AL, Lord CI, Ghim SH, Troup GM, Hughes AL, Letvin NL. New recombinat HLA-B a tribe of South American Amerindians indicate rapid evolution of MHC class I loci. Nature 357:329–333, 1992

Wislocki GB. Observations on twinning marmosets. Am J Anat 64:445–459, 1939

Yasutomi Y, McAdam SN, Boyson JE, Piekarczyk MS, Watkins ID, Letvin NL. A MHC class I B locus allele-restricted simian immunodeficiency virus envelope CTL epitope in rhesus monkey. J Immunol 154:2516–2522, 1995

Zemmour J, Koller BH, Ennis PD, Geraghty DE, Lawlor DA, Orr HT, Parham P. HLA-AR an inactivated antigen-presenting locus related to HLA-A. J Immunol 144:3619–3629, 1990

3 *Mhc* Class II Genes of Nonhuman Primates

The *DR* Loci

R.E. BONTROP

Back to the Roots of HLA-D/DR

In humans the HLA-D region was originally discovered by use of the mixed lymphocyte culture (MLC) test. This assay is based on coculturing lymphocyte suspensions of two unrelated individuals and actually measures the proliferative capacity of T lymphocytes. Normally one uses a so-called one way stimulation, meaning that one of the two cell populations has been inactivated, for instance, by irradiation. As a result the treated cells are unable to proliferate but still have the capacity to stimulate T cells of the second individual. Lymphocytes from HLA-identical siblings generally do not stimulate each other in the MLC whereas lymphocytes from HLA nonidentical individuals do. Since serologically typed HLA-A, -B and -C identical individuals could stimulate each others' cells, the incompatibility measured in the MLC was assigned to the *HLA-D* locus on chromosome 6. MHC antigens are inherited in a codominant fashion and as a consequence a given individual may be heterozygous for its *HLA-D* region products. Lymphocytes derived from donors descending from consanguineous offspring may be truly homozygous for the HLA region. Such homozygous typing cells (HTCs) were initially used to investigate and to inventory the polymorphism of *HLA-D* locus products. By now, more than 20 different HLA-D specificities have been identified in the human population. In parallel, MLC stimulating determinants have been documented for the chimpanzee and rhesus macaque (Jonker and Balner 1980). An apparently new set of antigens was defined using sera from multiparous women from which all HLA-A, -B and -C reactivity had been

Molecular Biology and Evolution of Blood Group
and MHC Antigens in Primates
Blancher/Klein/Socha (Eds.)
© Springer-Verlag Berlin Heidelberg 1997

depleted. These allo-antisera reactions were found to have good concordance with *HLA-D* alleles but, moreover, were able to inhibit MLC reactivity. The serologically defined B lymphocyte cell surface antigens detected in this way were designated HLA-DR, standing for HLA-D related.

The introduction of monoclonal antibodies which recognize MHC class II determinants was a great help in elucidating the various sorts of relationships that exist between different types of antigenic determinants. Immunoprecipitation studies demonstrated that the HLA-DR determinants are carried by a bimolecular complex consisting of an α (heavy) and a β (light) chain. According to population studies, the HLA-DR α-chain appeared to be invariant whereas the β-chain is polymorphic (Kaufman et al. 1984). Moreover, molecular and biochemical analysis provided evidence that the number of *HLA-DRB* genes and gene products expressed per haplotype is not constant (Böhme et al. 1985; Bontrop et al. 1986a).

In view of the fact that a close relationship appears to exist between HLA-D and -DR specificities it was postulated that both determinants are encoded by one locus (Termijtelen et al. 1982). The situation turned out to be even more complex since some HLA-D specificities were found to be included in one single HLA-DR serotype. For example, the HLA-DR4 specificity has been reported to be associated with multiple different cellularly defined clusters, notably HLA-Dw4, -Dw10, -Dw13, -Dw14 and -Dw15. Subsequent electrophoretic analyses demonstrated that two types of DR molecules could be isolated from HLA-DR4-positive B cells. One DR molecule turned out to be invariant and is shared between HLA-DR4, -DR7 and -DR9 positive individuals. The HLA-D typing disparities, however, correlated with the β-chain isoelectric point differences of a more polymorphic type of DR molecule (Nepom et al. 1983; Cairns et al. 1985; Bontrop et al. 1986b). Thus, antibodies that are present in HLA-DR4 allo-antisera may recognize a shared determinant on the DR molecules, whereas the MLC assay detected additional polymorphic T cell determinants on a different set of DR molecules. Similar observations were made for other HLA-D/DR specificities but also for Mamu-D/DR antigens in the rhesus macaque (Slierendregt et al. 1995).

HLA class II nucleotide sequence analyses and mapping studies indeed demonstrated that the *HLA-DR* subregion contains one α- and a differential number of β-chain genes. In brief, the HLA-DR8 family of haplotypes possesses only one *HLA-DRB* gene whereas two *-DRB* genes were found for the HLA-DR1 and -DR10 haplotypes. The HLA-DR2, -DR3, -DR5, -DR6 and the HLA-DR4, -DR7 and -DR9 group of haplotypes were found to possess three or four types of *HLA-DRB* genes, respectively. The various types of *DRB* genes have been denoted *HLA-DRB1–DRB9*. By convention the *HLA-DRB1* gene is present on all haplotypes whereas the other types of *-DRB* genes exhibit a more restricted haplotype distribution. Extensive sequencing studies and subsequent phylogenetic analysis demonstrated that the *Mhc-DRB* nucleotide sequences can be grouped into clusters or lineages. In humans, at least 124 *HLA-DRB1*, four *-DRB3*, five *-DRB4*, five *-DRB5*, three *-DRB6* and two *-DRB7* have been defined (Bodmer et al. 1995). From these data it is evident that the *HLA-DRB1* locus displays an extensive degree of polymorphism.

Allelic Diversity at the *Mhc-DRB* Loci in Apes

At present, the clade of the great apes comprises two species of chimpanzees, the gorillas and the orangutans, whereas the different species of gibbons are classified into the so called lesser apes. The common chimpanzee (*Pan troglodytes*) represents the ape species whose *Mhc* has been studied at considerable length. The ChLA system, now named *MhcPatr* (Klein et al. 1991), was discovered by means of serological typing reactions in the early 1970s (Balner et al. 1974). In contrast to MHC class I, allo-antisera that can be used for typing MHC class II polymorphisms in chimpanzees turned out to be difficult to generate. For that reason, first evidence that chimpanzees indeed have a *DR* region was based on cross-reactions observed with HLA-DR typing sera (Balner et al. 1977). The characterization of three full length *Patr-DRB* cDNAs clearly demonstrated that humans and chimpanzees share highly related alleles that group into lineages predating their speciation (Fan et al. 1989). As observed in humans, population studies manifested that the chimpanzee *DR* α-chain appears to be invariant, whereas the β-chain carries the allelic polymorphism (Bontrop et al. 1990a,b). By now, the most polymorphic unit; notably exon 2 of various *Patr-DRB* class II genes, has been studied extensively at the nucleotide level allowing the detection of chimpanzee orthologues for the *HLA-DRB1*02*, *-DRB1*03*, *-DRB1*07* and *-DRB*10* allelic lineages but also for the *HLA-DRB3*, *-DRB4*, *-DRB5*, *-DRB6* and *-DRB7* loci (Gyllensten et al. 1991; Kenter et al. 1992; Mayer et al. 1992). As found for the HLA system, the *Patr-DRB1* locus seems to be the most polymorphic class II locus in chimpanzees and by now at least 26 different alleles have been documented (O'hUigin et al. 1993). In addition, *Patr-DRB* alleles were detected for which no apparent human equivalents are known and these have been designated accordingly *Patr-DRB*W801* and *-DRB*W901–05*, respectively (O'hUigin et al. 1993). Equivalents of the *Mhc-DRB*W8* lineage are also present in gorillas whereas the *Mhc-DRB*W9* lineage may be unique to chimpanzees. Phylogenetic analyses showed that both the *Mhc-DRB*W8* and *W9* lineage are highly related to the *Mhc-DR3* lineage (Slierendregt and Bontrop 1994). These results not only emphasize that species may lose or gain lineages but also indicate that at a certain stage chimpanzees and gorillas shared a common ancestor.

Some loci which are oligomorphic in humans appear to display an extensive degree of polymorphism in chimpanzees. Only four and five alleles at the *HLA-DRB3* and *-DRB5* loci have been detected in humans whereas 17 *Patr-DRB3* and 15 *Patr-DRB5* equivalents have been described for the common chimpanzee (Table 1). This is even more striking if one realizes that less than 100 individuals have been analyzed. This marvel may be a reflection of the fact that the chimpanzee is an older species than modern humans or, alternatively, this high number of alleles may have been generated due to intensified selection constraints. In principle it is even possible that both factors had an impact.

At this stage a relatively large sample of chimpanzees originating from West Africa has been tested. The West African chimpanzee population shares many lineages with humans but seems to have lost the equivalents of the *HLA-DRB1*04* and *-DRB1*08* lineages (Bontrop et al. 1993). Alleles belonging to the

Table 1. Distribution of *Mhc-DRB* lineages and number of alleles among different primate species

Mhc-DRB	HLA	Patr/ Papa	Gogo	Popy	Mamu/ Mane	Camo	Caja	Ceap	Saoe	Sasc	Aotr	psm[a]
B₁*01/10	5	1	1		4							
B₁*02	11	13/2	1									
B₁*03[b]	85	10/2	8	2	14	4	1		5		2	+
B₁*04	22	–			5							+
B₁*07	1	2			1							
B₁*09	2	–	–									
B2	+	+	+									
B3	4	17	8		5	2			17		1	+
B4	5	7			1							
B5	5	15	7	4	4				1			
B6	3	10/3	7	2	11							+
B7	2	2										
B8	+	+	+									
B*W1	–				1/2							+
B*W2	–				1/2							
B*W3	–				6							+
B*W4/5	–				1							
B*W6	–				3							
B*W7	–				2							
B*W8	–	1	2									
B*W9	–	5										
B*W10	–		1									
B11	–					1	1		10			
B*W12	–						1		4	4		+
B*W13	–							3			2	
B*W14	–					2				3		
B*W15	–						2					
B*W16	–						3					
B*W17	–					1						
B*W18	–										1	
B*W19	–									3		
B*W20	–				2							+
B*W21	–				1							+
B*W22	–								9			

+, gene present; –, gene absent.

[a] The last column (psm) summarizes all the data available on prosimians (Figueroa et al. 1994).

[b] In humans the *Mhc-DRB1*03* lineage comprises the HLA-DR3, -DR5, -DR6 and -DR8 specificities.

*Mhc-DRB1*04* lineage have not been detected in any great ape species whereas highly related equivalents are present in humans and several Old World monkeys (O'hUigin et al. 1993). This may indicate that the *Mhc-DRB1*04* lineage was not only lost by chimpanzees of West African origin but more dramatically also by the other great ape species. Based on this limited sample size it is of course extremely difficult to make firm conclusions on the loss of allelic lineages. In contrast to humans, most great ape species have been studied only to a rather limited extent for their *DR* region variability. In the case of the pygmy chimpanzee (*Pan paniscus*) only a few samples have been analyzed and alleles were detected for the *Papa-DRB1*02*, *Papa-DRB1*03*, and *-DRB6* lineages (Table 1). The loss of particular *Patr-DR* lineages such as *Mhc-DRB1*04* is paralleled by the observation that both chimpanzee species seem to have lost a considerable segment of the ancestral MHC *A* locus lineages (McAdam et al. 1995).

A reasonable number of lowland gorillas has been studied by now (Gyllensten et al. 1991; Küpfermann et al. 1992; Kenter et al. 1993) and alleles were detected for the *Gogo-DRB1*02*, *-DRB1*03*, *-DRB1*10*, *-DRB3*, *-DRB5*, *-DRB6*, *-DRB8*, *-DRB*W8* and *-DRBW10* lineages (Table 1). As observed for chimpanzees, the gorilla equivalents of the *HLA-DRB3* and *-DRB5* loci encode many alleles. In the case of the gorilla it is not evident whether the *Gogo-DRB1* locus displays the most polymorphism. Again, in the case of the orangutan (*Pongo pygmaeus*) only a few animals have been analyzed (Schönbach et al. 1993), whereas there are currently no data available on the *DR* region of any of the gibbon species.

Allelic Diversity at the *Mhc-DRB* Loci in Old and New World Monkeys

The Old World monkey species whose MHC has been studied most completely is the rhesus macaque (*Macaca mulatta*). From other Old World monkey species only rather limited information is available at the population level (Heise et al. 1987; O'hUigin et al. 1993). For that reason the *MhcMamu* system located on chromosome 2 is taken as a reference (Balner et al. 1971). Allo-antisera can be generated in macaques by active immunizations. Typing sera allow the detection of at least ten Mamu-DR specificities which can be easily subdivided by restriction fragment length polymorphism (RFLP) techniques when *DR* β-chain gene-specific probes are used (Slierendregt et al. 1991). As found in humans and chimpanzees, immunoprecipitation studies demonstrated that the DR α-chain is invariant whereas the β-chain is polymorphic (Slierendregt et al. 1995; Lekutis and Letvin 1995). Thus far, 72 *Mamu-DRB* alleles have been reported in the international scientific literature and this number seems to represent the tip of the iceberg (Gyllensten et al. 1991; Slierendregt et al. 1992, 1994). Most of these alleles belong to already known lineages or loci that are also present in humans and apes and cluster within the *Mhc-DRB1*03*, *-DRB1*04*, *-DRB1*07*, *-DRB1*10*, *-DRB3*, *-DRB4*, *-DRB5* and *-DRB6* lineages (Table 1). Since humans, apes and Old World monkeys shared a common ancestor these lineages are at least 35 million years old. Some other allelic lineages have been designated *Mamu-DRB*W1–DRB*W7* and *-DRB*W20* and *-DRBW21*. For these *Mamu-DRB* alleles that are designated by a workshop number, no human equivalents have been detected. At this stage, it is not known whether they belong to different loci or not. Highly related equivalents have been detected, however, in other monkey species such as the pigtail macaque (Zhu et al. 1991). As one would expect to find based upon genetic distances, the number of *DRB*W* lineages is larger in Old World monkeys than in apes. The rhesus macaque equivalent of the HLA-DR3, -DR5, -DR6 cluster appears to be rather successful because at least 14 alleles have been identified for the *Mhc-DRB1*03* lineage (Table 1). In a recent report it was suggested that this may be due to functional constraints (Geluk et al. 1993). Another peculiarity is the high number of *Mamu-DRB6* alleles that has been found in rhesus monkeys.

The New and Old World monkey lineages radiated from each other about 55 million years ago. The New World monkeys are divided into two families, namely, the Callitridae (marmosets and tamarins) and Cebidae (sakis, owl and titi-, squirrel and capuchin-, spider and woolly-, howler- and capuchin monkeys). Sequencing studies demonstrated that at least three loci (*Mhc-DRB1*03*, -DRB3 and -DRB5)* are shared between Catarrhini (hominoids and Old World monkeys) and Platyrrhini (New World monkeys). These loci are occupied by functional genes in the Catarrhini and mostly by pseudogenes in the New World monkeys (Trtková et al. 1993). The New World monkey seems to have developed a complete new set of apparently functional genes clustering in the *Mhc-DRB*11*, -DRB*W12–DRB*W19* and -DRB*W22* lineages (Grahovac et al. 1992; Trtková et al 1993; Gyllensten et al. 1994). The *Mhc-DRB*W13*, -DRB*W14*, -DRB*W15*, -DRB*W17*, -DRB*W18* and -DRB*W19* seem to be restricted to the Cebidae family whereas the *Mhc-DRB*W16* and -DRBW22* thus far have been found only in the Callitrichidae family.

The best characterized New World monkey MHC system belongs to the cotton-top tamarin (*Saguinus oedipus*). Analysis of animals from captive and wild ranging populations demonstrated the existence of a high number of alleles at the following lineages *Saoe-DRB1*03*, -DRB3, -DRB5, -DRB11, -DRB*W12* and -DRB*W22*. Thus extensive polymorphism was found at all four apparently functional *DRB* loci. This is in sharp contrast with the observations done on the limited degree of variability seen within the tamarin MHC class I loci (Watkins et al. 1991). As found for the MHC class II genes in rodents and humans (Hughes and Nei 1989), the number of nonsynonymous substitutions at the codons specifying the contact residues of the peptide binding site was found to exceed the number of synonymous changes at three of the tamarin *Saoe-DRB* loci. This observation indicates that the polymorphism at *Saoe-DRB11, -DRB*W12* and -DRB*W22* is maintained by positive selection. There is no such indication for the *Saoe-DRB3* locus (Gyllensten et al. 1994). Similar findings have reported for *DR* lineages in other species (Bergström and Gyllensten 1995).

Allelic Diversity at the *Mhc-DRB* Loci in Prosimians

The lemurs, loris and tarsiers are commonly refered to as prosimians on the grounds that they are generally more primitive in morphological criteria than the simians. It is believed that simians and prosimians shared a common ancestor which lived about 85 million years ago. MHC class II genes of the *DRB* family were partially sequenced from ten individuals representing six species (Figueroa et al. 1994). Comparative analysis of the 41 different prosimian *Mhc-DRB* genes resulted in the following conclusions. Evidence was found that allelic lineages at the *Mhc-DRB1*03*, *-DRB1*04*, *-DRB3* and *-DRB6* are shared with the simians and thus are at least 85 million years old. In the ringtail lemur (*Lemur catta*) all *DRB* genes seem to have been lost, indicating that the function of antigen presentation by class II molecules was taken over by another locus.

Are All Primate *Mhc-DRB* Genes Functional?

A considerable number of the *DRB* loci in humans are known to be occupied by pseudogenes. As we will see some equivalents can be traced back in other primate species. The first example is provided by the *HLA-DRB2* gene which is present in humans on the DR52 family of haplotypes encoding the serological HLA-DR3, -DR5 and -DR6 specificities (Rollini et al. 1985). An equivalent of this gene has been described for chimpanzees (Brändle et al. 1992) and gorillas (Kasahara et al. 1992; Kenter et al. 1993). These *HLA-*, *Patr-* and *Gogo-DRB2* genes share, apart from 95% sequence similarity, a 20 base pair deletion in exon 3 which shifts the reading frame; they also lack exon 2. These results suggest that the *Mhc-DRB2* pseudogene has been inactivated more than 8 million years ago.

All HLA-DR53 related haplotypes harbor the *DRB4* locus which encodes the serologically defined DR53 determinants. In fact, two types of *HLA-DRB4* genes have been described, one functional and one nonfunctional which is present on HLA-DR7/Dw11 haplotypes and contains a so-called altered splice site (Sutton et al. 1989). An equivalent of this latter pseudogene was detected in chimpanzees (Kenter et al. 1992).

The *HLA-DRB6* gene lacks exon 1 which encodes the leader peptide and the first four amino acids of the protein. Additionally it also lacks its promotor. Chimpanzee transformed B cell lines, however, were found to express low amounts of *Patr-DRB6* message (Corell et al. 1992). This phenomenon is due to a retroviral insertion that took place more than 23 million years ago. The 3' long terminal repeat of the virus provided an open reading frame for a new stretch of mostly hydrophobic amino acids that seems to have taken over the function as leader (Mayer et al. 1993). In simians and prosimians all identified members of the *DRB6* locus seem to be pseudogenes, characterized either by the presence of stop codons or by deletions of several base pairs in the coding part of exon 2 (Figueroa et al. 1991, 1994; Corell et al. 1992; Kenter et al. 1992; Mayer et al. 1992; Slierendregt et al. 1992). For example, the *Mamu-DRB6* alleles that have been found in rhesus macaques all have features that render them as pseudogenes. The *Mamu-DRB6*0101* allele has two stop codons (TAG) at nucleotide positions 220 and 247, corresponding with amino acid positions 74 and 82, respectively. Four other alleles, *Mamu-DRB6* 0102*, *0103*, *0104*, *0105*, and *0106*, share the same deletion of 62 nucleotides spanning position 181–242 (Slierendregt et al. 1994). In addition to this deletion, the *DRB6*0102* allele has a four nucleo-tide insert at nucleotide position 101, while the *DRB6*0104* allele has an 81 nucleotide insertion at the same position. The four nucleotides that are inserted in the *DRB6*0102* allele are identical to the first four nucleotides of the 81 nucleo-tide insertion of the *DRB6*0104* allele, suggesting they have been part of an insertion that was later lost. A data bank search showed that the 81 nucleotide insertion in the *DRB6*0104* allele belongs to the family of *Alu* insertions (Slier-endregt et al. 1994). Thus far, no *HLA-DRB6* orthologue has been found in New World monkeys. This may indicate that this gene was lost during their emergence.

In humans the *HLA-DRB7* gene is restricted to the family of DR53 haplotypes (Larhammer et al. 1985). It contains, in addition to frame shift mutations, prema-ture stop codons and defective splice signals. A deletion of 56 base pairs is pre-sent in the 3' proximity of intron 1. The same deletion was found in the *HLA-DRB1*0401* gene but seems to be absent in other DR53 related *HLA-DRB1* genes (Andersson et al. 1987) This suggests that both genes shared a common ancestral gene or that either the *HLA-DRB1*04* or the -*DRB7* gene arose from a recombi-nation event. The *HLA-DRB7* gene differs by only two base pairs out of 246 sequenced from its chimpanzee equivalent (Kenter et al. 1992; Mayer et al. 1992).

The *Gogo-DRB8* pseudogene lacks exons 1 and 2 and also shares with its human counterpart the presence of two *Alu* repeats at identical positions (Klein et al. 1991; Kenter et al. 1993). As a result, these loci will not be detected after amplification with MHC *DRB* exon 2-specific primers. It is envisaged that the *Mhc-DRB8* has been a pseudogene since the divergence of Old World mon-keys and hominoids.

The *HLA-DRB9* locus is an isolated and truncated *DRB* exon 2 sequence (Meunier et al. 1986). Based on Southern blot hybridizations it was concluded that this sequence basically is nonpolymorphic and found on most haplotypes. Currently, it is not clear whether equivalents are present in other primate species.

In contrast to the situations described above, some *Mhc-DRB* genes that are functional in humans have become pseudogenes in other primate species. An example is provided by the *Mhc-DRB3* genes in some New World monkeys and prosimians that have been inactivated (Trtková et al. 1993; Figueroa et al. 1994).

Duplications, Duplications and Recombinations

Theoretically, all contemporary primate *Mhc-DRB* genes must originate from a shared ancestral gene and as such several rounds of duplications are necessary to explain the existence of the contemporary *Mhc-DRB* genes. The initial dupli-cations probably predate separation of the prosimians and simians as is evi-denced by the presence of an *Alu-J* element in all known *HLA-DRB1–DRB8* genes that integrated into the primate genome more than 65 million years ago

(Mňuková-Fajdelová et al. 1994). Thus at a certain point there must have been minimally two types of primate -DR DRB genes. One lineage probably comprised the highly related Mhc-DRB2, -DRB4 and -DRB6 genes that share a retroviral element absent in all other primate DRB genes (Klein and O'hUigin 1995). Moreover this is supported by nucleotide sequencing data demonstrating that the DRB6 gene is indeed present in contemporary prosimian and cattherrhine primate species (Table 1). Another founder gene was probably the ancestor of the HLA-DRB1, -DRB5, -DR7 and -DR8 genes, all of which are earmarked by the presence of two elements of the Alu-Sc family that integrated about 45 million years ago (Klein and O'hUigin 1995). About 15 million years later, yet another duplication round produced the primate Mhc-DRB3 and DRB*W8 genes. These events illustrate that a number of independent duplications is needed to explain the presence of the contemporary types of Mhc-DRB genes. Evidence for other more recent type of duplications can be found in the genomes of primates that are alive today. For example, multiplication of the Mhc-DRB3 pseudogene has been described for Otolemur garnetti (Figueroa et al. 1994), whereas the DRB5 locus was found to be duplicated in the orangutan (Schönbach et al. 1993). Another rather peculiar example comprises the duplication of Mamu-DRB6 and -DRB*W6 genes in rhesus macaques (Slierendregt et al. 1994). The most simple explanation for the presence of two members of both loci is that originally the DRB6 and DRB*W6 genes were located next to each other and duplicated in tandem. One of the DRB6 alleles is, however, characterized by an Alu insertion that is absent in its sister gene. This insertion must have taken place after the emergence of the Old World monkeys since it is not found in hominoid species. It also would mean that the duplication event concerning DRB6 is likely to have taken place before the insertion, because it is found in only one of the DRB6 alleles The alternative possibility, that the insertion in the other DRB6 allele was removed during evolution, seems unlikely since that would leave certain base pairs behind as a footprint. Calculations demonstrated that the DRB6 and DRBW6 genes arose from two independent duplication events.

The presence of multiple highly related genes may facilitate recombination-like processes. At least two examples show that lineages are generated de novo by recombination. Nucleotide sequencing studies, in concert with phylogenetic analyses, revealed that the HLA-DRB1*0901 gene, which is unique to humans, is a fusion product of an Mhc-DRB1*07 lineage member with an allele of the Mhc-DRB5 locus (Kenter et al. 1992; Mayer et al. 1992). The other precedent is provided by the HLA-DRB1*08 specificity which was generated by a recombination between an Mhc-DRB1*03 lineage member and an Mhc-DRB3 gene (Gorski 1989; Kasahara et al. 1992). Such type of recombinations underline that the origin and the duplication pathway of Mhc-DRB genes is a rather complicated subject to study. Similar conclusions were drawn when exon 2 and 3' untranslated (UT) sequences of the same gene were analyzed (Kenter et al. 1993).

The Organization of *DR* Subregions in Primates

In humans, chimpanzees and rhesus macaques there is a strong correlation between the number of *Taq* I RFLP fragments found upon hybridization with an *HLA-DRB* 3' UT-specific probe and the number of *DRB* genes present (Bontrop et al. 1990b). MHC class II variation studies at the population level demonstrated that the number of *Patr-DRB* gene associated fragments is not constant and varies from three to five per haplotype. This suggests that some chimpanzee class II haplotypes may harbor more MHC *DRB* genes than their human equivalents. Similar types of results were obtained in gorillas and rhesus macaque populations and in these species the *DRB* gene content per haplotype appears to vary from two to six (Kenter et al. 1993; Slierendregt et al. 1994).

The *DR* subregion of chimpanzee Hugo has been studied in great detail. From an Epstein-Barr virus (EBV) transformed B cell line, cDNA and cosmid libraries were made and the *Patr-DR* loci were sequenced and mapped. The organization of this chimpanzee's *DR* region was revealed by analyzing two nonoverlapping contigs. One contig contains the *Patr-DRB3*0201*, *-DRB6*0105* and *-DRB5*0301* loci, while the other contig contains the *Patr-DRB1*0201* and *-DRB2*0101* loci (Figueroa et al. 1991; Brändle et al. 1992). Pulsed field gel electrophoresis showed that both clusters are separated by approximately 250 kb and that the order of the genes is *Patr-DRB1*0201-DRB2*0101-DRB3*0101-DRB6*0105-DRB5*0301*. Most other chimpanzee haplotypes have been deduced based upon segregation studies. As can be seen in Fig. 1, in chimpanzees there is at least one haplotype that corresponds to a human equivalent, namely the HLA-DR7 haplotype (Kenter et al. 1992; Mayer et al. 1992; Klein et al. 1993; Slierendregt et al. 1993).

Two *Gogo-DRB* haplotypes were defined from gorilla Sylvia by contig mapping. One haplotype was shown to contain the *Gogo-DRB1*0301*, *-DRB2*0101* and *-DRB3*0402* genes and resembles the organization of the human DR52 family of haplotypes The other haplotype minimally contains the *Gogo-DRB6*0202*, *-DRB5*0201*, *-DRB8*0101* and *-DRB3*0101* loci (Kasahara et al. 1990, 1992). Studies in other gorillas provided evidence that this latter haplotype also may harbor *Gogo-DRB2 *0101* and *-DRB1*03* genes (Kenter et al. 1993). The haplotype obtained from gorilla Josephine is reported to contain *Gogo-DRB1*03*, *-DRB6* and *-DRB5* genes (Klein et al. 1993). In orangutans two rather similar haplotypes containing three *Popy-DRB5* genes and a *-DRB1*03* locus have been defined (Schönbach et al. 1993). For the rhesus macaque all haplotype information has been obtained by analyzing the gene content of animals that are of consanguineous origin (Slierendregt et al. 1994). It has to be stressed that the actual order or organization of the different loci in the genome may be different than is shown in Fig. 1. Despite this, some rhesus macaque haplotypes are found that harbor two alleles orthologous to different *HLA-DRB1* lineages. In some haplotypes a combination is found of *Mamu-DRB1*03* and *-DRB1*10* alleles while in others the combination of a *-DRB1*04* with a *-DRB1*10* allele is found. The schematic representation of haplotypes that is depicted in Fig. 1 makes it clear that, although allelic lineages are inherited in a *trans*-species mode of evolution, the haplotype generation is merely a postspeciation event. The question arises as

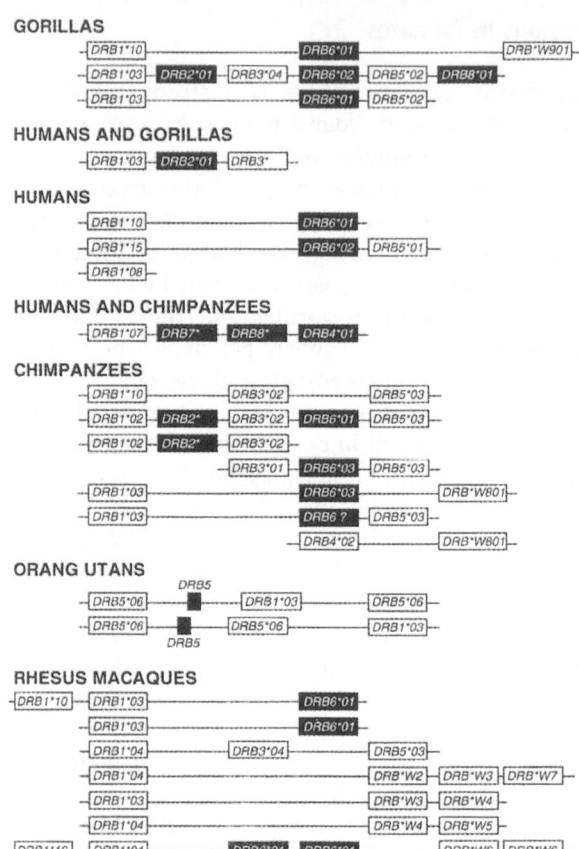

Fig. 1. The *Mhc-DR* regions in primates. Conventional nomenclature is used (Klein et al. 1991). Pseudogenes are depicted by *black boxes* whereas functional genes are represented by *white boxes*. Most haplotypes that are obtained from chimpanzees and rhesus macaques have been deduced by segregation studies whereas most other were defined by contig mapping. For more details and references see text

to how these haplotypes were generated. Unequal crossing over might be responsible for the generation of haplotypes that contain a different number of genes. As a consequence some haplotypes may lack a *DRB1* locus, as has been found for chimpanzees (Fig. 1). Furthermore, homologous recombinations and duplications seem to alter the organization of the primate *DR* region in time.

References

Andersson G, Larhammar D, Widmark E, Servenius B, Peterson PA, Rask L. Class II genes of the human major histocompatibility complex: organization and evolutionary relationship of the DRβ genes. J Biol Chem 262:8748–8758, 1987

Balner H, Gabb BW, Dersjant H, van Vreeswijk W, van Rood JJ. Major histocompatibility locus of rhesus monkeys (RhL-A). Nature 230: 177–180, 1971

Balner H, Gabb BW, D'Amaro J, van Vreeswijk W, Visser TP. Evidence for two linked loci controlling the serologically defined leukocyte antigens of chimpanzees (ChL-A). Tissue Antigens 4:313–328, 1974

Balner H, van Vreeswijk W, D'Amaro J, Roger J, Schreuder I, van Rood JJ. Chimpanzees share B cell as well as SD antigens with man. In: Bodmer WF, Batchelor JR, Bodmer JG, Festenstein H, Morris PJ (eds) Histocompatibility testing 1977. Munksgaard, Copenhagen, p 412, 1977

Bergström T, Gyllensten U: Evolution of Mhc class II polymorphism: the rise and fall of class II gene fuction in primates. Immunol Rev 143:13–32, 1995

Bodmer JG, Marsh SGE, Albert ED, Bodmer WF, Bontrop RE, Charron D, Dupont B, Erlich HA, Mach B, Mayr WR, Parham P, Sasazuki T, Schreuder GMT, Strominger JL, Svejgaard A, Terasaki P. Nomenclature for factors of the HLA system 1995. Tissue Antigens 46:1–18, 1995

Böhme J, Andersson M, Andersson G, Moller E, Peterson PA, Rask L. HLA-DR beta genes vary in number between different DR⁺ specificities whereas the number of DQ beta genes is constant. J Immunol 135:2149–2155, 1985

Bontrop RE, Schreuder GMT, Elferink BG, Mikulski MMA, Geerse R, Giphart MJ. Molecular serologic and functional evidence for an apparent HLA-DR triplet. J Immunol 137:211–216, 1986a

Bontrop RE, Schreuder GMT, Mikulski EMA, van Miltenburg RT, Giphart MJ. Polymorphisms within the HLA-DR4 haplotypes; various DQ subtypes detected with monoclonal antibodies. Tissue Antigens 27:22–31, 1986b

Bontrop RE, Elferink BG, Otting N, Jonker M, de Vries RRP. Major histocompatibility complex class II restricted antigen presentation across a species barrier: conservation of restriction determinants in evolution. J Exp Med 172:53–59, 1990a

Bontrop RE, Broos LAM, Pham K, Bakas RM, Otting Nand Jonker M. The chimpanzee major histocompatibility complex class II DR subregion contains an unexpectedly high number of beta-chain genes. Immunogenetics 32:272–280, 1990b

Bontrop RE, Kenter M, Otting N, Jonker M. Major histocompatibility complex polymorphisms in humans and chimpanzees. J Med Primatol 22:50–59, 1993

Bontrop RE, Otting N, Slierendregt BL, Lanchbury JS. Evolution of major histocompatibility complex polymorphisms and T-cell receptor diversity in primates. Immunol Rev 143:32–62, 1995

Brändle U, Ono H, Vincek V, Klein D, Golubic M, Grahovac B, Klein J. Trans species evolution of Mhc-DRB haplotype polymorphism in primates: organization of DRB genes in the chimpanzee. Immunogenetics 36:39–48, 1992

Cairns JS, Curtsinger JM, Darl CA, Freeman S, Alter BJ, Bach FH. Sequence polymorphism of HLA-DRB1 alleles relating to T cell recognized determinants. Nature 317:166–168, 1985

Corell A, Morales P, Varela P, Paz-Artal E, Martin-Villa JM, Martinez-Laso J, Arnaiz-Villena A. Allelic diversity at the primate major histocompatibility complex DRB6 locus. Immunogenetics 36:33–38, 1992

Fan WM, Kasahara M, Gutknecht J, Klein D, Mayer WE, Jonker M, Klein J. Shared class II MHC polymorphisms between humans and chimpanzees. Hum Immunol 26:107–121, 1989

Figueroa F, O'hUigin C, Inoko H, Klein J. Primate DRB6 pseudogenes–clue to the evolutionary origin of the HLA-DR2 haplotype. Immunogenetics 34:324–337, 1991

Figueroa F, O'hUigin C, Tichy H, Klein J. The origin of the primate Mhc-DRB genes and allelic lineages as deduced from the study of prosimians. J Immunol 152:4455–4465, 1994

Geluk A, Elferink BG, Slierendregt BL, van Meijgaarden KE, de Vries RRP, Ottenhoff THM, Bontrop RE. Evolutionary conservation of major histocompatibility complex- DR/peptide/T cell interactions in primates. J Exp Med 177:979–984, 1993

Gorski J. The HLA-DRw8 lineage was generated by a deletion in the DRB region followed by first domain diversification. J Immunol 142:4041–4045, 1989

Grahovac B, Mayer WE, Vincek V, Figueroa F, O'hUigin C, Tichy H, Klein J. Major histocompatibility complex DRB genes of a New-World monkey, the cottontop tamarin (Saguinus oedipus). Mol Biol Evol 9:403–416, 1992

Gyllensten U, Sundvall M, Ezcurra I, Erlich HA. Genetic diversity at class II DRB loci of the primate MHC. J Immunol 146:4368–4376, 1991

Gyllensten U, Bergstrom T, Josefson A, Sundvall M, Savage A, Blumer ES, Giraldo LH, Soto LH, Watkins DI. The cotton-top tamarin revisted: Mhc class I polymorphism of wild tamarins and polymorphism and allelic diversity of the class II DQA1, DQB1 and DRB loci. Immunogenetics 40:167–176, 1994

Heise ER, Cooke DJ, Schepart BS, Manning CH, McMahan MR, Chedid M, Keever CA. The major histocompatibility complex of primates. Genetica 73:53–68, 1987

Hughes AL, Nei M. Nucleotide substitution at major histocompatibility complex class II loci: evidence for overdominant selection. Proc Natl Acad Sci USA 86:958–962, 1989

Jonker M, Balner H. Current knowledge of the D/DR region of the major histocompatibility complex of rhesus monkeys and chimpanzees. Hum Immunol 1:305–316, 1980

Kasahara M, Klein D, Fan W, Gutknecht J. Evolution of the class II major histocompatibility complex alleles in higher primates. Immunol Rev 113:65–82, 1990

Kasahara M, Klein D, Vincek V, Sarapata DE, Klein J. Comparative anatomy of the primate major histocompatibility complex DR subregion: evidence for combinations of DRB genes conserved across species. Genomics 14:340–349, 1992

Kaufman JF, Auffray C, Korman AJ, Shackelford DA, Strominger JL. The class II molecules of the major histocompatibility complex. Cell 36:1–13, 1984

Kenter M, Otting N, Anholts J, Jonker M, Schipper R, Bontrop RE. Mhc-DRB diversity of the chimpanzee (Pan troglodytes). Immunogenetics 37:1–11, 1992

Kenter M, Otting N, de Weers M, Anholts J, Reiter C, Jonker M, Bontrop RE. Mhc- DRB and -DQA1 nucleotide sequences of three lowland gorillas: implications for the evolution of primate Mhc class II haplotypes. Hum Immunol 36:205–218, 1993

Klein D, Vincek V, Kasahara M, Shönbach C, O'hUigin C, Klein J. Gorilla major histocompatibility complex-DRB pseudogene orthologous to HLA-DRVIII. Hum Immunol 32:211–220, 1991

Klein J, O'hUigin C. Class II B Mhc motifs in an evolutionary perspective. Immunol Rev 143:89–111, 1995

Klein J, Bontrop RE, Dawkins RL, Erlich HA, Gyllensten UB, Heise ER, Jones PP, Parham P, Wakeland EK, Watkins DI. Nomenclature for the major histocompatibility complexes of different species: a proposal. Immunogenetics 31:217–219, 1990

Klein J, O'hUigin C, Figueroa F, Mayer WE, Klein D. Different modes of evolution in primates. Mol Biol Evol 10:48–59, 1993

Küpfermann H, Mayer WE, O'hUigin C, Klein D, Klein J. Shared polymorphism between gorilla and human major histocompatibility complex DRB loci. Hum Immunol 34:267–278, 1992

Larhammar D, Servenius B, Rask L, Peterson PA. Characterization of an HLA-DRβ pseudogene. Proc Natl Acad Sci USA 82:1475–1479, 1985

Lekutis C, Letvin NL. Biochemical and molecular characterization of rhesus monkey major histocompatibility complex class II DR. Hum Immunol 43:72–80, 1995

Mayer WE, O'hUigin C, Zaleska RZ, Klein J. Trans-species origin of Mhc-DRB polymorphism in the chimpanzee. Immunogenetics 37:12–23, 1992

Mayer WE, O'hUigin C, Klein J. Resolution of the HLA-DRB6 puzzle: a case of grafting a de novo-generated exon on an existing gene. Proc Natl Acad Sci USA 90:10720–10724, 1993

McAdam SN, Boyson JE, Liu X, Garber TL, Hughes AL, Bontrop RE, Watkins DI. Chimpanzee MHC class I A locus alleles are related to only one of the six families of human A locus alleles. J Immunol 154:6421–6429, 1995

Meunier HF, Carson S, Bodmer WF, Trowsdale J. An isolated β1 exon next to the DRα gene in the HLA-D region. Immunogenetics 23:172–180, 1986

Mňuková-Fajdelová F, Satta Y, O'hUigin C, Mayer W, Figueroa F, Klein J. *Alu* elements of the primate major histocompatibility complex. Mammalian Genome 5:405–415, 1994

Nepom BS, Nepom GT, Mickelson E, Antonelli P, Hansen JA. Electrophoretic analysis of human HLA-DR antigens from HLA-DR4 homozygous cell lines: correlation between β chain diversity and HLA-D. Proc Natl Acad Sci USA 80:6962–6966, 1983

O'hUigin C, Bontrop R, Klein J. Nonhuman primate Mhc-DRB sequences: a compilation. Immunogenetics 38:165–183, 1993

Rollini P, Mach B, Gorski J. Linkage map of three HLA-DR β- chain genes: evidence for a recent duplication event. Proc Natl Acad Sci USA 82:7179–7183, 1985

Schönbach C, Vincek V, Mayer WE, Golubic M, O'hUigin C, Klein J. Multiplication of Mhc-DRB5 loci in the orangutan: implications for the evolution of DRB haplotypes. Mammalian Genome 4:159–170, 1993

Slierendregt BL, Bontrop RE. Current knowledge on the major histocompatibility complex class II region in nonhuman primates. Eur J Immunogenet 21:391–402, 1994

Slierendregt BL, Otting N, Jonker M, Bontrop RE. RFLP analysis of the rhesus monkey MHC class II DR subregion. Hum Immunol 30:11–17, 1991

Slierendregt BL, van Noort JT, Bakas RM, Otting N, Jonker M, Bontrop RE. Evolutionary stability of trans-species major histocompatibility complex class II DRB lineages in humans and rhesus monkeys. Hum Immunol 35:29–39, 1992

Slierendregt BL, Kenter M, Otting N, Anholts J, Jonker M, Bontrop RE. Major histocompatibility complex class II haplotypes in a breeding colony of chimpanzees (Pan troglodytes). Tissue Antigens 42:55–61, 1993

Slierendregt BL, Otting N, Van Besouw N, Jonker M, Bontrop RE. Expansion and contraction of rhesus macaque DRB regions by duplication and deletion. J Immunol 154:2298–2307, 1994

Slierendregt BL, Otting N, Jonker M, Bontrop RE. Gel electrophoretic analysis of rhesus macaque major histocompatibility complex class II DR molecules. Hum Immunol 40:33–40, 1995

Sutton VR, Kienzle BK, Knowles RW. An altered splice site is found in the DRB4 gene that is not expressed in HLA-DR7Dw11 individuals. Immunogenetics 29:317–322, 1989

Termijtelen A, van Leeuwen A, van Rood JJ. HLA-linked lymphocyte activating determinants. Immunol Rev 66: 79–101, 1982

Trtková K, Kupfermann H, Grahovac B, Mayer WE, O'hUigin C, Tichy H, Bontrop R, Klein J. Mhc-DRB genes of platyrrhine primates. Immunogenetics 38:210–222, 1993

Watkins DI, Garber TL, Chen ZW, Toukatly G, Hughes AL, Letvin NL. Unusually limited nucleotide sequence variation of the expressed major histocompatibility complex class I genes of New World primate species (Saguinus oedipus). Immunogenetics 33:79–89, 1991

Zhu Z, Vincek V, Figueroa F, Schönbach C, Klein J. Mhc-DRB genes of the pigtail macaque (Macaca nemestrina): implications for the evolution of human DRB genes. Mol Biol Evol 8:563–578, 1991

Evolution of Length Variation in the Primate Mhc DR Subregion

J. Klein, Y. Satta, and R. Gongora

The length of the primate Mhc varies not only between species but also between individuals of the same species. The length differences can be as great as several hundred kilobase pairs (kb) (see chapter by Hanini et al., this volume). The most variable in this regard are the *DR* subregion of the class II region and the class III region. The length variability of the class III region is discussed elsewhere in this volume (see Figueroa 1997); here we shall focus on the variability of the *DR* subregion.

We define the *DR* subregion as the chromosomal segment harboring the loci that code for the α- and β-chains of the DR molecule. The subregion extends from the 5' end of the *DRB1* locus to the 3' end of the *DRA* locus (see also chapter by Bontrop) and is intercalated between the *DQA1* locus of the *DQ* subregion and the *G18* locus of the class III subregion. (The *G18* gene is at present the most proximal marker of the class III subregion relative to the centromere and to the *DRA* locus, but there are >200 kb of unexplored sequence between *G18* and *DRA* which may contain more proximal, as yet undetected loci; see chapter by Hanini et al., this volume). The *DRB1* and *DRA* loci form the borders of the *DR* subregion, which is present in all catarrhine primates studied thus far; the situation in non-catarrhine primates remains obscure. There is only one other segment of the *DR* subregion that appears to be relatively constant at least in apes, namely, the segment harboring the *DRB9* gene fragment; the rest of the region is variable.

The length variants are referred to as *haplotypes*. Individual haplotypes differ, however, not only in their length, but also in the composition of the loci residing in the subregion and in the alleles occupying these loci. The haplotypes can therefore be divided into *groups* (each group having a characteristic length and constellation of loci), *major haplotypes* (within each group loci are occupied by alleles that differ from corresponding alleles at loci in other haplotypes by a large number of substitutions), and *minor haplotypes* (alleles are distinguished by one or a few substitutions only).

Nine *DRB* loci (or gene fragments) and one *DRA* locus have been identified in the *HLA* complex and designated *DRB1–DRB9* (see J. Klein et al. 1991). The distribution of the loci in nonhuman primates, as far as it is known, is given in Table 1. Only four loci of the *DRB* loci (*DRB1*, *DRB3*, *DRB4*, and *DRB5*) are not defective and are presumed to be functional; the remaining loci are occupied by pseudogenes.

Molecular Biology and Evolution of Blood Group
and MHC Antigens in Primates
Blancher/Klein/Socha (Eds.)
© Springer-Verlag Berlin Heidelberg 1997

Table 1. Distribution of *DRB* loci among primate species

| Species | DRB locus | | | | | | | | |
	DRB1	DRB2	DRB3	DRB4	DRB5	DRB6	DRB7	DRB8	DRB9
Homo sapiens	+	+	+	+	+	+	+	+	+
Pan troglodytes	+	+	+	+	+	+	+	+	
Pan paniscus	+					+			
Gorilla gorilla	+	+	+		+	+		+	
Pongo pygmaeus	+				+	+			
Mandrillus leucopheus	+					+			
Macaca mulatta	+		+	+	+	+			
Macaca nemestrina	+								
Papio hamadryas	+								
Saguinus oedipus	+		+		+				
Aotus trivirgatus	+		+						
Callithrix jaccus	+								
Callicebus moloch	+		+						
Galago senegalensis	+					+			
Galago moholi	+								
Otolemur garnettii		+							

The symbol "+" stands for the presence of the locus.

Human *DR* haplotypes fall into five groups (Fig. 1): *HLA-DR8* (one major haplotype), *HLA-DR1* (major haplotypes *DR1* and *DR10*), *HLA-DR51* (major haplotype *DR2*), *HLA-DR52* (major haplotypes *DR3*, *DR5*, and *DR6*), and *HLA-DR53* (major haplotypes *DR4*, *DR7*, and *DR9*). The five groups represent the entire endowment of the major haplotypes in the human species. How many different haplotypes exist in nonhuman primates is not known for any species and may, in fact, never be known for the simple reason that haplotype identification is a laborious undertaking. Haplotype organization can be determined either by molecular or by classical genetics methods; however, both approaches are quite time-consuming. It is therefore unlikely that anybody will be tempted to characterize the chimpanzee, the gorilla or any other primate to the same extent as the international network of collaborating laboratories has done for the human species. Fortunately, it is not necessary to have a complete knowledge of all the genes and haplotypes in the various nonhuman primates species in order to be able to draw conclusions about the evolutionary origins of these genes and haplotypes. In fact, a great deal of evolutionary information can be extracted from the characterization of the human MHC alone. In what follows, we describe the origin of the human *DRB* subregion, the roots of which apparently go back to the origin of the primates some 80 million years (my) ago.

Earlier studies have concentrated on exon 2 of the *DRB* genes which codes for the β1 domain of the polypeptide chain and is the most variable part of the gene (Klein and Figueroa 1986). The exon contains sites that specify the peptide-bind-

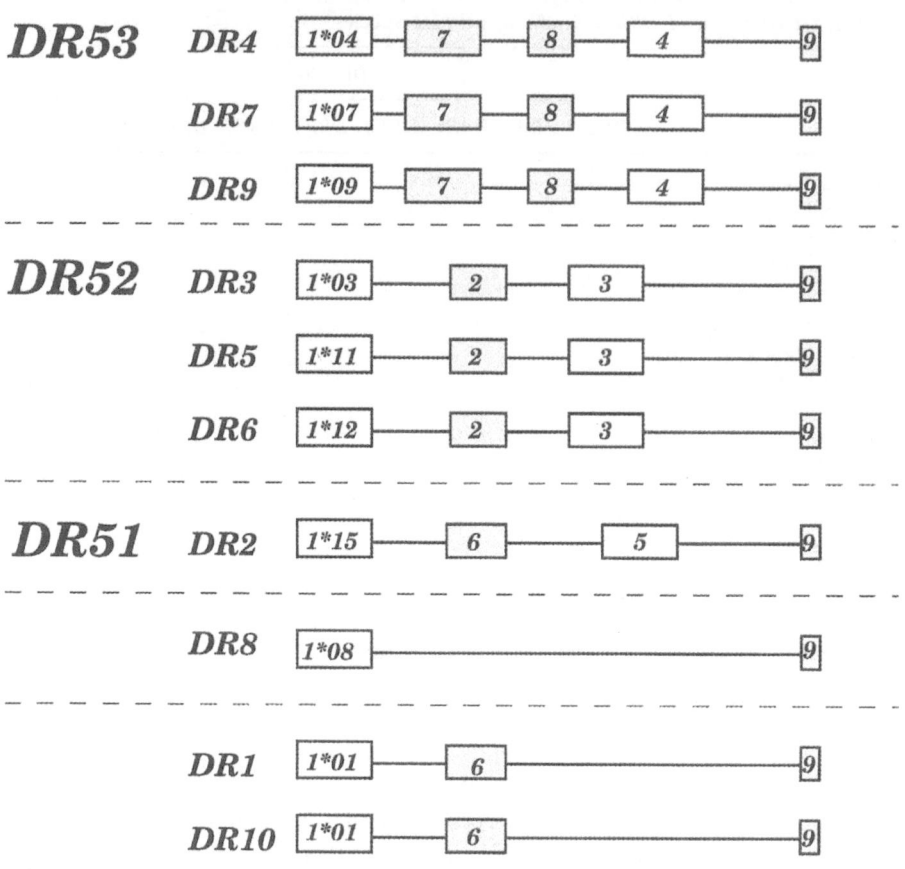

Fig. 1. Human *DR* haplotypes. Functional and nonfunctional *DRB* genes are represented by *open* and *shaded rectangles*, respectively. The diagram is not drawn to scale

ing region (PBR) of the protein – the residues that contact the peptide and others that are contacted by the T cell receptor. Several hundred exon 2 sequences have been accumulated from a dozen or so primate species representing major primate taxa (see chapter by Bontrop). The information has been used to construct phylogenetic trees of the genes in an attempt to deduce their origin. The attempt has been only a partial success. The major obstacle in these evolutionary deductions has proved to be parallel (convergent) evolution which produces similar sequences by independent pathways (O'hUigin 1995). Thus, for example, exon 2 sequences have revealed the presence of similar motifs at the PBR sites in prosimians, Platyrrhini and Catarrhini suggesting that some of the *DRB1* allelic lineages have persisted as polymorphisms for the entire period of primate evolution (Figueroa et al. 1994). This conclusion, however, is contradicted by other data obtained subsequently (Trtková et al. 1995).

Exon 2 provides misleading information because its PBR-encoding sites evolved under selection pressure from parasites, whose peptides the PBR presents to T lymphocytes (Hughes and Nei 1988; Takahata et al. 1992). The selection seems to favor certain substitutions over others and leads to their preferential incorporation into persisting allelic lineages. To obtain evolutionary information not disturbed by natural selection, it is necessary to turn from exons to introns, which are believed to evolve free of selection pressure. A typical DRB gene has five introns varying in length according to their position in the gene, the particular gene, and the species. The human DRB1 gene, for example, contains altogether approximately 13 kb of intron sequence; intron 1 alone is about 8.5 kb long. Approximately 5000 base pairs (bp) of intron 1 have been sequenced from three HLA-DRB1 alleleles (DRB1*03, 1*04, and 1*15) and five HLA-DRB genes (DRB2, 3, 4, 5, and 7; Andersson et al. 1987; Satta et al. 1996b) totaling about 40 kb of sequence information. Similarly, the entire introns 4 and 5 and part of the 3′ untranslated (3′UT) region have been sequenced from HLA-DRB1*15, 1*03, 1*04, DRB2, 3, 4, 5, 6, 7, and 8 genes (approximately 1 kb from each gene, a total of 10 kb of sequence information; Andersson et al. 1987; Svensson et al. 1995, 1996; Satta et al. 1996a).

Using introns rather than exons for evolutionary analysis has another advantage. Introns often contain inserts of various repetitive elements such as the Alu, L1, and MER repeats as well as retroviruses or their remnants. All these and other elements have been found to be present in primate DRB introns. The Alu elements are the most common (Mnuková-Fajdelová et al. 1994), followed by L1 repeats and retroviral elements. To facilitate communication, we have numbered the Alu elements found in primate Mhc genes consecutively in the order of their discovery (Schönbach and Klein 1991). The elements fall into several groups according to their presumed age (time of insertion; Kapitonov and Jurka 1996) and hence can be used to date the divergence of the genes which bear them. The distribution of the various repetitive elements among the human DRB genes is given in Fig. 2. The elements serve as useful cladistic markers because sharing of the same insert implies origin of gene segments from the same ancestor. The underlying assumption of this use is that insertion of a repetitive element is a unique event and that loss of an insert can be ascertained by comparison of nucleotide sequences from the relevant gene segments.

The question of the origin of the MHC genes can therefore be approached by two intron-based methods. First, the intronic nucleotide sequences can be used to construct phylogenetic trees according to genetic distance, parsimony or some other principle; second, the inserts of repetitive elements provide material for producing cladograms presumably reflecting evolutionary relationships among the DRB genes. Alternatively, the insertion markers can be superimposed on the phylogenetic trees to seek support for a particular branching pattern.

Before describing the origin of the nine DRB genes present in the human population, we must mention certain peculiarities of some of these genes, specifically HLA-DRB8, -DRB9, -DRB2, and DRB5. The DRB8 gene has been

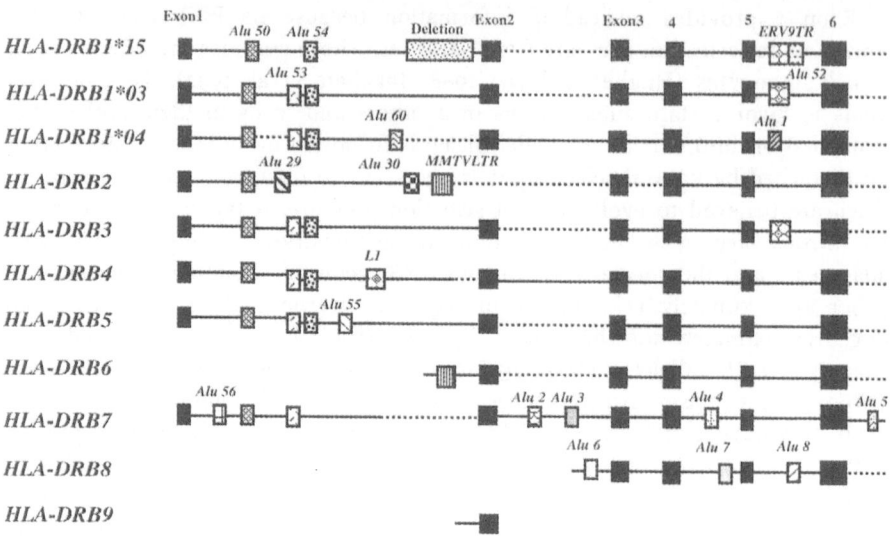

Fig. 2. Distribution of known insertions/deletions in nine *HLA-DRB* genes. Homologous insertions are indicated by the same type of *shading*. *Dotted lines* indicate regions that have not been sequenced. The *rectangle* in *DRB1*15* indicates a deletion of approximately 1.2 kb

found in humans, chimpanzees, and gorillas (Andersson et al. 1987; D. Klein et al. 1991; Slierendregt et al. 1993). In these three species it is a pseudogene that lacks the 5′ end, including exon 1, intron 1, and exon 2. This fact disqualifies the *DRB8* gene from comparisons involving intron 1. Analyses based on intron 4 and 5 sequences suggest the *DRB8* gene to be >50 my old (Satta et al. 1996a); whether it has been a pseudogene for most of this time is not known.

HLA-DRB9 is a gene fragment consisting only of a 3′ half of intron 1 exon 2, and 5′ end of intron 2 (Meunier et al. 1986; Haas et al. 1987; Gongora et al. 1996). Upstream from the 3′ half of intron 1 is an exon 1 and a 5′ half of intron 1, but the two halves of intron 1 are separated by a series of repeats which complicate the interpretation of the rearrangements in this region (Gongora et al. 1997). The two intron segments could indeed be the two halves of the same intron 1 that has been interrupted by several insertion events. In this case the exon 1 would be part of the *HLA-DRB9* gene. Alternatively, the exon 1 and the 5′ part of intron 1 could be the remains of another gene that was once a neighbor of *DRB9* but was later obliterated by a large deletion that cut out all of the gene except the most 5′ end and also removed the exon 1 and the 5′ half of intron 1 from the *DRB9* gene. (Another deletion removed the rest of the *DRB9* gene downstream of exon 2). Because of its fragmentary nature, the *DRB9* gene is difficult to place into the scheme of relationships of the human *DRB* genes. The gene has been found in humans, chimpanzees, gorillas, orangutans, and macaques and its age has been estimated to be more than 50 my (Gongora et al. 1996).

HLA-DRB2 and HLA-DRB6, as their designations indicate, are generally considered to be genes at two loci. It is, however, rather likely that they are, in fact, alleles at a single locus or, if they do occupy two loci, that they have duplicated recently. Both HLA-DRB2 and DRB6 are pseudogenes. HLA-DRB2 lacks exon 2 but has exon 1 (Rollini et al. 1985; Vincek et al. 1992), whereas the opposite is true of HLA-DRB6 (Kawai et al. 1989; Figueroa et al. 1991). Although both genes have the remaining exons, they each have additional defects that make them nonfunctional, though DRB6 is transcribed by using the promoter and exon 1 sequence of a retroviral insert (Mayer et al. 1993). Both are apparently present not only in humans, but also in the chimpanzee and the gorilla (Vincek et al. 1992; Figueroa et al. 1991; Kasahara et al. 1992). The chimpanzee DRB2 gene is, however, unusual in several respects (e.g., it lacks both exon 1 and exon 2 and it lacks the 20 bp deletion in exon 3 that characterizes the human and the gorilla DRB2 genes), so that its orthology with the human and gorilla genes remains in doubt (Vincek et al. 1992). The presence of a gene referred to as DRB6 has been reported in orangutans (Corell et al. 1992), macaques (Slierendregt et al. 1993; Mayer et al. 1993), and the prosimian Galago senegalensis (Figueroa et al. 1994); the DRB2 gene has not been reported for any primates other than human, chimpanzee, and gorilla. The divergence time of the HLA-DRB2 and HLA-DRB6 genes has been estimated to be approximately 20 my (Satta et al. 1996a). It seems, therefore, that before 20 my only one of the two genes existed and whether this ancestor should be designated as DRB2 or DRB6 is a matter of convention. At least three arguments can be made in support of the notion that DRB2 and DRB6 are alleles. First, the genetic distances between the introns of the DRB2 and DRB6 genes are comparable to those between two alleles at the DRB1 locus (Satta et al. 1996b). Second, the DRB2 and DRB6 genes share the Alu30 and mouse mammary tumor virus (MMTV) long terminal repeat (LTR) elements which are absent in other DRB genes (Mnuková-Fajdelová et al. 1994; Satta et al. 1996a,b). Third, the two genes occupy corresponding positions in the human DR3 and DR2 haplotypes, respectively, and are located at approximately the same distance from HLA-DRB1 (Rollini et al. 1985; Kawai et al. 1989). The only argument in favor of two loci is the observation that chimpanzees have a haplotype with both the DRB2 and DRB6 genes on the same chromosome (Brändle et al. 1992; Slierendregt et al. 1993). However, as mentioned above, the orthology of the chimpanzee DRB2 with the human DRB2 gene is in doubt, and even if it were real, the presence of the two genes in the same haplotype would not invalidate their allelism in other haplotypes. The composition of the chimpanzee haplotype suggests that it arose by unequal crossing-over from two other haplotypes (Brändle et al. 1992). If so, in most other haplotypes the two genes could still be alleles. Whether separate loci or alleles, however, it seems that only one of the two genes existed before the separation of apes and Old World monkeys. We shall refer to this ancestral gene as DRB2.

Two genes are the most likely candidates for being closest to the ancestor of all the nine human DRB genes – DRB1*04 and DRB2; however, DRB8 and DRB9 must also be considered. HLA-DRB1*04 and -DRB2 are the only two of the nine genes that originally lacked Alu53, an old repetitive element that was probably

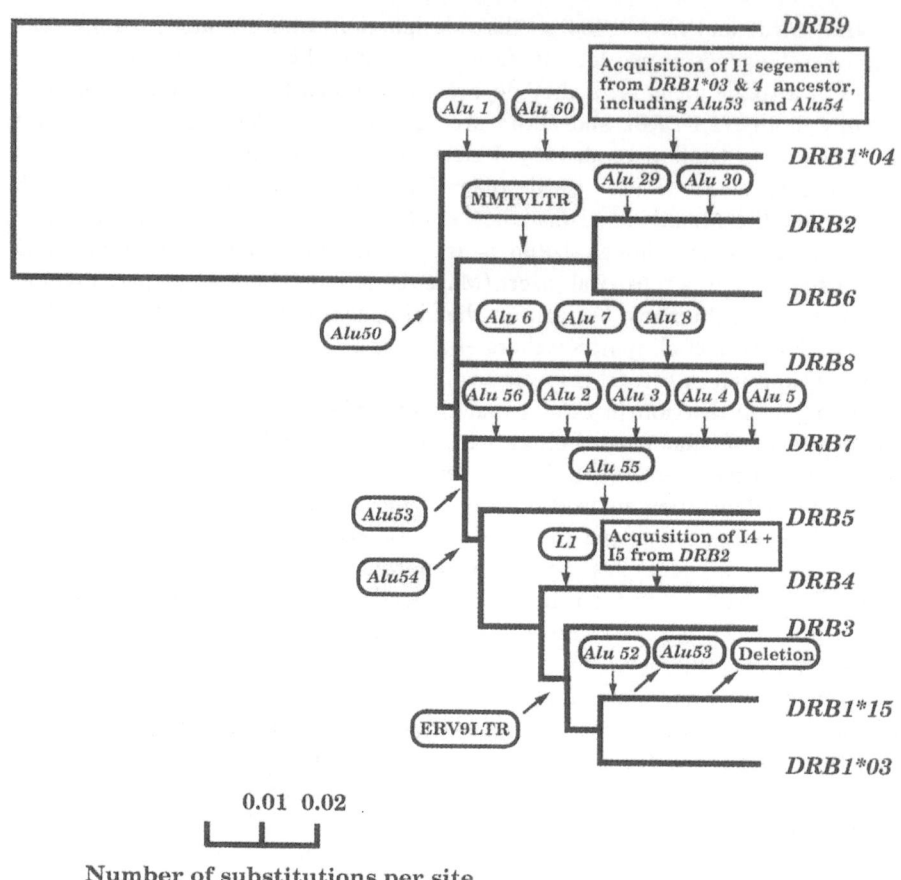

Fig. 3. The phylogenetic tree based on nucleotide substitutions and phylogenetically informative markers in introns 1, 4, and 5. The topology of the tree is determined by the phylogenetically informative markers and the neighbor joining method based on the number of nucleotide substitutions in the intron sequences. The branch length was determined by taking the average of the number of substitutions in pairs involving this particular gene and genes which comprise a cluster. Internodal branch length is the difference between the neighboring two branches. Possible exchanged sequences are excluded from the calculations. Insertions and excisions of Alu and other elements are indicated by arrows. Phylogenetically informative makers are shaded

inserted into the Mhc region >35 my ago (Mnuková-Fajdelová et al. 1994). Whether DRB8 once had Alu53 can no longer be ascertained because the gene now lacks intron 1, the place of Alu53 insertion; genetic comparisons indicate that it is an old gene (O'hUigin and Klein 1991; Satta et al. 1996a). The DRB9 gene is also old (Gongora et al. 1996).

Of the two most likely candidates, *DRB1*04* is favored over *DRB2* by genetic distance data (Fig. 3). The bootstrap test, however, fails to support the candidacy of *DRB1*04* statistically. By contrast, the *DRB1* lineage has remained functional throughout primate evolution, whereas *DRB2* was apparently inactivated early in its evolution (O'hUigin and Klein 1991). On the strength of these two arguments we therefore assume that a gene related to *DRB1*04* was the "mother of all the *HLA-DRB* genes" and that all the other genes arose from it by gene duplications, directly or indirectly (Fig. 4). The data suggest that there were three main duplications in close succession (Satta et al. 1996a,b). The first duplication occurred 58 my ago and produced the *DRB2* gene which became inactivated either during the duplication or subsequent to it. The ancestral *DRB1*04* gene contained *Alu50*, a member of the oldest family (J) of *Alu* elements (Kapitonov and Jurka 1996), and consequently *Alu50* is now present in the intron 1 of all nine *HLA-DRB* genes (Satta et al. 1996b). It, however, lacked all other insertion elements thus far identified in introns 1, 4, and 5 of the *HLA-DRB* genes. The original copy of the *DRB1*04* gene evolved into the extant *HLA-DRB1*04* allelic lineage but along the way was modified by at least three insertion events: It acquired the *Alu60* element in intron 1 more than 50 my ago, *Alu1* in intron 5 some 38 my ago, and later it procured part of intron 1 from a gene that was the ancestor of the *DRB3* and *DRB4* genes. By that time the ancestor had already acquired the *Alu53* and *Alu54* elements which the present-day *DRB1*04* gene still possesses. Sequence comparison with intron 1 of other *DRB* genes, however, clearly indicates that the segment of intron 1 bearing the *Alu53* and *Alu54* elements in *HLA-DRB1*04* is a late-date acquisition, presumably through recombination or transposition, from another gene (Satta et al. 1996b). Since the "foreign" segment in intron 1 is delineated by the two *Alu* elements, they may have actually mediated the acquisition. Following the first round of duplication, the ancestral *DRB2* gene acquired the *Alu29*, *Alu 30* and MMTV LTR elements 24, 18, and >20 my ago, respectively (Mnuková-Fajdelová et al. 1994; Mayer et al. 1993), before it diverged into *DRB2* and *DRB6* 20 my ago, (O'hUigin and Klein 1991; Satta et al. 1996a). Somewhere along the line the *DRB8* gene arose by duplication of either the primordial *DRB1*04* or the *DRB2* genes and later became the target of *Alu7* and *Alu8* insertions in introns 4 and 5, respectively.

In the meantime, the ancestral *DRB1*04* acquired the *Alu53* element in intron 1, then duplicated again some 56 my ago (Satta et al. 1996b) and the extra copy became the ancestor of the *DRB7* gene. Here again, the *DRB7* gene was inactivated either at its birth or shortly afterwards (O'hUigin and Klein 1991; Satta et al. 1996b). The gene also acquired the *Alu56* element in intron 1 >50 my ago (Satta et al. 1996b) and the *Alu4* element in intron 4 some 18 my ago (Mnuková-Fajdelová et al. 1994; Satta et al. 1996a).

The third round of *DRB1*04* duplication, which was preceded by the insertion of the *Alu54* element, occurred approximately 53 my ago and produced the ancestor of the *HLA-DRB5* gene which then acquired the *Alu55* element. The remaining *DRB* genes all arose from the ancestral *DRB1*04* gene carrying the *Alu50*, *Alu53* and the *Alu54* inserts in intron 1. The *DRB4* gene arose some 42 my ago, probably by a deletion of the 3' end of *DRB1*04* and 5' end of a neighboring *DRB2*

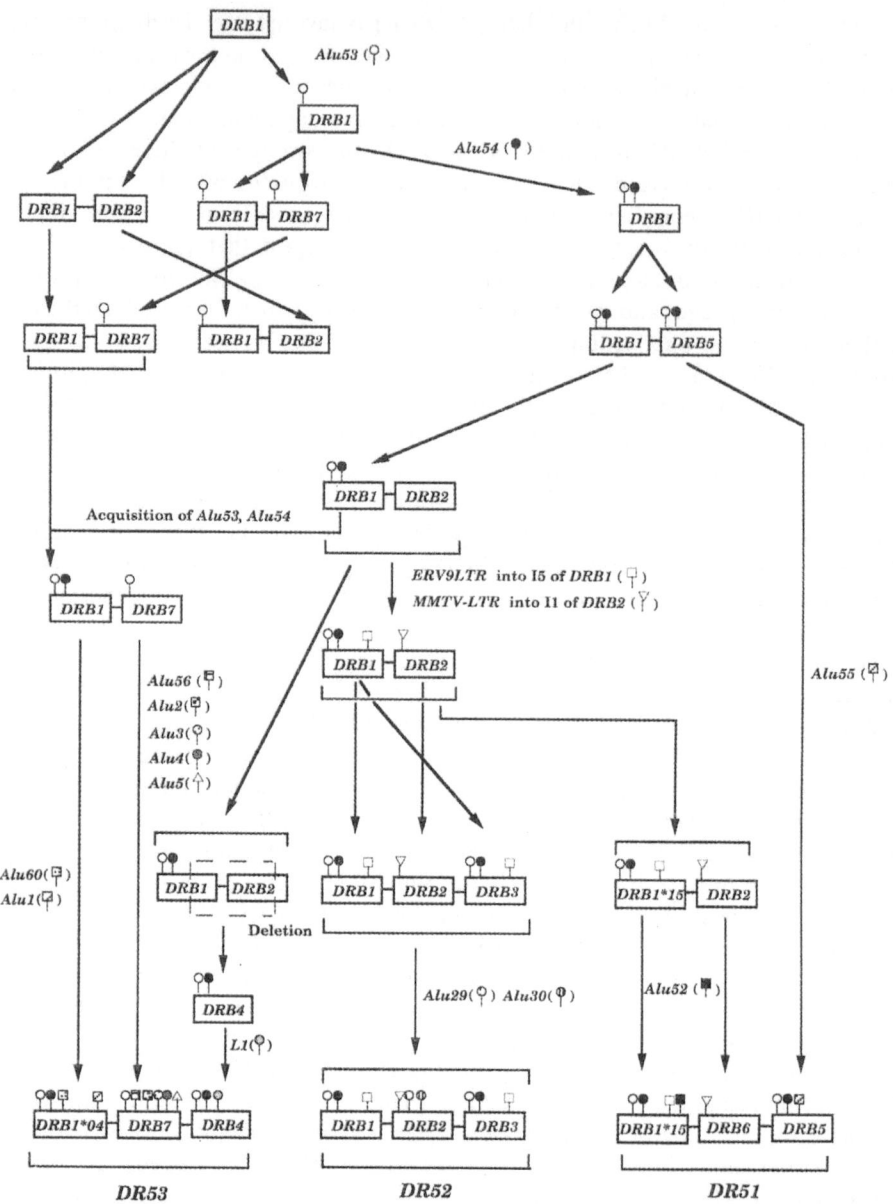

Fig. 4. Hypothetical scheme of *DR* haplotype evolution. Insertions and deletions observed in *DRB* genes are shown

gene, accompanied by the fusion of the 5' end of DRB1*04 with the 3' end of DRB2 (Satta et al. 1996a,b). The DRB4 gene then acquired the L1 element it now possess in intron 1. The DRB3 gene arose from DRB1*04 by the duplication of the DRB1-DRB2 doublet some 36 my ago (Rollini et al. 1985; Satta et al. 1996b). Before this duplication, the ancestral DRB1 gene acquired the retroviral insert ERV9LTR in intron 5 (Svensson et al. 1995, 1996; Satta et al. 1996a) and the DRB2 gene acquired another retroviral insert, MMTVLTR, in its intron 1 (Satta et al. 1996b). The DRB1 gene then diverged into two allelic lineages, DRB1*03 and DRB1*15. The latter lineage subsequently lost the Alu53 insert; evidence that the element was once present in the ancestor of HLA-DRB1*15 can be found in the intron 1 sequence of this gene (Satta et al. 1996b). The DRB1*15 lineage acquired another repeat instead, the Alu52 element in intron 5.

The proposed order of primate DRB gene divergence also provides clues to the identity of the primordial DR haplotypes (Fig. 4). The first haplotype created by the duplication of the ancestral DRB1*04 gene, followed by the divergence of the two copies, was presumably of the type DRB1-DRB2. The second and third duplications generated the primordial haplotypes DRB1-DRB7 and DRB1-DRB5, respectively. The duplication of the DRB1-DRB2 primordial haplotype, followed by the deletion of one DRB2 copy and divergence, generated the DRB1-DRB2 – DRB3 haplotype which is still present in the human population. The other extant HLA-DR haplotypes were presumably generated by more complex rearrangements involving unequal crossing-over, duplications, deletions, transpositions, and other mechanisms (Fig. 4). Because many such rearrangements probably occurred during the >60 my evolution since the inception of the primordial haplotypes, it is no longer possible to decide whether, for example, the combination DRB1*03-DRB2 found in some of the extant haplotypes is a remnant of the primordial haplotype or a recreation of the original combination.

It is, however, possible to estimate the minimal age of the DR haplotypes now present in the human population. This estimate can be based on two sources of information – the presence of similar haplotypes in different primate species and the mutual genetic relationship of the genes constituting the extant haplotypes. Limited haplotyping of nonhuman primates (Brändle et al. 1992; Kasahara et al. 1992; Slierendregt et al. 1993; Schönbach et al. 1993) has thus far revealed the presence of two human haplotypes, DRB1*03-DRB2-DRB3 and DRB1*07-DRB7-DRB8-DRB4, in nonhuman primates, the former in the gorilla, the latter in the chimpanzee. Furthermore, parts of human haplotypes have also been found in the chimpanzee and the gorilla (Fig. 5).

Genetic analysis of the haplotype-constituting genes can also provide information about the persistence of haplotypes. As long as a haplotype remains intact, its genes will coevolve and genetic distances between alleles at one locus should correlate with distances between alleles at another locus within the same haplotype group. When this is observed, the neutrally evolving parts of the haplotypes can be used to estimate how long the genes constituting the haplotypes have remained together. Recombination between haplotypes belonging to different groups disturbs the correlation, and if rearrangements were frequent in the evolution of haplotypes, no correlation between genetic distances of

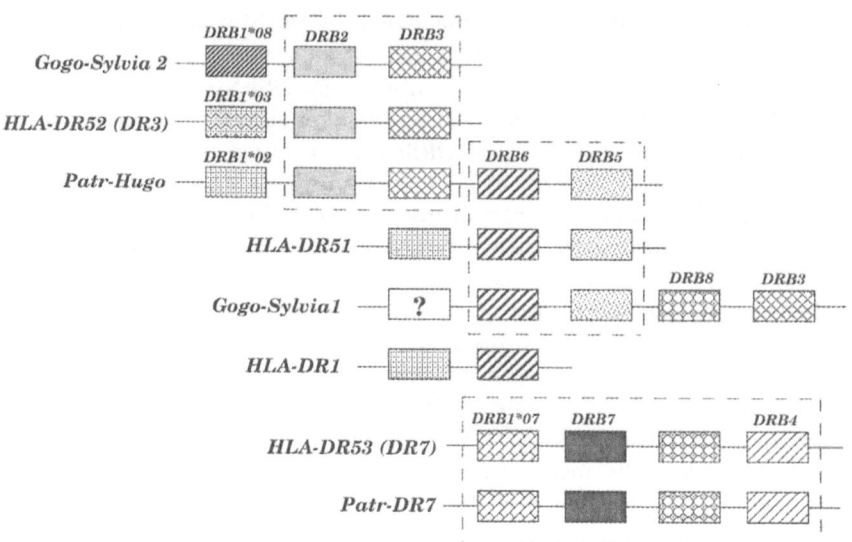

Fig. 5. The comparison of human (*HLA*), gorilla (*Gogo*), and chimpanzee (*Patr*) haplotypes. Homologous regions are indicated by *broken line rectangles*. An orthologous gene is indicated by the same *shading* pattern

alleles at one locus with those at another locus would be expected. These predictions have been put to test by using the assembly of human haplotype groups. The original classification into groups was based on the degree of similarity between alleles at the *DRB1* locus. Only later it was realized that each of the groups was also characterized by a unique constellation of loci. Recently, sequencing of a stretch of about 1200 bp at the *DRB9* locus from the major human haplotypes has revealed that the allelic forms fall into groups that correspond to those defined by the *DRB1* locus (Gongora et al. 1996). These data indicate that the gene combinations characterizing the groups have remained "frozen" (i.e., without recombination) for as long as it took the alleles at the constituting loci to diverge. Since DRB9 is a pseudogene and hence evolves free of selection pressure, the sequences could be used to estimate the divergence times of the alleles. The estimates indicate that the *DR53* and *DR52* groups have been frozen for 5 and 10 my, respectively. This conclusion is consistent with the reported presence of the *DR7* and *DR3* haplotypes in the chimpanzee and gorilla, respectively (Slierendregt et al. 1993).

The co-evolution of *DRB1* and *DRB9* genes does not, however, preclude certain types of rearrangement. Deletions, for example, can occur within a haplotype without disturbing this correlation. A case in point is the *DR8* haplotype, which contains two genes only, *DRB1*08* and *DRB9*. There is good evidence that the *DRB1*08* gene arose from the *DR3* haplotype by the deletion of *DRB2*, followed by the fusion of the 5' part of *DRB1*03* with the 3' part of DRB3 (Gorski

1989). The "before fusion" haplotype has actually been found in the gorilla (Kasahara et al. 1992). The *DRB9* sequencing places the *DR8* – associated allele into the *DR52* group, thus supporting the deletion hypothesis (Gongora et al. 1996). The deletion apparently occurred without the recombination of flanking markers.

The demonstrated conservation of haplotypes implies that not only alleles but also groups of haplotypes are passed on from ancestral to emerging species during speciation (Klein 1987). Apparently, when the genus *Homo* emerged some 3.5 my ago, all the extant major haplotypes were present in the founding population already. The coevolution of the *DRB1* and *DRB9* genes suggest that, in fact, recombination leading to reshuffling of genes within the *DR* haplotypes is extremely rare.

The reasons for the persisting linkage disequilibrium remain obscure. An obvious but probably least likely explanation is that the combinations are maintained by selection. The difficulty with this explanation is that there is no obvious advantage in keeping together alleles of a functional gene (*DRB1*) with a pseudogene (*DRB9*) or an invariant gene (*DRA*). The same argument applies to most *DRB* loci which are occupied by either pseudogenes or genes of doubtful functional significance. There is also no evidence for the presence of a non-*DR* gene in the *DR* subregion which could, like *DRB1*, be under selection pressure.

Another possible explanation for the rarity of recombination in the *DR* subregion is that any attempts at crossing-over are aborted because the haplotypes and genes are too divergent for the recombination mechanism to function properly. While this postulate may explain the lack of recombination between haplotypes of different groups and in the *DRB1* gene segment, it does not explain the paucity of recombination between haplotypes of the same group and in the *DRB9* gene segment.

Finally, the third and perhaps the most likely explanation is that the *DR* subregion contains "cold spots" of recombination – sequences that decrease the probability of crossing-over.

Whatever the correct explanation might be, the *trans*-species haplotype polymorphism offers excellent opportunities to study the genetic history of past populations, human or other. Like the allelic *trans*-species polymorphism, it makes it possible to pose questions that might be unanswerable by other genetic systems and other approaches.

Acknowledgements. We thank Ms. Moira Burghoffer for editorial assistance.

References

Andersson G, Larhammer D, Widmark E, Servenius B, Peterson PA, Rask L. Class II genes of the human major histocompatibility complex. Organization and evolutionary relationships of the DRβ genes. J Biol Chem 262:8748–8758, 1987

Bontrop R.E. *Mhc* class II genes of nonhuman primates. In Blancher, A., Klein, J. and Socha, W. (eds.) *Molecular Biology and Evolution of Blood Group and MHC antigens in Primates*, pp. 358–371, Springer-Verlag, Heidelberg

Brändle U, Ono H, Vincek V, Klein D, Golubic M, Grahovac B, Klein J. Trans-species evolution of *Mhc-DRB* haplotype polymorphism in primates: organization of DRB genes in the chimpanzee. Immunogenetics 36:39–48, 1992

Corell A, Morales P, Varela P, Paz-Artal E, Martin-Villa JM, Martinez-Laso A, Arnaiz-Villena A. Allelic diversity at the primate major histocompatibility complex *DRB6* locus. Immunogenetics 36:33–38, 1992

Figueroa F. The primate class III MHC region encoding complement components and other genes. In Blancher, A., Klein, J., and Socha, W. (eds.) *Molecular Biology and Evolution of Blood Group and MHC Antigens in Primates*, pp. 433–448, Springer-Verlag, Heidelberg

Figueroa F, O'hUigin C, Inoki H, Klein J. Primate *DRB6* pseudogenes: clue to the evolutionary origin of the *HLA-DR2* haplotype. Immunogenetics 34:324–337, 1991

Figueroa F, O'hUigin C, Tichy H, Klein J. The origin of primate *Mhc-DRB* genes and allelic lineages as deduced from the study of prosimians. J Immunol 152:4455–4465, 1994

Gongora R, Figueroa F, Klein J. The HLA-DRB9 gene and the origin of human DR haplotypes Hum Immunol 51:23–31, 1996a

Gongora R, Figueroa F, O'hUigin C, Klein J. Possible remnant of an ancient functional *DRB* region. Scand J Immunol, in press 1997

Gorski J. The HLA-DRw8 lineage was generated by a deletion in the DRB region followed by first-domain diversification. J Immunol 142:4041–4045, 1989

Haas DA, Boss JM, Strominger JL, Spies T. A highly diverged β1 exon in the *DR* region of the human *MHC*: sequence and evolutionary implications. Immunogenetics 25:15–20, 1987

Hanini RT, Henry J, Pontarotti P. Genes in the human major histocompatibility complex geography and history, or can we predict the MHC organization in nonhuman primates. In Blancher, A., Klein, J. and Socha, W. (eds.) *Molecular Biology and Evolution of Blood Group and MHC Antigens in Primates*, pp. 325–339, Springer-Verlag, Heidelberg 1997

Hughes AL, Nei M. Pattern of nucleotide substitution at major histocompatibility complex class I loci reveals overdominant selection. Nature 355:167–170, 1988

Kapitonov V, Jurka J. The age of Alu subfamilies. J Mol Evol 42:59–65, 1996

Kasahara M, Klein D, Vincek V, Sarapata DE, Klein J. Comparative anatomy of the primate major histocompatibility complex DR subregion: evidence for combinations of DRB genes conserved across species. Genomics 14:340–349, 1992

Kawai J, Ando A, Sato T, Nakatsuji T, Tsuji K, Inoko H. Analysis of gene structure and antigen determinants of DR2 antigens using DR gene transfer into mouse L cells. J Immunol 142:312–317, 1989

Klein D, Vincek V, Kasahara M, Schönbach C, O'hUigin C, Klein J. Gorilla major histocompatibility complex DRB pseudogene orthologous to HLA-DRBVIII. Hum Immunol 32:211–220, 1991

Klein J. Origin of major histocompatibility complex polymorphism: the trans-species hypothesis. Hum Immunol 19:155–162, 1987

Klein J, Figueroa F. Evolution of the major histocompatibility complex. CRC Crit Rev Immunol 6:295–386, 1986

Klein J, O'hUigin C, Kasahara M, Vincek V, Klein D, Figueroa F. Frozen haplotypes in *Mhc* evolution. In: Klein J, Klein D (eds) Molecular evolution of the major histocompatibility complex. Springer, Berlin Heidelberg New York, pp 261–286, 1991

Mayer WE, O'hUigin C, Klein J. Resolution of the HLA-DRB6 puzzle: a case of grafting a de novo generated exon on an existing gene. Proc Natl Acad Sci USA 90:10720–10724, 1993

Meunier HF, Carson S, Bodmer WF, Trowsdale J. An isolated β₁ exon next to the DRα gene in the *HLA-D* region. Immunogenetics 23:172–180, 1986

Mnuková-Fajdelová M, Satta Y, O'hUigin C, Mayer WE, Figueroa F, Klein J. Alu elements of the primate major histocompatibility complex. Mamm Genome 5:405–415, 1994

O'hUigin C. Quantifying the degree of convergence in primate *Mhc-DRB* genes. Immunol Rev 143:123–140, 1995

O'hUigin C, Klein J. The age and evolution of the *DRB* pseudogenes. In: Klein J, Klein D (eds) Molecular evolution of the major histocompatibility complex. Springer, Berlin Heidelberg New York, pp 287–297, 1991

Rollini P, Mach B, Gorski J. Linkage map of three HLA-DR β-chain genes: evidence for a recent duplication event. Proc Natl Acad Sci USA 82:7197–7201, 1985

Satta Y, Mayer WE, Klein J. Evolutionary relationships of *HLA-DRB* genes inferred from intron sequences. J Mol Evol 42:648–657, 1996a

Satta Y, Mayer WE, Klein J. HLA-DRB intron 1 sequences: implications for the evolution of HLA-DRB genes and haplotypes. Hum Immunol 51:1–12, 1996b

Schönbach C, Klein J. The *Alu* repeats of the primate *DRB* genes. In: Klein J, Klein D (eds) Molecular evolution of the major histocompatibility complex. Springer, Berlin Heidelberg New York, pp 243–255, 1991

Schönbach C, Vincek V, Mayer WE, Golubic M, O'hUigin C, Klein J. Multiplication of *Mhc-DRB5* loci in the orangutan: implications for the evolution of *DRB* haplotypes. Mamm Genome 4:159–170, 1993

Slierendregt BL, Kenter M, Otting N, Anholts J, Jonker M, Bontrop RE. Major histocompatibility complex class II haplotypes in a breeding colony of chimpanzees (*Pan troglodytes*). Tissue Antigens 42: 55–61, 1993

Svensson AC, Setterbald N, Sigurdardottir S, Rask L, Andersson G. Primate *DRB* genes from the *DR3* and *DR8* haplotypes contain *ERV9LTR* elements at identical positions. Immunogenetics 41:74–82, 1995

Svensson AC, Setterbald N, Philgre U, Rask L, Andersson G. Evolutionary relationship between human major histocompatibility complex *HLA-DR* haplotypes. Immunogenetics 43:304–314, 1996

Takahata N, Satta Y, Klein J. Polymorphism and balancing selection at major histocompatibility complex loci. Genetics 130:925–938, 1992

Trtková K, Mayer WE, O'hUigin C, Klein J. *Mhc-DRB* genes and the origin of New World monkeys. Mol Phylogenet Evol 4:408–419, 1995

Vincek V, Klein D, Figueroa F, Hauptfeld V, Kasahara M, O'hUigin C, Mach B, Klein J. The evolutionary origin of the *HLA-DR3* haplotype. Immunogenetics 35:263–271, 1992

The *DQ* Loci

U. GYLLENSTEN, T. BERGSTRÖM, AND H.A. ERLICH

Introduction

The MHC class II DQ molecule is composed of an α-chain, encoded by the *DQA1* locus and a β-chain, encoded by the *DQB1* locus. In contrast to the DR and DP molecules, both the α-chain and the β-chain are polymorphic among individuals, resulting in a vast combinatorial diversity at the antigen level. As is the case for other class II MHC molecules, the DQ antigen is displayed on the surface of antigen-presenting cells.

DQA1 and *DQA2*

Polymorphism

Extent of Polymorphism

In humans, 15 *DQA1* and three *DQA2* alleles have been found. As with the other polymorphic loci, virtually all of the sequence diversity is localized to the second exon. By comparison with *DRA1* (two alleles), and *DPA1* (eight alleles), the *DQA1*

Molecular Biology and Evolution of Blood Group
and MHC Antigens in Primates
Blancher/Klein/Socha (Eds.)
© Springer-Verlag Berlin Heidelberg 1997

locus is by far the most polymorphic of the class II α-chain loci (Marsh and Bod-mer 1995). In nonhuman primates, the *DQA1* locus has also been found to be polymorphic; the number of known alleles for the various species are: chimpan-zee, 5; gorilla, 6; orangutan, 3; gibbon, 4; rhesus monkey, 12; cotton-top tamarin, 2 (Bontrop 1994). With the exception of the rhesus monkey, there appears to be fewer *DQA1* alleles in nonhuman primates than in humans. This apparent differ-ence could, of course, reflect the number of samples analyzed in the different spe-cies. Given comparable sample sizes in the various species, it is likely that the number of *DQA1* alleles would be comparable and reasonably high. By contrast, only one or two different *DQA2* sequences have been found in these primate spe-cies (Bontrop 1994; Gyllensten and Erlich 1989). Although no mutations have been found in the coding sequences of the *DQA2* locus that would formally ren-der it a pseudogene, no evidence that this locus is transcribed has been reported.

Age of Alleles

Initial comparisons of *HLA-DQA1* sequences showed the existence of four main allelic types (usually referred to as allelic lineages), namely, *01*, *02*, *03* and *04–06*. Of these four lineages, *01*, *03* and *04–06* were present in nonhuman hominoids such as gorillas and chimpanzees and consequently must predate the separation of these species, 5–10 million years ago (Gyllensten and Erlich 1988, 1989). The *DQA1*01* lineage and the *DQA1*04–06* lineage could even be traced back prior to the separation between hominoids and Old World monkeys, i.e., langurs, baboons, and macaques (Gyllensten and Erlich 1989, 1990). Later, more extensive surveys also extended the age of most of the allelic lineages to a time predating the separation between hominoids and Old World monkeys (Kenter et al. 1992). Only for the *HLA-DQA1*0201* allele is there evidence, based on sequence compar-isons and its exclusive presence in the human species, that it could have arisen after separation of the hominoids. The ancient origins of the *DQA1* allelic lineages were originally postulated on the basis of the similarity of the second exon nucleotide sequences between species (Gyllensten and Erlich 1989; Kenter et al. 1992). These arguments could, in principle, be complicated by the possibi-lity of strong convergent evolution, resulting in similar protein sequences from different ancestral origins. Although this possibility was not supported by the analysis of the synonymous substitutions in the second exon, it remained a pos-sibility (Gyllensten and Erlich, 1989). Recently, however, analysis of sequences from intron 1 of the *HLA-DQA1* alleles provided additional support for the ancient origin of three *DQA1* allelic lineages (McGinnis et al. 1994). These intron-1 sequence analyses were also consistent with the hypothesis, based on exon 2 sequences, that the *DQA1*0201* allele was derived from the *0301* lineage (Gyllensten and Erlich 1989).

In order to assess the average age of the contemporary polymorphism, i.e., the average age of the alleles, the number of synonymous changes (silent nucleo-tide substitutions) per synonymous site can be used. We employed the model of Nei and Gojobori (1986) to estimate the number of synonymous (d_s) and nonsynonymous changes (d_n) (nucleotide substitutions resulting in an amino acid change) per synonymous and nonsynonymous site, respectively, for putative antigen recognition site (ARS) codons (Brown et al. 1988, 1993) and the rest of the second exon by using the MEGA computer program (Kumar et al. 1994; Rzhetsky and Nei 1994). As has been pointed out, this method of calculating the substitution pattern does not allow correction for deviations from equal nucleotide frequencies (Satta 1993). The d_n/d_s ratio in the absence of selection is expected to be approximately one (Kimura 1983). As nonsynonymous changes often are deleterious in functional genes, the nonsynonymous substitution rate is usually lower than the synonymous substitution rate (Kimura 1977). Most genes thus have a (d_n/d_s) ratio of less than one. However, if a gene or a region of a gene is subjected to positive selection, promoting amino acid substitution, nonsynonymous substitutions would increase rapidly and become more prevalent than synonymous substitutions in a population (Maruyama and Nei 1981). Our analysis shows that the ratios of nonsynonymous to synonymous changes (d_n/d_s) at codons not implicated as ARS (non ARS) are, with a few exceptions, less than one, indicating that these codons are subject to weak or conserving selection (Bergström and Gyllensten 1995).

Due to the presumed weak selection pressure, the number of synonymous substitutions for these codons can be taken to reflect the time since common ancestry of the contemporary polymorphism. Comparing the number of synonymous changes at nonARS codons for the class II loci reveals that *DPB1* has the lowest amount of synonymous substitutions and *DQA1* the highest amount. Thus, this analysis indicates that the average age of the polymorphism at class II loci increases in the following order *DPB1* <*DRB3* <*DRB5*, *DRB1*, *DRB4* <*DQB1* <*DQA1* (Bergström and Gyllensten 1995). The mean number of synonymous changes among the three hominoids examined is, for these seven loci, 1.9, 2.0, 4.5, 6.3, 6.4, 8.6 and 20.1, respectively. Thus, there appears to be an inverse relationship between the amount of polymorphism (estimated as the number of alleles) and the average age of the polymorphism (Bergström and Gyllensten 1995). The most polymorphic class II loci (*DPB1* and *DRB1*) are, in fact, also those with the youngest polymorphisms. At the other extreme, the *DQA1* locus, with a low or moderate amount of polymorphism, appears to have the most ancient polymorphism. It should be noted that these estimates do not reflect the age of the locus but the average age of the contemporary polymorphism. Furthermore, it should be emphasized that a set of polymorphic alleles at a locus or within an allelic lineage with an average age for the polymorphism of a million years (myr) could contain alleles as old as several myr or as young (recent) as 10–20 000 years, as indicated by the class II alleles found exclusively in South American Indian tribes (Titus-Trachtenberg 1994) The difference between the degree of polymorphism at ARS codons and the number of synonymous substitutions at nonARS codons is consistent with an increase in the selection allelic diversification at some class II loci, such as the *DPB1* and *DRB1*, at a more recent time than that for the *DQA1* and *DQB1*.

Nature, Generation and Maintenance of Polymorphism

The persistence of a number of allelic forms, such as at the *DQA1*, over considerable evolutionary time periods is inconsistent with the polymorphism being neutral. The existence of selection on the polymorphism can be detected by comparing the substitution pattern at codons assumed to participate in antigen binding (ARS) with the pattern for the remaining codons of exon 2 (Hughes and Nei 1988, 1989). At most expressed loci synonymous nucleotide changes among alleles predominate, presumably due to conserving selection (Li and Graur 1991). However, at the ARS codons of class I and class II loci, nonsynonymous substitutions are in excess, consistent with positive selection for variability (Hughes and Nei 1988, 1989). These studies provided evidence for positive selection at ARS codons but were unable to distinguish between two types of selection, pathogen-driven, frequency dependent selection and overdominance (heterozygote advantage). These two models of selection need not be mutually exclusive and it is plausible that both of these selective forces, as well as others (see below), may be involved in maintaining polymorphism at the HLA loci.

At *DQA1*, an excess of nonsynonymous changes is found at codons postulated to encode residues involved in antigen binding (Table 1). This pattern is consistent with the allelic variation being under selection. The ARS of *DQA1* appears to be under positive selection in all but three species: two New World monkeys and one Old World monkey (Table 1). This indicates a possible difference in the selection pattern between species from the two groups. The *DQA1* and *DQB1* polymorphism in some New World monkeys has also previously been shown to be more limited than in Old World monkeys (Gyllensten et al. 1994). A shift in the importance of the polymorphism at the *DQA1* may have occurred after the separation of the two primate groups and either involved relaxed selection in New World monkeys or intensified selection in Old World monkeys. Except for this difference, the ARS appears to have been under positive selection during the evolution of the primates examined. By contrast, for *DQA2*, which is essentially monomorphic within species and identical on a protein level between species, comparisons between species do not indicate selection for polymorphism (Table 1). However, there is evidence for conserving selection, consistent with the conservation of the locus sequence over long evolutionary time periods.

A number of mechanisms responsible for the maintenance of MHC polymorphism have been suggested, including viability selection, maternal-fetal interactions and nonrandom mating (Hedrick 1994). Among the viability selection models, the heterozygote advantage (overdominance) and frequency-dependent selection models have received the most attention. Although an excess of heterozygotes has been reported at MHC loci in one study (Markow et al. 1993), consistent with a selective maintenance of the polymorphism, the available data do not permit the two types of viability models to be distinguished (Hedrick 1994).

The putative mechanism(s) for generation of MHC polymorphism have been the subject of much controversy and speculation (Trowsdale et al. 1985; Kappes and Strominger 1988). Earlier proposals for a mechanism generating the allelic

Table 1. Nucleotide substitution pattern in the *DQA* locus of primates

Comparison[a]	ARS (N = 48)[b]			Non-ARS (N = 138)[c]		
	d_s	d_n	d_n/d_s	d_s	d_n	d_n/d_s
Homo sapiens (9)	3.94 ± 4.57	16.34 ± 4.75	4.1	16.05 ± 5.06	8.75 ± 2.01	0.5
Pan troglodytes (7)	11.61 ± 7.99	16.03 ± 4.91	1.4	23.37 ± 7.42	8.87 ± 2.08	0.4
Gorilla gorilla (6)	8.95 ± 6.54	17.75 ± 5.09	2.0	20.89 ± 6.32	9.80 ± 2.18	0.5
Pongo pygmaeus (3)	9.00 ± 8.13	23.49 ± 7.34	2.6	22.19 ± 7.66	16.37 ± 3.43	0.7
Hylobates lar (4)	4.32 ± 7.53	12.23 ± 4.26	2.8	5.93 ± 3.26	6.86 ± 1.91	1.2
Saginus oedipus (2)	11.13 ± 11.36	2.66 ± 2.67	0.2	13.44 ± 6.89	2.90 ± 1.68	0.2
Macaca mulatta (1.4)	9.43 ± 5.43	17.68 ± 3.92	1.9	16.43 ± 4.07	8.78 ± 1.60	0.5
Macaca arctoides (2)	0.00 ± 0.00	5.46 ± 3.90	>1.0	26.69 ± 10.67	2.86 ± 1.66	0.1
Macaca fascicularis (3)	7.16 ± 7.34	11.57 ± 4.87	1.6	24.29 ± 8.25	8.28 ± 2.36	0.3
Cercopithecus aethiops (3)	15.51 ± 11.44	11.82 ± 5.00	0.8	12.56 ± 5.80	10.54 ± 2.73	0.8
Papio hamadryas (3)	20.28 ± 15.08	22.71 ± 7.00	1.1	27.98 ± 8.73	16.05 ± 3.41	0.6
Papio leucophaeus (2)	12.03 ± 12.03	17.22 ± 7.22	1.4	32.17 ± 12.20	13.20 ± 3.75	0.4
Cebus capucinus (2)	17.73 ± 14.99	4.00 ± 3.29	0.2	6.42 ± 4.59	4.90 ± 2.21	0.8
Callithrix jaccus (2)	0.00 ± 0.00	5.44 ± 3.89	>1.0	17.57 ± 8.13	3.87 ± 1.95	0.2
DQA2 (between species)[d] (8)	5.98 ± 6.21	2.49 ± 1.75	0.4	4.63 ± 2.51	1.48 ± 0.78	0.3

Mean numbers and standard errors are expressed as percentage of synonymous substitutions per synonymous site (d_s) and nonsynonymous substitutions per nonsynonymous site (d_n) between sequences of modern human (*Homo sapiens*), common chimpanzee (*Pan troglodytes*), gorilla (*Gorilla gorilla*) orangutan (*Pongo pygmaeus*), common gibbon (*Hylobates lar*), cotton-top tamarin (*Saginus oedipus*), rhesus monkey (*Macaca mulatta*), stump-tailed or bear macaca monkey (*Macaca arctoides fascicularis*), Savanna or African green monkey (*Cercopithecus aethiops*), hamadryas baboon (*Papio hamadryas*), common baboon (*Papio leucophaeus*), capuchin monkey (*Cebus capucinus*) and common marmoset (*Callithrix jaccus*).

[a] Number of sequences compared in parantheses.

[b] ARS (antigen recognition sites) are codons 26, 28, 35, 41, 51, 53, 54, 56, 57, 62, 66, 69, 70, 73, 76 and 77.

[c] Non-ARS in the above comparison are codons 18–79, excludin the ARS codons.

[d] Species and number of alleles in the *DQA2* analysis: *H. sapiens* (2), *P. troglodytes* (2), *G. gorilla* (2), *P. pygmeus* (1), *G. gorilla* (1), *P. pygmeus* (1) and *H. lar* (2).

polymorphism emphasized the importance of putative sequence exchanges (gene conversion) based on a few human *DRB* sequences (Gorski and Mach 1986; Wu et al. 1986). Since most of the *DQA1* allelic lineages have remained unchanged over long periods, there is little evidence of sequence exchange. Instead, the differences seen among the members of an allelic lineage could be explained almost exclusively by the accumulation of single point mutations occurring on ancestral allelic types. Only for the *HLA-DQA1*0201* allele is there an indication, based on sequence comparisons and its exclusive presence in humans, that it could have been generated after the divergence of the hominoids by recombination (Gyllensten and Erlich 1989).

DQB1 and DQB2

Polymorphism

Extent of Polymorphism

In humans, 25 *DQB1* and three *DQB2* alleles have been found. By comparison with other class II β-chain loci, however, the *DQB1* locus is among those with the lowest level of polymorphism (Marsh et al. 1995). In nonhuman primates, the *DQB1* locus has been found to be polymorphic as well, with a variable number of reported alleles, e.g., chimpanzee, 12; pygmy chimpanzee, 5; gorilla, 10; orangutan, 6; gibbon, 2; rhesus monkey, 19; cotton-top tamarin, 2 (Bontrop 1994). As with the *DQA2* locus, only one or two different *DQB2* sequences have been found in any of these species (Bontrop 1994).

Age of Alleles

At the *DQB1* locus, initial studies showed that several of the allelic lineages (then denoted B1-B4) found in the human species must have origins that predate separation of the hominoids (Gyllensten et al. 1990). Some of these lineages were detected in Old World monkeys, implying an even more ancient origin. Subsequent studies, encompassing a larger sets of individuals and species and based on the analysis of an extended number of codons, identified by phylogenetic analysis a total of eight allelic lineages (I-VIII) (Otting et al. 1992; Gaur et al. 1992a,b). All of these could be found in more than one species, reinforcing the view that many of the lineages predate the separation of species (Otting et al. 1992). However, the distribution of allelic lineages present in the different species appears to vary considerably (Bontrop 1994). In the most extreme of these cases, the bonobo (*Pan paniscus*) was found to carry a number of different alleles, but all related to one allelic lineage, B3 (Gyllensten et al. 1990). This species is likely to have arisen with only a single *DQB1* allelic lineage and to subsequently have accumulated mutational differences on this lineage. In general, the extent of *DQB1* polymorphism appears to be much lower in New World monkeys than in Old World monkeys. However, since the *DQB1* lineages have evolved beyond recognition, shared lineages between Old and New World monkeys cannot be identified.

In the phylogenetic analyses of the *DQB1* polymorphism, sequences cluster to a high degree by species at the terminal twigs, suggesting that, unlike the *DQA1* alleles, the *DQB1* allelic forms have accumulated many additional changes subsequent to the separation of the different hominoid species (Gyllensten et al. 1990; Otting et al. 1992). This postspeciation diversification among *DQB1* alleles suggests that continued modification of the basic allelic types is occurring. Many of these changes probably involved point mutations (nucleotide substitutions) but some may have been generated by interallelic segmental exchange (gene conversion-like events). For example, the human *DQB1*0504* allele appears to have been generated by a polymorphic sequence motif "donated" from *DQB1*0401* to the "recipient" *DQB1*0502*.

Nature, Generation and Maintenance of Polymorphism

Analysis of the substitution pattern of the *DQB1* sequences shows that an excess of nonsynonymous substitutions can be detected at the ARS codons in most primates (Table 2). In general, the d_n/d_s ratio for the *DQB1* is somewhat lower than for *DQA1*. In the context of the phylogenetic relationships among primates, there are no indications of changes in selection during primate evolution. By contrast, the polymorphism at the ARS of *DQB2* does not appear to be under positive selection (Gaur et al. 1993; Bergström and Gyllensten 1995). Also, transcripts from the *DQB2* locus have been detected in a number of species (Gaur et al. 1991).

Additional evidence for selective forces operating on the *DQA1* and *DQB1* polymorphism, as well as on the allelic diversity at other HLA loci, comes from studies of population genetics. Not surprisingly, the primate species with the most population data is the human. Although the frequency distributions of *DQA1* and *DQB1* alleles vary significantly among different human populations, virtually all populations studied thus far have a similar property, namely, a relatively even allele frequency distribution. This distribution, as measured by the Ewens-Waterson F-statistic for homozygosity (Ewens 1972; Watterson 1978), differs significantly from that expected for a neutral polymorphism and has been interpreted as evidence of symmetric balancing selection (Begovich et al. 1992). Balancing selection is expected to maintain multiple polymorphic alleles in a population at appreciable frequencies due to forces such as overdominance (heterozygote advantage) or frequency-dependent selection. The theroretical expectation for multiple neutral alleles is that the population would contain one frequent allele with many rare alleles. The difference between the observed population distribution for *DQA1* and *DQB1* alleles and the neutrality expectation

Table 2. Nucleotide substitution pattern in the MHC class II *DQB* locus in primates

Comparison[a]		ARS (N = 30)[b]			Non-ARS (N = 141)[c]		
		d_s	d_n	d_n/d_s	d_s	d_n	d_n/d_s
Homo sapiens	(14)	14.53 ± 9.96	29.67 ± 9.96	2.0	9.65 ± 2.95	4.74 ± 1.21	0.5
Pan troglodytes	(10)	16.34 ± 11.03	13.64 ± 5.21	0.8	7.35 ± 2.78	4.00 ± 1.21	0.5
Pan paniscus	(5)	9.35 ± 9.73	2.69 ± 2.71	0.3	3.38 ± 2.18	3.35 ± 1.28	1.0
Gorilla gorilla	(9)	17.04 ± 11.48	39.28 ± 11.44	2.3	8.72 ± 3.28	6.14 ± 1.48	0.7
Pongo pygmaeus	(5)	4.17 ± 6.59	24.15 ± 7.88	5.8	18.27 ± 5.46	7.60 ± 1.86	0.4
Hylobates lar	(2)	0.00 ± 0.00	25.24 ± 11.89	> 1.0	8.84 ± 4.95	7.09 ± 2.72	0.8
Cercopithecus aethiops	(3)	14.96 ± 17.16	60.52 ± 18.79	4.0	11.61 ± 4.90	9.34 ± 2.54	0.8
Papio hamadryas	(3)	28.86 ± 22.21	37.53 ± 12.17	1.4	9.42 ± 4.37	12.66 ± 3.00	1.3
Saginus oedipus	(2)	34.90 ± 26.68	25.90 ± 12.22	0.8	13.11 ± 6.33	8.69 ± 3.03	0.7
Macaca mulatta	(17)	18.70 ± 10.30	33.44 ± 6.58	1.8	13.21 ± 3.16	8.37 ± 1.35	0.6
Macaca arctoides	(3)	7.65 ± 10.48	17.26 ± 7.70	> 2.3	10.60 ± 4.66	9.26 ± 2.51	0.9
Macaca fascicularis	(7)	28.44 ± 19.70	30.45 ± 9.56	1.1	13.01 ± 4.08	8.01 ± 1.90	0.6
Presbytis entellus	(3)	17.07 ± 15.32	49.52 ± 17.25	2.9	12.28 ± 5.22	6.84 ± 2.20	0.6
Callithrix jaccus	(2)	36.53 ± 28.06	31.91 ± 13.97	> 0.9	14.79 ± 6.80	7.07 ± 2.71	0.5

Mean numbers and standard errors are expressed as percentage of synonymous substitutions per synonymous site (d_s) and nonsynonymous substitutions per nonsynonymous site (d_n) between sequences of modern human (*Homo sapiens*), common chimpanzee (*Pan tryglodytes*), pygmy chimpanzee (*Pan pansicus*), gorilla (*Gorilla gorilla*) orangutan (*Pongo pygmaeus*), common gibbon (*Hylobates lar*), Savanna or African green monkey (*Cercopithecus aethiops*), hamadryas baboon (*Papio hamadryas*), cotton-top tamarin (*Saginus oedipus*), rhesus monkey (*Macaca mulatta*), stump-tailed or bear macaque (*Macaca arctoides*), crab-eating monkey (*Macaca fascicularis*), langur (*Presbytis entellus*) and common marmoset (*Callithrix jaccus*).

[a] Number of sequences compared in parantheses.

[b] ARS (antigen recognition sites) are codons 28, 30, 37, 38, 57, 61, 67, 70, 71 and 74.

[c] Non-ARS in the above comparison are codons 21–77, excludin the ARS codons.

is statistically significant in all populations studied thus far. This observation has been reported in populations containing many *DQA1* and *DQB1* alleles and in those groups, such as Native Americans, with a significant reduction in allelic diversity due, presumably, to population bottlenecks (Titus-Trachtenberg et al. 1995).

That balancing selection is operating at individual sites (polymorphic residues) can be inferred from inspection of the *DQB1* sequences. At position 57 of the *DQ* β-chain, a residue implicated in the formation of a potential salt bridge in the class II structure (Brown et al. 1988; Stern et al.) and in disease susceptibility (Todd et al. 1987; Horn et al. 1988), only the amino acid residues, Ala, Val, Ser, or Asp are observed. (This same constraint on the residues at position 57 is also evident in other class II β-chain loci.) There are many *DQB1* alleles that differ only at this position: *0501 (Val) vs 0502 (Ser), vs 0503 (Asp); *0301 (Asp) vs *0304 (Ala); *0302 (Ala) vs *0303 (Asp), and *0201 (Ala) vs *0203 (Asp). Balancing selection for the presence of both Asp-57 and Ala,Val, or Ser-57 alleles may be responsible for the patchwork pattern of polymorphism at this position.

Phylogenetic analysis of *DQB1* sequences also implicates position 57 as a site of high homoplasy (Erlich and Gyllensten 1991). Maximum parsimony phylogenetic trees allow the tracing of mutational changes of a single residue from the putative ancestral gene sequence to the sequences derived from contemporary species. This kind of analysis for position 57 of *DQB1* reveals a pattern in which an inferred ancestral Asp-57 allele diversifies in many different branches into the contemporary alleles containing Ala, Val, and Ser, consistent with convergent evolution and balancing selection at this site (Erlich and Gyllensten 1991).

Population studies of various human groups also reveal a characteristic property of HLA diversity that has been interpreted as evidence of selection at another level, namely, linkage disequilibrium, the nonrandom association of particular alleles at linked loci. This phenomenon, also known as gametic association, has been observed for many different combinations of HLA class I and class II loci but is particularly strong for specific alleles at the *DQA1* and *DQB1* loci. Although, in principle, linkage disequilibrium could reflect a lack of crossing over between the loci, or recent population admixture, selection for particular combinations of alleles is, in our view, the most plausible explanation. One possible explanation for *DQA1* and *DQB1* linkage disequilibrium is based on preferential pairing of some α- and some β-chains. Experimental evidence for preferential pairing in the formation of *DQA1/DQB1* encoded heterodimers has been reported in the analysis of cells containing transfected *DQA1* and *DQB1* genes (Kwok et al. 1989).

Origin and Age of *DQA* and *DQB* Loci

To generate the contemporary set of primate *DQ* α- and β-chain loci, either the whole or parts of the *DQ* region must have participated in several rounds of duplication (Hughes and Nei 1990). The simplest scheme predicts that a first

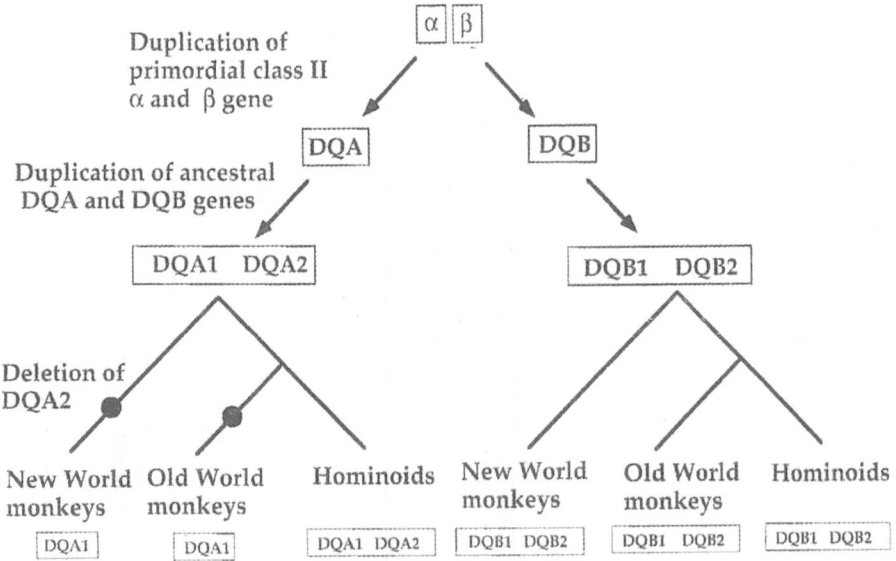

Fig. 1. Scenario for the evolution of *DQA* and *DQB* locus diversity

duplication generated the progenitor pair of *DQ* α-chain (*DQA*) and β-chain (*DQB*) loci from the primordial class II A and B loci. In a second event, the region carrying the ancestral DQ α- and β-chain loci was duplicated, resulting in the set of four loci (*DQA1, DQA2, DQB1, DQB2*) present in most primate species. The second duplication event was initially believed to have occurred in the hominoid lineage, based on the apparently exclusive presence of the *DQA2* and *DQB2* in these species (Gyllensten and Erlich 1989; Gyllensten et al. 1990). Subsequent findings of the *DQA2* locus in the orangutan and gibbon positioned the second duplication before the separation of the Old World monkeys and hominoids (20–30 myr ago) (Kenter et al. 1991, 1992). To explain the lack of the *DQA2* in Old World monkeys, we have to assume that this locus was subsequently deleted in the progenitor species of Old World monkeys, but this deletion did not involve the hominoid lineage. However, subsequent studies have also detected the presence of *DQB2* in a number of other primate species outside the hominoids (Gaur et al. 1992a, b; Otting et al. 1992; Gyllensten et al. 1994), indicating that the duplication resulting in generation of *DQB1* and *DQB2* must have occurred before separation of New World and Old World primates (Fig. 1). This implies that the deletion of the *DQA2* locus must have occurred independently in both the New World monkey lineage and the Old World monkey lineage and was retained only on the primate lineage giving rise to contemporary hominoids. Additional genomic rearrangement events must have taken place in certain species, since additional loci (such as the nonfunctional *DQB3* locus, which lacks exons 1 and 2) have been found in some species.

Table 3. Nucleotide substitution pattern in the allelic lineages of the *DQA1* locus in hominoids with and without humans

Comparison[a]		ARS (N = 48)[b]			Non-ARS (N = 138)[c]		
		d_s	d_n	d_n/d_s	d_s	d_n	d_n/d_s
A1 (all hominoids)	(6)	0.00 ± 0.00	2.24 ± 1.64	> 1.0	2.02 ± 2.03	1.13 ± 0.67	0.6
A3 (all hominoids)	(7)	12.67 ± 8.01	13.17 ± 4.95	1.0	14.18 ± 4.95	4.50 ± 1.35	0.3
A4 (all hominoids)	(6)	0.00 ± 0.00	3.58 ± 2.10	> 1.0	8.89 ± 3.82	1.67 ± 0.76	0.2
A1 (without *H. sapiens*)	(3)	0.00 ± 0.00	3.46 ± 2.47	1.0	0.00 ± 0.00	1.26 ± 0.89	> 1.0
A3 (without *H. sapiens*)	(4)	16.38 ± 10.13	13.41 ± 4.61	0.8	9.74 ± 4.19	3.40 ± 1.30	0.3
A4 (without *H. sapiens*)	(6)	0.00 ± 0.00	1.65 ± 1.66	> 1.0	7.09 ± 3.70	1.48 ± 0.76	0.2

Mean numbers and standard errors are expressed as percentage of synonymous substitutions per synonymous site (d_s) and nonsynonymous substitutions per nonsynonymous site (d_n) between sequences of three hominoid species: modern human (*Homo sapiens*), common chimpanzee (*Pan tryglodytes*) and gorilla (*Gorilla gorilla*).

[a] Number of sequences compared in parantheses.

[b] ARS (antigen recognition sites) are codons 26, 28, 35, 41, 51, 53, 54, 56, 57, 62, 66, 69, 70, 73, 76 and 77.

[c] Non-ARS in the above comparison are codons 18–79, excluding the ARS codons.

Molecular Coevolution of *DQA1* and *DQB1* Alleles

As noted above, the *DQA1* allelic polymorphism can be divided into the four main lineages *01*, *02* (unique to humans), *03* and *04–06*. Preliminary analysis of the dn/ds ratios for the hominoid alleles indicated different evolutionary rates (or different selective pressures) for these lineages (Gyllensten and Erlich 1989). This issue was reexamined using an extended set of sequences and an analysis of the substitution pattern (Bergström and Gyllensten 1995). In comparisons between the three hominoids, *Homo*, *Pan* and *Gorilla*, the three main allelic lineages showed evidence of differences in selection intensity and constraints. Consistent with the early findings (Gyllensten and Erlich 1988, 1989), the *01* and *04–06* both appear to be under positive selection at the ARS, while the *03* lineage does not show a similar excess of nonsynonymous changes (Table 3). To investigate whether the differences in selection on *DQA1* allele lineages is exclusive to humans, we also compared the d_n/d_s ratio for the different lineages using only chimpanzee and gorilla sequences (Table 3). Values similar to those found with all hominoids were obtained for the different allelic lineages in gorillas and chimpanzees.

Unlike the DR heterodimer formed by interaction of a highly polymorphic β-chain with a monomorphic α-chain, the DQ molecule consists of polymorphic α- and β-chains. Selective constraints imposed by preferential pairing may be reflected in the d_n/d_s ratios. Since the ARS codons themselves are unlikely to be involved in the interaction of α- and β-chains, analysis of these codons may not be representative for the overall pattern for the *DQA1* gene. A comparison of the d_n/d_s ratio for nonARS codons for second exon sequences of all hominoids and a comparison involving only gorillas and chimpanzees also indicate a higher ratio for the *01* lineage than for the *03* and *04* lineages. The d_s at the non-ARS codons indicates that the polymorphism at the *03* and *04* lineages is older than that of the *01* lineage.

To explain this variation in the evolutionary rate or in the selective constraints between lineages, it has been postulated that the amount of polymorphism within each *DQA1* allelic lineage is related to the diversity of the *DQB1* encoded β-chains with which the α-chain interacts to form stable dimers. In humans, the haplotype combinations of *DQA1* and *DQB1* are highly restricted (Gyllensten and Erlich 1989, 1990). In particular, the *01* lineage encoded chains can form dimers with a very restricted set of β-chain alleles encoded in *cis*, while the *03* lineage α-chains have to be able to interact with a wide variety of β-chains encoded in *cis* (Gyllensten and Erlich 1989). The existing haplotype combinations thus suggest limitations in the potential pairing of the α-and β-chains encoded by the various *DQA1* and *DQB1* alleles. Direct studies of the interaction of α- and β-chains have shown the inability of the products of the *DQA1*01* alleles to form stable dimers with *DQB1* products, except those encoded by alleles of the *DQB1*05–06* lineage (Kwok et al. 1989, 1993). A wider ability for dimer formation with a variety of β-chains has been shown by alleles of the *DQA1*03* lineage. The strong haplotype restriction of combinations of *DQA1* and *DQB1* alleles could thus be due to structural requirements for dimer formation. Thus, one might expect that the corresponding *DQB1* allelic lineages show some variation in evolutionary rate or selective constraints. Indeed, the evolutionary rate for the different *DQB1* lineages appears to vary considerably (Otting et al. 1992). Comparisons of the evolutionary rate of the *DQA1* and *DQB1* lineages revealed that the rate for *DQB1* alleles in the *B1* lineage (like the *DQB1*05* and *DQB1*06* alleles) that have to interact with *DQA1*01* is generally higher than for *DQB1* alleles in the *B2* (*DQB1*02*) or *B3* (*DQB1*03*) groups (Otting et al. 1992). These results support an hypothesis of reciprocal constraints on the evolution of the *DQA1* and *DQB1* alleles.

Summary

The *DQA1* and *DQB1* loci are polymorphic in all of the primate species studied to date. In general, the *DQA1* locus is the most polymorphic of all class II A loci while the *DQB1* locus is much less polymorphic than the *DRB1* and *DPB1* loci. With the exception of the human *DQA1*0201* allele, virtually all of the primate *DQA1* alleles and lineages predate separation of the hominoid species, consistent with the *trans*-species evolution hypothesis (Klein and Figueroa 1986). The *DQB1* lineages also predate divergence of the hominoid species, however, the alleles within these lineages show phylogenetic clustering by species and therefore have diversified considerably following speciation. Population genetics studies provide evidence for balancing selection for both *DQA1* and *DQB1* alleles and analysis of closely related *DQB1* sequences as well as phylogenetic analysis implicates individual amino acid residues (e.g., position 57) as sites of strong balancing selection. Analysis of the ratio of nonsynonymous to synonymous changes (d_n/d_s) in the antigen recognition sites (ARS) of *DQA1* and *DQB1* provides strong evidence for positive selection for diversification, whereas the d_n/d_s ratios for the non-ARS sequences as well as those for the nonexpressed *DQA2* and *DQB2* loci do not support the presence of balancing selection. Comparison of the synonymous substitution rate (d_s) for the alleles at different class II loci, used to estimate the relative age of the polymorphism at these various loci, reveals that the allelic diversity at *DQA1* is the oldest, followed by the *DQB1* polymorphism and that the polymorphism at *DPB1* is the youngest. Thus, there appears to be an inverse relationship between the number of alleles at a locus and the average age of the polymorphism.

References

Begovich AB, McClure GR, Suraj VC, Helmuth RC, Fildes N, Bugawan TL, Erlich HA et al. Polymorphism, recombination, and linkage disequilibrium within the HLA class II region. J Immunol 148:249–258, 1992

Bergström T, Gyllensten U. Evolution of Mhc class II polymorphism: the rise and fall of class II gene function in primates. Immunol Rev 143:13–31, 1995

Bontrop RE. Nonhuman primate Mhc-DQA and -DQB second exon nucleotide sequences: a compilation. Immunogenetics 39:81–92, 1994

Brown JH, Jardetzky TS, Benirschke K, Andersson JM, Brownhill LC, Saper MA, Samaraoui B, Bjorkman PJ, Wiley DC. A hypothetical model of the foreign antigen binding site of class II histocompatibility molecules. Nature 332:845–850, 1988

Brown JH, Jardetzky TS, Gorga JC, Stern LJ, Urban RG, Strominger JL, Wiley DC. Three-dimensional structure of the human class II histocompability antigen HLA- DR1. Nature 364:33–39, 1993

Erlich HA, Gyllenstern UB. Shared epitopes among HLA class II alleles: gene conversion common ancestry and balancing selection. Immunol Today 12 (11):411–414, 1991

Ewens WJ. The sampling theory of selectively neutral alleles. Theor Popul Biol 3:87–112, 1972

Gaur LK, Heise ER, Thurtle PS, Nepom GT. Is DQB2 functional among nonhuman primates? In: Klein J, Klein D (eds) Molecular evolution of the major histocompatibility complex. Springer, Berlin Heidelberg New York, pp 221–230 (NATO ASI series, vol 59), 1991

Gaur LK, Heise ER, Thurtle PS, Nepom GT. Conservation of the HLA-DQB2 locus in nonhuman primates. J Immunol 148:943–948, 1992a

Gaur LK, Heise ER, Thurtle PS, Nepom GT. Conservation of the HLA-DQB2 locus in nonhuman primates (published erratum). J Immunol 149:2530–2531, 1992b

Gaur LK, Hughes AL, Heise ER, Gutknecht J. Maintenance of DQB1 polymorphisms in primates. Mol Biol Evol 9:599–609, 1993

Gorski J, Mach B. Polymorphism of human Ia antigens: gene conversion between two DRb loci results in a HLA-D/DR specificity. Nature 322:67–70, 1986

Gyllensten U, Erlich HA. Generation of single-stranded DNA by the polymerase chain reaction and its application to direct sequencing of the HLA-DQA locus. Proc Natl Acad Sci USA 85:7652–7656, 1988

Gyllensten U, Erlich HA. Ancient roots for polymorphism at the DQA locus of primates. Proc Natl Acad Sci USA 86:9986–9990, 1989

Gyllensten U, Erlich HA. Evolution of HLA class II polymorphism in primates: the DQA locus. Immunol Res 9:223–233, 1990

Gyllensten U, Lashkari D, Erlich HA. Allelic diversification at the class II DQB locus of the mammalian major histocompability complex. Proc Natl Acad Sci USA 87:1835–1839, 1990

Gyllensten U, Bergström T, Josefsson A, Sundvall M, Savage A, Blumer ES, Humberto Giraldo L, Soto LH, Watkins DI. The cotton-top tamarin revisited: Mhc class I polymorphism of wild tamarins and polymorphism and allelic diversity of the class II DQA1 DQB1 and DRB loci. Immunogenetics 40:167–176, 1994

Hedrick P. Evolutionary genetics of the major histocompatibility complex. Am Natural 143:945, 1994

Horn GT, Bugawan TL, Long CM, Erlich HA. Allelic sequence variation of the HLA-DQ loci: relationship to serology aand to insulin-dependent diabetes susceptibility. Proc Natl Acad Sci USA 855:6012–6016

Hughes AL, Nei M. Pattern of nucleotide substitution at major histocompatibility complex class I loci reveals overdominant selection. Nature 335:167–170, 1988

Hughes A, Nei M. Nucleotide substitution at major histocompability complex class II loci: evidence for overdominance selection. Proc Natl Acad Sci USA 86:958–962, 1989

Hughes AL, Nei M. Evolutionary relationships of class II major-histocompatibility-complex genes in mammals. Mol Biol Evol 7 (6):491–514, 1990

Kappes D, Strominger JL. Human class II major histocompatibility complex genes and proteins. Annu Rev Biochem 57:991–1028, 1988

Kenter M, Anholts J, Ruff G, Otting N, Bontrop RE. Selective inactivation of the primate Mhc-DQA2 locus. In: Klein J, Klein D (eds) Molecular evolution of the major histocompatibility complex. Springer, Berlin Heidelberg New York, pp 213–220 (NATO ASI series, vol 59), 1991

Kenter M, Otting N, Anholts J, Leunissen J, Jonker M, Bontrop RE. Evolutionary relationships among the primate Mhc-DQA1 and DQA2 alleles. Immunogenetics 36:71–78, 1992

Kimura M. Preponderance of synonymous changes as evidence for the neutral theory of molecular evolution. Nature 267:275–276, 1977

Kimura M. The neutral theory of molecular evolution. Cambridge University Press, Cambridge, 1983

Klein J, Figueroa F. Evolution of the major histocompatibility complex. CRC Crit Rev Immunol 6:295–389, 1986

Kumar S, Tamura K, Nei M. MEGA: Molecular Evolutionary Genetics Analysis, version 10. Pennsylvania State University University Park. Comput Appl Biosci 10:189–191, 1994

Kwok WW, Thurtle P, Nepom GT. A genetically controlled pairing anomaly between HLA-DQα and HLA-DQβ chains. J Immunol 143:3598–3601, 1989

Kwok WW, Kovats S, Thurtle P, Nepom GT. HLA-DQ allelic polymorphisms constrain patters of class II heterodimer formation. J Immunol 150:2263–2272, 1993

Li W-H, Graur D. Fundamentals in molecular evolution. Sinauer, Sunderland, 1991

Markow TP, Hendrick PW, Zuerlein K, Danilovs J, Martin J, Vyvial T, Armstrong C. HLA polymorphism in the Havasupai: evidence for balancing selection. Am J Hum Genet 53:943–952, 1993

Marsh SG, Bodmer JG. HLA class II region nucleotide sequences. Tissue Antigens 46:258–280, 1995

Maruyama T, Nei M. Genetic variability maintained by mutations and overdominant selection in finite populations. Genetics 98:441–459, 1981

McGinnis MD, Lebo RV, Quinn DL, Simons MJ. Ancient, highly polymorphic human major histocompatibility complex DQA1 intron sequences. Am J Med Genet 52:438–444, 1994

Nei M, Gojobori T. Simple methods for estimating the number of synonymous and non-synonymous nucleotide substitutions. Mol Biol Evol 3:418–426, 1986

Otting N, Kenter M, van Weeren P, Jonker M, Bontrop RE. Mhc-DQB repertoire variation in hominoid and Old World primate species. J Immunol 149:461–470, 1992

Rzhetsky A, Nei M. Unbiased estimates of the number of nucleotide substitutions when substitution rate varies among different sites. J Mol Evol 38:295–299, 1994

Satta Y. How the ratio of nonsynonymous to synonymous pseudogene substitutions can be less than one. Immunogenetics 38:450–454, 1993

Stern LJ, Brown JH, Jardetzky TS, Gorga JC, Urban RG, Strominger JL, Wiley DC. Crystal structure of the human class II MHC protein HLA-DR! complexed with an influenza virus peptide. Nature 368:215–221, 1994

Titus-Trachtenberg EA, Rickards O, DeStefano GF, Erlich HA. Analysis of HLA class II haplotypes in the Cayapa Indians of Ecuador: a novel DRB1 allele reveals evidence for convergent evolution and balancing selection at position 86. Am J Hum Genet 5:160–167, 1994

Titus-Trachtenberg EA, Erlich HA, Rickards O, DeStefano GF, Klitz W. HLA class II linkage disequilibrium and haplotype evolution in the Cayapa Indians of Ecuador. Am J Hum Genet 57:415–424, 1995

Todd JA, Bell JI. HLa-DQβ gene contributes to susceptibility and resistance to insulin-dependent diabetes mellitus. Nature

Trowsdale J, Young JAT, Kelly AP, Austin PJ, Carson S, Meunier H, So A, Erlich HA, Spielman R, Bodmer J, Bodmer WF. Stucture sequence and polymorphism in the HLA-D region. Immunol Rev 85:5–43, 1985

Watterson GA. The homozygosity test of neutrality. Genetics 88:405–417, 1978

Wu S, Saunders T, Bach F. Polymorphism of human Ia antigens generated by reciprocal exchange between two DRb loci. Nature 324:676–679, 1986

Nonhuman Primate *Mhc-DP* Genes

B. GRAHOVAC

Introduction

The primate MHC class II region consists of six gene families, *DP, DN, DM, DO, DQ,* and *DR*, arranged respectively from centromere to telomere (Trowsdale et al. 1991). The *DM, DP, DQ,* and *DR* families each contain genes coding for the α- and β-chains of the class II molecules (the *A* and *B* genes, respectively), whereas in *DN* only one *A* gene and in *DO* only one *B* gene have thus far been identified. The MHC class II genes code for cell surface glycoprotein receptors which bind peptides derived from processed antigens and present them to helper T cells (Klein 1986; Bjorkman et al. 1990; Brown et al. 1993). The class II molecules exhibit differential tissue distribution; their constitutive expression is restricted primarily to cells of the immune system such as B cells, activated T lymphocytes, macrophages, and dendritic cells (Klein 1986).

The human *DP* region contains four loci: *DPB2, DPA2, DPB1,* and *DPA1*, situated on one chromosomal segment approximately 70 kilobases (kb) long. The genes are arranged respectively, with *DPB2* the most centromeric of the four. In contrast to other class II genes, which are oriented in a head-to-tail fashion, the *DP* genes are tail-to-tail oriented (Fig. 1; Trowsdale et al. 1984; Gorski et al. 1984; Servenius et al. 1984; Gustafsson et al. 1987). *DPA2* and *DPB2* are pseudogenes containing several defects which hinder their expression. No transcript of *HLA-DPA2* or *-DPB2* could be found in cDNA libraries or by northern blotting. *HLA-DPA1* and *-DPB1* are functional genes coding for the α and β polypeptide chains, respectively, which assemble at the cell surface to form the αβ heterodimeric DP molecules (Boss et al. 1985; Kappes and Strominger 1986).

Molecular Biology and Evolution of Blood Group
and MHC Antigens in Primates
Blancher/Klein/Socha (Eds.)
© Springer-Verlag Berlin Heidelberg 1997

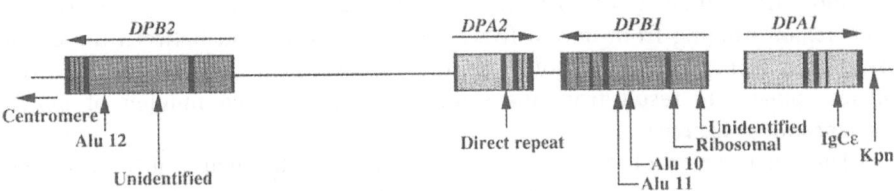

Fig. 1. *HLA-DP* region organization. Single loci are represented by *shaded boxes, vertical bars* represent exons, and *horizontal arrows* above the boxes indicate the transcriptional orientation of the genes. The approximate position of the various inserts are indicated by the *vertical arrows.* (From Klein et al. 1993)

Several insertions have been identified in the human *HLA-DP* genes (Fig. 1). Two *Alu* elements are present in intron 2 of *HLA-DPB1* and one *Alu* element resides in intron 2 of *HLA-DPB2* (Gustafsson et al. 1987; Young and Trowsdale 1985; Kelly and Trowsdale 1985). Intron 1 of the *HLA-DPB1* gene also shelters two other inserts: a processed pseudogene specifying the ribosomal protein of the L32 family and an unidentified element flanked by 18 bp direct repeats (Gustafsson et al. 1987; Kelly and Trowsdale 1985). In intron 4 of the *HLA-DPA1* gene there is a sequence strikingly similar to another found upstream of the immunoglobulin (Ig) Cε pseudogene and the IgCε functional gene. In addition, the 3′ untranslated region of this gene contains a *Kpn* repeat (Lawrance et al. 1985). The *HLA-DPB2* gene harbors an unidentified 5 kb element in intron 2 (Kappes and Strominger 1986) and the *HLA-DPA2* gene possesses a 110 bp direct repeat encompassing the 3′ end of exon 2 and the 5′ end of intron 2 (Servenius et al. 1984).

Serologic reagents specific for DP molecules have been difficult to generate and the molecules elicit a weak response only in primary mixed-lymphocyte culture (reviewed by Sanchez-Perez and Shaw 1986). These two observations have been interpreted as an indication that the *HLA-DPB1* and *HLA-DPA1* genes are expressed on the B lymphocyte surface at a level lower than that of other functional class II genes. The DP molecules, however, stimulate a strong secondary response of specifically primed T lymphocytes; hence, the primed lymphocyte test (PLT) has been the main method of allomorph detection.

HLA-DP allelic polymorphism was first described by Shaw and co-workers (1980), who used PLT to define six specificities (DPw1–DPw6). The cloning of *DPA* and *DPB* genes made possible the use of restriction fragment length polymorphism (RFLP) analysis to define ten variants (Hyldig-Nielsen et al. 1987; Mitsuishi et al. 1987). Later, the polymerase chain reaction (PCR) method permitted extensive study of *DPA* and *DPB* gene polymorphism (Bugawan et al. 1988; de Koster et al. 1991; Moonsamy et al. 1992, 1994). This study has largely been restricted to the second exon. The analysis of 65 reported *HLA-DPB1* allelic sequences (Marsh and Bodmer 1995) reveals an unusual polymorphic pattern. At both the amino acid and nucleotide levels, polymorphism within the second exon is localized almost exclusively to six variability regions with a limited number of polymorphic residues ($n=2-3$) at each position (Moonsamy et al. 1992). Within each region ranging from one to five amino acids in length, between three and five polymorphic sequence motifs exist, none of which is allele-specific. The alleles apppear to result from the shuffling of this limited number of motifs (Moonsamy et al. 1994).

The functionality of the DP molecules is indicated by their ability to present antigen-derived peptides to T lymphocytes. It has been shown that the HLA-DP molecules can present influenza and herpes virus peptides (Eckels et al. 1983) as well as peptides derived from other HLA molecules (de Koster et al. 1989; Essaket et al. 1990; Falk et al. 1994). Several reports demonstrate that HLA-DP molecules can serve as targets in the rejection of transplanted tissue (Bonneville et al. 1988) and induce graft-vs-host disease (Odum et al. 1987a; Santamaria et al. 1994). Certain HLA-DP specificities have been found to be associated with diseases such as juvenile chronic arthritis (Begowich et al. 1989), severe aplastic anemia (Odum et al. 1987b), acute lymphoblastic leukemia (Pawelec et al. 1988), and celiac disease (Kagnoff et al. 1989). The total number of known human DPB1 alleles is more than 60, rendering the DPB1 locus the second most polymorphic class II gene in this species (Marsh and Bodmer 1995).

A comparison of nucleotide sequences revealed that most polymorphisms of the MHC class II molecules are encoded in exon 2 of the functional Mhc-DRB,-DQB, and -DPB genes (Klein et al. 1993) – specifically in the β-pleated sheet of the peptide binding region (Bjorkman et al. 1987; Brown et al. 1993). MHC class II A genes are either invariant (DRA) or display a modest degree of polymorphism (DQA1 and DPA1); (Gyllensten and Erlich 1989; Kenter et al. 1992b; Marsh and Bodmer 1995).

Studies of nonhuman primate equivalents of HLA class II genes have shown that in many instances MHC polymorphism is extremely old and actually pre-dates the speciation of contemporary living primates (Gyllensten et al. 1990a,b; Mayer et al. 1992; Fan et al. 1989; Kenter et al. 1992a; Grahovac et al. 1991; Trtkova et al. 1993). These results support the trans-species hypothesis of MHC poly-morphism, formulated to explain why alleles from different species are more related to each other than are certain alleles within a species (Klein 1987).

Conservative Evolution of the Mhc-DP Region in Nonhuman Primates

The primate Mhc-DP region, in contrast to the DR and DQ regions, is genetically very stable with regard to gene content and genomic organization (Grahovac et al. 1993). The human type DP region organization was established before the Catarrhini-Platyrrhini split and has changed very little since then (Fig. 2). Conser-vation of the DP region organization is indicated on the basis of several obser-vations. First, the length of the DP region in the species tested (cotton-top

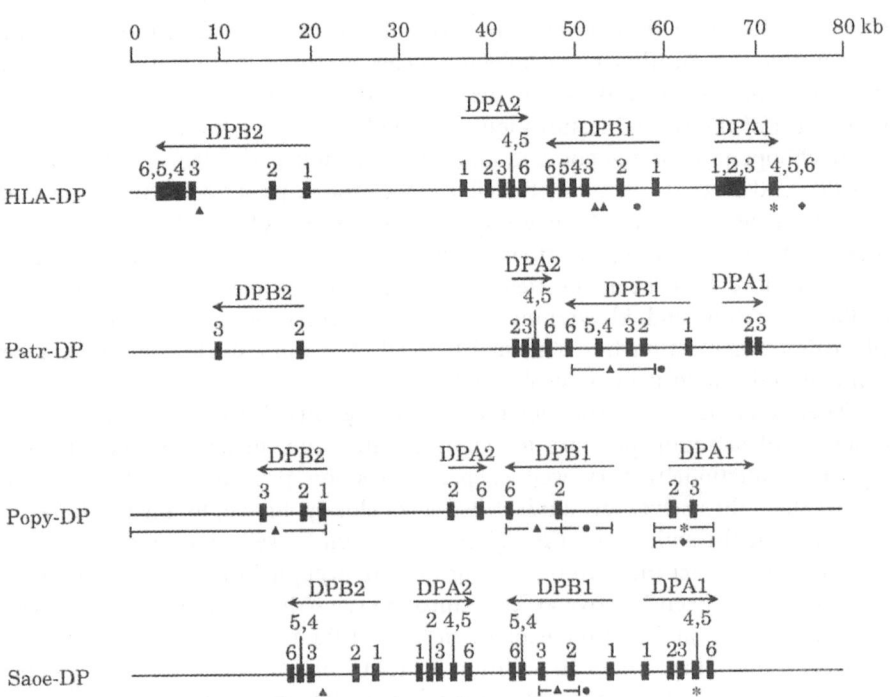

Fig. 2. The *DP* region in human (*HLA-DP*), chimpanzee (*Patr-DP*), orangutan (*Popy-DP*) and tamarin (*Saoe-DP*). The *top line* gives the length in kilobase pairs (kb). The exons are represented by *full boxes* numbered from 1 to 6. The transcriptional orientation of the genes is shown by *arrows*; the sites of the insertions of repetitive or other elements are indicated by the following symbols: Δ, *Alu* element; ◯, ribosomal pseudogene; ∗, Ig Cε gene; and ◇, the *Kpn* I repeat (uncertainty of location is indicated by a *line segment*). The map of the human *DP* region was adapted from Okada et al. (1985) and information about the location of the insertions in the *HLA-DP* region was taken from Young and Trowsdale (1985), Kelly and Trowsdale (1985), Gustafsson et al. (1987), and Lawrance et al. (1985). All other maps are from Grahovac et al. (1993)

tamarin, *Saguinus oedipus*; chimpanzee, *Pan troglodytes*; and orangutan, *Pongo pygmeus*) is between 60 and 75 kb, indicating that major deletions and insertions occurred rarely in the region's 37 million year (my) history. One such indel occurred between the *DPA2* and *DPB2* genes, which is 15 kb shorter in the tamarin than in humans and the chimpanzee. Second, restriction maps of the human, chimpanzee, and orangutan are highly similar, and even in the tamarin map several of the restriction sites appear to be in the same position as those in the human, chimpanzee, and orangutan. This constancy again shows that there have been singularly few major rearrangements in the *DP* region during its evolution from the Catarrhini and Platyrrhini common ancestor. Third, in each of the four species tested, the *DP* region contains four genes – two *DPA* and two *DPB* – which are not only arranged in the same order

(*DPB2...DPA2...DPB1...DPA1*), but are also all tail-to-tail oriented. Fourth, the exon-intron organization of orthologous genes is the same and the lengths of the corresponding introns closely resemble those of the four primate species. Fifth, most of the special features found in the human *DP* region are also present in the chimpanzee, orangutan, and tamarin *DP* regions: *Alu* elements in intron 2 of *DPB2* and intron 2 of *DPB1*, the ribosomal protein pseudogene in intron 1 of the *DPB1* gene, and the IgCε-like sequence in intron 4 of the *DPA1* gene are all present at corresponding positions in the four species. Each of these elements must therefore have been inserted into the *DP* region prior to the divergence of the Catarrhini and Platyrrhini lines. These observations point out that the *DP* region organization has remained basically the same during the more than 37 my of anthropoid primate evolution.

There may be two reasons for the region's stability. First, it could be a manifestation of selection pressure to retain a status quo, presumably because the region is functionally very well adapted. Selection pressure targets could be the *DP* loci themselves or unidentified loci closely linked to them. However, the notion of the anthropoid *DP* region evolving under selection pressure is contradicted by the fact that two of the four loci are occupied by pseudogenes which have apparently been inactive at least since the Catarrhini-Platyrrhini split (Grahovac et al. 1993) and that the expression level of the two active loci is low (Sanchez-Perez and Shaw 1986). In addition, in terms of nucleotide diversity (Marsh and Bodmer 1995), the polymorphism is relatively low in comparison to the *HLA-DRB* genes. Second, evolutionary stability in the *DP* region can be interpreted as a reflection of its waning significance: the region may simply be slowly phasing out in a functional sense. In certain taxa, such as some mouse-like rodents, the *DP* region has virtually been obliterated (Klein 1986), while in others (exemplified by the mole-rat, *Spalax ehrenbergi*) it has expanded considerably, and the DP-like molecules appear to be the main functional class II elements (Nizetic et al. 1987).

Allelic Diversity at the *Mhc-DP* Loci in Nonhuman Primates

The *DPA1* Locus

At present, only limited nucleotide sequence data on *Mhc-DP* alleles of nonhuman primates are available (Hashiba et al. 1993; Bidwell et al. 1994; Otting and Bontrop 1995; Slierendregt et al. 1995). The *DPA1* alleles have been determined in three great ape species – chimpanzee (*Pan troglodytes*), lowland gorilla (*Gorilla gorilla*), and orangutan (*Pongo pygmeus*). They have also been found in four representatives of Old World monkeys, the hamadryas baboon (*Papio hamadryas*), rhesus macaque (*Macaca mulatta*), cynomolgus macaque (*Macaca fascicularis*), and stump-tailed macaque (*Macaca arctoides*), and in one representative of New World monkeys – the squirrel monkey (*Saimiri sciureus*). The 15 nonhu-

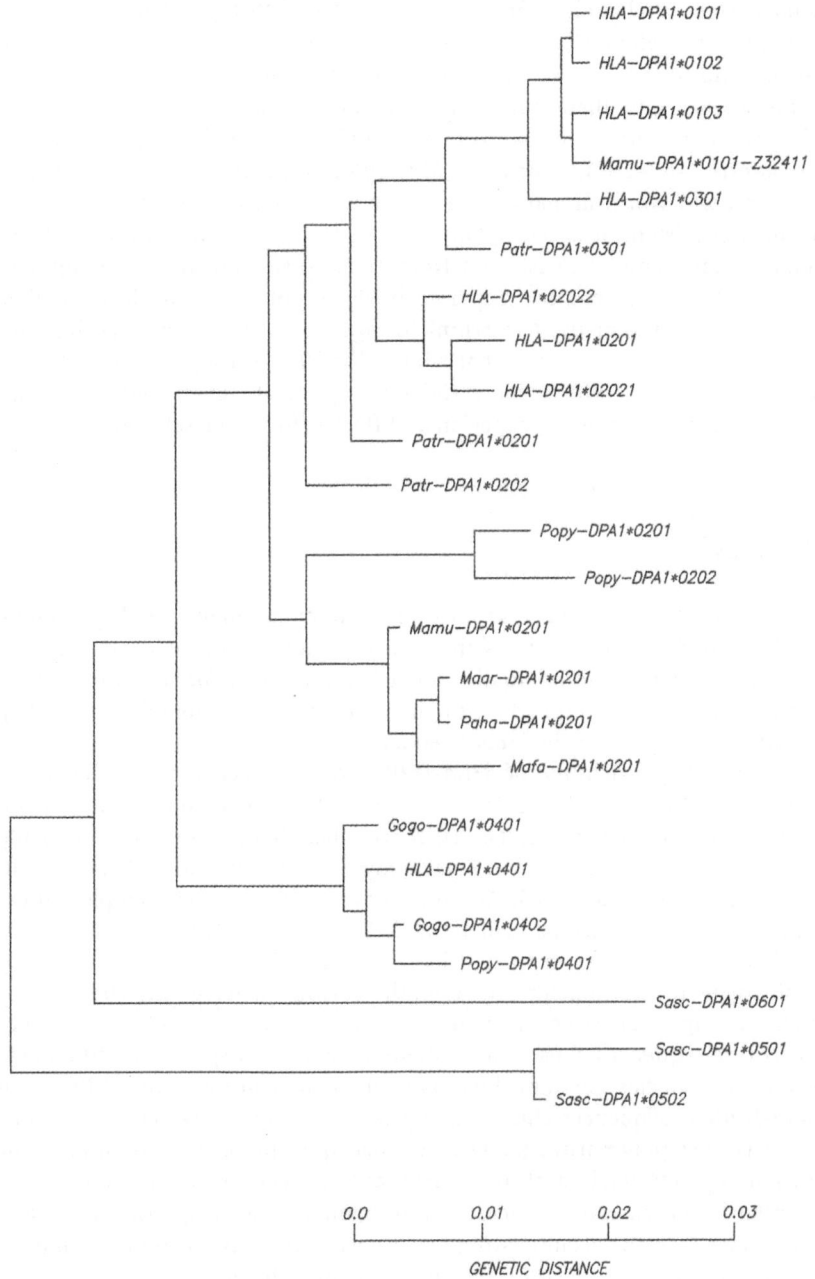

Fig. 3. Phylogenetic tree of the exon 2 of the primate *Mhc-DPA1* alleles constructed using the neighbor-joining method (Saitou and Nei 1987). The sequences are from Marsh and Bodmer (1995) and Otting and Bontrop (1995)

man primate *DPA1* alleles identified in these studies (Otting and Bontrop 1995) can be aligned with eight *HLA-DPA1* homologues and the *HLA-DPA2* paralogue and grouped into lineages based upon nucleotide sequence similarity. A comparison of the alleles with their human homologues revealed several differences, with all variations found within exon 2 probably having been generated by point mutations. At least five primate *Mhc-DPA1* lineages could be identified. Great apes and Old World monkeys share three lineages with humans, indicating that they have persisted for at least 35 my (Fig. 3). Sequences most divergent from their human equivalents were isolated from New World monkeys. The squirrel monkey alleles grouped into two separate lineages but it is not clear whether these are related to any of the Catarrhini lineages (Fig. 3). However, Otting and Bontrop (1995) have demonstrated that some *Mhc-DPA1* lineages are maintained much longer than *Mhc-DPB1* lineages (Slierendregt et al. 1995), suggesting that the *Mhc-DPA1* and *-DPB1* genes experience different forms of selection.

The *DPB1* Locus

Mhc-DPB1 alleles have only been reported for the cynomolgus monkey, cottontop tamarin, and rhesus macaque, with the exon 2 nucleotide sequence analysis of 19 rhesus macaques leading to the identification of 13 *Mamu-DPB1* alleles. These were able to be aligned with human (*HLA*), cynomolgus monkey (*Mafa*), and cotton-top tamarin (*Saoe*) sequences.

In humans, the variability of *HLA-DPB1* exon 2 nucleotide sequences is restricted to six major variable regions also present in other nonhuman primate species, often with similar motifs. The DNA stretches that separate these variable regions are conserved in humans but some show variation in Old and New World monkeys (Slierendregt et al. 1995). At amino acid position 9, for example, different codons are found in various primate *Mhc-DPB1* alleles (TAC, CAC, CTC, and TTC). Such variation can be explained by point mutations.

The phylogenetic relationships between the various primate *Mhc-DPB1* alleles are depicted in Fig. 4. The tree shows that the rhesus macaque and human *Mhc-DPB1* nucleotide sequences cluster into different clades in a species-specific manner and no *trans*-species polymorphism is evident. By contrast, *Mafa-DPB1*M09* and *Mamu-DPB1*13* sequences cluster tightly together and are evidently an example of *trans*-species polymorphism. The two macaque species shared an ancestor less than 1 my ago (Melnick et al. 1993). Humans and rhesus macaques are more distantly related and shared an ancestor approximately 36 my ago (Martin 1993). Within this time span the accumulating genetic variation has apparently removed the *trans*-species character of the *Mhc-DPB1* polymorphism.

Satta and co-workers (1994) have demonstrated that the selection coefficient of the human *HLA-DPB1* genes is much lower than that of the *HLA-DRB1*, *-DQA1*, and *-DQB1* genes. This difference correlates well with differences in the degree of polymorphism among various *HLA* loci and with the *trans*-specific retention of MHC polymorphism. The *DRB1* locus is more polymorphic than the other class

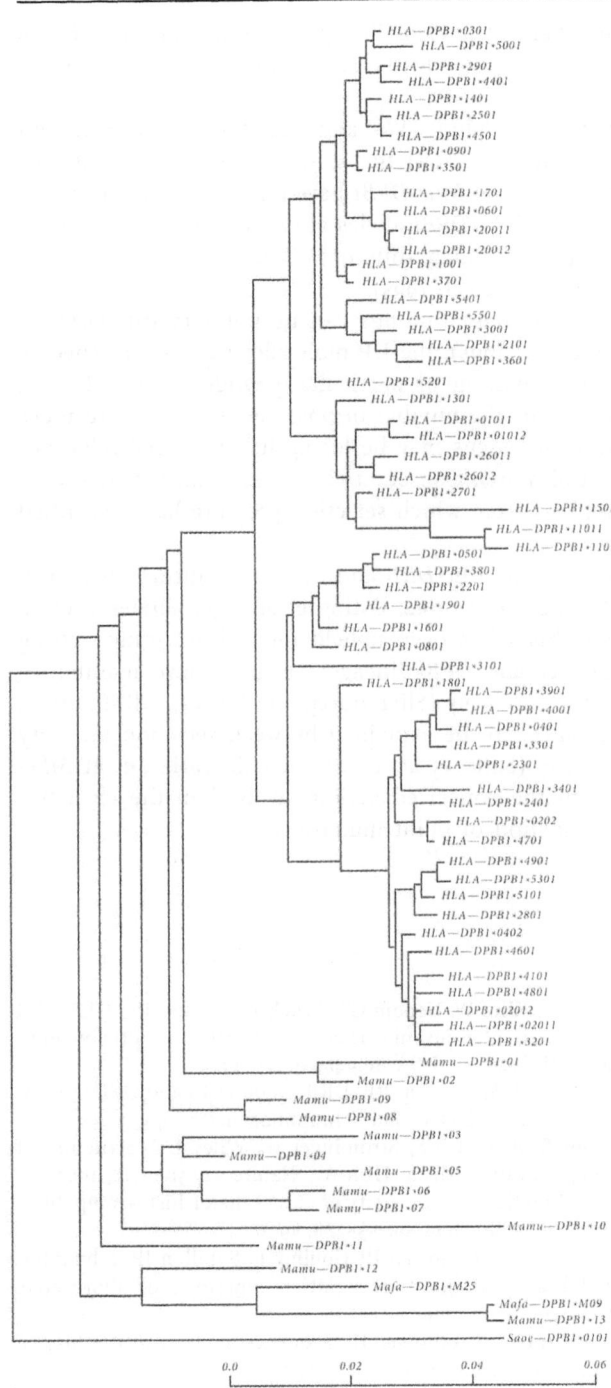

Fig. 4. Phylogenetic tree of the exon 2 of the primate *Mhc-DPB1* alleles constructed using the neighbor-joining method (Saitou and Nei 1987). The *Saoe-DPB1*0101* sequence was used to root the tree. The sequences are from Marsh and Bodmer (1995), Hashiba et al. (1993), Bidwell et al. (1994), and Slierendregt et al. (1995)

II loci and its alleles are members of ancient lineages. The human and rhesus macaque *Mhc-DPB1* alleles, by contrast, cluster separately in the phylogenetic tree (Fig. 4).

A comparison of the variability plots (Wu and Kabat 1970) of *HLA* and *Mamu-DRB*, *-DQB*, and *-DPB* sequences shows that most of the residues that contribute to the peptide-binding region of *Mhc-DPB1* have a lower variability index than those of *DRB1* and *DQB1*. The *Mhc-DPB1* alleles are more restricted than the other two loci in their ability to accumulate different amino acid residues at the peptide binding region (Slierendregt et al. 1995).

Two possible explanations for these findings can be put forward. First, as demonstrated by Slierendregt et al. (1995), the DP molecules have a very specific (restricted) function in the presentation of particular peptides to the T cells, which by some selective force keeps the number of possible amino acid residues at these positions low. Second, this locus may be losing its functional relevance (Grahovac et al. 1993; Klein et al. 1993). Conservative evolution may therefore be a characteristic of class II regions from which selection pressure has been lifted (Klein et al. 1993).

The *trans*-species origin of some *Mhc-DRB* and *Mhc-DQB* lineages has been maintained for over 58 my (Kenter et al. 1992; Otting et al. 1992; Figueroa et al. 1994) and *Mhc-DPA1* polymorphism has been upheld for at least 35 my (Otting et al. 1995). In contrast, the primate *DPB1* lineages seem to have maintained their *trans*-species character for only 5 my (Slierendregt et al. 1995; R.E. Bontrop, unpublished data). This difference can be explained by weak selection intensity operating at the *Mhc-DPB1* locus (Satta et al. 1994). Genetic variation at *Mhc-DPB1* may be due primarily to the concerted action involved in the exchange of sequence motifs and accumulation of point mutations.

References

Begovich AB, Bugawan TL, Nepon BS, Klitz W, Nepom GT, Erlich HA. A specific HLA-DPB allele is associated with pauciarticular juvenile rheumatoid arthritis but not adult rheumatoid arthritis. Proc Natl Acad Sci USA 86:9489–9493, 1989

Bidwell JL, Lu P, Wang Y, Zhou K, Clay TM, Bontrop RE. DRB, DQA, DQB and DPB nucleotide sequences of Sanguinus oedipus B95-8. Eur J Immunogenet 21:67–77, 1994

Bjorkman PJ, Saper MA, Samraoui B, Bennet WS, Strominger JL, Wiley DC. Structure of the human class I histocompatibility antigen, HLA-A2. Nature 329:506–512, 1987

Bjorkman PJ, Parham P. Structure, function and diversity of class I major histo-compatibility complex molecules. Annu Rev Biochem 59:253–288, 1990

Bonneville M, Moreau JF, Blokland E, Pool J, Moisan JP, Goulmy E, Souillon JP. T lymphocyte cloning from rejected kidney allograft: recognition repertoire of alloreactive T cell clones. J Immunol 141:4187–4195, 1988

Boss JM, Mengler R, Okada K, Auffray C, Strominger JL. Sequence analysis of the human major histocompatibility gene SX α. Mol Cell Biol 5:2677–2683, 1985

Brown JH, Jardetzky TS, Gorga JC, Stern LJ, Urban RG, Strominger JL, Wiley DC. Three-dimensional structure of the human class II histocompatibility antigen HLA-DR1. Nature 364:33–39, 1993

Bugawan TL, Horn GT, Long CM, Mickelson E, Hansen J, Ferrara GB, Angelini G, Erlich HA. Analysis of HLA-DP allelic sequence polymorphism using the in vitro enzymatic DNA amplification of DPα and DPβ loci. J Immunol 141:4020–4030, 1988

De Koster HS, Anderson DC, Termijtelen A. T cells sensitized to synthetic HLA-DR3 peptide give evidence of continuous presentation of denatured HLA-DR3 molecules by HLA-DP. J Exp Med 169:1191–1196, 1989

De Koster HS, Kenter MJ, D'Amaro J, Luiten RM, Schroeijers WE, Giphart MJ, Termijtelen A. Positive correlation between oligonucleotide typing and T-cell recognition of HLA-DP molecules. Immunogenetics 34:12–22, 1991

Eckels DD, Lake P, Lamb JR, Johnson AH, Shaw S, Woody JN, Hartzman RJ. SB-restricted presentation of influenza and herpes simplex virus antigens to human T-lymphocyte clones. Nature 301:716–718, 1983

Essaket S, Fabron J, de Preval C, Thomsen M. Corecognition of HLA-A1 and DPw3 by human CD4+ alloreactive T cell clone. J Exp Med 172:387–390, 1990

Falk K, Rötzschke O, Stevanovic S, Jung G, Rammensee H-G. Pool sequencing of natural HLA-DR DQ and DP ligands reveals detailed peptide motif constraints of proccesing and general rules. Immunogenetics 39:230–242, 1994

Fan WM, Kasahara M, Gutknecht J, Klein D, Mayer WE, Jonker M, Klein J. Shared class II MHC polymorphisms between humans and chimpanzees. Hum Immunol 26:107–121, 1989

Figueroa F, O'hUigin C, Tichy H, Klein J. The origin of the primate Mhc-DRB genes and allelic lineages as deduced from the study of Prosimians. J Immunol 152:4455–4465, 1994

Gorski J, Rollini P, Long P, Mach B. Molecular organization of the HLA-SB region of the human major histocompatibility complex and evidence for two SBβ-chain genes. Proc Natl Acad Sci USA 81:3934–3938, 1984

Grahovac B, O'hUigin C, Tichy H, Mayer WE, Klein J. Mhc class II genes of New World monkeys and their relationship to human genes. In: Klein J, Klein D (eds) Molecular evolution of the major histocompatibility complex. Springer, Berlin Heidelberg New York, pp 201–121, 1991

Grahovac B, Schönbach C, Brändle U, Mayer WE, Golubic M, Figueroa F, Trowsdale J, Klein J. Conservative evolution of the Mhc-DP region in anthropoid primates. Hum Immunol 38:75–84, 1993

Gustafsson K, Widmark E, Jonsson A-K, Servenius B, Sachs DH, Larhammar D, Rask L, Peterson PA. Class II genes of the human major histocompatibility complex: evolution of the DP region is deduced from nucleotide sequences of the four genes. J Biol Chem 262:8778–8786, 1987

Gyllensten UB, Erlich HA. Ancient roots for polymorphism at the HLA-DQ alpha locus in primates. Proc Natl Acad Sci USA 86:9986–9990, 1989

Gyllensten UB, Erlich HA. Evolution of HLA class II polymorphism in primates: the DQA locus. Immunol Res 9:223–233, 1990a

Gyllensten UB, Lashkari D, Erlich HA. Allelic diversification at the class II DQB locus of the mammalian major histocompatibility complex. Proc Natl Acad Sci USA 87:1835–1839, 1990b

Hashiba K, Kuwata S, Tokunaga K, Juji T, Noguchi A. Sequence analysis of DPB1- like genes in cynomolgus monkeys (Macaca fascicularis). Immunogenetics 38:462, 1993

Hyldig-Nielsen JJ, Morling N, Odum N, Ryder LP, Platz P, Jacobsen B, Svejgaard A. Restriction fragment length polymorphism of the HLA-DP subregion and correlation of HLA-DP phenotypes. Proc Natl Acad Sci USA 84:1644–1649, 1987

Kagnoff MF, Harwood JI, Bugawan TL, Erlich HA. Structural analysis of the HLA-DR, -DQ, and -DP alleles on the celiac disease-associated HLA-DR3 (DRw17) haplotype. Proc Natl Acad Sci USA 86:6274–6278, 1989

Kappes D, Strominger JL. Structure and evolution of the HLA class II SXβ gene. Immunogenetics 24:1–7, 1986

Kelly A, Trowsdale J. Complete nucleotide sequence of a functional HLA-DPβ gene and the region between the DPβ1 and DP 1 genes: comparison of the 5' ends of HLA-class II genes. Nucleic Acids Res 13:7515–7528, 1985

Kenter M, Otting N, Anholts J, Jonker M, Schipper R, Bontrop RE. Mhc-DRB diversity of chimpanzee (Pan troglodytes). Immunogenetics 37:1–11, 1992a

Kenter M, Otting N, de Weers M, Anholts J, Leunissen J, Jonker M, Bontrop RE. Evolutionary relationships among the primate Mhc-DQA1 andDQA2 alleles. Immunogenetics 36:71–78, 1992b

Klein J. Natural history of the major histocompatibility complex. Wiley, New York, 1986

Klein J. Origin of major histocompatibility complex polymorphism: the trans-species hypothesis. Hum Immunol 19:155–162, 1987

Klein J, Kasahara M, Gutknecht J, Schönbach C. Phylogeny of primate major histocompatibility complex DRB alleles. Hum Immunol 31:28–33, 1991a

Klein J, O'hUigin C, Kasahara M, Vincek V, Klein D, Figueroa F. Frozen haplotype in Mhc evolution. In: Klein J, Klein D (eds) Molecular evolution of the major histocompatibility complex. Springer, Berlin Heidelberg New York, pp 261–268, 1991b

Klein J, O'hUigin C, Figueroa F, Mayer WE, Klein D. Different modes of Mhc evolution in primates. Mol Biol Evol 10:49–59, 1993

Lawrance SK, Das HK, Pan J, Weissman SM. The genetic organization and nucleotide sequence of HLA-SB (DP) α genes. Nucleic Acids Res 13:7515–7528, 1985

Marsh SGE, Bodmer JG. HLA class II region nucleotide sequences 1995. Tissue Antigens 45:258–280, 1995

Martin RD. Primate origins: plugging the gaps. Nature 363:223–234, 1993

Mayer WE, O'hUigin C, Zaleska-Rutczynska Z, Klein J. Trans-species origin of Mhc-DRB polymorphism in the chimpanzee. Immunogenetics 37:12–23, 1992

Melnick DJ, Hoelzer GA, Absher R, Ashley MV. mtDNA diversity in rhesus monkeys revealed overestimates of divergence time and paraphyly with neighboring species. Mol Biol Evol 10:282–295, 1993

Mitsuishi Y, Urlacher A, Tongio MM, Mayer S. Restriction fragment length polymorphism of DP genes defines three new alleles. Immunogenetics 26:383–388, 1987

Moonsamy PV, Suraj VC, Bugawan TL, Saiki RK, Stoneking M, Rondier J, Magzoub MM, Hill AV, Begovich AB. Genetic diverity within the class II region: ten new DPB1 alleles and their population distribution. Tissue Antigens 40:153–157, 1992

Moonsamy PV, Aldrich CL, Petersdorf EW, Hill AV, Begovich AB. Seven new DPB1 alleles and their population distribution. Tissue Antigens 43:249–252, 1994

Nizetic D, Figueroa F, Dembic Z, Nevo E, Klein J. Major histocompatibility complex gene organization in the mole rat Spalax ehrenbergi: evidence for transfer of function between class II genes. Proc Natl Acad Sci USA 84:5828–5832, 1987

Odum N, Platz P, Jakobsen BK, Munck-Petersen C, Jakobsen N, Möller J, Ryder LP, Lamm L, Svejgaard A. HLA-DP and bone marrow transplantation: DP-incompatibility and severe acute graft-versus-host disease. Tissue Antigens 30:213–216, 1987a

Odum N, Platz P, Morling N, Jakobsen N, Jakobsen BK, Ryder LP, Svejgaard A. Increased frequencies of HLA-DPw3 in severe aplastic anemia. Tissue Antigens 29:184–185, 1987b

Okada K, Prentice HL, Boss JM, Levy DJ, Kappes D, Spies T, Raghupathy R, Mengler RA, Auffray C, Strominger JL. SB subregion of the human major histocompatibility complex: gene organization allelic polymorphism and expression in transformed cells. EMBO J 4:739–748, 1985

Otting N, Bontrop RE. Evolution of the major histocompatibility complex DPA1 locus in primates. Hum Immunol 42:184–187, 1995

Otting N, Kenter M, van Weeren P, Jonker M, Bontrop RE. Mhc-DQB repertoire variation in hominoid and Old World primate species. J Immunol 149:461–470, 1992

Pawelec G, Ehninger G, Müller C, Blaurock M, Schneider EM, Wernet P. Human leucocyte antigen – DP in leukemia. Cancer 61:475–477, 1988

Saitou N, Nei M. The neighbor-joining method: a new method for reconstructing phylogenetic trees. Mol Biol Evol 4:406–425, 1987

Sanchez-Perez M, Shaw S. HLA-DP: current status. In: Solheim BG, Möller E, Ferrone S (eds) HLA-class II antigens: a comprehensive review of structure and function. Springer, Berlin Heidelberg New York, pp 83–108, 1986

Santamaria P, Reinsmoen NL, Lindstrom AL, Boyce-Jacino MT, Barbosa JJ, Faras AJ, McGlave PB, Rich SS. Frequent HLA class I and DP sequence mismatches in serologically (HLA-A, HLA-B, HLA-DR) and molecularly (HLA-DRB1, HLA-DQA1, HLA-DQB1) HLA – identical unrelated bone marrow transplant pairs. Blood 83:280–287, 1994

Satta Y, O'hUigin C, Takahata N, Klein J. Intensity of natural selection at the major histocompatibility complex loci. Proc Natl Acad Sci USA 91:7184–7188, 1994

Servenius B, Gustafsson K, Widmark E, Emmoth E, Andersson G, Larhammar D Rask L and Peterson PA Molecular map of the human HLA-SB (HLA-DP) region and sequence of an SBA (DPA) pseudogene. EMBO 3:3209–3214, 1984

Shaw S, Jonhson AH, Shearer GM. Evidence for a new segregant of B cell antigens that are encoded in the HLA-D region and that stimulate secondary allogeneic proliferative and cytotoxic responses. J Exp Med 152:565–571, 1980

Slierendregt BL, Otting N, Jonker M, Bontrop RE. Major histocompatibility complex class II DQ diversity in rhesus macaques. Tissue Antigens 41:178–185, 1993

Slierendregt BL, Otting N, Kenter M, Bontrop RE. Allelic diversity at the MHC-DP locus in rhesus macaque (Macaca mulatta). Immunogenetics 41:29–37, 1995

Trowsdale J, Kelly A, Lee J, Carson S, Austin P, Travers P. Linkage map of two HLA-SBβ and two HLA-SBα- related genes: an intron in one of the SBβ genes contains a processed pseudogene. Cell 38:241–249, 1984

Trowsdale J, Ragoussis J, Campbell RD. Map of the human Mhc. Immunol Today 12:443–446, 1991

Trtkova K, Kupfermann H, Grahovac B, Mayer WE, O'hUigin C, Tichy H, Bontrop RE, Klein J. Mhc-DRB genes of platyrrhine primates. Immunogenetics 38:210–222, 1993

Young JAT, Trowsdale J. A processed pseudogene in an intron of the HLA-DPβ1 is a member of the ribosomal protein L32 gene family. Nucleic Acids Res 13:8883–8891, 1985

Wu TT, Kabat EA. An analysis of the sequences of the variable regions of Bence Jones proteins and myeloma light chain and their implications for antibody complementarity. J Exp Med 132:211–250, 1970

The *DM* and *DN/DO* Loci

C. O'hUigin and F. Figueroa

The *DM* Loci

Among the array of class II genes families known in eutherian mammals, the *DM* loci are the most recent to be described (Cho et al. 1991; Kelly et al. 1991). The sequences, determined in mouse and human, gave an indication of why the loci were not detected in rapid succession to the other class II loci. Many features are unconventional and indicate that *DM* has long diverged from the other known eutherian families. Although the unusual history of *DM* genes renders them of interest in the study of MHC evolution, to date only the human and mouse genes have been studied in detail. Characterization of the bovine *DMA* and *DMB* loci is under way (Niimi et al. 1995). Partial sequence data available from lemur (this report) give some indication of sequence conservation within primates. However, the organization and function of *DM* genes in nonhuman primates must be inferred by analysis of the human and mouse genes. Thus this section will focus on features of *DM* common to these species and compare and contrast the *DM* genes to the other class II gene families in order to ascertain what can be learned from the nonhuman primate *DM* genes.

Molecular Biology and Evolution of Blood Group
and MHC Antigens in Primates
Blancher/Klein/Socha (Eds.)
© Springer-Verlag Berlin Heidelberg 1997

DM Structure and Expression

In common with DR (H-2E in mouse), DQ (H-2A), DP (H-2P) and DN/DO (H-2O) proteins, DM (H-2M) is a heterodimer, consisting of α- and β-chains of the generalized class II structure and genetic organization. The γ-interferon inducibility and tissue specificity of expression are consistent with findings reported for other class II genes. Five domains are identified in the mature proteins, namely, membrane distal and proximal, connecting peptide, transmembrane and cytoplasmic tail. The genomic organization of the *DM* loci has been described in humans by Radley et al. (1994) and, taken together with the primary descriptions of Cho et al. (1991) and Kelly et al. (1991), it is possible to build up a picture of how the *DM* loci look.

DM shares several features with other class II families. In humans the α- and β-chain genes lie within 8 kb of each other in the region between the *DNA* and *DOB*. In the mouse the genes lie in a homologous region, the *Oa* to *Ob* interval. The β-chain locus is duplicated in mouse, with the *Mb1* β-chain distal to *Mb2* and the *Ma* α-chain gene. The α-chain has the five exons and the β-chain the six exons common to the other class II families. In both chains, a leader peptide-binding and immunoglubulin-like domains are present, encoded by exons 1–3. Connecting peptide, transmembrane and cytoplasmic domains of similar size and disposition to the known class II structures are also present. The characteristic disulfide bridges are present in the extracellular domains of the prosposed protein structure. The α-1 domain also contains a potential disulfide bridge, a characteristic known for the α-chains of lower vertebrates (Sültmann et al. 1993). A typical class II promoter region, with conserved X, Y, J and CCAAT elements, is present in both *DMA* and *DMB* (Radley et al. 1994).

The DM proteins also differ uniquely in several ways from other class II proteins. The size of the immunoglobulin-like domain is one amino acid longer and one amino acid shorter, respectively, in DMB and DMA when compared to other class II genes. The spacing between the promoter X and Y elements is slightly different in *DMB*, being 1 nucleotide longer than in other class II β molecules. In both DMA and DMB, potential glycosylation sites do not correspond with those found conserved between other class II chains. Polymorphisms, a defining feature of classical MHC genes, are rare in the *DM* genes (Carrington et al. 1993; Sanderson et al. 1994). Finally, a very high degree of sequence divergence occurs between the *DM* genes and their classical class II counterparts. Both the mouse and human gene sequences are strikingly different from other class II families. Even in those better conserved immunoglobulin-like domains only 30%–40% amino acid identity is found against 60%–80% found between the other class II families.

Figure 1 shows a nucleic acid alignment of *DMA* coding sequences in three species, human, mouse and lemur (partial). Figure 2 shows a protein alignment of these sequences with other class II sequences. Figure 3 shows the nucleic acid alignment of *DMB* coding sequences, again in human mouse and lemur (partial), while Fig. 4 shows the protein alignment with other class II chains. For both *DMA* (Fig. 1) and *DMB* (Fig. 3) the lemur sequences appear closest to

```
          107                                                     127
CONSENSUS ACG CTG AAG CCC CTG GAG TTT GGC AAG CCC AAC ACG TTG GTC TGT TTT GTC AGT AAT CTC TTC CCA CCC ACG CTG
Pefu-DMA  --- --C --- --- --- --- --- --- --- --- --- --- C-- --- --- --- --- --- --- --- --- --- T-- --- ---
HLA-DMA   --- --- --- --- --- --- --- --- --- --- --- --T --- --- --- --- --- --- --- --- --- -T- --- ---
H-2Ma     --A --- --- --- --- --- --- --- --- --- --- --T --- --C A-- --C --- --C --- --T --- --G --C --- ---
PROTEIN    T   L   K   P   L   E   F   G   K   P   N   T   L   V   C   F   .   S   N   L   F   P   P   .   L

          132                                                     152
CONSENSUS ACA GTG AAC TGG CAG CAT CAC TCC GAC CCT GTG GAA GGA GTT CCC ACT TTC GTC TCA GCC GTC GAT GGA CTC
Pefu-DMA  --G --- --- --- --- --- --- --- C-- --C N-- --- --- --- --- --- --- --- --G A-T --C --- ---
HLA-DMA   --- --- --C G-- --T --T --- -T- --- --- T-- --- --- --- -T- --- --- --T --- --- ---
H-2Ma     --C --- --- -T- --- -G- -C- --G -C- --- --G --- --CC A-C --A -A C- A-- --C A-- --- --G --G
PROTEIN    T   V   N   W   .   H   S   .   P   V   E   G   .   P   T   .   S   A   .   D   G   L

          157                          167                          177
CONSENSUS ACC TTC CAG GCC TTT TCT TAC TTA AAC TTC ACA CCG GAA CCC TCT GAC ATT TTC TCC TGC ATT GTG ACA
Pefu-DMA  --- --- --- --- --- --- --- --- --- --- --C --- --- --- --- --- --- --- --- --- --- ---
HLA-DMA   -G- --- --- --- --- --- --- --- --- --A --- --- --T --- --- --T --- --- --C --- --T
H-2Ma     --- --- --- -T- --C --- --T --- --- C-- -T- --- C-- --A- --C- --- --- ---
PROTEIN    .   F   Q   A   F   S   Y   L   N   F   T   P   E   P   D   .   S   C   .   V   T
```

Fig. 1. Alignment of lemur DMA coding regions. A *dash* indicates identity with the consensus sequence, determined by simple majority. Numbering follows the scheme of Kelly et al. (1991). Conserved amino acid residues are indicated using the standard single letter code. The lemur, *Petterus fulvus* (Pefu), sequence was obtained by PCR amplification from a cDNA library (Figueroa et al. 1994) using primers based on human and mouse sequence to amplify a central portion of the gene

Fig. 2. Alignment of MHC class IIα protein sequences. A *dash* indicates identity with the consensus sequence, determined by simple majority. An *asterisk* indicates an indel introduced to improve alignment, while a *dot* indicates unavailable sequence. The numbering follows that of the HLA-DMA protein (Kelly et al. 1991). The species designations follow the standard usage for the MHC and include human (*HLA*), mouse (*H-2*), lemur (*Pefu*), sheep (*Ovar*), wallaby (*Maru*) and shark (*Gici*).

```
                -25        -18        -8          2          12         22         32         42         52         62
Consensus  MALEGALFSI GAVVLGALTL AVLLSPSWAV LGASGAIKAD HVGIYGFLHT AFYQSYDPSG QFTFEFPGDE LFSVDLKKKE TVWRLPEFGD FASFDAQGGA
HLA-DMA    -GH-Q***NQ --AL-QM-P- LW--PH---- PE-PTPMWP- DLQNHT---- VYC-DGS--V GLSEAY-E-Q --FF-FSQNT R-P---A- W-QEQGDAP-
Pefu-DMA   -EH-Q****S --L--RL-R- LW--PH---- E-TPVLW-    DPQNHY-R-- L-C-DGI-NI GLSETY-E-- --F-FSQNT   R-P--D-AE W-QGQGDAS-
H-2Ma      --RAG-***  ***:--FH-  MT--:-**** *QEA--T-   -M-S-:-**P -------GA- -------H--E EQ ---S--A-  --R--P--*
HLA-DNA    .:-FVE-***  ***:-PVL--V MSF--:**** *RGVR----- -M-S-:-***P -------A-- ------YD--E Q ---R----N --Y--N--*
Ovar-DNA   .:-SRV-***  ***I-RT-S- V-----***** *-T-QS-E-- -M-S-:-***P -------ES- ------Y-Q-E S -I--SH--   -S--:--S-*
H-20a      --IS-*V***  **P---RFFII ---M-A****  *QE-W----- TI--QA****  E--LNP-Q-- ------E-M-D I--H-MA---  -E--A*
Maru-DNA   --TI---***  ****-RFFFI  ---M-S**** *QK-W----E TI--QA****  E--LLP-KR- ------E-M-D I--H-IE-S   -E--A*
HLA-DRA    -TSNKS-***  ****I---FI- S--G-W**** *--*R----EN -AS--**** E---THE-    ------E-M-D I--H--N---  -E--A*
H-2Ea      -I-NK-***   ****L---A  TTVM--C**** *-G*ED-V-- --AS-****V NL---G----  -----Y-H-- Q-Y--GR-    --C--VLRQ
Maru-DRA   -RP-DRM-H-  R--I-R--S- -F--L*****  *-RGA----- -ST-A****  *-V-THR-T-  ------M-M- E- M--Y--D-  --H-E--Q
HLA-DQA1   -EARN*Y-*   ****V-V*   -FLV-IQG*** *-WA-KYIY  FTQV-*****  *-V-QRS-EK H-DVME-    ------I-YM-FNL-K
HLA-DPA1                                                                                                        E-A-I---AH
Gici-DAA                                                                                                        LYMQGGEA-*
```

```
                72         82         92         102        112        122        132        142        151        161
Consensus  ILAANIAVIK ANLDILVKRS NRTPATNVPP EVTVFPKSPV ELGQPNTLIC FVDNIFPPVL NVTWLRNGQP VTEGVSETSF LARDDHSFRK FHYLFVPSA
HLA-DMA    --FDKEFCEW MIQQ-GP-LD GKI-VSRGF- IAE--TLK-L -F-K----V- --S-L---M- -T-N-HDHSV --FGP-FV   S-Y-GL--QA S--N-T-EP
Pefu-DMA   -AFGKSFCEM LMREVSF-LE GQI-VSRGLS VAE--TIK-L -F-K----V- --IS-L---T -T-N-QLHSA --*--GP-FV S--GLT-QA  S--N-T-EP
H-2Ma      *--*G--R-- H--EV--E-- --SR-I----  R--L--R-   Q--E-V----  --D------I -----A--H-  --*---AQ-  YSQP--L--
HLA-DNA    *--VS--M--  H--EV--E-- --G-R-P---  R--L--R-   --E-I----   --I-------  T---H----   --*--TQ-   YSQP-R---
Ovar-DNA   *--M*S-SM-  -R-EK-TC-- --SR-IS---  L--TR----   -SE------   --L-----V-  -IK-----V   --I-K--AQ- YSQPN-R--
H-20a      *--*---I--  Y--I------ --Y--I----  --SR-IS    -LTN-----   --RE----V-  --T------K  --I-T----D YS-P--K--
Maru-DNA   *--*---D-- --VMKE---- -N--DA--A-  --LSR----   -G----  --N-E---I-  --IK------  --T-------V Y--P-L-NT
HLA-DRA    *--*-L--D- --E-MM---- -N--N-I---  -----S-     -----  --T-------  --I-R-----  --I-D--F--V --P-S--A-
H-2Ea      *--T*---T- H--N-NI--- -S-A---EV-  -----E-     -----  --I-KFS---  --T---H--V  --P-S--A-  --I----
Maru-DRA   *--*---ILN N--NT-IQ-- -H-Q--D---  -----E      -----  --HI-KP---  --I--C-EL   --A-SL---  IS----
HLA-DQA1   *--IS---IV- N--KVVMNL- GG--EPK---  S-YSEDL    -W--L      --A-GFY--HI TMK-R--NE-  M-D-DNI-E  YIK-FTY-R  S--SI---P
Gici-DAA
```

```
                171        181        191        200        210        220        230        235
Consensus  EDFVDCKVEH WGLDQPLLKH WEPQVPTPLP ETTETVVCAL GLALGLVGII VGTVLIIKGT RSSNASAGRR QGYPPL
HLA-DMA    S-IFS-I-T- -V-RNAL-*S -V-NAL-*S   DLL-N-L-GV AFG-VL---  -I----YFR  KPC****S-D.
Pefu-DMA   S-IFS-I-T- EIDRYTALAY -V-NAL-*S  DLL-NAL-GV AP---VL-T-  -I-I-FFLCSQ -PC***S-D.
H-2Ma      F-L-S-T-T- EIDRYTALAY -L---I-P-   DAM--L--- --I---FL   YV-**-VP-.
HLA-DNA    D-V------- ----T--FQ- ----V----   D---LI-G   --I--G-FL  CL-**--P-.
Ovar-DNA   -V-------- ----E----- ----L-S-V-  D---LI-G    -V------  L--M-T-    -RP**-IR-.
H-20a      --V----R-- ----R----- ---FDA-S-V- ---N-M---   --V-----  I--I-R-M   -K--*RFQH   R-**-*
Maru-DNA   D-T----E-D ----EE--R-T -FEEK-L-- ---N-M---    -FV----V I--F--M-I  KKR*NVVE- R-***A
HLA-DRA    D-T-Y----- ---E--VV-- ---E-R-I--  -L------   -SV----V   -I-P-R-L  V**G-S-H  -*T
H-2Ea      DEI------- ----R----- --EI-A-MS   -L-----    --V---  --N-T-R-GS R-***
Maru-DRA   G-M-S-H--- SS-QD-VTVF -DQG--EEKS  G*PG-II-   -T--IISAV  --II-L--ER QRLQ-QQHGT
HLA-DQA1
HLA-DPA1
Gici-DAA
```

Fig. 3. Alignment of lemur DMB coding regions. A *dash* indicates identity with the consensus sequence, determined by simple majority. Numbering follows the scheme of Kelly et al. (1989). Conserved amino acid residues are indicated using the standard single-letter code. The lemur, *Petterus fulvus* (Pefu), sequence was obtained by PCR amplification from a cDNA library (Figueroa et al. 1994) using primers based on human and mouse sequence to amplify the central portion of the gene

Block 54 (54 · 64 · 74):

	54										64										74				
CONSENSUS	TTG	GCC	AAT	AAC	CTG	TCA	AAG	AAC	CTC	AAC	CAA	AAA	GAC	ACC	CTG	CTC	CAG	CGC	TTG	CGC	AAT	GGG	CTT	CAG	GAC
Pefu-DMB	---	---	---	--G	TG-	---	G-T	T--	---	---	--G	C--	---	---	---	---	---	---	C--	---	-G-	---	--C	---	---
HLA-DMB	--G	---	GT-	--C	C--	---	C--	C--	---	---	---	---	---	A-G	---	---	---	-T-	---	-AA	---	---	-A-	-C-	A-T
H-2Mb	---	G-A	-T	T-T	G-A	-G-	-T	-G-	A--	G--	-GT	-T	--G	-T	-G-	-G	G--	-G	-C-	A--	G-	-C-		
PROTEIN	L	A	·	·	·	S	·	·	L	N	·	·	·	·	L	·	Q	R	L	·	G	L	·	A	·

Block 79 (79 · 89 · 99):

	79										89										99				
CONSENSUS	TGT	GCC	ACA	CAC	ACC	CCC	TTC	TGG	GGA	TCA	CTG	ACC	CAC	AGG	ACA	CGG	CCA	CCA	TCT	GTG	CAA	GTA	GCC	CAA	
Pefu-DMB	---	--G	---	---	---	---	---	---	--N	---	---	---	A--	---	---	---	A--	---	---	---	---	---	---
HLA-DMB	---	---	---	T-C	---	---	---	---	---	A--	---	A--	---	-A	-A-A	-G	---	--C	-G-	-C	-G	---	A--	---
H-2Mb	T-C	---	---	---	AAT	G-G	---	---	---	---	---	---	---	---	---	---	---	---	---	---	---			
PROTEIN	C	A	·	H	T	Q	P	F	W	·	L	T	·	R	T	R	P	P	·	V	·	V	A	·

Block 104 (104 · 114 · 124):

	104										114										124				
CONSENSUS	ACC	ACT	CCT	AAC	ACG	AGG	GAG	ATG	CTG	GTG	CCT	GTG	TGC	TAC	GTG	TGG	GGC	TTC	TAT	CCA	GCA	GAT	GTG	ACC	
Pefu-DMB	---	---	---	---	---	---	---	---	---	---	---	---	---	---	---	---	---	---	---	--C	---	---	---	
HLA-DMB	---	---	---	---	---	---	---	---	---	---	---	-T	---	-T	---	---	---	---	---	---	---	-A-	---	-T
H-2Mb	---	--A	---	--A	---	---	---	---	---	---	---	-C	---	--C	---	---	---	---	---	---	--G	---	---	-T
PROTEIN	T	T	P	N	T	R	E	P	V	M	L	A	C	Y	V	V	W	G	F	Y	P	A	·	V	T

Block 129 (129):

	129						
CONSENSUS	ATC	ACG	TGG	AAG	AAT	GGG	CAG	C
Pefu-DMB	---	---	---	---	---	---	---	
HLA-DMB	---	---	---	-G-	---	--C	A--	
H-2Mb	---	--A	---	---	---	---	---	·
PROTEIN	I	T	W	·	K	N	G	·

```
                -18        -8         3          13         23         29         34         44         49         59
CONSENSUS       MAAMVGLGLG GAPWVVALLV NLMVLDSSVA LGRDSPEDFV IQAKAECYFT NGTLERVRFL VRYIYNQEEY ARFDSDVGEF VAVTELGEPD AEYWNSQKDL
HLA-DMB         -ITFLP-L-- LSLGCTGAGG FVAHVE-TCL -DDAGTPKDF TY****-ISF -KD-*****- TCWDPEENKM -PCE*****- GVLNS-ANVL SQHL-QKDT-
Pefu-DMB        .......... .......... .......... .......... .......... .......... .......... .......... .....-ANCL SD-L-Q-DT-
H-2Mb           ---LWL-L-V LSLHCMGAGG FVAHVE-TCV -DDAGTPQDF TY****-VSF -KD-*****- ACWDPIVGKI VPCE*****- GVLYP-A-NF SRIL-KEES-
HLA-DOB         ......M-S- WV-------- --TR----MT Q-T------- -----D---- ---*-K-Q-V --F-F-L--- V-------M- --L-K--Q-- --Q---RL--
Patr-DOB        ......M-S- WV-------- --TR----MT Q-T------- -----D---- ---*-K-Q-V --F-F-L--- V-------M- --L-K--Q-- --Q---RL--
H-2Ob           ......M-A- R--------- ---R---FMI E------N-- -----D---- ---*-K-HL- --F-F-L--- LH----L-M- --L------- -DQ--KRL--
Orcu-DOB        ......MDSK WV-------- --IR--Y-MT Q-*------- -----D---- ---*-Q--VV --F-F-L--- -H------M- --L----Q-- --S--NRP-I
HLA-DRB1        ...--C-K-P -GSCMA--T- T----S-PL- -AG-TRPR-L E---S--H-F ---*------ D--F------ V--------Y R------R-- ----------
Pefu-DRB        ...--C-WFP -GS-TA--T- I--M-SPPL- ----TRPR-L E---S--H-Y ---*------ D--FH-R--- ---------- R------GDA EYWNSLKNI-
Susc-DQB1       -SG--A-R-P RGL-TA--T- M-V--GAP-- E-----Q--- F-F-G----Y ---*Q---GV A--------H L--------- R---P--R-E -DS------V
HLA-DPB1        ...-MV-QVS A--RT---TA L----LT--V Q--AT--NYL F-GRQ---AF ---*Q**--- E-----R--F ---------- R------R-A ---------I
Maru-DAB1       .......... ........A- I-L--T-Q-- A--HA-K--L W---S--H-E ---*QH---M D--F--RQ-A L--------Y ---S---RLQ --L--KREEI
Gaga-BL-BIIB    ....M-R-VP --GA-LVA-- A-GARPARTS SEPRRDGF-Q WTF----HYL ---*--A--- E-H---RQQF MH------KY --D-P---RQ --I---NAEI
Xela-DAB        .......... ...SG-SVR- VSVL-TL--C -CSSL---Y- F-W-SQ--YR ---*DN--Y- EG-AN----- -Y-------Y K-KNDF--VQ -K--------
Sasa-c144       PV-V-P-*-S QQRNMSMSIF CVSLTLVLSI FSGTDG**YF YHMMTQ-RYS SKD-QGIELI TS-VF--A-N I--N-T--K- -GY--H-VKN --A--KGPE*
Auha-231a       .......... ......SSFL C**LSLLFIS -STADG**-L SYSVDR-Q-N STE-KDIEYI RSVY--KL-I F--S-SL-K- -GY--Y-VKQ -D-R-ND-AI
Dare-DAB1*01    .......... ..VKMYL-IL F-AI-ML-TF T-TADG**YY QYTML--IYS TSDYSDMVL- ESGSF-KVVD VQYN-T--KY -GY--Q-VIF -RNF-KNQAY
Gici-DAB*11     -LSLG-AASR SLWIRIPF-I AVL-IFNGER DSGAGA**HS EISLHR-V-N STG*DW-*-- KQQV-D--LI -YY-YNQRK- Y--KAWMKSN VDR--REGA*

                69         78         88         98         108        118        128        138        148        158
CONSENSUS       LERSRAAVDT VCRHNYQLGE PFTVQRRVQP EVTVYPTKTP PLQHHNLLVC SVTGFYPGDI KVTWFRNGQL EETAGVVSTG LIRNGDWTYQ TLVMLEMTPE
HLA-DMB         MQ-L-NGLQN *-ATHT-PFW GSLTN-TRP- S-Q-AK-TPF NTREPVM-A- Y-W----AEV TI--RK--K- VMPHSSAHKT AQP------- --SH-AL--S
Pefu-DMB        -Q-L-SGLQD *-ATHT-PFW GSLTH-TRP- T-Q-AQ-TPF NTREPVM-A- Y-W----A-V TI--MK---. .......... .......... ..........
H-2Mb           -Q-LQNGLPD *-ASHT-PFW NALTH-TRP- S-R-AQ-TPF NTREPVM-A- Y-W----A-V TI--MK---- VPSHSNKEKT AQP------- -VSY-AL--S
HLA-DOB         -----Q---G ------R--A ----G-K--- ------ER-- L-HQ----H- ---------- -IK--L---* --R--M---- P-------F- -V--------
Patr-DOB        -----Q---G ------R--A ----G-K--- ------ER-- L-HQ----H- ---------- -IR--L---* --R-R-M--- P-------F- -V--------
H-2Ob           --T-----NM ---QK-K--A ----E-N-P- ------ER-- L--Q----L- ---------- S-K------* --RS--M--- -V------F- -T-----I--
Orcu-DOB        -----RS--F L--R--R--A ----G-K-P- -G--I-ER-- VCGQQ\--L- -A----\--- R-R-L----* --R--IM--- --------F- -TE-------
HLA-DRB1        --QR--E--- Y-----GVV- S--------- K-----S--Q ---------- ---S-----S- E-R------* --K------- --Q-----F- ------TV-R
Pefu-DRB        DDERA-VDTV *-----GII- S-------E- K-----A--Q ---------- ---N-----N- E-K-L---R* --E-AED--- --Q-----F- ------IV-Q
Susc-DQB1       --QM--E--R --K----IE- GT-L------ T--IS-S-AE A-N------- A--D---SQV --Q------* ---------P ---------- V------NLQ
HLA-DPB1        --EK--VP-R M-----E--G -M-L------ R-N-S-S-KG ---------- H--D----S- Q-R--L---* ---------N --------F- I--------Q
Maru-DAB1       --DA------ L-----EIL- R-L-P---E- ------S-LA --G------- ---------V E-R--L---* ---------- -VS------- M--------K
Gaga-BL-BIIB    --DEMN---- F-----GV-- S-----S*E- K-R-SALQSG S-PETDR-A- Y------PE- E-K--L--R* ---ER----D VMQ------- V--V--TV-R
Xela-DAB        -DQK--V--- --------YK -Y--D-KS-- D-KIVN---L D-E-E--IT- I-D--F-PM- ----LK--I* --REQ-T-SA -LK-----FE IH----T-IK
Sasa-c144       -AGELGVLER Y-KF-APIDY SAILDKT-E- H-RLSSVAP- SGR-PAM-M- -AYD-F-KP- R---L-D-R* -VKSD-T--E ELA----Y-- IHSH--Y--R
Auha-231a       -SSMK-QKE- Y-HN-IGVDY SAILSKS--- S-KLHSM-PA SSH-PAM--- --YD---SE- --S-L-D-K* -I-SD-T--E EMAD---Y-- -HSH--Y--R
Dare-DAB1*01    -QQRK-E-ES F----A-ISD SAVRDKA-K- K--IQSVMQA EGK-PAM-L- DAYE---KK- -MS-L-DDK* VV-SD-T--I EMA--N-Y-- IHSH--Y--K
Gici-DAB*11     *-EQYESGKA Y-E--IPIVY ESALA-Q-E- K--IRTKEST YPGPSAI-I- YAV----AK- S---LK---* KVSDAD-TVE -LS------- VRQY-QYE-V

                168        178        188        197        207        217        226        236        245        245
CONSENSUS       SGDVYTCQVE HSSLTSPVSV EWRAQGLSES ARSKMLSGAG GFVLGLIFLL VGLIIYLRAQ KGSVRTQLSG SLVSRAVLLP QSCKAGPGHV T
HLA-DMB         Y--T---V-- -IGAPE-ILR D-TP*---PM QTL-VSVS-V TLG----IFS L-V-SW*-RA GH-SY-P-P- -NY-EGWHIS .......... .
H-2Mb           Y------V-Q --GTSE-IRG D-TP*---PI QTV-VSVS-A TLG--F-IFC --FFRW*-KS HS-SY-P--- -TYPEGRH.. .......... .
HLA-DOB         L-H----L-D ----L----- -----**--Y SWR-----IA A-L------- --IV-Q---- --Y----M-- NE-------- ---....... .
Patr-DOB        L-H----L-D ----L----- -----**--Y SWK-----IA A-L------- --IV-Q---- --Y----M-- NE-------- ---....... .
H-2Ob           L--I-S-L-- -PG-LR---- A-M--**--Y SWK-I----A V-L----VF- --VV-H-K-- -A--E--PGN ESR--MMERL TKF------- -
Orcu-DOB        L------L-D -P--L----- -YARP**QSN IWA-----VA --LF------ --T--HI--W -ICGD-A-CC C......... .......... .
HLA-DRB1        --E------- -P-V---LT- ----R**--- -Q------V- ------L--G A--F--F-N- --HSGL-PT- F-S....... .......... .
Pefu-DRB        --E------- -P-RM--LT- -----**--- -Q------V- ------L--G ---FV-F-N- -DTLDFSQQD -......... .......... .
Susc-DQB1       R------R-- ----QN-IL- -----**--- -Q------V- ---------G L--F-RH-S- --L--..... .......... .......... .
HLA-DPB1        Q--------- -T--D---T- --K--**-D- ----T-T--- -------ICG --IFMHR-SK -VQRGSA... .......... .......... .
Maru-DAB1       H--------- ----QK--IL D-K--**--- -Q------V- -L------FG ----V-K-N- --NRGS-PT- L-S....... .......... .
Gaga-BL-BIIB    R--S-V-R-- -A--RQ-I-Q A-EPP**ADA G---L-T-V- ------V--A L--FVF--G- --RPVAAAP- M-N....... .......... .
Xela-DAB        H--TF----- ----QQ---- N-EPD*V--- --N---T-IV -----S--II ---VV---SK -TMT-FSVVQ NENLMS.... .......... .
Sasa-c144       --EKIS-M-- -I---E-MVY H-DPS*-P-A E-N-IAI--S -L---A-LA- A---Y-KKKS S-VL...... .......... .
Auha-231a       --EKIS-V-- -A--KE-LKT D-DPS*MP-- E-NNVAI--S -LI----LS- A-F-Y-K-KA R-RILVPSN. .......... .
Dare-DAB1*01    --EKIQ-V-- -A-S-Q-ITK --NPH*I--- D-N-FAI--S -L---I-IAI A---Y-KKKS T-RILVPN.. .......... .
Gici-DAB*11     Y--K---H-- -------M-- D-EVEST-K- EKT-IIV-VL --GF---I-- A-V-MR-KNA -AILDSNHGP R*LMGPAVS. .......... .
```

Fig. 4. Alignment of MHC class IIB protein sequences. A *dash* indicates identity with the consensus sequence, determined by simple majority. An *asterisk* indicates an indel introduced to improve alignment, while a *dot* indicates unavailable sequence, and a *backslash* indicates a frameshifting indel. The numbering follows that of the HLA-DMB protein (Kelly et al. 1991). The species designations follow the standard usage for the MHC and include human (*HLA*), mouse (*H-2*), lemur (*Pefu*) chimpanzee (*Patr*), rabbit (*Orcu*), swine (*Susc*), wallaby (*Maru*), fowl (*Gaga*), clawed toad (*Xela*), salmon (*Sasa*), cichlid (*Auha*), zebrafish (*Dare*), shark (*Gici*)

human. Residue conservation between human and mouse is generally confirmed by the lemur sequences. Important cysteine residues (positions 79 and 117 of DMB; positions 121 and 176 of DMA) involved in disulfide bridge formation are present. N-linked glycolsylation sites in the DMA sequences are conserved. However, the potential glycosylation site identified at position 92–94 of human DMB (Kelly et al. 1991) is not conserved in lemur. The absence of endoglycosidase H sites in H-2Mb (Karlsson et al. 1994) and lack of conservation of the potential site in humans (Fig. 3) suggest that glycosylation is not an esssential feature of DMB. Both DMA and H-2Ma are endoglycosidase H sensitive (Sanderson et al. 1994; Karlsson et al. 1994), tallying with conservation of potential sites in lemur.

DM Function

DM is thought to function during the maturation of class II molecules before cell surface expression. In the endoplasmic reticulum (ER), conventional class II molecules bind endogenous invariant chain protein, forming a nine subunit complex of threefold α, β and invariant chain. This complex migrates to the Golgi apparatus and thence takes the endocytic route to the cell surface. Invariant chain is thought to enable transportation out of the ER and prevent premature antigen binding.

During maturation, the invariant chain is proteolytically degraded, leaving a central core of 24 amino acids called CLIP (for class II associated invariant chain peptide), held inside the peptide-binding domain (PBR) of the class II molecule (Ghosh et al. 1995). CLIP is removed and replaced by internalized degraded proteins in a process mediated by DM. Evidence from mutant human B cell lines indicate that DM is essential for normal class II maturation (Mellins et al. 1990, 1991; Ceman et al. 1992, 1994). DM accumulates in the intracellular compartment where exogenous peptide loading takes place and little if any DM is found on the cell surface (Sanderson et al. 1994; Karlsson et al. 1994). Appearance of unstable and CLIP, containing class II heterodimers on the cell surface of various DM mutants indicate the intimate involvement of DM with maturation. CLIP is replaced with exogenous peptide before migration of conventional class II molecules to the cell surface. The exact details of the replacement of CLIP are unclear, but may involve stabilization of transient conformation of the class II molecule which promotes CLIP dissociation, the loading of peptide into the PBR of conventional class II molecules by DM or simply chaperoning of class II by DM (Sanderson et al. 1994; Karlsson et al. 1994)

DM Phylogeny

Clearly, several features set *DM* apart from other class II loci. The divergence is most apparent on the sequence level. Conservation of key features such as the disulfide-bridge cysteines, the length, and exon and domain boundary conserva-

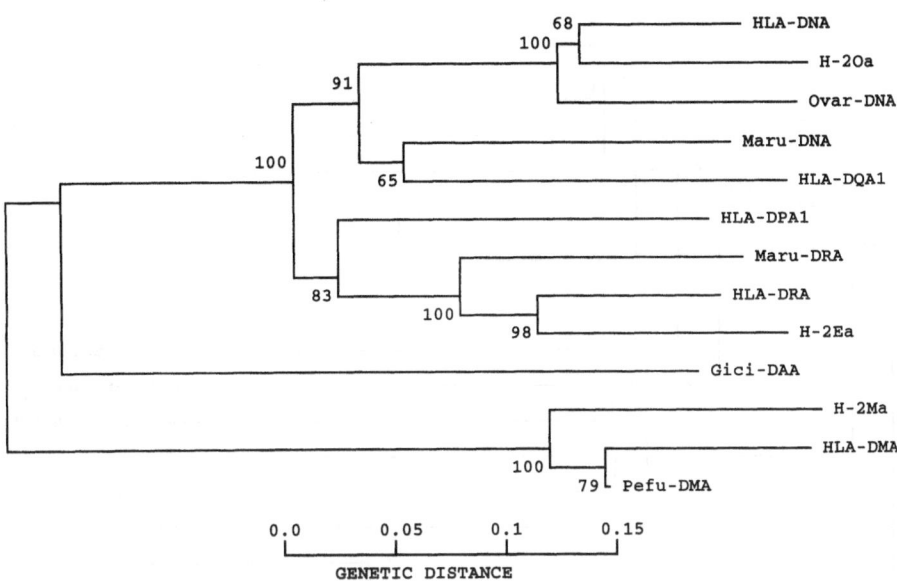

Fig. 5. Phylogenetic tree of vertebrate MHC class II A sequences. The sequences shown in Fig. 2 are shown in a dendrogram following estimation of pairwise distances based on protein identity and application of the neighbor-joining algorithm. *Numbers* on the nodes indicate the percentage recovery of that node in 500 bootstrap replicates

tion assist in determining the otherwise difficult alignment of *DM* with other class II molecules.

Representative phylogenetic trees of *DMA* and *DMB* chains from several species along with other class II chains are shown in Figs. 5 and 6. The *DM* loci, both α and β, are at the base of their respective trees. Sequences group according to family with *DMA* and *DMB* genes from various species forming single clades. A shark sequence divides the *DMA* chains from the other mammalian families (Fig. 5). Likewise, shark and bony fish genes are interposed between the *DMB* chains and the better characterized mammalian families (Fig. 6). This is a strong indication that the familial divergence of both *DM* genes is ancient.

There is no apparent rate change in substitutions in *DM* lineages. This would manifest itself in longer branch lengths connecting *DM* genes. The branch lengths found between human and mouse *DM* sequences are similar to those found between mouse and human in other families (compare branch lengths of *DNA* to *DMA* in Fig. 5 and of *DOB* to *DMB* in Fig. 6). This is an indication that *DM* evolves at a similar rate to the other genes. This trend is maintained in comparisons involving marsupial *DM* genes (O'hUigin et al. 1996, in preparation). Table 1 shows substitutional distances measured between various class II sequences in mouse and humans. The data indicate that there is little difference in nonsynonymous evolutionary rates between either the A or B genes. All values lie in the range of 14%–17% and are, within error, equal. The synonymous rates

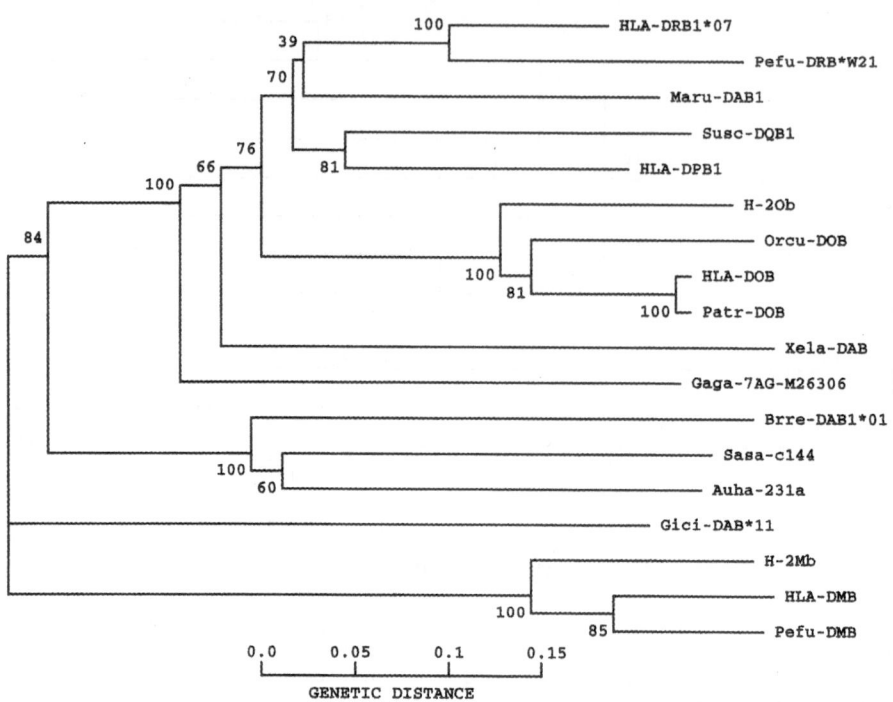

Fig. 6. Phylogenetic tree of vertebrate MHC class II B sequences. The sequences shown in Fig. 4. are arranged in a dendrogram following estimation of pairwise distances based on protein identity and application of the neighbor-joining algorithm. *Numbers* on the nodes indicate the percentage recovery of that node in 500 bootstrap replicates

do show stronger fluctuations, ranging from 43 % to 72 %. Much of this variation however may be due to saturation effects (Satta et al. 1991) and again provides little evidence of systematic rate changes for particular gene lineages.

The *DO/DN* Loci

With the exception of *DM*, the most atypical class II family in mammals are the *DNA* and *DOB* loci. Unusual features include separation of the loci constituent to the putative heterodimer, lack of polymorphism, restricted expression and early divergence from other class II families. It is not clear which role the DN/DO products play and if their function is the same in all mammals. The characterization of these genes has been primarily on the genetic level. Only in the mouse has the protein heterodimer been described (Karlsson et al. 1991). It remains to be seen whether the genetic evidence of function is reflected in a cellular role.

Table 1. Substitution distances for *MHC* class II loci in human-mouse sequence comparisons

	K_s	L_s	K_a	L_a	GC(3) %
α-Chains					
DMA	0.72 ± 0.09	184	0.14 ± 0.02	596	63.7
DNA	0.54 ± 0.07	179	0.14 ± 0.02	568	76.5
DQA	0.53 ± 0.07	180	0.16 ± 0.02	579	60.9
DRA	0.72 ± 0.09	167	0.14 ± 0.02	591	58.5
β-Chains					
DMB	0.65 ± 0.09	189	0.17 ± 0.02	591	64.8
DOB	0.58 ± 0.08	189	0.14 ± 0.02	621	63.5
DQB	0.46 ± 0.06	178	0.16 ± 0.02	581	80.2
DRB	0.43 ± 0.06	188	0.15 ± 0.02	601	71.8

Substitutional distance is measured using the method of Li (1993), in which K_s indicates synonymous and K_a nonsynonymous distances. L_s and L_a measure the number of sites used in each estimate, respectively. GC(3) % shows the third position GC content.

DN/DO Structure and Expression

The DNA and DOB loci lie towards the center of the MHC class II region. In humans, the loci are separated by some 160 kb, within which lie the *DM* loci, a cluster of genes involved in class I antigen processing and several genes unrelated to immune function (Trowsdale 1995). The *DNA* and *DOB* genes have a similar orientation in humans (Okamoto et al. 1991), in contrast to the tail-to-tail orientation of the *DR* and *DQ* gene pairs. In the rabbit, however, the *DNA* and *DOB* genes have a tail-to-tail arrangement, indicating that the region has undergone rearrangements in the course of eutherian history.

To date *DOB* genomic sequences have been obtained in mouse (Larhammar et al. 1985; originally called *H-2AB2*, now *H-2Ob*), humans (Servenius et al. 1987; Beck et al. 1992), sheep (a partial sequence: Wright et al. 1995a) and lemur (partial: this report). The cDNA has been sequenced in humans (Tonelle et al. 1985; Jonsson and Rask 1989), chimpanzee (Kasahara et al. 1989), rabbit (Chouchane et al. 1993). The DNA gene has been sequenced in humans (Trowsdale and Kelly 1985), mouse (Karlsson and Peterson 1992), tammar and redneck wallabies (partial: Slade et al. 1994, 1995) and sheep (partial: Wright et al. 1995b).

The *DOB* cDNA sequences indicate that the gene is transcribed in humans, chimpanzee, mouse and rabbit. That the DNA locus too is transcribed is indicated by cDNAs obtained in humans (Inoko et al. 1985; Jonsson and Rask 1989), mouse (Karlsson and Peterson 1992) and tammar wallaby (Slade et al. 1994). However, *DOB* transcripts were not detected in cattle when using the human cDNA probe (Stone and Muggli-Cockett 1993). The relative abundance of the *DOB* and *DNA* transcripts suggests that the expression level of the gene is an order of magnitude lower than that of the classical class II genes, *DR*, *DQ* and *DP*. Furthermore *HLA-DOB* transcripts are found only in limited set of lymphoid cells (DR, DQ and DP are expressed in pre-B cells while DOB is limited to mature B cells at low levels and is absent from the thymic cortex). Unlike *DP*, *DR* and *DQ*, the DOB transcript is not γ-interferon inducible. Both *DNA* and *DOB* display a low level of polymorphism (Jonsson and Rask 1989) with no indication of selective maintenance thereof. The aberrant expression and low polymorphism levels are characteristics of nonclassical MHC genes and pseudogenes (Klein and O'hUigin 1994). Nevertheless, no obvious defects which would impair function are seen in the coding regions of *DOB* sequences thus far obtained (Fig. 7, but note alignment alterations to rabbit sequence). The sequences contain the four conserved cysteine residues, two each in the B1 and B2 domains capable of forming a disulfide bridge as found in the b-chains of functional class II molecules (Fig. 4). The asparagine residues involved in glycosylation are present and conserved features of the B2 domains such as the amino acids FYP in positions 122–125 and GDW at positions 152–154 are retained. (In the rabbit sequence, the FYP is not found in the appropriate reading frame.) Nor are defects seen in DNA coding sequences (Fig. 8). In comparisions with other α-chain genes, (Fig. 2) the conserved cysteine residues involved in the disulfide bridge (Fig. 2, position 121 and 176) and the N-linked glycosylation sites (Fig. 2, positions 92–94 and 132–134) are present.

The *DOB* gene of humans (Servenius et al. 1987) and mice (Larhammar et al. 1985) has the intron/exon organization common to other MHC class II β-chains. No major rearrangements occur and known insertion elements are found only in the mouse gene (a possible B1 element in the sheep gene may be a cloning artefact). The intron-exon junctions follow the GT-AG rule. In the promoter region of the human, the mouse and the sheep genes, the characteristic Z, X and Y boxes are present. These show considerable differences when compared to promoter elements of other class II genes. There is evidence that the changes in the Z box (four substitutions in seven nucleotides when compared to DRA promoter sequences in humans) are responsible for the altered expression of *DOB* relative to other class II genes (Voliva et al. 1993). A short region of the *DOB* gene has been sequenced in *Lemur catta*. Figure 9 shows the extent of conservation of exon 2 and its flanking intron sequences in those species for which sequence is available. While lemur and human sequence are alignable, the alignment of mouse and sheep intron is tenuous outside the immediate flank of exon 2. Surprisingly, at about 300 bases in the 5' direction lies an intronic segment which shows a high degree of conservation between human and mouse. The reason for this conservation is unclear, but some functional constraint seems probable, as a neutral segment of DNA which has evolved over the 80–100 million years since the separation of these species would be approaching substitutional saturation.

DN/DO Function

The extensive molecular characterization of the *DN/DO* loci has not been accompanied by elucidation of possible function. The sequence conservation and prob-

```
              -26                               -16                                -6
CONSENSUS     ATG GGT TCT GGG TGG GTC CCC TGG GTG GTG GCT CTG CTA GTG AAT CTG ACC AGA CTG GAT TCC TCC ATG ACT CAA
HLA-DOB       --- --- --- --- --- --- --- --- --- --- --- --- --- --- --- --- CA- --- --- --- --- --- --- --- ---
Patr-DOB      --- --- --- --- --- --- --- --- --- --- --- --- --- --- --- --- C-- --- --- --- --- --- --- --- ---
H-20b         --- --C G-- --- A-- --C- --- --- --- --- --- --- T-G --- --C --C -TG --G --- --- --- -T- --- -T G--
Orcu-DOB      --- --A- --C AA- --- --T --- --- --- --- --- --- --G --- --- --T- --G --- --- --- -A- --- --- ---
PROTEIN       M   .   .   .   .   P   W   V   V   A   L   L   V   N   L   .   L   D   .   .   M   .   .

              -1 1                              10                                 20
CONSENSUS     GGC ACA GAC TCT CCA GAA GAT TTT GTG ATT CAG GCA AAG GCT GAC TGT TAC TTC ACC AAC GGG ACA GAA AAG GTG
HLA-DOB       --- --- --- --- --- --- --- --- --- --- --- --- --- --- --- --- --- --- --- --- --- --- --- --- ---
Patr-DOB      --- --- --- --- --- --- --- --- --- --- --- --- --- --- --- --- --- --- --- --- --- --- --- --- ---
H-20b         --- -G- --- --C --- --G A-- --- --- --- --- --- --- --G --- --- --- --T --T --- --T --- --- --- --T
Orcu-DOB      --A *** --- --- --- --A A-- --- --- --- --- --- --- --- --C --- --- --- --- --- --G C-- --- ---
PROTEIN       G   .   D   S   P   E   .   F   V   I   Q   A   K   A   D   C   Y   F   T   N   G   T   E   .   V

              25                               35                                 45
CONSENSUS     CAG TTT GTG GTC AGA TTC ATC TTT AAC TTG GAG GAG TAT GTA CAC TTC GAC AGT GAC GTG GGG ATG TTT GTG GCA
HLA-DOB       --- --- --- --- --- --- --- --- --- --- --- --- -GT --- --- --- --T --- --- --- --- --- --- --- ---
Patr-DOB      --- --- --- --- --- --- --- --- --- --- --- --- -GT --- --- --- --T --- --- --- --- --- --- --- ---
H-20b         --C --G C-- --- --G --- --C --- C-- --- --- --- T-- --- --C --T --- --C --- --- C-- --- --- --T
Orcu-DOB      -GT G-- --- --- --G --T --- --- --T --- --- T-- --- -C- --- --T --- --C --- --- --- --- --- ---
PROTEIN       .   .   V   R   F   I   F   N   L   E   E   Y   .   F   D   S   D   .   G   M   F   V   A

              50                               60                                 70
CONSENSUS     CTG ACG AAG CTG GGG CAG CCA GAT GCT GAG CAG TGG AAC AAC CGG CTG GAT CTC CTG GAG AGG AGC AGA CAG GCC
HLA-DOB       T-- --C --- --- --- --- --- --- --- --- --- --G- --- --- --- --T --- --- --- --- --- --- --- ---
Patr-DOB      T-- --C --- --- --- --- --- --- --- --- --- --G- --- --- --- --T --- --- --- --- --- --- --- ---
H-20b         --- --- G-- --- --- G-A --T --- --C --- --- --- --A --C --- --- --- --- --- --C --- --- GCT -T
Orcu-DOB      --- --- G-- --- --- --T --- --- --- TC- --- --T --- --CC A-- --- --- --- --- -G- T-T
PROTEIN       L   T   .   L   G   .   P   D   A   .   W   N   .   R   .   D   .   L   E   .   S   R   .

              75                               85                                 95
CONSENSUS     GTG GAT GGG GTC TGT AGA CAC AAC TAC AGG CTG GGG GCA CCC TTC ACT GTG GGG AGA AAA GTG CAA CCA GAG GTG
HLA-DOB       --- --- --- --- --- --- --- --- --- --- --C --- --- --- --- --- --- --- --- --- --- --- --- --- ---
Patr-DOB      --- --- --- --- --- --- --- --- --- --- --C --- --- --- --- --- --- --- --- --- --- --- --- --- ---
H-20b         --- A-C AT- --- --C --G --G --G --- -A- --- --- --T --- --- --- --A --- --T --- --CT --- --- ---
Orcu-DOB      --- --- TTT C-- --- --G- --- --- --- --T --- --- --- --- --- --- --C --- --- --GT
PROTEIN       V   .   .   .   C   R   .   Y   .   L   G   A   P   F   T   V   .   R   .   V   .   P   E   .

              100                              110                                120
CONSENSUS     ACA GTG TAC CCA GAG AGG ACC CCA CTC CTG CAC CAG CAT AAC CTG CTG CAC TGC TCT GTG ACA GGC TTC TAT CCA
HLA-DOB       --- --- --- --- --- --- --- --- --- --- --- --T --- --- --- --- --- --- --- --- --- --- --- --- ---
Patr-DOB      --- --- --T --- --- --- --- --- --- --- --- --T --- --- --- --- --- --- --- --- --- --- --- --C --C
H-20b         --- --- --- --- --- --- --- --G T-G --- --G --- --C --- --T --- --- --- --C- --- --G --- --- --C --C
Orcu-DOB      --- --- AT- --- --- --- --- --G G-- TGC GG- --A ** --- --T- --- --- -C- --- --- --- *G
PROTEIN       T   V   .   P   E   R   T   V   .   .   Q   .   N   L   L   .   C   S   .   T   G   F   Y   P

              125                              135                                145
CONSENSUS     GGG GAC ATC AAG ATC AAG TGG TTC CGG GAT GGG CAG GAG GAG AGA GCT GGG GTC ATG TCC ACT GGC CCT ATC AGG
HLA-DOB       --- --T --- --- --- --- --- --- -T- --- --- --- --- --- --- --- --- --- --- --C --- --C --- ---
Patr-DOB      --- --T --- --- --- --- --- --- -T- --- --- --- --- --- --- --C --- --- --- --- --C --- ---
H-20b         --- --- --A -GT G-- --- --- --- --- --A --- --- --G T-- --- --- --- --- --A --- --- --T- G-T ---
Orcu-DOB      --- --- -GA G-- -G- --- C-G --- --- --- --- --- --A- --- --- --- --- --T- --- --A
PROTEIN       G   D   I   .   .   W   .   N   G   Q   E   E   R   .   .   M   S   T   G   .   R


              150                              160                                170
CONSENSUS     AAT GGA GAC TGG ACC TTT CAG ACT ACG GTG ATG CTA GAA ATG ACT CCT GAA CTT GGA GAC GTC TAC ACC TGC CTC
HLA-DOB       --- --- --- --- --- --- --- -GT- --- --- --- --- --- --- --- --- --- --- --- --- C-T --- --- --- --T
Patr-DOB      --- --- --- --- --- --- --- -GT- --- --- --- --- --- --- --- --- --- --- --- --- C-T --- --- --- --T
H-20b         --- --- --- --- --- --C --- --A --A --- --G --- --- --- TC --A --G --- --T --- A-- --- --G- --- ---
Orcu-DOB      --C --- --- --- --- --- --- --C --- -A- --- --G --- --- --- --- --G --- --- --- --- --T -T- ---
PROTEIN       N   G   D   W   T   F   Q   T   .   .   M   L   E   M   .   P   E   L   G   .   .   Y   .   C   L

              175                              185                                195
CONSENSUS     GTG GAT CAC CCC AGC CTG CTG AGC CCT GTT TCT GTG GAG TGG AGA GCT CAG TCT GAA TAT TCC TGG AAA AAG ATG
HLA-DOB       --C --- --- --T-- --- --- --- --- --- --- --- --- --- --- --- --- --- --- --T --- --G- --- --- ---
Patr-DOB      --C --- --- --T-- --- --- --- --- --- --- --- --- --- --- --- --- --- --- --T --- --G- --- --- --A
H-20b         --- --G --- --- G-- --C --- --A --- --- --- --- --- CA --- --TG --- --- --- --- --- --- --- --A
Orcu-DOB      --- --- --- --- --- --- --- --- --C --- --- --- -AC GCG AGG -CT CAG TCT A-- AT- --- GC- ---
PROTEIN       V   .   H   .   L   L   .   P   V   S   V   .   .   .   .   .   .   .   .   W   .   K   .

              200                              210                                220
CONSENSUS     CTG AGT GGA ATT GCA GCC TTC CTA CTT GGG CTA ATC TTC CTT CTG GTG GGA ATC GTC ATC CAG CTA AGG GCT CAG
HLA-DOB       --- --- --C --- --- --- --- --- --- --- --- --- --- --- --- --- --- --- --- --- --- --- --- ---
Patr-DOB      --- --- --C --- --- --- --- --- --- --- --- --- --- --- --- --- --- --- --- --- --- --- --- ---
H-20b         --- --- --- GC- --- -TG --- --G --- --- --G --T G-- T-C --- --- --G G-T --T --- --T --C -A-- --- ---
Orcu-DOB      --A --- G-- --- -G- --- -G T-- --- -G --- --- --C- A-- --T -T A-C --- --- A-- TG-
PROTEIN       L   S   G   .   A   .   F   L   .   G   L   I   .   .   L   V   G   .   I   .   A   .

              225                              235                                245
CONSENSUS     AAA GGA TAT GTG AAG ACG CAG ATG TGC GGT AAT GAG GTC TCA AGA GCT GTT CTG CTC CCT GGA TCA TGC TAA GCT
HLA-DOB       --- --- --- --- --- -G- --- --- -CT --- --- --- --- --- --- --- --- --- --- --- --- --- ... ...
Patr-DOB      --- --- --- --- --- -G- --- --- -CT --- --- --- --- --- --- --- --- --- --- --- --- --- ... ...
H-20b         --C- --C- --- G-- --T --- CCT G-- AA- G-G AGT AGG ...-C C-G ATG A-G GA- -GG -TA ACC AAG -T- A-G ---
Orcu-DOB      --G AT- -G- -GA G-C --A GCT C-C --- T-C TGC TGA
PROTEIN       K   .   .   .   .   .   T   .   .   .   .   .   .   .   .   S   R   .   .   .   .   .   .   A

              250
CONSENSUS     GGA CCG GGA CAT GTC ACA TGA
PROTEIN       G   P   G   H   V   T   .
```

Fig. 7. Alignment of *DOB* sequences. A *dash* indicates identity with the consensus sequence, determined by simple majority. An *asterisk* indicates an indel introduced to improve alignment, while a *dot* indicates unavailable sequence. For the Orcu-DOB sequence, indels at positions 113 and 124 are included to enhance the degree of conservation. Numbering follows the scheme of Kasahara et al. (1989). Conserved amino acid residues are indicated using the standard single-letter code

```
Intron 1
            1          11         21         31         41         51         61         71         81         91
CONSENSUS   GAATAACATA CCCAGCATGC ACACAGCATG TGTGTAATGA TTGGCAACAG TGCAAGGACC TCTCAACTCC TCACTGGTCT CCGTTCCCAT TCTTTCCACT
HLA-DOB     -TT---TT-- ----A*G--T T-------CA -*------T- ------GT-- ------ACT- -----T---T --T--T-C-- ------T--- ---C---C--
Leca-DOB    -GG----C-- -A-----G-- ---------- -*------CC -----**--- -T------*- ----TC---- --C--C---- ---------- -------T--
H-2Ob       -----C---T --A--TT--- CTT--A---- A-----G--T -CA-A----T A-A-T--G-- GAAAG-TGG- ---G----TA AA-**--AC- -GC-G---AA
            101        111        121        131        141        151        161        171        181        191
CONSENSUS   CCTTAGAGAA TCCATTCTAA ACATATCACG TATGTGTCTG AAAAAGAGAG TAGATTTCAG AAATATGTAT CTTTATTACT TTTTTTGTTT TTTAAAAATT
HLA-DOB     --C-G----- ---------- --G-G--TG- -----C---A -GT-T----- -G-C---T-- ---------G -A-----G-- C---A---G- ---T------
Leca-DOB    ---------- --------G- ----G--T-- -----C-T-C *********- --TT----T- ***C---C-- T-CA---TT- -C------A- --------G-
Ovar-DOB    ********** ****A-GG-C -G-C-C---C --A-C-A--- --CT*****- C-T---C--T ***-G-A--- G-G--CC--- -------T-- ------G-A-
H-2Ob       ---G-C-ACT -GAG---A-T C-C--GG-AC C-C----TA- -T---C--GA -G--A-GGGA T-GAC-TAC- -C-G-AA--- G----CC--G --C-C-TT--
            201        211        221        231        241        251        261        271        281        291
CONSENSUS   AAATAAATGT CATTTGACAG GTTAAATCAC ATTAAGTGTC TTTGCTCACC TGACACTATG CTTGTCGAGC ATACTGAGGT CAAAGGACCT CCTCTCAAAT
HLA-DOB     T--------- -----CT-T- ---G----C- ---CT--T-- ---T-----T -------G-- ---T-T---- ---------- ---G-***** -----T-G--
Leca-DOB    T--------- ---GG-CAT- -***---TG- ---**--T-- ---T------ -------G-- -----*---- ----A----C ----A----- ----------
Ovar-DOB    -G--GTC--- TTGC---G-- -***--C--G GG-**---G- CCA-A-G-G- A--GG---C* T-----AGAT T--------- -----C*--- ------C---
H-2Ob       ---------C --A-AAG--T T**----G-- --------A- ----*--T-T -A-TGTC--T TAA--*--CT T-GTCTC**- T--T--TGGG -T-T-G-G-A
            301        311        321        331        341        351        361        371        381
CONSENSUS   CCATGTGGTC TGACGTAATT TACCAGGTAT GGGTTTTCCC AGAGGTGGGG GCTGGTTTCA TAGTCTTGGT TTTGGTTTTC CAG
HLA-DOB     ---------- ---------- -------C-- ---------- -----A---- -------*-- -G--T----- ---------- ---
Leca-DOB    ---C---T-- --C-ACC-C- ------A-G- -C-------- ---------- ---------- CG-----TC- ---------- ---
Ovar-DOB    ----A----G CATT-----A ---------* ---G-----T -A--A----- -------*-- -------**- ---------- ---
H-2Ob       -GT--CTA-- ----T--G-G -G------GC A--C------ -------CA- *------*-C --A--C-**- --AT-G---- ---

Exon 2
              5                                           15                                          25
CONSENSUS    AA GAT TTT GTG ATT CAG GCA AAG GCT GAC TGT TAC TTC ACC AAC GGG ACA GAA AAG GTG CAG TTT GTG GTC AGA
HLA-DOB      -- --- --- --- --- --- --- --- --- --- --- --- --- --- --- --- --- --- --- --- --- --- --- --- ---
Patr-DOB     -- --- --- --- --- --- --- --- --- --- --- --- --- --- --- --- --- --- --- --- --- --- --- --- ---
Leca-DOB     -- --- --- --- --- --- --- --- --- --- --- --- --- --- --T --- --G --- --- --- -GC --- --- --- ---
Ovar-DOB     -- --- --- --- -CC --- --- --- --- --- --- --- --- --- --T --- --- --- --A --- -G- --- --- --- ---
Orcu-DOB     -- --- --- --- --- --- --- --- --- --- --C --- --- --- --- --- --- --G C-- --- -GT G-- --- --- --G
H-2Ob        -G A-- --- --- --- --- --- --- --G --- --- --- --- --- --T --T --- --- --- --T --C --G C-- --- --G
 PROTEIN         .   F   V   .   Q   A   K   A   D   C   Y   F   T   N   G   T   E   .   V   .   .   .   V   R
             30                                          40                                          50
CONSENSUS    TTC ATC TTT AAC TTG GAG GAG TAT GCA CGT TTC GAC AGT GAC GTG GGG ATG TTT GTG GCA CTG ACG GAG CTG GGG
HLA-DOB      --- --- --- --- --- --- --- --- -T- --- --- --- --- --T --- --- --- --- --- --- T-- --C A-- --- ---
Patr-DOB     --- --- --- --- --- --- --- --- -T- --- --- --- --- --T --- --- --- --- --- --- T-- --C A-- --- ---
Leca-DOB     --- --- -A- --- --- --- --- --- --- --- --- --- --- --- -C- --- G-- --- --- -NC --- --- -C- --- ---
Ovar-DOB     --- --- --C --- C-- --- --- --- --- --- --- --- --C --- T-- --A --- --- --- --C T-- --- --- --- ---
Orcu-DOB     --T --- --- --T --- --- --- --- --- -AC --T --- --C --- --- --- --- --- --- --- --- --- --- --- ---
H-2Ob        --- --- --C --- C-- --- --- --- TT- -AC --- --- --- --- C-- --- --- --- --- --T --- --- --- --- ---
 PROTEIN      F   I   .   N   L   E   E   Y   .   .   F   D   S   D   .   G   .   F   V   A   L   T   .   L   G
             55                                          65                                          75
CONSENSUS    CAG CCA GAT GCT GAG CAG TGG AAC AAC CGG CCG GAT ATC CTG GAG AGG AGC AGA GCG GCC GTG GAT GTG GTC TGC
HLA-DOB      --- --- --- --- --- --- --- --- -G- --- -T- --- C-- T-- --- --- --- --- CA- --- --- --- -G- --- --T
Patr-DOB     --- --- --- --- --- --- --- --- -G- --- -T- --- C-- T-- --- --- --- --- CA- --- --- --- -G- --- --T
Leca-DOB     --- --C --C --C --- --- --- --N GC- --- G-- --- --A --- --- --- --- --- --- T-- --- --- -C- C-- ---
Ovar-DOB     A-- --C --- --- --- -T- --- --- --T --- --- --- --T --- -C- --- --T --- --C T-T --- --C A-- C-T ---
Orcu-DOB     --- --T --- --- --- TC- --- --- --T --- --C --C --- --- --- --- --- --- CG- T-T --- --- T-T C-- --T
H-2Ob        G-A --T --- --- --C --- --- --- --A --C -T- --- C-- --- --- -C- --- --- --T --T --- A-C A-- --- ---
 PROTEIN      .   P   D   A   .   .   W   N   .   R   .   D   .   L   .   .   S   R   .   .   V   .   .   .   C
             80                                          90
CONSENSUS    AGA CAC AAC TAC AAG CTG GGC GCA CCC TTC ACT GTG GGG AGA AAA G
HLA-DOB      --- --- --- --- -G- --- --- --- --- --- --- --- --- --- --- -
Patr-DOB     --- --- --- --- -G- --- --- --- --- --- --- --- --- --- --- -
Leca-DOB     --- -G- --- --- G-- --- --- --- --- --- --G --- --- --- -C- -
Ovar-DOB     --- -G- --- --- --- --- --T --- --- --- --C --- --- --G -G- -
Orcu-DOB     --- -G- --- --- -G- --- --G --T --- --- --- --- --- --- --- -
H-2Ob        --G --G --G --- --- --- --G --C --- --- --- --- -A- --- --T -
 PROTEIN      R   .   .   Y   .   L   G   A   P   F   T   V   .   R   .
```

Fig. 8. Alignment of *DNA* coding regions. A *dash* indicates identity with the consensus sequence, determined by simple majority. Numbering follows the scheme of Trowsdale and Kelly (1985). Conserved amino acid residues are indicated using the standard single-letter code

able structural similarity to other class II proteins suggest that any function of *DN/DO* will be related to the functions of other class II genes. That *H-2O* is known to form an α-β heterodimer and is generally expressed in tissues related to immune function also supports this suggestion. The aberrant expression pattern of the *DO* genes points to differences in the temporal expression and/or the localization of such function. It has been suggested that the absence of *H-2O* from thymic cortex implies its lack of involvement in positive selection of T cells, whereas its presence in medullary epithelial cells allows involvement in negative selection of AB receptors (Karlsson et al. 1991). The balance of evidence indicates that *DN* and *DO* do, as other class II genes, participate in immune function in a manner which at present evades elucidation because of their abberant expression.

DN/DO Phylogeny

Figures 5 and 6 indicate that, unlike their *DM* counterparts, the *DN/DO* genes fit firmly into the radiation of mammalian MHC gene families. There is evidence that this radiation predates the marsupial-eutherian split, the evidence being that *DN*-like and *DR*-like genes occur in marsupials. However, there is no suggestion from the phylogenetic trees (Fig. 6) that such families will be found in birds, and they may also be absent in monotremes.

There is some indication that *DN/DO* is among the oldest of the classical mammalian class II gene families. The *DOB* sequences group outside the other mammalian gene families (*DMB* excepted) with 70 % bootstrap support (Fig. 6). Ours is not the only tree where the *DOB* gene shows this tendency (Hughes and Nei 1990; Kasahara et al. 1992, Sato et al. 1993). Evolutionary rate estimates (Table 1) indicate that *DOB* evolves at the same rate as other class II genes. Thus its position in the tree likely reflects that it was one of the first branches in the expansion of the mammalian MHC gene families.

Conclusions

Both the *DM* and *DN/DO* genes are as yet poorly characterized, and little research has been undertaken on the evolution of these genes in nonhuman primates. It is clear however that their evolutionary histories must be different from those of their polymorphic counterparts. They lie in the center of the class II region, generally oligomorphic and evolving in parallel with, but apparently unaffected by, the widespread distal and proximal variation. They are therefore interesting contrasts to the classical story of the MHC, polymorphism and variation. The *DM* and *DN/DO* genes may thus provide insight into how new evolutionary function can be gained within the restraints of an ancient structure and into how many functions the same basic structure can house.

Intron 1

CONSENSUS		
HLA-DOB		
Leca-DOB		
H-20b		

Exon 2

CONSENSUS		
HLA-DOB		
Patr-DOB		
Leca-DOB		
Ovar-DOB		
Orcu-DOB		
H-20b		
PROTEIN		

```
Intron 2
                1          11         21         31         41         51         61         71         81         91
CONSENSUS   GTGAGCTGGA AGATGAGGTC TGGCGGGGCT CAGGAACGTC CTCCATGTGA ACCTTGGCGT AATCTTCTTC CTTACGAGAA GCAATTTTCT GCTTCTGGAT
HLA-DOB     |||||||||| |||||||||| |||||||||| |||||||-T| |-C||||||| ||-A-||-** **|||||||T |||||||||| |||T|||||| |||T-A-GA
Leca-DOB    ||||-G|||| ||-G-G-CC| ||||||-T-| ||||||-T|| T||||||||| ||-G-C|-** GGC||||||| |||||||||| ||||G|||C|| |||-G|||
Ovar-DOB    ||||-G||-G ||-T-G|--| ||-CCTA-*** -A-G--TT-- TG--G-*** **-TGG---* ||****-||-** **||-GTCC*GG |GTCC*GG |||G||C||| |||-A|||
H-20b       ||-G-T**-|| CA--A--A-| ||||||-T-- |-T-TG-GAA- -TGTAA-AG G-A--TAA** A--||-*****| A--||-*****| |-|--CA--| ATC-G-CTG-

                101        111        121        131        141
CONSENSUS   AAAAGATGA TGAGCGGAGA AGTCCGGGCA CGAGCTGTGT CC
HLA-DOB     T--T-G-TG -CT-R-T--C -C-||||-** **
Leca-DOB    -|C-|||-G ||-T-T-C-| -T-T-T-C|-| ||
Ovar-DOB    -TCT-TCC*| -C--G-TCAG *--G-A--GG A--C****** **
```

Fig. 9. Alignment of *DOB* exon 2 and intron flanks. For explanation of sequence names see Fig. 4. A *dash* indicates identity with the consensus sequence, determined by simple majority. An *asterisk* indicates an indel introduced to improve alignment while a *dot* indicates unavailable sequence. Numbering of exonic codons follows Kasahara et al. (1989). Conserved amino acid residues are indicated using the standard single-letter code. A seperate numbering is used for the introns. The lemur sequence (*Leca-DOB*) was obtained by screening a pWE15 genomic library with the chimpanzee DRB cDNA clone c4-2 (Fan et al. 1989). One of the positive clones was further digested with appropriate restriction enzymes and selected fragments subcloned into pUC18 prior to sequencing

References

Beck S, Kelly A, Radley E, Khurshid F, Alderton RP, Trowsdale J. DNA sequence analysis of 66kb of the human MHC class II region encoding a cluster of genes for antigen processing. J Mol Biol 228:433–441, 1992

Carrington M, Yeager M, Mann D. Characterization of HLA-DMB polymorphism. Immunogenetics 38:446–449, 1993

Ceman S, Rudersdorf RR, Long EO, DeMars R. MHC class II deletion mutant expresses normal levels of transgene encoded class II molecules that have an abnormal confromation and impaired antigen presentation ability. J Immunol 149:754–761, 1992

Ceman S, Petersen JW, Pinet V, DeMars R. Gene required for normal MHC class II expression and function is localized to approximately 45kb of DNA in the class II region of the MHC. J Immunol 152:2865–2873, 1994

Cho S, Attaya M, Monaco J. New class II-like genes in the murine MHC. Nature 353:573–576, 1991

Chouchane L, Brown TJ, Kindt TJ. Structure and expression of a nonpolymorphic rabbit class II gene with homology to HLA-DOB. Immunogenetics 38:64–66, 1993

Fan W, Kasahara L, Gutknecht J, Klein D, Mayer WE, Jonker M, Klein J. Shared class II Mhc polymorphisms between human and chimpanzees. Hum Immunol 26:107–121, 1989

Figueroa F, O'hUigin C, Tichy H, Klein J. The origin of primate Mhc-DRB genes and allelic lineages as deduced from the study of prosimians. J Immunol 152:4455–4465, 1994

Ghosh P, Amaya M, Mellins E, Wiley DC. The structure of an intermediate in class II MHC maturation: CLIP bound to HLA-DR3. Nature 378:457–462, 1995

Hermel EJY, Monaco JJ. Characterization of polymorphism within the H2-M MHC class II loci. Immunogenetics 42:136–142, 1995

Hughes AL, Nei M. Evolutionary relationships of class II major histocompatibility complex genes in mammals. Mol Biol Evol 7:491–514, 1990

Inoko H, Ando A, Kimura M, Tsuji K. Isolation and characterization of the cDNA clone and genomic clones of a new HLA class II antigen heavy chain, DOa. J Immunol 135:2156–2159, 1985

Jonsson A-K, Rask L. Human class II DNA and DOB genes display low sequence variability. Immunogenetics 29:411–413, 1989

Karlsson L, Surh CD, Sprent J, Peterson PA. A novel class II MHC molecule with unusual tissue distribution. Nature 351:485–488, 1991

Karlsson L, Peterson PA. The a chain gene of H-2O has an unexpected location in the major histocompatability complex. J Exp Med 176:477–483, 1992

Karlsson L, Peleraux A, Lindstedt R, Liljedahl M, Peterson PA. Reconstitution of an operational MHC class II compartment in Nonantigen-presenting cells. Science 266:1569–1572, 1994

Kasahara M, Klein D, Klein J. Nucleotide sequence of a chimpanzee DOB cDNA clone. Immunogenetics 30:66–68, 1989

Kasahara M, Vasquez M, Sato K, McKinney EC, Flajnik M. Evolution of the major histocompatibility complex: Isolation of class II cDNA clones from the cartilaginous fish. Proc Natl Acad Sci USA 89:6688–6692, 1992

Kelly AP, Monaco JJ, Cho SS, Trowsdale J. A new human HLA class II-related locus DM. Nature 353:571–573, 1991

Klein J, O'hUigin C. The conundrum of nonclassical major histocompatibility complex genes. Proc Natl Acad Sci USA 91:6251–6252, 1994

Larhammar D, Hammerling U, Rask L, Peterson PA, Sequence of Gene and cDNA encoding murine major histocompatability complex class II gene Ab2. J Biol Chem 260:14111–14119, 1985

Li W. Unbiased estimation of the rates of synonymous and nonsynonymous substitution. J Mol Evol 36:96–99, 1993

Mellins E, Smith L, Arp B, Cotner TEC, Pious D. Defective processing and presentation of exogenous antigens in mutants with nornal HLA class II genes. Nature 343:71–74, 1990

Mellins E, Kempin S, Smith L, Monji T, Pious D. A gene required for class II-restricted antigen presentation maps to the major histocompatability complex. J Exp Med 174:1607–1615, 1991

Niimi M, Nakai Y, Aida Y. Nucleotide sequences and the molecular evolution of the DMA and DMB genes of the bovine major histocompatability complex. Biochem Biophys Res Commun 217:522–528, 1995

Okamoto N, Ando A, Kawai J, Yoshiwara T, Tsuji K, Inoko H. Orientation of HLA-DNA gene and identification of a CpG island associated gene adjacent to DNA in human major histocompatability complex class II region. Hum Immunol 32:221–228, 1991

Radley E, Alderton RP, Kelly A, Trowsdale J, Beck S. Genomic organization of HLA-DMA and HLA-DMB. J Biol Chem 269:18834–18838, 1994

Sanderson F, Kleijmeer MJ, Kelly A, Verwoerd D, Tulp A, Neefjes JJ, Geuze HJ, Trowsdale J. Accumulation of HLA-DM a regulator of antigen presentation in MHC class II compartments. Science 266:1566–1572, 1994

Sato K, Flajnik MF, Du Pasquier L, Katagiri M, Kasahara M. Evolution of the MHC: isolation of class II b-chain cDNA clones from the amphibian Xenopus laevis. J Immunol 150:2831–2843, 1993

Satta Y, Takahata N, Schönbach C, Gutknecht J, Klein J. Calibrating evolutionary rates at major histocompatibility complex loci. In: Klein J, Klein D (eds) Molecular evolution of the major histocompatibility complex. Springer, Berlin Heidelberg New York, pp 51–62, 1991

Servenius B, Rask L, Peterson PA. Class II genes of the human major histocompatability complex. J Biol Chem 262:8759–8766, 1987

Slade RW, Hale PT, Francis DI, Graves JAM, Ra S. The marsupial MHC: the tammar wallaby, Macropus eugenii, contains and expressed DNA-like gene on chromosome one. J Mol Evol 38:496–505, 1994

Slade RW, Mayer WE. The expressed class II a-chain genes of the marsupial major histocompatibility complex belong to eutherian mammal gene families. Mol Biol Evol 12:441–450, 1995

Stone RT, Muggli-Cockett NE. BoLA-DIB: species distribution, linkage with DOB and northern analysis. Anim Genet 24:41–45, 1993

Sültmann H, Mayer WE, Figueroa F, O'hUigin C, Klein J. Zebrafish Mhc class II a chain-encoding genes: polymorphism, expression, and function. Immunogenetics 38:408–420, 1993

Tonelle C, DeMars R, Long EO. DOB: a new B chain gene in HLA-D with a distinct regulation of expression. EMBO J 4:2839–2847, 1985

Trowsdale J. Both man and bird and beast: comparative organization of MHC genes. Immunogenetics 41:1–17, 1995

Trowsdale J, Kelly A. The human HLA class II a chain gene DZa is dintinct from genes in the DP DQ and DR subregions. EMBO J 4:2231–2237, 1985

Voliva CF, Tsang S, Peterlin BM. Mapping cis-acting defects in promoters of transcriptionally silent DWA2 DQBs and DOB genes. Proc Natl Acad Sci USA 90:3408–3412, 1993

Wright H, Redmond J, Wright F, Ballingall KT. The nucleotide sequence of the sheep MHC class II DNA gene. Immunogenetics 41:131–133, 1995a

Wright H, Redmond J, Ballingall KT. The sheep orthologue of the HLA-DOB gene. Immunogenetics 43:76–79, 1995b

4 The Primate Class III MHC Region Encoding Complement Components and Other Genes

F. FIGUEROA

Introduction

The class III gene region, located between the loci encoding class I and class II MHC molecules, has been the subject of close scrutiny on the part of immunogeneticists for the last 20 years. As a result it has developed as one of the best genetically characterized regions in the entire vertebrate genome. In humans the class III region encompasses 1100 kilobases (kb) of DNA and is located on the short arm of chromosome 6 (Campbell and Trowsdale 1993). Its centromeric side is flanked by the class II MHC loci, specifically by the locus encoding the α-chain of the HLA-DR molecule (DRA locus). The telomeric end is bordered by the class I loci, specifically the HLA-B locus.

CpG islands, known to be frequently associated with expressed genes (Bird 1986; Lindsay and Bird 1987), characterize the region, and, in keeping with this, an average density of about one gene per less than 30 kb DNA length has been found. In all but one of the vertebrate classes the region is conserved in that it is always located between the class I and class II MHC regions (Trowsdale 1995). The exception is the teleost Danio rerio, in which homologs of the human class III

Molecular Biology and Evolution of Blood Group
and MHC Antigens in Primates
Blancher/Klein/Socha (Eds.)
© Springer-Verlag Berlin Heidelberg 1997

region map to chromosomes different from those encoding MHC molecules (J. Bingulac-Popovic, J. et al 1997.

Concordance is observed when human and mouse class III MHC regions are compared, and it is likely that in all nonhuman primates the class III region is also located between the class I and class II MHC regions, that the gene order is similar, and that perhaps only minor differences exist due to indels, especially in noncoding regions. A description of the human class III MHC region and its corresponding nonhuman primate counterparts follows. Only those genes characterized in nonhuman primates are discussed.

The Human TECYC4RP Module

About 400 kb telomeric from the class II *DRA* gene lies a complex cluster of four genes referred to as tenascin-X (*X*-gene), 21-hydroxylase (*CYP21* gene), complement component 4 (*C4* gene), and RP molecule (*RP* gene) (Fig. 1). The most rele-

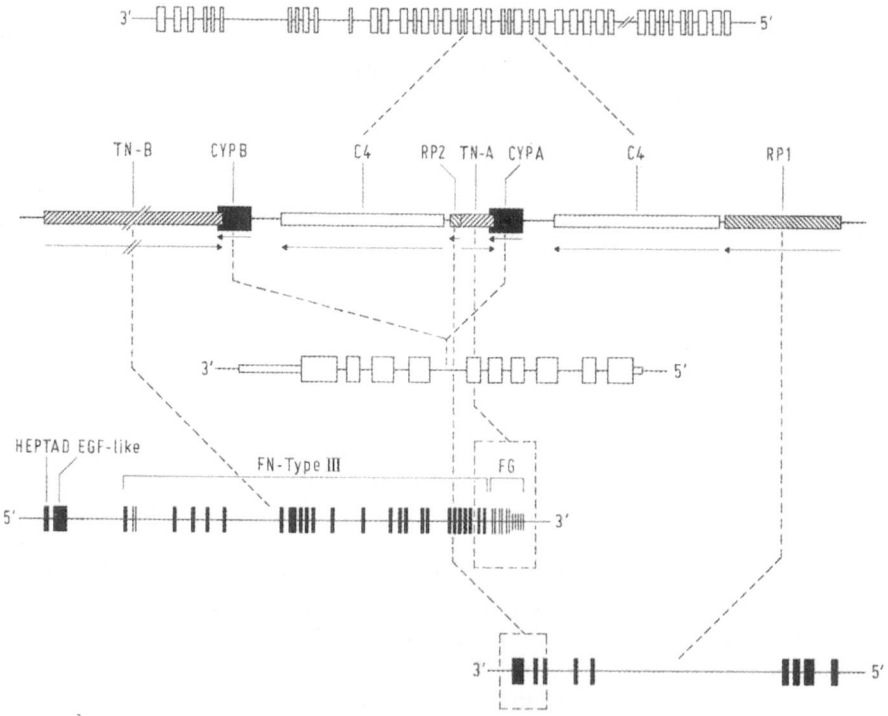

Fig. 1. Position and intron-exon organization of genes in the human TECYC4RP module: *TN-B* and *TN-A*, complete and truncated versions, respectively, of tenascin genes; *CYPA* and *CYPB*, 21-hydroxylase genes; C4, complement component *C4* genes; RP1 and RP2, complete and truncated versions, respectively, of the *RP* genes. *Arrows* transcriptional orientations

vant characteristic of the cluster is that it exists in one or more copies of a primogenial unit or module.

The Tenascin-X gene

Tenascin hexabrachion is a large glycoprotein component of the extracellular
matrix (Erickson and Bourdon 1989) composed of six disulfide-bound monomers
varying in sizes from M_r 220 000 to 230 000. Tenascin-X belongs to a family of
related proteins which also includes tenascin-C or cytotastin, tenascin-R or
restrictin, and tenascin-Y.

Tenascin glycoproteins appear to mediate interactions between the cell and
the extracellular matrix. A tenascin monomer is composed of four domains:
the NH_2-terminal or heptad domain is involved in the polymerization of the
polypeptide into oligomers or tenascin hexabrachion; the second domain is composed of a series of epidermal growth factor (EGF)-like repeats; the third domain
comprises a series of fibronectin type III (FnIII) repeats; and, finally, the COOH-
terminal or globular domain is similar to fibrinogen β- and γ-chains (Gulcher et
al. 1991).

There are two tenascin-X genes (TN in Fig. 1) in the module (Bristow et al.
1993; Morel et al. 1989; Matsumoto et al. 1992). A complete tenascin-X gene is
about 65 kb long, maps to the XB locus, encodes a 12 kb mRNA, and consists
of 39 exons interrupted by introns. Its exon 1 encodes the leader peptide and
the heptad domain, exon 2 encodes the EGF-like repeats, exons 3–34 encode
the 29 FnIII repeats, and exons 35–39 encode the fibrinogen-like domain. All
together 3816 amino acids are encoded by the 39 exons, but fairly large regions
of the gene remain unsequenced and additional exons encoding Fn repeats may
remain undiscovered. The second gene (XA) is truncated, about 4.5 kb long, contains 13 exons, and encodes a 2.6 kb adrenal gland-specific transcript. It encodes
a 673 amino acid protein that is identical to the 673 COOH-terminal residues of
the XB gene; the protein consists of five COOH-terminal FnIII repeats and the
fibrinogen-like domain. The promotor region of the XA gene is identical to
sequences present in intron 26 of the tenascin XB gene (Gitelman et al 1992;
Tee et al. 1995).

The CYP21 Gene

The human CYP21 gene encodes a key enzyme in the biosynthesis of glucocorti-
coid and mineralcorticoid hormones. The enzyme, the steroid 21-hydroxylase
(EC 1.14.99.10), also known as P450(C21), 21-OHase, steroid 21-monooxygenase,
steroid, or hydrogen-donor:oxygen oxidoreductase (21-hydroxylating), belongs
to the cytochrome P-450 superfamily and is involved in the conversion of 17-
hydroxyprogesterone to 11-deoxycortisol and of progesterone to 11-deoxycorti-
sone (Miller 1988; New and Levine 1984). There are two copies of CYP21 in
the class III region. CYP21B is the functional gene, while CYP21A is a pseudogene

(Carrol et al. 1985; White et al. 1985). Several kinds of genetic flaws can render the *CYP21B* gene defective, resulting in the recessive inherited disorder known as the congenital adrenal hyperplasia syndrome (CAH; New and Levine 1984; Speiser and New 1985; White et al. 1987; Miller 1988). Clinical symptoms include salt-wasting or the inability to conserve dietary sodium as a consequence of defective aldosterone production. Simple virilizing and nonclassical forms, characterized by the late development of symptoms, are less severe but quite common. The frequency of CAH as a result of deficiencies in *CYP21B* is about 1 in 10 000 births.

The human *CYP21B* gene is about 3.4 kb long, is composed of 10 exons and their corresponding intervening introns, and codes for 494 amino acid residues (Higashi et al. 1986). Molecular characterization of the functional *CYP21B* gene in patients with CAH of either the salt-wasting or the simple virilizing type has revealed several defects which impair the activity of steroid 21-hydroxylase (Higashi et al. 1988; Kohn et al. 1995; White et al. 1994; Tusie-Luna et al. 1991). The human *CYP21A* pseudogene carries three deleterious mutations compared with *CYP21B*: an 8 base pair (bp) deletion in exon 3, a single nucleotide (T) insertion in exon 7, and a 1 bp substitution in codon 318 of exon 8 (Higashi et al. 1986). Restriction fragment length polymorphisms (RFLPs) have shown human *CYP21* genes to be either 3.2 kb or 3.7 kb long on the basis of the size of *Taq*I-produced fragments. The 3.7 kb fragment appears to be characteristic of the functional *CYP21* gene and the 3.2 kb fragment is diagnostic of the *CYP21* pseudogene.

The Complement Component C4 Gene

The complement component C4 is synthesized as a single-chain precursor or pre-pro-C4 molecule with an M_r of about 200 000 and encompassing 1744 amino acid residues; the precursor is processed by proteolytic cleavage to a structure consisting of three disulfide-linked chains, β, α, and γ, with an M_r of 75 000, 98 000, and 30 000, respectively (Schreiber and Muller-Eberhard 1974). The respective lengths of the β, α and γ chains are 656, 770, and 292 amino acid residues. The activated C1 clips off a 77 amino acid fragment (C4a) from the NH_2-terminal of the α-chain, which diffuses into the fluid phase to become active as a weak anaphylatoxin (Hugli 1986). The removal of the C4a fragment unmasks the reactive site located in the α-chain of the remainder of the C4b molecule. The action of factor I in the presence of cofactors like the C4-binding protein and complement receptor 1 (CR1) further cleaves the C4b molecule at two sites in the α-chain, producing the C4d and C4c fragments. The *C4* gene spans approximately 21 kb, encompassing 41 exons and the intervening introns. It codes for 1744 amino acids of the pre-pro-C4 molecule and is located about 3 kb from each *CYP21* gene.

The C4 component exists in two isotypic forms – C4A and C4B – differing in the amino acid sequence at position 1106 of the molecule (Belt et al. 1984, 1985; Carroll et al. 1990; Yu et al. 1986). The human C4A isotype has Asp at this position and is characterized by a high binding affinity for amino groups of IgG immune aggregates. The C4B isotype has His at this position and is characterized by a high affinity for hydroxyl groups of carbohydrate targets (Yu et al. 1986; Carroll et al. 1990; Dodds and Law 1990). Both isotypes can be distinguished by hemolytic assays in which C4B isotypes display a three- to fourfold greater hemolytic activity than the C4A type (Law et al. 1984).

The C4 component is responsible for the difference between the Rg and Ch blood groups, which were defined on the basis of erythrocyte reactivity to anti-bodies originally present in the serum of two individuals named Rodgers and Chido, respectively. Each of these sera reacted with more than 95 % of the samples tested. Rg and Ch antigens are C4d fragments released from C4A and C4B isotypes, respectively, and passively deposited on erythrocytes (O'Neill et al. 1978; Tilley et al. 1978). Six Ch, three Rg, and one hybrid Ch/Rg epitope have been described. Four are sequential epitopes in that they depend on a specific amino acid sequence at defined positions (Rg1, Ch1, Ch4, Ch5, and Ch6). The others are conformational epitopes resulting from the interaction of two non-contiguous sites (Rg2, Ch2, Ch3, and WH). Eight polymorphic sites have been proposed as being responsible for the Rg and Ch groups, and they include positions 1054, 1101, 1102, 1105, 1106, 1157, 1188, and 1191 (Yu et al. 1988; Barba et al. 1994).

C4 molecules can be distinguished according to the electrophoretic mobility of the intact molecule or its subunits. At least 19 and 22 allotypes can be distinguished among the C4A and C4B isotypes, respectively, including the null alleles (i.e., those characterized by the absence of gene products; see Mauff el al. 1991). Depending on the ethnic groups, each allotype is present in the populations at different frequencies. The C4A3 (range 40 %-94 %) and C4B1 (range 36 %-80 %) are the most frequent, while C4A4 and C4B2 are present in populations at intermediate frequencies. Null alleles, on the C4A and C4B isotypes, occur at frequencies of 0 %-33 % and 0 %-21 %, respectively, depending on the population tested. Some but not all null alleles are due to deletions at the C4 loci (Braun et al. 1990).

At the genetic level two differents forms of C4 are known, namely, the C4 long (21 kb) and C4 short (14.5 kb) genes (Yu et al. 1986; Prentice et al. 1986). The two forms differ by the presence of a 6.5 kb element in intron 9 of the C4 long gene and its absence in the C4 short gene. The insert has recently been identified as a defective retrovirus, a member of the group HERV-II, a family dispersed over several human chromosomes (Dangel et al. 1994; Tassabehji et al. 1994). This insert is present in all C4A and about two thirds of the C4B isotypes. Long and short versions of the C4 gene can be identified by RFLP using BamHI- or KpnI-digested genomic DNA hybridized to labeled probes specific for the 5'end of the C4 genes. The resulting 4.8 kb BamHI and 7.5 kb KpnI fragments or 3.3 kb BamHI and the 8.5 kb KpnI fragments are diagnostic markers for the long and short C4 genes, respectively. Additional RFLPs which produce 7.0 or 6.0 kb fragments from the long C4 gene and either 6.4 or 5.4 kb fragments from the C4 short gene (Carrol et al. 1984; Yu et al. 1986; Schneider et al. 1986) can be detected upon digestion with the TaqI enzyme.

The *RP* Gene

The basic and highly hydrophilic RP molecule consists of 364 amino acid residues and may be a nuclear protein whose function has not been elucidated (Sargent et al. 1994; Shen et al. 1994). The protein is encoded by the *RP* gene (*G11*; Sargent et al. 1989) located about 600 bp upstream of the *C4* gene. The gene encompasses about 12 kb of DNA and consists of nine exons and introns. A series of nine *Alu* elements, one of them telescoped into the other, is present in the fourth intron. Three of the elements belong to the *Alu-J* family and four to the *Alu-S* family. In addition, between the third and fourth *Alu* elements a tandem repeat of 21 copies of a GC-rich 35–45 bp long sequence is present. The tandem repeat sequences seem to be related to a group of nonviral retroelements. The *RP* gene appears to be ubiquitously expressed with a major transcript size of about 1.6–1.8 kb and to exist in two forms. The *RP1* or complete version is located upstream of the human *C4A* gene, and the *RP2* or truncated version is located upstream of the human *C4B* gene. The *RP2* gene encompasses only exons 7–9, compared with the *RP1* gene (Shen et al. 1994).

Multicopies of the TECYC4RP Module

The number of copies of the *TECYC4RP* gene cluster per chromosome varies (McLean et al. 1988; Collier et al. 1989; Zhang et al. 1990). Human chromosomes with one, two, three, or four copies of the module exist (Fig. 2). The original

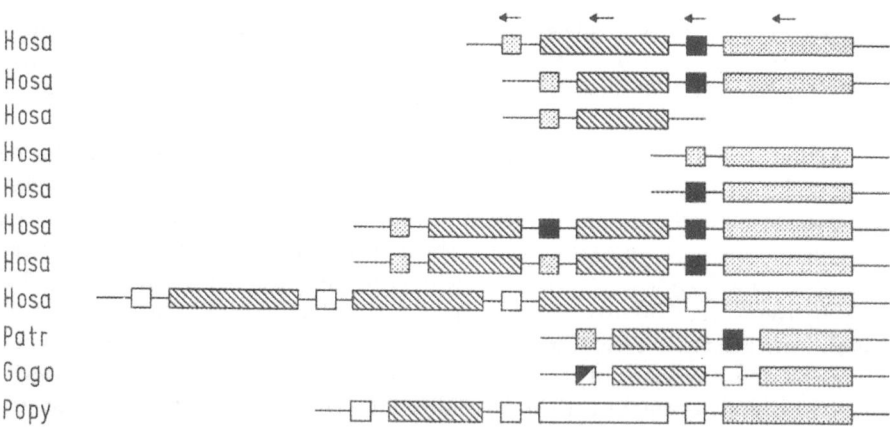

Fig. 2. Common *TECYC4RP* haplotypes of human (*Hosa*), chimpanzee (*Patr*), gorilla (*Gogo*) and orangutan (*Popy*). *Rectangles* and *squares*, *C4* and *CYP21* genes, respectively; *dotted rectangles*, C4A isotypes; *slashed rectangles*, C4B isotypes; *black* or *half-black squares*, CYP21 pseudogenes; *dotted squares*, functional CYP21 genes; *long* and *short rectangles*, long and short *C4* genes, respectively

event that led to this situation may have been a partial duplication resulting from a rare unequal crossing over involving exon 7 of the *RP* gene in one chromosome and intron 26 of the tenascin gene in the homologous chromosome (Gitelman et al. 1992; Bristow et al. 1993). As a consequence of this genetic recombination, these two genes became truncated in the duplicate, whereas complete copies of the *CYP21* and *C4* genes were generated (Fig. 3). The emergence of bimodular versions had the important evolutionary consequence that it created conditions for further expansion and contractions of the region through additional crossing over events.

The TECYC4RP Module in Nonhuman Primates

A systematic investigation of the *TECYC4RP* region has been carried out mainly in the great apes, but partial information about other nonhuman primates is also available. These studies have provided evidence for the region's instability and for the concerted evolution of some of the genes in the cluster.

Number of TECYC4RP Modules

To define the number of TECYC4RP modules in nonhuman primates, two approaches have been used. The first consisted of cloning the whole region in the chimpanzee, gorilla, and orangutan (Kawaguchi et al. 1990; Kawaguchi and Klein 1992). Two clusters of cosmid clones, each about 100 kb in length, and one 100 kb cluster and one 150 kb cluster were isolated from chimpanzee, gorilla, and orangutan, respectively. Two copies of the module were found to be present in the chimpanzee and gorilla clusters and three copies were observed in the single orangutan cosmid cluster (Fig. 2). Each module contained both the *C4* and the *CYP21* genes. The second approach consisted of digesting genomic DNA, transferring the fragments to filters, hybridizing them to labeled probes, and estimating the number of copies of the modules from the number of hybridizing bands. The subsequent typing of more than 60 related and unrelated individuals indicated the bimodular haplotype to be present in most chimpanzees (Bontrop et al. 1990; Christiansen et al. 1991; Zhang et al. 1993). Two gorillas have been tested in this manner, and they appeared to contain the single-module version (Bontrop et al. 1990; Zhang et al. 1993). In the orangutan, both single- and multi-module versions seem to exist, the latter with up to four modules per haplotype. Multimodular versions have also been observed in rhesus monkey (*Macaca mulatta*; Bontrop et al. 1990; Mevag et al. 1983). Thus, like humans, nonhuman primates possess duplicated versions of the *TECYC4RP* regions and, at least in the gorilla and the orangutan, haplotype polymorphism is present.

The presence of a multiple module in nonhuman primates suggests that the primate primogenial modular duplication may have occurred at least 23 my ago, before the divergence of Old World monkeys and apes. This notion is further

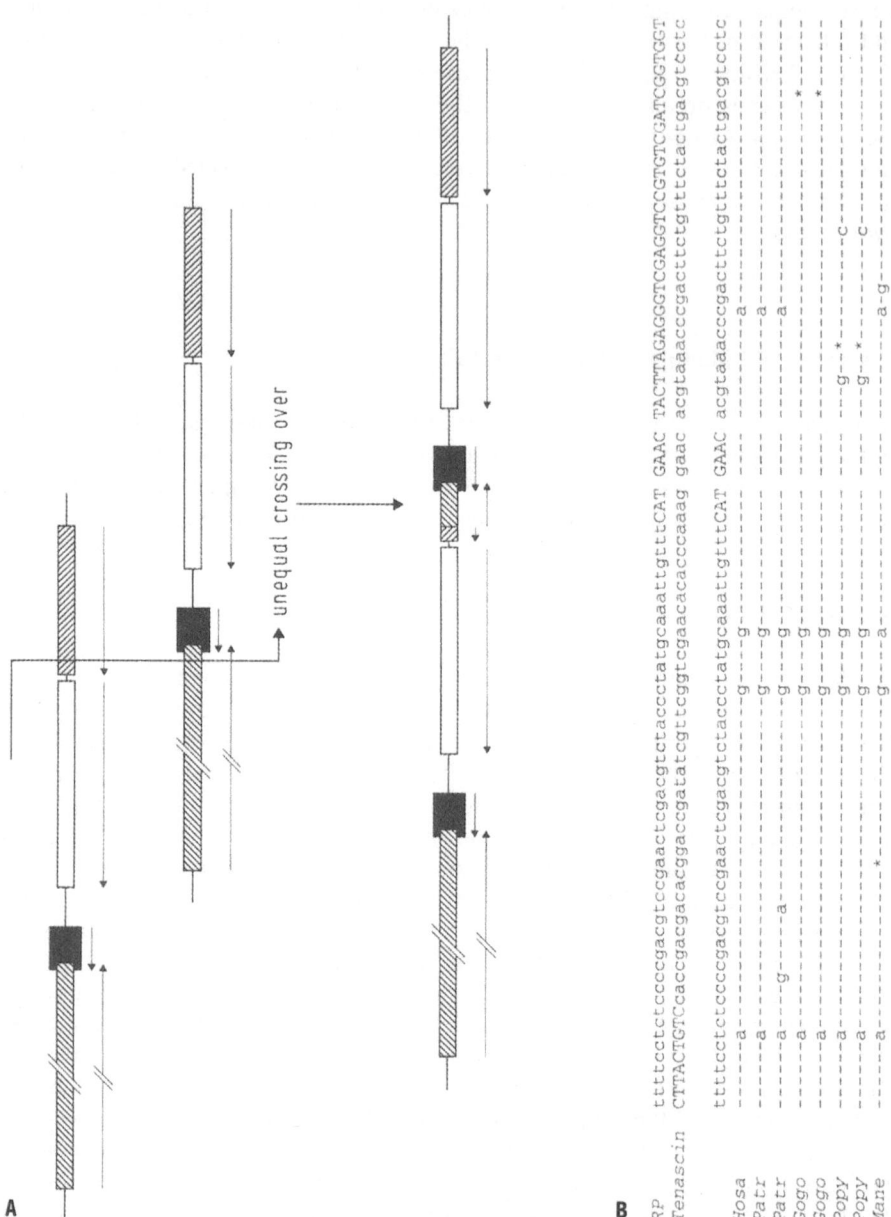

Fig. 3 A,B. A The unequal crossing over giving origin to the primigenial duplication of the TECYC4RP module and **B** the nucleotide sequence around the crossing over break point from human (*Hosa*), chimpanzee (*Patr*), gorilla (*Gogo*), orangutan (*Popy*), and pigtail macaque (*Mane*) C4 intergenic regions. *Capital* and *small letters*, exon and intron sequences, respectively

supported by the finding that in apes and in macaque, as a representative of Old World monkeys, the multiple TECYC4RP modules derive from one original module (Horiuchi et al 1993). The sequence of the flanking regions of the TECYC4RP modules and C4 regions from two independent chimpanzee gene clusters, one gorilla region, and C4 intergenic regions from a three-module orangutan haplotype, as well as part of the C4 region of the pigtail macaque, indicate that the duplication break point in all these species occurred at exactly the same position as that found previously in humans (i.e., in the *tenascin* and the *RP* genes; see Fig. 3).

Several other mammal species, such as mice (Roos et al. 1978), rats (Tosi et al. 1985), cattle (Yoshioka et al. 1986; Chung et al. 1986), pigs (Kirszenbaum et al. 1985), and horses (Kay et al. 1987), also possess duplicated TECYC4RP regions, but at least for rodents it has been suggested that these derive from a duplication event independent of the primogenial duplication in primates (Meo and Tosi 1985).

C4 Isotypes

Two approaches have also been used in defining isotypes of the C4 genes carried by nonhuman primates, namely, hybridization with isotype- specific oligonucleotides and sequencing of C4d portions of the genes. Both methods have revealed the presence of C4A- and C4B-like molecules in apes and Old World monkeys (Kawaguchi et al. 1990, 1992a; Kawaguchi and Klein 1992; Paz-Artal et al. 1994). As mentioned earlier, in humans, the A or B isotypes of the C4 molecules are distinguished by the presence of the Asp or His amino acid residues at position 1106 of the polypeptide. Sequences of C4d fragments derived from eight different nonhuman primate species have revealed that the same variation at position 1106 also exists in nonhuman primates (Kawaguchi et al. 1992b; Paz-Artal et al. 1994). Thus, of six C4 genes from two chimpanzee species (*Pan paniscus* and *Pan troglodytes*), five had the Asp residue at this position and therefore could be assigned to the C4A isotype, while the remaining gene had His at this position and hence represented the C4B isotype. Similarly, both C4A- and C4B-like genes can be found in gorilla, orangutan, and Old World monkeys (*Cercopithecus aethiops, Macaca fascicularis,* and *M. mulatta*). In New World monkeys, however, in four unrelated cotton-top tamarins (*Saguinus oedipus*) only C4B-like genes have been identified (Paz-Artal et al. 1994). Whether C4A isotypes are absent in New World monkeys will need to await further sampling of other species, but if they are indeed absent, one might postulate that C4B represents the ancestral form of C4. The fact that mouse C4 and C4-Slp molecules have His at position 1106 could favor this interpretation. These results indicate that isotype polymorphism in primates is at least 25 my old and has evolved *trans*-specifically. In humans, three other positions differentiate the A or B isotypes: 1101, 1102, and 1105. In nonhuman primates these positions are also variable and are occupied by either one or the other of the human-specific residues, but in different combinations. Thus, in addition to the human-specific PCLD and LSIH combinations,

LCLD, LSLD, PCIH, and PCLH combinations have been found in nonhuman primates. It may therefore be that at each of these positions only one of two residues is allowed because of functional constraints, but that the combinations of residues at different positions are not functionally constrained. In the mouse, orthologous genes also contain one or the other residue at these positions.

Long and Short *C4* Genes

The characterization by Kawaguchi and co-workers (1990, 1992a) of the chimpanzee, gorilla, and orangutan *C4* genes indicates that the long 21 kb vs short 14.5 kb *C4* length polymorphism was passed on *trans*-specifically. Although only short (*C4s*) genes were found to be present in chimpanzee and gorilla, both C4 long (*C4l*) and *C4s* genes were found in orangutan. In chimpanzee, the genes contain, as do the orthologous human genes, the 3.3 kb *Bam*HI diagnostic fragment and the 6.4 or 5.4 kb *Taq* fragments characteristic of the short versions of the genes. Both gorilla genes contain the 6.4 kb *Taq*I fragment and are therefore of the *C4s* type. In orangutan, long genes carry either the diagnostic human 7.0 kb or 6.0 kb *Taq*I fragments or a new 4.3 kb *Taq*I fragment. The orangutan *C4s* gene contains the characteristic 5.4 kb *Taq*I fragment.

Other studies based on RFLP analysis of a large sample of nonhuman primates confirm that apparently only *C4s* genes are present in chimpanzee and the gorilla and that both types of genes are common in orangutan (Zhang et al. 1993; Bontrop et al. 1991). In the rhesus monkey, additional RFLPs have been identified. The presence of 6.4, 5.4, and 4.3 kb *Taq*I fragments appears to indicate that both *C4l* and *C4s* genes occur in Old World monkeys. Thus, taken together the data suggest dichotomy of the long and short *C4* genes to be at least 25 my old and the retroviral insertion in intron 9 of the *C4* gene must also be at least that old.

The *CYP21* Genes

The characterization by Kawaguchi and co-workers (1990, 1992a) of the TECYC4RP modules from chimpanzee, gorilla, and orangutan has demonstrated that each module bears a corresponding duplicate of the *CYP21* gene. Chimpanzee *CYP21* can be found in two versions, one characterized by the presence of a 3.2 kb *Taq*I fragment and the second by a 3.7 kb fragment. The 3.2 kb *Taq*I fragment is derived from the intergenic *C4* region. In humans, the 3.2 kb version is associated with the presence of the *CYP21* pseudogene. In the gorilla, all four copies of the *CYP21* genes are of the 3.2 kb *Taq*I type. In the orangutan one copy produced the 3.2 kb fragment after *Taq*I digestion but the other three copies showed a new RFLP type characterized by a 5.3 kb *Taq*I fragment. Additional typing of a large panel of chimpanzees has revealed that all their *CYP21* genes are either of the 3.2 kb or the 3.7 kb type; no other RFLP pattern has been identified. A similar conclusion has been reached for gorillas. In the orangutan, by contrast, no 3.7 kb *Taq*I pattern has been found, and the identified genes are either of the 3.2 kb *Taq*I type or reveal a new polymorphism characterized by a 5.3 kb *Taq*I fragment (Bontrop et al. 1991; Zhang et al. 1993). In rhesus monkeys a high polymorphism has been observed, with at least nine different RFLP patterns detected (Bontrop et al. 1991).

Kawaguchi and co-workers (1992b) partially sequenced the *CYP21* genes present in four chimpanzee, three gorilla, and four orangutan modules. The nucleotide sequences covering five complete exons and one partial and four complete introns revealed that only one of the defects rendering the human gene a pseudogene was present in these nonhuman primates – the 8 bp deletion found in the 3.2 kb versions of the chimpanzee *CYP21* genes. Neither the 8 bp deletion nor the other human pseudogene defects were present in gorilla or orangutan genes. However, the intron 2 CG nucleotide substitution leading to aberrant mRNA splicing and present in the functional human *CYP21* in some CAH patients was found in one of the gorilla genes, and a previously undetected defect, a stop codon in exon 7, was found in another gorilla gene. No obvious defects were found in the available sequence of orangutan *CYP21* genes.

Evolution of *C4* and *CYP21* Primate Genes

As described above, the overall organization of the TECYC4RP module is known from four primate species: human, chimpanzee, gorilla, and orangutan. Sequencing information from *C4* and *CYP21* genes borne by a number of primate species, including humans, apes, Old World monkeys, and to lesser extent New World monkeys, is also available. Futhermore, specific gene characteristics, such as *C4* length variability, isotypes, and the presence or absence of *CYP21* functional genes and pseudogenes, are also known. These data enable us to ask questions about the origin of modular organization and the evolution of genes in the modules through primate history. What follows is a description of an evolutionary scenario based on the available data.

The variability in number of modules per chromosome has its origins in the primogenial duplication that occurred at least 25 my ago (Horiuchi et al. 1993). The existence of different primate haplotypes with a varying number of modules indicates that subsequent to this rare primogenial duplication, frequent secondary rounds of duplication have been occurring repeatedly. It has been documented that the secondary duplications occur by homologous but unequal crossing over between misaligned modules (Sinnott et al. 1990). Dendrograms based on partial sequences of human, gorilla, chimpanzee, and orangutan genes reveal that at least in the orangutan lineage the three *C4* genes borne by a single chromosome derive from a single ancestral gene different from that of the gorilla, chimpanzee, and human ancestral *C4* genes (Kawaguchi et al. 1991, 1992a). The same is also true for the *CYP21* genes of these primate species (Kawaguchi et al. 1992b). Unequal crossing over can contract a multimodular haplotype, render it monomodular, and then expand it to become multimodular again (Kawaguchi et al. 1991). The region thus apparently undergoes repeated contractions and expansions, with the consequence that the sequences are continually homogenized within each species. Unequal crossing over is probably the main mechanism of concerted evolution in the *TECYC4RP* region. (Kawaguchi et al. 1991).

Homogenization by unequal crossing over is also required to explain certain other features of the primate TECYC4RP modules. First, although in each species the *C4* genes are more similar to one onother than they are to genes in other species, the differentiation into *A*- and *B*-like genes is retained in all primate species (Kawaguchi et al. 1992a). Thus, different regions of the *C4* genes seem to display different phylogenies: The *A*- and *B*-specifying regions appear to have diverged before the divergence of the orangutan and the African ape lineage more than 12 my ago, whereas the rest of the *C4d* region of the genes diverged after their formation separately in each of the four species. These contradictory phylogenies can be reconciled by assuming that most of the *C4d* region, but not its *A*- and *B*-specifying part, has been homogenized.

Second, the *C4* genes occur in two versions, long and short, differentiated by the presence or absence of a 6.5 kb long defective retrovirus (Dangel et al. 1994; Tassabehji et al. 1994). In humans, all the *C4A* genes are of the long variety, whereas the *C4B* genes can be either long or short. In chimpanzee (Kawaguchi et al. 1990) and gorilla (Kawaguchi and Klein 1992), both the *C4A* and *C4B* genes are short; in orangutan (Kawaguchi and Klein 1992), the *C4 A* genes are long, whereas the *C4B* genes are short. The original *C4* gene must have been of the short type and the retrovirus insertion must have occurred before the divergence of the orangutan and the African ape lineages but after the primogenial duplication in one of the two *C4* genes. The transfer of the insert from one copy to its duplicate must again have resulted from unequal crossing over.

Third, the human and chimpanzee *CYP21P* genes contain a characteristic 8 bp deletion in exon 3, which might have been the original mutation silencing them (White et al. 1986; Higashi et al. 1988; Rodriguez et al. 1987; Kawaguchi et al. 1990, 1992b). This deletion is absent in gorilla and orangutan *CYP21* genes (Kawaguchi et al. 1992b), signaling that it occurred after separation of the gorilla/orangutan from the chimpanzee/human lineages, but before separation of the human from the chimpanzee lineage. One could therefore expect the *Hosa-CYP21P* and *Patr-CYP21P* genes to have derived from a common ancestor. In fact, this is demonstrated by the dendrogam based on overall *CYP21P* sequence similarity (Kawaguchi et al. 1992b). Yet, the *C4* genes borne by the same haplotypes as the *CYP21* genes show a different phylogeny (Kawaguchi et al. 1991). Once again, these contradictory findings are reconciled by the postulate of unequal crossing over and the resulting homogenization of the *C4* genes.

Acknowledgements. I thank Ms Donna Devine for editorial assistance and Dr Jan Klein for critical reading of the manuscript.

References

Barba GMR, Braun-Heimer L, Rittner C, Schneider PM. A new PCR- based typing of the Rodgers and Chido antigenic determinants of the fourth component of human complement. Eur J Immunogenet 21:325, 1994

Belt KT, Carroll MC, Porter RR. The structural basis of the multiple forms of human complement component C4. Cell 36:907–914, 1984

Belt KT, Yu CY, Carroll MC, Porter RR. Polymorphism of human complement component C4. Immunogenetics 21:173–180, 1985

Bingulac-Popouic, J, JH Postlethwait, A Sato, WS Talbot, SL Johnson, M Gates, F Figueroa, J Klein. Mapping of Mhc class I and class II to different chromosomes in the zebrafish, Danio rerio Immunogenetics (in press)

Bird AP, CpG-rich islands and the function of DNA methylation. Nature 321:209–213, 1986

Bontrop RE, Broos LAM, Otting N, Jonker MJ. Polymorphism of C4 and CYP21 genes in various primate species. Tissue Antigens 37:145–151, 1991

Braun L, Schneider PM, Giles CM, Bertrans J, Rittner C. Null alleles of human complement C4: evidence for pseudogenes at the C4A locus and for gene conversion at the C4B locus. J Exp Med 171:129–140, 1990

Bristow J, Tee MK, Gitelman SE, Mellon SH, Miller WL. Tenascin-X: a novel extracellular matrix protein encoded by the human XB gene overlapping P450c21B. J Cell Biol 122:265–278, 1993

Campbell RD, Trowsdale J. Map of the human major histocompatibility complex. Immunol Today 14:349–352, 1993

Carroll MC, Belt T, Palsdottir A, Porter RR. Structure and organization of the C4 genes. Philos Trans R Soc B 306:379–388, 1984

Carroll MC, Campbell RD, Porter RR. Mapping of steroid 21-hydrolase genes adjacent to complement component C4 genes in HLA, the major histocompatibility complex in man. Proc Natl Acad Sci USA 82:521–525, 1985

Carroll MC, Fathallah DM, Bergamaschini L, Alicot EM, Isenman DE. Substitutions of a single aminoacid (aspartic acid for hystidine) converts the functional activity of human complement C4B to C4A. Proc Natl Acad Sci USA 87:6868–6872, 1990

Christiansen FT, Bontrop R, Giphart MH, Cameron PU, Zhang WJ, Townend D, Jonker M, Dawkins RL. Major histocompatibility complex haplotype in the chimpanzee: identification using C4 haplotyping. Hum Immunol 31:34–39, 1991

Chung B, Matteson KJ, Miller W. Cloning and characterization of the bovine gene for steroid 21-hydroxylase. DNA 4:211–219, 1986

Collier S, Sinnott PJ, Dyer PA, Price DA, Harris R, Strachan T. Pulsed field gel electrophoresis identifies a high degree of variability in the number of tandem 21-hydrolase and complement C4 genes repeats in the 21-hydrolase deficiency haplotypes. EMBO J 8:1393–1402, 1989

Dangel AW, Mendoza AR, Baker BJ, Daniel CM, Carroll MC, Wu L-C, Yu CY. The dichotomous size variation of human complement C4 genes is mediated by a novel family of endogenous retroviruses which also establishes species-specific genomic patterns among Old World primates. Immunogenetics 40:425–436, 1994

Dodds AW, Law ASK. The complement component C4 of mammals. Biochem J 265:495–502, 1990

EricksonHP, Bourdon MA. Tenascin: an extracellular matrix protein prominent in specialized embryonic tissues and tumors. Annu Rev Cell Biol 5:71–92, 1989

Gitelman SE, Bristow J, Miller WL. Mechanism and consequences of the duplication of the human C4/P450c21/geneX locus. Mol Cell Biol 12:2124–2134, 1992

Gulcher JR, Nies DE, Alexakos MJ, Ravikant NA, Sturgill ME, Marton LS, Stefanson K. Structure of the human hexabrachion (tenascin) gene. Proc Natl Acad Sci USA 88:9438–9442, 1991

Higashi Y, Yoshioka H, Yamane M, Gotoh O, Fujii-Kuriyama Y. Complete nucleotide sequence of the two steroid 21-hydrolase genes tandemly arranged in human chromosome: a pseudogene and a genuine gene. Proc Natl Acad Sci USA 83:2841–2845, 1986

Higashi Y, Tanae A, Inoue H, Hiromasa T, Fujii-Kuriyama Y. Aberrant splicing and missence mutation causing steroid 21-hydrolase (P-450c21) deficiency in humans: possible gene conversion products. Proc Natl Acad Sci USA 85:7486–7490, 1988

Horiuchi Y, Kawaguchi H, Figueroa F, O'hUigin C, Klein J. Dating the primigenial C4-CYP21 duplication in primates. Genetics 134:331–339, 1993

Hugli TE. Biochemistry and biology of anaphylatoxins. Complement 3:111, 1986

Kawaguchi H, Klein J. Organization of C4 and CYP21 loci in gorilla and orangutan. Hum Immunol 33:153–162, 1992

Kawaguchi H, Golubic M, Figueroa F, Klein J. Organization of the chimpanzee C4-CYP21 region: implications for the evolution of human genes. Eur J Immunol 20:739–749, 1990

Kawaguchi H, O'hUigin C, Klein J. Evolution of primate C4 and CYP21 genes. In: Klein J, Klein D (eds) Molecular evolution of the major histocompatibility complex. Springer, Berlin Heidelberg New York, pp 357–383, 1991

Kawaguchi H, Zaleska-Rutczynska Z, Figueroa F, Klein J. C4 genes of the chimpanzee gorilla and orangutan: evidence for extensive homogenization. Immunogenetics 35:16–23, 1992a

Kawaguchi H, O'hUigin C, Klein J. Evolutionary origin of mutations in the human cytochrome P450c21 gene. Am J Hum Genet 50:766–780, 1992b

Kay PH, Dawkins RL, Bowling AT, Bernovo D. Heterogeneity and linkage of equine C4 and steroid 21-hydroxylase genes. J Immunogenet 14:247–253, 1987

Kirszenbaum M, Renard C, Geffrontin C, Chardon P, Vaiman M. Evidence for mapping C4 gene(s) within the pig major histocompatibility complex (SLA). Anim Blood Groups Biochem Genet 16:65–68, 1985

Kohn B, Day D, Alemzadeh R, Enerio D, Patel SV, Pelczar JV, Speiser PW. Splicing mutation in CYP21 associated with delayed presentation of salt-wasting congenital adrenal hyplerplasia. Am J Med Genet 57:450–454, 1995

Law SKA, Dodds AW, Porter RR. A comparison of the properties of two classes, C4A and C4B, of the human complement component C4. EMBO J 3:1819–1823, 1984

Lindsay S, Bird AP. Use of restriction enzymes to detect potential gene sequences in mammalian DNA. Nature 327: 336–338 1987

Matsumoto K, Ishihara N, Ando A, Inoko H, Ikemura T. Extracellular matrix protein tenascin-like gene found in human MHC class III region. Immunogenetics 36:400–403, 1992

Mauff G, Suzuki K, Tokunaga K, Hauptman G, Brenden M, Hasbach H. C4 reference report. In: Tsuji K, Aizawa M, Sasazuki T (eds) HLA 1991. Oxford Scientific, Oxford, pp 954–957, 1991

McLean RH, Donohoue PA, Jospe N, Biase WB, van Dorp C, Migeon CJ. Restriction fragment analysis of duplication of the fourth component of complement (C4A). Genomics 2:76–85, 1988

Meo T, Tosi M. Molecular genetics of the S region of the murine H-2 major histocompatibility complex. Ann Inst Past 136C:225–243, 1985

Mevag B, Olaisen B, Teisberg P, Smith DG. Two C4 loci in macaca monkeys 10th international complemenet workshop, Mainz. Immunobiology 164:1 (abstract), 1983

Miller WL. Molecular biology of steroid hormone synthesis. Endocr Rev 9:295–318, 1988

Morel Y, Bristow J, Gitelman SE, Miller WL. Transcript encoded on the opposite strand of the human steroid 21-hydroxylase/complement component/C4 gene locus. Proc Natl Acad Sci USA 86:6582–6586, 1989

New MI, Levine LS. Recent advances in 21-hydroxylase deficiency. Annu Rev Med 35:649–663, 1984

O'Neill GJ, Yang SW, Tegoli J, Berger J, Dupont B. Chido and Rodgers blood groups are distinct antigenic components of human complement C4. Nature 273:668–670, 1978

Paz-Artal E, Corell A, Alvarez M, Varela P, Allende L, Madrono A, Rosal M, Arnaiz-Villena A. C4 gene polymorphism in primates: evolution, generation, and Chido and Rodgers antigenicity. Immunogenetics 40:381–396, 1994

Prentice HL, Schneider PM, Strominger JL. C4B polymorphism detected in a human cosmid clone. Immunogenetics 23:274–276, 1986

Rodrigues NR, Dunham I, Yu Y, Carroll MC, Porter RR, Campbell RD. Molecular characterization of the HLA-linked steroid 21- hydroxylase B gene from an individual with congenical adrenal hyperplasia. EMBO J 6:1753–1661, 1987

Roos MH, Atkinson JP, Shreffler DC. Molecular characterization of the Ss and Slp (C4) proteins of the mouse H-2 complex: subunit composition, chain size polymorphism, and an intracellular (Pro-Ss) precursor. J Immunol 121:1106–1115, 1978

Sargent CA, Anderson MJ, Hsieh SL, Kendall E, Gomez-Escobar N, Campbell RD. Characterization of the novel gene G11 lying adjacent to the complement C4A gene in the human major histocompatibility complex. Hum Mol Genet 3:481–488, 1994

Schneider PM, Carroll MC, Alper CA, Rittner C, Whitehead AS, Colten HR. Polymorphism of the human complement C4 and steroid 21- hydroxylase genes. Restriction fragment length polymorphisms revealing structural deletions, homoduplications and size variants. J Clin Invest 78:650–657, 1986

Schreiber RD, Muller-Eberhard HJ. Fourth component of human complement: description of a three chain structure. J Exp Med 140:1324–1335, 1974

Shen L, Wu L, Sanlioglu S, Chen R, Mendoza AR, Dangel AW, Carroll MC, Zipf WB, Yu CY. Structure and genetics of the partially duplicated gene RP located immediately upstream of the complement C4A and the C4B genes in the HLA class III region. J Biol Chem 269:8466–8476, 1994

Sinnott P, Collier S, Costigan C, Dyer PA, Harris R, Strachan T. Genesis by meiotic unequal crossing-over of a novo deletion that contributed to steroid 21-hydroxilase deficiency. Proc Natl Acad Sci USA 87:2107–2111, 1990

Speiser PW, New MI. Genetics of steroid 21-hydroxylase deficiency. Trends Genet 1:275–278, 1985

Tassabehji M, Strachan T, Anderson M, Campbell RD, Collier S, Lako M. Identification of a novel family of human endogenous retroviruses and characterization of one family member HERV-K (C4), located in the complement C4 gene cluster. Nucleid Acids Res 22:5211–5217, 1994

Tee M, Thompson AA, Bristow J, Miller WL. Sequences promoting the transcription of the human XA gene overlapping P450c21A correctly predict the presence of a novel, adrenal-specific, truncated form of tenascin-X. Genomics 28:171–178, 1995

Tilley CA, Romans DG, Crookston MC. Localization of Chido and Rodgers antigenic determinants to the C4d fragment of human C4. Nature 276:713–715, 1978

Tosi M, Levi-Strauss M, Georgatson E, Amor M, Meo T. Duplication of complement and non complement genes of the H-2S region: evolutionary aspects of the C4 isotypes and molecular analysis of their expression variants. Immunol Rev 87:141–183, 1985

Trowsdale J. Both man & bird & beast: a comparative organization of MHC genes. Immunogenetics 41:1–17, 1995

Tusie-Luna MT, Speiser PW, Dumic M, New MI, White PC. A mutation (Pro-30 to Leu) in CYP21 represents a potential nonclassical steroid 21-hydroxylase deficiency allele. Mol Endrocrinol 5:685–692, 1991

White PC, Grossberger D, Onufer BJ, Chaplin DD, New MI. Dupont B, Strominger JL. Two genes encoding steroid 21-hydroxylase are located near the genes encoding the fourth component in man. Proc Natl Acad Sci USA 82:1089–1093, 1985

White PC, New MI, Dupont B. Structure of human 21-hydroxylase genes. Proc Natl Acad Sci USA 83:5111–5115, 1986

White PC, New MI, Dupont B. Congenital adrenal hyperplasia. N Engl J Med 316:1519–1524, 1987

White PC, Tusie-Luna MT, New MI, Speiser PW. Mutations in steroid 21-hydroxylase (CYP21). Hum Mutat 3:373–378, 1994

Yoshioka H, Morohashi K, Sogawa K, Ymane M, Komonami S, Takemori S, Okada Y, Fuji-Kuriyama Y. Structural analysis of cloned cDNA for mRNA of microsomal cytochrome P-450 (c21) which catalyzes steroid 21-hydroxylation in bovine adrenal cortex. J Biol Chem 261:4106–4109, 1986

Yu CY, Campbell RD, Porter RR. A structural model for the location of the Rodgers and Chido antigenic determinants and their correlation with the human complement component C4A/C4B isotypes. Immunogenetics 27:399–405, 1988

Yu CY, Belt KT, Giles CM, Campbell RD, Porter RR. Structural basis of the polymorphism of human complement components C4A and C4B: gene size reactivity and antigenicity. EMBO J 5:2873–2881, 1986

Yu CY The complete exon-intron sequence of a human complement component C4A gene. DNA sequence, polymorphism and linkage to the 21-hydroxylase gene. J Immunol 146:1057–1066, 1991

Zhang WJ, Degli-Esposi MA, Cobain TJ, Cameron PU, Christiansen FT, Dawkins R. Differences in gene copy number carried by different MHC ancestral haplotypes. J Exp Med 171:2101–2114, 1990

Zhang WJ, Christiansen FT, Wu X, Abraham LJ, Giphart M, Dawkins RL. Organization and evolution of C4 and CYP21 genes in primates: importance of genomic segments. Immunogenetics 37:170–176, 1993

5 MHC Genes, Immune Response, and Vaccines

R.E. BONTROP

Structure and Function of MHC Molecules

In primates, two major types of cell surface MHC molecules have been characterized, designated class I and class II. Both the *Mhc* class I and II regions encode multiple products and as a consequence have been divided into subregions. The MHC class I molecules are cell surface structures composed of a membrane spanning heavy chain of about 45 000 M_r which forms a noncovalent complex with a light chain of 12 000 M_r. The latter polypeptide, also known as β2-microglobulin, is encoded outside the *Mhc* region. The classical MHC class I molecules, also known as transplantation antigens, are normally found on the cell surface of nucleated cells. Some of the nonclassical MHC class I molecules, however, show a more differential type of tissue distribution.

MHC class II region products are heterodimers composed of two noncovalently associated sialoglycoproteins, designated α and β, of approximately 34 000 and 29 000 M_r, respectively (Kaufman et al. 1984). Both MHC class II chains span the membrane and the corresponding sequences map within the *Mhc* region. In contrast to MHC class I, class II molecules display a limited tissue distribution and are primarily expressed on white blood lineage cells.

One of the tasks of the immune system is to protect the host from danger, for instance, resulting from infections with pathogens such as viruses and bacteria which may cause damage to infected tissues. MHC molecules play a key role

Molecular Biology and Evolution of Blood Group
and MHC Antigens in Primates
Blancher/Klein/Socha (Eds.)
© Springer-Verlag Berlin Heidelberg 1997

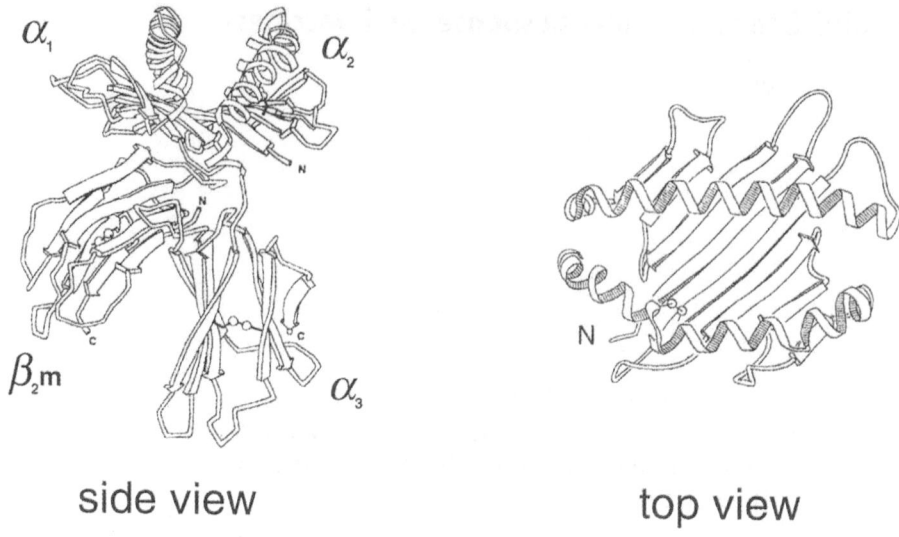

side view   top view

Fig. 1. The structure of the MHC class I molecule showing the α1, α2 and α3 domains in association with β2-microglobulin (side view), as well as the top view illustrating the peptide binding site formed by β-pleated sheets and two α-helices

in the initiation of an antigen-specific immune response. Transfection studies have demonstrated that only MHC class I or II molecules are required to convert any nonpresenting cell effectively into an antigen presenting cell.

The actual binding of immunogenic peptides to MHC molecules was first documented about a decade ago (Babitt et al. 1985; Guillet et al. 1986; Buus et al. 1987). Another major breakthrough was achieved when the crystal structure of MHC class I and II gene products was resolved documenting the existence of a so-called peptide binding site (Bjorkman et al. 1987; Brown et al. 1993). The peptide binding site of MHC molecules is formed by two parallel α-helices on top of anti-parallel oriented β-strands (Fig. 1). In the case of MHC class I gene products, the peptide binding site exhibits a closed configuration whereas the peptide binding site of MHC class II molecules is "open-ended". MHC class I molecules generally "select" for binding peptides that are 8–11 amino acids long whereas the peptides that bind to MHC class II molecules are generally longer but also more heterogeneous in size (Schumacher et al. 1991; Rudensky et al. 1991). In principle, MHC molecules assemble with peptides inside the cell and are then transported to the cell surface. The complex of the MHC molecule and peptide can be recognized by the T cell receptor (TCR) and this event can be considered as the actual start of an antigen-specific immune reaction. Thus with regard to biological function, MHC molecules can be considered as peptide receptors that provide the context for recognition of degraded antigen segments by T cells. At the cell surface, stable MHC structures are found complexed with peptide. In the normal situation, MHC associated peptides are derived from self proteins. During ontogeny, T cells that are reactive with MHC molecules and self peptides are deleted from the repertoire or become tolerant to such complexes (Von Boehmer 1992; Matzinger 1994). As soon as the MHC molecules present new peptides, which can originate from pathogens, they can become potential targets for T cells. In order to be presented by MHC molecules, antigens must be properly handled. The collection of events leading to the denaturation and cleavage of native antigens into immunogenic peptides which subsequently may bind to MHC molecules is called processing. The antigen processing pathways of MHC class I and class II are different. Most of the recent data have been obtained in rodent or human model systems. The situation in humans is considered to resemble the situation encountered in other primates. For that reason it is taken as a reference system and will be discussed here in some detail.

Antigen Processing in the Context of MHC Class I Molecules

In the endoplasmatic reticulum (ER), newly synthesized MHC class I heavy chains bind to calnexin, a so-called chaperone molecule. After dissociation, the heavy chain complexes with β2-microglobulin and TAP (transporter associated with antigen processing) which seems to facilitate the binding of peptide (Sugati and Brenner 1994). In humans, two TAP genes have been identified, designated TAP1 and TAP2, which map within the HLA class II region. The corresponding gene products form a heterodimer that is equipped with two ATP binding cassettes. In its proper configuration, this complex generates a peptide pump controlling the translocation of peptides from the cytosol to the ER (Neefjes and Momburg 1993). TAP gene products exhibit some allelic variation at the population level and the specificity of TAP genes may define which peptides become available for antigen presentation. It is thought that cytosolic proteins are partially degraded by the endopeptidase activity of the proteasome, a multi-component structure (Rock et al. 1994). The fine specificity of the endopeptidase activity, however, is influenced by two proteasome units, LMP2 and LMP7. Both genes map within the MHC class II region and, like MHC class II genes, their expression levels can be influenced by interferon (IFN)-γ.

Thus, MHC class I molecules transport peptide fragments from endogenously expressed cytosolic and nuclear proteins from the ER to the cell surface (Rammensee et al. 1993). Edman degradation of unfractionated mixtures of peptides isolated from MHC class I molecules elucidated the existence of so-called peptide binding motifs and anchor residues that are often allele-specific (Rötzschke et al. 1990). These type of studies indicate that the peptide binding site is promiscuous and can accommodate different sorts of peptides. In contrast, some MHC class I molecules may be involved in performing highly specialized functions. The classic example is the mouse H-2M3 molecule which presents N-formylated peptides of bacterial and mitochondrial origin (Fischer-Lindahl et al. 1991). Equivalents of such MHC class I molecules have not yet been discovered in primates.

Antigen Processing in the Context of MHC Class II Molecules

Early in their biosynthesis MHC class II molecules associate with Ii, the invariant chain (Anderson and Miller 1992). A short internal segment of the Ii chain, called CLIP (class II invariant chain peptide), associates with the MHC class II binding site, thereby avoiding binding of free peptides from cytosolic proteins that are potential ligands for class I (Rogmagnoli and Germain 1994; Ghosh et al. 1995). The second function of Ii is to direct the MHC class II heterodimer to the class II loading compartment. Just before arriving at these specialized endocytic compartments, the Ii chain is dissociated from the MHC class II dimer by proteolytic cleavage, thereby restoring the peptide binding capacity (Neefjes and Ploegh 1992; Morton et al. 1995). Recently HLA-DMA and -DMB genes were identified which map within the class II region and show an intermediate similarity with MHC class I and II genes (Kelly et al. 1991). In the absence of DM, MHC class II molecules are not loaded with foreign peptides, but appear at the cell surface with CLIP (Riberdy et al. 1992; Sette et al. 1992). Exposure to low pH conditions and proteases apparently is not sufficient to remove CLIP (Fling et al. 1994). According to the latest reports, DM products help to liberate the MHC class II antigen binding site and thus facilitate loading with peptides (Sloan et al. 1995; Denzin and Cresswell 1995).

Peptides that bind to MHC class II molecules do not have a restricted size. The open-ended groove of MHC class II molecules can accommodate peptides with lengths varying from 12 to 25 residues (Rudensky et al. 1992). Pool sequencing studies of eluted MHC class II binding peptides suggested that such peptides bear a motif of two or three anchor positions in the core region of the peptide, dependent on allelic variation (Rudensky et al. 1991; Chicz et al. 1992; Verreck et al. 1994). Epitope prediction of potential MHC class II ligands within a protein is not as easy as with MHC class I, since the anchor residues are more degenerate with regard to specificity.

According to the immunological dogma, MHC class II molecules can bind peptides derived from exogenous antigens that are taken up by antigen presenting cells through endocytosis or phagocytosis. It should be noted, however, that although MHC class I molecules present peptides derived from endogenous proteins and MHC class II molecules present peptides derived from internalized exogenous proteins, many exceptions to the rule exist (Moore et al. 1988; Nuchtern et al. 1990; Rock et al. 1990).

Something About T Cell Receptors

The majority of primate T cells are equipped with a CD3-TCR αb complex which recognizes enzymatically degraded peptide samples in the context of self MHC structures. This phenomenon, known as MHC restriction, is explained at the molecular level by the assumption that the compound TCR complex interacts both with the MHC molecule and its peptide. Although promiscuous T cells are known, the specificity of this interaction is generally very high. Fundamental to the generation of TCR recognition capacity is the existence in germline of multiple, tandemly duplicated, variable region genes. By convention these are classified into families and 33 $V\alpha$ and 26 $V\beta$ families have been described in humans. *TCR* diversity is achieved by the somatic recombination of germline segments. Each rearranged TCR α- or β-chain is composed of a variable (V), diversity (D, not TCR α), joining (J) and constant (C) region (Toyonaga and Mak 1987).

Recently substantial amounts of data have become available on *TCR* gene segments or functionally rearranged receptors in nonhuman primates. In the case of the rhesus macaque and chimpanzee, which are the only primates to have been extensively studied apart from humans, the material has comprised peripheral blood T cells and antigen-specific T cell clones (Levinson et al. 1992; Chen et al. 1992; Jaeger et al. 1994). In retrospect, each of the $V\beta$ segments isolated thus far can be assigned to one of the 26 known human families. Alignment of $V\beta$ region sequences showed that key residues which are conserved among variable regions of TCR and which have been suggested to be responsible for the domain framework are conserved in primates. For the rhesus macaque detailed descriptions of $V\beta$ segments belonging to 19 families have been reported whereas for the chimpanzee members of 14 families have been characterized so far (Levinson et al. 1992; Chen et al. 1992; Jaeger et al. 1994). Other reports indicate that similar findings exist for the $V\alpha$ segments (Thiel et al. 1996). Based upon such results, it seems justified to conclude that most of the contemporary primate species share a very similar *TCR* system, both in quality and quantity (Bontrop et al. 1995; Charmley et al. 1995).

In rodents and humans, predominant usage of TCR $V\alpha$ or $V\beta$ gene segments has been reported for several type of disease related immune responses. A similar type of repertoire skewing has been observed for rhesus monkey cytotoxic T lymphocyte (CTL) responses directed to simian immunodefiency virus (SIV) gag antigen (Chen et al. 1992).

Apart from the αb TCR, some human T lymphocytes are equipped with a so-called $\gamma\delta$ receptor. Recently it was documented that $\gamma\delta$ T lymphocytes may recognize nonpeptide antigens such as monoalkyl phosphates (Tanaka et al. 1995). Like humans, apes and Old World monkeys have $\gamma\delta$-positive T lymphocytes. Human and chimpanzee $\gamma\delta$-positive T lymphocytes even seem to recognize identical or similar targets (Sturm et al. 1992).

Signal 1 and 2

The complex formation of MHC-peptide and TCR-CD3 complex is normally referred to as signal 1. The CD3 molecular complex is involved in the intracellular signal transduction. To induce an antigen-specific T cell response, another costimulatory event (signal 2) is required that is provided by several receptor ligand interactions such as CD2-LFA3 or CD28-B7.

Immunosurveillance of virus-infected cells or malignant cells is mainly performed by CD8-positive CTLs which may recognize a peptide derived from endogenously synthesized viral or oncogenic peptides in the context of cell surface MHC class I molecules. β2- microglobulin deficient mice that do not express MHC class I molecules show delayed clearance of viruses and are susceptible to a variety of pathogens (Zijlstra et al. 1992; Bender et al. 1992). The CD8 glycoprotein is composed of an α- and β-chain and acts as a coreceptor. The CD8 α-chain binds to a nonpolymorphic site on the α3 domain of the MHC class I molecule which strengthens the interaction with the TCR/MHC but also provides an intracellular signal (Julius et al. 1993). The gene encoding the CD8 β-chain gene is duplicated in humans, chimpanzee, and gorillas and only one of the copies appears to be functional (DeLarbre et al. 1993). For effective lysis by a MHC class I restricted CTL, about 200 cell surface peptide-MHC class I complexes are needed (Christinck et al. 1991). It should be taken into account, however, that one CTL has the capacity to lyse multiple target cells.

MHC class II molecule-peptide complexes are generally recognized by the TCR on CD4-positive T helper (Th) cells resulting in the secretion of immunoregulatory cytokines. The CD4 molecule performs a similar function as the CD8 molecules in MHC class I restricted responses (Julius et al. 1993). Expression of MHC class II molecules is restricted to a rather small group of cells that possess an intralysosomal system facilitating breakdown of internalized antigens. Dendritic cells seem to be very potent antigen presenting cells since they express MHC class II molecules constitutively. It is generally assumed that dendritic cells pick up the exogenous antigens, process them and transport them to the T cell areas of draining lymph nodes where they can activate antigen-specific T cells. Expression of MHC class II can be up-regulated or even induced by cytokines such as interleukin (IL)-4 and IFN-γ, which are secreted by the activated T cells. At present two different subsets of Th cells can be discriminated. Both cell types produce the lymphokines granulocyte/macrophage colony-stimulating factor (GM-CSF), tumor necrosis factor (TNF)-α and IL-3. The Th1 cells also produce the cytokines; IFN-γ, IL-2 and TNF-β whereas Th2 cells secrete IL-4, IL-5, IL-6 and IL-10 (Mosmann and Moore 1989). As a consequence Th1 cells have stronger macrophage-activating potentials than Th2 cells, due to the production of IFN-γ but they selectively may increase the secretion of IgG2α (Stevens et al. 1988). Th2 cells are excellent helpers for B cells and regulate specific humoral responses such as isotype switching to IgE and IgA (Mosmann and Moore 1991). On average, the cytokine genes of primates have been conserved in evolution to a high degree, hinting at their importance (Villinger et al. 1995).

Biological Relevance of MHC Polymorphism

In the animal kingdom the *Mhc* is probably the genetic system that is characterized by the highest degree of polymorphism. On top of that, the repertoire itself is broadened due to duplications and this is reflected by the presence of multiple loci encoding gene products with similar functions. MHC polymorphism may be regarded as a sort of biological live assurance policy since it minimizes the chance that a given species or population is exterminated by one single pathogen. The cotton-top tamarin (*Saguinus oedipus*), a New World primate species, exhibits unusually low levels of MHC class I polymorphism. Indeed, these animals are extremely susceptible to fatal infections with a variety of pathogens (Watkins et al. 1988).

Due to codominant expression of the MHC an individual may express two alleles encoded by the same locus. Heterozygous advantage may promote the distribution of alleles associated with resistance to various types of diseases. MHC polymorphism, with special emphasis on the invariant residues that contribute to the configuration of the peptide binding site, has been the result of positive Darwinian selection (Hughes and Nei 1988, 1989). This polymorphism directly influences the types of peptides selected for binding to the MHC molecules and subsequent T cell activation. Hence, the contemporary repertoire of primate MHC alleles is probably at least in part a collection of "winners" since many MHC lineages are of considerable age and predate speciation (Bontrop et al. 1995).

Since humans, apes and monkeys possess highly similar MHC class II molecules that are the end products of natural selection, it is possible to dissect to what extent differences and similarities in these biovariants influence the immune response. To sort out whether structural similarities could be translated into function, the capacity of humans and chimpanzees to respond to purified protein derivative of *Mycobacterium tuberculosis* (PPD) was investigated (Bontrop et al. 1990). Some chimpanzee T cell lines were identified that recognized PPD in a *Mhc-DRB3* restricted fashion. Antigen presentation experiments showed that chimpanzee antigen presenting cells were able to present the peptide to *HLA-DRB3* restricted, PPD reactive T cell lines (and vice versa). Thus, human and chimpanzee MHC molecules seem to select the same peptide from a crude PPD extract for activation of the immune response. This suggests that not only the MHCs but also the enzymatic degradation pathways and TCR systems of both species are highly similar.

Humans, chimpanzees, gorillas and rhesus monkeys share *Mhc-DRB1*03* lineage members that have the capacity to bind the p3–13 peptide of the 65 kDa heat shock protein (hsp) of *M. tuberculosis/leprae* (Geluk et al. 1993). Positive binding was due to the presence of a shared ancestral motif that is characteristic of the *Mhc-DRB1*03* lineage and maps to the floor of the antigen binding site. Subsequently it was demonstrated that cells of some individuals could operate effectively as antigen presenting cells, employing an *HLA-DRB1*0301* 65 kDa hsp restricted human T cell line as a readout system. Presentation was observed for the autologous and allogeneically matched donor cells and some rhesus monkey antigen presenting cells but not for cells of chimpanzee and gorilla origin. Variation in the groove may still permit binding, but could disrupt the orientation of the peptide or more likely the configuration of the MHC-peptide complex and therefore recognition by the T cell line. Positive binding by the chimpanzee and gorilla cells but failure to induce T cell proliferation in a xenogeneic way is probably due to differences in the α-helix segments that prohibit successful T cell recognition by the *HLA-DRB1*0301* restricted 65 kDa hsp antigen-specific T cell clone. Due to the large TCR repertoire, it is likely that p3–13 peptide-MHC complexes will be successfully recognized by rearranged receptors of other clones. For this reason, we believe that the most important selective constraint is for maintenance of variation in the hypervariable pleated sheets folding into the floor of the antigen binding site. According to this view, the α-helix, which primarily interacts with the components of a very large TCR repertoire, should have much greater freedom to accumulate variation than the hypervariable pleated sheets. As a consequence, one can consider that the exon encoding the antigen binding site of MHC class II β chains is divided into two subregions with different evolutionary histories (Gyllensten et al. 1991; Sigurdardottir et al. 1992; Elferink et al. 1993). Recently similar types of findings were documented on myelin basic protein (MBP) reactive human T cell clones which can be activated by rhesus monkey antigen presenting cells that are positive for the *Mhc-DRB1*03* lineage (Meinl et al. 1995).

Theoretically, individuals that possess too many copies of *Mhc* genes may start to mimic the degree of MHC polymorphism present at the population level. Such a population would be at severe risk of being wiped out by a single pathogenic agent. T cells bearing high affinity receptors specific for MHC molecules loaded with self peptides are deleted in the thymus. Theoretically, this would imply that individuals who possess too many copies of functional *Mhc* genes may become susceptible to parasitic infections because their T cell compartment is ineffective due to deletion of too many receptors. Another reason for so-called nonresponsiveness may be the fact that potentially reactive T cells have been inactivated (tolerance or anergy). Alternatively, nonresponders may lack the ability to load their MHC antigens with the relevant peptide. In theory, such a phenomenon would only be observed when one is dealing with small or repetitive antigens which may generate only a few epitopes for MHC binding. For all these reasons it is evident that the evolution of the MHC and TCR systems must have had a great impact on each another.

T Helper Cells, MHC Class II Epitopes in Nonhuman Primates

At the level of MHC class II, relatively few epitopes have been defined thus far in nonhuman primates. As has been mentioned before, the p3–13 peptide of 65 kDa hsp of *M. tuberculosis/leprae* successfully binds to HLA-DR3-like molecules in nonhuman primates (Geluk et al. 1993; Elferink et al. 1993).

More recent data have become available on the recognition MBP in rhesus macaques in the context of experimental autoimmune encephalomyelitis (EAE), an autoimmune disorder that in rodents is known to be influenced by genetic background, specifically the *Mhc* class II region. Immunization of a group of outbred rhesus macaques with bovine brain homogenate resulted in induction of the disease in about 65 % of the animals (Slierendregt et al. 1995). No clear association between the *Mamu-DR* or *-DQ* subregion of the rhesus macaque

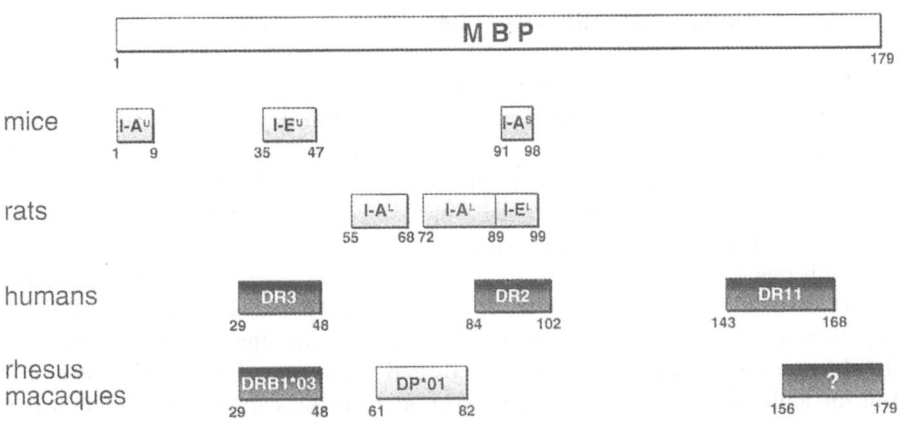

Fig. 2. Myelin basic protein (MBP) and epitopes as they are recognized by MHC class II restriction elements in different species of rodents and primates

MHC and susceptibility or resistance to the disease was apparent. The MHC class II restriction profile of an antigen-specific, $CD4^+$ Th cell line, isolated from an animal diagnosed with EAE, was investigated. This cell line recognized a peptide including residues 61–82 of the MBP molecule. The antigen-specific proliferative response was inhibited by a monoclonal antibody reactive with DP molecules. Molecular analysis of the *Mamu-DP* region, in concert with allogeneic antigen presentation studies, elucidated that the *Mamu-DPB1*01* gene product functions as the restriction element for MBP peptide presentation. Retrospective analyses showed that this particular allele is frequently found in the group of EAE susceptible animals but is absent in the resistant animals ($p < 0.01$). As a consequence, the *Mamu-DPB1*01* allele may represent one of the risk factors involved in determining susceptibility to EAE in an outbred population of rhesus macaques. It was also observed that some of the animals that lack the *Mamu-DPB1*01* allele developed EAE. This indicates that other immune response genes may contribute to EAE susceptibility in rhesus monkeys. By now, some other epitopes have been mapped that may play a role in the onset of EAE in rhesus macaques (Meinl et al. 1995). A schematic representation of the MBP epitopes as they are bound by MHC class II molecules in different species is depicted in Fig. 2. As can be seen, in rhesus macaques one of these epitopes is recognized by *Mamu-DPB1*01* gene products whereas another epitope is bound by HLA-DR3-like molecules. At least one other epitope has been mapped in rhesus monkeys and its corresponding restriction element is under investigation. These type of studies illustrate that susceptibility to autoimmune diseases in an outbred population of primates is of a complex character and as such difficult to sort out.

Cytotoxic T Lymphocytes, MHC Class I Restriction Elements and Vaccines

Nonhuman primates are often used to study the efficacy and safety of vaccines at a preclinical stage. Despite the important relationship that exists between the initiation of an immune response and the MHC, relatively little is known on this issue in nonhuman primates. This is at least partly due to the fact that most nonhuman primate colonies used in biomedical research are rather poorly characterized for their MHC antigens.

With regard to infectious diseases, most data have been obtained in hepatitis or SIV/HIV-infected subjects. More than a decade ago, successful replication of hepatitis A virus (HAV) under cell culture conditions opened the way for the development of live attenuated and inactivated vaccine candidates. The introduction of molecular biological methods also enabled the construction of so-called recombinant vaccines. Detailed studies on the immune response in primates as chimpanzees and human volunteers demonstrated the availability of a safe and potent HAV vaccine. One group of investigators reported that antigen-specific CD8-positive CTLs could be isolated from chimpanzee livers during acute and chronic stages of hepatitis C virus (HCV) infection (Erickson et al. 1993). Analysis of the epitopes that are recognized by these CTLs indicated that both structural and nonstructural proteins of HCV are recognized in the context of MHC class I molecules. The results, however, also indicated that virus-specific CTL populations persisted in the liver for months but were unable to resolve chronic HCV infection. Later on it was found that persistent HCV infections in chimpanzees turned out to be associated with the emergence of a CTL escape variant (Weiner et al. 1995). These observations, which were made in apes, have important implications for the construction and design of future vaccines against HCV

Infection of rhesus macaques with SIV results in the induction of a disease similar to AIDS in humans. For that reason the macaque model is extensively used to test the effectiveness of various AIDS vaccine approaches. In one population of rhesus monkeys, gag-specific CTL were mainly restricted by the Mamu-A*o1 antigen, an MHC molecule expressed at a high frequency and homologous to HLA-A molecules (Miller et al. 1991). This demonstration of a predictable gag-specific CD8-positive immune response recognizing a defined epitope (CTPY-DINQML; residues 182–190) in Mamu-A*o1-positive monkeys provided a good model system for vaccine research. Indeed, Mamu-A*o1-positive animals immunized with a live SIV recombinant vaccinia virus generated a CTL response specific for the relevant epitope (Shen et al. 1991). Also, immunizations with attenuated BCG-SIV constructs elicited SIV-specific CTL responses in rhesus monkeys (Yasutomi et al. 1993). By now a variety of MHC class I restricted responses to SIV or HIV antigens, such as gag, nef, and gp120, have been documented in rhesus macaques (Bourgault et al. 1994; Erickson and Walker 1994; Watanabe et al. 1994; Voss et al. 1995; Yasutomi et al. 1995).

Vaccine protection from SIV-infected peripheral blood mononuclear cells (PBMCs) derived from a rhesus macaque diagnosed with AIDS is dependent on sharing of a particular MHC class I allele (Heeney et al. 1994). In this study, all animals were challenged with 10 MID of the 1XC stock of infected

PBMCs. The donor of the infected PBMCs, animal 1XC, was serotyped as being Mamu-A14/A26; -B10/-B10; -DR3/-DR3-positive. It should be noted that the Mamu-A34 specificity is serologically included in the Mamu-A11 and -26 serotypes (Bontrop et al. 1995). For that reason, monkey 1XC potentially may express an additional MHC class I molecule that is recognized by Mamu-A34 sera. The establishment of intravenous SIV infection was found to be dose-dependent but no apparent effect of MHC sharing (or differences) in establishment of infection with infected donor cells in naive recipients was found. Only four of the eight SIV immunized animals developed an SIV infection. These latter four animals remained negative for plasma antigen and SIV provirus in PBMCs, lymph nodes and bone marrow. The Mamu-A26 allele was found to be shared between the donor (1XC) cells and all the protected and one of the four infected vaccinated animals. Nucleotide sequence determination revealed that this infected vaccinated animal from Burma expresses MHC molecules of subtypes that are hardly seen in monkeys of Indian origin. Thus there appears to exist an apparent effect of Mamu-A26 sharing on resistance to infection with infected donor cells in immunized recipients. Subsequent cellular studies demonstrated that SIV gp120-reactive CTLs were detected in three out of four protected vs one out of four unprotected animals and this suggests that MHC class I restricted CTLs may play a crucial role in protection. Mamu-A26 related protection was not observed when other Mamu-A26 positive vaccinated animals were rechallenged with cell-free virus grown SIV_{mac}. Hence it is concluded that the sharing of Mamu-A26 between infected donor cells and host may elicit a protective response.

References

Anderson MS, Miller J. Invariant chain can function as a chaperone protein for class II major histocompatibility complex molecules. Proc Natl Acad Sci USA 89:2282–2286, 1992

Babbitt BP, Allen PM, Matsueda G, Haber E, Unanue E. Binding of immunogenic peptides to Ia histocompatibility molecules. Nature 317:359–360, 1985

Bender BS, Croghan T, Zhang L, Small PA Jr. Transgenic mice lacking class I major histocompatibility complex-restricted T cells have delayed viral clearance and increased mortality after influenza virus challenge. J Exp Med 175:1143–1145, 1992

Bjorkman PJ, Saper MA, Samraoui B, Bennet WS, Strominger JL, Wiley DC. Structure of the human class I histocompatibility antigen HLA-A2. Nature 329:506–512, 1987

Bontrop RE, Elferink DG, Otting N, Jonker M, de Vries RRP. Major histocompatibility complex class II-restricted antigen presentation across a species barrier: conservation of restriction determinants in evolution. J Exp Med 172:53–59, 1990

Bontrop RE, Otting N, Slierendregt BL, Lanchbury JS. Evolution of Major histocompatibility complex polymorphisms and T cell receptor diversity in primates. Immunol Rev 143:33–62, 1995

Bourgault I, Chirat F, Tartar A, Levy JP, Guillet JG, Venet A. Simian Immunodeficiency virus as a model for vaccination against HIV Induction of GAG or NEF- specific cytotoxic T cell lymphocytes by lipopeptides. J Immunol 152:2530–2537, 1994

Brown JH, Jardetzky TS, Gorga JC, Stern LJ, Urban RG, Strominger JL, Wiley DC. Three-dimensional structure of the human class II histocompatibility antigen HLA-DR1. Nature 364:33–39, 1993

Buus S, Sette A, Grey HM. The interaction between protein-derived immunogenic peptides. Immunol Rev 98:115–134, 1987

Charmley P, Keretan E, Snyder K, Clark EA, Concannon P. Relative size and evolution of the germline repertoire of T cell receptor beta-chain segments in nonhuman primates. Genomics 25:150–156, 1995

Chen ZW, Yamamoto H, Watkins DI, Levinson G, Letvin NL. Predominant use of a T cell receptor Vβ gene family in Simian Immunodeficiency virus gag-specific cytotoxic T lymphocytes in a rhesus monkey. J Virol 66:3913–3917, 1992

Chen ZW, Kou Z-C, Shen L, Reimann KA, Letvin NL. Conserved T-cell receptor repertoire in simian immunodeficiency virus-infected rhesus monkeys. J Immunol 151:2177–2187, 1993

Chicz RM, Urban RG, Lane WS, Gorga JC, Stern LJ, Vignal DAA, Strominger JL. Predominant naturally processed peptides bound to HLA-DR1 are derived from MHC-related molecules and are heterogeneous in size. Nature 358:764–768, 1992

Christinck ER, Luscher MA, Barber BH, Williams DB. Peptide binding to class I MHC on living cells and quantitation of complexes required for CTL lysis. Nature 352:67–70, 1991

DeLarbre C, Nakauchi H, Bontrop R, Kourilsky PH, Gachelin G. Duplication of the CD8 β chain gene as a marker of the man-gorilla-chimpanzee clade. Proc Natl Acad Sci USA 90:7049–7053, 1993

Denzin LK, Cresswell P. HLA-DM induces clip dissociation from MHC class II alpha beta dimers and facilitates peptide loading. Cell 82:155–165, 1995

Elferink BG, Geluk A, Otting N, Slierendregt BL, van Meijgaarden KE, de Vries RRP, Ottenhof THM, Bontrop RE. The biologic importance of conserved major histocompatibility complex class II motifs in primates. Hum Immunol 38:201–205, 1993

Erickson AL, Houghton M, Choo QL, Weiner AJ, Ralston R, Muchmore E, Walker CM. Hepatitis C virus specific responses in the liver of chimpanzees with acute and chronic hepatitis C. J Immunol 151:4189–4199, 1993

Erickson AL, Walker CM. An epitope in the V1 domain of the Simian Immunodefiency virus (SIV) gp120 protein is recognized by CD8 positive cytotoxic T lymphocytes from an SIV infected rhesus macaque. J Virol 68:2756–2759, 1994

Fischer-Lindahl KF, Hermel E, Loveland BE, Wang CR. Maternally transmitted antigen of mice: a model transplantation antigen. Annu Rev Immunol 9:351–372, 1991

Fling SP, Arp B, Pious D. HLA-DMA and DMB genes both required for MHC class II-peptide complex formation in antigen presenting cells. Nature 368:554–558, 1994

Geluk A, Elferink DG, Slierendregt BL, van Meijgaarden KE, de Vries RRP, Ottenhoff THM, Bontrop RE. Evolutionary conservation of major histocompatibility complex-DR/peptide/T cell interactions in primates. J Exp Med 177:979–987, 1993

Ghosh P, Amaya M, Mellins E, Wiley DC. The structure of an intermediate in MHC maturation: CLIP bound to HLA-DR3. Nature 378:457–462, 1995

Guillet JG, Lai M, Briner TJ, Smith JA, Gefter ML. Interaction of peptide antigens and class II MHC antigens. Nature 324:260–262, 1986

Gyllensten U, Sundvall M, Ezcurra I, Erlich HA. Genetic diversity at class II DRB loci of the primate MHC. J Immunol 146:4368–4376, 1991

Heeney JL, van Els C, de Vries P, ten Haaft P, Otting N, Koornstra W, Boes J, Dubbes R, Niphuis H, Dings M, Cranage M, Norley S, Jonker M, Bontrop RE, Osterhaus A. Major Histocompatibilitry complex class I associated vaccine production from simian immunodeficiency virus infected peripheral blood cells. J Exp Med 180:769–774, 1994

Hughes AL, Nei M. Pattern of nucleotide substitution at major histocompatibility complex class I loci reveals overdominant selection. Nature 335:167–170, 1988

Hughes AL, Nei M. Nucleotide substitution at major histocompatibility complex class II loci: evidence for overdominant selection. Proc Natl Acad Sci USA 86:958–962, 1989

Jaeger EEM, Bontrop RE, Lanchbury JS. Structure diversity and evolution of T cell receptor VB gene repertoire in primates. Immunogenetics 40:184–191, 1994

Julius M, Maroun CR, Haugn L. Distinct roles for CD4 and CD8 as co-stimulatory signals. Immunol Today 14:177–183, 1993

Kaufman JF, Auffray C, Korman AJ, Shackelford DA, Strominger JL. The class II molecules of the human and murine major histocompatibility complex. Cell 36:1–13, 1984

Kelly AP, Monaco JJ, Cho S, Trowsdale J. A new HLA class II locus DM. Nature 353:571–573, 1991

Levinson G, Hughes AL, Letvin NL. Sequence and diversity of rhesus monkey T-cell receptor beta chain genes. Immunogenetics 35:75–88, 1992

Matzinger P. Tolerance danger and the extended family. Annu Rev Immunol 12:991–1045, 1994

Meinl E, 't Hart B, Bontrop RE, Hoch RM, Iglesias A, de Waal Malefyt R, Fickensher H, Muller-Fleckenstein I, Fleckenstein B, Wekerle H, Hohlfeld R, Jonker M. Activation of a myelin basic protein-specific human T cell clone by antigen-presenting cells from rhesus monkeys. Int Immunol 7:1489–1495, 1995

Miller MD, Yamamoto H, Hughes AL, Watkins DI, Letvin NL. Definition of an epitope and MHC class I molecule recognized by gag-specific cytotoxic T lymphocytes in SIVmac-infected rhesus monkeys. J Immunol 147:320–329, 1991

Moore MW, Carbone FR, Bevan MJ. Introductioof soluble protein into the class I pathway of antigen processing and presentation. Cell 54:777–785, 1988

Morton PA, Zacheis ML, Giacoletto KS, Manning JA, Schwartz BD. Delivery of nascent MHC class II invariant chain complexes to lysosomal compartments and proteolysis of invariant chain by cysteine proteases precedes peptide binding in B-lymphoblastoid cells. J Immunol 154:137–150, 1995

Mosmann TR, Moore KW. TH1 and TH2 cells: different patterns of lymphokine secretion lead to different functional properties. Annu Rev Immunol 7:145–162, 1989

Neefjes JJ, Momburg F. Cell biology of antigen presentation. Curr Opin Immunol 5:27–34, 1993

Neefjes JJ, Ploegh HL. Inhibition of endosomal proeolytic activity by leupeptin blocks surface expression of MHC class II molecules and their conversion to SDS resistant alpha-beta heterodimers in endosomes. EMBO J 11:411–416, 1992

Nuchtern JG, Biddeson WE, Klausner RD. Class II MHC molecules can use the endogeneous pathway of antigen presentation. Nature 343:74–76, 1990

Rammensee HG, Falk K, Rötschke O. Peptides naturally presented by MHC class I molecules. Annu Rev Immunol 11:213–244, 1993

Riberdy JM, Newcomb JR, Surman MJ, Barbosa JA, Cresswell P. HLA-DR molecules from an antigen-processing mutant cell line are associated with invariant chain peptides. Nature 360:474–476, 1992

Rock KL, Gamble S, Rothstein L. Presentation of exogenous antigen with Class I MHC molecules. Science 249:918–921, 1990

Rock KL, Gramm C, Rothstein L, Clark K, Stain R, Dick L, Hwang D, Goldberg AL. Inhibitors of the proteasome block the degradation of most cell proteins and the generation of peptides presented by class I molecules. Cell 78:761–771, 1994

Rogmagnoli P, Germain RN. The CLIP region of the invariant chain plays a critical role in regulating major histocompatibility complex class II folding transport and peptide occupancy. J Exp Med 180:1107–1113, 1994

Rötzscke O, Falk K, Deres K, Schild H, Norda M, Metzger J, Jung G, Rammensee H-G. Isolation and analysis of naturally processed viral peptides as recognized by cytotoxic T cells. Nature 348:252–254, 1990

Rudensky AY, Preston-Hurlburt P, Hong S-C, Barlow A, Janeway CA Jr. Sequence analysis of peptides bound to MHC class II molecules. Nature 353:622–627, 1991

Rudensky AY, Preston-Hurlburt P, Al-Ramadi BK, Rothbard J, Janeway CA. Truncation variants of peptides isolated from MHC class II molecules suggests sequence motifs. Nature 359:429–431, 1992

Schumacher TN, de Bruijn ML, Vernie LN, Kast WM, Melief CJ, Neefjes JJ, Ploegh HL. Peptide selection by MHC class I molecules. Nature 350:703–706, 1991

Sette AS, Ceman RT, Kubo K, Sakaguchi E, Apella DF, Hunt TA, Davis H, Michel J, Shabanowitz R, Rudersdorf HM, Grey HM, DeMars R. Invariant chain peptides in most HLA-DR molecules of an antigen processing mutant. Science 258:1801–1804, 1992

Shen L, Chen ZW, Miller MD, Stallerd V, Mazzara GP, Panicali DL, Letvin NL. Induction of simian immunodeficiency virus specific CD8 positive responses cytotoxic T lymphocytes following vaccination with a recombinant vaccine. Science 252:440–443, 1991

Sigurdardottir S, Borsch C, Gustafsson K, Andersson L. Exon encoding the antigen binding site of MHC class II beta-chains is divided into two subregions with different evolutionary histories. J Immunol 148:968–973, 1992

Slierendregt BL, Hall M, 't Hart B, Otting N, Anholts J, Verduin W, Claas F, Jonker M, Lanchbury J, Bontrop RE. Identification of an Mhc-DPB1 allele involved in susceptibility to experimental autoimmune encephalomyelitis in rhesus macaques. Int Immunol 7:1671–1679, 1995

Sloan VS, Cameron P, Porter G, Gammon M, Amaya M, Mellins E, Zaller DM. Mediation by HLA-DM of dissociation of peptides from HLA-DR. Nature 375:802–806, 1995

Stevens TL, Bossie A, Sanders VM, Fernandez-Botran R, Coffman RL, Mosmann TR, Vitetta ES. Subsets of antigen specific helper T cells regulate isotype secretion by antigen-specific B cells. Nature 334:255–258, 1988

Sturm E, Bontrop RE, Vreugdenhil RJ, Otting N, Bolhuis RLH. T-cell receptor gamma/delta: comparison of gene configurations between humans and chimpanzees. Immunogenetics 36:294–301, 1992

Sugati M, Brenner ML. An unstable β2-microglobulin major histocompatibility complex class I heavy chain intermediate dissociates from calnexin and then is stabilized by binding peptide. J Exp Med 180:2163–2169, 1994

Tanaka Y, Morita CT, Tanaka Y, Nieves E, Brenner MB, Bloom BR. Natural and synthetic non-peptide antigens recognized by human γδ T cells. Nature 375:155–158, 1995

Thiel C, Otting N, Bontrop RE, Lanchbury JS. Generation and reactivation of T cell receptor A joining region pseudogenes in primates. Immunogenetics 43:57–62, 1996

Toyonaga B, Mak TW. Genes of the T-cell antigen receptor in normal and malignant T cells. Annu Rev Immunol 5:585–620, 1987

Verreck FAW, Van de Poel A, Termijtelen A, Amons R, Drijfhout JW, Koning F. Identification of an HLA-DQ2 peptide binding motif and HLA-DPw3 bound self peptides by pool sequencing. Eur J Immunol 24:375–379, 1994

Villinger F, Brar SS, Mayne A, Chikkala N, Ansari AA. Comparative sequence analysis of cytokine genes from human and nonhuman primates. J Immunol 155:3946–3954, 1995

Von Boehmer H. Thymic selection a matter of life and death. Immunol Today 13:454–458, 1992

Voss G, Li J, Manson K, Wyand M, Sodroski J, Letvin NL. Human Immunodeficiency virus type 1 envelope glycoprotein-specific cytotoxic T lymphocytes in simian human immunodeficiency virus infected rhesus monkeys. Virology 208:770–775, 1995

Watanabe N, McAdam SN, Boyson JE, Piekarczyk MS, Yasutomi Y, Watkins DI, Letvin NL. A simian Immunodeficiency virus envelope V3 cytotoxic epitope in rhesus monkeys and its restricting major histocompatibility complex class I molecule Mamu-A*02. J Virol 68:6690–6696, 1994

Watkins DI, Hodi FS, Letvin NL. A primate species with limited major histocompatibility complex classs I polymorphism. Proc Natl Acad Sci USA 85:7714–7718, 1988

Weiner A, Erickson AL, Kansopon J, Crawford K, Muchmore E, Hughes AL, Houghton M, Walker CM. Persistent hepatitis C virus infection in a chimpanzee is associated with emergence of a cytotoxic T lymphocyte escape variant. Proc Natl Acad Sci USA 92:2755–2759, 1995

Yasutomi Y, Koenig S, Haun SS, Stover CK, Jackson RK, Conard P, Conley AJ, Emini EA, Fuerst TR, Letvin NL. Immunization with recombinant BCG-SIV elicits SIV specific cytotoxic T lymphocytes in rhesus monkeys. J Immunol 150:3101–3107, 1993

Yasutomi Y, McAdam SN, Boyson JE, Piekarczyk MS, Watkins DI, Letvin NL. MHC class I B locus allelel restricted simian immunodeficiency virus envelope CTL epitope in rhesus monkeys. J Immunol 154:2516–2522, 1995

Zijlstra M, Bix M, Simister NE, Loring JM, Raulet DH, Jaenisch R. β2- microglobulin deficient mice lack CD4-8+ cytolytic cells. Nature 344:885–893, 1992

6 MHC and Disease Associations in Nonhuman Primates

S. Gaudieri, J.K. Kulski, and R.L. Dawkins

Introduction

Disease and Evolution

Diseases can be seen as a reflection of the interaction between the genome and the prevailing environment and therefore provide an approach to the study of evolution. The potential of this approach is exciting but limited by current understanding of the nature and extent of contemporaneous changes in the genome and environment. Here we take advantage of recent advances in the description of the differences in the MHC of several primates and relate these to the available examples of relevant diseases. A minimalist view might be that the genomic differences are minor and largely irrelevant and that the lack of knowledge of diseases in nonhuman primates (NHPs) simply reflects inadequate data and small population sizes. Alternatively, we will argue that the admittedly

Molecular Biology and Evolution of Blood Group and MHC Antigens in Primates
Blancher/Klein/Socha (Eds.)
© Springer-Verlag Berlin Heidelberg 1997

limited information suggests some important possibilities and potentially valuable hypotheses. Certainly, it is appropriate to address the paradox confronting paleopathology: disease is said to be important in selection (Haldane 1949) but the diseases of all primates are said to be similar (Lovell 1990).

MHC and Disease

The MHC can be defined in many different ways (Klein 1985; Bodmer et al. 1986; Dawkins et al. 1989; Klein and Takahata 1990), but in the context of primate disease we refer to the region of the genome which contains the genes relevant to transplantation rejection, antigen presentation, autoimmunity, complement activation and immunoregulation mediated by tumor necrosis factors (TNFs).

In humans, the MHC consists of about four megabases which encode at least several functional HLA class I, HLA class II, complement and TNF genes together with more than 100 other genes or pseudogenes. A striking feature of the region is the propensity for duplication of genes and, more particularly, blocks of genes (Leelayuwat et al. 1995; Zhang et al. 1993). The precise structure of the MHC varies between different humans depending upon their mix of ancestral haplotypes (Tokunaga et al. 1988; Zhang et al. 1990). In fact, the MHC could also be defined in terms of its extreme polymorphism.

The functions of many of the component genes and their alleles remain uncertain and may be best approached by considering the disease susceptibility genes which have been mapped to the region. Indeed, the MHC could be defined as that region of the genome which controls susceptibility for, or severity of, hundreds of distinct and diverse diseases. As described below, some of these diseases are infectious or postinfectious and some are autoimmune but many are idiopathic or apparently metabolic in pathogenesis. Only a few can be explained in terms of a specific genetic lesion; obviously there is much to learn of the functions of genes within the MHC.

Potential Importance of MHC Associated Diseases in Primates

It is our prejudice that similarities and differences between primates may be informative in terms of the following issues.

Role of MHC Genes

The precise function of many MHC genes remains unclear. A major difficulty is due to the fact that there are so many potentially important genes in such close proximity. These are inherited en bloc and cannot be dissected through relatively short evolutionary periods. Conceivably, it might be possible to isolate the effects of components by comparing different primates separated by sufficient generations to rearrange the component genes without substantially altering the protein products of each gene. A key assumption is that most of the differences found in primates are due to genomic changes (e.g. insertions and deletions (indels) rather than mutations affecting coding regions.

Pathogenesis of Disease

Any differences between the diseases of different primates may provide a means of understanding the mechanism or pathogenesis of disease. For example, primates differ dramatically in the diseases follow HIV/SIV infection. Comparative genomics provides a potential means of determining the role of viral receptors, cofactors, antiviral responses, endogenous retroviruses and especially the relevance of MHC gene products.

Evolutionary Bottlenecks and Selection

Presumably, successive expansions and contractions of populations reflect, inter alia, the interactions between the genome and the prevailing environment. Differences between primates may relate to environmental changes over time, may suggest the nature of insults which have led to bottlenecks and could help to identify the selective forces which have operated periodically during primate evolution.

Specificity of Infectious Agents

As will be discussed below, many parasites including bacteria do not appear to discriminate between primates whereas some viruses exhibit exquisite species specificity. Indeed, HIV/SIV distinguishes between subspecies and races (Myers et al. 1994b). In humans, HIV has greater or lesser effects depending upon the MHC of the host (Cameron et al. 1992; Itescu et al. 1991; Just 1995) and possibly other factors. It is well known that inbred strains of mice vary in their susceptibility to viruses, again largely depending upon their mix of MHC genes. Another factor which deserves further study is the potential protection conferred by endogenous retroviruses. Sequence analysis of endogenous and exogenous retroviruses reveals a close relationship between certain human endogenous retroviruses and exogenous retroviruses such as the Moloney viruses (Doolittle et al. 1990; Li et al. 1995; Gaudieri et al. 1996), and yet Moloney retroviruses are not known to cause disease in humans. As one of several factors, integration and vertical transmission could be protective.

Potential Transmission of Infectious Agents

Nonhuman primates can serve as a reservoir for infection of humans and vice versa. It is presumed that species-hopping has occurred periodically in the

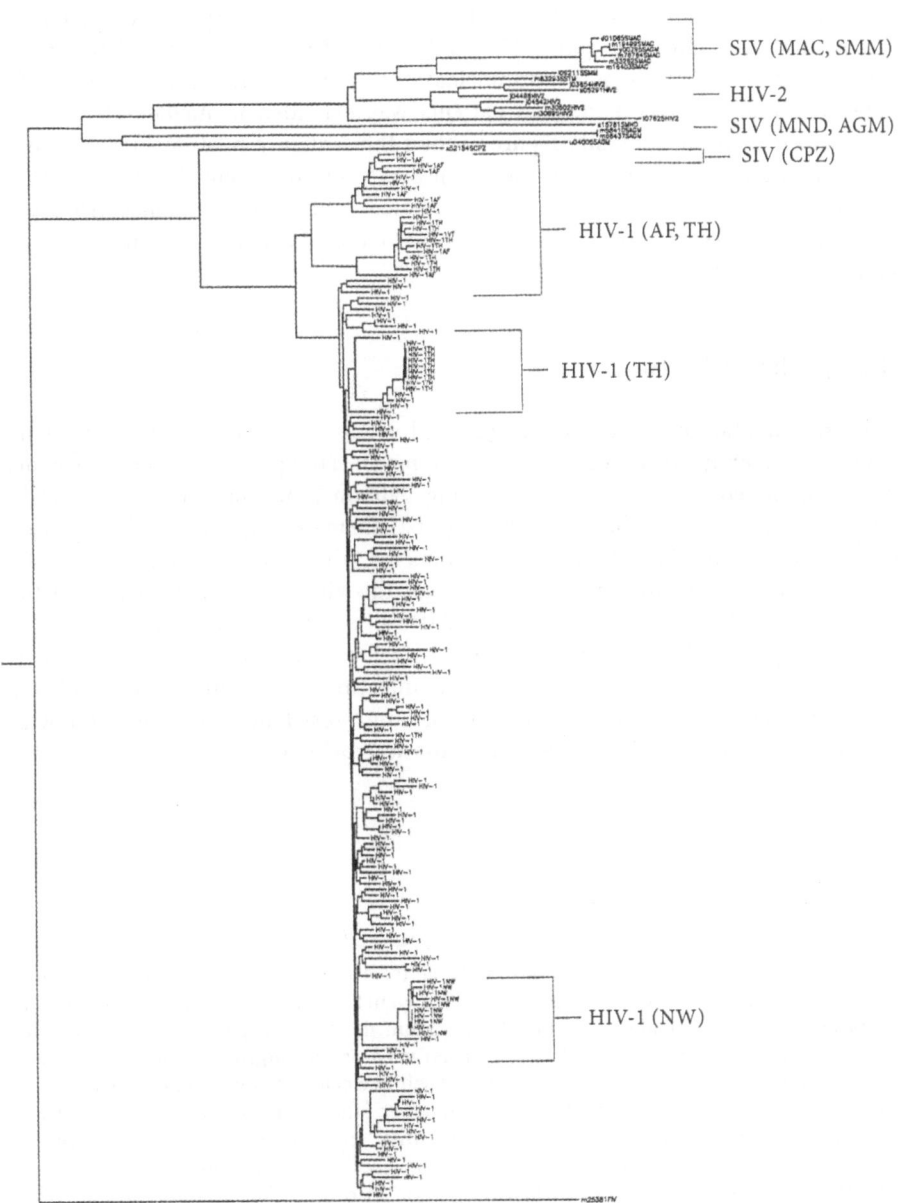

past and anticipated that the same will occur in the future. In fact, it has been said that the greatest risk to the survival of NHPs will be the infections transmitted by humans. The magnitude of the threat may be inversely related to the severity of the infection in the donor species.

Interspecies transmission from NHPs to humans has previously been illustrated for HIV and SIV and HTLV and STLV (Crandell 1996; Seibert et al. 1995). Sequence analysis of 497 bp nucleotides of the reverse transcriptase (RT) gene found in our HIV-1 infected patients was extended to include HIV-2 and SIV sequences from GenBank. Figure 1 shows the sequence relationships and suggests that there have been two cross-species transmission events leading to HIV-1 and HIV-2. The HIV-1 clade is closely related to the SIV$_{CPZ}$ isolate from the chimpanzee. However, HIV-2 is very close to the SIV isolates from the Asian and African monkeys.

Therapy of Human Disease

Xenotransplantations between humans and NHPs have been performed using baboon and chimpanzee organs (O'Brien 1994). Though these xenotransplants have had mixed results, a new trial using baboon bone marrow transplants in HIV-1 infected patients has recently begun (Lehrman 1995). This trial is based on the observation that baboons can be infected with a chimeric HIV/SIV isolate but do not progress to disease even though the virus actively replicates in the baboon cells (Allen et al. 1995) and an AIDS-like condition can occur in HIV-2 infected baboons (Barnett et al. 1994). Therefore, there are two NHP models for long-term survival following HIV-1 infection: the chimpanzee and the baboon. However, due to availability, cost and resistance to reinfection with HIV-1, the baboon has been selected as the test primate.

Fig. 1. Phylogenetic analysis of a 497 bp sequence from the reverse transcriptase (RT) gene in HIV-1, HIV-2 and various SIV isolates. A phylogenetic analysis of 204 RT sequences was performed using the neighbor-joining method from the PHYLIP package (Saitou and Nei 1987) with FIV (feline immunodeficiency virus) used as an outgroup. Sequences below SIV$_{CPZ}$ are locally derived and were compared with selected examples obtained from GenBank and denoted by GenBank accession number followed by: S for SIV; HIV-2; or FIV. The different isolates of SIV we designated as: *MAC*, macaque monkey; *SMM*, sooty mangebey monkey; *MND*, mandrill; *AGM*, african green monkey; and *CPZ*, chimpanzee. The HIV-1 isolates have been described in relation to origin, that is, where the infection was obtained or from where the HIV-1 infected contact had obtained the infection. The initials following HIV-1 designate the geographical location: *AF*, Africa; *VT*, Vietnam; *NW*, Northwest Australia; *TH*, Thailand. The isolates designated as HIV-1 alone are from the metropolitan area in Perth, Western Australia. From these relationships it would appear that HIV-1 and HIV-2 are from two separate cross-species transmission events, with HIV-1 grouping with SIV from the chimpanzee and HIV-2 grouping with the monkey SIV isolates. In addition, the HIV-1 samples form a separate cluster but also show a wide range of diversity within the clade. The geographical origin correlates well with the placement of these strains within the tree.

Approach to Comparative Pathology and Genomics

For the purposes of comparison, we will describe the diseases associated with MHC genes in humans and ask which of these have been observed in NHPs. We will then compare and contrast the content of the MHC and discuss differences in the context of the pathogenic mechanisms which are thought to operate in humans.

The promise of this approach has yet to be realised in full. Until recently most information on diseases of NHPs came from paleopathology and was therefore limited to evidence contained within skeletal remains. Further difficulties in studying NHP diseases arise in relation to population size, age-dependent penetrance and enviromental cofactors. Determination of genetic susceptibility will depend on adequate evidence of sufficient exposure and this will often be difficult to estimate. Assumptions in relation to some infectious agents may be reasonable but minor changes in virulence may be important.

MHC and Disease

MHC and Associated Diseases

Here we discuss those spontaneous diseases which have been associated with MHC genes. In humans, many thousands of studies have been reported; only those with reasonable design, high corrected p values and confirmatory data are accepted. By contrast, in other species less rigorous criteria are possible.

Humans

In humans several hundred diseases are associated with specific alleles at one or more MHC loci. Some of these are listed in Table 1, which, however, shows only HLA loci, whereas in some cases it is widely believed that other MHC loci may be of more fundamental importance. For example in systemic lupus erythematosus (SLE) there are sound reasons to believe that the association is largely with haplotypes containing reduced numbers of C4 genes – so-called C4 null alleles (Christiansen et al. 1983, 1991b).

There are at least two different categories of MHC associated diseases. As shown in Table 1, in some instances there is an association with a particular allele at a particular locus and this is true in different racial groups even when the specific allele is carried by an ethnic-specific haplotype. It can be concluded that, in the case of ankylosing spondylitis (AS), B27 is found in all racial groups, irrespective of haplotype, and is either directly involved or an excellent marker for another closely linked gene. Other diseases are also associated with B27, including psoriatic spondyloarthropathy, but if these disorders are excluded

Table 1. HLA associations with disease in humans (after Tiwari and Terasaki 1985)

Disease	HLA association	Race	Disease	HLA association	Race
Acute anterior uveitis	B27	C	Juvenile rheumatoid arthritis	B27	C
				DR5	C
Ankylosing spondylitis	B27	C, O, N	Multiple sclerosis	B7	C
				DR2	C
Behcet's disease	B5	C, O	Myasthenia gravis	B8	C
Celiac disease	B8	C	Narcolepsy	DR2	C, O
	DR3	C	Pemphigus vulgaris	A26	C
	DR7	C		B38	C
Dermatitis Herpetiformis	B8	C		DR4	C
	DR3	C	Psoriasis vulgaris	B13	C, O
Goodpasture's syndrome	DR2	C		B17	C, O
				B37	C, O
Grave's disease	B8	C		Cw6	C, O
	DR3	C		DR7	C, O
	B35	O	Reiter's disease	B27	C
Idiopathic hemochromatosis	A3	C	Rheumatoid arthritis	B27	C
	B7	C		DR4	C, O, N
	B14	C		Bw54	O
Juvenile diabetes mellitus	B8	C, N	Sjogren's syndrome	B8	C
	B15	C, N		Dw3	C
	DR3	C, N, O	Systemic lupus Erythematosus	B8	C
	DR4	C, N, O		DR3	C
	Bw54	O			

C, Caucasian; O, Oriental; N, Negroid.

and only "true" AS considered, it can be shown that the B27 marker is a more or less essential requirement (sine qua non) although clearly not sufficient by itself. Although some 5 %–10 % of Caucasoids have B27, only a few percent of these will develop AS and many do not do so until late in adult life (Christiansen et al. 1977, 1980). A second gene may be involved possibly explaining the higher penetrance in males than females (Hollingsworth et al. 1982). In many respects psoriasis is similar to AS except that the main association is with HLA Cw6 rather than B27. Narcolepsy can also be classified in this group.

A second category of association (Table 2) is illustrated by autoimmune diseases such as myasthenia gravis (MG), insulin-dependent diabetes mellitus (IDDM) and SLE. Here the association differs in different racial groups and involves one or more blocks of an ancestral haplotype. It seems likely that more than one MHC gene is responsible. In MG, there is evidence for one gene within the β block marked by HLA B8 in Caucasoids (but other HLA-B alleles in other racial groups) and a second gene within the δ block which, how-

Table 2. MHC functions revealed by disease associations

	Disease association	Immunological
HLA class I alleles		
B27	Ankylosing spondilitis	?
Cw6	Psoriasis	No
HLA class II alleles		
DQβ1 0602	Narcolepsy	No
DR9	Seronegative myasthenia gravis	?
Ancestral haplotypes		
8.1	Multiple autoimmune disorders	Yes
7.1	Multiple sclerosis	?
	Hemochromatosis	No
47.1	21-Hydroxylase deficiency	No
18.1	C2 deficiency	No

Ancestral haplotyes are denoted as: 8.1-A1, Cw7, B8, C4Q0, C4B1, DR3, DQ2; 7.1-A3, Cw7, B7, C4A3, C4B1, DR15, DQ6; 47.1-A3, Cw6, B47, C4A1, C4BQ0, DR7, DQ2; 18.1-A25, B18, C4A4, C4B2, DR15, DQ6.

ever, appears to vary depending upon the precipitating factor (Degli-Esposti et al. 1992b). In IDDM, there is also evidence for at least two genes. In humans, there is at least one gene in the β block and another in the δ block (Degli-Esposti et al. 1992a). The second of these appears to prevent the development of disease although it does not protect against the production of autoantibodies (Hollingsworth et al., unpublished). In the NOD mouse, there are also at least two genes: one is in the class II region (equivalent to the human δ block) and the other is probably close to K (Hattori et al., personal communication). These and other data raise the interesting possibility that autoimmunity may be under the control of multiple immunoregulatory genes within the MHC. It is certainly clear that: (1) autoimmunity, as reflected by autoantibodies, is extremely common in humans and is *not* critically dependent on any single HLA allele and (2) disease is a reflection of excessive or up-regulated autoimmune responses rather than an all-or-nothing phenomenon. Support for this interpretation is provided by considering the influence of the MHC on infectious disease. The degree of response to hepatitis B virus is dependent upon the haplotype present (Alper et al. 1989). Similarly, the MHC appears to play a major role in determining the form of leprosy which develops after *Mycobacterium leprae* infection (Mehra et al. 1994). Indeed, responses to streptococcal antigens are influenced by MHC antigens (Christiansen et al. 1978). However, probably the most relevant example is the effect of the MHC on outcomes after HIV infection (reviewed by Just 1995). Two Caucasoid ancestral haplotypes are associated with more rapid loss of CD4 cells and more rapid progression to AIDS (Cameron et al. 1992). The nature of complications is also affected by the MHC (Mallal et al. 1994). It is less

Table 3. Rate of substitutions and ambiguities in the reverse transcription gene sequences in sequential samples from the same patient

Patient	8.1 AH	Substitutions per year	Ambiguities per year
1	+	0	0.018
2	−	0.004	0.009
3	−	0.007	0.021
4	−	0.006	0.016
5	+	0	0.012
6	−	0.008	0.022
7	+	0	0.005

Patients with the 8.1 ancestral haplotype (A1, Cw6, B8, C4AQ0, DR3) have a lower rate of change.

clear whether the MHC affects susceptibility to actual infection but there is increasing evidence that the MHC does influence the evolution/selection of the virus after infection (Itescu et al. 1995). We have shown that there is less variation in the RT gene sequence within patients with a particular haplotype than in patients as a whole (Table 3).

The MHC also appears to play a role in the development of some cancers that have been associated with viral infections. For example, HLA haplotypes have been associated with an increased risk of genital cancer in women who were infected with oncogenic types of human papillomaviruses (HPVs) (Wank et al. 1991; David et al. 1992; VanDenVelde et al. 1993; Apple et al. 1994) and may affect the variability of the host response to HPV infections (Han et al. 1992; Coleman et al. 1994). Women with HLA-DQw3 antigen are 2.5 times more likely to develop cervical intraepithelial neoplasia and 7.1 times more likely to develop invasive squamous cell carcinoma (Wank et al. 1991; David et al. 1992; VanDenVelde et al. 1993), although the reliability of these results has been questioned (Glew et al. 1993). In a study of Hispanic patients with HPV-16 associated invasive cervical cancer, DRB1*1501-DQB1*062 were significantly correlated with susceptibility to cervical cancer, whereas DR13 appeared to confer a protective effect against cervical cancer (Apple et al. 1994). The inflammatory cytokines TNF and interleukin (IL-1) have been shown to interact with a non-coding control element of HPV-16 to suppress the expression of the viral E6 and E7 oncogenes (Kyo et al. 1994), and may in part help to retard tumor development. By contrast, a single nucleotide mutation in the HPV-16 E6 gene sequence which was identified in some tumors has been reported to influence the evolution/selection of HPV after infection and contribute to the development of carcinoma in HLA-B7 individuals (Ellis et al. 1995). In this case, a mutation in an HPV epitope may alter the conformation of the presenting viral peptide-HLA-B7 binding complex to prevent interaction with T cells and contribute to the development of carcinoma by escaping from immunosurveillance of HPV-16. Other mechanisms may result in a loss or down-regulation

Table 4. HLA associated diseases which occur spontaneously in nonhuman primates

Disease	Nonhuman primate	HLA association
Ankylosing spondylitis	Gorilla	B27
	Rhesus monkey	
	Gibbon	
	Baboon	
Psoriasis vulgaris	Chimpanzee	?
Systemic lupus erythematosus	Rhesus monkey	?

of major HLA class I expression (Garrido et al. 1993). In this regard, the loss of HLA-B7/B40 expression has been significantly correlated with metastatic spread of cervical carcinoma and a poor outcome for patients (Honma et al. 1994).

The explanations for all these associations are far from clear. However there are two situations which can be understood on the basis of specific genetic changes. C2 deficiency is due to a sequence change affecting expression of C2: the relevant sequence is carried by the 18.1 ancestral haplotype. Also, 21-hydroxylase deficiency comes in several different forms, depending upon the ancestral haplotype; but haplospecific sequence differences appear to result in a deficiency of the product.

Nonhuman Primates

Table 4 lists the HLA associated diseases which occur spontaneously in NHPs. In most cases the influence of the MHC has not been investigated as yet.

Infectious Diseases

Parasites
Numerous parasites have been observed in NHPs but there is an impression that there is little consequential disease in the wild and in the absence of other contributing factors. In relation to plasmodium, the symptoms of malaria are said to be mild (Lovell 1990).

Bacteria
Treponemal diseases are thought to be similar in primates. Apparently yaws is "very prevalent" among wild gorillas, chimps and Old World monkeys but detailed laboratory data are lacking (Lovell 1990). NHPs appear to be highly susceptible to mycobacteria although tuberculosis is not found in the wild (Roberts

1995). Naturally acquired and experimentally induced leprosy has been reported in NHPs (Meyers et al. 1991).

Viruses

In general, viruses exhibit more species and strain specificity than bacteria and parasites. However, many viruses which are pathogenic to humans also infect NHPs – sometimes with devastating consequences. Factors such as previous exposure, herd immunity, genetic susceptibility (including MHC and non-MHC immunogenetics,viral receptors,vertical transmission of endogenous retrovirus), viral drift, nonspecific reactivation and contemporaneous environmental cofactors are clearly important and will be illustrated below.

Infection of NHPs with yellow fever results in various signs and outcomes which are dependent on the species of monkey infected. Howler monkeys (*Alouatta* sp), spider monkeys (*Ateles* sp), squirrel monkeys (*Saimiri* sp) and owl monkeys (*Aotus* sp) all develop fatal infections from yellow fever virus (Monath 1990). In contrast, the capuchin monkeys (*Cebus* sp) and wooly monkeys (*Lagothrix* sp) do not develop clinical signs of disease following infection.

Measles has not been reported to have occurred in NHPs in their natural environment. However, several NHPs have been infected with the measles virus experimentally and after exposure to humans and other infected monkeys. The rhesus (*Macaca mulatta*) and cynomolgus monkeys (*Macaca fascicularis*) show a variety of signs following infection, ranging from fever and mild respiratory signs to conjunctivitis and maculopapular rashes. Chimpanzees (*Pan troglodytes*) can also be experimentally infected with the measles virus but lack the typical clinical signs observed in humans and the macaque monkeys. In contrast, the marmoset (*Saguinus mystax* and *Saguinus labiatus*) progresses to a fatal disease (Van Binnendijk et al. 1995).

Experimental paralytic poliomyelitis has been documented in the bonnet monkey (*Macaca radiata*) (Samuel et al. 1993). Limb paralysis and virus isolation from the spinal cord was reported in these monkeys following inoculation with the poliovirus type 1 isolate (Samuel et al. 1993).

NHPs develop papillomas (warts) and neoplasias that correspond to a range of papillomavirus-induced lesions observed in humans (Kloster et al. 1988; Reszka et al. 1991). A sexually transmitted rhesus monkey papillomavirus type 1 (RhPV-1) infection has been strongly associated with the development of genital malignancies in a mating cohort of rhesus monkeys (Ostrow et al. 1990) in a manner similar to what has been proposed for humans (Herrington 1994). Although RhPV-1 DNA was frequently (71 %) detected in a mating cohort of 32 rhesus monkeys, including a male with penile carcinoma and lymph node metastasis, about 9 % of monkeys had malignant tumors of the genital tract with integrated papillomavirus DNA, and another 35 % had some pathological evidence of papillomavirus infection such as koilocytosis, mild dysplasia, and acetowhitening (Ostrow et al. 1990). The genomic sequence of RhPV-1 (Ostrow et al. 1991) is closer to the genital human papilloma virus type 16 (HPV-16) than to any of the other 60 or more mucosal-cutaneous HPV types that have been partially or fully sequenced (Myers et al. 1994a; Chan et al. 1995). This supports the view that a related group of papillomaviruses are involved in the development of genital neoplasias in primates.

Pygmy chimp papillomavirus type 1 (PCPV-1) is another NHP papillomavirus type which has been completely sequenced after isolation from a pygmy chimp *(Pan paniscus)* in a colony with a high incidence of focal epithelial hyperplasia (FEH) (Van Ranst et al. 1992). The PCPV-1 has 85% sequence similarity to HPV-13 which also is strongly associated with oral FEH, a benign lesion frequently found among Indians in Central and South America and in Eskimos in Greenland and Alaska but rarely in Caucasians or Asians. Some cases of FEH have been reported in the chimpanzee, although the papillomavirus type in this species has not yet been identified (Van Ranst et al. 1992). Nevertheless, the closeness of RhPV and PCPV sequence to HPV-16 and HPV-13, respectively, suggests a cross-species transmission which has parallels to the evolutionary history of the SIV-HIV groups (Chan et al. 1995). Most aspects of the papillomavirus infection in NHPs, including transmission, latency, integration and the strong association with the development of benign and malignant neoplasia, are similar to the observations in humans. Therefore, NHPs offer an excellent animal model for understanding the host immune response, immunogenetics and pathogenesis associated with human papillomavirus infection.

Fortunately and not surprisingly, HIV/SIV is now receiving serious attention. However it is not obvious that the key issues have been considered. Many authors discuss HIV and SIV as though there are two distinct viruses, but, as shown in Fig. 1, these terms refer to the source of the isolate rather than the nature of the virus. There are at least two major groups which could be referred to as PIV-1 and PIV-2, where P refers to primate to avoid confusion. PIV-1 includes HIV-1 and SIV cpz, whereas PIV-2 refers to most SIV and HIV-2. A systematic study of the entire sequences of equal numbers of isolates from different primates in diverse environments would be valuable.

Much of the literature is based on the assumptions that infection only occurs in the native host and that genetic susceptibility can be deduced from the severity of the clinical disease. However there are several complexities to consider. Prior and even asymptomatic infection could influence susceptibility; undoubtedly there is a relationship between the degree of sequence difference of the two strains and the consequences of infection. The potential importance of repeated infection/exposure is often ignored.

To date we have not been able to undertake a meaningful comparison of NHPs in terms of genomic sequences related to retroviruses. However, by amplifying human, chimp and rhesus genomic DNA with HIV RT primers it is obvious that there must be substantial differences between these species. The sequence relationship between an infecting virus and the host content of endogenous retroviruses also requires further study.

In spite of the shortcomings, we have attempted to review available data on the relationship between HIV and SIV. Numerous natural isolates of SIV have been characterised from NHPs such as SIV_{MAC} from rhesus monkey (*Macaca mulatta*), SIV_{AGM} from African green monkey (*Cercopithecus aethiops*) and

SIV$_{CPZ}$ from chimpanzee (*Pan troglodytes*) (Huet et al. 1990; Johnson and Hirsch 1992). As described above, there are close sequence relationships between SIV, HIV-1 and HIV-2. However, HIV-1 and HIV-2 in humans are pathogenic, whereas SIV in the natural host does not appear to induce disease even though it can be pathogenic in another monkey host. Rhesus macaques can be infected with a SIV isolate from the sooty mangabey and the induced infection most closely resembles the acquired immunodeficiency syndrome associated with HIV infection in humans (Lewis et al. 1992; Simon et al. 1992).

There does appear to be a difference between the disease progession of African and Asian monkeys infected with SIV. African monkeys such as the African green monkey, chimpanzee and the baboon do not progress to disease if infected with their natural isolate of SIV (Allan et al. 1995). However, as discussed above, Asian monkeys, such as the macaque, progress rapidly to disease if infected with SIV. Therefore an hypothesis might assume that an ancestral SIV infected an African monkey species after the split between the African and Asian monkeys. Following this primary infection there was an expansion in the diversity of this virus as it was passaged through serial hosts. The population of this African species would have gone through a bottleneck with selection pressures favoring a "resistant" gene(s) and/or integration of this retrovirus into the genome thereby conferring resistance (Doolittle et al. 1990; O'Brien 1995). Transmission of this virus to other African species occurred when the diversity of the virus was sufficient to allow crossing to another species. This process of virus diversification, disease, a resulting population bottleneck and selection of host and virus is repeated, eventually permitting back transmission to the previous species (Fig. 2). Therefore, one could expect that the African monkeys have a "resistant" gene(s) in common and/or an endogenous retrovirus which is conferring resistance to the virus of that particular host. If a gene(s) is involved in conferring resistance, then it could be present in the Asian monkeys, but at a lesser frequency. If the resistance is due to an endogenous retrovirus, then the Asian monkeys should be negative.

Chimpanzees provide an excellent model to investigate HIV-1 infection, as humans and chimpanzees share 98% identity at the DNA level (Diamond 1988; Leelayuwat et al. 1993). Chimpanzees can be actively infected with the HIV-1 virus and mount an immune response (Goudsmit et al. 1988) but do not progress to disease (Watanabe et al. 1991; Warren et al. 1990). The region between HLA-B and TNF, which has been implicated in rapid progression to AIDS following HIV-1 infection in humans (Cameron et al. 1992), has been shown to have a 50% length difference between the two species (Leelayuwat et al. 1993). Therefore, this region which has been deleted in the chimpanzee can be used to evaluate possible candidate genes characterised in this region.

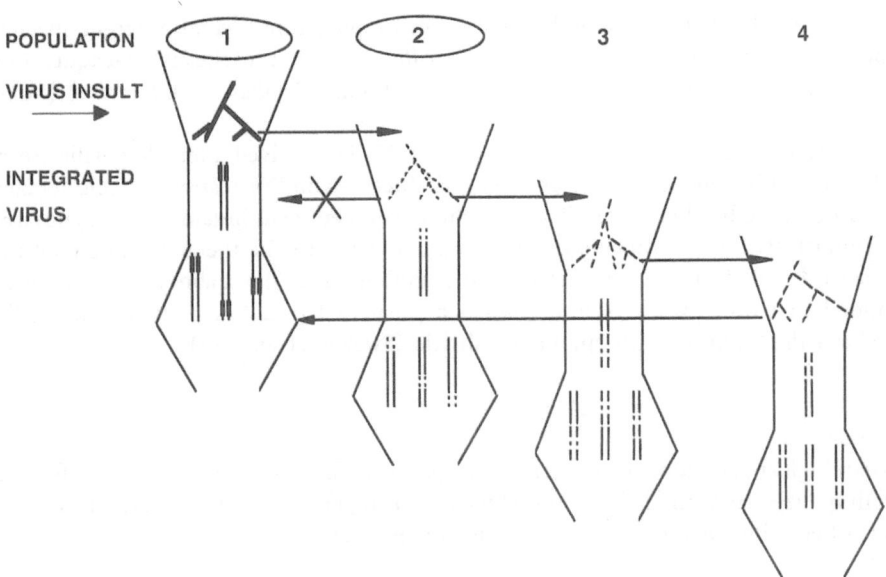

Fig. 2. Retroviral evolution in primate species. A virus insult on population 1 causes a bottleneck phenomenon in this species but also genetic diversity of the invading virus, represented by the *expanding tree*. Resistance is conferred when that virus has integrated into the host genome, represented by the *two vertical lines*. This integrated virus is transfered vertically with resulting expansion of the species population. An *arrow* indicates a cross-species transfer of this virus. The same pattern is repeated in species 2. The *arrow* from species 2 to species 1 indicates a block in a back-transmission of the virus. The integrated virus in species 1 can still confer resistance to the virus from species 2. However, when the virus has diverged considerably from the integrated virus, that species population may not be resistant. This is shown by the *arrow* extending from species 4 to species 1

Idiopathic Diseases

Arthitides

Diseases of bone and joints are probably the most useful for our present purpose simply because there is an enduring record. Lovell (1990) describes pathology which must be largely degenerative in nature but, in view of the presence of lumbar and sacroiliac fusion, AS must be a possibility. In view of the features of AS and B27 described above, it would not be surprising if NHPs were affected and if B27 or a B27-like allele were associated. Indeed there are some case reports which support these possibilities. Unfortunately the HLA typing is not conclusive and sequence data do not appear to be available.

Classical ankylosing spondylitis has been reported in several different species of nonhuman primates, the gorilla (*Gorilla gorilla*), the rhesus macaque, the baboon (*Mandrillus leucophaeus*), and the gibbon (*Hylobates syndactylus*) (Swezey et al. 1991).

The first reported case of classical AS in NHPs involved a female gorilla from the Memphis zoo (Adams et al. 1987). Histocompatibility testing on this female was not done but her male offspring who developed local inflammatory synovitis demonstrated B27 antigenicity using human allosera. In the publication citing classical ankylosing spondylitis in the gibbon and the baboon, a note was added that other workers have shown a positive B27 antigenicity in mandrills (*Mandrillus sphinx*) with spondyloarthritis (Swezey et al. 1991).

Psoriasis

Histological symptoms characteristic of psoriasis has been described in a female chimpanzee from the Leipzig zoo (Biella et al. 1991). Psoriasiform dermatoses has also been documented in other species of monkeys.

Autoimmune Diseases

Systemic Lupus Erythematosus

Naturally occurring SLE has been documented in a rhesus macaque (Anderson and Klein 1993). Characteristic features such as antinuclear antibodies, hemolytic anemia and glomerulonephritis have been observed in this case report.

Insulin-Dependent Diabetes Mellitus

Naturally occurring insulin-dependent diabetes does occur in NHPs but is not always clear whether the disease is type 1 or 2. However, we have demonstrated islet cell antibodies in rhesus monkeys with diabetes (Pummer et al., unpublished).

A model for IDDM in humans has been found in the chimpanzee. Chimpanzees do not appear to develop type I autoimmune diabetes (IDDM) but have been reported to develop type II nonautoimmune diabetes (NIDDM) (Rosenblum et al. 1981; Hamilton and Ciaccia 1978). It has been shown by Degli-Esposti and colleagues that the region between HLA-B and TNF in the central region of the MHC carries a susceptibility gene(s) to the autoimmune diseases IDDM and MG (Degli-Esposti et al. 1992a,b). As described earlier, this region is 50% shorter in chimpanzees than in humans. Again, this region can be used to evaluate candidate genes that are characterised in this region.

Table 5. Experimentally induced diseases in nonhuman primates that model HLA associated diseases in humans

Human disease	Induced model	Nonhuman primate	HLA association
Juvenile diabetes mellitus	Streptozotocin	Rhesus monkey	
Rheumatoid arthritis	CIA	Rhesus monkey	Mamu-A26
Multiple sclerosis	EAE	Rhesus monkey	Mamu-DR8 Mamu-DR2
	EAE	Common marmoset	
Myasthenia gravis	EAMG	Rhesus monkey	

CIA, type II collagen induced arthritis; EAE, experimental allergic encephalomyelitis in rhesus monkey and experimantal autoimmune encephalomyelitis in common marmoset; EAMG, experimental autoimmune myastenia gravis.

Experimentally Induced Diseases

Table 5 lists the experimentally induced diseases in NHPs.

Type II Collagen-Induced Arthritis in the Rhesus Monkey

Type II collagen-induced arthritis (CIA) is an experimentally induced autoimmune disease characterised by systemic polyarthritis and is used as an animal model for human rheumatoid arthritis. The first evidence of induced arthritis from immunisation with bovine type II collagen occurred in the squirrel monkey (*Saimiri sciureus*) (Cathcart et al. 1986). In this case all six squirrel monkeys that were immunised with the known antigen progressed to soft tissue swelling of the peripheral joints. Though no histocompatibility typing was completed on these monkeys, evidence for probable resistance due to genetics was observed when three cebus monkeys (*Cebus albifrons*) given the same antigen did not show any of the pathological features observed in the squirrel monkeys. A later analysis of MHC association with CIA and the rhesus monkey (*Macaca mulatta*) showed no evidence between disease onset and MHC DR types or restriction fragment length polymorphisms (RFLPs) of DRß, DQ and DQß (Jonker et al. 1991). Admittedly the sample group, which contained nine diseased monkeys from a total of 14, was quite small and no definite associations could be determined. Further analysis using the rhesus monkey has shown a strong association between resistance to CIA and Mamu-A26 (Bakker et al. 1992). This analysis was based on the distinction of 13 *Mamu-A* and 14 *Mamu-B* alleles and greater than ten DR allelic specificities. Of the 24 monkeys immunised with native bovine type II collagen, 16 developed CIA. There was no detection of class II DR association with susceptibility or resistance in the CIA monkeys. Of the eight resistant monkeys, seven shared the class I allele *Mamu-A26* ($p<0.00002$). The resistance due to *Mamu-A26* diminished with age and may be due to the inability of the immune system to tolerate autoantigens (Bakker et al. 1992). This association has also been linked to a lower level of anti-type II collagen IgM autoantibodies in the circulation (Jonker et al. 1991; Bakker et al. 1992; 't Hart et al. 1993).

A later study by 't Hart and colleagues (1983) showed that heat denaturation of the antigen type II collagen destroyed its arthritogenic capacity and B cell epitopes, and therefore resistance could be attained in Mamu-A26–monkeys. The authors suggested that resistance or susceptibility to CIA is determined by which Th cell response is instigated by the *Mhc-Mamu* alleles. Further evidence from other animal models of CIA suggest that the interaction between T and B cells via CD40 and its ligand gp39 are critical in the induction of arthritis (Durie et al. 1994). However it remains to be determined which *Mhc* genes are involved directly. The evidence presented above supports a role for class I but class II (Nepom et al. 1989), TNF and other genes deserve further attention.

Interleukin 3-Induced Arthritis in Juvenile Rhesus Monkey

Interleukin-3 administration to juvenile rhesus monkeys (*Macaca mulatta*) produced an increased hematopoiesis of multiple cell lineages but also arthritis (Van Gils et al. 1993). Four of 20 monkeys developed swollen joints and extremities characteristic of arthritis, and upon analysis of their typing it was found that there was an association between MHC alleles *B9* and *DR5* ($p=0.03$) and disease onset. However, there was no association between resistance to disease onset and *Mamu-A26* as had been previously reported.

Experimental Allergic Encephalomyelitis in Rhesus Monkeys

Experimental allergic encephalomyelitis (EAE) is an induced autoimmune disease that resembles human multiple sclerosis. In humans, one of the candidate autoantigens is myelin basic protein, which can induce EAE (Hafler and Weiner 1995). The disease has been induced in rhesus monkeys (*Macaca mulatta*) using bovine brain homogenate (Jonker et al. 1991). In this analysis of 22 immunised monkeys, 14 developed EAE. Consequences included neurological features such as motor weakness and tremors. Investigation of a possible association between MHC alleles and disease onset showed that *Mamu-DR8* occurred at a higher frequency in EAE-susceptible monkeys while *Mamu-DR2* was found in resistant monkeys. However, further analysis of the DR allelic specificities using RFLP showed no significant associations with susceptibility or resistance to EAE (Jonker et al. 1991). The DR specificities of the rhesus monkey do not correlate with those occurring in humans; HLA-DR2 is associated with multiple sclerosis in humans. This model can help determine how self-peptides can induce autoimmunity in unrelated individuals via a particular MHC class II molecule (Jonker et al. 1991).

As an approach to use human cells in a primate model of multiple sclerosis, a recent review by Hohlfeld and colleagues from a meeting on "T-cell autoimmunity in multiple sclerosis", reports that a group ('t Hart et al. 1993) has been able to show that a CD4+ HLA-DRB1*0301-restricted human T cell clone can recognise myelin basic protein in vitro from different rhesus monkeys expressing two very closely related MHC alleles *Mamu-DRB1*0305* and *Mamu-DRB1*0306* (Hohlfeld et al. 1995). Further analysis in NHPs may be facilitated by a recent model of EAE in the common marmoset (*Callthrix jacchus*) (Massacesi et al. 1995).

Simian Immunodeficiency Virus Vaccine Production

In an attempt to develop safe vaccines for HIV infection, SIV epitopes associated with MHC class I molecules were identified and used as vaccines in NHPs to elicit MHC class I-restricted CD8+ cytotoxic T lymphocytes (CTLs) (Miller et al. 1992). A nine amino acid epitope from 182–190 of SIV_{MAC} p27 restricted by an HLA-A homolog Mamu-A*01 (Miller et al. 1991; Letvin et al. 1993) was shown to produce a gag-specific $CD8^+$ CTL response, and when used in a subunit vaccine produced a virus-specific MHC class I-restricted CTL response in rhesus monkeys (Miller et al. 1992). Using similar procedures, another HLA-A homolog, Mamu-A*02, was found to bind to an eight amino acid peptide covering residues 306–313 of SIV_{MAC} envelope V3 region and reacted with $CD8^+$ CTLs in rhesus monkeys (Watanabe et al. 1994). An HLA-B homolog in the rhesus monkey, Mamu B*01, was also found to bind to an eight amino acid peptide from residues 502–510 of both the CD4-binding and V4 region in SIV_{MAC} (Yasutomi et al. 1995).

In addition, the effectiveness of an SIV vaccine to protect rhesus monkeys from infection was observed in animals vaccinated with whole inactivated SIV and sharing the *Mamu-A26* allele with the donor (infecting) monkey (Heeney et al. 1994). The detection of CTLs specific for SIV envelope proteins suggested a role for MHC class I-restricted CTL in protection. The association between SIV vaccine protection and the sharing of *Mamu-A26* was found to be significant ($p<0.005$). However, *Mamu-A26* was not associated in prolonged survival after infection nor to resistance in general.

Streptozocin Induced Diabetes

Streptozocin induced diabetes has been studied in the rhesus monkey and the baboon (Jonasson et al. 1985; Bagdade et al. 1994). These primates have been used as a model for IDDM in humans. Alterations in plasma lipoprotein and apolipoprotein and accelerated cholesteryl ester transfer have been observed in these primates (Bagdade et al. 1994).

Comparative Genomics of Primates

The MHC genomic size and gene orientation in the primates is relatively similar and sequence comparisons of class I and class II alleles reveal a *trans*-species mode of evolution (Trowsdale 1995). In addition, some endogenous retroviruses isolated from the human MHC are also present in the NHPs. Using fluorescent in situ hybridization, the chromosomal position of the MHC in primates has been shown to differ between the lower primates and apes. For example, the chimpanzee MHC is situated on chromosome 5p21.3 in the exact position as in humans on 6p21.3. However, the MHC in macaque monkeys is positioned on the long arm of chromosome 5 (Hirai et al. 1991).

HLA Class II (δ Block)

The MHC δ block contains several membrane bound glycoprotein antigens such as DR, DQ and DP. These polymorphic antigens have been shown to exist in the NHPs (Bontrop et al. 1990; Kenter et al. 1992; Otting and Bontrop 1995; Slierendregt et al. 1995) and sequence comparisons reveal their mode of evolution. However, there is evidence to suggest that the δ block in primates is subject to duplications and subtractions (Slierendregt et al. 1994) which may involve genes or other informative DNA that could play a critical role in various diseases associated with this region. Diseases of potential interest in relation to these genomic variations include rheumatoid arthritis, narcolepsy and perhaps multiple sclerosis but we are not aware of relevant data.

Complement (γ Block)

The γ block of the MHC includes the *C4*, *CYP21* loci and the other members of the complement module. In NHPs the gene organisation and copy numbers of the *C4* and *CYP21* loci have been shown to be essentially the same (Christiansen et al. 1991a; Kawaguchi et al. 1990; Zhang et al. 1993). In fact, complotypes/haplotypes described in humans can also be described in the NHPs (Christiansen et al. 1991a). The size of this genomic region in NHPs appears to be similar with differences relating mainly to the size of the *C4* genes. For example, chimpanzees and gorillas appear to have only the short form of the *C4* genes, and therefore, have a genomic size corresponding to its human equivalent (Zhang et al. 1993). The orangutan appears to have extra copies of the *C4* and *CYP21* genomic segments. Typing of any NHPs with SLE would be informative.

A retroposon and an endogenous retroviruses have been described in the γ block. SINE-R.C2 and HERV-K(C4) are located within the third intron of *C2* and long intron 9 of *C4*, respectively (Zhu et al. 1994; Dangel et al. 1995). As chimpanzees and gorillas only have a short form of *C4* they do not have the *C4* endogenous retrovirus. Other primates which exhibit both forms of *C4* such as orangutans, Old World and New World monkeys contain the endogenous

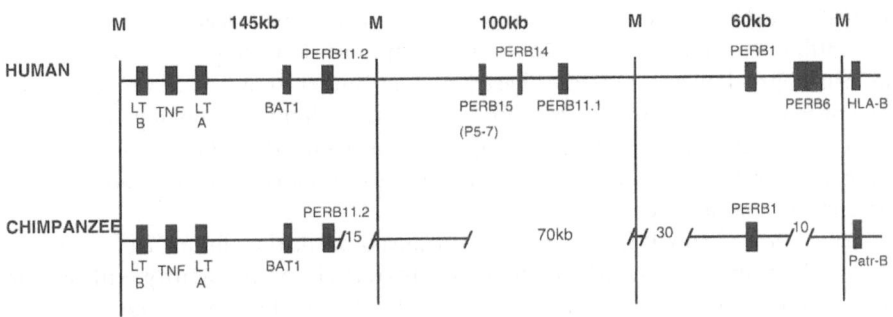

Fig. 3. Comparison of the central MHC in primates reveals a large deletion in chimpanzees

retrovirus in the exact position as in humans. The C2 endogenous retroposon SINE-R.C2 is found only in humans. The endogenous retroviruses not only provide an evolutionary tool to investigate divergence times and phylogenetic clades, but they may be involved in the kinetics of exogenous retroviruses in the individual. Studies of these genomic variations in relation to HIV/SIV infection are awaited with interest.

Tumor Necrosis Factor and HLA-B (β Block)

This region of the MHC between HLA-B and TNF termed the β block has been associated with susceptibility to autoimmune diseases such as IDDM and MG (Degli-Esposti et al. 1992a,b) and rapid progression to AIDS following HIV-1 infection (Cameron et al. 1992). By pulse field gel electrophoresis (PFGE) this region varies in humans between 270 and 300kb. However, in chimpanzees the size is approximately 170 kb. Figure 3 shows an updated map of this region between the two species, highlighting the regions of deletion/translocation in the chimpanzee. Many genes and gene fragments have been identified in this region and include a newly described gene family, PERB11 (Marshall et al. 1993; Leelayuwat et al. 1994), which has also been alternately termed MIC by an independent group (Bahram et al. 1994) and includes an FCRn-like gene (PERB11.1). The chimpanzee appears to have fewer copies of this gene family; one gene resides near BAT1. Other deletions/translocations in this region include the X locus and approximately 10 kb of a highly polymorphic and duplicated region in humans (Leelayuwat et al. 1993). These deletions/translocations in the chimpanzee may influence autoimmune diseases such as IDDM and MG and progression to AIDS following HIV-1 infection. Therefore, the chimpanzee provides an excellent model in which to evaluate possible candidate genes as described previously. As more sequence becomes available in this region the identification of the candidates gene(s) involved in the autoimmune processes and in HIV-1 disease progression will enable a better understanding of the mechanisms involved in their pathogenesis. In addition, the rhesus macaque does develop autoantibodies and develops AIDS-like symptoms following SIV infection as described previously. Therefore, based on the model above, one would expect the macaque to contain the insertions present in humans and deleted in the chimpanzee. Preliminary data on the macaque show that the region between the genes for HLA-B and TNF is comparable to humans if not expanded (Gaudieri et al., unpublished).

The X locus, which has been deleted/transposed in the chimpanzee, contains a retrovirus-like element which shares approximately 90 % identity with a new class of retrovirus-like sequences termed *HERK-L* (Gaudieri et al. 1996; Cordonnier et al. 1996). At the amino acid level, the sequence shares homology with the POL sequence of foamy and Moloney retroviruses. In addition, it has been shown using transgenic mice that human foamy virus can transactivate HIV-1 in vivo (Marino et al. 1995). Therefore, a potential retroviral element may be critically involved in the kinetics of HIV.

In addition, Nco1 and microsatellite typing of the four AOH workshop panels revealed new TNF alleles in chimpanzees and orangutans that had not previously been described in the human population (Abraham et al. 1993). As TNF is a critical cytokine in the immune response, further investigation of this gene in primates would be invaluable.

HLA-A (α Block)

Sequence analysis of the α block in primates reveals not only several *HLA-A* alleles in all the NHPs but also several of the pseudogenes described in this region (Lawlor et al. 1988, 1990; Otting and Bontrop 1993).

In humans the α block is associated with hemochromatosis. The apparent failure to observe this disease in NHPs may be due to the low penetrance and the late age of onset especially in females. However NHPs appear to be efficient at absorbing iron and direct analysis is awaited.

Acknowledgements. We are grateful to Dr. Ronald Bontrop and colleagues at the Biomedical Primate Research Centre in Rijswijk the Netherlands and Dr. Jan Klein and colleagues at the Max Planck Institute for Biology, Department of Immunogenetics, Tubingen, Germany for material and discussions. We acknowledge the suggestions of Dr. Paul Cameron, Dr. Wenjie Zhang and Dr. Chanvit Leelayuwat. We thank Dr. Rosemary Markham at the Western Australian Zoo. This work is supported by NH&MRC, Immunogenetics Research Foundation and the Lotteries Commission of Western Australia. Silvana Gaudieri is the recipient of an APRA Award and Scholarship. Roger Dawkins was supported by the Humboldt Foundation whilst visiting the Max Planck Institute in Tubingen, Germany 1994. Publication number 9507.

References

Abraham LJ, Marley J, Nedospasov SA, Cambon-Thomsen A, Dawkins RL, Giphart M. Microsatellite, RFLP and SSO typing of the TNF region: comparisons on the 4AOH cell panel. Hum Immunol 38:17–23, 1993

Adams RF, Flinn GS, Douglas M. Ankylosing spondylitis in a nonhuman primate: a monkey tale. Arthritis Rheum 30:956–957, 1987

Allan JS, Ray P, Broussard S, Whitehead E, Hubbard G, Butler T, Brasky K, Luciw P, Cheng-Mayer C, Levy JA, Steimer K, Li J, Sodroski J, Garcia-Moll M. Infection of baboons with simian/human immunodeficiency viruses. J Acquir Immune Defic Syndr Hum Retrovirol 9:429–441, 1995

Alper CA, Kruskall MS, Marcus-Bagley D, Craven DE, Katz AJ, Brink SJ, Dienstag JL, Awdeh Z, Yunis EJ. Genetic prediction of nonresponse to hepatitis B vaccine. N Engl J Med 321:708–712, 1989

Anderson ST, Klein EC. Systemic lupus erythematosus in a rhesus macaque. Arthritis Rheum 36:1739–1742, 1993

Apple RJ, Erlich HA, Klitz W, Manos MM, Becker TM, Wheeler CM. HLA DR-DQ associations with cervical carcinoma show papillomavirus-type specificity. Nature Genet 6:157–162, 1994

Bagdade JD, Koerker DJ, Ritter MC, Weigle DS, Goodner CJ. Accelerated cholesteryl ester transfer in baboons with insulin-requiring diabetes mellitus. Horm Metab Res 27:70–75, 1994

Bahram S, Bresnahan M, Geraghty DE, Spies T. A second lineage of mammalian major histocompatibility complex class I genes. Proc Natl Acad Sci USA 91:6259–6263, 1994

Bakker NPM, Van Erck MGM, Otting N, Lardy NM, Noort RC, 't Hart B, Jonker M, Bontrop RE Resistance to collagen-induced arthritis in a nonhuman primate species maps to the major histocompatibility complex class I region. J Exp Med 175:933–937, 1992

Barnett SW, Murthy KK, Herndier BG, Levy JA. An AIDS-like condition induced in baboons by HIV-2. Science 266:642–646, 1994

Biella U, Haustein UF, Seifert S, Adler J, Schuppel KF, Eulenberger K. Psoriasis in a female chimpanzee. Hautarzt 42:322–323, 1991

Bodmer WF, Trowsdale J, Young J, Bodmer J. Gene clusters and the evolution of the major histocompatibility system. Philos Trans R Soc Lond B Biol Sci 312:3030, 1986

Bontrop RE, Broos LAM, Pham K, Bakas RM, Otting N, Jonker M. The chimpanzee major histocompatibility complex class II DR subregion contains an unexpectedly high number of beta-chain genes. Immunogenetics 32:272–280, 1990

Cameron PU, Mallal SA, French MAH, Dawkins RL. Central MHC genes between HLA-B and complement C4 confer risk for HIV-1 disease progression. In: Tsuji K, Aizawa M, Sasazuki T (eds) HLA 1991, vol 2, Oxford University Press, New York, pp 544–547, 1992

Cathcart ES, Hayes KC, Gonnerman WA, Lazzari AA, Franzblau C. Experimental arthritis in a nonhuman primate. Lab Invest 54:26–31, 1986

Chan SY, Delius H, Halpern AL, Bernard HU. Analysis of genomic sequences of 95 papillomavirus types: uniting typing, phylogeny, and taxonomy. J Virol 69:3074–3083, 1995

Christiansen FT, Owen ET, Dawkins RL, Hanrahan P. Symptoms and signs among relatives of patients with HLA B27 ankylosing spondylitis: correlation between back pain, spinal movement, sacroiliitis, and HLA antigens. J Rheumatol Suppl 3:11–17, 1977

Christiansen FT, Hawkins BR, Dawkins RL. Immune function in ankylosing spondylitics and their relatives: influence of disease and HLA B27. Clin Exp Immunol 33:270–275, 1978

Christiansen FT, Dawkins RL, Hawkins BR, Owen ET. On the reassessment of the prevalence of as among B27 positive normal individuals. J Rheumatol 7:577–578, 1980

Christiansen FT, Dawkins RL, Uko G, McCluskey J, Kay PH, Zilko PJ. Complement allotyping in SLE: association with C4A null. Aust NZ J Med 13:483–488, 1983

Christiansen FT, Bontrop R, Giphart MJ, Cameron PU, Zhang WJ, Townend D, Jonker M, Dawkins RL. Major histocompatibility complex ancestral haplotypes in the chimpanzee: identification using C4 allotyping. Hum Immunol 31:34–39, 1991a

Christiansen FT, Zhang WJ, Griffiths M, Mallal SA, Dawkins RL. Major histocompatibility complex (MHC) complement deficiency, ancestral haplotypes and systemic lupus erythematosus (SLE): C4 deficiency explains some but not all of the influence of the MHC. J Rheumatol 18:9:1350–1358, 1991b

Coleman N, Birley HDL, Renton AM, Hanna NF, Ryait BK, Byrne M, Taylor-Robinson D, Stanley M. Immunological events in regressing genital warts. Ann Path 102:768–774, 1994

Cordonnier A, Casella JF, Heidmann T. Isolation of novel human endogenous retrovirus-like elements with foamy virus-related pol sequence. J Virol 69:1475–1479, 1996

Crandell KA. Multiple interspecies transmissions of human and simian t-cell leukemia/lymphoma virus type I sequences. Mol Biol Evol 13:115–131, 1996

Dangel A, Baker B, Mendoza A, Yu C. Complement component C4 gene intron 9 as a phylogenetic marker for primates: long terminal repeats of the endogenous retrovirus ERV-K(C4) are a molecular clock of evolution. Immunogenetics 42:41–52, 1995

David ALM, Taylor GM, Aplin JD, Seif MW, Tindal VR. HLA-DQB1*03 and cervical intraepithelial neoplasia type III. Lancet 340:52, 1992

Dawkins RL, Martin E, Andreas-Zietz A et al. Linkage disequilibrium, interlocus association and ancestral haplotypes. In: Dupont B (ed) Immunobiology of HLA, vol 1. Springer, Berlin Heidelberg New York, pp 891–892, 1989

Degli-Esposti MA, Abraham LJ, McCann V, Spies T, Christiansen FT, Dawkins RL. Ancestral haplotypes reveal the role of the central MHC in the immunogenetics of IDDM. Immunogenetics 36:345–356, 1992a

Degli-Esposti MA, Andreas A, Christiansen FT, Schalke B, Albert E, Dawkins RL. An approach to the localization of the susceptibility genes for generalised myasthenia gravis by mapping recombinant ancestral haplotypes. Immunogenetics 35:355–364, 1992b

Diamond JM. DNA based phylogenies of the three chimpanzees. Nature 332:685–686, 1988

Doolittle RF, Feng DF, McClure MA, Johnson MS. Retrovirus phylogeny and evolution. Curr Top Microbiol Immunol 157:1–18, 1990

Durie FH, Fava RA, Noelle RJ. Collagen-induced arthritis as a model of rheumatoid arthritis. Clin Immunol Immunopathol 73:11–18, 1994

Ellis J, Keating P, Baird J, Hounsell E, Renouf D, Rowe M, Hopkins D, Duggan-Keen M, Bartholomew J, Young L, Stern P. The association of an HPV16 oncogene variant with HLA-B7 has implications for vaccine design in cervical cancer. Nature Med 1:464–470, 1995

Garrido F, Cabrera T, Concha A, Glew S, Ruiz-Cabello F, Stern PL. Natural history of HLA expression during tumour development. Immunol Today 14:491–499, 1993

Gaudieri S, Kulski JK, Dawkins RL. The central region of the major histocompatibility complex contains a motif of amino acid sequence with homology for the pol gene of Moloney retroviruses. Immunogenetics 44:157–158, 1996

Glew SS, Duggan-Keen M, Gosh AK, Ivinson A, Sinnott P, Davidson J, Dyer PA, Stern PL. Lack of association of HLA polymorphisms with human papillomavirus-related cervical cancer. Hum Immunol 37:157–164, 1993

Goudsmit J, Debouck C, Heloen RH, Smit L, Bakker M, Ashers DM, Wolff AV, Gibbs CJ, Gajdusek DC. Human immunodeficiency virus type 1 neutralizaiton epitope with conserved architecture elicits early type-specific antibodies in experimentally infected chimpanzees. Proc Natl Acad Sci USA 85:4478–4482, 1988

Hafler DA, Weiner HL. Immunologic mechanisms and therapy in multiple sclerosis. Immunol Rev 144:75–108, 1995

Haldane JBS. Disease and evolution. Ric Scien [Suppl] 19:68–76, 1949

Hamilton CL, Ciaccia P. The course of development of glucose intolerance in the monkey. J Med Primatol 7:165–173, 1978

Han R, Breitburd F, Marche PN, Orth G. Linkage of regression and malignant conversion of rabbit viral papillomas to MHC class II genes. Nature 356:66–68, 1992

Heeney JL, Van Els C, De Vries P, Ten Haaft P, Otting N, Koornstra W, Boes J, Dubbes R, Niphuis H, Dings M, Cranage M, Norley S, Jonker M, Bontrop RE, Osterhaus A. Major histocompatibility complex class I-associated vaccine protection from simian immunodeficiency virus-infected peripheral blood cells. J Exp Med 180:769–774, 1994

Herrington CS. Human papillomaviruses and cervical neoplasia. I. Classification, virology, pathology, and epidemiology. J Clin Pathol 47:1066–1072, 1994

Hirai M, Takahashi E, Ishida T, Hori T. Chromosomal localization of the major histocompatibility complex (MHC) in the rhesus monkey and chimpanzee by fluorescence in situ hybridization. Cytogenet Cell Genet 57:204–205, 1991

Hohlfeld R, Londei M, Massacesi L, Salvetti M. T-cell autoimmunity in multiple sclerosis. Immunol Today 16:259–261, 1995

Hollingsworth PN, Dawkins RL, Christiansen FT, Owen ET. Ankylosing spondylitis: B27, sacroiliitis and ankylosing spondylitis in caucasians. In: Dawkins RL, Christiansen FT, Zilko PJ (eds) Immunogenetics in rheumatology: muscoskeletal disease and D-penicillamine, pp 179–183, Excerpta Medica, Amsterdam, 1982

Honma S, Tsukada S, Honda S, Nakamura M, Takakuwa K, Maruhashi T, Kodama S, Kanazawa K, Takahashi T, Tanaka K. Biological-clinical significance of selective loss of HLA-class-I allelic product expression in squamous-cell carcinoma of the uterine cervix. Int J Cancer 57:650–655, 1994

Huet T, Cheynier R, Meyerhans A, Roelants G, Wain-Hobson S. Genetic organization of a chimpanzee lentivirus related to HIV-1. Nature 345:356–359, 1990

Itescu S, Mathur-Wagh U, Skovron ML, Brancato LJ, Marmor M, Zelenuich-Jacquotte A, Winchester R. HLA-B35 is associated with accelerated progression to AIDS. J Acquir Immune Defic Syndr 5:37–45, 1991

Itescu S, Rose S, Dwyer E, Winchester R. Grouping HLA-B locus serologic specificities according to shared structural motifs suggests that different peptide-anchoring pockets may have contrasting influences on the course of HIV-1 infection. Hum Immunol 42:81–89, 1995

Johnson PR, Hirsch VM. Genetic variation of simian immunodeficiency viruses in nonhuman primates. AIDS Res Hum Retroviruses 8:367–371, 1992

Jonasson O, Jones CW, Bauman A, John E, Manaligod J, Tso MOM. The pathophysiology of experimental insulin-deficient diabetes in the monkey. Annu Surg 201:27–35, 1985

Jonker M, Bakker K, Slierendregt B, 't Hart B, Bontrop R. Autoimmunity in non-human primates: the role of major histocompatibility complex and T cells, implications for therapy. Hum Immunol 32:31–40, 1991

Just JJ. Genetic predisposition to HIV-1 infection and acquired immune deficiency virus syndrome. Hum Immunol 44:156–169, 1995

Kawaguchi H, Golubic M, Figueroa F, Klein J. Organization of the chimpanzee C4-CYP21 region: implications for the evolution of human genes. Eur J Immunol 20:739–745, 1990

Kenter M, Otting N, Anholts J, Leunissen J, Jonker M, Bontrop RE. Evolutionary relationships among the primate Mhc-DQAI and DQA2 alleles. Immunogenetics 36:71–78, 1992

Klein J. Is it possible to formulate a unified concept for the biological function of the major histocompatibility complex (MHC). Vox Sang 49:354–367, 1985

Klein J, Takahata N. The major histocompatibility complex and the quest for origins. Immunol Rev 113:5–25, 1990

Kloster BE, Manias DA, Ostrow RS, Shaver MK, McPherson SC, Rangen SRS, Uno H, Jaras AJ. Molecular cloning and characterization of the DNA of two papillomaviruses from monkeys. Virology 166:30–40, 1988

Kyo S, Inoue M, Hayasaka N, Inoue T, Yutsudo M, Tanizawa O, Hakura A. Regulation of early gene expression of human papillomavirus type 16 by inflammatory cytokines. Virology 200:130–139, 1994

Lawlor DA, Ward FE, Ennis PD, Jackson AP, Parham P. HLA-A and B polymorphisms predate the divergence of humans and chimpanzees. Nature 335:268–271, 1988

Lawlor DA, Warren E, Ward FE, Parham P. Comparison of class I MHC alleles in humans and apes. Immunol Rev 113:147–185, 1990

Leelayuwat C, Zhang WJ, Townend DC, Gaudieri S, Dawkins RL. Differences in the central MHC between man and chimpanzee: implications for autoimmunity and acquired immune deficiency development. Hum Immunol 38:30–41, 1993

Leelayuwat C, Townend DC, Degli-Esposti MA, Abraham LJ, Dawkins RL. A new polymorphic and multicopy MHC gene family related to nonmammalian class I. Immunogenetics 40:339–351, 1994

Leelayuwat C, Pinelli M, Dawkins R. Clustering of diverse replicated sequences in the major histocompatibility complex (MHC): Evidence for en bloc duplication. J Immunol 155:692–698, 1995

Lehrman S. AIDS patient given baboon bone marrow. Nature 378:756, 1995

Letvin NL, Miller MD, Shen L, Chen ZW, Yasutomi Y. Simian immunodeficiency virus-specific cytotoxic t lymphocytes in rhesus monkeys: characterization and vaccine induction. Immunology 5:215–223, 1993

Lewis MG, Zack PM, Elkins WR, Jahrling PB. Infection of rhesus and cynomolgus macaques with a rapidly fatal SIV (SIVsmm/pbj) isolate from sooty mangabeys. AIDS Res Hum Retroviruses 8:1631–1633, 1992

Li MD, Bronson DL, Lemke TD, Faras AJ. Phylogenetic analyses of 55 retroelements on the basis of the nucleotide and product amino acid sequences of the pol gene. Mol Biol Evol 12:657–670, 1995

Lovell NC. Patterns of injury and illness in great apes. Smithsonian Institute, Washington, 1990

Mallal SJ, James IR, French MAH. Detection of subclinical Mycobacterium avium intracellulare complex infection in immunodeficient HIV-infected patients treated with zidovudine. AIDS 8:1263–1269, 1994

Marino S, Kretschmer C, Brandner S, Cavard C, Zider A, Briand P, Isenmann S, Wagner EF, Aguzzi A. Activation of HIV transcription by human foamy virus in transgenic mice. Lab Invest 73:103, 1995

Marshall B, Leelayuwat C, Degli-Esposti MA, Pinelli M, Abraham LJ, Dawkins RL. New MHC genes. Hum Immunol 38:24–29, 1993

Massacesi L, Genain CP, Lee-Parritz D, Letvin NL, Canfield D, Hauser SL. Active and passively induced experimental autoimmune encephalomyelites in common marmosets: a new model for multiple sclerosis. Ann Neurol 37:519–530, 1995

Mehra NK, Rajalingam R, Mitra DK, Taneja V, Giphart MJ. Variants of HLA-DR2/DR51 group haplotypes and susceptability to tuberculoid leprosy and pulmonary tuberculosis in asian indians. Int J Leprosy 63:241–248, 1994

Meyers WM, Gormus BJ, Walsh GP, Baskin GB, Hubbard GB. Naturally acquired and experimental leprosy in nonhuman primates. Am J Trop Med Hyg 44:24–27, 1991

Miller MD, Yamamoto H, Hughes AL, Watkins DI, Letvin NL. Definition of an epitope and mhc class 1 molecule infected rhesus monkeys. J Immunol 147:320–329, 1991

Miller MD, Gould-Fogerite S, Shen L, Woods RM, Koenig S, Mannino RJ, Letvin NL. Vaccination of rhesus monkeys and synthetic peptide in a fusogenic proteoliposome elicits simian immunodeficiency virus-specific cd8+ cytotoxic T lymphocytes. J Exp Med 176:1739–1744, 1992

Monath TP. Flaviviruses. Virology 2:763–814, 1990

Myers G, Bernard HU, Delius H et al. Human papillomaviruses 1994. A compilation and analysis of nucleic and amino acid sequences. Los Alamos National Laboratory, Los Alamos NM, 1994a

Myers G, Wain-Hobson S, Henderson LE, Korber B, Jeang KT, Pavlakis GN. Human retroviruses and AIDS 1994. Theor Biol Biophy 1994b

Nepom GT, Byers P, Seyfried C, Healey LA, Stage D, Nepom BS. HLA genes associated with rheumatoid arthritis. Arthritis Rheum 32:15–21, 1989

O'Brien C. Xenotransplants set to resume. Science 266:1148–1151, 1994

O'Brien S. Genomic prospecting. The importance of preserving biodiversity extends beyond the discovery of new drugs to understanding how other species have dealt with medical problems we currently face Nature Medicine 1:742–744, 1995

Ostrow RS, McGlennen RC, Shaver MK, Kloster BE, Houser D, Faras AJ. A rhesus monkey model for sexual transmission of a papillomavirus isolated from a squamous cell carcinoma. Proc Natl Acad Sci USA 87:8170–8174, 1990

Ostrow RS, LaBresh KV, Faras AJ. Characterization of the complete RhPV 1 genomic sequence and an integration locus from a metastatic tumor Virology 181:424–429, 1991

Otting N, Bontrop RE. Characterization of the rhesus macaque (Macaca mulatta) equivalent of HLA-F. Immunogenetics 38:141–145, 1993

Otting N, Bontrop RE. Evolution of the major histocompatibility complex DPA1 locus in primates. Hum Immunol 42:184–187, 1995

Reszka AA, Sundberg JP, Reichmann ME. Transformation and molecular characterization of colobus monkey venereal papillomavirus DNA. Virology 181:787–792, 1991

Roberts JA. Occupational health concerns with nonhuman primates in zoological gardens. J Zoo Wildlife Med 26:10–23, 1995

Rosenblum IY, Barbolt TA, Howard CFJ. Diabetes mellitus in the chimpanzee (Pan troglodytes). J Med Primatol 10:93–101, 1981

Saitou N, Nei M. The neighbor-joining method: a new method for reconstructing phylogenetic trees. Mol Biol Evol 4:408, 1987

Samuel BU, Ponnuraj E, Rajasingh J, John TJ. Experimental poliomyelitis in bonnet monkey. Clinical features, virology and pathology. Dev Biol Stand 78:71–78, 1993

Seibert SA, Howell CY, Hughes MK, Hughes AL. Natural selection on the gag, pol, and env genes of human immunodeficiency virus 1 (HIV-1). Mol Biol Evol 12:803–813, 1995

Simon MA, Chalifoux LV, Ringler DJ. Pathologic features of SIV-induced disease and the association of macrophage infection with disease evolution. AIDS Res Hum Retroviruses 8:327–335, 1992

Slierendregt BL, Otting N, Kenter M, Bontrop RE. Allelic diversity at the Mhc-DP locus in rhesus macaques (Macaca mulatta). Immunogenetics 41:29–37, 1995

Slierendregt EL, Otting N, van Besouw N, Jonker M, Bontrop RE. Expansion and contraction of rhesus macaque DRB regions by duplication and deletion. J Immunol 152:2298–2307, 1994

Swezey RL, Cox C, Gonzales B. Ankylosing spondylitis in nonhuman primates: the drill and the siamang. Semin Arthritis Rheum 21:170–174, 1991

't Hart B, Bakker NPM, Jonker M, Bontrop RE. Resistance to collagen-induced arthritis in rats and rhesus monkeys after immunization with attenuated type II collagen. Eur J Immunol 23:1588–1594, 1993

Tokunaga K, Saueracker GC, Kay PH, Christiansen FT, Anand R, Dawkins RL. Extensive deletions and insertions in different MHC supratypes detected by pulsed field gel electrophoresis. J Exp Med 168:933–940, 1988

Trowsdale J. Both man and bird and beast: comparative organization of MHC genes. Immunogenetics 41:1–17, 1995

Van Binnendijk RS, Van Der Heijden RWJ, Osterhaus ADME. Monkeys in measles research. Curr Top Microbiol Immunol 191:135–148, 1995

VanDenVelde C, De Foor M, Van Beers D. HLA-DQB1*03 and cervical intraepithelial neoplasia grade I-III. Lancet 341:442–443, 1993

Van Gils FCJM, Mulder AH, Van Den Bos C, Burger H, Van Leen RW, Wagemaker G. Acute side effects of homologous interleukin-3 in rhesus monkeys. Am J Pathol 143 6:1621–1633, 1993

Van Ranst M, Fuse A, Fiten P, Beuken E, Pfister H, Burk RD, Opdenakker G. Human papillomavirus type 13 and pygmy chimpanzee papillomavirus type 1: comparison of the genome organizations. Virology 190:587–596, 1992

Wank R, Thomssen C. High risk of squamous cell carcinoma of the cervix for woman with HLA-DQw3. Nature 352:723–725, 1991

Warren RQ, Wolf H, Suler KR, Eichberg JW, Zajac RA, Boswell RN, Kanda P, Kennedy RC. Synthetic peptides define the fine specificity of the human immunodeficiency virus (HIV) gp160 humoral immune response in HIV Type 1-infected chimpanzees. J Virol 64:486–492, 1990

Watanabe M, Ringler DJ, Fultz PN, MacKey JJ, Boyson JE, Levine CG, Letvin NL. A chimpanzee-passaged human immunodeficiency virus isolate is cytopathic for chimpanzee cells but does not induce disease. J Virol 65:3344–3348, 1991

Watanabe N, McAdam SN, Boyson JE, Piekarczyk MS, Yasutomi Y, Watkins DI, Letvin NL. A simian immunodeficiency virus envelope v3 cytotoxic t-lymphocyte epitope in rhesus monkeys and its restricting major histocompatibility complex class 1 molecule mamu-a*02. J Virol 68:6690–6696, 1994

Yasutomi Y, McAdam SN, Boyson JE, Piekarczyk MS, Watkins DI, Letvin NL. A mhc class 1 b locus allele-restricted simian immunodeficiency virus envelope ctl epitope in rhesus monkeys. J of Immunol 154:2516–2522, 1995

Zhang WJ, Degli-Esposti MA, Cobain TJ, Cameron PU, Christiansen FT, Dawkins RL. Differences in gene copy number carried by different MHC ancestral haplotypes: quantitation after physical separation of haplotypes by pulsed field gel electrophoresis. J Exp Med 171:2101–2114, 1990

Zhang WJ, Christiansen FT, Wu X, Abraham LJ, Giphart MJ, Dawkins RL. Organization and evolution of C4 and CYP21 genes in primates: importance of genomic segments. Immunogenetics 37:170–176, 1993

Zhu ZB, Jian B, Volanakis JE. Ancestry of SINE-RC2 a human-specific retroposon. Hum Genet 93:545–551, 1994

7 Nonhuman Primate MHC Class I Sequences: 1997

COMPILED BY C. SHUFFLEBOTHAM AND D.I. WATKINS

Introduction

Sequences used in this compilation are from: Chen et al. 1992, Chen et al. 1993, Lawlor et al. 1988, Lawlor et al. 1990, Lawlor et al. 1991, Lawlor et al. 1995, Klein et al. 1990, Mayer et al. 1988, Miller et al. 1991, McAdam et al. 1994a,1994b, Watanabe et al. 1994, Yasutomi et al. 1995, Otting et al. 1993, Corell et al. 1994, Boyson et al. 1995, Watkins et al. 1990a, 1990b, 1991a, 1991b. Gogo-A3 and Gogo-A4 (Watkins et al. 1991b) are identical to *Gogo-OKO* and *Gogo-A*0401*, respectively (Lawlor et al. 1991). The cotton-top tamarin MHC class I cDNAs (Watkins et al. 1990a, 1990b, 1991a) have been renamed (Klein et al. 1990) as follows; *Saoe-F*01* (So-3), *Saoe-G*01* (So-4), *Saoe-G*02* (So-5), *Saoe-G*03* (So-6), *Saoe-G*04* (So-8), *Saoe-G*05* (So-16), *Saoe-G*06* (So-17), *Saoe-G*07* (So-21), *Saoe-G*08* (So-24), *Saoe-G*09* (So-32), *Saoe-G*10* (So-46), *Saoe-G*11* (So-47). Identity to the consensus sequence is indicated by a hyphen (-) and gaps in the sequence inserted to improve alignment are indicated by a period (.).

Molecular Biology and Evolution of Blood Group
and MHC Antigens in Primates
Blancher/Klein/Socha (Eds.)
© Springer-Verlag Berlin Heidelberg 1997

```
Exon 1
                    10        20        30        40        50        60        70
                     *    *    *    *    *    *    *    *    *    *    *    *    *    *
Consensus   ATGCCGGTCATGGCGCCCCGAACCCTCCTCCTGCTGCTCTCGGGGGCCCTGGCCCTGACCGAGACCTGGGCCC

Patr-A*01   ---G-C------C----------T----------A-----------------------C----------G-
Patr-A*02   ---G-C------C----------T----------A-----------------------C----------G-
Patr-A*03   G---C------C----------T----------A-----------------------C----------G-
Patr-A*04   ---G-C--------------------------A-----------------------C----------G-
Patr-A*05                           ---G-------A-----------------------C----------G-
Patr-A*06                           ---G-------A-----------------------C----------G-
Patr-A*07                           ---G-------A-----------------------C----------G-
Patr-A*08                           ---G-------A-----------------------C----------A-
Patr-A*09                           ---G-------A-----------------------C----------G-
Patr-A*10                           ---G-------A-----------------------C----------A-
Patr-A*11                           ---G-------A-----------------------C----------A-
Patr-A*12                           ---G-------A-----------------------C----------G-
Patr-A*13                           ---G-------A-----------------------C----------G-
Patr-A*14                           ---G-------A-----------------------C----------G-
Papa-A*01   ---G-C---C--C----------T----------A-----------------------C----------A-
Papa-A*02   ---G-C------------------G----------A-----------------------C----------A-
Papa-A*03                           --------------A---------A-----------C----------G-
Papa-A*04                           --------------A-----------------------C----------G-
Papa-A*05                           --------------A-----------------------C----------G-
Gogo-A*0101 ---G-C--------------------G----------A-----------------------C----------G-
Gogo-A*0201 ---G-C--------------------A---T-------------------------C----------G-
Gogo-A*0401 ---G-C--------------------G----------A-----------------------C----------G-
Gogo-A*0501 ---G-----G--------------------A-----------------------C----------G-
Gogo-OKO    ---G-----G--------------------A---------A-----------CG---G----G-
Popy-A*01   ---G--A-------------C----------A-----------------------C----------G-
Popy-A*02                           --------------A-----------------------C----------G-
Popy-A*03                           --------------A-----------------------C----------G-
Hyla-A*01                           --------------A---------A-----------C----------A-
Hyla-A*02                           --------------A---------A-----------C----------A-
Mamu-A*01                  --T------------------G------A---------T----A---------G-
Mamu-A*02                           --------G------A-----------------TC----C----A-
Patr-B*01   ----A-----C--------------G--------------C----------------------
Patr-B*02        C-------------------G--------------C----------------------
Patr-B*03   ----A-----C--------------G--------------C----------------------
Patr-B*04
Patr-B*05   ----A-----C--------------G--------------C----------------------
Patr-B*06                           G------------------------C----------------------
Patr-B*07                           G------------------------C----------------------
Patr-B*08                           ------------------G------AG----------------------
Patr-B*09                           G------------------------C----------------------
Patr-B*10                           G------------------------C----------------------
Patr-B*11
Patr-B*12
Patr-B*13
Patr-B*14
Patr-B*15
Papa-B*01                           G---------------------C--------------G----------
Papa-B*02                           G---------------------C--------------G----------
Papa-B*03                           G---------------------C--------------G----------
Papa-B*04                           G---------------------C--------------G----------
Gogo-B*0101 ----G-----C--------------------------C----------------------
Gogo-B*0102 ----G-----C--------------------------C----------------------
Gogo-B*0103 ----G-----C--------------------------C----------------------
Gogo-B*0201 ----A-----C--------------------------C----------------------
Popy-B*01                           ---------------------C----------------------
Popy-B*02                           ---------------------C--------C---------------
Popy-B*03                           -----------------A--T-----------------------
Hyla-B*01                           ---------------------C----------------G-----G-
Mamu-B*01                           -------------------------C-------------
Patr-C*01
Gogo-C*0101 ----G---------T------------A-----------A--A---------------
Gogo-C*0102 ----G---------T------------A-----------A--A---------------
Gogo-C*0201 ----G---------------------A----C----------A---------------
Gogo-C*0202 ----G---------------------A--------------A---------------
Gogo-C*0203 ----G---------------------A--------------A---------------
Popy-E*01                           -G--------TT-A--C-------A------A----T-------------G-
Mamu-E*01
Mamu-E*02
Mamu-E*03
Mamu-E*04
Mafa-E*01                           --A--..----CT-CT-T-.------T---A---------G-
Mafa-E*02                           --A--..----CT-CT-T-.------T---A---------G-
Patr-F*01   T--GG-----------------G--------------T---A-----------------T----A-
Mamu-F      T-.GG-------------------------G------A-----------------------G-
Saoe-F*01                           -----------------T--------A------------------G-
Patr-G*I
Gogo-G*I
Saoe-G*01                           ---------T-----C------------T-----T----------G-
Saoe-G*02                           ---------T------------------T-----T----------G-
Saoe-G*03                           ---------T------------------T-----T----------G-
Saoe-G*04                           ---------T------------------T-----T----------G-
Saoe-G*05                           ---------T------------------T-----T----------
Saoe-G*06   C--A----------T------------------------T-----T----------G-
Saoe-G*07                           ---------T------------------T-----T----------G-
Saoe-G*08                           ---------T------------------T-----T----------G-
Saoe-G*09                           ---------T------------------T--G--T----------G-
Saoe-G*10                           ---------T------------------T-----T----------G-
Saoe-G*11                           ------T------------------T-----T--C--------G-
Saoe-N1
Saoe-N3

Consensus   ATGCCGGTCATGGCGCCCCGAACCCTCCTCCTGCTGCTCTCGGGGGCCCTGGCCCTGACCGAGACCTGGGCCG
```

```
Exon 2

                 10        20        30        40        50        60        70        80        90       100
                 *    *    *    *    *    *    *    *    *    *    *    *    *    *    *    *    *    *    *    *
Consensus        GCTCCCACTCCATGAGGTATTTCTACACCTCCGTGTCCCGGCCCGGCCGCGGGGAGCCCCGCTTCATCGCCGTGGGCTACGTGGACGACACGCAGTTCGT
Patr-A*01        ---------------------T---A----------------------------------------------------------------------------
Patr-A*02        ---------------------T---A----------------------------------------------------------------------------
Patr-A*03        ---------------------C---A---G------------------------------------------------------------------------
Patr-A*04        ---------------------C---A---G------------------------------------------------------------------------
Patr-A*05        ---------------------C---A---G------------------------------------------------------------------------
Patr-A*06        ---------------------C---A---G------------------------------------------------------------------------
Patr-A*07        ---------------------C---A---G----------------------------------------------------A--------------------
Patr-A*08        ---------------------C---A---G----------------------------------------------------A--------------------
Patr-A*09        ---------------------C---A---G------------------------------------------------------------------------
Patr-A*10        ---------------------C---A---G----------------------------------------------------A--------------------
Patr-A*11        -----------------------------------------------------------------------------------------------------
Patr-A*12        ---------------------C---A---G------------------------------------------------------------------------
Patr-A*13        ---------------------C---A---G------------------------------------------------------------------------
Patr-A*14        ---------------------C---A---G------------------------------------------------------------------------
Papa-A*01        ---------------------T--------------------------------------------------------------------------------
Papa-A*02        ---------------------C---A---G----------------------------------------------------A--------------------
Papa-A*03        ---------------------T---A----------------------------------------------------------------------------
Papa-A*04        ---------------------T--------------------------------------------------------------------------------
Papa-A*05        ---------------------C---A----------------------------------------------------------------------------
Gogo-A*0101      ---------------------C---A----------------------------------------------------------------------------
Gogo-A*0201      ---------------------C---A----------------------------------------------------------------------------
Gogo-A*0401      -----------------------------------------------------------------------------------------------------
Gogo-A*0501      ---------------------C---A---------------T-----------------------A------------------------------------
Gogo-OKO         --------------------A---A---------------------------------------T---C---------------T------------------
Popy-A*01        -------------------AG----------------------------------------------A----------------------------------
Popy-A*02        ----------T--------------------------------------------------------A----------------------------------
Popy-A*03        ---------------------T----G------------------------------------------A----------------------------------
Hyla-A*01        -------------------AG----A----------------A--T----------G-A-C---AA--T--------------------------------
Hyla-A*02        -------------------AG----A----------------A-AT---------G-A-C---AA--C---------------------------------
Mamu-A*01        ----T---------A----------A-----------------------C----------------------------------------C----------
Mamu-A*02        ----T---------------A----------------------T----------------------------------------------C----------
Patr-B*01        -----------------A----------------------------------------A--------------------------------C----------
Patr-B*02        -------------------------G------------------------------T-A-------------------------------------------
Patr-B*03        -----------------G----------------------------------------A--------------------------------C----------
Patr-B*04        ------------------------------------------------------------------------------------------------   --
Patr-B*05        ----------A----------------G---------------------------T-A--------------------------------------------
Patr-B*06        -------------------------G------------------------------T-A--------------------------------C----------
Patr-B*07        ----------A-----TA--A--------------------------------------A--------------------------------C----------
Patr-B*08        ---------------G--A-------------G---------------------A--A----T---------C---------------------------
Patr-B*09        ----------------------------------------------------------A--------------------------------C----------
Patr-B*10        ----------------------------------------------------------A--------------------------------C----------
Patr-B*11        -----------------------------------------------------------------------------------------------------
Patr-B*12        -----------------------------------------------------------------------------------------------------
Patr-B*13        -----------------------------------------------------------------------------------------------------
Patr-B*14        -----------------------------------------------------------------------------------------------------
Patr-B*15        ----------------------------------------------------------T-A-----------------------------------------
Papa-B*01        ---------------------------------------------------------T-A-----------------------------------------
Papa-B*02        --------------------------------------GC---------------T-A-----------------------------------------
Papa-B*03        -----------------G----------G-------------------GC---------A--------------------------------------------
Papa-B*04        ----------------------------------------------------------T-A-----------------------------------------
Gogo-B*0101      -----------------G----------G-----------------------------A--------------------------------C----------
Gogo-B*0102      -----------------G----------G-----------------------------A--------------------------------C----------
Gogo-B*0103      -----------------G----------G-----------------------------A--------------------------------C----------
Gogo-B*0201      ---------------------C------G--A-----------------A--------A------------------------------------------
Popy-B*01        ---------------A--------GGG--A--------------------------T----------------------------------------------
Popy-B*02        ----------A------------GGG--A--------------------------T----------------------------------------------
Popy-B*03        -----------------G-------------------------------------------A-----------------------T----------------
Hyla-B*01        ----------------------------------------------------------A--------------------------------C----------
Mamu-B*01        ------G-------------------------------------------------T--------------------------------------   ----
Patr-C*01        -----------------------------------------------------------------------------------------------------
Gogo-C*0101      -----------------T----G-----G---C----------A-------------A--------------------------------------------
Gogo-C*0102      -----------------T----G-----G---C----------A-------------A--------------------------------------------
Gogo-C*0201      ----------------------------------------------------------T-A-----------------------------------------
Gogo-C*0202      ----------------------------------------------------------T-A-----------------------------------------
Gogo-C*0203      ----------------------------------------------------------T-A-----------------------------------------
Popy-E*01        -----------T---A--------C----T-----------------------------T-T----------------------------------------
Mamu-E*01        ------------------C----T------------------------------G-----T----------------------------C------------
Mamu-E*02        ------------------C----T----------------G------------------T----------------------------C------------
Mamu-E*03        ------------------C----T------------------------------G-----T----------------------------C------------
Mamu-E*04        ------------------C----T-----------------------------------T----------------------------C------------
Mafa-E*01        -----------T---A--------C----T-----------------------------G-----T----------------------------C------------
Mafa-E*02        -----------T---A--------C----T-----------------------------G-----T----------------------------C------------
Patr-F*01        -----------T----------AG---G-T----G-----------------------G-A----------AG----A-----------------C-
                                                   AGTACA
                                                     |
Mamu-F           -----------T--C-------TAG---TG-T-----G----------A---------T--G-A----G---A--------------------------C-
                                                   GATACA
                                                     |
Saoe-F*01        -----------T----------AGT--TG--A----------T-----------------G-A-----GA-------------T------T-----
Patr-G*I         ------------------G---------------------------_------------------------A-----------------------------
Gogo-G*I         ------------------G--------------------------------------------------A--------------------------------
Saoe-G*01        ------------------AG----G------------T-----A--------------AA----------------------------------------
Saoe-G*02        ------------------G-A------------------------------------G-A---T-------------------------------------
Saoe-G*03        ------------------A-A--------------------------------------G-A---T-------------------------------------
Saoe-G*04        ------------------G-A------------------------------------G-A---T-------------------------------------
Saoe-G*05        ------------------A-A----------------------------T-----------------------------------------------------
Saoe-G*06        ------------------AG----G--------------G-------A---------G-A-----AA--C-------------------------------
Saoe-G*07        ------------------AG---A--------------------------------G-A-----AA--C-------------------------------
Saoe-G*08        ------------------AG---A-----------T--------------------G-A-----AA--C-------------------------------
Saoe-G*09        ------------------A-A--------------------------------------G-A---T------------------------------------
Saoe-G*10        ------------------AG---G--A-----------G-----------------G-A-----AA--C-------------------------------
Saoe-G*11        ------------------G-A------------------------------------G-A-----AA--C-------------------------------
Saoe-N1          ------------------G----T----------------------T------------------T-------------------------------------
Saoe-N3          ----------GT-----------------------------T-------------G-A-----AA--C----------------T-----------

Consensus        GCTCCCACTCCATGAGGTATTTCTACACCTCCGTGTCCCGGCCCGGCCGCGGGGAGCCCCGCTTCATCGCCGTGGGCTACGTGGACGACACGCAGTTCGT
```

```
                 110       120       130       140       150       160       170       180       190       200
                  *    *    *    *    *    *    *    *    *    *    *    *    *    *    *    *    *    *    *
Consensus        GCGGTTCGACAGCGACGCCGCGAGTCCGAGGATGGAGCCGCGGGCGCCGTGGATAGAGCAGGAGGGGCCGGAGTATTGGGACCGGGAGACACAGAACTCC

Patr-A*01        --------------------C-A--------------------------------------T--------------A-------G--GTG-G
Patr-A*02        --------------------C-A--------------------------------------T--------------GA------G--GTG-G
Patr-A*03        --------------------C-A--------------------------------------T--------------A-------G--TATG
Patr-A*04        --------------------C-A--------------------------------------T--------------GA------G--GTGTG
Patr-A*05        --------------------C-A--------------------------------------T--------------GA------G--GTGTG
Patr-A*06        --------------------C-A--------------------------------------T--------------A-------G--GTGTG
Patr-A*07        --------------------C-A----------------A-------------------T-----------------------G--GTGTG
Patr-A*08        --------------------C-A-----------------A-------------------------------------A----C-G--G-GT-
Patr-A*09        --------------------C-A--------------------------------------T--------------GA------G--GTGTG
Patr-A*10        --------------------C-A--------------------------------------T--------------A-------G--TATG
Patr-A*11        --------------------C-A--------------------------------------T--------------A-------G--TATG
Patr-A*12        --------------------C-A--------------------------------------T--------------GA------G--GTGTG
Patr-A*13        --------------------C-A--------------------------------------T--------------GA------G--GTGTG
Patr-A*14        --------------------C-A--------------------------------------T--------------GA------G--GTGTG
Papa-A*01        --------------------C-A--------------------------------------T--------------A-------G--TGTG
Papa-A*02        --------------------C-A----------------A---------------------T--------------A----C-G--G-GT-
Papa-A*03        --------------------C-A--------------------------------------T--------------A-------G--GTGTG
Papa-A*04        --------------------C-A--------------------------------------T--------------A----C-G--G-GT-
Papa-A*05        --------------------C-A--------------------------------------T--------------A-------G--GTGTG
Gogo-A*0101      --------------------C-A--------------------------------------------------A-C----G----G--GTG
Gogo-A*0201      ------T-------------C-A---------------------------------------A---------------G------G--AGTG
Gogo-A*0401      --------------------C-A---------------------------------------A---------------G------G--TATG
Gogo-A*0501      --------------------C-A--------------------G-------------------A---------------C------T----
Gogo-OKO         ---------------A-----------AGA---------A--------G---G-----------------------A-C------T--A-
Popy-A*01        --------------------C-A---------------A----------------------------------------G-G-GTG
Popy-A*02        --------------------C--------------------A-----------------------------------G--G-G-G
Popy-A*03        --------------------C--------------------A-----------------------------------G--G-G-G
Hyla-A*01        ------------T---T------C-------------------------T----G-G--------------------A-A-C---G------
Hyla-A*02        ---------------T------C---------------------T----G-G--------------------A-C---------
Mamu-A*01        ------------------C-AA------------------------G-G-------------------------------G---ATG
Mamu-A*02        --------------------C-A----------------------G-G---------------T--A-------------G---ATG
Patr-B*01        -A-----------------------------------------C-----A--------------------------------G--TATG
Patr-B*02        -T----------------------AGA------------------A--------------------------A---------T----
Patr-B*03        -A-------------------------------------C-----A------------------------------------G--GTG
Patr-B*04        -T------------------------------------A---A------------------------------A---------T----
Patr-B*05        -T----------------------AGA------------------A------------------------------------T----
Patr-B*06        -A-----------------------------------C-------A--------------------------------G--GTG
Patr-B*07        -A-----------------------------------C-------A--------------------------------G--GTG
Patr-B*08        -A--------------T-------------------------A--------------------------------G--TATG
Patr-B*09        -A-----------------------------------C-------A--------------------------------G--TATG
Patr-B*10        -A-----------------------------------C-------A--------------------------------G--TATG
Patr-B*11        --------------------------AGA----------------G-----------------T-----------A-C------T-A-
Patr-B*12        --------------------------AGA----------------G-----------------T-----------A-C------T-A-
Patr-B*13        -----------------CAA------------------G------A---------A---------------A-----------T----
Patr-B*14        ------------------T---------------------A-----------------------------------A-C----GG-AT-
Patr-B*15        --------------------------AGA----------------A--------------------------A-----------T----
Papa-B*01        -A------------------------AGA----------------G-----------------T-----------A-C------T-G-
Papa-B*02        -A------------------------AGA----------------G-----------------T-----------A-C------T-G-
Papa-B*03        -A------------------------AGA----------------G-----------------T-----------A-C------T-G-
Papa-B*04        -A------------------------AGA----------------G-----------------T-----------A-C------T-A-
Gogo-B*0101      -A-----------------------------------C-------A--------------------------------A--------C----
Gogo-B*0102      -A-----------------------------------C-------A--------------------------------A--------C----
Gogo-B*0103      -A-----------------------------------C-------A--------------------------------A--------C----
Gogo-B*0201      -T--------------------------A--------A-------A--------------------------------A--------T----
Popy-B*01        -T--------------------------A----------------A--------------------------------A-C------T-G-
Popy-B*02        -T----------------T-------AGA----------------A--------------------------A-C------T-A-
Popy-B*03        ----------------------------------------T--A-C---------------------------A---------AG--
Hyla-B*01        ----------------------------------------------------------------------A-C----------T----
Mamu-B*01        ---A---------------A----G-------------------C----------------------------------G---GG--
Patr-C*01        ----------------------A-AGG---------------G-G-------------------C------------------G-A-
Gogo-C*0101      ----------------A-A---AGG---------------G-G----------------------------------------G-A-
Gogo-C*0102      ----------------A-A---AGG---------------G-G----------------------------------------G-A-
Gogo-C*0201      ----------------A-A---AGG---------------G-G----------------------------------------G-A-
Gogo-C*0202      ----------------A-A---AGG---------------G-G----------------------------------------G-A-
Gogo-C*0203      ----------------A-A---AGG---------------G-G----------------------------------------G-A-
Popy-E*01        ---C--T----A-------------T--------A---------A---------------A-------------------G--G-G--
Mamu-E*01        ----AT----------------A----------------G-G-------------------A--------------A-------G--G-G--
Mamu-E*02        ----AT----------------G---------------T-----GA-------------------------------A-------G--G-G--
Mamu-E*03        ----A----------------A--------------T-----GA-------------------------------A-------G--G-G--
Mamu-E*04        ----A----------------A--------------T-----GA-------------------------------A-------G--G-G--
Mafa-E*01        ----AT----------------A----------------G-G-------------------A--------------A-------G--G-G--
Mafa-E*02        ----AT----------------A----------------G-G-------------------A--------------A-------G--G-G--
Patr-F*01        -----------------T--------------T---A--G-G----A-------------------------------A--------G-G--
Mamu-F           ------------------T----------------A---G-G--A-----A---------C-------A---ACC---GG-T--G-G
Saoe-F*01        -A-------------A-C-------------------G-G--A----------A-----------GGACC------GG-T--G-G
Patr-G*I         -T-----------------T-G---T--------------G-G------------------------AGA------G----A--
Gogo-G*I         -------------------T-G---T--------------G-G------------------------AG-------G----A--
Saoe-G*01        ---------------C--------G----------------G-------------------GGAA------T---AG--
Saoe-G*02        ---------------------A-----------------G----------------------GGA------G---AGG-
Saoe-G*03        ----------------A-------------------G--------------T-------GGA------G--CAG--
Saoe-G*04        ----------------A-------------------G----------------------GGA------G---AGG-
Saoe-G*05        ------------------G----------------G----------------------GGA------G---AGT-
Saoe-G*06        --------------------------------G--A--G-G---A---------G----------GGA------GAG--
Saoe-G*07        ---------T---A-----------------------G--------------------GGA------GAG--
Saoe-G*08        ------------------------------G--------------------------T--T-C-----G-------
Saoe-G*09        ----------------A-------------------G---------------------C-----GAA------G--CAG--
Saoe-G*10        ------------------------------G--G---A---------------G----------GGA------GAG--
Saoe-G*11        ----------------------------------G-------------------C-----GAA------G--GAG--
Saoe-N1          --------------T----------------AGA-----------A---G-T---------------------------T-C--------T----
Saoe-N3          ---------------A---A-A--------------------G-G---A-----------G---------------A-T---G--GAG--

Consensus        GCGGTTCGACAGCGACGCCGCGAGTCCGAGGATGGAGCCGCGGGCGCCGTGGATAGAGCAGGAGGGGCCGGAGTATTGGGACCGGGAGACACAGAACTCC
```

```
                 210       220       230       240       250       260       270
                  *    *    *    *    *    *    *    *    *    *    *    *    *
Consensus     AAGGCCCACGCACAGACTGACCGAGTGAACCTGCGGACCCTGCGCGGCTACTACAACCAGAGCGAGGCCG


Patr-A*01     ---------T----------------G----G----------------------------------A--
Patr-A*02     ---------T----------------G----G----------------------------------A--
Patr-A*03     ------TC------------------G----G----------------------------------A--
Patr-A*04     ------TC------------------G----G----------------------------------A--
Patr-A*05     ------TT------------------G----G----------------------------------A--
Patr-A*06     ------TC------------------G----------------------------------------A--
Patr-A*07     ------GG--------TCT-------C----G----------------------------------A--
Patr-A*08     ------GG--------TCT-------C----G----------------------------------A--
Patr-A*09     ------TC------------------G----G----------------------------------A--
Patr-A*10     ------TT--------TCT-------C----G----------------------------------A--
Patr-A*11     ------TC------------------C----G----------------------------------A--
Patr-A*12     ------TC------------------C----G----------------------------------A--
Patr-A*13     ------TT------------------C----G----------------------------------A--
Patr-A*14     ------TT------------------C----G----------------------------------A--
Papa-A*01     ---------T----------------G----------------------------------------A--
Papa-A*02     ------GG--------TCA-------C----G----------------------------------A--
Papa-A*03     ------G---T---------------G----G----------------------------------A--
Papa-A*04     ------GG--------TCT-------C----G----------------------------------A--
Papa-A*05     ---------T----------------G----G----------------------------------A--
Gogo-A*0101   ---------T----------------G----G-----------------------------------A--
Gogo-A*0201   ---------T----------------G----G--------------------------------------
Gogo-A*0401   ------G-T-----------------G----G----------------------------------A--
Gogo-A*0501   ---A--A---------T---G-T--A--G-------T-GC--T-C-------T--------------A--
Gogo-OKO      ---------A-----------------G-T--GA--------------------------------G--
Popy-A*01     ------------------A---------G----G---G------------------------C-G--
Popy-A*02     --------------------------G----G---G-------------------------C-G--
Popy-A*03     --------------------------G----G--A-------------------------C-G--
Hyla-A*01     ------GG------CT----------G----AG-------------------------------
Hyla-A*02     ------GG------CT----------G----AG-----------C-------------------
Mamu-A*01     ---A--G-GA-----A--C--C----------------T-C----------------------
Mamu-A*02     ------G-GA-----A--C--C----------------A------------------------
Patr-B*01     ------TC------------------A----------T-GC--T-C-----------------
Patr-B*02     ---A--A--------T------A--G-------------T-C---------------------
Patr-B*03     ------TC------------------A----------T-GC--T-C-----------------
Patr-B*04     ---A--A--------T------A--GG------------T-C---------------------
Patr-B*05     ---A--A--------T------A------A--------------------------------
Patr-B*06     ------TC-----------------G--G----GC-A--------------------------
Patr-B*07     ------TC---------T------A-----------T-GC--T-C----T---TT----------A--
Patr-B*08     ---A--A--------T------A--GG------------T-C---------------------
Patr-B*09     ------TC------------------A----------T-GC--T-C-----------------
Patr-B*10     ------TC------------------A----------T-GC--T-C-----------------
Patr-B*11     ---------G---------------GG----G--A----------------------------
Patr-B*12     --------G----------------G--G----G--A--------------------------
Patr-B*13     --A-T-T---------A---T---------G----------T---------------------
Patr-B*14     ---A-G--G----------------A------------T-C----------------------
Patr-B*15     ---A--A--------T------A------A--------------------------------
Papa-B*01     --------G----------------A--G----G--A--------------------------
Papa-B*02     --------G----------------A----------T-GC--T-C------------------
Papa-B*03     --------G----------------A----------T-GC--T-C------------------
Papa-B*04     --------G----------------A--G--G---A--------------------------
Gogo-B*0101   -----T--G----------------A----------T-GC--T-C-----------------
Gogo-B*0102   -----T--G----------------A----------T-GC--T-C-----------------
Gogo-B*0103   -----T--G----------------A----------T-GC--T-C-----------------
Gogo-B*0201   ---A--A--A-------T------GG----G--------------------------------A--
Popy-B*01     ---A--A--A-------T------A--GG-----T-----------------A-------A--
Popy-B*02     ------A--A-------T------A--G--A------T-----------------------A--
Popy-B*03     ------TC------------------G--------------A-------------------A--
Hyla-B*01     ---A--A------------------A----------T-GC--T-C------------------A--
Mamu-B*01     ---G-A-------------------A----------T-GC--T-A----------------A-G--
Patr-C*01     ---CG---G-------G---------G--------A---------------------------A--
Gogo-C*0101   ---CG---G----------------G--------AA-------------------------A--
Gogo-C*0102   ---CG---G----------------G--------AA-------------------------A--
Gogo-C*0201   ---CG---G------------------------AA-------------------------A--
Gogo-C*0202   ---CG---G------------------------AA-------------------------A--
Gogo-C*0203   ---CG---G------------------------AA-------------------------A--
Popy-E*01     -G--A-AC----------TT--------------------------T----C---------
Mamu-E*01     -G--A-AC----------TT--------GA--------------------------------
Mamu-E*02     -G----AC----------TT--------GA--------------------------------
Mamu-E*03     -G--A-AC----------TT--------GA--------------------------------
Mamu-E*04     -G--A-AC----------TT--------AA--------------------------------
Mafa-E*01     -G--A-AC----------TT--------GA--------------------------------
Mafa-E*02     -G--A-AC----------TT--------GA--------------------------------
Patr-F*01     ------A---------------GC----A--A----T-C--CG-------------------
Mamu-F        ------A-------G-------GC----A--AG---T-CT-CG-------------------
Saoe-F*01     ------A-------------------A----AA----T-C---------T--------G--
Patr-G*I      -----------------------A--A-A----A---------------------------
Gogo-G*I      ---------------------A--A-------A-----------------------------
Saoe-G*01     ----ATTGG---------T-------------A----T--T--------------------
Saoe-G*02     ------G-----------TT------GG------A-------T------------------
Saoe-G*03     ------GC----------TT------------A--------T-------------------
Saoe-G*04     ------G-----------TT------G-------A-------T------------------
Saoe-G*05     -----------------T------A-G------A--------------------C----A-
Saoe-G*06     ------TT----------TT--------C-A----GC--T---------------------
Saoe-G*07     ------TTG---------TT-------------A----T-C------------G--T--
Saoe-G*08     ------TC----------TT-------------A----T--T-------------------
Saoe-G*09     ------GC----------TT-------------A--------T------------------
Saoe-G*10     ------TT----------------------A----GC--T---------------------
Saoe-G*11     ------GC----------TT-------------A--------T------------------
Saoe-N1       ------AT----------C-G--TA--------A--------------G------------A-
Saoe-N3       ------A-----------CT------GG----A----------------------------

Consensus     AAGGCCCACGCACAGACTGACCGAGTGAACCTGCGGACCCTGCGCGGCTACTACAACCAGAGCGAGGCCG
```

Exon 3

```
                 10        20        30        40        50        60        70        80        90       100
                 *    *    *    *    *    *    *    *    *    *    *    *    *    *    *    *    *  .  *  .  *
Consensus     GGTCTCACACCATCCAGAGGATGTATGGCTGCGACGTGGGGCCGGACGGGCGCCTCCTCCGCGGGTATCACCAGTACGCCTACGACGGCAAGGATTACAT

Patr-A*01     -T---------------TA----------------T----------T-----------C-GG---G-------------------------------
Patr-A*02     -T---------------TA----------------T----------T-----------C-GG---G-------------------------------
Patr-A*03     -T---------------TA----------------T----------T-----------C-GG---G-------------------------------
Patr-A*04     -T--------------TTA----T-----------T----------T-----------C-GG---G-------------------------------
Patr-A*05     -T--------------TTA----------------T----------T-----------C-GG---G-------------------------------
Patr-A*06     -T---------------TA----------------T----------T-----------C-GG---G-------------------------------
Patr-A*07     -T--------------TTA----T-----------T----------T-----------G-A---CT-------------------------------
Patr-A*08     -T---------------TA----T-----------T----------T-----------G-A---CT-------------------------------
Patr-A*09     -T---------------TA----------------T----------T-----------C-GG-----------------------------------
Patr-A*10     -T--------------TTA----T-----------T----------T-----------G-A-----C-------------------------------
Patr-A*11     -T---------------TA----------------T----------T-----------C-GG---G-------------------------------
Patr-A*12     -T--------------TTA----T-----------T----------T-----------G-A-----C-------------------------------
Patr-A*13     -T--------------TTA----------------T----------T-----------C-GG-----------------------------------
Patr-A*14     -T--------------TTA----T-----------T----------T-----------G-A-----C-------------------------------
Papa-A*01     -T---------------TA----------------T----------T-----------C-GG---G-------------------------------
Papa-A*02     -T---------------TA----T-----------T----------T-----------G-G-----T-------------------------------
Papa-A*03     -T---------------TA----------------T----------T-----------C-GG---G-------------------------------
Papa-A*04     -T--------------TTA----T-----------T----------T-----------C-GG---G-------------------------------
Papa-A*05     -T--------------TTA----------------T----------T-----------C-GG---G-------------------------------
Gogo-A*0101   -T---------------------------------T----------T-----------C-G---G-------------------------------
Gogo-A*0201   -T------------------A--------------T----------T-----------C-G---G-------------------------------
Gogo-A*0401   -T---------------------------------T----------T-----------C-G---G-------------------------------
Gogo-A*0501   -T---------------------------------T----------T-----------C-G---G-------------------------------
Gogo-OKO      -----------------A------------A---------------T-----------C-TG---G---T----------------------------
Popy-A*01     -T--------------------T------------C----------T-----------G-A---C-------------------------------
Popy-A*02     -T---------------------------------C----------T-----------C-GG-----------------------------------
Popy-A*03     -T---------------------------------C----------T-----------CTGG-----------------------------------
Hyla-A*01     -T--------------TTA---------T------C----------T-----------C-GG---G----------T---------------------
Hyla-A*02     -T----T----------TA--------A-------C----------T-----------C-GG---G----------T---------------------
Mamu-A*01     -----------C-----------GT------C--------C-----------------G-A-------------------------------------
Mamu-A*02     -----------------------C------C----------------------------C-------------------------------------
Patr-B*01     ----------TGG-----C--------A------------------------------GG--------------------------------------
Patr-B*02     -T------T--------C--------A-------------------------------G--------------------------------------
Patr-B*03     --------TC-----C-----------------------T------------------C-GG---T--------------------------------
Patr-B*04     --------TGG-----C-----T----A------------------------------C-GG---T--------------------------------
Patr-B*05     -T-------T-------C----------------A-----------------------G-G-------------------------------------
Patr-B*06     --------TGG-----C---------A-------------------------------G---------------------------------------
Patr-B*07     --------TGG-----C-----T----A------------------------------T---------------------------------------
Patr-B*08     --------TGG-----C---------A-------------------------------C-GG---T--------------------------------
Patr-B*09     -T------T-------C---------A-------------------------------T---------------------------------------
Patr-B*10     --------TGG-----C---------A-------------------------------GG--------------------------------------
Patr-B*11     --------TGG-----C-----------------------------------------G---------------------------------------
Patr-B*12     ----------C-------------------------------------------------T---G---------------------------------
Patr-B*13     -T-----------TTA-----------------C-----A---T---T----------C-GG------------------------------------
Patr-B*14     --------TC-----TC-----------A-----------------------------G---------------------------------------
Patr-B*15     -T------T-------C---------A-------------------------------G-G-------------------------------------
Papa-B*01     ----------C-----C-------------------------------T---------C-GG---T--------------------------------
Papa-B*02     ----------C-----C---------------------------------------T-C-GG---T--------------------------------
Papa-B*03     ----------C-----C--------------------------------------AG----C-----------------------------------
Papa-B*04     ----------C-----C---------------------------------------T-G-G-----T-------------------------------
Gogo-B*0101   ----------------------T----------------------------------AG----C-----------------------------------
Gogo-B*0102   ----------T------------T----------------------------------AG----C-----------------------------------
Gogo-B*0103   ----------------T--------------A--------------------------AG----C-----------------------------------
Gogo-B*0201   ----------------------------A----------------------------AG-----TG---------------------------C-
Popy-B*01     ----------TGG---------------------C---------------------G-----T-----------------------------------
Popy-B*02     ----------TGG----C-----T----------C---------------------G-----T-----------------------------------
Popy-B*03     ----------TGG---------T----------C----------------------G-----------------------------------------
Hyla-B*01     ----------C-----C----------C------------------------------G---------------------------------------
Mamu-B*01     ----------C-----T----C-C----C-----A--C---------------------T---CG---------------------------------
Patr-C*01     ----------TC----------------------C----------------------G---------------T------------------------
Gogo-C*0101   ----------T------------------------C----------AA----------G---TG---------T------------------------
Gogo-C*0102   ----------T------------------------C----------AA----------G---TG---------T------------------------
Gogo-C*0201   ----------TC----TC-----------------C----------------------AG----T--------------------------------
Gogo-C*0202   ----------TC----TC-----------------C----------------------AG----T--------------------------------
Gogo-C*0203   ----------TC----TC-----------------C----------------------AG----T--------------------------------
Popy-E*01     ----------C-G---T-----C------------C-----------T----------G-A---T--------------------------TC-
Mamu-E*01     -------T---C-----T-----C------------C-----------T----------G-A---T--------------------------TC-
Mamu-E*02     -------T---C-----T-----C------------C-----------T----------G-A---T--------------------------TC-
Mamu-E*03     -------T---C-----T-----C------------C-----------T----------G-A---T--------------------------TC-
Mamu-E*04     -------T---TC----T-----C------------C-------------T--T-----G-A---TT------------------------TC-
Mafa-E*01     -------T---C-----T-----C------------C----AT-C-------------G-A---T--------------------------TC-
Mafa-E*02     -------T---C-----T-----C------------C----AT-C-------------G-A---T--------------------------TC-
Patr-F*01     ----------C-----G-A---A------------A------C-----A---------T----------G---------------------------
Mamu-F        ----------C-----G-A---A-C-----------A------C-----A-------------------------C-----------------------
Saoe-F*01     ----------C-----G-A-C-A-----------T----X--C-T-A--T--------A---------C--------------------C-----
Patr-G*I
Gogo-G*I
Saoe-G*01     ----------TT-G---T------C---------C-----------------------G-G----------------------------C-----
Saoe-G*02     --------------T------C---------C--------------------------G----C--------------------------C-----
Saoe-G*03     --------------T------C---------C--------------------------G----C--------------------------C-----
Saoe-G*04     --------------T-----------------C-------------------------G----C--------------------------C-----
Saoe-G*05     --------------GT-----------------C------------------------G----C--------------------------C-----
Saoe-G*06     --------T------T-----------------C------------------------G----C--------------------------C-----
Saoe-G*07     --------------T------C------T--C--------------------------T----C--------------------------C-----
Saoe-G*08     ---T----------T------C---------C--------------------------T----C--------------------------C-----
Saoe-G*09     --------------T------C---------C--------------------------G-G--C--------------------------C-----
Saoe-G*10     --------------T------C-----------------------T------------G-A---TA------------------------C-----
Saoe-G*11     --------------T------C---------C--------------------------G-G--C--------------------------C-----
Saoe-N1       -----------TA-------------------C------T-C---A---------------AG-----------TT--T--------C----C
Saoe-N3       A-----TT----------TA-C---C--------------.-G-----T------T-----C-GA---G--------T---------------------

Consensus     GGTCTCACACCATCCAGAGGATGTATGGCTGCGACGTGGGGCCGGACGGGCGCCTCCTCCGCGGGTATCACCAGTACGCCTACGACGGCAAGGATTACAT
```

```
                 110       120       130       140       150       160       170       180       190       200
                 *    *    *    *    *    *    *    *    *    *    *    *    *    *    *    *    *    *    *    *
Consensus        CGCCCTGAACGAGGACCTGCGCTCCTGGACCGCGGCGGACACGGCGGCTCAGATCACCCAGCGCAAGTGGGAGGCGGCCCGTGTGGCGGAGCAGCTGAGA

Patr-A*01        ----------------------T--------------T---A----------A---------------------A--C---------A----
Patr-A*02        ---T------------------T--------------T---A----------A---------------------A--C---------G----
Patr-A*03        ----------------------T----------C--T--A----------A---------------------A--C---------A----
Patr-A*04        ----------------------T--------------T------------A---------------------A--C---------A----
Patr-A*05        ----------------------T--------------T-----------------------------------A--C---------TG---
Patr-A*06        ----------------------T--------------T-----------------------------------A--C---------TG---
Patr-A*07        ----------------------T--------------T--------------------------------T--A--C---------A----
Patr-A*08        --------T-------------T--------------T-------------T-----------------------A--C---------A----
Patr-A*09        ----------------------T--------------T-----------------------------------A--C-------G------
Patr-A*10        ----------------------T--------------T-----------------------------------A--C---------TG---
Patr-A*11        ----------------------T--------------T---A----------A---------------------A--C---------G----
Patr-A*12        ----------------------T--------------T-----------------------------------A--C---------A----
Patr-A*13        ----------------------T--------------T-----------------------------------A--C---------A----
Patr-A*14        ----------------------T--------------T--------------T--------------------A--C---------A----
Papa-A*01        ----------------------T--------------T---A----------A---------------------A--C---------G----
Papa-A*02        ----------------------T--------------T-------------T---------------------A--C---------AC---
Papa-A*03        ----------------------T--------------T-------------T---------------------A--A---------TG---
Papa-A*04        ----------------------T--------------T-------------T---------------------A--A---------AC---
Papa-A*05        ----------------------T--------------T---A----------A---------------------A--A---------G----
Gogo-A*0101      ---T-----------------T--------------T--A---G---T--A--------------------A-T-T--------T------
Gogo-A*0201      ---T-----------------T--------------T--A----------------------------------A-T-T--------T------
Gogo-A*0401      ---T-----------------T--------------T--A----------------------------------A--A--------T------
Gogo-A*0501      ---TC----------------T--------------T--A---G---T--A--------------------A-T-T--------T------
Gogo-OKO         -A-------------------T--------A----T-------------------------------------A---------G-T----
Popy-A*01        ------A--------------------------------T---------------------------------G---C---------GAC--
Popy-A*02        ------A---G----------------------------T---------------------------------G---A---------G----
Popy-A*03        ------A---G----------------------------T-------------------------T----G---C---------GAC--
Hyla-A*01        -T-----------------T----G------GT-----------A--------------------------G---------------TAC--
Hyla-A*02        -T-----------------T-------------T----------A--------------------------G--------------GAC--
Mamu-A*01        ------------------------------GT----------A-----------------------GGA---------AGCA--
Mamu-A*02        ------------------C-------T----------A---------------------------GG--A---------AC---
Patr-B*01        -----------------A-----------C-----------------------------------------TG-----
Patr-B*02        -----------------A-----------C-------------------------------------C----
Patr-B*03        -----------------A---------------------------------------------------C----
Patr-B*04        ----------C------A-----------C---------------------------------------------
Patr-B*05        ----------C------A-----------C-----------------------------------------TG----
Patr-B*06        -----------------A---------------------------------------------------TG---
Patr-B*07        -----------------A-----------------------------------------------------AC---
Patr-B*08        --------A--------A-----------C---C-----------------------------TG---C------
Patr-B*09        -----------------A-----------C-----------------------------------------G----
Patr-B*10        -----------------A-----------C---T----------------------------------A-----G----
Patr-B*11        -----------------A-----------------------------------------------------A------G----
Patr-B*12        -----------------A-----------------------------------------------------AC---
Patr-B*13        ------A----------A-----------T------G------A----------------------CG--------T----
Patr-B*14        --------A--------A-----------C-----------------------------------------G----
Patr-B*15        ----------C------A-----------C-----------------------------------------TG----
Papa-B*01        -----------------A-----------------------------------------------A--A---------A----
Papa-B*02        -----------------A-----------------------------------------------------A----
Papa-B*03        -----------------A----------------------------------------------------------
Papa-B*04        -----------------A-----------------------T----------T--------------C-----------TG---
Gogo-B*0101      -----------------A-----------C-----------------------------------------A----
Gogo-B*0102      -----------------A-----------C-----------------------------------------A----
Gogo-B*0103      -----------------A-----------C-----------------------------------------A----
Gogo-B*0201      -----------------A-----------C-----------------------------------C---------GA---
Popy-B*01        -----------------A----C------C----------------------------------------A------T-C--
Popy-B*02        -----------------A----C------C-----------------------------------G--TG--------T-C--
Popy-B*03        -----------------A----T------C---------------------------------CG---TG---------T----
Hyla-B*01        -----------------A----C------C--------------------------------------A--A------A----
Mamu-B*01        --------------A-----C------TT---------A------------G---------------G------A----G----
Patr-C*01        -----------------C----------------------------------T-----------------C----
Gogo-C*0101      -----------------C-------------------------------------------A---TG-------G--A---
Gogo-C*0102      -----------------C--------------------------------------A---TG-------G--A---
Gogo-C*0201      -----------------C-------------G--A--------------------------------------GA---
Gogo-C*0202      -----------------C--------------------------T-----------------C---------A----
Gogo-C*0203      -----------------C-------------G--A--------------------------------------GA---
Popy-E*01        -A------T----------------T------------------T-G-----A---CAA-T--AT---T---A---------C-A---
Mamu-E*01        -A------T----------T------T------------A---T-G-----AA---CAA-T-AT-G-TC--A---T-----C-A---
Mamu-E*02        -A------T----------T------T------------A---T-G-----AA---CAA-T-AT-G-TC--A---T-----C-A---
Mamu-E*03        -A------T----------T------T------------A---T-G-----AA---CAA-T-AT-G-TC--A---T-----C-A---
Mamu-E*04        -A------T----------T------T------------A---T-G-----AA---CAA-T-AT-G-TC--A---T-----C-A---
Mafa-E*01        -A------T----------T------T------------A---T-G-----AA---CAA-T-AT-G-TC--A---T-----C-A---
Mafa-E*02        -A------T----------T------T------------A---T-G-----AA---CAA-T-AT-G-TC--A---T-----C-A---
Patr-F*01        -T--------------------------------C--T-------------------TTC-AT-----A-AGGAATAT--A---G--T-C--G
Mamu-F           -T--------------------------C---------TA---G------------TTC-AT-----A-AGGAATAT--A---G--T-C--G
Saoe-F*01        -T------------------------A----A--------------------T--AT---T-A-AGAAATAT--A---G--T-C--C
Patr-G*I
Gogo-G*I
Saoe-G*01        ---------------------G--C-----GT------------------------------AA--C---T---GG-A----
Saoe-G*02        ---------------------G--C-----GT------------------------------AA--A---T---AG-AC---
Saoe-G*03        ---------------------G--C-----TG-----------------------C--A---AA--C---T---GG-A----
Saoe-G*04        ---------------------G--CA----GT------------------------------AA--A---T---AG-AC---
Saoe-G*05        ---------------------G--C-----GT------------------------------AA--C---T---AG-AC---
Saoe-G*06        -T-------------------G--C-----GT------------------------------AA--A---T---AG-AC---
Saoe-G*07        ---------------------G--C-----TT------------------------------AA--C---T---G-ATG---
Saoe-G*08        ---------------------G--C-----TG------------------------C-A---AA--C---T---GG-A----
Saoe-G*09        ---------------------G--C-----GT------------------------------AA-TC--T---AG-AC---
Saoe-G*10        --------------T------G--C-----GT------------------------------AA--C---T---G-TG---
Saoe-G*11        --------------T------G--C-----GT------------------------------AA--C---T---AG-AC---
Saoe-N1          ----------------------.,..,.---..-T-----...........--A---AAC-A---T---AG-A---A-
Saoe-N3          ----------A-----------------C-------T---------G-A-A-----C-A-----AA--A---T---AG-A---C-

Consensus        CGCCCTGAACGAGGACCTGCGCTCCTGGACCGCGGCGGACACGGCGGCTCAGATCACCCAGCGCAAGTGGGAGGCGGCCCGTGTGGCGGAGCAGCTGAGA
```

```
                   210       220       230       240       250       260       270
               *   *   *   *   *   *   *   *   *   *   *   *   *   *   *
Consensus      GCCTACCTGGAGGGCACGTGCGTGGAGTGGCTCCGCAGATACCTGGAGAACGGGAAGGAGACGCTGCAGCGCGCGG

Patr-A*01      -------------------------------------------------------------------A---
Patr-A*02      -------------------------------------------------------------------A---
Patr-A*03      --------------CG-----------------------------------------------A---
Patr-A*04      -------------------------------------------------------------------A-T-
Patr-A*05      ---------------------CG------------------------------------A-T-
Patr-A*06      ---------------------CG------------------------------------A---
Patr-A*07      --------------CG--------C--------------------------------------A---
Patr-A*08      -------------------------------------------------------------------A-T-
Patr-A*09      -------------------------------------------------------------------A-T-
Patr-A*10      --------------CG---------CG------------------------------------A---
Patr-A*11      -------------------------CG------------------------------------A---
Patr-A*12      -------------------------------------------------------------------A-T-
Patr-A*13      -------------------------------------------------------------------A-T-
Patr-A*14      -------------------------------------------------------------------A-T-
Papa-A*01      --------------CT---------CG------------------------------------A---
Papa-A*02      ----C--------------------------------------------------------------A-T-
Papa-A*03      --------------CT---------------------------------------------------A---
Papa-A*04      ----C--------------------------------------------------------------A-T-
Papa-A*05      --------------CG---------CG------------------------------------A-T-
Gogo-A*0101    --------------------------------C-----------------------------A---
Gogo-A*0201    --------------GA-----------------------------------------------A---
Gogo-A*0401    -------------------------------------------------------------T-A---
Gogo-A*0501    -------------------------------------------------------------------A---
Gogo-OKO       ------A-------------------------C-----------------------------A---
Popy-A*01      --------------CT---------C-----------------------------T-------A---
Popy-A*02      -------------------------------------------------------T-------A---
Popy-A*03      --------------GA---------------------------------------T-------A---
Hyla-A*01      ------T------G----------A-------------------TA---------A---
Hyla-A*02      -----T------A-----A--------C---------------------------A---
Mamu-A*01      --------------CA---------C----------------G-------------------A---
Mamu-A*02      A-------------GA---C---------------------------------------------
Patr-B*01      -----------------------------------------------------------------
Patr-B*02      -----------------------------------------------------------------
Patr-B*03      -----------------------------------------------------------------
Patr-B*04      -----------------------------------------------------------------
Patr-B*05      -----------------------------------------------------------------
Patr-B*06      --------------CT-------------------------------------------------
Patr-B*07      --------------CT-------------------------------------------------
Patr-B*08      -----------------------------------------------------------------
Patr-B*09      -----------------------------------------------------------------
Patr-B*10      ---------------------------------------------------A------------
Patr-B*11      --------------CT---------------------------------
Patr-B*12      --------------CT---------------------------------
Patr-B*13      -T-------A-----CG---------C--------------G-
Patr-B*14      ------------------------------------------------
Patr-B*15      ------------------------------------------------
Papa-B*01      --------------CT---------------------------------A------------
Papa-B*02      --------------CT---------------------------------A------------
Papa-B*03      --------------CT-------------------------------------------------
Papa-B*04      --------------CT---------------------------------A------------
Gogo-B*0101    ----------------------------------------------GA----------------
Gogo-B*0102    ----------------------------------------------GA----------------
Gogo-B*0103    ----------------------------------------------GA----------------
Gogo-B*0201    --------------CT-------------------------------------------------
Popy-B*01      --------------CT---------C---------------------------------------
Popy-B*02      --------------CG---------C---------------------------------------
Popy-B*03      --------------CG-----C-------------------------------------------
Hyla-B*01      --------------CG---C-------------------------------T-----------
Mamu-B*01      --------------CG----------------------------G-------A-----------
Patr-C*01      -----------------------------------------------------------------
Gogo-C*0101    --------------CT-------------------------------T--------------A-
Gogo-C*0102    --------------CT-------------------------------T--------------A-
Gogo-C*0201    --------------CT-------------------------------------------------A-
Gogo-C*0202    --------------CT---------C---------------------------------------
Gogo-C*0203    --------------CT-------------------------------------------------
Popy-E*01      ----------A-A--A----------------------A------------G------------TT-A-CT--
Mamu-E*01      ----------A-A--A----------------A-----------T
Mamu-E*02      ----------A-A--A------------------------------T
Mamu-E*03      ----------A-A--A----------------A-----------T
Mamu-E*04      ----------A-A--A------------------------------T
Mafa-E*01      ----------A-A--A----------------A--------------T------------T-A-
Mafa-E*02      ----------A-A--A----------------A--------------T------------T-A-
Patr-F*01      A-------------GA----C------T---------------T-----------------A------A--
Mamu-F         A-------------GA----C------T---------------------------------A------A-
Saoe-F*01      A--------A----GG-------------------A---------------------------T-
Patr-G*I
Gogo-G*I
Saoe-G*01      -------------------------------A------------------------------C-
Saoe-G*02      -------------------------------A------------------------------C-
Saoe-G*03      --------------GT---------------T------------------------------C-
Saoe-G*04      -------------------------------T------------------------------C-
Saoe-G*05      ---------------T---------------T------------------------------C-
Saoe-G*06      -------------------------------T------------------------------C-
Saoe-G*07      -------------------------------A------------------------------C-
Saoe-G*08      ----------A--------------------C-----------------T------------C-
Saoe-G*09      ------------T---------------T------------------------------C-
Saoe-G*10      -------------------------------A------------------C----------C-
Saoe-G*11      -------------------------------A------------------------------C-
Saoe-N1        ------G---------G----C------C-----T----------------
Saoe-N3        ------T----------C--------------A---TG---C-----------

Consensus      GCCTACCTGGAGGGCACGTGCGTGGAGTGGCTCCGCAGATACCTGGAGAACGGGAAGGAGACGCTGCAGCGCGCGG
```

Exon 4

```
                        10        20        30        40        50        60        70        80        90       100
                  *     *    *    *    *    *    *    *    *    *    *    *    *    *    *    *    *    *    *    *
Consensus         ACCCCCCAAAGACACATGTGACCCACCACCCCATCTCTGACCATGAGGCCACCCTGAGGTGCTGGGCCCTGGGCTTCTACCCTGCGGAGATCACACTGAC

Patr-A*01         ------C---------A-----------------------------------------------------------------------------------
Patr-A*02         ------C---------A----------------------------------------------------------------------T------------
Patr-A*03         ------C---------A---------------------------------------------------------------------------------- 
Patr-A*04         ------C---------A---------------C---G---------------------------------------------------------------
Patr-A*05         ------C---------A---------------C---G---------------------------------------------------------------
Patr-A*06         ------C---------A---------------C---G----------------------------C----------------------------------
Patr-A*07         ------C---------A---------------C---G---------------------------------------------------------------
Patr-A*08         ------C---------A---------------C---G---------------------------------------------------------------
Patr-A*09         ------C---------A---------------C---G---------------------------------------------------------------
Patr-A*10         ------C---------A---------------C---G---------------------------------------------------------------
Patr-A*11         ------C---------A---------------------------------------------------------------------------------- 
Patr-A*12         ------C---------A---------------C---G---------------------------------------------------------------
Patr-A*13         ------C---------A---------------C---G---------------------------------------------------------------
Patr-A*14         ------C---------A---------------C---G---------------------------------------------------------------
Papa-A*01         ------C---------A---------------------------------------------------------------------------------- 
Papa-A*02         ------C---------A---------------C---G---------------------------------------------------------------
Papa-A*03         ------C---------A---------------------------------------------------------------------------------- 
Papa-A*04         ------C---------A---------------C---G---------------------------------------------------------------
Papa-A*05         ------C---------A---------------------------------------------------------------------------------- 
Gogo-A*0101       --G----C-----G--A----T------G-TG----------------T---------------------------A-----------------------
Gogo-A*0201       --G----C-----G--A----T------G-TG----------------T---------------------------A-----------------------
Gogo-A*0401       --G----C-----G--A----T------TG------------------T---------------------------A-----------------------
Gogo-A*0501       --G----C-----G--AC---T----AG-TG-----------------------A---------------------A-----------------------
Gogo-OKO          ------C---------A----T------TG----------------------------------------------------------------------
Popy-A*01         --G----C---------A----T------TG----------------------------------------------------------------------
Popy-A*02         -TG----C---------A----T------G-----------------------------------T-----------------------------------
Popy-A*03         -TG----C---------A----T------G-----------------------------------T-----------------------------------
Hyla-A*01         ------C---------A----T-------------G-------------------------------------A-------------------T----
Hyla-A*02         ------C---------A----T------G-----------------------------------A------------A------------TG----
Mamu-A*01         ------C---------A----G------------------------------------------------------------------------------
Mamu-A*02         ------C---------A----G-------------A----------------------------------------------------------------
Patr-B*01         ---------------C-----------------------------------------------T------------------------------------
Patr-B*02         ---------------C-----------------------------------------------T------------------------------------
Patr-B*03         ---------------CA-----------------------------------------------T------------------------------------
Patr-B*04         ---------------C-----------------------------------------------T------------------------------------
Patr-B*05         ---------------C------------------------------------------------------------------------------------
Patr-B*06         ---------------C-----------------------------------------------T------------------------------------
Patr-B*07         ---------------C------------------------------------------------------------------------------------
Patr-B*08         ---------------C-----------------------------------------------T------------------------------------
Patr-B*09         ---------------C------------------------------------------------------------------------------------
Patr-B*10         ---------------C------------------------------------------------------------------------------------
Patr-B*11         
Patr-B*12         
Patr-B*13         
Patr-B*14         
Patr-B*15         
Papa-B*01         ---------------C------------------------------------------------------------------------------------
Papa-B*02         ---------------C------------------------------------------------------------------------------------
Papa-B*03         ---------------C------------------------------------------------------------------------------------
Papa-B*04         ---------------C------------------------------------------------------------------------------------
Gogo-B*0101       --A------------C-------------G----------------------------------------------------------------------
Gogo-B*0102       --A------------C-------------G----------------------------------------------------------------------
Gogo-B*0103       --A------------C-------------G----------------------------------------------------------------------
Gogo-B*0201       ---------------C------------------------------------------------------------------------------------
Popy-B*01         -G------G-----CA---------T-----------------A-T--------------------------------------------------------
Popy-B*02         -G------G-----CA---------T-----------------A-T--------------------------------------------------------
Popy-B*03         ---------------C----------A---------------A-----------------------------------------------------------
Hyla-B*01         -G------G------C------------------------------------------------------------------------------------
Mamu-B*01         ---------------C---------TG----------------A---T------------------------------------------------------
Patr-C*01         -A-A-----------C----T--G-----------------------------------------------------------------------------
Gogo-C*0101       ---------------C------------------------------------------------------------------------------------
Gogo-C*0102       ---------------C------------------------------------------------------------------------------------
Gogo-C*0201       -A-------------C---------C--------------------------------------------------------------------------
Gogo-C*0202       -A-------------C---------C--------------------------------------------------------------------------
Gogo-C*0203       -A-------------C---------C--------------------------------------------------------------------------
Popy-E*01         -T-------------------------G------------------------------------------------------------------------
Mamu-E*01         
Mamu-E*02         
Mamu-E*03         
Mamu-E*04         
Mafa-E*01         -A-------------C-------------------G------T------------------------------------T-------------G----
Mafa-E*02         -A-------------C-------------------G------T------------------------------------T-------------G----
Patr-F*01         -T--T------G---CA-TG--------------------------------------------------------------------G-----
Mamu-F            -T--T------G---CC-TG---------G-----------------------------------------------------------AC------------
Saoe-F*01         TT--T------G---C--TG--------------------------T-------------------------------------------T-TG-----
Patr-G*I          
Gogo-G*I          
Saoe-G*01         -G-----T--------C---------TG------------------------------------------------------------------G----
Saoe-G*02         -G-----T--------C---------T--TG--------------------------------------------------------------G----
Saoe-G*03         -G-----T--------C---------TG---------------------------------C-----T---------------------------------
Saoe-G*04         -G-----T--------C---------T--TG------------------------------------------------------------G----
Saoe-G*05         -G-----T--------CA--------TG------------------------------------------------------------------------
Saoe-G*06         -G-----T--------C---------TG------------------------------------------------------------------------
Saoe-G*07         -G-----T--------C---------TG------------------------T-------------------------------------------------
Saoe-G*08         -G-----T--------C---------TG------------------------------------------------------------------------
Saoe-G*09         -G-----T--------C---------TG---------------------------------C-----T---------------------------------
Saoe-G*10         -G-----T--------C---------T-------------------------------------------------------------------------
Saoe-G*11         -G-----T--------C---------T--TG-----------------------------------------------------------------------
Saoe-N1           
Saoe-N3           

Consensus         ACCCCCCAAAGACACATGTGACCCACCACCCCATCTCTGACCATGAGGCCACCCTGAGGTGCTGGGCCCTGGGCTTCTACCCTGCGGAGATCACACTGAC
```

```
              110      120      130      140      150      160      170      180      190      200
              *    *    *    *    *    *    *    *    *    *    *    *    *    *    *    *    *    *    *
Consensus     CTGGCAGCGGGATGGCGAGGACCAAACTCAGGACACGGAGCTTGTGGAGACCAGGCCAGCAGGAGATGGAACCTTCCAGAAGTGGGCAGCTGTGGTGGTG

Patr-A*01     ---------------G--------G--C---------------C----------C-----G-------------------------G----------A
Patr-A*02     ---------------G--------G--C---------------C----------C-----G-------------------------G----------A
Patr-A*03     ---------------G--------G--C---------------C----------C-----G-------------------------G----------A
Patr-A*04     ---------------G--------G--C---------------C----------T-----G-------------------------G----------A
Patr-A*05     ---------------G--------G--C---------------C----------T-----G-------------------------G----------A
Patr-A*06     ---------------G--------G--C---------------C----------T-----G-------------------------G----------A
Patr-A*07     ---------------G--------G--C---------------C----------T-----G-------------------------G----------A
Patr-A*08     ---------------G--------G--C---------------C----------T-----G-------------------------G----------A
Patr-A*09     ---------------G--------G--C---------------C----------T-----G-------------------------G----------A
Patr-A*10     ---------------G--------G--C---------------C----------T-----G-------------------------G----------A
Patr-A*11     ---------------G--------G--C---------------C----.,-.-C---------------------------------G----------A
Patr-A*12     ---------------G--------G--C---------------C----------T-----G-------------------------G----------A
Patr-A*13     ---------------G--------G--C---------------C----------T-----G-------------------------G----------A
Patr-A*14     ---------------G--------G--C---------------C----------T-----G-------------------------G----------A
Papa-A*01     ---------------G--------G--C---------------C----------T-----G-------------------------G----------A
Papa-A*02     ---------------G--------G--C---------------C----------T-----G-------------------------G----------A
Papa-A*03     ---------------G--------G--C---------------C----------T-----G-------------------------G----------A
Papa-A*04     ---------------G--------G--C---------------C----------T-----G-------------------------G----------A
Papa-A*05     ---------------G--------G--C---------------C----------T-----G-------------------------G----------A
Gogo-A*0101   ---------------G--------G--C---------------C----------T-----G-------------------------G----------
Gogo-A*0201   ---------------G--------G--C---------------C----------T-----G-------------------------G----------
Gogo-A*0401   ---------------G--------G--C---------------C----------T-----G-------------------------G----------
Gogo-A*0501   ---------------G--------G--C---------------C----------T-----G-------------------------G----------
Gogo-OKO      ---------------A--------G--C---------------C----------T-G----G-------G-----------------G----------
Popy-A*01     ---------------G--------G--C---------------C----------T-----G-------------------------G----------
Popy-A*02     ---------------G--------G--C---------------C----------T-----G-------------------------G----------
Popy-A*03     ---------------G--------G--C---------------C----------T-----G-------------------------G----------
Hyla-A*01     ---------------G--------G--C------------------------------G-------G-----------------------
Hyla-A*02     ---------------G--------G--C------------------------------G-------G-----------------------
Mamu-A*01     ---------------G-------------------C----------T-----------------------------------G----------
Mamu-A*02     ---------------A-------------------C----------T-G--C--G-------------------------------G----------
Patr-B*01     -----------------------------C-------------------A------A-------------------------
Patr-B*02     -----------------------------C-----------A------A-------------------------
Patr-B*03     -----------------------------C-------------------A------A-------------------------
Patr-B*04     -----------------------------C-------------------A------A-------------------------
Patr-B*05     -----------------------------C-------------------A------A-------------------------
Patr-B*06     -----------------------------T-------------------------A-------------------------
Patr-B*07     ---------------G-------------T-------------------------A-------------------------
Patr-B*08     -----------------------------C-------------------A------A-------------------------
Patr-B*09     -----------------------------C-------------------A------A-------------------------
Patr-B*10     -----------------------------C-------------------A------A-------------------------
Patr-B*11
Patr-B*12
Patr-B*13
Patr-B*14
Patr-B*15
Papa-B*01     ---------------G-----------------------T-----------------A-------------------------
Papa-B*02     ---------------G-----------------------T-----------------A-------------------------
Papa-B*03     ---------------G-----------------------T-----------------A-------------------------
Papa-B*04     ---------------G-----------------------T-----------------A-------------------------
Gogo-B*0101   -----------------------------C-------------------------------------------
Gogo-B*0102   -----------------------------C-------------------------------------------
Gogo-B*0103   -----------------------------C-------------------------------------------
Gogo-B*0201   ----------A------------------C-----------------------A-------------------------
Popy-B*01     ---------------A-G-------TG--C--------C--------------------C-----------------
Popy-B*02     ---------------A-G-------G--C--------C--------------------C-----------------
Popy-B*03     ---------------G--------G--C-------------------------------------
Hyla-B*01     ---------------G-----------------------T-------------------------CA-------------T-------
Mamu-B*01     ---------------G-----------------------T---------------------A---G--------
Patr-C*01     ---------------G-------------C----------------------------------------
Gogo-C*0101   ---------------G--------G---------TC----------------------A--------
Gogo-C*0102   ---------------G--------G---------TC----------------------A--------
Gogo-C*0201   ---------------G--------G--C----------------------------------
Gogo-C*0202   ---------------G--------G--C----------------------------------
Gogo-C*0203   ---------------G--------G--C----------------------------------
Popy-E*01     ---------------G--------G-AC---T----A--------------------T--C--G--------
Mamu-E*01
Mamu-E*02
Mamu-E*03
Mamu-E*04
Mafa-E*01     ---------------G-------G--C--------------------------T-----G-------T--------------------A
Mafa-E*02     ---------------G-------G--C---------------T-----------T-----G-------T--------------------A
Patr-F*01     ---------------G------A--G--C--------A------------T-----G---------A------------------T------
Mamu-F        ----------C-G------A--G--C------------------------T-----G--------
Saoe-F*01     -------A--------A----T--G--C--------A--A--------------T-----G--------
Patr-G*I
Gogo-G*I
Saoe-G*01     ---------------G--------G--C-------T------C--A---------------G--------------------G--------
Saoe-G*02     ---------------G--------G--A--------T------C--A---------------G--------------------G--------
Saoe-G*03     ---------------G--------G--A--------T------C--A---------------G--------------------G--------
Saoe-G*04     ---------------G--------G--A--------T------C--A---------------G--------------------G--------
Saoe-G*05     ---------------G--------G--A--------T------C--A---------------G--------------------G--------
Saoe-G*06     ---------------G--------G--A--------T------C--A---------CA----GA-------------------G--------
Saoe-G*07     ---------------G--------G--A--------T------C--A---------CA---GA-------------------G--------
Saoe-G*08     ---------------G--------GG-A--------T------C--A---------CA-G--GA-------------------G--------
Saoe-G*09     ---------------G--------G--A--------T------C--A---------------G--------------------G--------
Saoe-G*10     ---------------G--------G--A--------T------C--A---------------G--------------------G--------
Saoe-G*11     -------T-------G--------G--A--------T------C--A---------------G--------------------G--------
Saoe-N1
Saoe-N3

Consensus     CTGGCAGCGGGATGGCGAGGACCAAACTCAGGACACGGAGCTTGTGGAGACCAGGCCAGCAGGAGATGGAACCTTCCAGAAGTGGGCAGCTGTGGTGGTG
```

```
              210        220       230       240       250       260       270
              *    *     *    *    *    *    *    *    *    *    *    *    *    *
Consensus     CCTTCTGGAGAAGAGCAGAGATACACATGCCATGTGCAGCATGAGGGGCTGCCGAAGCCCCTCACCCTGAGATGGG

Patr-A*01     -----------G-------------C--------------------T----C-----------------------
Patr-A*02     -----------G-------------C--------------------T----C-----------------------
Patr-A*03     -----------G-------------C--------------------T----C-----------------------
Patr-A*04     -----------G-------------C--------------------T----C-----------------------
Patr-A*05     -----------G-------------C--------------------T----C-----------------------
Patr-A*06     -----------G-------------C--------------------T----C-----------------------
Patr-A*07     -----------G-------------C--------------------T----C-----------------------
Patr-A*08     -----------G-------------C--------------------T---TC-----------------------
Patr-A*09     -----------G-------------C--------------------T----C-----------------------
Patr-A*10     -----------G-------------C---------------------T---TC----------------------
Patr-A*11     -----------G-------------C--------------------T----C-----------------------
Patr-A*12     -----------G-------------C--------------------T----C-----------------------
Patr-A*13     -----------G-------------C--------------------T----C-----------------------
Patr-A*14     -----------G-------------C--------------------T----C-----------------------
Papa-A*01     -----------G-------------C--------------------T--T-C-----------------------
Papa-A*02     -----------G-------------C--------------------T---TC-----------------------
Papa-A*03     -----------G-------------C--------------------T--T-C-----------------------
Papa-A*04     -----------G-------------C--------------------T---TC-----------------------
Papa-A*05     -----------G-------------C--------------------T----C-----------------------
Gogo-A*0101   ---------C-G-------------C--------------------T----CG----------------------
Gogo-A*0201   ---------C-G-------------C------------------A-T----C-----------------------
Gogo-A*0401   ---------C-G-------------C--------------------T----C-----------------------
Gogo-A*0501   ---------C-G-------------C--------------------T----C-----------------------
Gogo-OKO      ---------A-G-------------C--------------------T----C-----------------------
Popy-A*01     ---------A-G-------------C--------------------T----CG----------------------
Popy-A*02     ---------A-G-------------C----T---------------T----TG----------------------
Popy-A*03     ---------A-G-------------C----T---------------T----TG----------------------
Hyla-A*01     ---------A-G-------------C--------------------T----CG-----------------T-----
Hyla-A*02     ---------A-G-------------C--------------------T----CG----------------------
Mamu-A*01     ---------G---------------C--T-----------------T----C-----A-------A----------
Mamu-A*02     ---------A-G-------------C--T-----------------T----GTG---------------------
Patr-B*01     --------------------------------------------------------------------------
Patr-B*02     --------------------------------------A-----------------------------------
Patr-B*03     --------------------------------------------------------------------------
Patr-B*04     --------------------------------------A-----------------------------------
Patr-B*05     --------------------------------------A-----------------------------------
Patr-B*06     ---------C----------------------------------------T-G---------------------
Patr-B*07     ---------C-----------------------------------------G----------------------
Patr-B*08     --------------------------------------A-----------------------------------
Patr-B*09     --------------------------------------------------------------------------
Patr-B*10     --------------------------------------------------------------------------
Patr-B*11
Patr-B*12
Patr-B*13
Patr-B*14
Patr-B*15
Papa-B*01     ---------C-----------------------------------CG--G------------------------
Papa-B*02     ---------C----------------------------------------G----------------------
Papa-B*03     ---------C----------------------------------------G----------------------
Papa-B*04     ---------C----------------------------------------G----------------------
Gogo-B*0101   ------------G--------------------------------------------------------------
Gogo-B*0102   ------------G--------------------------------------------------------------
Gogo-B*0103   ------------G--------------------------------------------------------------
Gogo-B*0201   ------------A---------------------G----------C-----------G-------T---------
Popy-B*01     ---------C------------------------------------C-----------G----------------
Popy-B*02     ---------C------------------------------------C-----------G----------------
Popy-B*03     ---------C------------------------------------C-----------G---------------A
Hyla-B*01     ---------------------------------------------C-----------G----------------
Mamu-B*01     ----------------------G----------------------C-----------G----------------
Patr-C*01     --------------------------------------C----------C------------------------
Gogo-C*0101   ----------------------G--------------C----CA------T-G---------------------
Gogo-C*0102   ----------------------G--------------C----CA------T-G---------------------
Gogo-C*0201   ---------C------------G------A-------C-----------AG-----------------------
Gogo-C*0202   ---------C------------G------A-------C-----------AG-----------------------
Gogo-C*0203   ---------C------------G------A-------C-----------AG-----------------------
Popy-E*01     ------------G---------------G--------------------A--CG-A------------------
Mamu-E*01
Mamu-E*02
Mamu-E*03
Mamu-E*04
Mafa-E*01     -----------G---------------G--------------------A--CG------GAG------------
Mafa-E*02     -----------G---------------G--------------------A--CG---------------------
Patr-F*01     -----------G--A------------G---------------C-----CC-----------------------
Mamu-F        ----C----G------------------------A---A---------CC-----------------------
Saoe-F*01     -T------G--G-----C------C-----------C-----------AG--T---------------------
Patr-G*I
Gogo-G*I
Saoe-G*01     -T---------G-----C------------------C-----------CG-----------------------
Saoe-G*02     -T---------G-----C------------------C-----------CG-----------------------
Saoe-G*03     -T---------G--A--C------------------C-----------CG-----------------------
Saoe-G*04     -T---------G-----C------------------C-----------CG-----------------------
Saoe-G*05     -T---------G-----C------------------C-----------CG-----------------------
Saoe-G*06     -T---------G-----C------------------C-----------TG-----------------------
Saoe-G*07     -T---------G-----C-T----------------C-----------CG-----------------------
Saoe-G*08     -T---------G-----C-A----------------C-----------CG----T------------------
Saoe-G*09     -T---------G--A--C------------------C-----------CG-----------------------
Saoe-G*10     -T---------G-----C------------------C-----------CG-----------------------
Saoe-G*11     -T---------G-----C------------------C-----------CG-----------------------
Saoe-N1
Saoe-N3

Consensus     CCTTCTGGAGAAGAGCAGAGATACACATGCCATGTGCAGCATGAGGGGCTGCCGAAGCCCCTCACCCTGAGATGGG
```

Exon 5

```
                    10        20        30        40        50        60        70        80        90       100
                 *    *    *    *    *    *    *    *    *    *    *    *    *    *    *    *    *    *    *    *
Consensus    AGCCCTCTTCCCAGCCCACCATCCCCATCGTGGGCATCGTTGCTGGCCTGGCTGTCCTTGGAGTCCCTGTGGTCATTGGAGCTGTGGTCGCTGCTGTGAT

Patr-A*01    ----G---------------------T---------A-----------T-C---------.,-----A---C-----------------------C----
Patr-A*02    ----G---------------------T---------A-----------T-C---------.,-----A---C-----------------------C----
Patr-A*03    ----G---------------------T---------A-----------T-C---------.,-----A---C-----------------------C----
Patr-A*04    ----G---------------------T---------A-----------T-C---------.,-----A---C-----------------------C----
Patr-A*05    ----G-------------------------------A-----------T-C---------.,-----A---C-----------------------C----
Patr-A*06    ----G-------------------------------A-----------T-C---------.,-----A---C-----------------------C----
Patr-A*07    ----G-------------------------------A-----------T-C---------.,-----A---C-----------------------C----
Patr-A*08    ----G-------------------------------A-----------T-C---------.,-----A---C----------------------CA----
Patr-A*09    ----G-------------------------------A-----------T-C---------.,-----A---C-----------------------C----
Patr-A*10    ----G-------------------------------A-----------T-C---------.,-----A---C----------------------CA----
Patr-A*11    ----G---------------------T---------A-----------T-C---------.,-----A---C-----------------------C----
Patr-A*12    ----G-------------------------------A-----------T-C---------.,-----A---C-----------------------C----
Patr-A*13    ----G-------------------------------A-----------T-C---------.,-----A---C-----------------------C----
Patr-A*14    ----G-------------------------------A-----------T-C---------.,-----A---C-----------------------C----
Papa-A*01    ----G-------------------------------A-----------T-C---------.,-----A---C-----------------------C----
Papa-A*02    ----G-------------------------------A-----------T-C---------.,-----A---C----------------------CA----
Papa-A*03    ----G-------------------------------A-----------T-C---------.,-----A---C-----------------------C----
Papa-A*04    ----G-------------------------------A-----------T-C---------.,-----A---C----------------------CA----
Papa-A*05    ----G---------------------T---------A-----------T-C---------.,-----A---C-----------------------C----
Gogo-A*0101  ----A-------------------------------A-----------T-C--T------.,-----A-GC-----------------------------G
Gogo-A*0201  ----A-------------------------------A-----------T-C--T------.,-----A-GC--------A--------------------G
Gogo-A*0401  ----A-------------------------------A-----------T-C--T------.,-----A-GC-----------------------------G
Gogo-A*0501  ----A-------------------------------A-----------T-C---------.,-----A-GC-----------------------------G
Gogo-OKO     ----G-------------------------------A-----------T-C-T-------.,-----A---C---------------------A----.--
Popy-A*01    ---TG-------------------------------A-----------T-C-T-------.,-----A---C----------------------------
Popy-A*02    ----A------------------------------------------T-C-T-------.,-----A---C----------------------------
Popy-A*03    ----A------------------------------------------T-C-T-------.,-----A---C----------------------------
Hyla-A*01    ----G---------A---------------------TA----------T-C---------.,-----A---C----------------------------
Hyla-A*02    ----G---------------------T---------A-----------T-C---------.,-----A---C----------------------------
Mamu-A*01    ----G-T-------T------------G--------A-----------T-C---------.,---------C-----------------T----------
Mamu-A*02    ----G--------T----------T-----------A-----------T-C-------A--.,--------C----------------A-T---------
Patr-B*01    ----A---------T------------T--------T-----------TG---A-C--A-CTG-G-TC----------------------------
Patr-B*02    ----A---------T--------------------T-----------A-C---------.,--------C----------------------
Patr-B*03    ----A---------T------------T--------T-----------TG---A-C--A-CTG-G-TC------------------------T-
Patr-B*04    ----A---------T--------------------T-----------A-C---------.,--------C----------------------
Patr-B*05    ----A---------T--------------------T-----------A-C---------.,--------C----------------------
Patr-B*06    ----A---------T--------------------T-----------A-C--.T-------CC----------------------
Patr-B*07    ----A---------T--------------------T-----------A-C--.,-------CC----------------------
Patr-B*08    ----A---------T--------------------T-----------A-C---------.,--------C----------------------
Patr-B*09    ----A---------T------------T--------T-----------TG---A-C--A-CTG-G-TC------------------------T-
Patr-B*10    ----A---------T--------------------T-----------.,-------TG---A-C--A-CTG-G-TC----------------
Patr-B*11
Patr-B*12
Patr-B*13
Patr-B*14
Patr-B*15
Papa-B*01    ----A---------T--------------------T-----------A-C---.,------CC----------------------
Papa-B*02    ----A---------T--------------------T-----------A-C---.,------CC----------------------
Papa-B*03    ----A---------T--------------------T-----------A-C---.,------CC----------------------
Papa-B*04    ----A---------T--------------------T-----------A-C---.,------CC----------------------
Gogo-B*0101  ----A---------T--------------------T-----------A-C---.,------C----------------A---
Gogo-B*0102  ----A---------T--------------------T-----------A-C---.,------C----------------A---
Gogo-B*0103  ----A---------T--------------------T-----------A-C---.,------C----------------A---
Gogo-B*0201  ----A---------T--------------------T-----------TG---A-C--A-CTG-G-TC----------------
Popy-B*01    ----A---------T--------------------T-----------A-C---.,------C--AT----------------
Popy-B*02    ----A---------T--------------------T-----------A-C---.,------C--AT----------------
Popy-B*03    ----A---------T--------------------T-----------A-C---.,------C--AT----------------
Hyla-B*01    ----A---------T--------------------T-----------A-C---.,------CC------------T------
Mamu-B*01    ----A---------T--------------------T-----------A-C---.,------CC----------------------
Patr-C*01    G---G--------------------------C------------G-CT----TA-CT---C-A-----------G----T-----
Gogo-C*0101  ----G-------------------------------------------A-C---.,----T---C----A------
Gogo-C*0102  ----G------------------------------------------A-C---.,----T---C----A------
Gogo-C*0201  ----A-------------T---------C------------G-TT----TA-CT---C----------------A-C--A----
Gogo-C*0202  ----A-------------T----T-----------------G-TT----TA-CT---C-----------------A-C--A----
Gogo-C*0203  ----A-------------T---T------------------G-TT----TA-CT---C--------------A-C--A----

                                                                          CTA
                                                                           |
Popy-E*01    ----GG--------A-----------------T------------T-C----------.,------C------------T------
Mamu-E*01
Mamu-E*02
Mamu-E*03
Mamu-E*04
Mafa-E*01    ----G-------G-T---------------A-----------T-C---------.,---------C-----------T-T----
Mafa-E*02    ----G--------T---------------A-----------T-C---------.,---------C-----------T-T----
Patr-F*01    --AG---C--------------------------------T-T----------
Mamu-F       --T-G----T---------T--------------------A-C--.,-------T-C-----------------A--
Saoe-F*01    ----A---------------G---T----A------AC-----TG---T----------.,------C----.,............-----
Patr-G*I
Gogo-G*I
Saoe-G*01    ----AC-----------------A--------AT-T----A--------.,---A---C-------------
Saoe-G*02    ----GC-----------------A--------AT-T----A--------.,---A---C-------------
Saoe-G*03    ----AC-----------------A--------AT-T----A--------.,---A---C------------A-------
Saoe-G*04    ----GC-----------------A--------AT-T----A--------.,---A---C-------------
Saoe-G*05    ----AC-----------------A--------AT-T----A--------.,---A---C-------------
Saoe-G*06    ----GC-----------------A--------AT-T----A--------.,---A---C-------------
Saoe-G*07    ----AC-----------------A--------AT-T----A--------.,---A---C-------------
Saoe-G*08    ----GC-----------------A--------AT-T----A--------.,---A---C-------------
Saoe-G*09    ----AC-----------------A--------AT-T----A--------.,---A---C-------------
Saoe-G*10    ----AC-----------------A--------ATGT----A--------.,---A---C------------A-------
Saoe-G*11    ----GC-----------------A--------AT-T----A--------.,---A---C-------------
Saoe-N1
Saoe-N3

Consensus    AGCCCTCTTCCCAGCCCACCATCCCCATCGTGGGCATCGTTGCTGGCCTGGCTGTCCTTGGAGTCCCTGTGGTCATTGGAGCTGTGGTCGCTGCTGTGAT
```

```
Exon 5                                    Exon 6
                    110       120                10        20        30
               *      *      *      *        *      *      *      *      *      *
Consensus      GTGTAGGAGGAAGAGCTCAG       GTGGAAAAGGAGGGAGCTACTCTCAGGCTGCAT

Patr-A*01      ---G--                     
Patr-A*02      ---G----------------       A-A-----------------A----------A
Patr-A*03      ---G----------------       A-A-----------------A----------A
Patr-A*04      ---G----------------       A-A---------------------------A
Patr-A*05      ---G----------------       A-A---------------------------A
Patr-A*06      ---G----------------       A-A---------------------------A
Patr-A*07      ---G----------------       A-A---------------------------A
Patr-A*08      ---G----------------       A-A---------------------------A
Patr-A*09      ---G----------------       A-A---------------------------A
Patr-A*10      ---G----------------       A-A---------------------------A
Patr-A*11      ---G----------------       A-A-----------------A---------A
Patr-A*12      ---G----------------       A-A---------------------------A
Patr-A*13      ---G----------------       A-A---------------------------A
Patr-A*14      ---G----------------       A-A---------------------------A
Papa-A*01      ---G----------------       A-A-----------------A---------A
Papa-A*02      ---G----------------       A-A---------------------------A
Papa-A*03      ---G----------------       A-A-----------------A---------A
Papa-A*04      ---G----------------       A-A---------------------------A
Papa-A*05      ---G----------------       A-A---------------------------A
Gogo-A*0101    ---G----------------       A-A---------------------------A
Gogo-A*0201    ---G----------------       A-A--------------------------A
Gogo-A*0401    ---G----------------       A-A---------------------------A
Gogo-A*0501    ---G----------------       A-A---------------------------A
Gogo-OKO       ---G----A-----------       --A---------------------------A
Popy-A*01      ---G-------G--A-----       A-A---------------------------A
Popy-A*02      ---G-------G--------       A-A---------------------------A
Popy-A*03      ---G-------G--------       A-A---------------------------A
Hyla-A*01      ---G----------------       A-A---------------------------A
Hyla-A*02      ---G----------------       A-A---------------------------A
Mamu-A*01      ---G----------------       A-A---------------------------A
Mamu-A*02      A--G----------------       A-A---------------------------A
Patr-B*01      --------------------       ------------------------------
Patr-B*02      --------------------       ----------------------------G-
Patr-B*03      --------------------       ------------------------------
Patr-B*04      --------------------       ----------------------------G-
Patr-B*05      --------------------       ----------------------------G-
Patr-B*06      --------------------       -----------------C--------G-
Patr-B*07      --------------------       -----------------C--------G-
Patr-B*08      --------------------       ----------------------------G-
Patr-B*09      --------------------       ------------------------------
Patr-B*10      --------------------       ------------------------------
Patr-B*11
Patr-B*12
Patr-B*13
Patr-B*14
Patr-B*15
Papa-B*01      --------------------       -----------------C--------G-
Papa-B*02      --------------------       -----------------C--------G-
Papa-B*03      --------------------       -----------------C--------G-
Papa-B*04                                 -----------------C--------G-
Gogo-B*0101    A-------------------       ----------------------------G-
Gogo-B*0102    A-------------------       ----------------------------G-
Gogo-B*0103    A-------------------       ----------------------------G-
Gogo-B*0201    --------------------       ----------------------------G-
Popy-B*01      --------------------       ---G------------------------G-
Popy-B*02      --------------------       ---G------------------------G-
Popy-B*03      --------------------       ---G------------------------G-
Hyla-B*01      ---G------------T--         --------------A------G-
Mamu-B*01      ---G----------------       ----------------------------G-
Patr-C*01      --------------------       -------------G--------------
Gogo-C*0101    --------------------       -------------G------------G-
Gogo-C*0102    --------------------       -------------G------------G-
Gogo-C*0201    --------------------       -------------G------------G-
Gogo-C*0202    --------------------       -------------G------------G-
Gogo-C*0203    --------------------       -------------G------------G-
Popy-E*01      ---G----A-----------       --------------A------AG-
Mamu-E*01
Mamu-E*02
Mamu-E*03
Mamu-E*04
Mafa-E*01      ---G----------------       --A---------------------------T-G-
Mafa-E*02      ---G----------------       --A--------------------------TTG-
Patr-F*01      ---G----A-----------       A-A-----CA-------------------G
Mamu-F         ---G----A-----------       A-A-----CA----A---------------A
                         TGGGAAGGG
Saoe-F*01      ---G----G-----T--GGT       A-A-----CA-------------------A
Patr-G*I
Gogo-G*I
Saoe-G*01      ---G----A-----------       A-AA---------A-----------A------A
Saoe-G*02      ---G----A-----------       A-AA---------A-----------A------A
Saoe-G*03      ---G----A-----------       A-AA---------A---------------A
Saoe-G*04      ---G----A-----------       A-AA---------A---------------A
Saoe-G*05      ---G----A-----------       A-AA---------A---------------A
Saoe-G*06      ---G----A-----------       A-AA---------A---------------A
Saoe-G*07      ---G----A-----------       A-AA---------A---------------A
Saoe-G*08      ---G----A-----------       A-AA---------A---------------A
Saoe-G*09      ---G----A-----------       A-AA---------A---------------A
Saoe-G*10      ---G----A-----------       A-AA---------A---------------A
Saoe-G*11      ---G----A-----------       A-AA---------A---------------A
Saoe-N1
Saoe-N3

Consensus      GTGTAGGAGGAAGAGCTCAG       GTGGAAAAGGAGGGAGCTACTCTCAGGCTGCAT
```

Exon 7   Exon 8

```
                  10        20        30        40
                  *    *    *    *    *    *    *    *    *
Consensus    GCAGCGACAGTGCCCAGGGCTCTGATGTGTCTCTCACAGCTTGTAAAG      TGTGA

Patr-A*01
Patr-A*02    ----T------------------------------------------      -----
Patr-A*03    ----T------------------------------------------      -----
Patr-A*04    ----T------------------------------------------      -----
Patr-A*05    ----T------------------------------------------      -----
Patr-A*06    ----T------------------------------------------      -----
Patr-A*07    ----T------------------------------------------      -----
Patr-A*08    ----T-------------------------------
Patr-A*09    ----T------------------------------------------      -----
Patr-A*10    ----T------------------------------------------      -----
Patr-A*11    ----T------------------------------------------      -----
Patr-A*12    ----T------------------------------------------      -----
Patr-A*13    ----T------------------------------------------      -----
Patr-A*14    ----T------------------------------------------      -----
Papa-A*01    ----T------------------------------------------      -----
Papa-A*02    ----T------------------------------------------      -----
Papa-A*03    ----T------------------------------------------      -----
Papa-A*04    ----T------------------------------------------      -----
Papa-A*05    ----T------------------------------------------      -----
Gogo-A*0101  ----T------------------------------------------      -----
Gogo-A*0201  ----T------------------------------------------      -----
Gogo-A*0401  ----T------------------------------------------      -----
Gogo-A*0501  ----T------------------------------------------      -----
Gogo-OKO     --------------------------------------G-----A
Popy-A*01    ---AT------------------------------------------      -----
Popy-A*02    ---AT------------------------------------------      -----
Popy-A*03    ---AT------------------------------------------      -----
Hyla-A*01    ----T---------------------------------G--------      -----
Hyla-A*02    ----T---------------------------------G--------      -----
Mamu-A*01    ----T---------------------------------G--------      -----
Mamu-A*02    ----T---------------------------------G--------      -----
Patr-B*01    C-------------------------------------------A
Patr-B*02    C-------------------------------------------A
Patr-B*03    C-------------------------------------------A
Patr-B*04    C-------------------------------------------A
Patr-B*05    C-------------------------------------------A
Patr-B*06    C-------------------------------------------A
Patr-B*07    C-----------------------------------------
Patr-B*08    C-------------------------------------------A
Patr-B*09    C-------------------------------------------A
Patr-B*10    C-------------------------------------------A
Patr-B*11
Patr-B*12
Patr-B*13
Patr-B*14
Patr-B*15
Papa-B*01    C-------------------------------------------A
Papa-B*02    C-------------------------------------------A
Papa-B*03    C-------------------------------------------A
Papa-B*04    C-------------------------------------------A
Gogo-B*0101  C-------------------------------------------A
Gogo-B*0102  C-------------------------------------------A
Gogo-B*0103  C-------------------------------------------A
Gogo-B*0201  C-------------------------------------------A
Popy-B*01    ------------------------------------G-----A
Popy-B*02    ------------------------------------G-----A
Popy-B*03    C---------C-T-----------------------G-----A
Hyla-B*01    C----------------------------------G-----A
Mamu-B*01    C----------------------------------G-----AA
Patr-C*01    C-----A-----------------A--------TC----------     CC---
Gogo-C*0101  ------A-----------------A--------TC----------     CC---
Gogo-C*0102  ------A-----------------A--------TC----------     CC---
Gogo-C*0201  ------A-----------------A--------TCA---------     CC---
Gogo-C*0202  ------A-----------------A--------TCA---------     CC---
Gogo-C*0203  ------A-----------------A--------TCA---------     CC---
Popy-E*01    -G--T------------A-----...------------------     CC---
Mamu-E*01
Mamu-E*02
Mamu-E*03
Mamu-E*04
Mafa-E*01    -T---------A-------A-------A-----------------     CC---
Mafa-E*02    -T---------A-------A-------A-----------------     CC---
Patr-F*01    ...............................................     -----
Mamu-F       ...............................................     -----
Saoe-F*01    ...............................................     -----
Patr-G*I
Gogo-G*I
Saoe-G*01    -A--T---------------------------------G--------     C----
Saoe-G*02    -A--------------------------------------------     -----
Saoe-G*03    -A--------------------------------------------     -----
Saoe-G*04    -A--------------------------------------------     -----
Saoe-G*05    -A------------------------------------G--------     -A---
Saoe-G*06    -A------------------------------------G--------     -----
Saoe-G*07    -A------------------------------------G--------     -----
Saoe-G*08    -A------------------------------------G--------     -----
Saoe-G*09    -A------------------------------------G--------     -----
Saoe-G*10    -A------------------------------------G--------     -----
Saoe-G*11    -A------------------------------------G--------     -----
Saoe-N1
Saoe-N3

Consensus    GCAGCGACAGTGCCCAGGGCTCTGATGTGTCTCTCACAGCTTGTAAAG      TGTGA
```

Acknowledgements. This work was supported by grants from the National Institutes of Health (RR00167, HD34215, DK44886 and A132426) and a Biomedical Science Grant from the Arthritis Foundation.

References

Boyson JE, McAdam SN, Gallimore A, Golos TG, Lui X, Gotch FM, Hughes AL, Watkins DI. The MHC E locus in macaques is polymorphic and is conserved between macaques and humans. Immunogenetics 41:59–68, 1995

Chen ZW, McAdam SN, Hughes AL, Dogon AL, Letwin NL, Watkins DI. Molecular cloning of orangutan and gibbon MHC class I cDNAs: the HLA-A and B loci diverged over 30 million years ago. J Immunol 148:2547–2554, 1992

Chen ZW, Hughes AL, Ghim SH, Letvin NL, Watkins DI. Two more chimpanzee Patr-A locus alleles related to the HLA-A1/A3/A11 family. Immunogenetics 38:238–240, 1993

Corell A. Morales P, Martínez-Laso J, Martín-Villa JM, Varela P, Paz-Artal E, Allende LM, Rodríguez C, Arnaiz-Villena A. Species-specific alleles at the primate MHC-G locus. Hum Immunol 41:52–55, 1994

Klein J, Bontrop RE, Dawkins RL, Erlich HA, Gyllensten UB, Heise ER, Jones PP, Parham P, Wakeland EK, Watkins DI. Nomenclature for the major histocompatibility complexes of different species: a proposal. Immunogenetics 31:217–219, 1990

Lawlor DA, Ward FE, Ennis PD, Jackson AP, Parham P. HLA-A and B polymorphism predate the divergence of humans and chimpanzees. Nature 335:268–271, 1988

Lawlor DA, Warren E, Ward FE, Parham. Comparison of class I MHC alleles in humans and apes. Immunol Rev 113:147–185, 1990

Lawlor DA, Warren E, Taylor P, Parham P. Gorilla class I alleles: comparison to human and chimpanzee class I. J Exp Med 174:1491–1509, 1991

Lawlor DA, Edelson BT, Parham P. Mch-A locus molecules in pygmy chimpanzees: conservation of peptide pockets. Immunogenetics 42:291–295, 1995

Mayer WE, Jonker M, Klein D, Ivanyi P, van Seventer G, Klein J. Nucleotide sequences of chimpanzee MHC class I alleles: evidence for a trans-species mode of evolution. EMBO J 7:2765–2774, 1988

McAdam SN, Boyson JE, Xiaomin L, Garber TL, Hughes AL, Bontrop RE, Watkins DI. Chimpanzee major histocompatibility complex class I A locus alleles are related to only one of the six families of human A locus alleles. J Immunol 154:6421–6429, 1994a

McAdam SN, Boyson JE, Lui X, Garber TL, Hughes AL, Bontrop RE, Watkins DI. A uniquely high level of recombination at the HLA-B locus. Ptroc Natl Acad Sci 91;5893–5897, 1994b

Miller MD, Yamamoto H, Hughes AL, Watkins DI, Letvin NL. Definition of an epitope and an MHC class I molecule recognized by gag-specific cytotoxic T-lymphocytes in SIVmac-infected rehsus monkeys. J Immunol 147(1):320–329, 1991

Otting N, Bontrop RE. Characterisation of the rhesus macaque (Macaca mulatta) equivalent of HLA-F. Immunogenetics 38:141–145, 1993

Watanabe N, McAdam SN, Boyson JE, Piekarczyk MS, Yasutomi Y. Cytotoxic T-lymphocyte epitope in rhesus monkeys and its restricting major histocompatibility complex class I molecule Mamu-A*02. J Virol 68(10):6690–6696, 1994

Watkins DI, Letvin NL, Hughes AL, Tedder TF. Molecular cloning of cDNAs that enclode MHC class I molecules from a new world primate (Saguinus oedipus): natural selection acts at positions that may affect peptide presentation to T cells. J Immunol 144:1136–1143, 1990a

Watkins DI, CChen ZW, Hughes AL, Evans MG, Tedder TF, Letvin NL. Evolution of the MHC class I genes of a New World primate from ancestral homologues of human non-classical genes. Nature 346:60–63, 1990b

Watkins DI, Garber TL, Chen ZW, Toukatly G, Hughes AL, Letvin NL. Unusually limited nucleotide sequence variation of the expressed major histocompatibility complex class I genes of a New World primate species (Saguinus oedipus). Immunogenetics 33;79–89, 1991a

Watkins DI, Chen ZW, Garber TL, Hughes AL, Letvin NL. Segmental xchange between MHC class I genes in a higher primate. Recombination in the gorilla between the ancestor of a human non-functional gene and an A locus gene. Immunogenetics 34:185–191, 1991b

Yasutomi Y, McAdam SN, Boyson JE, Piekarcyzk MS, Watkins DI, Letvin NL. A MHC class I B locus allele-restricted simian immunodeficiency virus envelope CTL epitope in rhesus monkeys. J Immunol 154(5);2516–2522, 1995

8 Nonhuman Primate MHC Class II Sequences: A Compilation

C. O'hUigin

Introduction

The figures and tables that follow bring together the nonhuman primate class II sequences reported thus far. The compilation is in three parts, corresponding to separate *DR* (Table 2 and Fig. 2), *DQ* (Tables 3,4 and Fig. 3) and *DP* (Table 5 and Fig. 4) sections. The designations used are the standardized versions proposed by Klein et al. (1990). This compilation extends and updates those of O'hUigin et al. (1993) for nonhuman primate DR loci and of Bontrop (1994) for nonhuman primate DQ loci. The present and former compilations are limited to the variable second exon. References to species names are given in Table 1 and to synonyms and publications in Tables 2–5. Where differences were noted between GEN-BANK/EMBL/DDBJ entries and the sequences in original publications, the published version is used.

Table 1. Species names and their abbreviations

Abbreviation	Scientific name	Common name
Aotr	*Aotus trivirgatus*	Northern night (owl) monkey
Caja	*Callithrix jaccus*	Common marmoset
Camo	*Callicebus moloch*	Dusky titi
Ceae	*Cercopithecus aethiops*	African (vervet) green monkey
Ceap	*Cebus apella*	Brown capuchin
Gamo	*Galago moholi*	Southern lesser bushbaby
Gase	*Galago senegalensis*	Northern lesser bushbaby
Gogo	*Gorilla gorilla*	Gorilla
Hyla	*Hylobates lar*	Gibbon
Lota	*Loris tardigradus*	Slender loris
Maar	*Macaca arctoides*	Bear (stumptailed) macaque
Mafa	*Macaca fascicularis*	Crab eating macaque
Male	*Mandrillus leucophaeus*	Drill

Molecular Biology and Evolution of Blood Group
and MHC Antigens in Primates
Blancher/Klein/Socha (Eds.)
© Springer-Verlag Berlin Heidelberg 1997

Table 1. (cont.)

Abbreviation	Scientific name	Common name
Mamu	*Macaca mulatta*	Rhesus macaque (monkey)
Mane	*Macaca nemestrina*	Pigtail macaque
Otga	*Otolemur garnetti*	Greater bushbaby
Pren	*Presbytris entellus*	Entellus langur
Paha	*Papio hamadryas*	Hamadryas baboon
Papa	*Pan paniscus*	Pygmy chimpanzee
Patr	*Pan troglodytes*	Chimpanzee
Pef	*Petterus (Lemur) fulvus*	Brown lemur
Popy	*Pongo pygmaeus*	Orangutan
Sasc	*Saimiri sciureus*	Common squirrel monkey
Saoe	*Saguinus oedipus*	Cotton-top tamarin

Table 2. Sources of DRB sequences

Name	Synonym	Reference
*Patr-DRB1*1001*		5, 6
*Gogo-DRB1*1001*		8, 9
*Mamu-DRB1*1001*		10
*Mamu-DRB1*1002*		10
*Mamu-DRB1*1003*		10
*Mamu-DRB1*1004*		20
*Mamu-DRB1*1005*		20
*Mamu-DRB1*1006*		20
*Mamu-DRB1*1007*		20
*Patr-DRB1*0201*	*Patr-DRB1*02*	5, 6, 13
*Patr-DRB1*0202*		5
*Patr-DRB1*0203*		5
*Patr-DRB1*0204*		5, 6
*Patr-DRB1*0205*		5
*Patr-DRB1*0207*	PATRDRB*07	1
*Patr-DRB1*0208*	PATRDRB*08	1
*Patr-DRB1*0209*	PATRDRB*09	1
*Patr-DRB1*0210*	PATRDRB*05	1
*Patr-DRB1*0212*	PATRDRB*43	1
*Patr-DRB1*0213*		5
*Papa-DRB1*0201*	PAPADRB*12	1
*Papa-DRB1*0202*	PAPADRB*13	1
*Gogo-DRB1*0201*	GOGODRB*15	1
*Patr-DRB1*0301*	PATRDRB*01	1
*Patr-DRB1*0302*	PATRDRB*02	1, 5, 6
*Patr-DRB1*0303*	PATRDRB*03	1
*Patr-DRB1*0304*	PATRDRB*04	1
*Patr-DRB1*0305*		5, 6
*Patr-DRB1*0306*		5
*Patr-DRB1*0307*		5
*Patr-DRB1*0308*		5

Table 2. (cont.)

Name	Synonym	Reference
Patr-DRB1*0309		5, 6
Patr-DRB1*0310		5
Papa-DRB1*0301	PAPADRB*10	1
Papa-DRB1*0302	PAPADRB*11	1
Gogo-DRB1*0301	Gogo-DRB1*08	4, 9
Gogo-DRB1*0302	GOGODRB*14	1
Gogo-DRB1*0303	GOGODRB*47	1
Gogo-DRB1*0304		8
Gogo-DRB1*0305	Gogo-DRB1*0307	9
Gogo-DRB1*0306		8
Gogo-DRB1*0307		8
Gogo-DRB1*0308		8
Popy-DRB1*0301		12
Popy-DRB1*0302		12
Male-DRB1*0301	PALEDRB*20	1
Paha-DRB1*0301	Paha-DRB	14
Mane-DRB1*0301	Mane-DRB*01	2
Mamu-DRB1*0301	MAMUDRB*18	1
Mamu-DRB1*0302	MAMUDRB*19	1
Mamu-DRB1*0303		10
Mamu-DRB1*0304		10
Mamu-DRB1*0305		10
Mamu-DRB1*0306		10
Mamu-DRB1*0307		10
Mamu-DRB1*0308		10
Mamu-DRB1*0309		10
Mamu-DRB1*0310		20
Mamu-DRB1*0311		20
Mamu-DRB1*0312		20
Mamu-DRB1*0313		20
Mamu-DRB1*0314		20
Aotr-DRB1*0301		11
Aotr-DRB1*0302		11
Caja-DRB1*0301		11
Camo-DRB1*0301		11
Camo-DRB1*0302		11
Camo-DRB1*0303		11
Camo-DRB1*0304		11
Camo-DRB1*0305		11
Saoe-DRB1*0301	Saoe-DRB2*02	16
Saoe-DRB1*0302		11
Saoe-DRB1*0303		11
Saoe-DRB1*0304		11
Saoe-DRB1*0305		11
Gamo-DRB1*0301		17
Gase-DRB1*0301		17
Gase-DRB1*0302		17
Gase-DRB1*0303		17

Table 2. (cont.)

Name	Synonym	Reference
*Male-DRB1*0401*	*PALEDRB*21*	1
*Mamu-DRB1*0401*	*MAMUDRB*17*	1
*Mamu-DRB1*0402*		10
*Mamu-DRB1*0403*		10
*Mamu-DRB1*0404*		10
*Mamu-DRB1*0405*		20
*Gamo-DRB1*0401*		17
*Gamo-DRB1*0402*		17
*Gamo-DRB1*0403*		17
*Patr-DRB1*0701*		5
*Patr-DRB1*0702*		5, 6
*Mamu-DRB1*0701*		20
*Patr-DRB3*0102*		5, 6
*Patr-DRB3*0103*		5
*Gogo-DRB3*0101*	*Gogo-DRB3*01*	4
*Gogo-DRB3*0102*	*GOGODRB*45*	1
*Gogo-DRB3*0103*	*GOGODRB*46*	1
*Gogo-DRB3*0104*	*GOGODRB*48*	1
*Gogo-DRB3*0105*	*GOGODRB*49*	1
*Gogo-DRB3*0106*		8
*Gogo-DRB3*0107*		8
*Patr-DRB3*0201*	*Patr-DRB3*03*	6, 13
*Patr-DRB3*0202*	*PATRDRB*37*	1
*Patr-DRB3*0203*	*PATRDRB*38*	1
*Patr-DRB3*0204*	*PATRDRB*39*	1
*Patr-DRB3*0205*	*PATRDRB*42*	1
*Patr-DRB3*0206*	*PATRDRB*44*	1
*Patr-DRB3*0207*		5, 6
*Patr-DRB3*0208*		1, 5, 6
*Patr-DRB3*0209*		5
*Patr-DRB3*0210*		5
*Patr-DRB3*0211*		5
*Patr-DRB3*0212*		5
*Patr-DRB3*0213*		5
*Patr-DRB3*0214*		5
*Patr-DRB3*0215*		5
*Gogo-DRB3*0401*	*Gogo-DRB3*02*	4, 9
*Mamu-DRB3*0401*		10
*Mamu-DRB3*0402*		10
*Mamu-DRB3*0403*		20
*Mamu-DRB3*0404*		20
*Mamu-DRB3*0405*		20
*Camo-DRB3*0501*		11
*Camo-DRB3*0502*		11
*Saoe-DRB3*0501*		11
*Saoe-DRB3*0502*		11
*Saoe-DRB3*0503*	*Saoe-DRB3*	16
*Saoe-DRB3*0504*		18

Table 2. (cont.)

Name	Synonym	Reference
Saoe-DRB3*0505		19
Saoe-DRB3*0506		19
Saoe-DRB3*0507		19
Saoe-DRB3*0508		19
Saoe-DRB3*0509		19
Saoe-DRB3*0510		19
Saoe-DRB3*0511		19
Saoe-DRB3*0512		19
Saoe-DRB3*0513		19
Saoe-DRB3*0514		19
Saoe-DRB3*0515		19
Saoe-DRB3*0516		19
Saoe-DRB3*0517		19
Otga-DRB3*0501		17
Otga-DRB3*0502		17
Otga-DRB3*0503		17
Otga-DRB3*0504		17
Otga-DRB3*0505		17
Otga-DRB3*0506		17
Otga-DRB3*0507		17
Otga-DRB3*0508		17
Aotr-DRB3*0601		11
Camo-DRB3*0601		11
Camo-DRB3*0701		11
Patr-DRB4*0101	PATRDRB*50	1
Patr-DRB4*0102	PATRDRB*51	1
Patr-DRB4*0104		5, 6
Patr-DRB4*0105		6
Patr-DRB4*0106		5
Patr-DRB4*0107		5
Patr-DRB4*0201		5
Mamu-DRB4*0101		10
Gogo-DRB5*0101	Gogo-DRB5*01	4
Gogo-DRB5*0102	GOGODRB*60	1
Gogo-DRB5*0103		8
Gogo-DRB5*0104		9
Mamu-DRB5*0101	MAMUDRB*64	1
Patr-DRB5*0101		5
Patr-DRB5*0102		5, 6
Gogo-DRB5*0201	GOGODRB*62	1
Patr-DRB5*0301	Patr-DRB5*01	5, 6, 13
Patr-DRB5*0302	PATRDRB*53	1
Patr-DRB5*0303	PATRDRB*57	1
Patr-DRB5*0304		5
Patr-DRB5*0305		5
Patr-DRB5*0306	PATRDRB*54	1, 5
Patr-DRB5*0307		5
Patr-DRB5*0308		5

Table 2. (cont.)

Name	Synonym	Reference
Patr-DRB5*0309		6
Patr-DRB5*0310	PATRDRB*55	1, 5, 6
Patr-DRB5*0311		5
Mamu-DRB5*0301		10
Mamu-DRB5*0302		10
Mamu-DRB5*0303		20
Mamu-DRB*0304		20
Patr-DRB5*0401	PATRDRB*58	1
Patr-DRB5*0402	PATRDRB*59	1
Gogo-DRB5*0401	GOGODRB*63	1
Gogo-DRB5*0402		8
Gogo-DRB5*0403		9
Gogo-DRB5*0404		8
Gogo-DRB5*0501		8
Gogo-DRB5*0502		8
Gogo-DRB5*0503		8
Popy-DRB5*0601		12
Popy-DRB5*0602		12
Popy-DRB5*0603		12
Popy-DRB5*0604		12
Saoe-DRB5*0701	Saoe-DRB4	16
Patr-DRB6*0101	PATRDRB*22	1
Patr-DRB6*0102	PATRDRB*24	1
Patr-DRB6*0103	PATRDRB*25	1
Patr-DRB6*0104	PATRDRB*26	1
Patr-DRB6*0105	Patr-DRB6	5, 6, 15
Patr-DRB6*0106		7
Patr-DRB6*0107		7
Patr-DRB6*0108		7
Patr-DRB6*0109		5, 6
Patr-DRB6*0110		5
Papa-DRB6*0101	PAPADRB*29	1
Gogo-DRB6*0101	GOGODRB*31	1
Gogo-DRB6*0102		7, 8, 9
Gogo-DRB6*0103		9
Mamu-DRB6*0101		10
Mamu-DRB6*0102		20
Mamu-DRB6*0103		20
Mamu-DRB6*0104		20
Mamu-DRB6*0105		20
Mamu-DRB6*0106		20
Mamu-DRB6*0107		20
Mamu-DRB6*0108		20
Mamu-DRB6*0109		20
Mamu-DRB6*0110		20
Mamu-DRB6*0111		20
Papa-DRB6*0201	PAPADRB*30	1
Papa-DRB6*0202	PAPADRB*28	1

Table 2. (cont.)

Name	Synonym	Reference
*Gogo-DRB6*0201*	*GOGODRB*33*	1
*Gogo-DRB6*0202*	*Gogo-DRB6*	4, 9
*Gogo-DRB6*0203*		7
*Popy-DRB6*0201*		7
*Popy-DRB6*0202*		7
*Patr-DRB6*0301*	*PATRDRB*23*	1, 5, 7
*Patr-DRB6*0302*		5
*Patr-DRB6*0303*		5
*Patr-DRB6*0304*		5
*Patr-DRB6*0305*	*PATRDRB*23*	1, 5, 6
*Gogo-DRB6*0401*	*GOGODRB*34*	1
*Male-DRB6*0401*	*PALEDRB*35*	1
*Gase-DRB6*0401*		17
*Gase-DRB6*0402*		17
*Gase-DRB6*0403*		17
*Patr-DRB7*0101*		5, 6
*Mamu-DRB*W101*	*Mamu-DRBW1*0101*	10
*Mane-DRB*W101*	*Mane-DRB*05*	2
*Mane-DRB*W102*	*Mane-DRB*04*	2
*Gase-DRB*W101*		17
*Gamo-DRB*W101*		17
*Mamu-DRB*W201*	*Mamu-DRBW2*0101*	10
*Mane-DRB*W201*	*Mane-DRB*02*	2
*Mane-DRB*W202*	*Mane-DRB*03*	2
*Mamu-DRB*W301*	*Mane-DRBW3*0101*	10
*Mamu-DRB*W302*	*Mamu-DRBW3*0102*	10
*Mamu-DRB*W303*	*Mamu-DRBW3*0103*	10
*Mamu-DRB*W304*	*Mamu-DRBW3*0201*	10
*Mamu-DRB*W305*	*Mamu-DRBW3*0202*	10
*Mamu-DRB*W306*	*Mamu-DRBW3*0203*	10
*Mamu-DRB*W307*		20
*Mamu-DRB*W308*		20
*Mamu-DRB*W309*		20
*Gase-DRB*W301*		17
*Gase-DRB*W302*		17
*Mamu-DRB*W401*	*Mamu-DRBW4*0101*	10
*Mamu-DRB*W501*	*Mamu-DRBW5*0101*	10
*Mamu-DRB*W601*	*Mamu-DRBW6*0101*	10
*Mamu-DRB*W602*	*Mamu-DRBW6*0102*	10
*Mamu-DRB*W603*	*Mamu-DRBW6*0103*	10
*Mamu-DRB*W604*		20
*Mamu-DRB*W605*		20
*Mamu-DRB*W606*		20
*Mamu-DRB*W607*		20
*Mamu-DRB*W701*	*Mamu-DRBW7*0101*	10
*Mamu-DRB*W702*	*Mamu-DRBW7*0102*	10
*Patr-DRB*W801*	*Patr-DRB*Y0101*	5
*Gogo-DRB*W801*	*GOGODRB*16*	1

Table 2. (cont.)

Name	Synonym	Reference
Gogo-DRB*W802	Gogo-DRBY*01	8, 9
Patr-DRB*W901	Patr-DRB*X0101	5
Patr-DRB*W902	Patr-DRB*X02	6
Patr-DRB*W903	Patr-DRB*X03	6
Patr-DRB*W904	Patr-DRB*X0104	5
Patr-DRB*W905	Patr-DRB*X0105	5
Gogo-DRB*W1001	Gogo-DRB*Z01	8
Camo-DRB11*0101		11
Caja-DRB11*0101		11
Saoe-DRB11*0101	Saoe-DRB1*0101	3
Saoe-DRB11*0102		11
Saoe-DRB11*0103	Saoe-DRB1*0102	3
Saoe-DRB11*0104		11
Saoe-DRB11*0105		11
Saoe-DRB11*0106		19
Saoe-DRB11*0107		19
Saoe-DRB11*0108		19
Saoe-DRB11*0109-		19
Saoe-DRB11*0110-		19
Caja-DRB*W1201		11
Saoe-DRB*W1201	Saoe-DRB2	3
Saoe-DRB*W1202		11
Saoe-DRB*W1202		18
Saoe-DRB*W1203		19
Saoe-DRB*W1204		19
Saoe-DRB*W1205		19
Sasc-DRB*W1201		11
Sasc-DRB*W1202		11
Sasc-DRB*W1203		11
Sasc-DRB*W1204		11
Lota-DRB*W1201		17
Otga-DRB*W1201		17
Otga-DRB*W1202		17
Otga-DRB*W1203		17
Otga-DRB*W1204		17
Otga-DRB*W1205		17
Otga-DRB*W1206		17
Otga-DRB*W1207		17
Aotr-DRB*W1301		11
Aotr-DRB*W1302		11
Ceap-DRB*W1301		11
Ceap-DRB*W1302		11
Ceap-DRB*W1303		11
Camo-DRB*W1401		11
Camo-DRB*W1402		11
Sasc-DRB*W1401		11
Sasc-DRB*W1402		11
Sasc-DRB*W1403		11

Table 2. (cont.)

Name	Synonym	Reference
Sasc-DRB*W1404		11
Ceap-DRB*W1501		11
Ceap-DRB*W1502		11
Caja-DRB*W1601		11
Caja-DRB*W1602		11
Caja-DRB*W1603		11
Camo-DRB*W1701		11
Aotr-DRB*W1801		11
Sasc-DRB*W1901		11
Sasc-DRB*W1902		11
Sasc-DRB*W1903		11
Mamu-DRB*W2001		20
Mamu-DRB*W2101		20
Mamu-DRB*W2102		20
Saoe-DRB*W2201		19
Saoe-DRB*W2202		19
Saoe-DRB*W2203		19
Saoe-DRB*W2204		19
Saoe-DRB*W2205		19
Saoe-DRB*W2206		19
Saoe-DRB*W2207		19
Saoe-DRB*W2208		19
Saoe-DRB*W2209	Saoe-DRB*W1202	11
Lota-DRB*W2301	Lota-DRB*W2001	17
Lota-DRB*W2302	Lota-DRB*W2002	17
Lota-DRB*W2303	Lota-DRB*W2003	17
Lota-DRB*W2304	Lota-DRB*W2004	17
Lota-DRB*W2305	Lota-DRB*W2005	17
Lota-DRB*W2306	Lota-DRB*W2006	17
Lota-DRB*W2307	Lota-DRB*W2007	17
Lota-DRB*W2308	Lota-DRB*W2008	17
Lota-DRB*W2309	Lota-DRB*W2009	17
Pefu-DRB*W2401	Pefu-DRB*W2101	17
Pefu-DRB*W2402	Pefu-DRB*W2102	17

The references are; 1, Gyllensten et al. 1991; 2, Zhu et al. 1991; 3, Grahovac et al. 1991; 4, Kasahara et al. 1992; 5, Kenter et al. 1992a; 6, Mayer et al. 1992; 7, Corell et al. 1992; 8, Kupfermann et al. 1992; 9, Kenter et al. 1992b; 10, Slierendregt et al. 1992; 11, Trtková et al. 1992; 12, Schönbach et al. 1992; 13, Fan et al. 1989; 14, Riess et al. 1990; 15, Figueroa et al. 1991; 16, Grahovac et al., unpublished; 17, Figueroa et al. 1994; 18, Bidwell et al. 1994; 19, Gyllensten et al. 1994; 20, Slierendregt et al. 1994.

Table 3. Sources of DQA alleles

Name	Synonym	Reference
*Patr-DQA1*0101*	*Patr-DQA1*01*	1
*Patr-DQA1*0501*	*Patr-DQA1*03*	1
*Patr-DQA1*0502*	*Patr-DQA1*04*	1
*Patr-DQA1*2001*	*Patr-DQA1*02*	1
*Patr-DQA1*2002*	*Patr-DQA1*02*	2
*Patr-DQA2*0101*	*Patr-DQA2*01*	1
*Gogo-DQA1*0101*		3
*Gogo-DQA1*0301*		3
*Gogo-DQA1*0501*	*Gogo-DQA1*02*	1
*Gogo-DQA1*0502*		3
*Gogo-DQA1*0503*		3
*Gogo-DQA1*2001*	*Gogo-DQA1*01*	1
*Gogo-DQA2*0101*	*Gogo-DQA2*01*	1
*Gogo-DQA2*0102*	*Gogo-DQA2*02*	1
*Popy-DQA1*0101*	*Popy-DQA1*01*	1
*Popy-DQA1*0501*	Popy-DQA1*03	1
*Popy-DQA1*2101*	*Popy-DQA1*02*	1
*Popy-DQA2*0101*	*Popy-DQA2*01*	1
*Hyla-DQA1*0101*	*Hyla-DQA1*04*	1
*Hyla-DQA1*2201*	*Hyla-DQA1*01*	1
*Hyla-DQA1*2202*	*Hyla-DQA1*02*	1
*Hyla-DQA1*2203*	*Hyla-DQA1*03*	1
Hyla-DQA2+0101	*Hyla-DQA2*01*	1
*Hyla-DQA2*0102*	*Hyla-DQA2*02*	1
*Paha-DQA1*0101*	*Paha-DQA1*02*	1
*Paha-DQA1*0501*	*Paha-DQA1*03*	1
*Paha-DQA1*2501*	*Paha-DQA1*01*	1
*Ceae-DQA1*0501*	*Ceae-DQA1*03*	1
*Ceae-DQA1*2501*	*Ceae-DQA1*01*	1
*Ceae-DQA1*2502*	*Ceae-DQA1*02*	1
*Mamu-DQA1*0101*	*Mamu-DQA1*08*	1
*Mamu-DQA1*0102*	*Mamu-DQA1*09*	1
*Mamu-DQA1*0103*	*Mamu-DQA1*10*	1
*Mamu-DQA1*0104*	*Mamu-DQA1*11*	1
*Mamu-DQA1*0105*		6
*Mamu-DQA1*0106*		6
*Mamu-DQA1*0501*	*Mamu-DQA1*12*	1
*Mamu-DQA1*2301*	*Mamu-DQA1*01*	1
*Mamu-DQA1*2401*	*Mamu-DQA1*02*	1
*Mamu-DQA1*2402*	*Mamu-DQA1*03*	1
*Mamu-DQA1*2403*	*Mamu-DQA1*04*	1
*Mamu-DQA1*2501*	*Mamu-DQA1*05*	1
*Mamu-DQA1*2502*	*Mamu-DQA1*06*	1
*Mamu-DQA1*2503*	*Mamu-DQA1*07*	1
*Maar-DQA1*0101*	*Maar-DQA1*01*	1
*Maar-DQA1*0102*	*Maar-DQA1*02*	1
*Mafa-DQA1*0101*	*Mafa-DQA1*03*	1
*Mafa-DQA1*2401*	*Mafa-DQA1*01*	1

Table 3. (cont.)

Name	Synonym	Reference
Mafa-DQA1*2501	Mafa-DQA1*02	1
Saoe-DQA1*2501		4
Saoe-DQA1*2502		4
Saoe-DQA1*2503	Saoe-DQA1*02	5

The references are: 1, Kenter et al. 1992; 2, Kenter et al. 1993a; 3, Kenter et al 1993b; 4, Bidwell et al. 1993; 5, Gyllensten et al. 1994; 6, Sauermann et al. 1995.

Table 4. Sources of DQB alleles

Name	Synonym	Reference
Patr-DQB1*0601		2, 3
Patr-DQB1*0602		2
Patr-DQB1*0603		3
Patr-DQB1*0604		2, 3
Patr-DQB1*0605		3
Patr-DQB1*0606	Patr-DQB1*08	4
Patr-DQB1*0301		3
Patr-DQB1*0302		2
Patr-DQB1*0303		7
Patr-DQB1*1501		2
Patr-DQB1*1502	Patr-DQB1*09	4
Patr-DQB1*2101		3
Patr-DQB2*0101		3
Patr-DQB2*0102		3
Patr-DQB2*0103		5
Papa-DQB1*0301		3
Papa-DQB1*0302		3
Papa-DQB1*0303		3
Papa-DQB1*0304		3
Papa-DQB1*0305		3
Papa-DQB2*0101		3
Gogo-DQB1*0501		3
Gogo-DQB1*0502		2
Gogo-DQB1*0503		3
Gogo-DQB1*0601		3
Gogo-DQB1*0201		3
Gogo-DQB1*0202		2, 3
Gogo-DQB1*1901		3
Gogo-DQB1*1902		3
Gogo-DQB1*1903	Patr-DQB1*01	2
Gogo-DQB1*1904		5
Gogo-DQB2*0101		2, 4, 5
Gogo-DQB2*0102		3
Popy-DQB1*0601		2
Popy-DQB1*0602		2
Popy-DQB1*0603		2

Table 4. (cont.)

Name	Synonym	Reference
Popy-DQB1*1601		2
Popy-DQB1*1701		5
Popy-DQB2*0101		2
Hyla-DQB1*2001	*Hyla-DQB1*01*	4
Hyla-DQB1*2002		2
Hyla-DQB2*0101		2
Hyla-DQB2*0102		5
Ceae-DQB1*1501		2
Ceae-DQB1*1801		2
Ceae-DQB1*1802		2
Male-DQB1*1501		3
Paha-DQB1*0601		2
Paha-DQB1*0602		2
Paha-DQB1*1501		2
Mamu-DQB1*0601		2
Mamu-DQB1*0602		2
Mamu-DQB1*0603		2
Mamu-DQB1*0604		2
Mamu-DQB1*0605		2
Mamu-DQB1*1501		2
Mamu-DQB1*1502		6
Mamu-DQB1*1601		2
Mamu-DQB1*1602		2
Mamu-DQB1*1701		3
Mamu-DQB1*1702		2
Mamu-DQB1*1801		2, 6
Mamu-DQB1*1802		2
Mamu-DQB1*1803		2
Mamu-DQB1*1804		2
Mamu-DQB1*1805		2
Mamu-DQB1*1806		2
Mamu-DQB1*1807		2
Mamu-DQB1*1808		2
Mafa-DQB1*0601		2
Mafa-DQB1*0602		2
Mafa-DQB1*0603		2
Mafa-DQB1*0604		2
Mafa-DQB1*0605	*Mafa-DQB1*01*	5
Mafa-DQB1*1501		2
Mafa-DQB1*1502		2
Mafa-DQB1*1701		2
Mafa-DQB1*1702	*Mafa-DQB1*02*	5
Maar-DQB1*0601		2
Maar-DQB1*0602		2
Maar-DQB1*1701		2
Pren-DQB1*1601		3
Pren-DQB1*1801		3

Table 4. (cont.)

Name	Synonym	Reference
*Caja-DQB1*2201*		2
*Caja-DQB1*2301*		2
*Caja-DQB2*0101*		2
*Aotr-DQB2*0101*		4
*Aotr-DQB2*0102*		4
*Saoe-DQB1*2201*		1
*Saoe-DQB1*2202*	*Saoe-DQB1*02*	8
*Saoe-DQB1*2301*	*Saoe-DQB1*01*	1, 8
*Saoe-DQB2*0101*	*Saoe-DQB2*01*	1, 8

The references are: 1, Bidwell et al 1993; 2, Otting et al. 1992; 3, Gyllensten and Erlich 1990; 4, Gaur et al. 1992b; 5, Gaur et al. 1992a; 6, Slierendregt et al. 1993a; 7, Slierendregt et al. 1993b; 8, Gyllensten et al. 1994.

Table 5. Sources of *DP* alleles

Name	Synonym	Reference
DPA		
*Mamu-DPA1*0101*		EMBL Z32411
*Patr-DPA1*0201*		1
*Patr-DPA1*0202*		1
*Popy-DPA1*0201*		1
*Popy-DPA1*0202*		1
*Paha-DPA1*0201*		1
*Mafa-DPA1*0201*		1
*Mamu-DPA1*0201*		1
*Patr-DPA1*0301*		1
*Gogo-DPA1*0401*		1
*Gogo-DPA1*0402*		1
*Popy-DPA1*0401*		1
*Sasc-DPA1*0501*		1
*Sasc-DPA1*0502*		1
*Sasc-DPA1*0601*		1
DPB		
*Mamu-DPB1*01*		2
*Mamu-DPB1*02*		2
*Mamu-DPB1*03*		2
*Mamu-DPB1*04*		2
*Mamu-DPB1*05*		2
*Mamu-DPB1*06*		2
*Mamu-DPB1*07*		2
*Mamu-DPB1*08*		2
*Mamu-DPB1*09*		2
*Mamu-DPB1*10*		2
*Mamu-DPB1*11*		2

Table 5. (cont.)

Name	Synonym	Reference
*Mamu-DPB1*12*		2
*Mamu-DPB1*13*		2
Mafa-DPB1-M09		3
Mafa-DPB1-M25		3
*Saoe-DPB1*0101*		4

The references are: 1, Otting and Bontrop 1995; 2, Slierendregt et al. 1995; 3, Hashiba et al. 1993; 4, Bidwell et al. 1994.

Fig. 1. Alignment of exon 2 *Mhc-DRB* nucleotide sequences of nonhuman primates. Consensus nucleotide sequence is that used by O'hUigin et al. (1993) for DRB. A *dash* indicates identity with this consensus, an *asterisk* a deletion. Unavailability of sequence data is indicated by a *dot*. The numbering of codons is according to the amino acid position in the β1 domain. Frameshifting insertions are omitted from the alignment

Fig. 2. Alignment of exon 2 *Mhc-DQA* and *Mhc-DQB* nucleotide sequences of nonhuman primates. Consensus nucleotide sequence is that used by Bontrop (1994) for DQA and DQB. A *dash* indicates identity with this consensus, an *asterisk* a deletion. Unavailability of sequence data is indicated by a *dot*. The numbering of codons is according to the amino acid position in the α1 or β domain

Fig. 3. Alignment of exon 2 *Mhc-DPA* and *Mhc-DPB* nucleotide sequences of nonhuman primates. Consensus nucleotide sequence is determined by simple majority for a particular position. A *dash* indicates identity with this consensus, an *asterisk* a deletion. Unavailability of sequence data is indicated by a *dot*. The numbering of codons is according to the amino acid position in the α1 or β domain

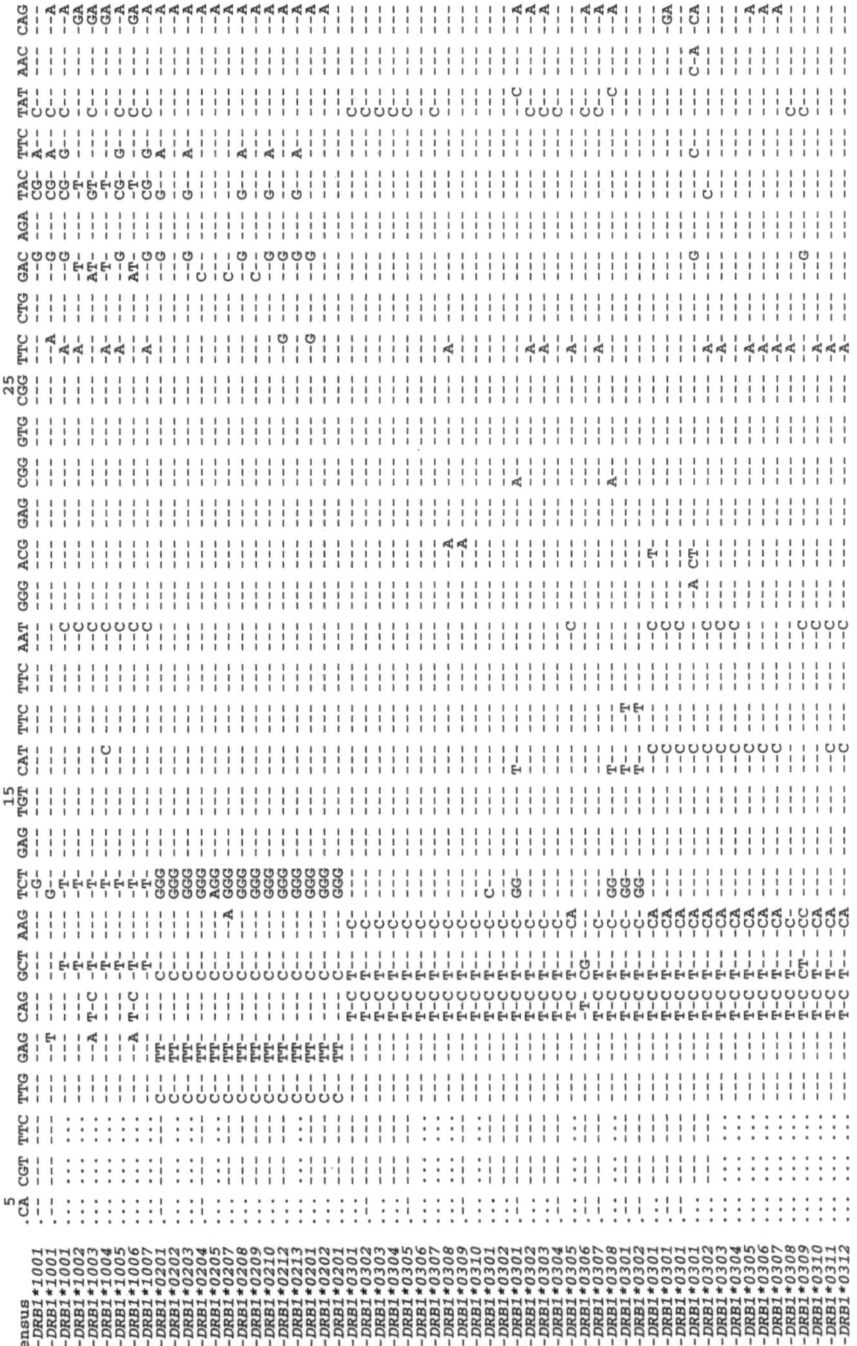

Fig. 1.

Fig. 1.

Fig. 1.

Fig. 1.

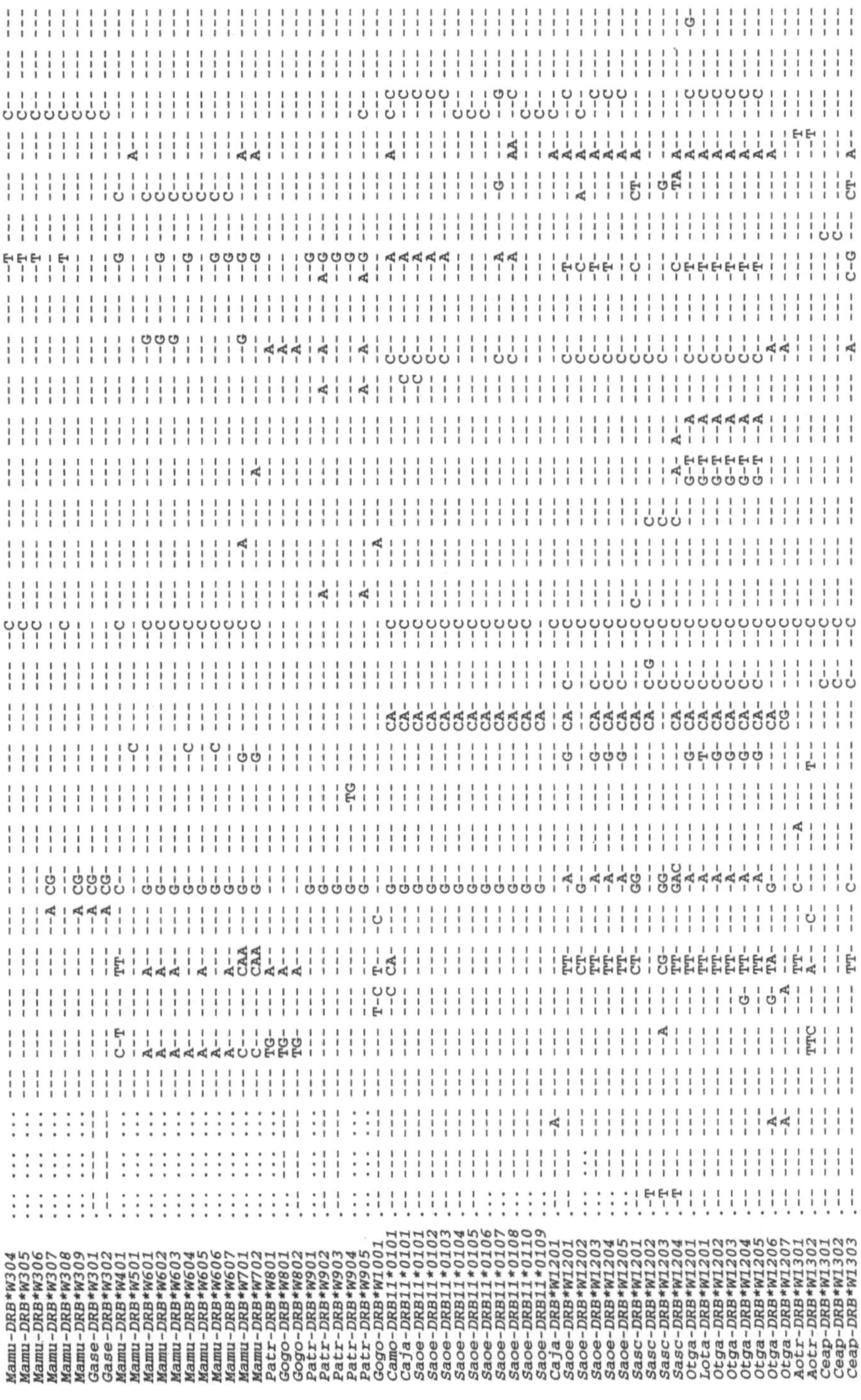

Fig. 1.

Fig. 1.

Fig. 1.

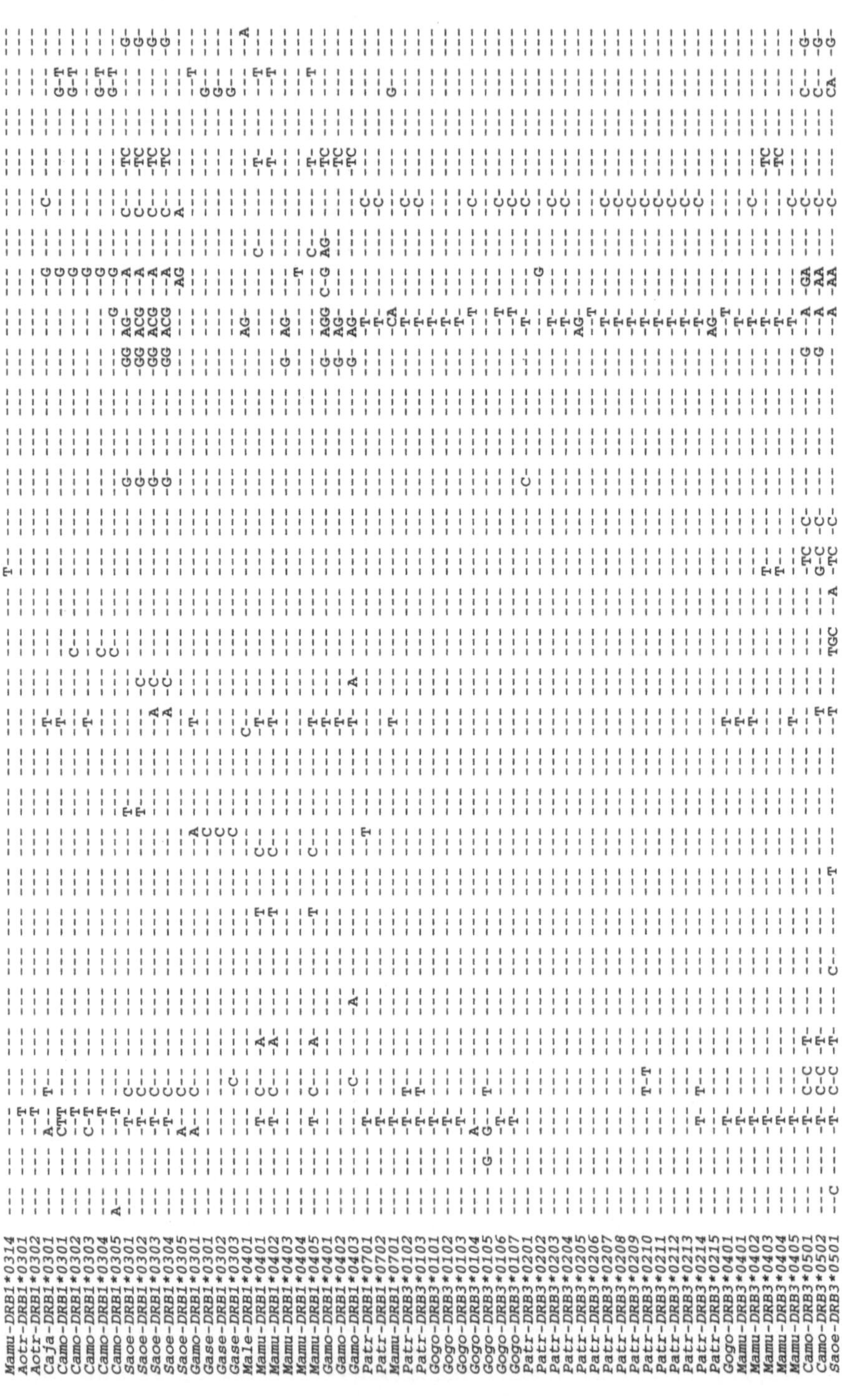

Fig. 1.

Fig. 1.

Fig. 1.

Fig. 1.

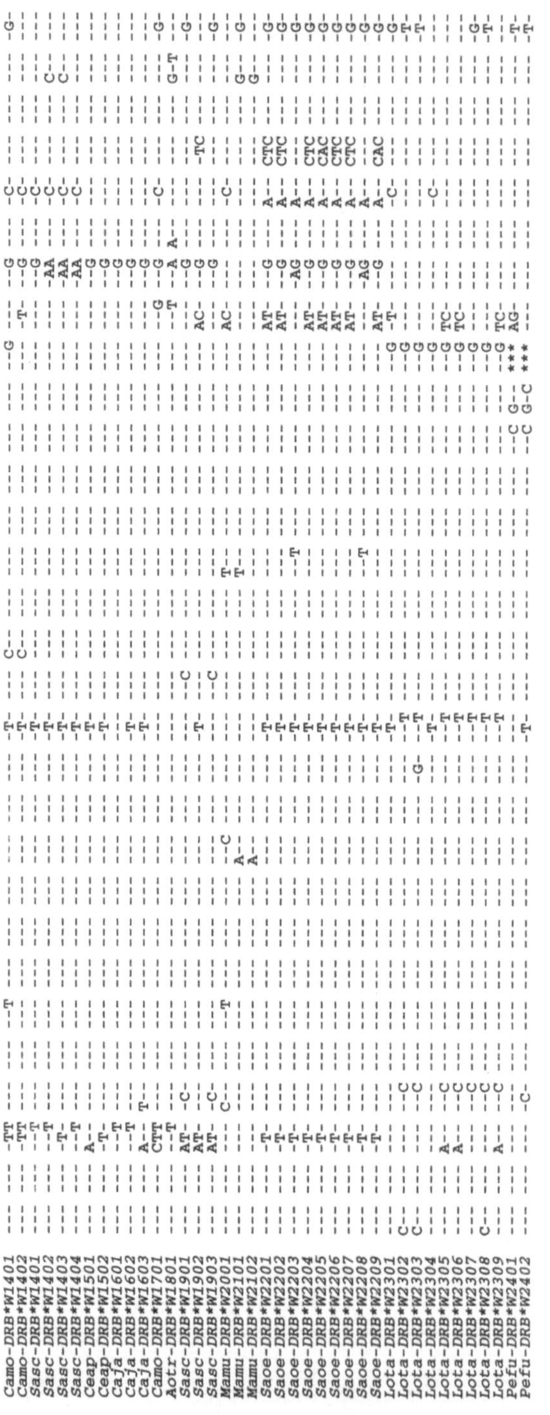

Camo-DRB*W1401
Camo-DRB*W1402
Sasc-DRB*W1401
Sasc-DRB*W1402
Sasc-DRB*W1403
Sasc-DRB*W1404
Ceap-DRB*W1501
Ceap-DRB*W1502
Caja-DRB*W1601
Caja-DRB*W1602
Caja-DRB*W1603
Camo-DRB*W1701
Aotr-DRB*W1801
Sasc-DRB*W1901
Sasc-DRB*W1902
Sasc-DRB*W1903
Mamu-DRB*W2001
Mamu-DRB*W2101
Mamu-DRB*W2102
Saoe-DRB*W2201
Saoe-DRB*W2202
Saoe-DRB*W2203
Saoe-DRB*W2204
Saoe-DRB*W2205
Saoe-DRB*W2206
Saoe-DRB*W2207
Saoe-DRB*W2208
Saoe-DRB*W2209
Lota-DRB*W2301
Lota-DRB*W2302
Lota-DRB*W2303
Lota-DRB*W2304
Lota-DRB*W2305
Lota-DRB*W2306
Lota-DRB*W2307
Lota-DRB*W2308
Lota-DRB*W2309
Pefu-DRB*W2401
Pefu-DRB*W2402

Fig. 1.

Fig. 1.

```
Mamu-DRB1*0314 --- --- --- --- --A G-C C-- --- --- TC- --- --- --- -T- --- --- --- --- --- C-- --- -CG --- --- --- ... ... ... ... ... .
Aotr-DRB1*0301 --- --- TA- G-- --- --- --- --- -G- CG- --- --- -A- --- --- --- --- --- --- --- --- -G- --- --- --- --- --- --- --- A-- -
Aotr-DRB1*0302 --- --- TA- G-- --- -G- --- --- -G- C-- --- --- -A- --- --- --- --- --- --- --- --- -G- --- --- --- --- --- --- --- A-- -
Caja-DRB1*0301 --- --- T-- --- --- G-C -G- --- --- --- --- --- --- GTG --- --- --- --- --- -A- A-- T-A --C --G --- CT- --- -C- --- A-- -
Camo-DRB1*0301 G-- --- T-- --- --- -G- --- --- -G- C-- --- --- -A- --- --- --- --- --- --- --- --- T-- --- --- --- --- --- --- --- A-- -
Camo-DRB1*0302 G-- --- T-- --- --- --- --- --- -G- C-- --- --- -A- --- --- --- --- --- --- -A- A-- TCA --- --G --- CT- --- -C- --- A-- -
Camo-DRB1*0303 --- --A C-- --- --- --- --- --- -G- C-- --- --- -A- --- --- --- --- --- --- -A- A-- T-A --- --G --- CT- --- -C- --- A-- -
Camo-DRB1*0304 G-- --- T-- --- --- --- --- --- -G- C-- --- --- -A- --- --- --- --- --- --- -A- A-- TCA --- --G --- CT- --- -C- --- A-- -
Camo-DRB1*0305 G-- --- T-- --- --- --- --- --- -G- C-- --- --- -A- --- --- --- --- --- --- -A- A-- T-A --- --G --- CT- --- -C- --- A-- -
Saoe-DRB1*0301 --- A-- TA- A-- --- G-T GC- --- --- -C- --- --- -GT *** --- --- --- --- --- -A- A-- TCA --- --G --- CT- --- -T- --- A-- .
Saoe-DRB1*0302 --- --A TA- A-- --- G-T GC- --- --- -C- --- --- -GT *** --- --- --- --- --- -A- A-- TCA --- --G --- CT- --- -T- --- A-- -
Saoe-DRB1*0303 --- --A TA- A-- --- G-T GC- --- --- -C- --- --- -GT *** --- --- --- --- --- -A- A-- TCA --- --G --- CT- --- -T- --- A-- -
Saoe-DRB1*0304 --- --A TA- A-- --- G-T GC- --- --- -C- --- --- -GT *** --- --- --- --- --- -A- A-- TCA --- --G --- CT- --- -T- --- A-- -
Saoe-DRB1*0305 --- --- --- --- --- --- --- --- --- -C- --- --- --- -T- --- --- --- --- --- -A- A-- TCA --- --G --- CT- --- -C- --- A-- -
Gamo-DRB1*0301 --- --- T-- --- --- --- -G- --- --- -C- --- --- --- --- --- --- --- --- --- --- -C- --G --- --- --- --- --- --- --- --- -
Gase-DRB1*0301 G-- --- --- --- --A G-C GC- --- --- -C- --- --- --- --- --- --- --- --- --- --- --- T-- --- --- --- --- --- --- --- --- -
Gase-DRB1*0302 G-- --- --- --- --A G-C GC- --- --- -C- --- --- --- --- --- --- --- --- --- --- --- T-- --- --- --- --- --- --- --- --- -
Gase-DRB1*0303 G-- --- --- --- --A G-C GC- --- --- -C- --- --- --- --- --- --- --- --- --- --- --- T-- --- --- --- --- --- --- --- --- -
Male-DRB1*0401 --- --- --- --- --A G-C C-- --- --- -C- --- --- --- --- ... ... ... ... ... ... ... ... ... ... ... ... ... ... ... ... .
Mamu-DRB1*0401 --- --- C-- --- --A G-C -G- --- --- CG- --- --- --- --- ... ... ... ... ... ... ... ... ... ... ... ... ... ... ... ... .
Mamu-DRB1*0402 --- --- C-- --- --A G-C -G- --- --- -G- --- --- --- --- --- --- --- --- --- --- --- -CG --- --- --- ... ... ... ... ... .
Mamu-DRB1*0403 --- --- --- --- --A G-C -G- --- --- AT- --- --- --- --- --- --- --- --- --- --- --- -G- --- --- --- ... ... ... ... ... .
Mamu-DRB1*0404 --- --- T-- --- --A G-C -G- --- --- C-- --- --- --- GTG --- --- --- --- --- --- --- --G --- --- --- ... ... ... ... ... .
Mamu-DRB1*0405 --- --- C-- --- --A G-C -G- --- --- -G- --- --- --- --- --- --- --- --- --- --- --- T-- --- --- --- ... ... ... ... ... .
Gamo-DRB1*0401 --- --- C-- --- --- --- --- --- --- -C- --- --- -AT --- --- --- --- --- --- --- --- --G --- --- --- --- --- --- --- --- -
Gamo-DRB1*0402 --- --- C-- --- --- --- --- --- --- -C- --- --- -AT --- --- --- --- --- --- --- --- --G --- --- --- --- --- --- --- --- -
Gamo-DRB1*0403 --- --- C-- --- --- --- --- --- --- -C- --- --- -AT --- --- --- --- --- --- --- -C- --G --- --- --- --- --- --- --- --- -
Patr-DRB1*0701 --- --- C-- --- --- G-C -G- --- -G- C-- --- --- --- GTG --- --- --- --- --- --- --- T-G --- --- --- ... ... ... ... ... .
Patr-DRB1*0702 --- --- C-- --- --- G-C -G- --- -G- C-- --- --- --- GTG --- --- --- --- --- --- --- -G- --- --- --- --- --- --- --- --- -
Mamu-DRB1*0701 --- --- T-- --- --- --- -G- --- --- TC- --- --- -AT --- --- --- --- --- --- A-- -G- --G --- --- --- ... ... ... ... ... .
Patr-DRB3*0102 --- --- TA- G-- --- --- --- --- -G- C-- --- --- -AT --- --- --- --- --- --- A-- --- -G- --- --- --- --- --- --- --- --- -
Patr-DRB3*0103 --- --- TA- G-- --- --- --- --- --- A-- --- --- -AT --- --- --- --- --- --- A-- --- -G- --- --- --- ... ... ... ... ... .
Gogo-DRB3*0101 --- --- C-- --- --- --- --- --- -G- CG- --- --- -AT --- --- --- --- --- --- --- --- --G --- --- --- --- --- --- --- --- -
Gogo-DRB3*0102 --- --- C-- --- --- --- --- --- -G- CG- --- --- -A- --- ... ... ... ... ... ... ... ... ... ... ... ... ... ... ... ... .
Gogo-DRB3*0103 --- --- C-- --- --- --- --- --- -G- CC- --- --- --- --- ... ... ... ... ... ... ... ... ... ... ... ... ... ... ... ... .
Gogo-DRB3*0104 --- --- C-- --- --- --- --- --- --- -C- --- --- -A- --- ... ... ... ... ... ... ... ... ... ... ... ... ... ... ... ... .
Gogo-DRB3*0105 --- --- C-- --- --- --- --- --- --- -C- --- --- -A- --- ... ... ... ... ... ... ... ... ... ... ... ... ... ... ... ... .
Gogo-DRB3*0106 --- --- C-- --- --- --- --- --- -G- C-- --- --- -AT --- --- --- --- --- --- --- --- -G- --- --- --- --- --- --- --- --- -
Gogo-DRB3*0107 --- --- C-- --- --- --- --- --- -G- C-- --- --- -AT --- --- --- --- --- --- --- --- -G- --- --- --- --- --- --- --- --- -
Patr-DRB3*0201 --- --- C-- --- --- --- --- --- -G- C-- --- --- -AT --- --- --- --- --- --- --- -C- --G --- --- --- --- --- --- --- --- -
Patr-DRB3*0202 --- --- --- --- --- -G- -G- --- --- -C- --- --- -AT --- ... ... ... ... ... ... ... ... ... ... ... ... ... ... ... ... .
Patr-DRB3*0203 --- -G- C-- --- --- --- --- --- -G- C-- --- --- -AT --- ... ... ... ... ... ... ... ... ... ... ... ... ... ... ... ... .
Patr-DRB3*0204 --- --- C-- --- --- --- --- --- -G- C-- -A- --- -AT --- ... ... ... ... ... ... ... ... ... ... ... ... ... ... ... ... .
Patr-DRB3*0205 --- --- TA- G-- --- --- --- --- -G- C-- --- --- -A- --- ... ... ... ... ... ... ... ... ... ... ... ... ... ... ... ... .
Patr-DRB3*0206 --- --- TA- G-- --- --- --- --- -G- C-- --- --- -A- --- ... ... ... ... ... ... ... ... ... ... ... ... ... ... ... ... .
Patr-DRB3*0207 --- --- C-- --- --- --- --- --- -G- C-- --- --- -AT --- --- --- --- --- --- --- -C- --G --- --- --- --- --- --- --- --- -
Patr-DRB3*0208 --- --- C-- --- --- --- --- --- -G- C-- --- --- -AT --- --- --- --- --- --- --- -C- --G --- --- --- --- --- --- --- --- -
Patr-DRB3*0209 --- --- C-- --- --- --- --- --- -G- C-- --- --- -AT --- --- --- --- --- --- A-- -C- --G --- --- --- ... ... ... ... ... .
Patr-DRB3*0210 --- --- C-- --- --- --- --- --- --- A-- --- --- -AT --- --- --- --- --- --- --- -C- --G --- --- --- ... ... ... ... ... .
Patr-DRB3*0211 --- --- C-- --- --- --- --- --- --- A-- --- --- -AT --- --- --- --- --- --- --- -C- --G --- --- --- ... ... ... ... ... .
Patr-DRB3*0212 --- --- C-- --- --- --- --- --- --- AG- --- --- -AT --- --- --- --- --- --- --- -C- --G --- --- --- ... ... ... ... ... .
Patr-DRB3*0213 --- --- C-- --- --- --- --- --- --- A-- --- --- -AT --- --- --- --- --- --- --- -C- --G --- --- --- ... ... ... ... ... .
Patr-DRB3*0214 --- --- C-A --- --- --- --- --- --- A-- --- --- -AT --- --- --- --- --- --- A-- --- -G- --- --- --- ... ... ... ... ... .
Patr-DRB3*0215 --- --- TA- G-- --- --- --- --- --- A-- --- --- -AT --- --- --- --- --- --- A-- --- -G- --- --- --- ... ... ... ... ... .
Gogo-DRB3*0401 --- --- --- --- --A G-C G-- --- --- TC- --- --- --- --T --G --- T-- --- --- --- A-- --G --- --- --- --- --- --- --- --- -
Mamu-DRB3*0401 --- --- T-- --- --- --- --- --- --- -C- --- --- -A- --- --- --- T-- --- --- --- --- --G --- --- --- ... ... ... ... ... .
Mamu-DRB3*0402 --- --- T-- --- --- --- --- --- --- -G- --- --- -A- --- --- --- T-- --- --- --- --- --G --- --- --- ... ... ... ... ... .
Mamu-DRB3*0403 --- --- TA- --- --- --- -G- --- --- C-- --- --- -A- --- --- --- T-- --- --- C-- --- -G- --- --- --- ... ... ... ... ... .
Mamu-DRB3*0404 --- --- TA- --- --- --- -G- --- --- AG- T-- --- -A- --- --- --- T-- --- --- C-- --- -G- --- --- --- ... ... ... ... ... .
Mamu-DRB3*0405 --- --- T-- --- --- --- --- --- --- -G- --- --- -A- --- --- --- T-- --- --T --- --- --G --- --- --- ... ... ... ... ... .
Camo-DRB3*0501 --- --A C-- --- --- -G- --- --- -G- C-- --- --- -AG --- --- --- --- --- --- A-- --- -G- --- --- --- --- --- --- --- --- -
Camo-DRB3*0502 --- --A C-- --- --- -G- --- --- -G- C-- --- --- -A- --- --- --- --- --- --- A-- --- -G- --- --- --- --- --- --- --- --- -
Saoe-DRB3*0501 --- --A C-- --- --- -T- --- -C- -G- C-- --A --- --- --- --- --- --- --- --- --- AC- -G* --- --- --- -T- --- --- --- T-- -
```

Fig. 1.

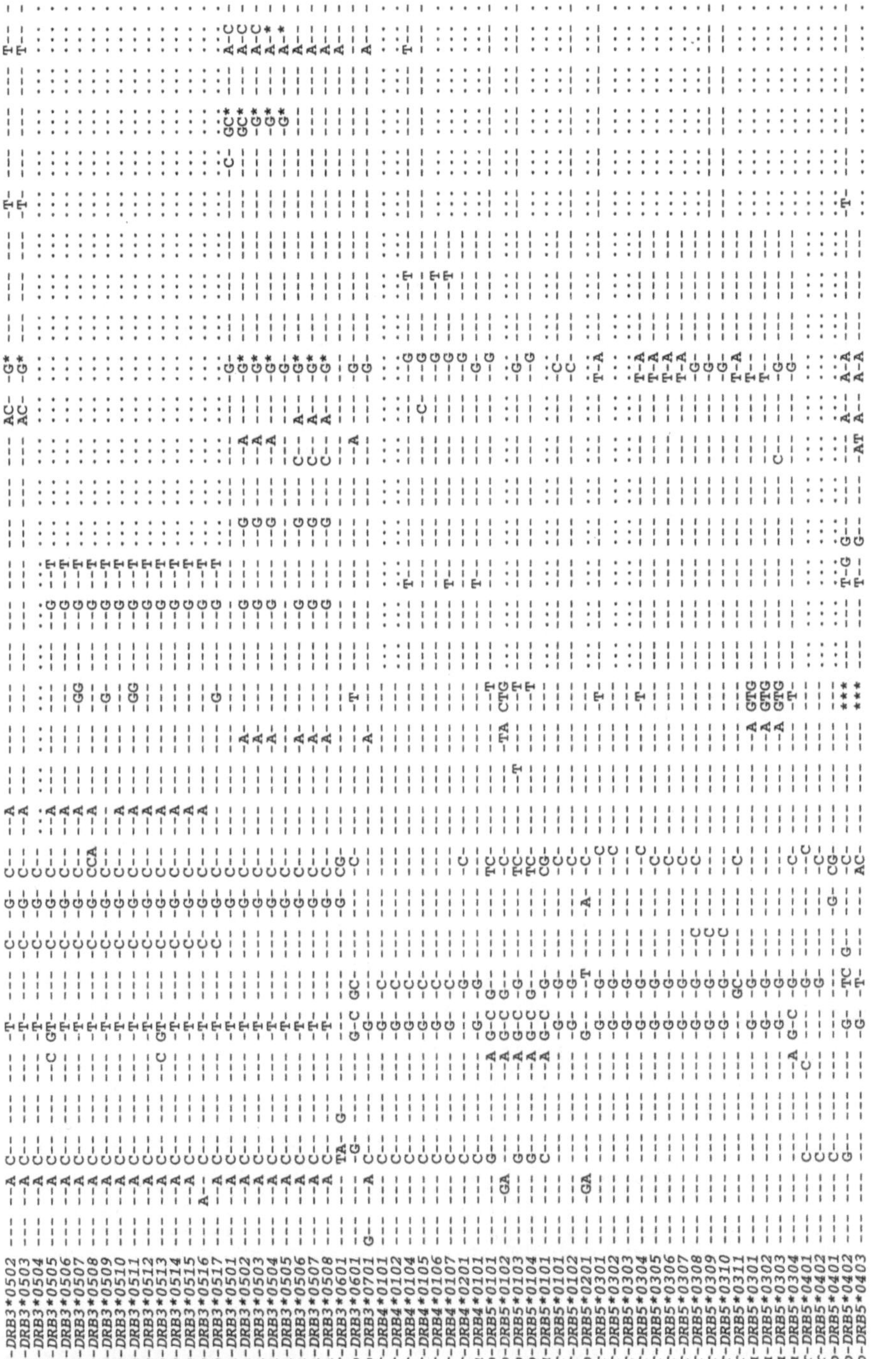

Saoe-DRB3*0502
Saoe-DRB3*0503
Saoe-DRB3*0504
Saoe-DRB3*0505
Saoe-DRB3*0506
Saoe-DRB3*0507
Saoe-DRB3*0508
Saoe-DRB3*0509
Saoe-DRB3*0510
Saoe-DRB3*0511
Saoe-DRB3*0512
Saoe-DRB3*0513
Saoe-DRB3*0514
Saoe-DRB3*0515
Saoe-DRB3*0516
Saoe-DRB3*0517
Otga-DRB3*0501
Otga-DRB3*0502
Otga-DRB3*0503
Otga-DRB3*0504
Otga-DRB3*0505
Otga-DRB3*0506
Otga-DRB3*0507
Otga-DRB3*0508
Aotr-DRB3*0601
Camo-DRB3*0701
Patr-DRB4*0101
Patr-DRB4*0102
Patr-DRB4*0104
Patr-DRB4*0105
Patr-DRB4*0106
Patr-DRB4*0107
Patr-DRB4*0201
Mamu-DRB4*0101
Gogo-DRB5*0101
Gogo-DRB5*0102
Gogo-DRB5*0103
Gogo-DRB5*0104
Mamu-DRB5*0101
Patr-DRB5*0102
Gogo-DRB5*0201
Patr-DRB5*0301
Patr-DRB5*0302
Patr-DRB5*0303
Patr-DRB5*0304
Patr-DRB5*0305
Patr-DRB5*0306
Patr-DRB5*0307
Patr-DRB5*0308
Patr-DRB5*0309
Patr-DRB5*0310
Patr-DRB5*0311
Mamu-DRB5*0301
Mamu-DRB5*0302
Mamu-DRB5*0303
Mamu-DRB5*0304
Patr-DRB5*0401
Patr-DRB5*0402
Gogo-DRB5*0401
Gogo-DRB5*0402
Gogo-DRB5*0403

Fig. 1.

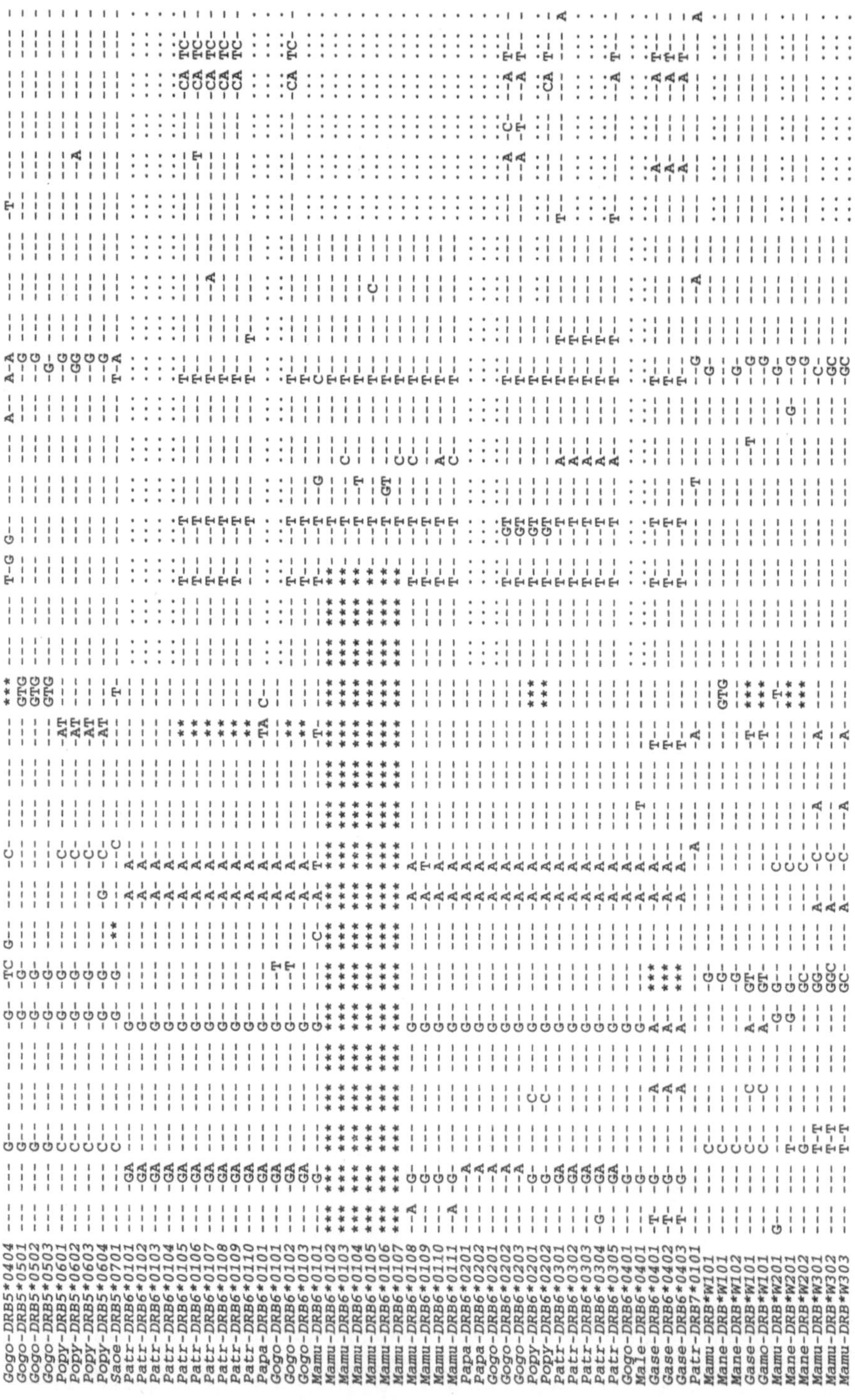

Gogo-DRB5*0404
Gogo-DRB5*0501
Gogo-DRB5*0502
Gogo-DRB5*0503
Popy-DRB5*0601
Popy-DRB5*0602
Popy-DRB5*0603
Popy-DRB5*0604
Saoe-DRB5*0701
Patr-DRB6*0101
Patr-DRB6*0112
Patr-DRB6*0103
Patr-DRB6*0104
Patr-DRB6*0105
Patr-DRB6*0106
Patr-DRB6*0107
Patr-DRB6*0108
Patr-DRB6*0109
Patr-DRB6*0110
Papa-DRB6*0101
Gogo-DRB6*0101
Gogo-DRB6*0102
Gogo-DRB6*0103
Mamu-DRB6*0101
Mamu-DRB6*0102
Mamu-DRB6*0103
Mamu-DRB6*0104
Mamu-DRB6*0105
Mamu-DRB6*0106
Mamu-DRB6*0107
Mamu-DRB6*0108
Mamu-DRB6*0109
Mamu-DRB6*0110
Mamu-DRB6*0111
Papa-DRB6*0201
Papa-DRB6*0201
Gogo-DRB6*0201
Gogo-DRB6*0203
Popy-DRB6*0201
Popy-DRB6*0202
Patr-DRB6*0301
Patr-DRB6*0302
Patr-DRB6*0303
Patr-DRB6*0304
Patr-DRB6*0305
Patr-DRB7*0101
Gogo-DRB6*0401
Mane-DRB6*0401
Gase-DRB6*0401
Gase-DRB6*0402
Gase-DRB6*0403
Gamo-DRB*W101
Mamu-DRB*W101
Mane-DRB*W101
Mane-DRB*W102
Gase-DRB*W101
Gamo-DRB*W101
Mamu-DRB*W201
Mane-DRB*W201
Mane-DRB*W202
Mamu-DRB*W301
Mamu-DRB*W302
Mamu-DRB*W303

Fig. 1.

Fig. 1.

```
Camo-DRB*W1401 G-- --- --- --- --- -G- --- --- --- --- --- --- --G ATG --- --- --- --- --- -A- --- TCA --- --G --- CT- --- -C- --- A-- -
Camo-DRB*W1402 --- --A C-- --- --- -G- --- --- --- --- --- --- --G GTG --- --- --- --- --- -A- T-- TCA --- --G A-- C.. ... ... ... ... .
Sasc-DRB*W1401 G-- --- --- --- --- --- -T- --- --- --- --- --- -AG *** --- --- TT- --- --- -A- --- T-A --- --G --- CTG --- -C- -A- A-- -
Sasc-DRB*W1402 G-- --- --- --- --- --- -T- --- --- --- --- --- -AG *** --- --- TT- --- --- --A --- T-A --- --G --- CTG --- -C- -A- A-- -
Sasc-DRB*W1403 --- --- C-- --- --- --- -T- --- --- --- --- --- -GG -T- --- --- --- --- --- -A- A-- TCA --- --G --- CTG --- -C- --- A-- -
Sasc-DRB*W1404 G-- --- --- --- --- --- -T- --- --- --- --- --- -AG *** --- --- TT- --- --- --A --- T-A --- --G --- CTG --- -C- -A- A-- -
Ceap-DRB*W1501 --- --- TA- A-- --- --- -G- --- --- -C- --- --- --- --- --- --- --- --- --- -A- A-- TCA --- --G --- CT- --- -C- --- A-- -
Ceap-DRB*W1502 --- --- TA- A-- --- --- -G- --- --- -C- --- --- --- --- --- --- --- --- --- -A- A-- TCA --- --G --- CT- --- -C- --- A-- -
Caja-DRB*W1601 --- --- --- --- --- --- --- --- --- -C- --- --- -GA *** --- --- --- --- --- -A- -G- T-C --- -CG --- CT- --- -C- --- A-- -
Caja-DRB*W1602 --- --- --- --- --- --- --- --- --- -C- --- --- -GA *** --- --- --- --- --- -A- -G- T-C --- -CG --- CT- --- -C- --- A-- -
Caja-DRB*W1603 --- --- --- --- --- --- --- --- --- -C- --- --- -GA *** --- --- --- --- --- -A- --- TCA --- --G --- CT- --- -C- --- A-- -
Camo-DRB*W1701 G-- --- --- --- --- G-C G-- --- --- -C- --- --- --- --- --- --- --- --- --- -A- A-- T-A --- --G --- CT- --- -C- --- A-- -
Aotr-DRB*W1801 C-- --- --- --- --- -T- --- --- -G- C-- --A --- -A- --- --- --- --- --- --- --- --- T-- --- --- --- --- --- --- --- A-- -
Sasc-DRB*W1901 --- --- TA- --- --- --- -G- --- --- --- --- --- GAG *** --- --- --- --- --- CA- A-- T-- --- --G --- CTG --- --- --- A-- -
Sasc-DRB*W1902 G-- --- T-- --- --- --- --- --- --- -C- --- --- --- GTG --- --G --- --- --- CA- A-- TCG --- --G --- CTG --- -CC --- A-- -
Sasc-DRB*W1903 --- --- TA- --- --- --- -G- --- --- -C- --- --- --- --- --- --- --- --- --- CA- A-- T-- --- --G --- CTG --- --- --- A-- -
Mamu-DRB*W2001 --- --- TA- --- --- --- --- --- --- --- --- --- -A- --- --- --- --- --- --- --- --- --- --- --- --- ... ... ... ... ... .
Mamu-DRB*W2101 --- -G- TA- --- --- --- --- --- -G- C-- --T --- -A- --- --- --- T-- --- --- C-- --- -G- --- --- --- ... ... ... ... ... .
Mamu-DRB*W2102 --- -G- TA- --- --- --- --- --- -G- C-- --T --- -A- --- --- --- T-- --- --- C-- --- -G- --- --- --- ... ... ... ... ... .
Saoe-DRB*W2201 --- --A G-- --- --- T-C -T- --- --- -C- --- --- --- -T- --- --G --T ... ... ... ... ... ... ... ... ... ... ... ... ... .
Saoe-DRB*W2202 --- --A G-- --- --- T-C -T- --- --- -C- --- --- --- -T- --- --G --T ... ... ... ... ... ... ... ... ... ... ... ... ... .
Saoe-DRB*W2203 --- --- --- --- --- --- -CA --- --- --- --- --- --- -T- --- --G --T ... ... ... ... ... ... ... ... ... ... ... ... ... .
Saoe-DRB*W2204 --- --- --- --- --- --- -CA --- --- -C- --- --- --- -T- --- --G --T ... ... ... ... ... ... ... ... ... ... ... ... ... .
Saoe-DRB*W2205 --- --A G-- --- --- T-C -T- --- --- -C- --- --- --- -T- --- --G --T ... ... ... ... ... ... ... ... ... ... ... ... ... .
Saoe-DRB*W2206 --- --A G-- --- --- T-C -T- --- --- -C- --- --- --- -T- --- --G --T ... ... ... ... ... ... ... ... ... ... ... ... ... .
Saoe-DRB*W2207 --- --A G-- --- --- T-C -T- --- --- -C- --- --- --- -T- --- --G --T ... ... ... ... ... ... ... ... ... ... ... ... ... .
Saoe-DRB*W2208 --- --A G-- --- --- --- -CA --- --- -C- --- --- --- -T- --- --G --T ... ... ... ... ... ... ... ... ... ... ... ... ... .
Saoe-DRB*W2209 --- --A G-- --- --- T-C -TC G-- --- -C- --- --- --- -T- --- --- --- --- --- AA- -G- T-- --- -TG --- CT- --- -C- --- A-- -
Lota-DRB*W2301 --- --- G-- --- --- G-- GC- --- --- -C- --- --- --G --T --- --- --- --- --- CA- A-C TCA --C G-- ... ... ... ... ... ... .
Lota-DRB*W2302 --- --- T-A A-- --- -G- --- --- A-G -C- --- --- TGG -T- --- --- GT- -G- --- CA- A-C CCA --C G-- ... ... ... ... ... ... .
Lota-DRB*W2303 --- --- T-A A-- --- -G- --- --- A-G -C- --- --- TGG -T- --- --- GT- -G- --- CA- A-C TCA --C G-- ... ... ... ... ... ... .
Lota-DRB*W2304 --- --- G-- --- --- G-- GC- --- --- -C- --- --- --G --T --- --- --- --- --- CA- A-C TCA --C G-- ... ... ... ... ... ... .
Lota-DRB*W2305 --- --- G-- --- --- G-- GC- --- --- -C- --- --- --G --T --- --- --- --- --- CA- A-C TCA --C G-- ... ... ... ... ... ... .
Lota-DRB*W2306 --- --- G-- --- --- G-- GC- --- --- -C- --- --- --G --T --- --- --- --- --- CA- A-C T-A --C G-- ... ... ... ... ... ... .
Lota-DRB*W2307 --- --- G-- --- --- G-- GC- --- --- -C- --- --- --G --T --- --- --- --- --- CA- A-C TCA --C G-- ... ... ... ... ... ... .
Lota-DRB*W2308 --- --- T-A A-- --- -G- --- --- A-G -C- --- --- TGG -T- --- --- GT- -G- --- CA- A-C TCA --C G-- ... ... ... ... ... ... .
Lota-DRB*W2309 --- --T G-- --- --- G-- GC- --- --G CG- TG- *-- --G --T --- --- GGG *** --- CA- C-C -AA TTC GTA ... ... ... ... ... ... .
Pefu-DRB*W2401 --- --- CG- --- --T T-C GGC --- --- -C- --- --- --G -T- --- --- T-- --- --- --- --C -GC --- --- --- --C --- --- --- --- -
Pefu-DRB*W2402 --- A-- --- --- --T G-C G-- --- --- -C- --- --- --G GTG --- --- --- --- --- --- A-C A-C --- --- --- --C --- --- --- --- -
```

Fig. 1.

Fig. 2.

Fig. 2.

Fig. 2.

```
                  5                                       15                                      25
Consensus        .GA GAC TCT CCC GAG GAT TTC GTG TAC CAG TTT AAG GGC ATG TGC TAC TTC ACC AAC GGG ACG GAG CGC GTG CGG GGT GTG ACC AGA TAC
Patr-DQB1*0601   ... ... ... ... ... ... ... ... -T- --- --- --- --- --- --- --- --- --- --- --- --- --- --- --- --T CT- --A --- --- ---
Patr-DQB1*0602   ... ... ... ... ... ... ... ... -T- --- --- --- --- --- --- --- --- --- --- --- --- --- --- --- --T CT- --A --- --- ---
Patr-DQB1*0603   ... ... ... ... ... ... ... ... ... ... ... ... ... ... ... ... ... ... ... ... --- --- --- --- --T CT- --A --- --- ---
Patr-DQB1*0604   ... ... ... ... ... ... ... ... -T- --- --- --- --- --- --- --- --- --- --- --- --- --- --- --- --T CT- --A --- --- ---
Patr-DQB1*0605   ... ... ... ... ... ... ... ... ... ... ... ... ... ... ... ... ... ... ... ... --- --- --- --- --T CT- --- --G --- ---
Patr-DQB1*0606   ... ... ... ... ... ... ... ... ... ... ... ... ... ... ... ... ... ... ... ... --- --- --- --- --T CT- --- --G --- ---
Patr-DQB1*0301   ... ... ... ... ... ... ... ... ... ... ... ... ... ... ... ... ... ... ... ... --- --- --- --- --T TA- --- --- --- ---
Patr-DQB1*0302   ... ... ... ... ... ... ... ... --- --- --- --- --- --- --- --- --- --- --- --- --- --- --- --- --T CT- --- --G --- ---
Patr-DQB1*0303   ... ... ... ... ... ... ... ... --- --- --- --- --- --- --- --- --- --- --- --- --- --- --- --- --T CT- --- --G --- ---
Patr-DQB1*1501   ... ... ... ... ... ... ... ... --- --- --- --- --- --- --- --- --- --- --- --- --A --- --- --- T-- A-- --A GA- --- ---
Patr-DQB1*1502   ... ... ... ... ... ... ... ... ... ... ... ... ... ... ... ... ... ... ... ... --A --- --- --- T-- A-- --A --G --- ---
Patr-DQB1*2101   ... ... ... ... ... ... ... ... ... ... ... ... ... ... ... ... ... ... ... ... --- --- --- --- --T TA- --- --- --- ---
Patr-DQB2*0101   ... ... ... ... ... ... ... ... ... ... ... ... ... ... ... ... ... ... ... ... --A --- --- --- --- --- --- G-- --- ---
Patr-DQB2*0102   ... ... ... ... ... ... ... ... ... ... ... ... ... ... ... ... ... ... ... ... --A --- --- --- --- --- --- G-- --- ---
Patr-DQB2*0103   ... ... ... ... ... ... ... ... ... ... ... ... ... ... ... ... ... ... ... ... --A --- --- --- --- --- --- G-- --- ---
Papa-DQB1*0301   ... ... ... ... ... ... ... ... ... ... ... ... ... ... ... ... ... ... ... ... --- --- --- --- --T TA- --- --- --- ---
Papa-DQB1*0302   ... ... ... ... ... ... ... ... ... ... ... ... ... ... ... ... ... ... ... ... --- --- --- --- --T CT- --- --G --- C--
Papa-DQB1*0303   ... ... ... ... ... ... ... ... ... ... ... ... ... ... ... ... ... ... ... ... --- --- --- --- --T CT- --- --- --- ---
Papa-DQB1*0304   ... ... ... ... ... ... ... ... ... ... ... ... ... ... ... ... ... ... ... ... --- --- --- --- --T CT- --- --G --- ---
Papa-DQB1*0305   ... ... ... ... ... ... ... ... ... ... ... ... ... ... ... ... ... ... ... ... --- --- --- --- --T CT- --- --G --- C--
Papa-DQB2*0101   ... ... ... ... ... ... ... ... ... ... ... ... ... ... ... ... ... ... ... ... --A --- --- --- --- --- --- G-- --- ---
Gogo-DQB1*0501   ... ... ... ... ... ... ... ... ... ... ... ... ... ... ... ... ... ... ... ... --- --- --- --- --- -C- --- --- --- C--
Gogo-DQB1*0502   ... ... ... ... ... ... ... ... --- --- --- --- --- C-- --- --- --- --- --- --- --- --- --- --- --- -C- --- --- --- C--
Gogo-DQB1*0503   ... ... ... ... ... ... ... ... ... ... ... ... ... ... ... ... ... ... ... ... --- --- --- --- --- -C- --- --- --- C--
Gogo-DQB1*0601   ... ... ... ... ... ... ... ... ... ... ... ... ... ... ... ... ... ... ... ... --- --- --- --- --T TA- --- --- --- ---
Gogo-DQB1*0201   ... ... ... ... ... ... ... ... ... ... ... ... ... ... ... ... ... ... ... ... --A --- --- --- --T CT- --- -G- --- A--
Gogo-DQB1*0202   ... ... ... ... ... ... ... ... --- --- --- --- --- --- --- --- --- --- --- --- --A --- --- --- --T CT- --- -G- --- A--
Gogo-DQB1*1901   ... ... ... ... ... ... ... ... ... ... ... ... ... ... ... ... ... ... ... ... --- --- --- --- --T CT- --- --G --- ---
Gogo-DQB1*1902   ... ... ... ... ... ... ... ... ... ... ... ... ... ... ... ... ... ... ... ... --- --- --- --- --T CT- --- --G --- ---
Gogo-DQB1*1903   ... ... ... ... ... ... ... ... --- --- --- --- --- --- --- --- --- --- --- --- --- --- --- --- --T -T- --- --G --- ---
Gogo-DQB1*1904   ... ... ... ... ... ... ... ... ... ... ... ... ... ... ... ... ... ... ... ... --- --- --- --- --T CT- --- --G --- ---
Gogo-DQB2*0101   ... ... ... ... ... ... ... ... GT- --- --- --- --- --- --- --- --- --- --- --- --A --- --- --- --- --- --- G-- --- C--
Gogo-DQB2*0102   ... ... ... ... ... ... ... ... ... ... ... ... ... ... ... ... ... ... ... ... --A --- --- --- --- --- --- G-- --- ---
Popy-DQB1*0601   ... ... ... ... ... ... ... ... --- --- --- --- --- C-- --- --- --- --- --- --- --- --- --- --- --- --- --- --- --- C--
Popy-DQB1*0602   ... ... ... ... ... ... ... ... --- --- --- --- --- C-- --- --- --- --- --- --- --- --- --- --- --- --- --- --- --- ---
Popy-DQB1*0603   ... ... ... ... ... ... ... ... --- --- --- --- --- C-- --- --- --- --- --- --- --- --- --- --- --T CT- --- --- --- ---
Popy-DQB1*1601   ... ... ... ... ... ... ... ... GT- --- --- --- --- --- --- --- --- --- --- --- --A --- --- --- --- TTG --A G-- --- C--
Popy-DQB1*1701   ... ... ... ... ... ... ... ... ... ... ... ... .-- --- --- --- --- --- --- --- --- --- --- --- -AT TA- --- --- --- ---
Popy-DQB2*0101   ... ... ... ... ... ... ... ... GT- --- --- --- --- --- --- --- --- --- --- --- --A --- --- --- --- --- --- G-- --- ---
Hyla-DQB1*2001   ... ... ... ... ... ... ... ... ... ... ... ... .-- --- --- --- --- --- --- --- --- --- --- --- --- --- --- --- --- C--
Hyla-DQB1*2002   ... ... ... ... ... ... ... ... --- --- --- --- --- --- --- --- --- --- --- --- --- --- --- --- --A --- --- G-- --- ---
Hyla-DQB2*0101   ... ... ... ... ... ... ... ... GT- --- --- --- --- --- --- --- --- --- --- --- --A --- --- --- --- --- --- G-- --- ---
Hyla-DQB2*0102   ... ... ... ... ... ... ... ... ... ... ... ... ... ... ... ... ... ... ... ... --A --- --- --- --- --- --- G-- --- ---
Ceae-DQB1*1501   ... ... ... ... ... ... ... ... -T- --- --- --- --- --- --- --- --- --- --- --- --- --- --- --- T-- A-- --A GA- --- ---
Ceae-DQB1*1801   ... ... ... ... ... ... ... ... --- --- --- --- --- C-- --- --- --T --- --- --- --- --- --- --- --T CT- --- --- --- C--
Ceae-DQB1*1802   ... ... ... ... ... ... ... ... --- --- --- --- --- C-- --- --- --- --- --- --- --- --- --- --- --- --- --- --- --- C--
Male-DQB1*1501   ... ... ... ... ... ... ... ... ... ... ... ... ... ... ... ... ... ... ... ... --- --- --- --- T-- A-- --A GA- --- ---
```

Fig. 2.

Fig. 2.

Fig. 2.

Fig. 2.

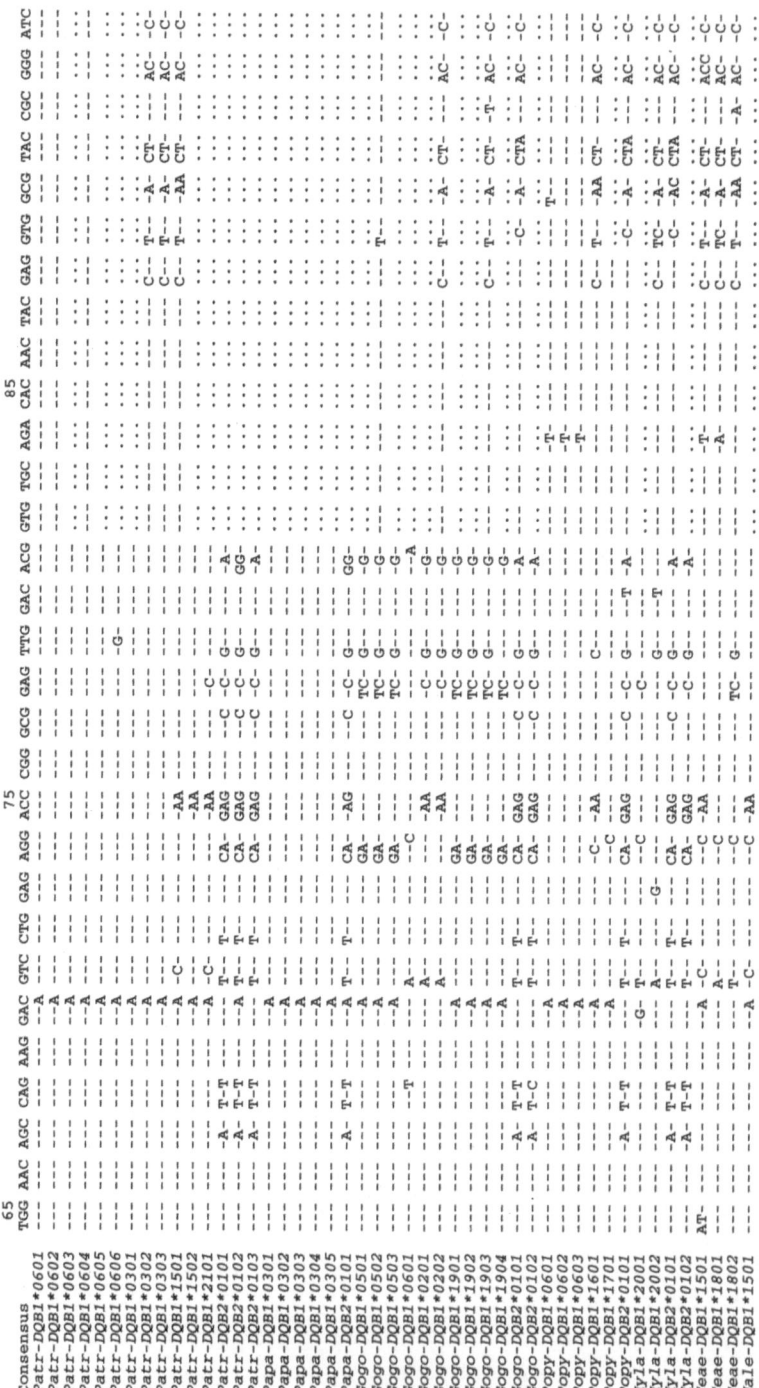

Fig. 2.

	65										75										85									
Consensus	TGG	AAC	AGC	CAG	AAG	GAC	GTC	CTG	GAG	AGG	ACC	CGG	GCG	GAG	TTG	GAC	ACG	GTG	TGC	AGA	CAC	AAC	TAC	GAG	GTG	GCG	TAC	CGC	GGG	ATC
*Paha-DQB1*0601*	---	---	---	---	---	---	---	---	---	--C	---	---	---	--A	---	---	---	---	---	-AG	---	---	---	---	---	---	-T-	---	---	---
*Paha-DQB1*0602*	---	---	---	---	---	---	A--	---	---	--C	-AA	---	---	---	---	---	-GT	---	---	---	---	---	---	---	---	--T	---	---	---	---
*Paha-DQB1*1501*	---	---	---	---	---	--A	-C-	---	---	--C	-AA	---	---	---	---	---	-G-	---	---	-T-	---	---	---	C--	T-A	-A-	CT-	G--	AC-	---
*Mamu-DQB1*0601*	---	---	---	---	---	---	--G	---	---	--C	---	---	---	--A	---	---	---	---	---	-A-	---	---	---	---	---	---	-T-	---	---	---
*Mamu-DQB1*0602*	---	---	---	---	---	---	A--	---	---	--C	---	---	---	---	---	---	---	---	---	-TG	---	---	---	---	---	---	---	---	---	---
*Mamu-DQB1*0603*	---	---	---	---	---	---	A--	---	---	--C	---	---	---	---	---	---	---	---	---	-TG	---	---	---	---	---	---	-G-	---	---	---
*Mamu-DQB1*0604*	---	---	---	---	---	---	A--	---	---	---	---	---	---	---	---	---	--T	---	---	-T-	---	---	---	---	---	-T-	---	---	---	---
*Mamu-DQB1*0605*	---	---	---	---	---	---	A--	---	---	---	---	---	---	---	---	---	--T	---	---	---	---	---	---	---	---	--T	---	---	---	---
*Mamu-DQB1*1501*	---	---	---	---	---	--A	-C-	---	---	--C	-AA	---	---	---	---	---	---	---	---	-T-	---	---	---	C--	T--	-A-	CT-	G--	AC-	-C-
*Mamu-DQB1*1502*	---	---	---	---	---	--A	-C-	---	---	--C	-AA	---	---	---	---	---	---	---	---	-T-	---	---	---	C--	T--	-A-	CT-	G--	AC-	-C-
*Mamu-DQB1*1601*	---	---	---	---	---	--A	---	---	---	---	-AA	---	---	---	---	---	---	---	---	---	---	---	---	C--	T--	-AA	CT-	---	TC-	-C-
*Mamu-DQB1*1602*	---	---	---	---	---	--A	---	---	---	--C	-AA	---	---	-C-	---	---	---	---	---	-T-	---	---	---	C--	T--	-AA	CT-	---	AC-	-C-
*Mamu-DQB1*1701*	---	---	---	---	---	---	---	---	---	---	---	---	---	--A	C--	---	---	...	...	...	...	...	...	...	...	...	...	...	...	...
*Mamu-DQB1*1702*	---	---	-A-	---	---	---	T--	---	---	--C	---	---	---	---	---	---	---	---	---	-A-	---	---	---	---	T--	-AA	CT-	-T-	AC-	-C-
*Mamu-DQB1*1801*	---	---	---	---	---	---	A--	---	---	---	---	---	---	---	---	---	---	---	---	-T-	---	---	---	C--	T--	-A-	CT-	---	ACC	-C-
*Mamu-DQB1*1802*	---	---	---	---	---	---	A--	---	---	---	---	---	---	--A	C--	---	---	---	---	---	---	---	---	C--	TCA	-A-	CT-	---	AC-	-C-
*Mamu-DQB1*1803*	---	---	---	---	---	---	T--	---	---	---	---	---	---	TC-	G--	---	---	---	---	---	---	---	---	C--	TC-	-A-	CT-	---	AC-	-C-
*Mamu-DQB1*1804*	---	---	---	---	---	---	A--	---	---	---	---	---	---	---	---	---	---	---	---	---	---	---	---	C--	TC-	-A-	CT-	---	AC-	-C-
*Mamu-DQB1*1805*	---	---	---	---	---	---	A--	---	---	--C	---	---	---	---	GC-	---	---	---	---	---	---	---	---	C--	TCA	-AC	CT-	---	AC-	-CG
*Mamu-DQB1*1806*	---	---	-A-	---	---	---	T--	---	---	---	---	---	---	TC-	G--	---	---	---	---	---	---	---	---	C--	T--	-AA	CT-	---	AC-	-C-
*Mamu-DQB1*1807*	---	---	---	---	---	---	A--	---	---	--C	---	---	---	---	GC-	---	---	---	---	---	---	---	---	C--	T--	-A-	CT-	---	AC-	-C-
*Mamu-DQB1*1808*	---	---	---	---	---	---	A--	---	---	--C	---	---	---	-T-	G--	---	---	---	---	---	---	---	---	C--	TC-	-AC	CT-	---	AC-	-C-
*Mafa-DQB1*0601*	---	---	---	---	---	---	--G	---	---	--C	---	---	---	--A	---	---	---	---	---	-A-	---	---	---	---	---	---	-T-	---	---	---
*Mafa-DQB1*0602*	---	---	---	---	---	---	A--	---	---	---	---	---	---	---	---	---	---	---	---	-TG	---	---	---	---	---	---	---	---	--A	---
*Mafa-DQB1*0603*	---	---	---	---	---	---	A--	---	---	---	---	---	---	---	---	---	---	---	---	---	---	---	---	---	---	---	---	---	---	---
*Mafa-DQB1*0604*	---	---	---	---	---	---	A--	---	---	---	---	---	---	---	---	---	--T	---	---	-T-	---	---	---	---	---	---	---	---	---	---
*Mafa-DQB1*0605*	---	---	---	---	---	---	A--	---	---	---	---	---	---	---	---	---	---	...	...	...	...	...	...	...	...	...	...	...	...	...
*Mafa-DQB1*1501*	---	---	---	---	---	--A	-C-	---	---	--C	-AA	---	---	---	---	---	---	---	---	-T-	---	---	---	C--	T--	-A-	CT-	G--	AC-	-C-
*Mafa-DQB1*1502*	---	---	---	---	---	--A	-C-	---	---	--C	-AA	---	---	---	---	---	---	---	---	-T-	---	---	---	C--	T--	-A-	CT-	G--	AC-	-C-
*Mafa-DQB1*1701*	---	---	---	---	---	---	T--	---	---	--C	---	---	---	---	---	---	---	---	---	---	---	---	---	C--	T--	-A-	CT-	---	AC-	---
*Mafa-DQB1*1702*	---	---	---	---	---	---	---	---	---	--C	---	---	---	---	---	---	---	...	...	...	...	...	...	...	...	...	...	...	...	...
*Maar-DQB1*0601*	---	---	---	---	---	---	A--	---	---	--C	---	---	---	---	---	---	---	---	---	-TG	---	---	---	---	---	---	---	---	---	---
*Maar-DQB1*0602*	-T-	---	---	---	---	---	--G	---	---	--C	---	---	---	---	---	---	---	---	---	-TG	---	---	---	---	---	A--	---	---	---	---
*Maar-DQB1*1701*	---	---	---	---	---	---	---	---	---	--C	---	---	---	---	---	---	---	---	---	-T-	---	---	---	C--	T--	-A-	CT-	---	AC-	---
*Pren-DQB1*1601*	---	---	---	---	---	--A	---	---	---	---	-AA	---	---	---	---	---	---	...	...	...	...	...	...	...	...	...	...	...	...	...
*Pren-DQB1*1801*	G--	---	---	---	---	---	A--	---	---	--C	---	---	---	---	GC-	---	---	...	...	...	...	...	...	...	...	...	...	...	...	...
*Caja-DQB1*2201*	-T-	---	---	---	---	---	T--	---	---	--A	---	---	---	---	C--	---	---	---	---	---	---	---	---	C--	CA-	-A-	-T-	-C-	-C-	---
*Caja-DQB1*2301*	---	---	---	---	---	---	A--	---	---	--C	G--	---	---	---	C--	---	---	---	---	-A-	---	---	---	C-A	T--	---	CT-	---	AC-	-C-
*Caja-DQB2*0101*	---	---	-A-	T-T	---	---	T--	T--	---	CA-	GAG	---	---	-C-	G--	---	---	---	---	-T-	---	---	---	---	-C-	-A-	CT-	---	AC-	-C-
*Aotr-DQB2*0101*	---	---	-A-	T-T	---	---	T--	T--	---	CA-	CAG	---	--C	-C-	G--	T--	---	...	...	...	...	...	...	...	...	...	...	...	...	...
*Aotr-DQB1*0102*	---	---	-A-	T-T	---	---	T--	T--	---	CA-	CAG	---	--C	CC-	G--	T--	---	...	...	...	...	...	...	...	...	...	...	...	...	...
*Saoe-DQB1*2201*	-T-	---	---	---	---	---	T--	---	---	---	---	---	---	--C	G--	---	---	...	...	...	...	...	...	...	...	...	...	...	...	...
*Saoe-DQB1*2202*	-T-	---	---	---	-CT	--A	---	---	---	---	---	---	---	---	C--	---	---	...	...	...	...	...	...	...	...	...	...	...	...	...
*Saoe-DQB1*2301*	---	---	---	---	---	---	A--	---	---	--C	---	---	---	---	C--	---	---	...	...	...	...	...	...	...	...	...	...	...	...	...
*Saoe-DQB2*0101*	---	---	-A-	T-T	---	---	T--	T--	---	CA-	GAG	---	--C	-C-	G--	T--	---	...	...	...	...	...	...	...	...	...	...	...	...	...

Fig. 2.

Fig. 3.

Fig. 3.

References

Bidwell JL, Lu P, Wang Y, Zhou K, Clay TM, Bontrop RE. DRB DQA DQB and DPB nucleotide sequences of Sanguinus oedipus B95-8. Eur J Immunogenet 21:67-77, 1994

Bontrop RE. Nonhuman primate Mhc-DQA and -DQB second exon nucleotide sequences: a compilation. Immunogenetics 39:81-92, 1994

Corell A, Morales P, Varela P, Paz-Artal E, Martin-Villa JM, Martinez-Laso A, Arnaiz-Villena A. Allelic diversity at the primate major histocompatibility comple DRB6 locus. Immunogenetics 36:33-38, 1992

Fan W, Kasahara L, Gutknecht J, Klein D, Mayer WE, Jonker M, Klein J. Shared class II Mhc polymorphisms between human and chimpanzees. Hum Immunol 26:107-121, 1989

Figueroa F, O'hUigin C, Tichy H, Klein J. The origin of primate Mhc-DRB genes and allelic lineages as deduced from the study of prosimians. J Immunol 152:4455-4465, 1994

Gaur LK, Heise ER, Thurtle PS, Nepom GT. Conservation of the HLA-DQB2 locus in nonhuman primates. J Immunol 148:943-948, 1992

Gaur L, Hughes AL, Heise ER, Gutknecht J. Maintenance of DQB1 polymorphism in primates. Mol Biol Evol 9:599-609, 1992

Grahovac B, Mayer WE, Vincek V, Figueroa F, O'hUigin C, Tichy H, Klein J. Major histocompatibility complex DRB genes of a New World monkey the cotton-top tamarin (Saguinus oedipus). Mol Biol Evol 9:403-416, 1992

Grahovac B, Schönbach C, Brändle U, Mayer WE, Golubic M, Figueroa F, Trowsdale J, Klein J. Conservative evolution of the Mhc-DP region in anthropoid primates. Hum Immunol 37:75-84, 1993

Gyllensten UB, Erlich HA. Ancient roots for polymorphism at the HLA-DQ alpha locus in primates. Proc Natl Acad Sci USA 86:9986-9990, 1989

Gyllensten UB, Erlich HA. Allelic diversification at the class II DQB locus of the mammalian major histocompatability complex. Proc Natl Acad Sci USA 87:1835-1839, 1990

Gyllensten UB, Sundvall M, Ezcurra I, Erlich HA. Genetic diversity at the class II DRB loci of the primate MHC. J Immunol 146:4368-4376, 1991a

Gyllensten UB, Sundvall M, Erlich HA. Generation of allelic polymorphism at the DRB1 locus of primates by exchange of polymorphic domains: a plausible hypothesis? In: Klein J, Klein D (eds) Molecular evolution of the major histocompatibility complex. Springer, Berlin Heidelberg New York, pp 111-118, 1991b

Gyllensten UB, Bergström T, Josefsson A, Sundvall M, Savage A, Blumer ES, Humberto Giraldo L, Soto LH, Watkins DI. The cotton-top tamarin revisited: Mhc class I polymorphism of wild tamarins and polymorphism and allelic diversity of the class II DQA1, DQB1, and DRB loci. Immunogenetics 40:167-176, 1994

Hashiba K, Kuwata S, Tokunaga K, Juji T, Noguchi A. Sequence analysis of DPB1-like genes in cynomolgus monkeys (Macaca fascicularis). Immunogenetics 38:462, 1993

Kenter M, Otting N, Anholts J, Jonker M, Schipper R, Bontrop RE. Mhc-DRB diversity of the chimpanzee. Immunogenetics 37:1-11, 1992

Kenter M, Otting N, Anholts J, Leunissen J, Jonker M, Bontrop RE. Evolutionary relationships among the primate Mhc-DQA1 and DQA2 alleles. Immunogenetics 36:71-78, 1992

Kenter M, Otting N, de Weers M, Anholts J, Reiter C, Jonker M, Bontrop RE. Mhc-DRB and -DQA1 nucleotide sequences of three lowland gorillas Implications for the evolution of primate Mhc class II haplotypes. Hum Immunol 36:205-218, 1993a

Klein J, Bontrop RE, Dawkins RL, Erlich HA, Gyllensten UB, Heise ER, Jones PP, Parham P, Wakeland EK, Watkins DI. Nomenclature for the major histocompatibility complexes of different species: a proposal. Immunogenetics 31:217-219, 1990

Kupfermann H, Mayer WE, O'hUigin C, Klein D, Klein J. Shared polymorphism between gorilla and human major histocompatibility complex DRB loci. Hum Immunol 34:267-278, 1992

Mayer WE, O'hUigin C, Zaleska-Rutczynska Z, Klein J. Trans-species origin of Mhc-DRB polymorphism in the chimpanzee. Immunogenetics 37:12–23, 1992

O'hUigin C, Bontrop R, Klein J. Nonhuman primate Mhc-DRB sequences: a compilation. Immunogenetics 38:165–183, 1993

Otting N, Bontrop RE. Evolution of the major histocompatability complex DPA1 locus in primates. Hum Immunol 42:184–187, 1995

Otting N, Kenter M, van Weeren P, Jonker M, Bontrop RE. Mhc-DQB repertoire variation in hominoid and Old World primate species. J Immunol 149:461–470, 1992

Riess O, Kammerbauer C, Roewer L, Steimle V, Andreas A, Albert E, Nagai T, Epplen JT. Hypervariability of intron simple (gt)n(ga)m repeats in HLA-DRB genes. Immunogenetics 32:110–116, 1990

Sauermann U, Christ R, Hunsmann G. Characterization of two novel Mamu-DQA1 alleles of rhesus monkeys. Tissue Antigens 46:408–410, 1995

Schönbach C, Vincek V, Mayer WE, Golubic M, O'hUigin C, Klein J: Multiplication of Mhc-DRB5 loci in the orangutan: implications for the evolution of DRB haplotypes. Mamm Genome 4:159–170, 1993

Slierendregt BL, Otting N, Jonker MJ, Re B. Major histocompatability complex class II DQ diversity in rhesus macaques. Tissue Antigens 41:178–185, 1993

Slierendregt BL, Otting N, van Besouw N, Jonker MJ, Re B. Expansion and contraction of rhesus macaque DRB regions by duplication and deletion. J Immunol 152:2298–2307, 1994

Slierendregt BL, Otting N, Kenter M, Re B. Allelic diversity at the Mhc-DP locus in rhesus macaques (Macaca mulatta). Immunogenetics 41:29–37, 1995

Trtková K, Kupfermann H, Grahovac B, Mayer WE, O'hUigin C, Tichy H, Bontrop RE, Klein J. Mhc-DRB genes of platyrrhine primates. Immunogenetics 38:210, 1993

Zhu Z, Vincek V, Figueroa F, Schönbach C, Klein J. Mhc-DRB genes of the pigtail macaque (Macaca nemestrina): implications for the evolution of human DRB genes. Mol Biol Evol 8:563–578, 1991

9 Nomenclature for the MHCs and Alleles of Different Nonhuman Primate Species

R.E. Bontrop and J. Klein

During the fall of 1989, a group of investigators attended the first meeting on The Evolution of the Major Histocompatibility Complexes, which was held in Oegstgeest, The Netherlands. One of the direct spin-offs of the meeting was that the participants formulated, accepted and published a proposal for the nomenclature of the MHCs of different species that is now in common use (Klein et al. 1990). For sake of clarity, the historical and practical reasons behind formulating such a proposal are highlighted, with special emphasis on the situation that is relevant to nonhuman primates. At first, the MHC was given different names in different species (Klein 1986). It was designated H-2 in the mouse, B in the domestic fowl, *RT1* in the rat, and Smh in the mole rat. In most other species that have been studied, the MHC has been referred to by the LA symbol (for lymphocyte or leukocyte antigen), prefixed by an abbreviation of the species' common name. Thus, it is called *HLA* in the human, RLA in the rabbit, *BOLA* in the domestic cattle, *SLA* in the pig, and so on. This practice has two problems associated with it. First, MHC products are expressed on many other tissues in addition to lymphocyte or leukocyte (and lymphocytes express many other antigens in addition to those controlled by the MHC) and their antigenicity is secondary to their biological function. Second, the use of common names to identify a species is a potential source of confusion. Common names are notoriously vague and imprecise. The designation "lemur", for example, can refer to any of the genera *Lemur, Hapalemur, Varecia, Lepilemur, Avahi, Propithecus, and Indri*, of which only the first four belong to the family Lemuridae; the last three are members of the family Indriidae. A "bushbaby" can be a *Galago, Otolemur, or Euoticus*. Obviously, common names not only fail to identify the species appropriately, they often do not even identify the genus or the family. If the trend in choosing common names for MHC symbols were to continue, chaos would soon ensue because we can expect MHCs in many different species to be identified in the future. For that reason the following rules were formulated and have been updated whenever necessary (It was not proposed, however, to change the name of well-established designations such as *HLA* and *H2*).

Molecular Biology and Evolution of Blood Group
and MHC Antigens in Primates
Blancher/Klein/Socha (Eds.)
© Springer-Verlag Berlin Heidelberg 1997

Table 1. Designations for nonhuman primate MHCs

Species name		*Mhc* designation	
Common	Scientific	New	Old
Apes			
Chimpanzee	*Pan troglodytes*	*Patr*	ChLA
Bonobo or pygmy chimpanzee	*Pan paniscus*	*Papa*	
Gorilla	*Gorilla gorilla*	*Gogo*	GoLA
Orangutan	*Pongo pygmaeus*	*Popy*	OrLA
Common gibbon	*Hylobates lar*	*Hyla*	
Old World monkeys			
African green monkey	*Cercopethicus aethiops*	*Ceae*	
Stump-tailed macaque	*Macaca arctoides*	*Maar*	
Cynomolgus monkey	*Macaca fascicularis*	*Mafa*	
Rhesus macaque	*Macaca mulatta*	*Mamu*	RhLA
Pig-tailed macaque	*Macaca nemestrina*	*Mane*	
Olive baboon	*Papio anubis*	*Paan*	
Hamadryas baboon	*Papio hamadryas*	*Paha*	
Drill	*Mandrillus leucophaeus*	*Male*	
Hanuman langur	*Presbytis entellus*	*Pren*	
New World Monkeys			
Northern night (owl) monkey	*Aotus trivirgatus*	*Aotr*	OmLA
Black spider monkey	*Ateles paniscus*	*Atpa*	
Common squirrel monkey	*Saimiri sciureus*	*Sasc*	
Cottop-top tamarin	*Saguinus oedipus*	*Saoe*	
Common marmoset	*Callithrix jacchus*	*Caja*	MaLA
Duski titi	*Callicebus moloch*	*Camo*	
Brown capuchin monkey	*Cebus apella*	*Ceap*	

1. Major histocompatibility complexes of the different nonhuman primate species are designated by the symbol *Mhc,* in which only the first letter of the abbreviation is capitalized. The abbreviation MHC is retained for general discussions, in which no commitments are made regarding genotype vs phenotype and/or a particular zoological species. The *Mhc* symbol, however, is a genetic designation used in combination with the species' name.
2. The *Mhc* symbol is followed by a four-letter abbreviation of the species' scientific name. The first two letters in this abbreviation are derived from the name of the genus, the last two letters from the name of the species. In both cases, the first two letters of the scientific names are chosen; where the letter combination has already been used to designate the MHC of another species, the next letters from the genus or species name are used. Only the first of the four letters is capitalized. For mammals, the scientific names to be used are those listed by Corbet and Hill (1986).
3. A register of MHC symbols will be maintained by the scientific journal *Immunogenetics.* For information contact the editor: Prof. Dr. Jan Klein,

Max-Planck-Institut für Biologie, Abteilung Immungenetik, Corrensstrasse 42, D-7400 Tübingen, FRG; FAX (**49) 70 71 60 04 37. To avoid duplications, all new symbols must be cleared through the register before they are used in publications. A list of accepted symbols for nonhuman primate MHCs already defined and in use appears in Table 1.

4. At the first mention of the MHC in a publication, the entire symbol (i. e., *Mhc* + abbreviation of the species' name) should be given; afterwards the *Mhc* symbol may be dropped and only the species part of the designation used. For example, in a publication dealing with gorilla (*Gorilla gorilla*) MHC, the symbol *MhcGogo* is first introduced but subsequently only the *Gogo* symbol is used.

5. The designation of the complex is followed by a hyphen ("dash") and then by a locus symbol. Loci are designated by capital letters and Arabic numerals. The letter D is used to designate class II loci; all other letters may be used for designations of class I loci. Families of class II loci are designated by a second capital letter following the D symbol (e.g. *DP, DQ, DR, DM*). The α- and β-chain loci are depicted by the letters A and B, respectively. For example, the MHC class II α-chain gene of the chimpanzee DR locus should be designated in publications as *Patr-DRA*.

6. Where orthologous relationships to human loci are obvious, symbols corresponding to human equivalents will be used; in situations in which the correspondence to human genes is doubtful or in which no human equivalent has been described, letters other than those used in the human MHC designations are introduced. The numbering of loci within each family (e.g., *DRB1, DRB2, DRB3*) should be sequential in the order of description. Again, homology to presumptive human equivalents will be taken into account whenever possible. This issue needs careful consideration since experience has shown that it is difficult to establish reliably which relationships between loci are of the orthologous and which are of the paralogous type.

7. The locus designation is followed by an on-line asterisk and then by an allelic symbol. Alleles are designated by Arabic numerals 01, 02, 03, etc., in the sequence of their description. Alleles in each gene family (e.g., *DP, DQ, DR*) are numbered separately. When orthologous relationships to human lineages are obvious, the same symbols will be used. Alleles at different loci within each family (e.g., *DRB1, DRB2, DRB3*) are numbered in the order of their discovery, irrespective of the loci. For example, the evolutionary equivalent of the *HLA-DRB1*0301* gene in rhesus monkeys is named *Mamu-DRB1*0301*. The first two digits identify the corresponding *HLA-DRB* lineage, whereas the next rhesus macaque allele discovered that belongs to the same lineage is called *Mamu-DRB1*0302*, and so. The MHC region has been subject to many duplications. Each series may therefore consist not only of true alleles, but also of pseudo-alleles. If one insisted on maintaining lists of pure allelic series, with the growth of information some designations may need to be shuffled around.

8. When the locus designation is evident, but the nonhuman primate sequences cluster into a lineage that is evidently absent in humans, than a novel serial

number may be introduced. In practice this number is chosen to be rather high in such a way that it may not interfere with HLA nomenclature politics. An example is provided by the *chimpanzee Patr-DQB1*1501* and rhesus macaque *Mamu-DQB1*1501* sequences (Bontrop 1994). Allelic variation at some primate *Mhc* loci is generated by means of frequent segmental exchange. Examples are the *HLA-A, -B* and *-DPB1* loci and their nonhuman primate equivalents. Although the locus definition is clear, experience has shown that the proper clustering of a relatively small number of sequences into lineages turns out to be difficult. In this case sequences will be numbered sequentially. For instance, the rhesus macaque equivalents of HLA-DP are named *Mamu-DPB1*01, Mamu-DPB1*02*, etc. If more information becomes available, such alleles will be grouped into lineages and subsequently renamed.

9. Allelic lineages without human counterparts and of unknown locus alliance are designated by the letter W followed by a two-digit serial number in the order of discovery, irrespective of the species. The W designations should be regarded as "workshop" designations and may be dropped when the locus they occupy is established. Thus, *Mamu-*and *Mane-DRB*W101* represent the names of macaque *DRB* sequences clustering into one lineage that are encoded by a locus for which the definition is not clear. By now, more than 20 *DRB*W* lineages have been defined (O'hUigin at al. 1993).

10. The entire genetic symbol is printed in italics (e. g., *MhcGogo-DRB1*01*). Alleles shared between species may be designated by similar symbols, although this is not a necessity.

11. New alleles receive official designations only when shown by sequencing to be different from all other alleles already described. In the case of doubtful results (recombinants), official allele designations may be handed out when the data are reported by two independent laboratories. For nonhuman primates, the MHC register is maintained by Dr. Ronald E. Bontrop, BPRC, Lange Kleiweg 151, 2288 GJ Rijswijk, The Netherlands; FAX (**31) 15 2843999.

References

Bontrop RE. Nonhuman primate Mhc-DQA and -DQB second exon nucleotide sequences: a compilation. Immunogenetics 39:81–92, 1994

Corbet GB, Hill JE. A world list of mammalian species, 2nd edn. British Museum (natural history), London, 1986

Klein J. Natural history of the major histocompatibility complex. Wiley, New York, 1986

Klein J, Bontrop RE, Dawkins RL, Erlich HA, Gyllensten UB, Heise ER, Jones PP, Parham P, Wakeland EK, Watkins DI. Nomenclature for the major histocompatibility complexes of different species: a proposal. Immunogenetics 31:217–219, 1990

O'hUigin C, Bontrop RE, Klein J. Nonhuman primate Mhc-DRB sequences: a compilation. Immunogenetics 38:165–183, 1993

Index

Molecular Biology and Evolution of Blood Group
and MHC Antigens in Primates
Blancher/Klein/Socha (Eds.)
© Springer-Verlag Berlin Heidelberg 1997